# Heat and Mass Transfer

Addison-Wesley Publishing Company
Reading, Massachusetts ■ Menlo Park, California ■ New York
Don Mills, Ontario ■ Wokingham, England ■ Amsterdam ■ Bonn
Sydney ■ Singapore ■ Tokyo ■ Madrid ■ Bogotá
Santiago ■ San Juan

# Heat and Mass Transfer

Frank M. White
University of Rhode Island

**To my parents,
Frank Mangrem
and
Dorothy Dorr White**

This book is in the **Addison-Wesley Series in Mechanical Engineering.**

**Library of Congress Cataloging-in-Publication Data**

White, Frank M.
Heat and mass transfer.

Includes bibliographies and index.
1. Heat — Transmission. 2. Mass transfer. I. Title.
I. Title.
TJ260.W47 1988 621.402′2 87-13081
ISBN 0-201-17099-X



ABCDEFGHIJ-HA-8987

# Preface

This book is a revision and extension of my earlier book on this subject, *Heat Transfer,* Addison-Wesley 1984. The two primary goals of this new text are to add the topic of mass transfer to the coverage, and to update and improve the original topics based on new literature and the suggestions of faculty reviewers.

The new Chapter 11 covers mass transfer in six sections: introduction and molecular concepts, definitions of mass transfer averaging, Fick's law, the equations of species conservation, diffusion in a stagnant medium, and mass transfer with forced convection. The convection coverage includes wall blowing effects for both laminar and turbulent flow and a special treatment of the naphthalene sublimation experimental technique that is now so popular.

Chapter 11, Mass Transfer, is self-contained and may be either added or omitted in a heat transfer course. The only other discussion of mass transfer in the text is a brief introduction in Chapter 1 of Fick's law and the concept of diffusion and its relation to heat conduction and viscous shear. There is no attempt whatever to thread mass transfer topics or tidbits throughout the other chapters, as I have found this tends to overload the brain circuits of both faculty and students. Heat transfer is a difficult and comprehensive subject that demands thorough coverage before going on to the related topic of mass transfer.

The new mass transfer chapter contains 32 references and 60 problem assignments. The prior edition had 284 references, and 43 more have been added for those topics. The grand total now is 359 references. Similarly, the previous edition had 642 problems, of which 225 were deleted and replaced by 231 new and more challenging problems. There are now 730 problem assignments.

A highlight of this new text is the addition of 22 new Special Design Projects, covering conduction, free and forced convection, radiation, condensation, boiling, and heat exchangers. These are challenging projects, and attempting a few of them during the course should give students a good feel for the broad scope of this subject.

The major improvements to the previous edition are in Chapters 2, 3, 6, and 7. But let us briefly outline the changes in each of the original ten chapters.

Chapter 1 has added Fick's law (page 15), the Schmidt number, and a new Section 1.7 on how to analyze, step by step, a general heat transfer problem. Two new examples are given that require iteration for an unknown temperature in a combined mode problem (pages 20 and 31). One of the iteration schemes diverges and fails — a good lesson for the reader new to numerical methods (pages 19–20).

Chapter 2 now begins with the general heat conduction equation, a topic that formerly was in Chapter 3. There is a very careful treatment of boundary conditions, an important concept that I feel is overlooked or too brief in most texts (page 52). The classical solutions for a slab, cylinder, and hollow sphere now precede the generalized solution for arbitrary area variation. The concept of overall coefficient $U$, formerly omitted, is now introduced and applied to a simple condensing heat exchanger problem (pages 73–74) — as an example, of course, of the fuller treatment in Chapter 9. The fin conduction formulas are now collected in a convenient table (page 93).

Chapter 3 now briefly reviews three–dimensional heat conduction equations and applies them to specific examples. A section on orthogonality of flux lines and isotherms has been added with a description of how to draw a flux plot (page 125). A new Section 3.4.5 has been added on the *boundary element* method, a very powerful and popular new approach to conduction. A single example of conduction through a hollow chimney is carried through the four techniques of flux plots, analytical formulas, finite differences, and boundary elements.

In Chapter 4, Unsteady Conduction, the Heisler/Gröber charts have been resurrected (pages 209, 211, and 212) for comparison with the one–term Fourier series method I had recommended in the earlier edition. Each example compares the charts with an analytical solution, and the chart is shown to be preferable when the surface heat transfer coefficient is unknown.

In Chapter 5, Principles of Convection, the boundary layer equations are developed as before but now are supplemented by the full Navier–Stokes and energy equations for a compressible, viscous, heat-conducting fluid (page 268). This gives the reader a sense of the full glory of convection analysis without the details covered so lovingly in graduate school. A special Example 5.1 is added, giving experimental data for a boundary layer temperature profile from which the heat transfer coefficient can be extracted (page 254). This data can be compared later with a theoretical temperature profile.

Chapter 6, Forced Convection, is completely rearranged so that, after development of the basic integral relations, laminar flows are considered in detail before any turbulent flows are introduced. The earlier edition had interspersed the two. Although the integral methods are still included for laminar flat plate flow, they are supplemented with the full Blasius solution for both velocity and temperature (page 290). The instructor may choose either approach or both. Turbulent plate and duct flows and eddy-coefficient correlations are now grouped together. The section on heat transfer on a cylinder in crossflow is now expanded to include new research on the effects of freestream turbulence, cylinder yaw, tip effects, root effects, and noncircular sections. The basic convection formulas are now collected in a convenient table at the end of the chapter.

Chapter 7, Free Convection, has only a few improvements. The old "King's law" for blocky shapes is now shown to be discredited and inaccurate. The older Elenbaas correlation for an array of vertical fins is now replaced with the newer Bar–Cohen/Rohsenow formulas (pages 407–409). And there is more discussion of mixed convection, including the assisting, opposing, and transverse modes (beginning on page 414).

Chapter 8, Radiation, is essentially unchanged except for 18 new problems and 4 new Special Design Projects.

Chapter 9, Convection with Phase Change, contains new formulas for vertical condensation in the laminar–wavy and turbulent–flow regimes (page 540). There are 12 new problems and 3 Design Projects.

Chapter 10, Heat Exchangers, is unchanged except for 12 new problems and 7 Design Projects.

As mentioned, Chapter 11 is a new stand–alone treatment of mass transfer, with and without convection effects. It has 60 new problems.

The sequence of topics is traditional and follows the approach used at the author's institution. Faculty may choose to include mass transfer or may not. Typical coverage in one semester is as follows.

| | | Hours of Course Coverage | |
|---|---|---|---|
| | *Chapter* | *With Mass Transfer* | *Without Mass Transfer* |
| Conduction | 1 | 3 | 3 |
| | 2 | 3 | 4 |
| | 3 | 5 | 6 |
| | 4 | 5 | 5 |
| Convection | 5 | 2 | 2 |
| | 6 | 5 | 6 |
| | 7 | 3 | 3 |
| Radiation | 8 | 6 | 6 |
| Phase Changes | 9 | 3 | 4 |
| Heat Exchangers | 10 | 3 | 3 |
| Mass Transfer | 11 | 4 | 0 |
| | | 42 hours | 42 hours |

These are only guidelines. The book is too rich in material to be covered completely in one semester; the instructor must choose some topics to be cut back or treated lightly. The sections that have asterisks in the Table of Contents can be omitted without loss of continuity.

The writer's intent is always to make the material scholarly and up to date, yet informal and readable. The coverage should be accurate and broad so that the instructor has leeway to make interesting excursions or variations in class, yet still be backed up by a solid textbook. There are now fourteen unusually complete Appendixes, three more having been added with mass transfer properties. It is hoped that this book will give its owner a lifetime of use in engineering practice.

The book is primarily presented in the SI system of units, but approximately 25% of the examples and problems are in the English system. Eventually SI units will probably be universal, but engineers of the 20th century must continue to be familiar with both systems.

Numerical methods are encountered at several places in the text, but no specific computer programs are presented. Engineering graduates now are generally expert computer programmers. They need only know the methods presented in this text and can easily program them to suit their own taste and their own specific desktop computer.

## Acknowledgments

In preparing this revision, the author is especially grateful for many stimulating discussions with colleagues at the University of Rhode Island: George Brown, Mohammad Faghri, Warren Hagist, Richard Lessmann, Martin Sadd, and Fred Test. Six reviewers read the entire manuscript and gave many helpful suggestions: John R. Biddle of California State Polytechnic University/Pomona, Robert D. Haberstroh of Colorado State University, William E. Lear Jr. of the University of Florida, Gregory V. Selby of Old Dominion University, Graig A. Spolek of Portland State University, and Ralph L. Webb of Pennsylvania State University. The continuing support of Addison-Wesley staff was gratifying, especially Barbara Rifkind, Tom Robbins, Laura Skinger, Karen Guardino, Sheila Bendikian, Sharon van Gundy, and Holly Wallace.

The introspective, self–absorbed, slowly moving activities of a writer can be very frustrating to others. Yet those who have to humor me the most — my wife, Jeanne, my daughters, Jennifer, Ellen, Amy, and Sarah, and my department chair, Tom Kim — have in fact made it a pleasure for me to be a writer.

*Kingston, Rhode Island* Frank M. White

# Contents

## 7

## Free Convection



## 8

## Radiation



## 9

## Heat Transfer with Phase Changes



## 10

## Heat Exchangers

## 11

## Mass Transfer



## Appendixes

# List of Symbols

## English Symbols

| | |
|---|---|
| $a$ | speed of sound |
| $A$ | area ($m^2$) |
| $C$ | heat capacity rate, $= \dot{m}c_p$ (W/K) |
| $C_D$ | dimensionless drag coefficient, $= \text{Drag}/(\frac{1}{2}\rho U^2 A)$ |
| $C_i$ | molar density of species $i$ (kmols/$m^3$) |
| $c_f$ | dimensionless friction coefficient, $= 2\tau/(\rho U^2)$ |
| $c_o$ | speed of light in a vacuum, ($2.9979 \times 10^8$ m/s) |
| $c_p$, $c_v$ | specific heats at constant pressure and volume (J/kg · K) |
| $C_{sf}$ | dimensionless nucleate boiling coefficient, Eq. (9.26) |
| $D$ | diameter (m) |
| $D_{AB}$ | molecular diffusion coefficient ($m^2$/s) |
| $D_h$ | hydraulic diameter, $= 4A/P$ (m) |
| $e$ | internal energy per unit mass (J/kg) |
| $E$ | system energy ($J$) |
| $E_b$ | blackbody radiation rate (W/$m^2$) |
| $E_s''$ | solar radiation received on earth, Eq. (8.79) (W/$m^2$) |
| $f_e$ | dimensionless fractional radiation function, Eq. (8.12) |
| $F$ | force ($N$) |
| $F_{i \to j}$ | radiation shape factor (dimensionless) |
| $g$ | acceleration of gravity (9.807 m/$s^2$) |
| $G$ | $= \dot{m}/A$, (kg/$m^2$ · s) — also irradiation in Sect. 8.2 (W/$m^2$) |
| $h$ | heat transfer coefficient (W/$m^2$ · K) |
| $h_{fg}$ | latent heat of vaporization ($J$/kg) |
| $h_m$ | mass transfer coefficient (m/s) |
| $i$ | enthalpy ($J$/kg) |
| **i, j, k** | unit cartesian vectors |
| $I$ | radiation intensity (W/$m^2$) |
| $J$ | radiosity, $= \text{E} + \rho\text{G}$ (W/$m^2$) |
| $j_i$ | diffusion mass flux of species $i$ (kg/$m^2$) |
| $k$ | thermal conductivity (W/m · K) |
| $L$ | body length (m) |
| $L_e$ | mean radiation beam length, Table 8.5 (m) |
| LMTD | log-mean temperature difference, Eq. (10.10) (K) |

| | |
|---|---|
| $m$ | mass (kg) |
| $\dot{m}$ | mass flow rate (kg/s) |
| $\dot{m}_i'''$ | rate of generation of species $i$ (kg/s · m$^3$) |
| $n$ | coordinate normal to the surface, Fig. 3.3 ($m$) |
| NTU | number of transfer units, $= UA/C_{min}$ (dimensionless) |
| $p$ | pressure ($N$/m$^2$) |
| $P$ | perimeter ($m$) |
| Pa | Pascal |
| $P_i$ | partial pressure of gas species i ($N$/m$^2$) |
| $\mathrm{Pr}_t$ | turbulent Prandtl number, $= \varepsilon_M/\varepsilon_H$ (dimensionless) |
| $Q$ | heat ($J$) |
| $q$ | heat transfer rate (W) |
| $q''$ | heat transfer per unit area (W/m$^2$) |
| $\dot{q}$ | heat generated per unit volume (W/m$^3$) |
| $r$, $z$ | cylindrical coordinates, Fig. 3.4 ($m$) |
| $R$ | thermal resistance, (K/W) — also gas constant in Sect. 6.6 |
| $S$ | conduction shape factor ($m$) |
| $t$ | time (s) |
| $T$ | temperature (K) |
| $U$ | freestream velocity (m/s) — also overall heat transfer coefficient in Chap. 10 (W/m$^2$ · K) |
| $V$ | resultant velocity, $= (u^2 + v^2 + w^2)^{1/2}$ (m/s) |
| $u$, $v$, $w$ | cartesian velocity components (m/s) |
| $v^*$ | friction velocity, $= (\tau_w/\rho)^{1/2}$, (m/s) |
| W | work (J) — also body width in Chap. 3 (m) |
| $x$, $y$, $z$ | cartesian coordinates (m) |
| $x_e$ | duct entry length, Fig. 6.7, (m) |
| $x_i$ | mole fraction of species $i$ (dimensionless) |
| $y^+$ | law-of-the-wall coordinate, $yv^*/\nu$, (dimensionless) |

## Greek Symbols

| | |
|---|---|
| $\alpha$ | thermal diffusivity, $= k/\rho c_p$ (m$^2$/s) — also absorptivity |
| $\beta$ | thermal expansion coefficient, Eq. (7.4), (K$^{-1}$) — also wall blowing parameter, Eq. (11.49) |
| $\gamma$ | $c_p/c_v$ (dimensionless) — also resistivity in Sect. 2.7.2 |
| $\delta$ | hydrodynamic boundary layer thickness (m) |
| $\Delta$ | thermal boundary layer thickness (m) |

| | |
|---|---|
| $\varepsilon$ | radiation emissivity (dimensionless) — also wall roughness in Chap. 6 and exchanger effectiveness in Chap. 10 |
| $\varepsilon_M$ | turbulent eddy viscosity, Eq. (5.14) ($m^2/s$) |
| $\varepsilon_H$ | turbulent eddy conductivity, Eq. (5.22), ($m^2/s$) |
| $\zeta$ | $y/\Delta$ (dimensionless)—also wall blowing parameter, Eq. (11.47) |
| $\eta$ | $y/\delta$ (dimensionless) — also fin efficiency in Sect. 2.7.3 — also equals $x/2(\alpha t)^{1/2}$ in Sect. 4.3 |
| $\theta$, $\phi$ | polar coordinate angles, Fig. 8.4 (dimensionless) |
| $\Theta$ | temperature difference variable |
| $\kappa$ | von Kármán's constant, Eq. (6.11b), $\doteq$ 0.41 |
| $\lambda$ | radiation wavelength (m) — also fin length scale in Sect. 2.7.2 — also equals $(h_0/k)(\alpha t)^{1/2}$ in Sect. 4.3 |
| $\mu$ | viscosity, (kg/m · s) |
| $\nu$ | kinematic viscosity, ($m^2/s$) |
| $\xi$ | unheated starting length in Fig. 6.6 (m) |
| $\pi$ | 3.14159 . . . |
| $\rho$ | density, ($kg/m^3$) — also reflectivity in Chap. 8 |
| $\sigma$ | Stefan-Boltzmann constant, $5.6696 \times 10^{-8}$ $W/m^2 \cdot K^4$ — also mesh Fourier number, $\alpha\Delta t/(\Delta x)^2$, in Chap. 4 |
| $\tau$ | shear stress ($N/m^2$) — also transmissivity in Chap. 8 |
| $\upsilon$ | volume ($m^3$) |
| $\Phi$ | conduction variable defined by Eq. (3.24) |
| $\omega$ | solid angle (dimensionless) — also frequency in Sect. 4.2.1 |
| $\omega_i$ | mass fraction of species $i$ (dimensionless) |
| $\Upsilon$ | surface tension coefficient, ($N/m$) |

## Subscripts

| | |
|---|---|
| b | base |
| crit | critical or transition value |
| *e* | entrance |
| *f* | of the liquid — also film or average value |
| *g* | of the vapor |
| *fg* | in the two-phase region |
| HFD | hydrodynamically fully developed |
| i | inner — also species $i$ in Chap. 11 |
| m | mixed mean |

| | |
|---|---|
| o | outer |
| TFD | thermally fully developed |
| tr | transition |
| w | wall |
| $x$ | in the $x$ direction |
| $\lambda$ | at a given wavelength |
| $\infty$ | ambient or freestream conditions |

## Superscripts

| | |
|---|---|
| * | dimensionless variable — also molar average in Chap. 11 |
| + | law-of-the-wall variable |
| — | overbar, denotes mean or average value |

## Fundamental Dimensionless Parameters

| | |
|---|---|
| Bi | Biot number, $hL/k_{\text{solid}}$ or $h\Delta x/k_{\text{solid}}$ |
| Bo | Bond number, $g\Delta\rho L^2/\Upsilon$ |
| Ec | Eckert number, $U^2/c_p\Delta T$ |
| Fo | Fourier number, $\alpha t/L^2$ |
| Gr | Grashof number, $\rho g\Delta\rho L^3/\mu^2$ or $g\beta\Delta \mathrm{T}L^3/\nu^2$ |
| Gr* | modified Grashof number, = GrNu |
| Gz | Graetz number, $(x/D)/(\mathrm{Re}_D\mathrm{Pr})$ |
| Ja | Jakob number, $c_p\Delta T/h_{fg}$ |
| Le | Lewis number, $\alpha/D_{\text{AB}}$ |
| Ma | Mach number, $U/a$ |
| Nu | Nusselt number, $hL/k_{\text{fluid}}$ |
| Pe | Peclet number, = RePr = $\rho c_p UL/k$ |
| Pr | Prandtl number, $\mu c_p/k$ |
| Ra | Rayleigh number, = GrPr |
| Re | Reynolds number, $\rho UL/\mu$ |
| Sc | Schmidt number, $\nu/D_{\text{AB}}$ |
| Sh | Sherwood number, $h_m\, L/D_{\text{AB}}$ |
| St | Stanton number, $h/(\rho U c_p)$ |

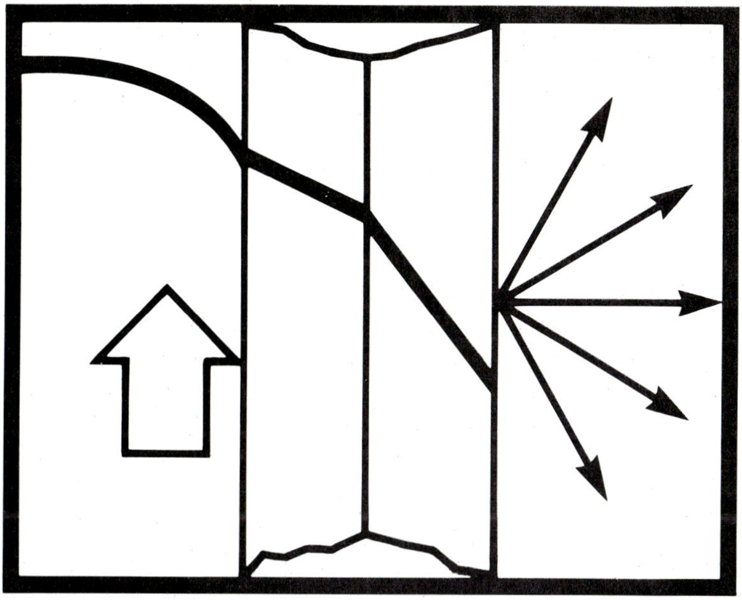

# Introduction

# 1

## 1.1 The Subject of Heat Transfer

*Heat transfer* is a branch of applied thermodynamics. It may be defined as the analysis of the rate at which heat is transferred across system boundaries that are subjected to specific temperature difference conditions. Whereas classical thermodynamics deals with the *amount* of heat transferred during a process, heat transfer estimates the *rate* at which heat transfers and the temperature distribution of the system during the process. We are interested in the transient, nonequilibrium exchange process, not just the static equilibrium states before and after the process.

Consider a hot sphere quenched by immersion in cold water. Thermodynamics tells us what the equilibrium states of sphere and water will be after the process. Heat transfer enables us to analyze the temperature distributions in both sphere and water and the transient heat exchange during the process. It may be a complex analysis — for example, the water may boil or be in streaming motion, the sphere may be anisotropic or layered — but nevertheless the time-varying temperatures and exchange rates are the *goals* of a heat transfer analysis.

The foundation blocks of heat transfer analysis are the first and second laws of thermodynamics. The second law tells us that heat flows from high to low temperature, that is, in the direction of decreasing temperature. Certain limitations on maximum and minimum temperatures and on overall system efficiency are also required by the second law. Otherwise the second law does not directly intrude upon a heat transfer analysis.

In contrast, the first law of thermodynamics is the fundamental relation behind every heat transfer analysis. It may be stated either for a closed *system* or for an open *control volume*. In system form, the total energy increase of the system equals the heat received plus the work received [1–3]:

$$\frac{dE}{dt} = \frac{dQ}{dt} + \frac{dW}{dt}. \tag{1.1}$$

Some authors define work received by the system as negative. For conduction in solids, the work term, $dW/dt$, is negligible. For convection in fluids, the work done by fluid pressure forces is important, but the changes in fluid kinetic and potential energy are usually small in comparison to the heat transfer rate, $dQ/dt$.

For steady flow at low speed through a fixed control volume, the first law of thermodynamics takes the form of a control surface integral

[4]:

$$\frac{dQ}{dt} = \int_{CS}\int i\rho V_n dA, \tag{1.2}$$

where $i = e + p/\rho$ is the fluid enthalpy and $V_n$ is the fluid velocity normal to the surface, taken as positive if exiting and negative if entering. We assume no shaft work in Eq. (1.2). In very high speed gas flows, the fluid kinetic energy and dissipative viscous work should be retained, as will be discussed in Section 6.6.

Equation (1.1) or (1.2) is the basis for practically every analysis in this text. The many different kinds of shaft work and shear work occurring in thermodynamics are rarely important here, so heat transfer is somewhat simpler in principle than thermodynamics, a fact that you may appreciate.

Both equations have a defect, however: They do not provide explicit information about the heat transfer rate, $dQ/dt$. It is the task of heat transfer analysis to provide additional equations, either theoretical or empirical, to relate $dQ/dt$ to system parameters such as geometry, material properties, flow rates, and environmental temperatures. The purpose of this text is thus to supplement the first law of thermodynamics with accurate and physically plausible heat transfer relationships. Additional data about the thermodynamic state are also required: density and specific heats and, if phase changes occur (see Chapter 9), the latent enthalpies of melting, vaporization, and sublimation. Problems involving fluid flow may require still other properties such as viscosity, surface tension, and coefficient of thermal expansion.

Heat transfer is concerned with temperature differences, and we live in a world full of such differences, owing to either natural or artificial causes. Thus an immense variety of heat transfer equipment has been created to deal with these differences: boilers, condensers, solar collectors, radiators, compact heat exchangers, combustion chambers, furnaces, driers, distillation columns, refrigerators, insulators, stoves — the list is almost endless. Each application requires the first law of thermodynamics plus material data and heat transfer relations that are appropriate to the specific system.

A familiar example is the truck radiator shown in Fig. 1.1. This is a type of compact heat exchanger. The fluid entering the radiator has been heated in another heat exchanger surrounding the combustion chambers of the truck engine. It must be cooled in the radiator† by

†Is the word *radiator* a misnomer? Does this design effectively "radiate" energy away from the coolant?

Figure 1.1 Cutaway of a finned, drawn-tank truck radiator. (Courtesy of Young Radiator Co.)

air passing over the multi-finned surfaces. The fins enhance the exchange by providing more hot surface for the air to cool (this is discussed more fully in Section 2.7). The cooler exit fluid then returns to the truck engine to absorb more heat from the combustion surfaces. It is the task of the heat transfer analyst to ensure that this radiator is economical and reliably efficient over all practical operating ranges of engine power, coolant liquid flow, and air flow. This text develops accurate and plausible methods of analyzing such problems.

## Example 1.1

Develop Eq. (1.2) into a form suitable for analyzing the heat transfer from air to the coolant fluid in the radiator of Fig. 1.1.

*Solution* Let the control volume surround the radiator and coolant, cutting through its entrance and exit connections. In steady flow with one entrance and one exit, Eq. (1.2) becomes

$$\left(\frac{dQ}{dt}\right)_{\text{air to coolant}} = (i\rho V_n A)_{\text{out}} - (i\rho V_n A)_{\text{in}}. \tag{1}$$

Assuming one-dimensional flow, the continuity equation [4] is

$$(\rho V_n A)_{\text{out}} = (\rho V_n A)_{\text{in}} = \dot{m}, \tag{2}$$

where $\dot{m}$ is the mass flow of coolant through the radiator. Equation (1) may thus be rewritten as

$$\frac{dQ}{dt} = \dot{m}(i_{\text{out}} - i_{\text{in}}). \tag{3}$$

It is a good approximation, made throughout this text, that for fluid flow with no phase changes, $c_p$ is approximately constant and $\Delta i \doteq c_p \Delta T$. Equation (3) thus becomes

$$\frac{dQ}{dt} = \dot{m}\, c_p(T_{\text{out}} - T_{\text{in}}). \tag{4}$$

This is the desired form of the first law for a heat exchanger. The temperature change $(T_{\text{out}} - T_{\text{in}})$ is negative; that is, heat is lost from the coolant to the air. Equation (4) is not the complete solution to the radiator analysis. We must also relate $dQ/dt$ to the flow properties and geometry of the exchanger. ■

## 1.2 Modes of Heat Transfer

Given the fact from the second law of thermodynamics that heat flows across temperature differences, how many ways can heat be transferred? As you probably know from experience, there are three modes of heat transfer: conduction, radiation, and convection. Conduction and radiation are fundamental physical mechanisms, while convection is really conduction as affected by fluid flow.

*Conduction* is an exchange of energy by direct interaction between molecules of a substance containing temperature differences. It occurs in gases, liquids, or solids and has a strong basis in the molecular kinetic theory of physics.

*Radiation* is a transfer of thermal energy in the form of electromagnetic waves emitted by atomic and subatomic agitation at the surface of a body. Like all electromagnetic waves (light, X-rays, microwaves), thermal radiation travels at the speed of light, passing most easily through a vacuum or a nearly "transparent" gas such as oxygen or nitrogen. Liquids, "participating" gases such as carbon dioxide and water vapor, and glasses transmit only a portion of incident radiation. Most other solids are essentially opaque to radiation. The analysis of thermal radiation (see Chapter 8) has a strong theoretical basis in physics, beginning with the work of Maxwell and of Planck [5].

*Convection* may be described as conduction in a fluid as enhanced by the motion of the fluid. It may not be a truly independent mode, but convection is the most heavily studied problem in heat transfer: More than three-quarters of all published heat transfer papers deal with convection. This is because convection is a difficult subject, being strongly influenced by geometry, turbulence, and fluid properties.

Let us look at each mode in somewhat more detail. A given problem may, of course, involve two or even all three modes.

## 1.3 Conduction

Conduction transfers energy from hot to cold regions of a substance by molecular interaction. In fluids, the exchange of energy is primarily by direct impact. In solids, the primary mechanism is relative lattice vibrations, enhanced in the case of metals by drift of free electrons through the lattice. Thus good electrical conductors are also good heat conductors. Both the molecular and the free-electron interactions are well founded in theoretical atomic physics.

As engineers, we wish to know on a *macroscopic* level how much heat is transferred by conduction. How can we express such a law using macroscopic variables such as temperature? The following homely example may help.

### 1.3.1 Fourier's Law of Conduction

Consider the insulation in the wall of a home. Neglect the plaster and wood frame, and let the thick insulation be the dominant effect on conduction through the wall. What affects heat loss from the warm room to the cold outside? First, we know that the cooler it is outside, the greater the heat loss. The bigger the house is, the greater the heat loss. But the thicker the insulation, the less the heat loss: Advertisements tell us that doubling the insulation halves the loss. Thus we deduce three effects characteristic of conduction heat flow and can write a qualitative expression:

$$\text{Conduction heat flow} \propto (\text{Wall area}) \frac{(\text{Temperature difference})}{(\text{Wall thickness})}. \tag{1.3}$$

The proportionality "constant" is a property of the material (in this case the insulation).

These deductions were made more than a century and a half ago by Joseph Fourier, a French mathematical physicist, and published in 1822 in his pioneering heat transfer book [6]. Fourier wrote Eq. (1.3) in limiting mathematical form, assuming that temperature $T$ varies in the $x$ direction, say, and using the short notation $q_x = (dQ/dt)$ in the $x$ direction:

$$q_x = -kA_x \frac{dT}{dx}$$

or **(1.4)**

$$q_x'' = q_x/A_x = -k\frac{dT}{dx}.$$

This is *Fourier's law of heat conduction;* it is valid for all common solids, liquids, and gases. The coefficient $k$ is a material transport property called the *thermal conductivity*. The double-prime notation, $q'' = q/A$, will be convenient in many analyses. The quantity $A_x$ is the area normal to the $x$ direction through which the heat flows.

If $T(x, y, z)$ is a multidimensional function, Fourier's law becomes a vector relation involving all three directions:

$$\mathbf{q}'' = \mathbf{i}\, q_x'' + \mathbf{j}\, q_y'' + \mathbf{k}\, q_z'' = -k\left(\mathbf{i}\frac{\partial T}{\partial x} + \mathbf{j}\frac{\partial T}{\partial y} + \mathbf{k}\frac{\partial T}{\partial z}\right)$$

or **(1.5)**

$$\mathbf{q}'' = -k\nabla T.$$

We assume that the material is *isotropic,* that is, that its conductivity $k$ does not vary with direction.

Note that the minus sign in Eqs. (1.4) and (1.5) is required by the second law of thermodynamics to ensure that heat flow is positive in the direction of decreasing temperature.

Conductivity $k$ is a thermodynamic property and varies, as will be shown in Section 1.3.2, with temperature, pressure, and composition of the substance.

Consider the simple example of steady conduction through a plane slab, as in Fig. 1.2. Let the control volume surround the wall. Since there is no enthalpy being carried by fluid flow across the boundaries, Eq. (1.2) predicts that $dQ/dt = 0$, or $q_x$ entering the left side equals $q_x$ leaving the right side. The wall area $A_x$ is constant, and for the small (2°C) temperature difference, $k$ will be constant. With $q_x$, $A_x$, and $k$ constant, Eq. (1.4) predicts that $dT/dx$ is constant. Thus for all materials, $T(x)$ is a straight line connecting the boundary temperatures of 21°C and 19°C.

If Fourier's general law, Eq. (1.5), is combined with the first law of thermodynamics, Eq. (1.1), and applied to a differential element of solid material, the result is the three-dimensional *heat conduction equation*, which will be derived and discussed in Section 2.1.

### Example 1.2

Suppose that the iron slab in Fig. 1.2 is 40 cm wide and 70 cm high, as shown in Fig. E1.2. Find an analytical formula for the temperature distribution and the numerical value of the total heat transfer $q_x$.

*Solution* In steady flow, the heat transfer through the slab is independent of position. From Eq. (1.4),

$$q_x'' = -k\frac{dT}{dx} = \text{constant} = -C_1.$$

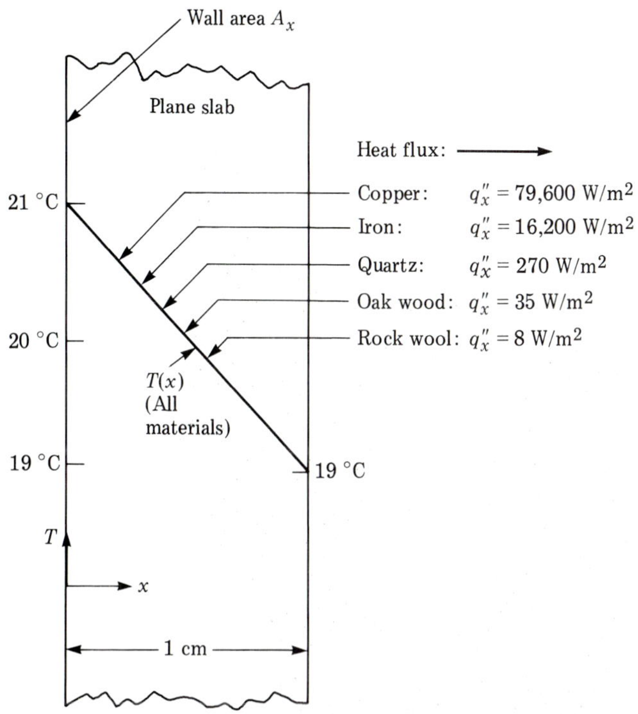


**Figure 1.2** **Temperature distribution and conduction heat flux in a plane slab for various solid materials.**

Integrate

$$T = C_1x + C_2. \tag{1}$$

The boundary conditions are

$$T(x_1) = T_1 = C_1x_1 + C_2,$$

$$T(x_2) = T_2 = C_1x_2 + C_2.$$

Solve for $C_1 = (T_1 - T_2)/(x_1 - x_2)$ and $C_2 = T_1 - C_1x_1$. Substitute into Eq. (1) and rearrange for the desired formula:

$$\frac{T - T_1}{T_2 - T_1} = \frac{x - x_1}{x_2 - x_1}. \quad [Ans.]$$

Differentiate and substitute into Fourier's law:

$$q_x'' = -k\frac{dT}{dx} = -k\frac{T_2 - T_1}{x_2 - x_1} = \frac{q_x}{A_x}. \tag{2}$$

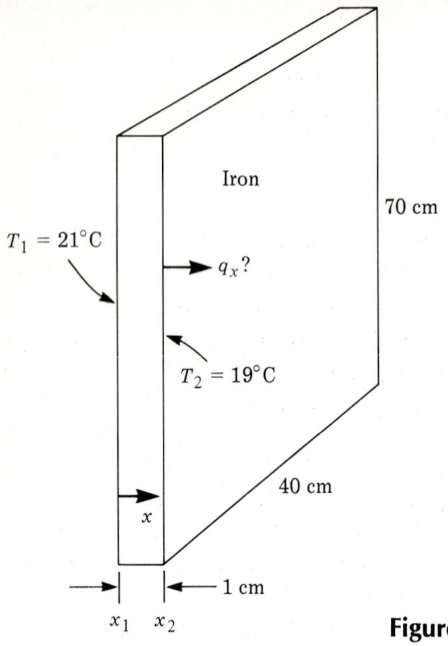


**Figure E1.2**

From Appendix C, for iron at 20°C, $k = 81$ W/(m · K). The total heat flux is thus

$$q_x = -\left(81\,\frac{\text{W}}{\text{m}\cdot\text{K}}\right)(0.4\text{ m})(0.7\text{ m})\left(\frac{19°\text{C} - 21°\text{C}}{0.01\text{ m}}\right)$$
$$= 4540\text{ W.} \quad [\textit{Ans.}]$$

Note that $T(x)$ is entirely independent of the material conductivity, as was stated in Fig. 1.2. ■

---

### 1.3.2 Thermal Conductivity of Various Materials

The differences in heat flux in Fig. 1.2 are entirely due to differences in thermal conductivity, $k$. At 20°C, copper ($k = 3.98$ W/cm · K) has ten thousand times greater conductivity than rock wool ($k = 0.00040$ W/cm · K). For a *homogeneous* substance like copper, $k$ is a transport property and varies with temperature and pressure. Rock wool, on the other hand, is a *heterogeneous* mixture of fibers and air spaces. For engineering purposes we measure and list its "apparent" conductivity $k$, which varies with temperature, density (or void fraction), and partial pressure of the air spaces. The uncertainty in measured conductivity of a substance such as rock wool can be rather high.

Figure 1.3 shows the *thermal conductivity* of various homogeneous substances. There is a five-orders-of-magnitude difference between the highest (diamonds) and the lowest (dense gases). Most metals have high values of $k$ because they allow energy transfer through drift of free electrons; thus there is a good correlation between electrical and thermal conductivity of materials. Gases, with low density and few molecular collisions, have very low conductivity. The kinetic theory

**Figure 1.3** **Thermal conductivity of various materials at low pressures.**

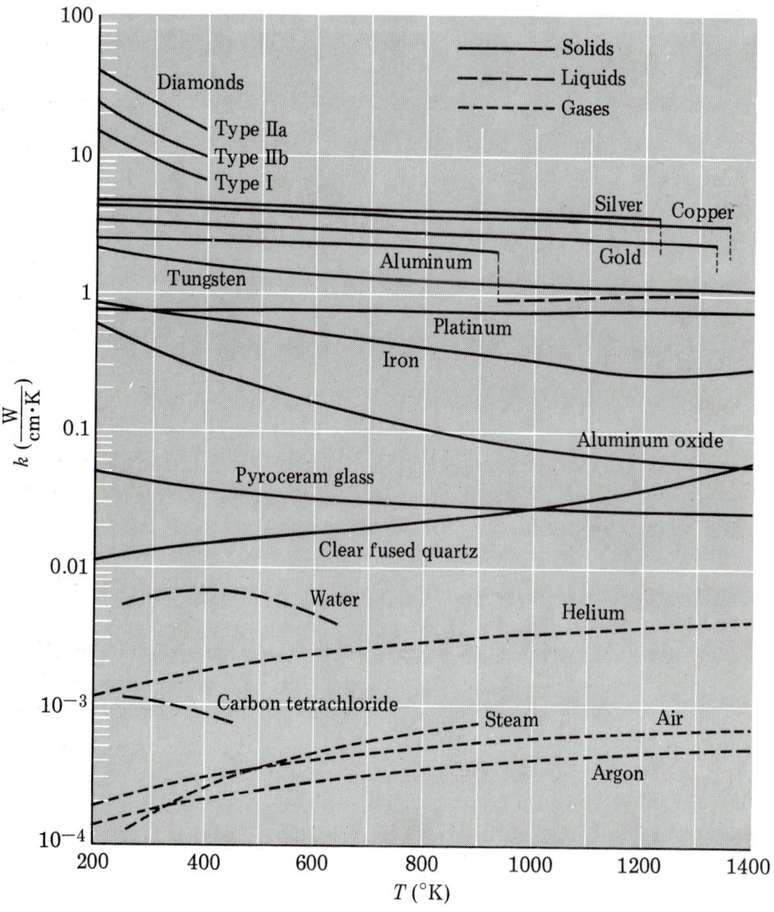

of gases [7] predicts that $k$ is inversely proportional to the square root of the molecular weight, so helium (M = 4.003) has higher conductivity than argon (M = 39.944).

Only the pure metals are shown in Fig. 1.3, and they all have relatively high thermal conductivities. Small amounts of alloy material tend to decrease conductivity. All of the common alloys of copper, aluminum, and iron (steel) have lower — sometimes much lower — conductivity than the pure metals in Fig. 1.3. For example, steel with 1% manganese has a conductivity 32% lower than that of pure iron. Further data on alloys and other metals are given in Appendix C.

The common insulating materials — glass fibers, rock wool, cellulose, cork, polymer foams, powders, sawdust — have apparent conductivity comparable to that of air: $< 10^{-3}$ W/cm · K. Still lower values can be attained by special designs.

### 1.3.3 Extreme Values of Apparent Conductivity

The conductivity values in Fig. 1.3 are representative of commercial materials of relatively uniform composition. Diamonds have the highest known conductivity; for this reason, diamond coverings, though expensive, are used as a heat sink for sensitive electronic semiconductors. Even higher apparent conductivity can be achieved by an artificial device called a *heat pipe* [8], which consists of an evaporator and a condenser connected by an insulated annular wick. The heat pipe allows very large heat fluxes comparable to an apparent system conductivity of 100 W/cm · K.

At normal pressures, gases and insulation materials have the lowest thermal conductivity. However, by an artificial construction using vacuum, multiple layers, and shiny (low-radiation) surfaces, insulating "walls" can achieve apparent conductivities as low as $0.3 \times 10^{-6}$ W/cm · K, or a thousand times less than air. These so-called superinsulations are used in cryogenic applications [9] (a few more details are given in Section 1.9).

### 1.3.4 Heat Storage Capacity

Complementing the conductivity of a material is its heat storage capacity: the amount of energy it absorbs per unit volume for each degree rise in temperature. Since almost all heat transfer applications involve free expansion of the material, the appropriate specific heat is the constant pressure value, $c_p$. The *heat capacity* is thus defined as $\rho c_p$ with units of J/(cm$^3$ · K) or Btu/(ft$^3$ · °F).

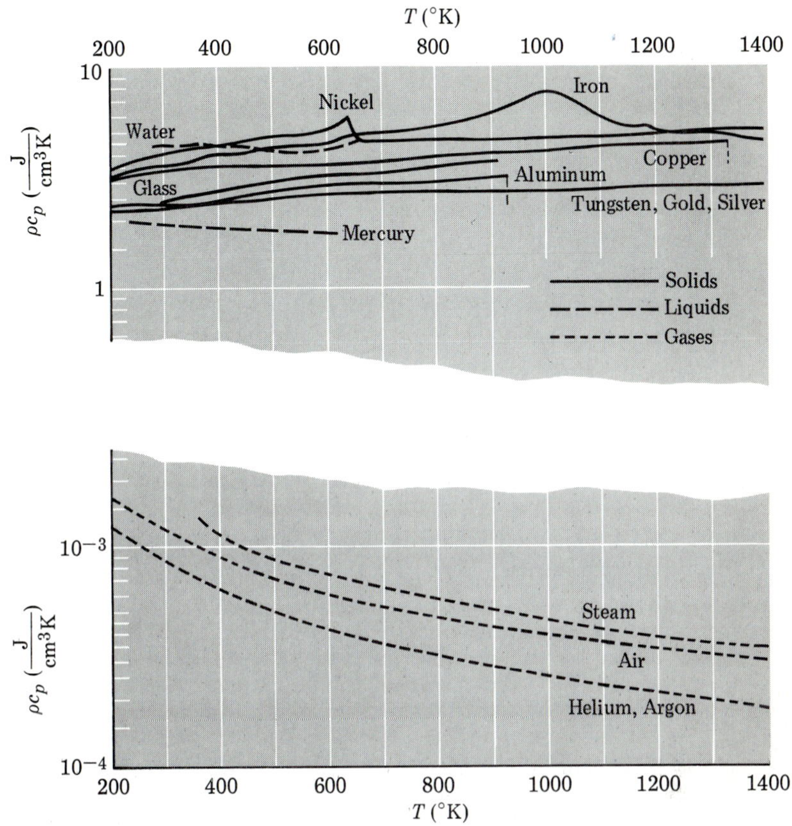


**Figure 1.4 Heat storage capacity of various materials. Solids and gases are at atmospheric pressure; liquids are at saturation pressure.**

Figure 1.4 shows the heat capacity of common materials. It turns out that substances of high density generally have low specific heat, so that most solids and liquids have comparable heat capacities. Nickel and iron show very high capacity at high temperatures, and water is a very good storage medium, though it has a narrow temperature range. Gases are hopelessly poor for storage because of their low densities.

### 1.3.5 Thermal Diffusivity

Since conductivity expresses the rate of heat flow into a substance, and thermal capacity denotes its ability to store this received energy, it follows that their *ratio* is a measure of the rate of change of temperature of the material. Substances with, for example, high conductivity and low capacity will react rapidly to transient external conditions. This important ratio is called the *thermal diffusivity*, $\alpha$:

$$\alpha = \frac{k}{\rho c_p}. \tag{1.6}$$

Energy and temperature units cancel in this ratio, so that $\alpha$ has dimensions of length-squared per time: $cm^2/s$ or $ft^2/s$.

Figure 1.5 completes our triad of heat transfer properties, showing the thermal diffusivity of various materials. We see that the fastest-reacting substances are the highly conducting metals and most gases, which have very low heat capacity. Liquids (even liquid metals), refractories such as aluminum oxide, and most insulation materials have low diffusivity. Diffusivity is the chief property affecting unsteady heat conduction, which is studied in detail in Chapter 4.

**Figure 1.5 Thermal diffusivity of various materials. Solids and gases are at atmospheric pressure; liquids are at saturation pressure.**

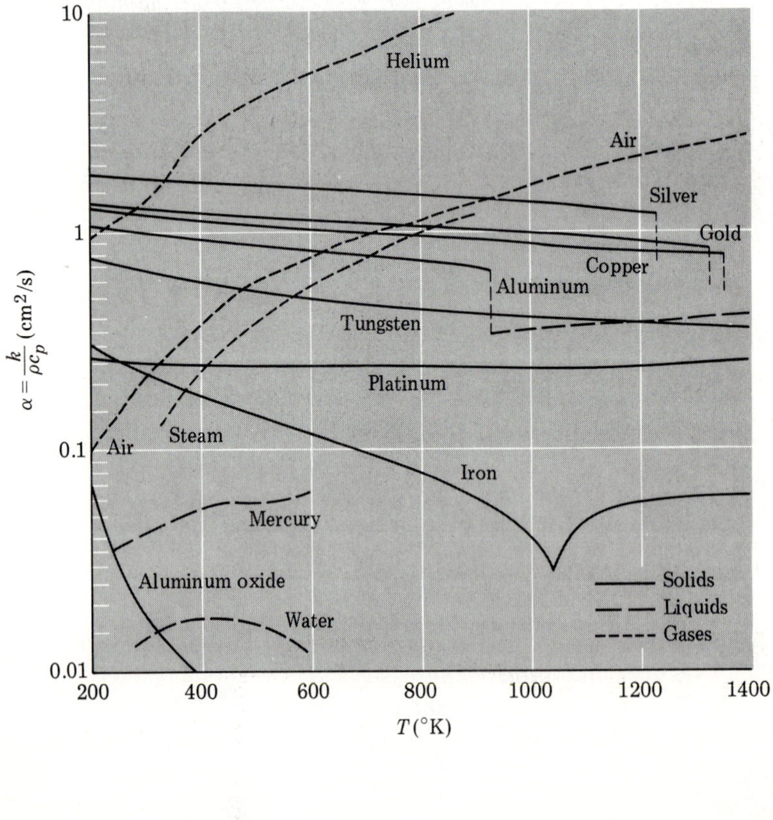

### 1.3.6 Fick's Law: The Heat–Mass–Momentum Analogy

Fourier's law, Eq. (1.4), relating heat flow to temperature gradient, resembles the gradient relation that occurs in mass transfer. Consider a mixture of two fluid species, A and B, with respective densities $\rho_A$ and $\rho_B$. Suppose that $\rho_A$ varies in the $x$ direction. Then there will be a *diffusion mass transfer* of species A in the direction that its density decreases:

$$j_{Ax} = -D_{AB}\frac{d\rho_A}{dx},$$

or in vector form,

$$\mathbf{j}_A = -D_{AB}\nabla\rho_A. \tag{1.7}$$

The units of $j_{Ax}$ are kg/(s · m$^2$). This is *Fick's law* of mass diffusion. The quantity $D_{AB}$ has units of m$^2$/s and is called the *molecular diffusion coefficient,* which varies with temperature, pressure, and mixture composition.

Note that Fourier's law, Eq. (1.5), and Fick's law, Eq. (1.7), are mathematically identical. Thus there is a strong analogy between heat transfer and mass transfer, and both theory and experiment from these two fields are often interchangeable. We will explain this analogy and the theory of mass transfer in Chapter 11.

Recall from the study of fluid mechanics [4] that a similar Newtonian law relates viscous shear to velocity gradient in a simple shear flow $u(y)$, where $u$ is the $x$ component of velocity:

$$\tau = \mu\frac{du}{dy}. \tag{1.8}$$

Recall also that $\mu$ is the fluid property known as the *coefficient of viscosity* or simply the *viscosity*. This is similar in form to Fourier's law, Eq. (1.5), and Fick's law, Eq. (1.7), so that, at least for simple shear flows, there is an analogy among the transports of momentum, heat, and mass. Special textbooks have been written [12] exploiting these and other analogous "transport phenomena."

But this seeming analogy is really quite weak. For one thing, the units of Eqs. (1.5) and (1.8) are not at all the same. And, in general, heat flux $q''$ is a *vector* (three components $q''_x$, $q''_y$, $q''_z$), while viscous stress $\tau$ is a *tensor* (nine components $\tau_{xx}$, $\tau_{xy}$, $\tau_{xz}$, $\tau_{yx}$, $\tau_{yy}$, $\tau_{yz}$, $\tau_{zx}$, $\tau_{zy}$, $\tau_{zz}$). Whatever that might mean, it certainly doesn't bode well for a multidimensional "transport analogy." In fact, it fails: Under three-dimensional conditions, Eqs. (1.5) and (1.8) take the following more general forms:

$$q''_y = -k\frac{\partial T}{\partial y},$$

$$\tau_{xy} = \mu\left(\frac{\partial u}{\partial y} + \frac{\partial v}{\partial x}\right).$$

The shear stress law has picked up an extra gradient, destroying the mathematical similarity. Thus one should take this widely cited analogy with a grain of salt: It is mostly a "one-dimensional" resemblance. In spite of this, certain unsteady viscous flow problems are indeed mathematically identical to unsteady conduction problems. Also, the viscous diffusivity or *kinematic viscosity,* $\nu$,

$$\nu = \mu/\rho,$$

has the same units as thermal diffusivity $\alpha$ ($cm^2/s$) and much the same meaning: It denotes the ability of a fluid to react to changes in stress conditions. The ratio of these two,

$$\text{Pr} = \frac{\nu}{\alpha} = \mu c_p/k, \tag{1.9}$$

is an important dimensionless fluid parameter called the *Prandtl number,* which has a strong effect on heat convection (Chapters 5–7).

In similar manner, the ratio of kinematic viscosity and mass diffusivity,

$$\text{Sc} = \frac{\nu}{D_{AB}}, \tag{1.10}$$

a dimensionless parameter called the *Schmidt number,* has a strong effect on convection mass transfer (Chapter 11).

In closing this section on material properties in heat transfer, we remark that additional data are given in this book in Appendices C–G. For more extensive data, the literature has recently been blessed with the monumental compilation by Y. S. Touloukian and co-workers [10], as well as a complementary set of Russian data [11].

## Example 1.3

The rock wool insulation in a house wall of area 140 $ft^2$ is subjected to a 60°F temperature difference. What thickness of insulation is needed to keep the heat loss below 800 Btu/hr?

*Solution* Assume that Fig. 1.2 applies and solve for thickness:

$$q = kA\frac{\Delta T}{\Delta x}, \qquad \text{or} \qquad \Delta x = \frac{kA\,\Delta T}{q}.$$

From Appendix D, $k$ for rock wool is 0.023 Btu/(hr · ft · °F). Then, for the given data, the minimum desired thickness is

$$\Delta x = \frac{(0.023\ \text{Btu/hr}\cdot\text{ft}\cdot{}^\circ\text{F})\,(140\ \text{ft}^2)(60^\circ\text{F})}{(800\ \text{Btu/hr})} = 0.242\ \text{ft} = 2.9\ \text{in.} \quad [Ans.].$$

They don't sell such an odd thickness. Buy three-inch batts. ■

## 1.4 Radiation

Unlike conduction, which requires a medium, radiation is an electromagnetic phenomenon and travels easily through a vacuum at the speed of light. Most gases transmit nearly all incident radiation, but liquids, even clean water, rapidly attenuate radiation. Most solids, except for glasses and clear plastics, are completely opaque to radiation.

All opaque surfaces emit thermal radiation and absorb or reflect incident radiation. A perfect or "blackbody" surface emits at a maximum rate and, correspondingly, absorbs *all* incident radiation. The term *blackbody* derives from the fact that a black surface absorbs all incident radiation in the visible range and thus reflects no colors to the eye. The net rate of heat flux from an opaque surface equals the total energy emitted and reflected minus the total energy absorbed from the surroundings.

Experiments by J. Stefan in 1879 and a theory by L. Boltzmann in 1884 showed that a black surface emits radiant energy at a rate that is proportional to the fourth power of the absolute temperature of the surface. If a black surface has area $A$ and temperature $T$, its radiant emission is given by

$$E_b = \sigma A T^4, \tag{1.11}$$

where $\sigma$ is a fundamental proportionality called the *Stefan-Boltzmann constant.* As we shall see in Chapter 8, $\sigma$ is related to Planck's constant and Boltzmann's constant and has a numerical value of

$$\sigma = 5.67 \times 10^{-8}\text{W/m}^2\cdot\text{K}^4 = 1.712 \times 10^{-9}\ \text{Btu/hr}\cdot\text{ft}^2\cdot\text{R}^4. \tag{1.12}$$

Note that $E_b$ in Eq. (1.11) has units of heat flux: W or Btu/hr.

Real surfaces are nonblack and emit radiation at a rate less than maximum. A convenient way to express this is to say that they emit at a fraction, $\varepsilon$, of the blackbody rate. For a real surface,

$$E = \varepsilon E_b = \varepsilon\sigma A T^4. \tag{1.13}$$

The dimensionless parameter $\varepsilon$ is called the *emissivity* of the surface and varies between zero and unity. Experiments (see Section 8.2) show that $\varepsilon$ varies with temperature and also with surface parameters: roughness, texture, color, degree of oxidation, and the presence of coatings. An idealized material with constant emissivity is called a *gray body*.

For some real surfaces, the variation $\varepsilon(T)$ can be quite wide over a short temperature range: as fast as $T^6$ and as slow as $T^{-2}$. Thus in these ranges, Eq. (1.13) predicts that $E$ could vary as rapidly as $T^{10}$ and as slowly as $T^2$. This is a warning that nonblack radiation emission may at times deviate considerably from the ideal fourth-power relation.

The analysis of radiant interchange between two or more surfaces can be a complex algebraic procedure. Some detailed analyses are given in Section 8.4. A common special case is when body 1 has temperature $T_1$ and constant emissivity $\varepsilon_1$ and is completely enclosed by a large surface, $A_2 >> A_1$, with temperature $T_2$ and emissivity $\varepsilon_2$. The net radiant heat transfer from the small body to the large enclosure is

$$q_{1\to 2} \doteq \varepsilon_1\sigma A_1(T_1^4 - T_2^4), \tag{1.14}$$

which is independent of the size and emissivity of the enclosure. Most multisurface radiation problems are more complex than this case.

Because of the fourth-power relationship, radiation becomes a dominant heat transfer mechanism at high temperatures: in combustion chambers, furnaces, incandescent filaments, and so on. At room temperature or lower, radiation is usually significant only if the conduction and convection modes happen to be small, as for example in a vacuum-bottle insulator. One should always begin a heat transfer analysis, however, by including radiation as a possible mode. For example, ice can form at night because of ground radiation losses even though the air temperature is above freezing.

### Example 1.4

A four-inch-diameter gray-body sphere, $\varepsilon_1 = 0.6$, with surface temperature $T_1 = 60°F$, is surrounded by a room whose walls are at $T_2 = 200°F$. How much net radiant energy is emitted by this sphere?

*Solution* For this case, Eq. (1.14) holds with $\varepsilon_1 = 0.6$. The sphere surface area is $A_1 = \pi D_1^2 = \pi(4/12 \text{ ft})^2 = 0.349 \text{ ft}^2$. Absolute temperatures are $T_1 = 60 + 460 = 520°\text{R}$ and $T_2 = 660°\text{R}$. Then

$$q_{1\to 2} = \varepsilon_1 \sigma A_1 (T_1^4 - T_2^4) = (0.6)(1.712 \times 10^{-9})(0.349)(520^4 - 660^4)$$
$$= -41.8 \text{ Btu/hr.} \quad [\textit{Ans.}]$$

The minus sign indicates that the net heat is *received* by the sphere, which is colder than the walls. ■

---

The next example illustrates a problem in combined conduction and radiation. Such problems combine first and fourth powers of temperature and thus often need to be solved by iteration. When this occurs, one must be careful to choose a *convergent* iterative process. The following example shows the folly of selecting an improper iterative scheme.

### Example 1.5

As shown in Fig. E1.5, a 2-cm-thick slab, with a conductivity of 4.0 W/m · K and surface emissivity of 0.8, has an inside temperature of 450°C. Its outside is exposed to a vacuum and black surrounding walls at 25°C. Estimate the temperature $T_o$ on the outside of the slab.

*Solution* In steady flow, the conduction heat transfer through the slab must equal the heat transfer radiated away:

$$q_{\text{cond}} = kA\left(\frac{T_i - T_o}{\Delta z}\right) = q_{\text{rad}} = \varepsilon\sigma A\left(T_o^4 - T_s^4\right).$$

Cancel the wall area $A$ and use absolute temperatures because they are required for radiation problems:

$$q = \left(4.0 \frac{\text{W}}{\text{m} \cdot \text{K}}\right)\left(\frac{723\text{K} - \text{T}_o}{0.02 \text{ m}}\right)$$
$$= 0.8\left(5.67 \times 10^{-8} \frac{\text{W}}{\text{m}^2 \cdot \text{K}^4}\right)(T_o^4 - 298^4\text{K}^4),$$

or

$$4.41 \times 10^9 (723 - T_o) = T_o^4 - 298^4. \quad (1)$$

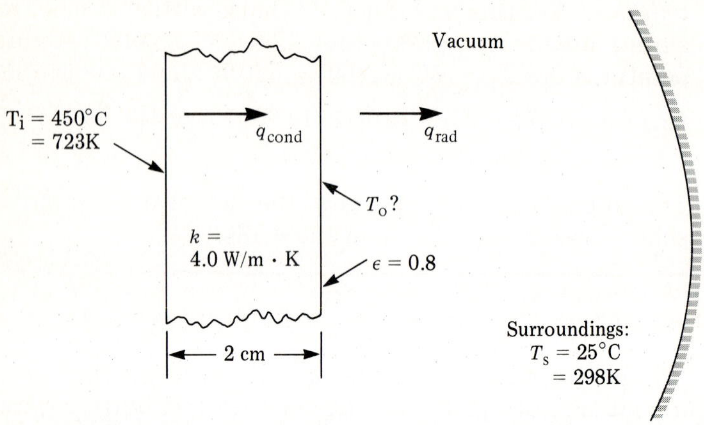


**Figure E1.5**

We may solve Eq. (1) by trial and error for the desired outside temperature:

$T_o = 677\text{K} = 404°\text{C}$. [*Ans.*]

There is an interesting facet to this process. Suppose we solve Eq. (1) by iteration on a computer or calculator. Depending upon the scheme we choose, the result may either converge to 677K or else diverge ridiculously. One convergent scheme is to solve the left-hand side of Eq. (1) for $T_o$:

(a) $T_o(\text{new}) = 723 - \{T_o^4(\text{old}) - 298^4\}/4.41 \times 10^9.$

Guessing, say, $T_o(\text{old}) = 600\text{K}$, the iteration leads to $T_o(\text{new}) = 695\text{K}$, then to $672\text{K} \rightarrow 679\text{K} \rightarrow 677\text{K}$ (converged).

Suppose, however, that we solve the right-hand side of (1) for $T_o$ as an alternate scheme:

(b) $T_o(\text{new}) = \{298^4 + 4.41 \times 10^9\,(723 - T_o(\text{old}))\}^{1/4}.$

Again guessing, say, $T_o(\text{old}) = 600\text{K}$, sequence (b) leads to $T_o(\text{new}) = 861\text{K}$, and then we get an error message (fourth root of a negative number). Even an *excellent guess* will diverge, for example, $675\text{K} \rightarrow 684\text{K} \rightarrow 649\text{K} \rightarrow 760\text{K} \rightarrow$ error message. Students who are familiar with algebraic iteration schemes will recognize the reason for the failure of scheme (b). We assign the explanation as Problem 1.45. ■

## 1.5 Convection

The dictionary defines *convection* as "the conveying of heat through a liquid or gas by motion of its parts." This is good, but how does it happen? What is the mechanism?

First, we require a fluid, which may be a liquid or gas or a mixture (for example, boiling or condensation). And the fluid must be in *motion,* or else things revert to a static conduction type of situation, such as in Fig. 1.2. For heat flux in Fig. 1.2, we could have added some static fluid examples such as water ($q''_x$ = 115 W/m$^2$) or air ($q''_x$ = 5 W/m$^2$). But we would have to justify that no motion ensues (see Chapter 7), because motion enhances the heat flux.

Second, we usually have a solid surface next to the fluid. Granted, there are cases of convection involving only fluids, such as a hot jet discharging into a cold reservoir, but the majority of practical problems involve a hot (or cold) surface adding (or subtracting) heat to (or from) the fluid.

Finally, we distinguish between forced versus free convection, as illustrated in Fig. 1.6. In *forced convection* (Fig. 1.6a), there is a dominant streaming motion far from the solid surface and of independent cause, such as a fan or a pump or the wind. In *free convection* (Fig. 1.6b), farfield streaming is negligible, and the only significant motion is caused by differential buoyancy adjacent to the hot (cold) surface. The heat flux transferred from the solid is usually much stronger for forced convection and increases with the magnitude of the streaming velocity. Finally, if the farfield streaming is relatively minor, such as the small fan in Fig. 1.6(c), the effects of streaming and buoyancy will be comparable. This is called *mixed convection,* and the convection heat flux will be intermediate or moderate in magnitude.

The mechanism of convection occurs in the *boundary layer* near the solid surface, as sketched in Fig. 1.7 for forced convection. As streaming velocity $U_\infty$ increases, both the velocity and thermal boundary layers become thinner, and the wall slopes of velocity and fluid temperature increase. The wall heat transfer and wall shear are proportional to these slopes:

$$\begin{aligned} \tau_w &= \mu_{\text{fluid}}\left(\frac{\partial u}{\partial y}\right)_{\text{wall}}, \\ q''_w &= -k_{\text{fluid}}\left(\frac{\partial T}{\partial y}\right)_{\text{wall}}. \end{aligned} \tag{1.15}$$

The no-slip condition enforces zero relative velocity at the wall, and

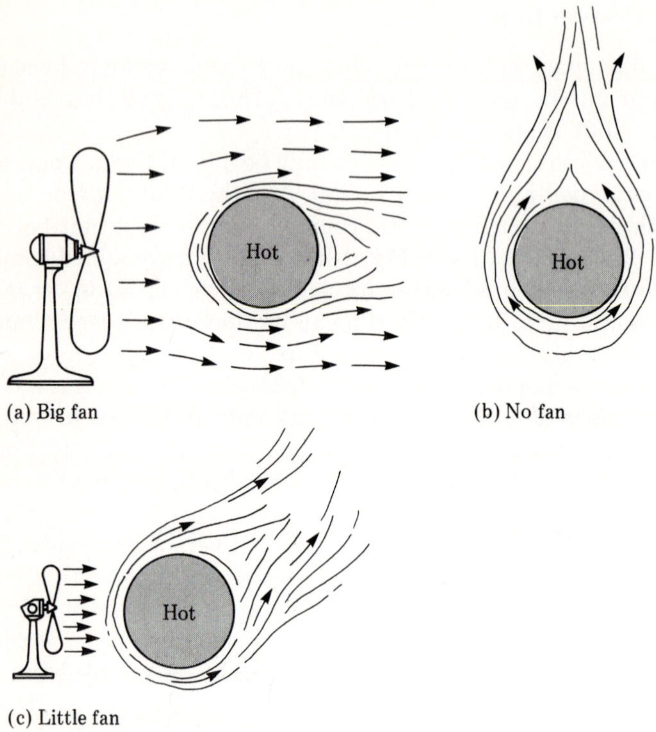


**Figure 1.6** **Convection regimes for flow near a hot sphere. (a) Forced convection, strong streaming motion. (b) Free convection due to buoyancy alone. (c) Mixed convection with both streaming and buoyancy important.**

thus wall heat transfer is due to pure conduction only. Since these wall slopes depend in turn on stream parameters such as $U_\infty$, body size, and fluid density, it is customary to correlate $\tau_w$ and $q''_w$ directly with stream parameters rather than use Eqs. (1.15).

### 1.5.1 The Heat Transfer Coefficient

For forced convection of a single-phase fluid with moderate temperature differences, a nice thing happens: The heat flux $q''_w$ is nearly proportional to the temperature difference $\Delta T = T_w - T_\infty$. This was discovered by Sir Isaac Newton, who experimented with small objects heated in a forge and then cooled outside in the wind. He found that the cooling was nearly exponential with time and from this inferred

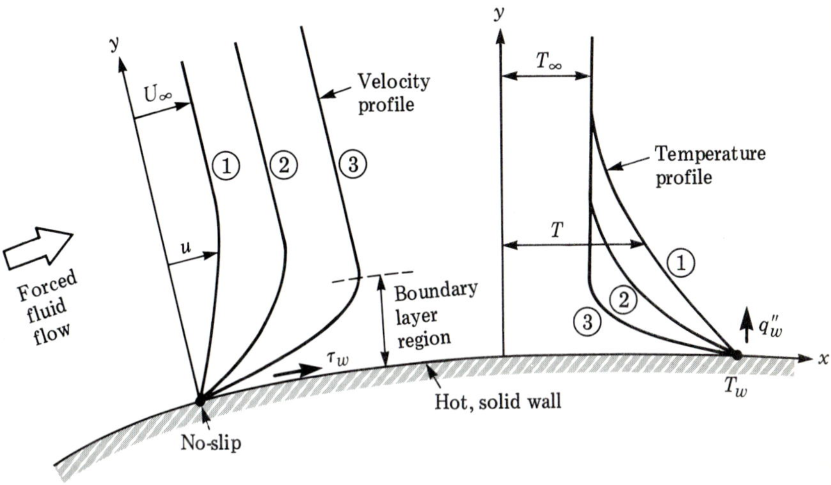


**Figure 1.7** Sketch illustrating how increasing forced-convection velocity thins both velocity and thermal boundary layers and increases their wall slopes. Profiles shown for low (1), medium (2), and high (3) velocity.

that $q''_w \propto \Delta T$ (see Example 1.6). The idea has stayed with us for three centuries, and it is customary to restate Eq. (1.15) in the form of Newton's law of cooling:

$$q''_w = h\,(T_w - T_\infty), \tag{1.16}$$

where $h$ is called the *heat transfer coefficient,* with units of $W/m^2 \cdot K$ or $Btu/hr \cdot ft^2 \cdot {}^\circ F$. We then correlate $h$ with stream parameters. The advantage, at least for forced convection, is that $\Delta T$ is eliminated as a variable, making the analysis more efficient. In many cases, however, $h$ actually varies with $\Delta T$: free convection, boiling, condensation, large temperature differences. This is quite annoying, since it belies the supposed proportionality and brings $\Delta T$ back into the analysis. Nevertheless, you will find that the use of $h$ in convection is widespread among heat transfer workers, even when it is quite inappropriate.

In most convection problems, $h$ varies around the surface of the body, so that an area-averaged value must be defined:

$$\bar{h} = \frac{1}{A}\int_A h\,dA. \tag{1.17}$$

We then correlate $\bar{h}$ as a measure of overall heat transfer from the body.

An example is shown in Fig. 1.8 for flow of various fluids normal to the axis of a long, smooth 5-cm-diameter cylinder. Since $h$ varies dramatically around the cylinder (see Chapter 6) we plot $\bar{h}$, from which the total heat transfer may be computed:

$$q_w = \bar{h}\, A_w(T_w - T_\infty), \qquad A_w = \pi DL. \tag{1.18}$$

We assume forced convection with moderate, room-temperature conditions: $T_w \simeq T_\infty \simeq 20°\text{C}$. No data are shown for $V$ less than 1 m/s — where free convection effects may be important. Note that the actual temperature difference is not specified, nor whether the body

**Figure 1.8 Effect of stream velocity on forced-convection heat transfer in crossflow past a 5-cm cylinder at near-room-temperature conditions. These curves have a 20% uncertainty.**

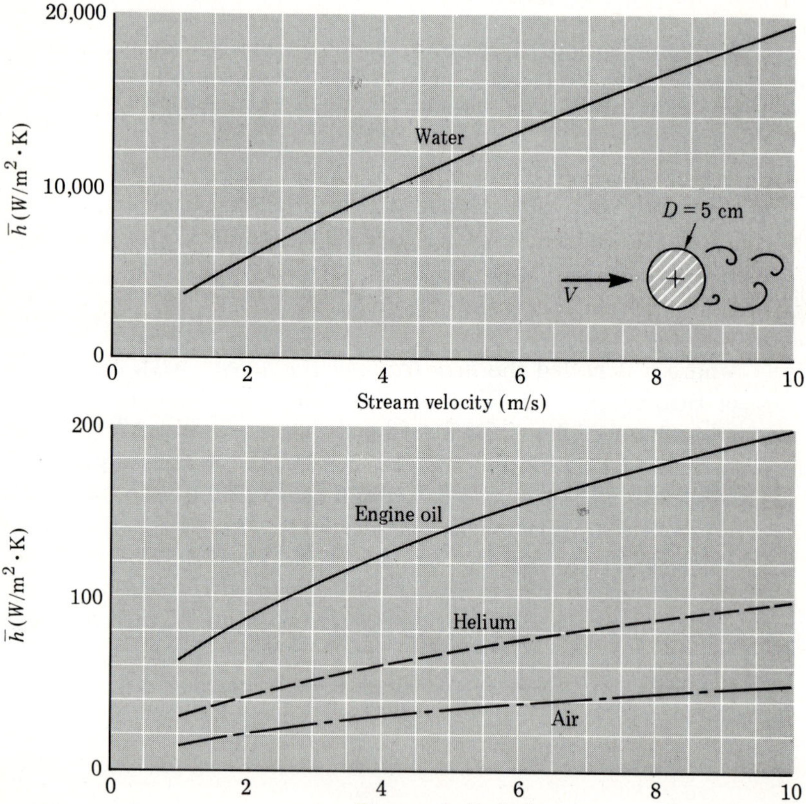

is hot or cold: Eq. (1.18) takes care of these two items through its structure.

Now look at the numbers in Fig. 1.8. Helium, which is much lighter than air, has twice the heat flux. Why is this so? Then engine oil, which is 700 times heavier than air, can provide only 4 times as much heat flux. Why is that so? Finally, water, only slightly heavier than engine oil, needs a whole new scale and produces 300 times as much heat flux as does air. Let us save the details for Chapter 6. It should be clear, though, that convection is a complex mixture of pro and con effects of fluid viscosity, thermal conductivity, density, and specific heat. Note in passing that $\bar{h}$ (or $q_w$) in Fig. 1.8 increases with stream velocity but at a somewhat less than linear rate.

One final note about Fig. 1.8: The cylinder *material* is not specified; nothing about the material affects $q_w$ except its surface roughness. Could you have deduced this from physical principles?

The curves in Fig. 1.8 are taken from data, not theory, and are subject to $\pm 20\%$ uncertainty. There is very little convection theory available for flow past complex body shapes, although textbooks are now beginning to present new digital computer simulations of convection problems [14].

If phase changes occur from condensation (boiling), the heat flux can be quite large because of the latent heat absorption (or release) and the droplets (or agitated bubbles). Some estimated ranges of heat transfer coefficient for the various types of convection are given in Table 1.1. Note our instinctive use of $\bar{h}$ even in the three inappropriate cases.

### Example 1.6

Show analytically how Newton might have deduced that exponential cooling of a small body in a steady wind would imply that $q_w \propto \Delta T$ for forced convection.

*Solution* If work is negligible, the first law, Eq. (1.1), states that convection heat loss $dQ/dt$ equals the decrease in body energy, $dE/dt$. If the body is small and highly conducting (a metal), its temperature $T$ will be nearly uniform and will vary only with time. The energy decrease of the body will be $dE/dt = -mc_p(dT/dt)$. Even if $\bar{h}$ is variable, the convective heat loss can be defined as $\bar{h}\,A_w(T - T_\infty)$, where $T_\infty$ is the air temperature, assumed to be

**Table 1.1** Typical convection heat transfer coefficients

| Type of Convection | $\bar{h}$(W/m · K)† |
|---|---|
| *Free Convection‡* | |
| Gases | 5–25 |
| Oils | 10–60 |
| Light liquids | 100–1000 |
| *Forced Convection* | |
| Gases | 10–300 |
| Oils | 50–2000 |
| Light liquids | 100–20,000 |
| Liquid metals | 5000–50,000 |
| *Boiling Water‡* | 1000–100,000 |
| *Condensing Steam‡* | 5000–100,000 |

† 1 W/m² · K = 0.1761 Btu/hr · ft² · °F.
‡ $\bar{h}$ varies with $\Delta T$ in these cases.

constant. Then the first law for a small cooling body reduces to

$$-mc_p\frac{dT}{dt} \doteq \bar{h}\,A_w(T - T_\infty),$$

where $A_w$ is the body surface area. If $\bar{h}$ is constant, the solution is an exponential decrease with time:

$$T = T_\infty + (T_0 - T_\infty)e^{-at}, \qquad a = \bar{h}\,A_w/mc_p,$$

where $T_0$ is the initial body temperature at $t = 0$. This is what Newton observed experimentally.

The assumption that $T$ varies with time only is the "lumped-mass" approach and is explained in Section 4.2. ■

## Example 1.7

For the truck radiator of Fig. 1.1, suppose that water enters at 180°F and leaves at 140°F with a flow rate of 7 gal/min (1 gal = 231 in³). If the ambient air is at 80°F and the air convection heat transfer coefficient is 12 Btu/hr · ft² · °F, how much radiator surface area is required?

*Solution* In steady flow, $dE/dt = 0$, and the heat lost by the water must be gained by the air. We found an expression for water heat loss in Eq. (3) of Example 1.1, so we equate

$$(dQ/dt)_{\text{water}} = (dQ/dt)_{\text{air}},$$

or (1)

$$\dot{m}\, c_p(T_{\text{in}} - T_{\text{out}}) = \bar{h}\, A_w(\bar{T}_w - T_\infty).$$

We know or can estimate every quantity in Eq. (1) except $A_w$. First assume that $\bar{T}_w$ equals the average water temperature, or $1/2(180° + 140°) = 160°\text{F}$. Then estimate the water properties at an average 160°F from Appendix F: $\rho = 978\ \text{kg/m}^3 = 61.0\ \text{lb}_m/\text{ft}^3$, $c_p = 4188\ \text{J/kg} \cdot \text{K} = 1.00\ \text{Btu/lb}_m \cdot °\text{F}$. The volume flux of water is 7 gal/min $= 0.0156\ \text{ft}^3/\text{s}$, which when multiplied by the density of water gives the mass flux:

$$\dot{m} = (61.0\ \text{lb}_m/\text{ft}^3)(0.0156\ \text{ft}^3/\text{s})(3600\ \text{s/hr}) = 3430\ \text{lb}_m/\text{hr}.$$

Then Eq. (1) may be evaluated as follows:

$$(3430\ \text{lb}_m/\text{hr})(1.00\ \text{Btu/lb}_m \cdot °\text{F})(180°\text{F} - 140°\text{F}) = 137{,}000\ \text{Btu/hr}$$
$$= (12\ \text{Btu/hr} \cdot \text{ft}^2 \cdot °\text{F})(A_w \text{ft}^2)(160°\text{F} - 80°\text{F}),$$

or

$$A_w = 143\ \text{ft}^2. \quad [\textit{Ans.}]$$

The only way that such a large area can be packed into such a small space is to use finned or "compact" heat exchanger surfaces, which the cutaway in Fig. 1.1 supposedly reveals. ■

## 1.6 Units and Dimensions

Look back again at Example 1.7. The use of common English terms such as gallons, Btu, and pounds of mass made the problem slightly cumbersome, somewhat messy, did it not? English terms, though familiar, are often inconsistent and require constant insertion of conversion factors. For example, in high-speed convection (discussed in Section 6.8), the additive term $(c_pT + V^2/2)$ occurs. In English units, $c_pT$ has units of $\text{Btu/lb}_m$ while $V^2$ would be $\text{ft}^2/\text{s}^2$. How can they be added? The answer lies — rather subtly, it seems — in the conversion factor $1\ \text{Btu/lb}_m = 25{,}040\ \text{ft}^2/\text{s}^2$.

A completely consistent set of units is the SI or *International System* of metric units, first adopted internationally in 1875 and now almost universally used by scientists. The fundamental units in this system are the kilogram of mass (kg), the meter, the second, and the newton of force (N). The energy unit is the newton-meter or joule (J). In this system, all equations that arise in heat transfer are totally consistent, that is, they need no conversion factors. For example, in SI units, the term $c_pT + V^2/2$ gives rise to identical units: 1 J/kg = 1 m$^2$/s$^2$. Except for the factors of 10 — centimeter versus meter, megawatt versus watt — the student can use the kg-m-s-N system in confidence that all terms in all equations will have consistent units.

In spite of the fact that the SI system is now required for publication by American professional societies such as ASME and AIChE, many engineers are still more comfortable with the English system of units. Many industries, especially the mature ones, retain the English system, and there may even be a bit of backlash presently against the adoption of SI units. This textbook thus gives emphasis to the SI system in presenting fundamentals but alternates SI and English units in the examples and problems. The two different systems are illustrated in Table 1.2.

### Example 1.8

In turbulent forced convection (see Chapter 6), a theoretical expression for wall heat flux is $q''_w = c_p\tau_w\Delta T/U_\infty$, where $\Delta T$ is the temperature difference across the boundary layer. How does this relation fare with SI or English units?

*Solutions* In SI units, the right-hand side has the units

$$\frac{(\mathrm{J/kg\cdot K})(\mathrm{N/m^2})(\mathrm{K})}{\mathrm{m/s}} = \frac{\mathrm{J\cdot N\cdot s}}{\mathrm{kg\cdot m^3}},$$

**Table 1.2 Two competing unit systems**

| Dimension | SI Unit | English Unit | Conversion |
|---|---|---|---|
| Mass | kg | $\mathrm{lb}_m$ | 1 kg = 2.2046 $\mathrm{lb}_m$ |
| Length | m | ft | 1 ft = 0.3048 m |
| Time | s | s | 1 s = 1 s |
| Force | N = kg · m/s$^2$ | $\mathrm{lb}_f$ | 1 $\mathrm{lb}_f$ = 4.4482 N |
| Energy | J = N · m | Btu | 1 Btu = 1055.1 J |

whereas we expected the SI flux unit of $W/m^2$. However, from Table 1.2, 1 N = 1 kg · $m/s^2$, which we use to eliminate the newton in our result above to obtain J/s · $m^2$ or $W/m^2$. Thus we should have had full confidence that our SI units would indeed lead to the proper heat flux unit, $W/m^2$.

In English units, however, the given equation leads to

$$\frac{(\text{Btu/lb}_m \cdot {}^\circ\text{F})(\text{lb}_f/\text{ft}^2)({}^\circ\text{F})}{\text{ft/s}} = \frac{\text{Btu} \cdot \text{lb}_f \cdot \text{s}}{\text{lb}_m \cdot \text{ft}^3},$$

whereas the traditional English heat flux unit is Btu/hr · $ft^2$. We cannot recover this result without introducing the rather ungainly conversion factor 1 $lb_f$ = 115,900 $lb_m$ · ft/s · hr. The possibility of serious error in applying such conversions is obvious, yet many engineers, the author included, feel quite comfortable with the English system. ■

## 1.7 Heat Transfer Analysis: Academic Versus Real

The chief purposes of an academic textbook are (1) to acquaint students with the fundamentals of a subject; (2) to give them practice in analyzing and solving relatively short problem assignments, sometimes aided by worked examples; and (3) to provide sufficient breadth and design information that the book becomes a lasting part of a person's library. A text can be used for self-teaching, of course, but mainly it serves as a supplement to a college professor's lectures. These are academic goals, not to be confused with *real industrial practice*.

As in most texts, the problems presented here are meant to be short assignments to be completed in about one hour. The student gains valuable experience in using the engineering approach to an unfamiliar problem:

1. *Read the problem carefully*. Determine what is known and what needs to be found. Writing the problem out again or paraphrasing it is helpful. It is astonishing how often students fail to give their full attention to this crucial first step.
2. *Make a sketch of the physical problem*. Be generous with the labeling and the details. You will gain insight, and people who read your work will be impressed with the professional appearance of your schematics.

3. *List all appropriate assumptions for the analysis.*
4. *Find and list the needed physical properties.* Heat transfer calculations require knowledge of densities, diffusivities, emissivities, viscosities, surface tensions, and latent heats. Many of these are tabulated in the appendices. Heaven forbid, you may even have to consult the primary sources [10, 11].
5. *Apply your analysis at an appropriate fundamental level.* Proceed as far as you can in symbolic form before inserting numerical values. Don't just "plug and chug."
6. *Present your final results clearly.* If possible, give remarks about the trends and principles illustrated. Estimate the uncertainty of your results and be sparing in your use of significant figures. If necessary, criticize the validity of your assumptions.
7. *Do your own work.* If you worked with others, cite them. If you found information in another reference, cite it. All colleges have strict regulations against plagiarism. Ignorance of such regulations is no excuse.

These seven steps are all part of the engineering approach to any problem. They work for a one-hour homework problem, and they work in real practice, when an industrial heat transfer analysis or design may take months and involve a team of engineers. The large scope of most real problems makes it imperative that your own contributions be clearly and professionally presented. Working with a team means cooperative efforts across disciplines, good interpersonal relations, and giving generous credit to others. A team effort requires prompt and timely contributions from each member. Both oral and written reports are required.

It is not possible in this academic text to simulate a major industrial project. However, in addition to the usual short problems, several lengthy and challenging Case Studies are presented as possible major class projects. The instructor may also present major assignments on other topics, with possible follow-up in engineering design courses.

## Example 1.9

Modify Example 1.5 to replace the vacuum on the outside by a stream of air at 25°C that provides a convection heat transfer coefficient of 75 $W/(m^2 \cdot K)$ to the outside surface. Estimate the surface temperature and the heat transfer rate.

*Solution* First note that dry air is "nonparticipating" and does not affect radiation to the surroundings. The conduction through the slab now equals the sum of radiated and convected heat transfer away from the outside wall:

$$q''_{cond} = q''_{rad} + q''_{conv},$$

or

$$k\frac{T_i - T_o}{\Delta z} = \varepsilon\,\sigma\,(T_o^4 - T_s^4) + \bar{h}(T_o - T_{air}).$$

Introducing the numerical data from Example 1.5, we obtain

$$4.0\left(\frac{723 - T_o}{0.02}\right) = 0.8(5.67 \times 10^{-8})\,(T_o^4 - 298^4) + 75(T_o - 298).$$

When cleaned up, this looks like

$$T_o + 1.65 \times 10^{-10}\,T_o^4 = 608.4. \qquad (1)$$

This must be solved by iteration if you are not comfortable with quartic polynomials. Recall from Example 1.5 that one iteration scheme diverges: If we solve for $T_o^4$ in Eq. (1) and iterate, the solution diverges no matter how excellent our first guess is. If we solve for $T_o$ in (1), the solution rapidly converges to

$T_o = 589\text{K} = 316°\text{C}.$ [*Ans.*]

The heat transfer rate is

$$q'' = q''_{cond} = 4.0\left(\frac{723 - 589}{0.02}\right) = 26{,}900\ \text{W/m}^2. \quad [\textit{Ans.}]$$

This is three times as much as the value $q'' = 9200\ \text{W/m}^2$ without convection in Example 1.5. This problem is ideal for the electric analogy technique to be developed in Section 2.4. ■

## †1.8 Complications Affecting Heat Transfer Analysis

If thoroughly studied, the material in this text should enable you to make reasonably sophisticated analyses of some very practical conduction, convection, and radiation heat transfer problems. Each chapter also cites additional literature for advanced or supplementary topics. But some remarks are in order to warn you of possible pitfalls or

†This section may be omitted without loss of continuity.

complications in heat transfer analysis. We may list some difficulties as follows:

1. Predictive uncertainty in both theory and data.
2. Deviations from (a) Fourier's law, (b) Newton's law of cooling, and (c) the fourth-power radiation law.
3. Anomalous physical properties near the critical point.

Of these three complications, predictive uncertainty is a constant problem. Thermophysical properties are not very well known. Consider, for example, the thermal conductivity of aluminum, a very common material. The data sets in [10] show a $\pm 20\%$ uncertainty in values of $k$ for aluminum. The same is true of other solids, especially if impurities are present.

Heat transfer experiments are difficult and uncertain. Typical measurement uncertainties are $\pm 40\%$ for emissivities, $\pm 50\%$ for convection coefficients, and $\pm 100\%$ for boiling and condensation correlations. Turbulent-flow theories are semi-empirical and may be inaccurate. Boundary conditions may be poorly known in all modes of heat transfer analysis. The engineer must act conservatively in using heat transfer analyses in design because of this uncertainty.

Our heat transfer mode formulas can each be inaccurate. At high heat fluxes, materials deviate from Fourier's law of conduction, Eq. (1.4), and have rate-sensitive effects. (See [16] for an example.) In certain convection problems — free convection, boiling, and condensation — the coefficient $h$ is a strong function of temperature difference, leading to nonlinear heat transfer correlations. Radiation properties of real surfaces can deviate widely from the classic blackbody fourth-power radiation law, Eq. (1.13). In design work, the engineer must often revise the analysis for these effects.

Finally, the physical properties of all fluids exhibit anomalous behavior near the critical point. Figure 1.9 shows the thermal conductivity of $CO_2$ near its critical point. The sharp rise in $k$ is not well formulated theoretically, and calculations near the critical point will be very uncertain.

## †1.9 Outline of Cryogenics

This text is intended to treat heat transfer problems that occur in the "commonly" encountered temperature range 200K–1200K. But there is a widely practiced subdivision of heat transfer, called *cryogenics*,

†This section may be omitted without loss of continuity.

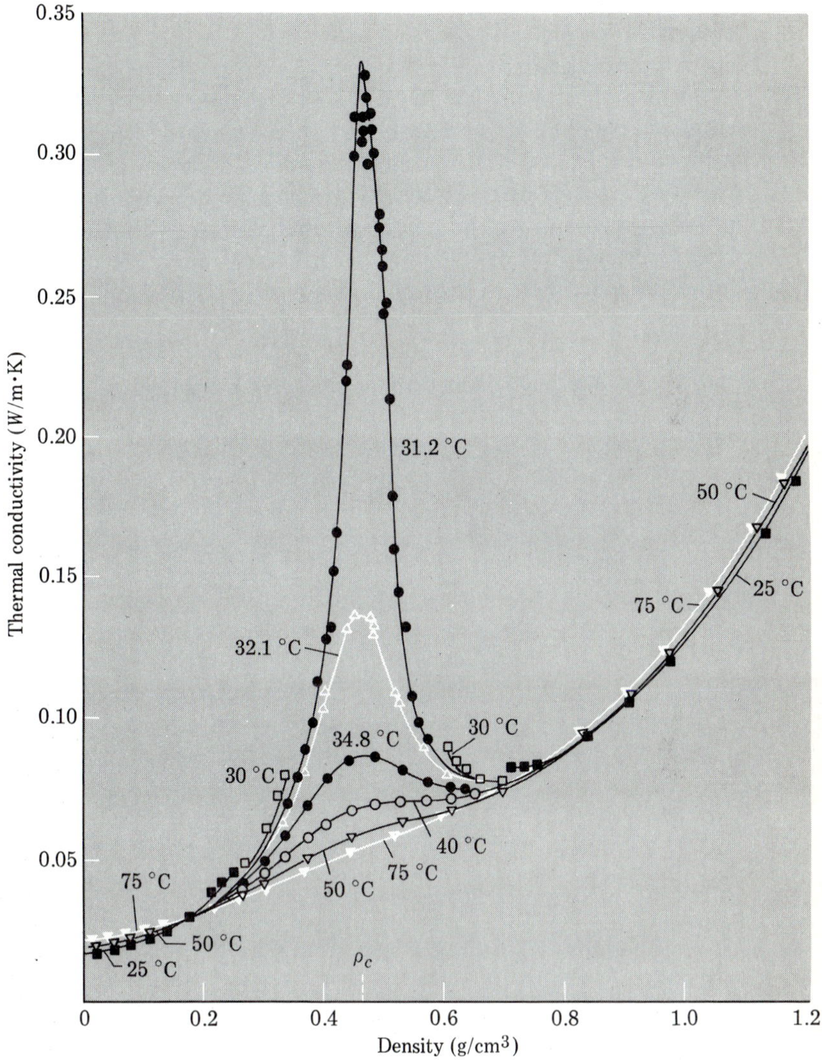


**Figure 1.9** **Thermal conductivity of carbon dioxide, showing the anomalous spike near the critical point, $T_c$ = 31°C, $p_c$ = 7.4 × $10^6$ Pa. (Data from [18].)**

dealing with temperatures below about 200K, where many of the common gases such as oxygen, nitrogen, argon, methane, helium, and hydrogen boil. Our basic modes of conduction, convection, and radiation are still valid at low temperatures, but thermal design problems associated with preparation, storage, and transport of cryogenic materials meet with several distinct complexities.

An analytical problem in cryogenics is the strong variation in physical properties at low temperatures. Figure 1.10 shows the thermal conductivity of selected materials in the cryogenic range. Metals show a marked increase and nonmetallic solids a striking decrease in conductivity with decreasing temperature. Many cryogenic solids become *anisotropic,* meaning that their conductivity and other properties vary with direction. At very low temperatures, of the order of 4K, quantum liquids appear, such as helium II, possessing "superconductivity" (nearly

**Figure 1.10** **Thermal conductivity at cryogenic temperatures.**

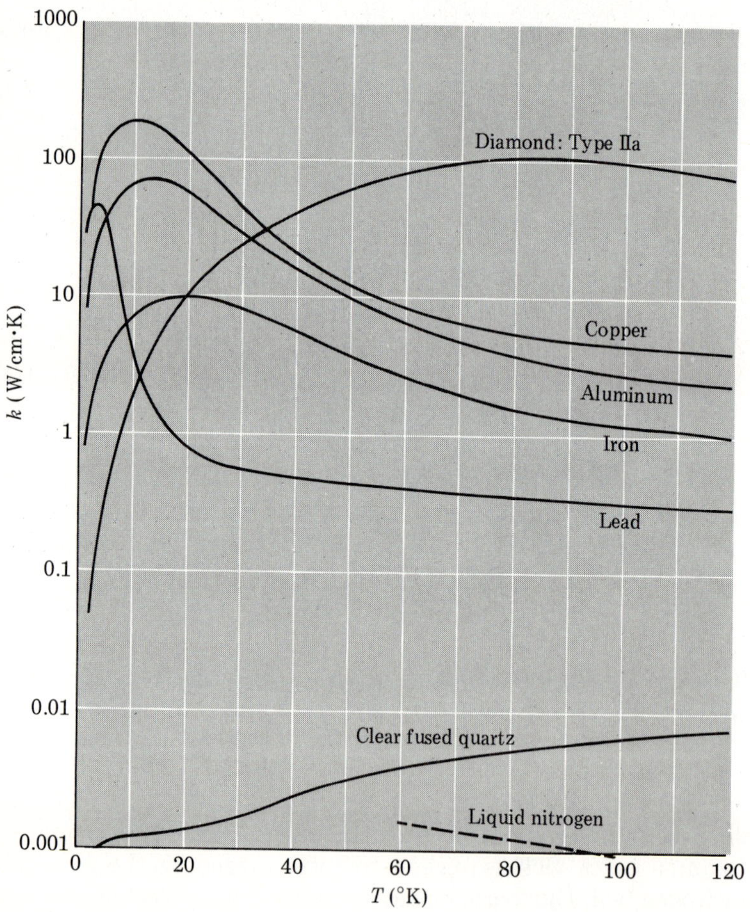

infinite $k$) and "superfluidity" (nearly zero viscosity). Boiling, condensation, and forced convection estimates are strongly dependent on the highly variable cryogenic properties of fluids.

The need to store cryogenic materials efficiently has led to ingenious insulation techniques because the common room-temperature materials such as glass fibers and rock wool are not good enough. Fundamental to all the cryogenic insulation designs is the use of high vacuum to eliminate gases from the insulation, thus minimizing conduction and convection. For best results, the pressure should be reduced below about 0.1 Pa. The apparent thermal conductivity achieved with the various designs is as follows:

| | $k^*$ (W/m · k) |
|---|---|
| 1. Unevacuated home insulation | 0.01–0.05 |
| 2. Evacuated foams | 0.001–0.02 |
| 3. Evacuated powders | 0.0003–0.002 |
| 4. Multilayer superinsulation | 0.00002–0.0001 |

*Apparent conductivity if nonhomogeneous.

The superinsulations are layers of highly reflective foils or aluminized sheets separated by insulation material such as glass fibers, all under very high vacuum. Optimum layer density is about 25 layers per centimeter.

Whole textbooks have been written about cryogenics [9, 19], and there are many periodicals, such as *Cryogenics, Cryogenic News,* and *Cryogenic Information Report,* plus an annual review article compilation, *Advances in Cryogenic Engineering*. The reader is referred to these sources for further study.

## 1.10 The Heat Transfer Literature

The literature of heat transfer is large and diffuse: More than 300 technical periodicals regularly publish papers relating to heat transfer. This is primarily because heat transfer affects so many disciplines, from metallurgy to chemical processing to nuclear fusion. Workers in many fields must be aware of heat transfer research.

The two most widely circulated and read heat transfer journals are *Journal of Heat Transfer* and *International Journal of Heat and Mass Transfer*. Several other specialized journals are also published: *Applied Heat Transfer, Heat Transfer Engineering, Numerical Heat*

*Transfer, Letters in Heat and Mass Transfer, Heat Transfer: Soviet Research, Heat Transfer: Japanese Research,* and *Heat and Fluid Flow.*

For a quick search of the hundreds of other relevant journals, one can peruse the weekly title journal *Current Contents: Engineering, Technology, and Applied Sciences* for technical articles. There is also a quarterly abstract journal with a brief discussion of nearly every published heat transfer paper: *Previews of Heat and Mass Transfer.*

Many of these journals are cited in the chapter references.

### Example 1.10

What type of heat transfer research is being published today? What are the most important topics now under study? Give quantitative estimates from a top journal.

*Solution* First pick out a top heat transfer journal. The author selects his favorite, the *Journal of Heat Transfer.* Study of the *JHT* for the three years 1984–1986 reveals the following breakdown of topic coverage:

| Topic | Number of Papers | Percent Coverage |
|---|---|---|
| 1. Free or mixed convection | 82 | 23 |
| 2. Forced convection | 77 | 22 |
| 3. Two-phase flow and boiling | 44 | 12 |
| 4. Heat exchangers | 33 | 9 |
| 5. Condensation or evaporation | 22 | 6 |
| 6. Freezing or melting | 18 | 5 |
| 7. Conduction | 23 | 7 |
| 8. Radiation (including solar) | 16 | 5 |
| 9. Mass transfer | 12 | 3 |
| 10. Combustion, fires, flames | 10 | 3 |
| 11. Environmental heat transfer | 8 | 2 |
| 12. Miscellaneous special topics | 9 | 3 |
| *Totals* | 354 | 100 |

Thus various aspects of convection, an experimental science, occupy four-fifths of this journal's offerings. Radiation and conduction constitute only 12% of this representative sample. ∎

## Summary

Heat transfer is the engineering science of estimating the rate of heat flow through various systems as a function of various environmental conditions. Since our world is filled with a vast assortment of both natural and artificial temperature differences, heat transfer considerations enter nearly every aspect of engineering.

One can distinguish three modes of heat transfer: conduction owing to molecular interaction, thermal radiation by electromagnetic waves, and convection or fluid conduction enhanced by the fluid velocity. In all modes of heat transfer, the thermophysical properties of materials are extremely important and can vary by many orders of magnitude among metals, nonmetallic solids, liquids, and gases. Often all three modes are important in a single problem.

The fundamental discussions in this text stress the SI metric system of units, but the examples and assigned problems alternate between SI and the English system, which is still popular among U.S. heat transfer workers.

Since heat transfer is a universal subject, its literature is vast and diffuse. Section 1.10 gives some hints for searching this literature for specific research.

## References

1. W. C. Reynolds and H. C. Perkins, *Engineering Thermodynamics,* 2nd ed., McGraw-Hill, New York, 1977.
2. G. J. Van Wylen and R. E. Sonntag, *Fundamentals of Classical Thermodynamics,* 3rd ed., John Wiley, New York, 1985.
3. K. Wark, *Thermodynamics,* 4th ed., McGraw-Hill, New York, 1983.
4. F. M. White, *Fluid Mechanics,* 2nd ed., McGraw-Hill, New York, 1986.
5. M. Planck, *The Theory of Heat Radiation,* Dover, New York, 1959.
6. J. B. Fourier, *Théorie Analytique de la Chaleur,* Paris, 1822 (trans.: Dover, New York, 1955).
7. R. C. Reid, J. M. Pravsnitz, and T. K. Sherwood, *The Properties of Gases and Liquids,* 3rd ed., McGraw-Hill, New York, 1977.
8. P. D. Dunn and D. A. Reay, *Heat Pipes,* 3rd ed., Pergamon Press, New York, 1977.
9. W. Frost (Ed.), *Heat Transfer at Low Temperatures,* Plenum Press, New York, 1975.
10. Y. S. Touloukian et al., *Thermophysical Properties of Matter,* 13 vols. plus index, IFI/Plenum, New York, 1970–1977.

11. N. B. Vargaftik, *Tables on the Thermophysical Properties of Liquids and Gases.* 2nd ed., Halsted Press, New York, 1975.
12. W. J. Beek and K. M. J. Muttzall, *Transport Phenomena,* Wiley-Interscience, New York, 1975.
13. R. Siegel and J. R. Howell, *Thermal Radiation Heat Transfer,* 2nd ed., McGraw-Hill, New York, 1980.
14. S. V. Patankar, *Numerical Heat Transfer and Fluid Flow,* Hemisphere Publishing, New York, 1980.
15. B. D. Coleman and V. J. Mizel, "Thermodynamics and Departures from Fourier's Law of Heat Conduction," *Arch. Rat. Mech. Anal.,* vol. 13, 1963, p. 245.
16. M. H. Sadd and J. E. Didlake, "Non-Fourier Melting of a Semi-Infinite Solid," *J. Heat Transfer,* Feb. 1977, vol. 99, no. 1, pp. 25–28.
17. E. F. Adiutori, *The New Heat Transfer,* Ventuno Press, Cincinnati, 1974.
18. J. V. Sengers, *Recent Advances in Engineering Science,* A. C. Eringen (Ed.), vol. 3, Gordon and Breach Science Publishers, New York, 1969, p. 153.
19. R. F. Barron, *Cryogenic Systems,* McGraw-Hill, New York, 1966.
20. J. P. Holman, *Experimental Methods for Engineers,* 3rd ed., McGraw-Hill, New York, 1978.

## Review Questions

1. What is the difference between heat transfer and thermodynamics?
2. What basic laws govern heat transfer analyses?
3. Can control volume analyses be used in heat transfer?
4. Define the three modes of heat transfer.
5. Give six samples of industries in which heat transfer is very important.
6. What is Fourier's law?
7. Which property of a material determines (a) the amount of energy it can store per unit volume, (b) the heat it can conduct under steady-state conditions, and (c) the rate at which it will react to transient temperature conditions?
8. Distinguish between homogeneous and heterogeneous materials.
9. Is conduction exactly analogous to viscous shear? Explain.
10. What is the Stefan-Boltzmann law?
11. What is a gray body?
12. Distinguish between forced and free convection.

13. What is Newton's law of cooling?
14. Name three cases in which the convection heat transfer coefficient varies with temperature difference.
15. As stream velocity increases, do boundary layers get thicker or thinner? Explain.
16. Make a comparison between the SI and the English unit systems, especially with regard to heat transfer.
17. Describe the specialized field of cryogenics.
18. Name the two top scholarly heat transfer journals.

## Problems

### Problem distribution by sections

**The Problem Assignments Are Organized as Follows:**

| Problems | Sections Covered | Topics Covered |
|---|---|---|
| 1.1–1.7 | 1.1, 1.2 | Introductory remarks |
| 1.8–1.22 | 1.3 | Conduction |
| 1.23–1.26 | 1.4 | Radiation |
| 1.27–1.34 | 1.5 | Convection |
| 1.35–1.37 | 1.6 | Units and dimensions |
| 1.38–1.41 | 1.7 | Methodology: combined modes |
| 1.42–1.43 | 1.8 | Messy complications |
| 1.44–1.45 | 1.10 | The literature |

1.1 A veteran professor once remarked to me that the only difference between thermodynamics and heat transfer is "time." Can you explain this statement?

1.2 Write Eq. (1.1) in the general form that applies to a fixed control volume (CV) with control surface (CS). What assumptions are needed to reduce this to Eq. (1.2)?

1.3 Why is the negative sign needed in Eq. (1.4)?

1.4 Is the "continuity" or mass-conservation equation from fluid mechanics [4, p. 126 or p. 201] ever necessary in a heat transfer analysis? Explain.

1.5 Does the truck radiator of Fig. 1.1 actually "radiate" energy? Explain.

1.6 Is the viscosity of a fluid ever important in a heat transfer analysis? Explain.

1.7 Since they occur across temperature differences, heat transfer processes are, by definition, *not* in equilibrium. How, then, can we justify using thermodynamics, a science of equilibrium states, as part of a heat transfer analysis?

1.8 When a slab of wood has inner and outer surface temperatures of 25°C and −5°C, respectively, its heat transfer rate is 30 W/m$^2$. What is the thermal conductivity of the wood if its thickness is 100 mm?

1.9 A plain glass window is 3 ft by 5 ft by 0.25 in. thick and has inside and outside temperatures of 70°F and 25°F, respectively. What is its conduction heat loss in Btu/hr?

1.10 A furnace wall 15 cm thick has an area of 23 m$^2$ and is made of fireclay brick ($k = 1.04$ W/m · K). Inside and outside temperatures are 525°C and 20°C, respectively. What is the heat loss through the wall? If the heating value of fuel oil is 19,000 Btu/lb$_m$, how many liters per day of fuel oil are needed to balance this loss?

1.11 A house has 3000 ft$^2$ of exposed area and is insulated with 3 in. of rock wool throughout. If the temperature difference is 68°F, what is the heat loss through the insulation? If the furnace delivers heat at 15,000 Btu/lb$_m$ of fuel oil, how many gallons of oil are needed per day?

1.12 It seems odd that the slab temperature distribution in Fig. 1.2 is independent of the type of material. Another example is the Couette shear flow between a fixed and a moving plate of Fig. P1.12. For laminar flow, the velocity distribution is linear, $u = Vy/H$, independent of the fluid viscosity. Are these two cases related? What explains the lack of dependence on material properties?

1.13 A foam plastic cooler box is 46 cm long, 38 cm high, and 31 cm wide, with a thickness of 1.5 cm. When the inner and outer surface temperatures are 32°C and −6°C, respectively, the heat loss is 70 W. Estimate the thermal conductivity of the foam.

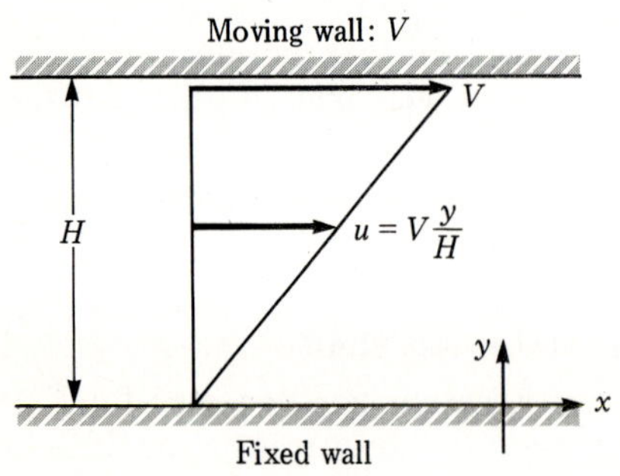


**Figure P1.12**

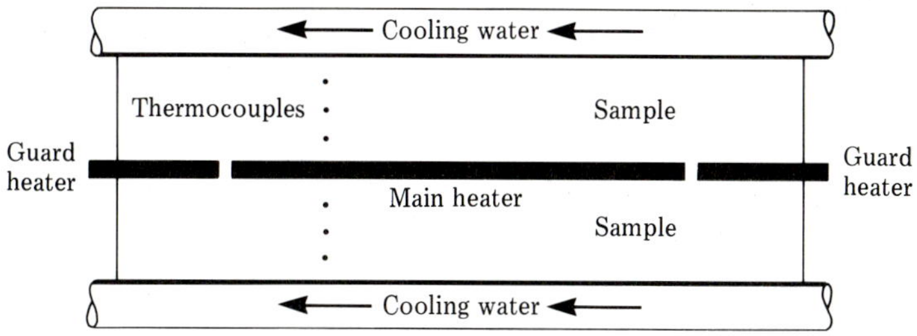


**Figure P1.14**

1.14 Figure P1.14 illustrates a *guarded hot-plate* device to measure thermal conductivity. The hot plate is sandwiched between two symmetric samples and two outer water-cooled plates. The guard heaters are adjusted to eliminate side losses. Suppose the hot plate temperature is 225°C while the water-cooled plates are at 20°C when the heater delivers 900 W/m$^2$. What is the measured thermal conductivity if each sample is 1.5 cm thick?

1.15 A sphere of diameter 3 m has 5-cm-thick walls of superinsulation ($k$ = $1.2 \times 10^{-4}$ W/m · K) and contains liquid oxygen at 90K. If the outside temperature is 20°C, how much liquid oxygen evaporates in g/day if the heat of vaporization is 213 J/g?

1.16 A stainless steel sphere of 0.5 in. thickness and 3 ft outer diameter loses heat at 500,000 Btu/hr. If its outside temperature is 68°F, estimate the inside temperature.

1.17 A solid slab 3 ft by 5 ft by 7 ft is to be heated from 60°F to 140°F. How much heat in Btu is required if the material is (a) aluminum, (b) cast iron, (c) type 304 stainless steel? For given boundary conditions, which material will heat up (a) the fastest and (b) the slowest?

1.18 Figure P1.18 illustrates a double-rod device to measure thermal conductivity of solids, as described by Holman [20]. If the thermocouple spacings are $\Delta z_A$ and $\Delta z_B$ and the measured temperature differences are $\Delta T_A$ and $\Delta T_B$, derive a formula for computing $k_B$ in terms of $k_A$.

1.19 The thermal diffusivity $\alpha$ combines with time $t$ and body size $L$ to make an important dimensionless transient conduction parameter. What is its form? What does it imply about the effect of size on the rate of change of body temperature if other similarity parameters are constant?

1.20 According to elementary kinetic theory of gases, the thermal conductivity of a gas increases as the square root of the absolute temperature. How does this prediction compare with the gas data in Fig. 1.3 and Appendix G? What is a more accurate power law?

1.21 A warehouse has 7500 ft$^2$ of exposed surface insulated with loose rock wool ($k$ = 0.039 Btu/hr · ft · °F). The maximum expected temperature

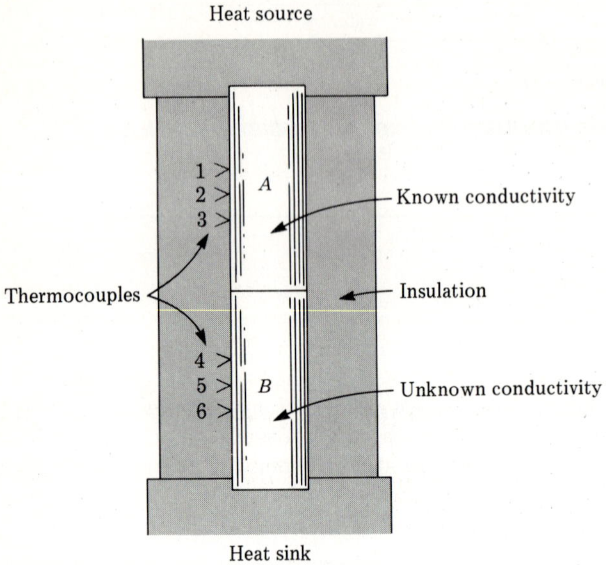


**Figure P1.18**

difference is 80°F. If the furnace can deliver 60,000 Btu/hr, what should the insulation thickness be?

1.22 For the double-rod device of Fig. P1.18, the thermocouple spacings are $z_A = 1$ cm and $z_B = 2$ cm. The measured temperatures are as follows:

| Thermocouple: | 1 | 2 | 3 | 4 | 5 | 6 |
|---|---|---|---|---|---|---|
| Temperature (°C): | 419 | 400 | 381 | 269 | 252 | 235 |

If rod A is 347 stainless steel, what is the thermal conductivity of alloy B?

1.23 A heater wire from a toaster is 1 mm in diameter and 30 cm long. The electric power delivered is 15 W. If the wire is a blackbody radiating to a cold environment and conduction and convection are neglected, what is the estimated wire temperature?

1.24 A thin copper sheet is in the evacuated space between two walls, as in Fig. P1.24. If all surfaces are black and heat flow is steady, what is the temperature of the sheet?

1.25 A gray-body cylinder, $\varepsilon = 0.7$, has a diameter of 3 cm and a length of 5 cm and is at 20°C inside a blackbody room whose walls are at 100°C. What is the net radiant heat transfer from the room to the cylinder?

1.26 If a stainless steel beaker of boiling water is placed on the kitchen table

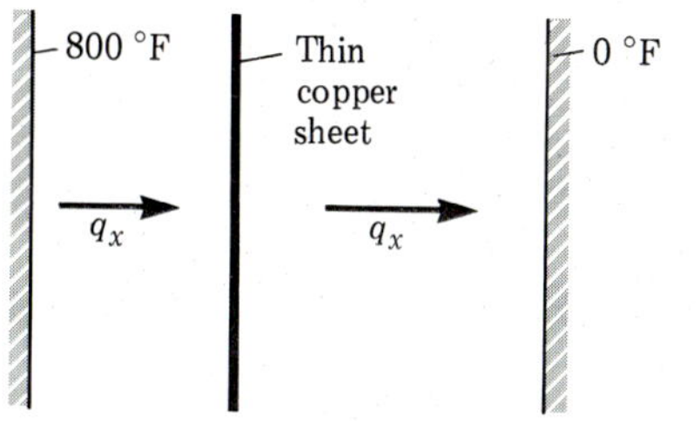


Figure P1.24

and wrapped with a thin layer of kraft-faced insulation, it may actually cool faster than if not wrapped. How can you explain this unexpected behavior?

1.27 An important dimensionless convection parameter can be formed from the heat transfer coefficient $h$, the body size $L$, and the fluid conductivity $k$. Find an expression for this parameter.

1.28 When exposed to a moving stream that is 15°C warmer, a 1-m-diameter sphere receives heat at the rate of 600 W. Find the heat transfer coefficient for this flow.

1.29 Consider the slab in Fig. P1.29, surrounded by convection conditions on each side. Show that the steady heat transfer through the slab is given by

$$q_x'' = \frac{T_1 - T_2}{1/h_1 + L_A/k_A + 1/h_2}.$$

1.30 A 5-cm-diameter cylinder that is 50 cm long is at 55°C and is to be cooled by crossflow immersion in a 10°C stream. From Fig. 1.8, if the desired initial cooling rate is 300 W, what should be the stream velocity if the fluid is (a) helium, (b) engine oil?

1.31 An immersion heater coil delivers 25 W and is 4 mm in diameter and 20 cm long. What is its surface temperature if immersed in (a) water at 20°C, $h = 80\ \mathrm{W/m^2 \cdot K}$; (b) air at 20°C, $h = 10\ \mathrm{W/m^2 \cdot K}$?

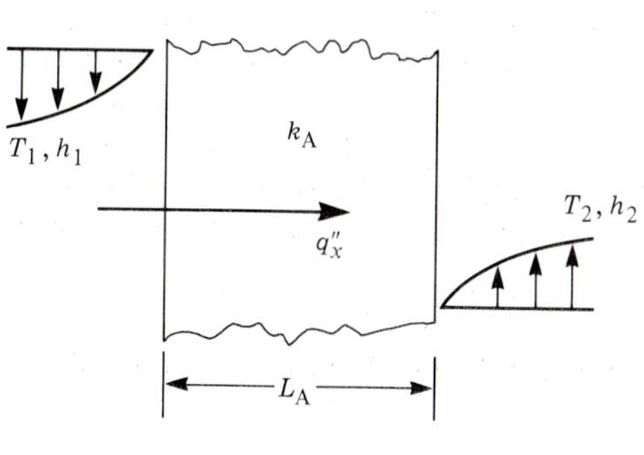


Figure P1.29

1.32 A steam pipe has surface temperature of 300°F and emissivity of 0.55. If immersed in room air at 68°F with $h = 6$ Btu/hr · ft$^2$ · °F, what percentage of its heat loss will be due to (a) convection, (b) radiation?

1.33 A thin plate has one side insulated and the other side exposed to a 20°C air flow such that $h = 35$ W/m$^2$ · K. The plate is electrically heated at a rate of 12,000 W/m$^2$. If radiation is neglected, find the equilibrium temperature of the plate.

1.34 Figure P1.34 illustrates a heated and guarded cylinder section used to measure convection heat transfer coefficients. (See Holman [20].) If the section has emissivity $\varepsilon$ and heater power $q$, derive an expression for heat transfer coefficient $h$ in terms of the parameters shown in the figure.

1.35 The two most important parameters in convection are the Reynolds number, $VL/\nu$, and the Prandtl number, $\nu/\alpha$. Compute these parameters for a flow with $\rho = 50$ lb$_m$/ft$^3$, $V = 15$ mi/hr, $L = 16$ in., $\mu = 3 \times 10^{-5}$ slug/ft · s, $k = 0.28$ Btu/hr · ft · °F, and $c_p = 0.4$ Btu/lb$_m$ · °F.

1.36 For laminar flow in a long tube [4, p. 235], viscous dissipation effects cause the centerline fluid temperature to rise by an amount $\mu V^2/4k$. Verify the dimensional consistency of this expression and compute the temperature rise in °C for engine oil at $V = 12$ mi/hr.

1.37 From the formula given in Problem 1.29, we may define the conduction "resistance" of a slab as the ratio of its thickness to its conductivity. In American home construction, insulation is quoted in "R" values of resistance per unit area. For example, 3.5 in. of glass-fiber insulation are listed as R-11, while 9 in. of rock wool are R-30. Are these values consistent with English units and the definition given above?

1.38 The fireclay brick slab in Fig. P1.38 is exposed to both convection and

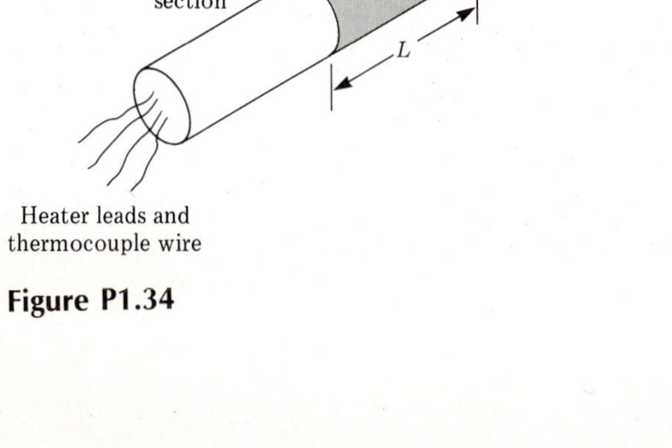


**Figure P1.34**

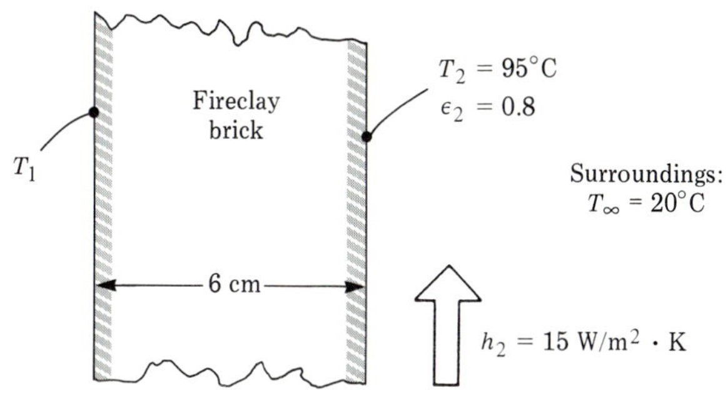


**Figure P1.38**

radiation on its right side. What is the total heat transfer rate in $W/m^2$? What is the left side temperature $T_1$ in °C?

1.39 Suppose that the heated cylinder test section in Fig. P1.34 has $D = 5$ cm and $L = 65$ cm and is exposed to an air flow of 7 m/s and $T_\infty =$ 20°C. If the surface emissivity is 0.7 and the measured surface temperature is 75°C, what is the estimated heat transfer coefficient for a heater power of 280 W? Suppose that the experimental parameters have the following uncertainties: emissivity = 15%, heat power = 5%, wall and air temperatures = 3%, air velocity = 2%. What is the overall uncertainty in the heat transfer coefficient? Compare your results with Fig. 1.8.

1.40 The wall of a house consists of 2 cm of plaster, 8 cm of rock wool, and 3 cm of oak wood. Inside and outside temperatures are 20°C and 0°C, respectively, while inside and outside heat transfer coefficients are 8 $W/m^2 \cdot K$ and 15 $W/m^2 \cdot K$, respectively. Compute the heat loss in $W/m^2$ and the temperature at the inside face of the insulation.

1.41 A heat exchanger consists of a one-quarter inch aluminum sheet separating a gas flow, $T_1 = 300°F$ and $h_1 = 25\ Btu/hr \cdot ft^2 \cdot °F$, from a liquid flow, $T_2 = 80°F$ and $h_2 = 450\ Btu/hr \cdot ft^2 \cdot °F$. What is the steady heat flux per unit area? What percent of total temperature drop does the gas boundary layer offer?

1.42 The non-Fourier analysis of [16] uses the following modification of Fourier's law, Eq. (1.5):

$$q'' = -k\frac{\partial T}{\partial x} - \Lambda\frac{\partial q''}{\partial t}.$$

Discuss the dimensions and physical significance of the added non-Fourier material property $\Lambda$.

1.43 Suppose it were necessary to make a conduction analysis of carbon dioxide near its critical point. Can you suggest a curve-fitting idea for approximating the data of Fig. 1.9 by a suitable algebraic expression?

1.44 Which of the top two journals, the *JHT* or the *IJHMT*, regularly publishes bibliographies of heat transfer papers from specific countries such as Japan, the Soviet Union, or Italy?

1.45 Explain why iteration scheme (a) in Example 1.5 converges while scheme (b) diverges. Can you cite an applied mathematics text which discusses this disparate behavior? Can you show graphically why it occurs?

# Heat Conduction: One-Dimensional Applications

# 2

## 2.1 The Heat Conduction Equation

In this text, the three modes of heat transfer are covered in the following order: conduction (Chapters 2–4), convection (Chapters 5–7), and radiation (Chapter 8). The present chapter derives and discusses the general heat conduction equation. This relation is then applied to a variety of simple but important cases of steady, one-dimensional heat conduction, including the possibility of internal heat sources.

In heat conduction, work effects are negligible, so that the first law of thermodynamics, Eq. (1.1), reduces to

$$\frac{dE}{dt} = \frac{dQ}{dt}. \tag{2.1}$$

Meanwhile, in heat convection, the work done by pressure and viscous forces may also be important, especially at high speeds (Section 6.6).

### 2.1.1 Fourier's Law

In Chapter 1, we considered some simple heat conduction problems involving the one-dimensional form of Fourier's law of conduction. Here we assume a general three-dimensional distribution $T(x, y, z, t)$. We treat only *isotropic* materials possessing a single thermal conductivity, $k$, independent of direction and varying only with temperature, pressure, and type of material. A temperature gradient in any direction $i$ causes a transfer $q_i''$ in that direction:

$$q_x'' = -k\frac{\partial T}{\partial x}; \qquad q_y'' = -k\frac{\partial T}{\partial y}; \qquad q_z'' = -k\frac{\partial T}{\partial z}. \tag{2.2}$$

For most common materials (as in Fig. 1.3), $k \doteq k(T)$ only, with little pressure dependence. Note that Eqs. (2.2) hold even if $T$ is time-dependent: The instantaneous temperature gradient determines the instantaneous heat transfer.

### 2.1.2 The Heat Conduction Equation

The basic differential equation of heat conduction is a combination of the first law of thermodynamics and Fourier's law, applied to a small element such as the one in Fig. 2.1. For any such element, the first law, in the absence of work terms, states that

$$\begin{matrix}\text{Net heat}\\ \text{conducted in}\end{matrix} + \begin{matrix}\text{Heat}\\ \text{generated within}\end{matrix} = \begin{matrix}\text{Element}\\ \text{energy increase.}\end{matrix}$$

Examining the six flux terms in Fig. 2.1, we see that the net heat conducted in is

$$dq_{\text{in}} = dq_x - dq_{x+\Delta x} + dq_y - dq_{y+\Delta y} + dq_z - dq_{z+\Delta z}. \tag{2.3}$$

Now, if we make the usual *continuum hypothesis* that all properties vary smoothly enough to use calculus, we may expand the outgoing flux terms in a first-order Taylor series:

$$dq_{x+\Delta x} = dq_x + \frac{\partial q_x}{\partial x}\Delta x,$$

and similarly for the $y$ and $z$ variations. Then Eq. (2.3) becomes

$$dq_{\text{in}} = -\frac{\partial q_x}{\partial x}\Delta x - \frac{\partial q_y}{\partial y}\Delta y - \frac{\partial q_z}{\partial z}\Delta z.$$

Finally, we introduce Fourier's law from Eqs. (2.2):

$$dq_{\text{in}} = \left[\frac{\partial}{\partial x}\left(k\frac{\partial T}{\partial x}\right) + \frac{\partial}{\partial y}\left(k\frac{\partial T}{\partial y}\right) + \frac{\partial}{\partial z}\left(k\frac{\partial T}{\partial z}\right)\right]\Delta x\Delta y\Delta z. \tag{2.4}$$

Thus the net conduction in is proportional to the element volume.

**Figure 2.1** Heat flux components on the six faces of a small Cartesian solid element.

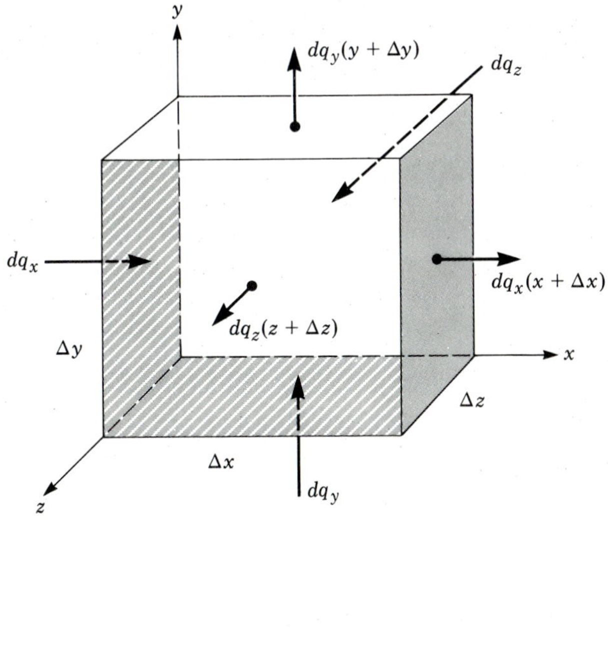

This is true also of the heat generated and the element energy increase:

$$\text{Heat generated} = \dot{q}\Delta x\Delta y\Delta z,$$
$$\text{Energy increase} = (\rho\Delta x\Delta y\Delta z)\, c_v \frac{\partial T}{\partial t}. \tag{2.5}$$

Substituting Eqs. (2.4) and (2.5) into the first law, Eq. (2.1), and canceling the element volume, we obtain

$$\frac{\partial}{\partial x}\left(k\frac{\partial T}{\partial x}\right) + \frac{\partial}{\partial y}\left(k\frac{\partial T}{\partial y}\right) + \frac{\partial}{\partial z}\left(k\frac{\partial T}{\partial z}\right) + \dot{q} = \rho c_v \frac{\partial T}{\partial t}. \tag{2.6}$$

Equation (2.6) is the general *heat conduction equation* for an isotropic solid continuum. It is to be solved for the temperature field $T(x, y, z, t)$ for various boundary conditions to be discussed. The physical properties $\rho$, $c_v$, and $k$ may all be functions of temperature. The heat generation $\dot{q}$ is assumed to be a known function of space and time and, possibly, temperature.

For constant $k$, $\rho$, and $c_v$, Eq. (2.6) takes the following simpler form:

$$\alpha\left(\frac{\partial^2 T}{\partial x^2} + \frac{\partial^2 T}{\partial y^2} + \frac{\partial^2 T}{\partial z^2}\right) + \frac{\dot{q}}{\rho c_v} = \frac{\partial T}{\partial t}, \tag{2.7}$$

where $\alpha = k/\rho c_v$ is the material thermal diffusivity (Fig. 1.5). This equation is very well behaved mathematically, and both analytical and numerical solutions techniques will be developed in Chapter 3.

Note, finally, that Eqs. (2.6) and (2.7), though complex in appearance, are merely statements of conservation of energy for a stationary material satisfying Fourier's law.

### 2.1.3 Cylindrical and Spherical Coordinates

Equation (2.6) is the Cartesian form of the heat conduction equation. Many conduction problems are better suited to a cylindrical or spherical geometry. For that reason, we give here two alternative forms of the conduction equation.

Cylindrical coordinates are defined in Fig. 2.2(a) by the transformations

$$x = r\cos\theta; \qquad y = r\sin\theta; \qquad z = z.$$

When these changes are made in Eq. (2.6), we obtain, after much algebra, the cylindrical heat conduction equation:

$$\frac{1}{r}\frac{\partial}{\partial r}\left(rk\frac{\partial T}{\partial r}\right) + \frac{1}{r^2}\frac{\partial}{\partial \theta}\left(k\frac{\partial T}{\partial \theta}\right) + \frac{\partial}{\partial z}\left(k\frac{\partial T}{\partial z}\right) + \dot{q} = \rho c_v \frac{\partial T}{\partial t}. \tag{2.8}$$

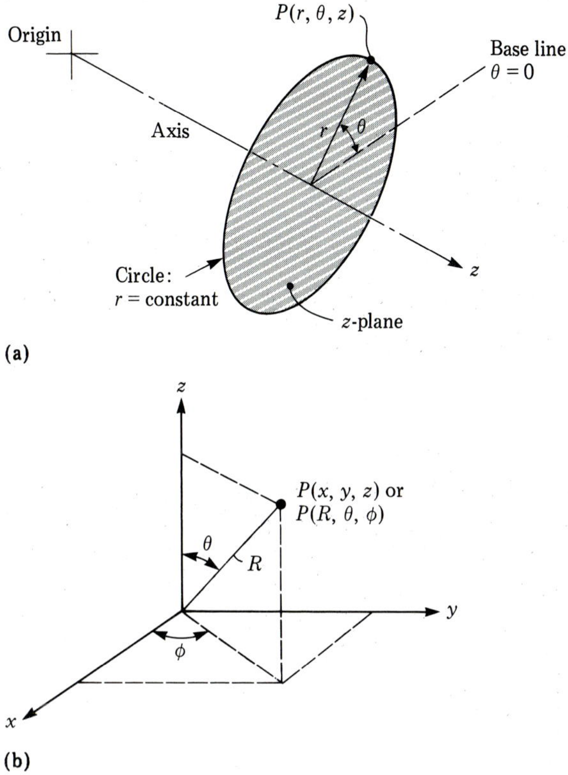


**Figure 2.2** **Definition sketch for (a) cylindrical coordinates and (b) spherical coordinates.**

Spherical coordinates are defined by a single radius $R$ and two angles $\theta$ and $\phi$, according to the transformation

$$x = R \sin \theta \cos \phi; \qquad y = R \sin \theta \sin \phi; \qquad z = R \cos \theta.$$

These coordinates are shown in Fig. 2.2(b). Equation (2.6) transforms, after a great deal of algebra, into the spherical heat conduction equation:

$$\frac{1}{R^2}\frac{\partial}{\partial R}\left(kR^2\frac{\partial T}{\partial R}\right) + \frac{1}{R^2 \sin \theta}\frac{\partial}{\partial \theta}\left(k \sin \theta \frac{\partial T}{\partial \theta}\right) + \frac{1}{R^2 \sin^2 \theta}\frac{\partial^2}{\partial \phi^2}(kT) + \dot{q} = \rho c_v \frac{\partial T}{\partial t}. \tag{2.9}$$

We shall encounter, in Chapters 2–4, several applications of these alternative orthogonal coordinate systems.

### 2.1.4 Boundary and Initial Conditions

The heat conduction equation, Eq. (2.6), being second order in space and first order in time, requires carefully specified boundary and initial conditions for any given solution. We briefly state them here and give more details in the applications of Chapters 3 and 4.

If the problem is time-dependent, $T = T(x, y, z, t)$, the entire temperature field must be specified as an *initial condition* at some particular time:

*At* $t = t_o$:

$$T = T_o(x, y, z). \tag{2.10}$$

The most common case is a uniform starting temperature, $T = T_o$ everywhere at $t = 0$.

Then, whether the problem is time-dependent or not, *boundary conditions* must be specified at every boundary point at all times. The four most common conditions are as follows, where $n$ denotes the outward-normal coordinate from the boundary and subscript $b$ denotes a boundary point.

1. Prescribed boundary temperature:

$$T_b = T_o = \text{known}. \tag{2.11a}$$

2. Insulated wall:

$$\left.\frac{\partial T}{\partial n}\right|_b = 0. \tag{2.11b}$$

3. Prescribed boundary heat flux:

$$-k_b \left.\frac{\partial T}{\partial n}\right|_b = q''_o = \text{known}. \tag{2.11c}$$

4. Convection to known surroundings $(T_\infty, h_\infty)$:

$$-k_b \left.\frac{\partial T}{\partial n}\right|_b = h_\infty(T_b - T_\infty). \tag{2.11d}$$

These four cases are illustrated in Fig. 2.3. In the latter three conditions, the boundary temperature $T_b$ is not known until after the solution is found.

**Example 2.1**

Write out the proper differential equation and boundary conditions for steady heat conduction in the two-dimensional body of Fig. E2.1. The conductivity is constant.

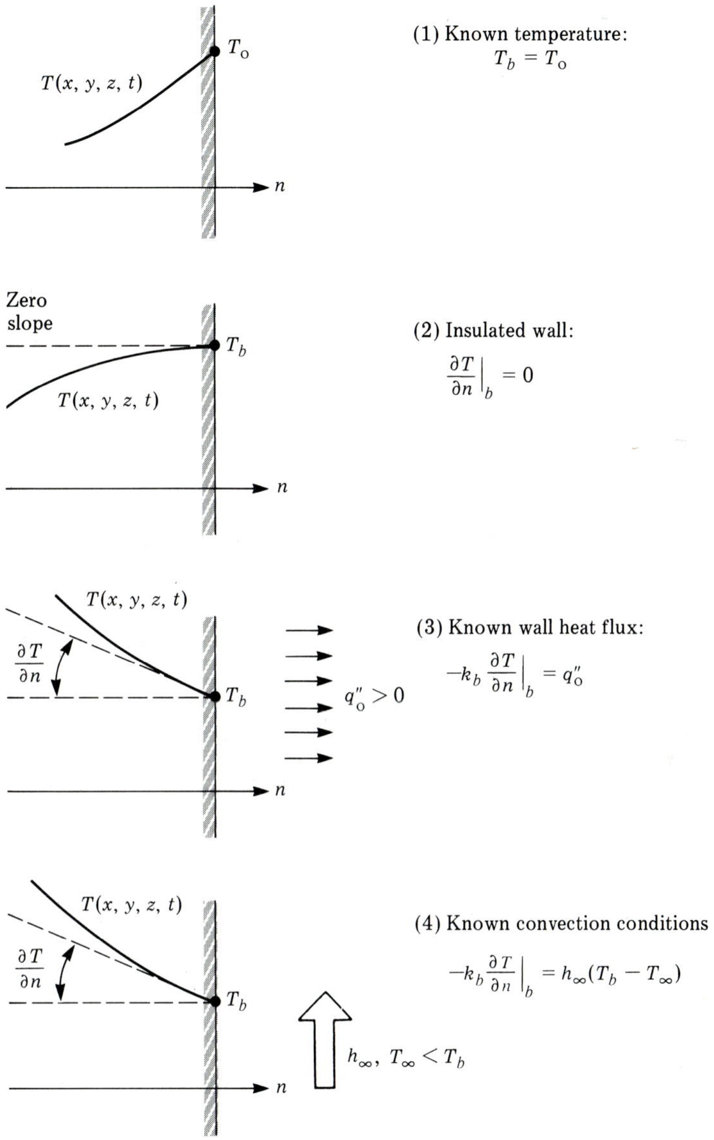


**Figure 2.3 Illustration of four different boundary conditions for the heat conduction equation.**

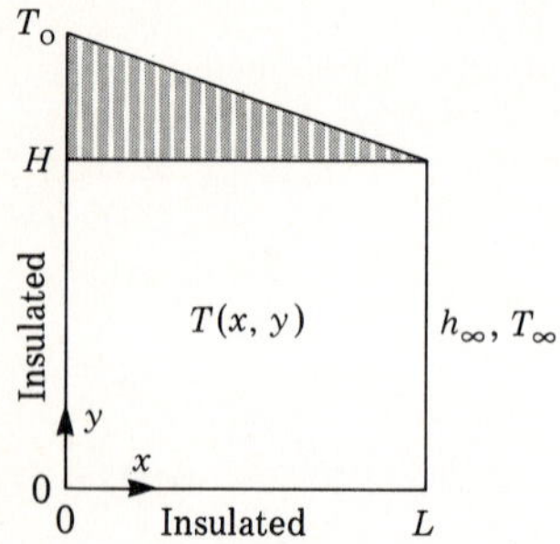


Figure E2.1

*Solution* By "steady" we mean $\partial T/\partial t \equiv 0$, and "two-dimensional" means $\partial T/\partial z \equiv 0$. Thus we assume $T = T(x, y)$ and constant $k$, with no internal heat generation. Equation (2.7) reduces to

$$\frac{\partial^2 T}{\partial x^2} + \frac{\partial^2 T}{\partial y^2} = 0.$$

This is Laplace's equation in the $x$-$y$ plane. The boundary conditions are of three types:

a) Known linear temperature distribution across the top:

$$T(x, H) = T_o(1 - x/L).$$

b) Insulated on the left and bottom sides:

$$\left.\frac{\partial T}{\partial x}\right|_{x=0} = 0, \qquad \left.\frac{\partial T}{\partial y}\right|_{y=0} = 0.$$

c) Convection to a known environment on the right-hand side ($n = x$):

$$\left(-k\frac{\partial T}{\partial x}\right)_{x=L} = h_\infty[T(y, L) - T_\infty].$$

Both the differential equation and the boundary conditions are linear. This problem is well posed and may be solved either analytically or numerically by the methods of Chapter 3. ∎

## 2.2 Steady One-Dimensional Conduction

There are many heat transfer applications that involve steady conduction through walls that are plane, cylindrical, or spherical, with known surface temperatures. These three geometries are shown in Table 2.1 and are special cases of the general heat conduction equation,

**Table 2.1** One-dimensional conductive resistances and temperatures

| | Thermal Resistance | Temperature Distribution* |
|---|---|---|
| 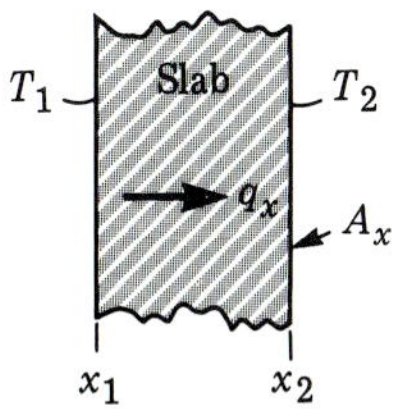 | $R_k = \dfrac{x_2 - x_1}{k_m A_x}$ | $\dfrac{T - T_1}{T_2 - T_1} = \dfrac{x - x_1}{x_2 - x_1}$ |
| 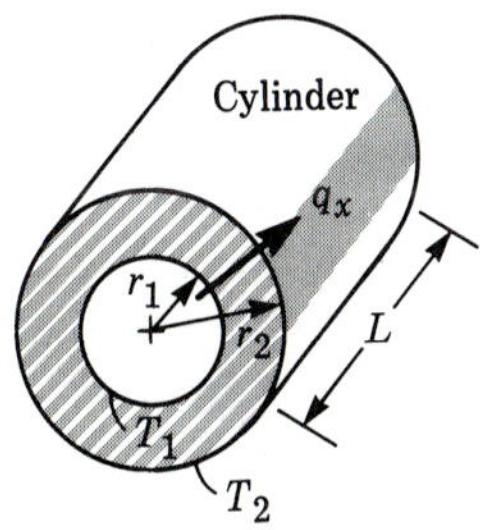 | $R_k = \dfrac{\ln(r_2/r_1)}{2\pi L k_m}$ | $\dfrac{T - T_1}{T_2 - T_1} = \dfrac{\ln(r/r_1)}{\ln(r_2/r_1)}$ |
| 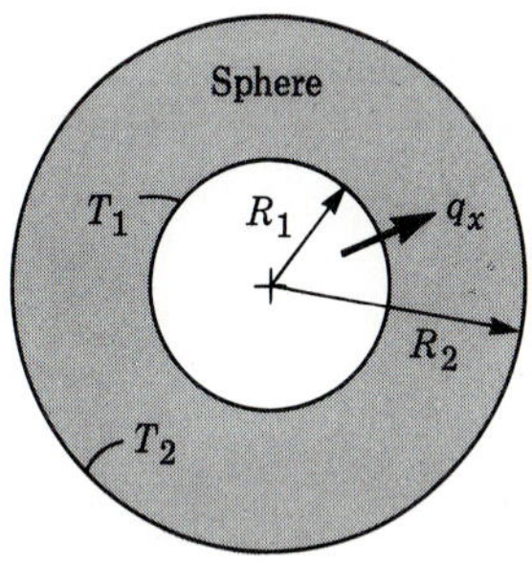 | $R_k = \dfrac{R_2 - R_1}{4\pi R_1 R_2 k_m}$ | $\dfrac{T - T_1}{T_2 - T_1} = \left(\dfrac{R - R_1}{R_2 - R_1}\right)\dfrac{R_2}{R}$ |

* for constant $k$.

Eq. (2.6). We assume no heat generation and variable conductivity, $k(T)$. Let us solve each case in turn.

### 2.2.1 The Plane Slab

As in Table 2.1, let $x$ be normal to the slab and let $A_x$ be the area through which heat transfer $q_x$ flows. Equation (2.6) reduces to

$$\frac{d}{dx}\left(k\frac{dT}{dx}\right) = 0,$$

or

$$k(T)\frac{dT}{dx} = \text{constant} = -\frac{q_x}{A_x} = -q_x''.$$

Integrate once again, using the conditions $T_1(x = x_1)$ and $T_2(x = x_2)$:

$$q_x''(x_2 - x_1) = -\int_{T_1}^{T_2} k(T)\, dT. \tag{2.12}$$

Define the *average* conductivity of the slab as

$$k_m = \frac{1}{T_2 - T_1}\int_{T_1}^{T_2} k(T)\, dT. \tag{2.13}$$

Then Eq. (2.12) may be rewritten in the convenient form

*Slab:*

$$q_x = k_m A_x \frac{T_1 - T_2}{x_2 - x_1}. \tag{2.14}$$

This is typical of plane conduction: The heat transfer is proportional to the average conductivity, the wall area, and the temperature difference and is inversely proportional to the wall thickness. Compare Eq. (2.14) with Eq. (1.4).

The temperature distribution $T(x)$ depends upon the variation $k(T)$. If $k$ is constant, $T(x)$ is linear, and Eq. (2.12) reduces to

*Slab:*

$$\frac{T - T_1}{T_2 - T_1} = \frac{x - x_1}{x_2 - x_1}. \tag{2.15}$$

This was shown earlier in Example 1.2. For variable $k(T)$, $T(x)$ is nonlinear, as shown in Fig. 2.4. If $k$ increases with $T$, the curve $T(x)$ is concave downward.

### 2.2.2 The Long Hollow Cylinder

If the cylinder in Table 2.1 is very long, $L \gg r_1, r_2$, with uniform inner and outer temperatures, then axial gradients near the ends can be neglected. The temperature equals $T(r)$ only, and Eq. (2.8) reduces to

$$\frac{d}{dr}\left(rk\frac{dT}{dr}\right) = 0,$$

or

$$rk(T)\frac{dT}{dr} = \text{constant} = -rq_x'' = -\frac{q_x}{2\pi L}.$$

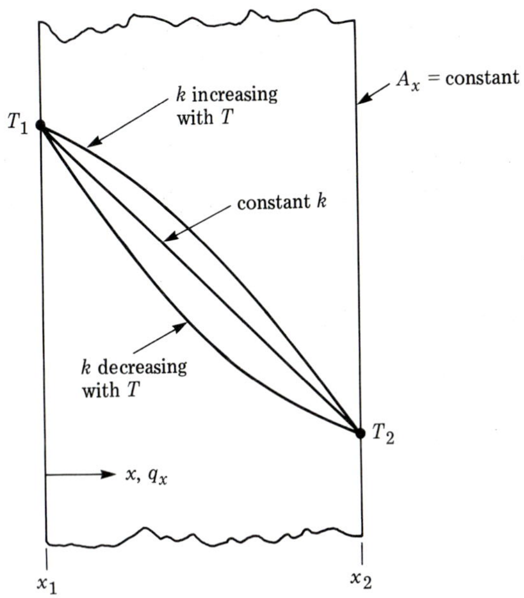


**Figure 2.4 Illustration of steady-flow temperature distributions in a plane slab with variable thermal conductivity $k(T)$.**

Note that, in this case, $q''_x = q_x/(2\pi rL)$ varies inversely with $r$. We may separate the variables and integrate again, using the concept of average conductivity from Eq. (2.13). The result is

*Cylinder:*

$$q_x = \frac{2\pi L k_m}{\ln(r_2/r_1)}(T_1 - T_2). \tag{2.16}$$

Note the sharp difference from the plane case, Eq. (2.14).

For constant $k$, the integration yields a logarithmic temperature distribution in the cylinder:

*Cylinder:*

$$\frac{T - T_1}{T_2 - T_1} = \frac{\ln(r/r_1)}{\ln(r_2/r_1)}. \tag{2.17}$$

### 2.2.3 The Hollow Sphere

For the hollow sphere in Table 2.1, the temperature is purely radial, $T = T(R)$. Equation (2.9) reduces to

$$\frac{d}{dR}\left(kR^2\frac{dT}{dR}\right) = 0,$$

or

$$R^2k(T)\frac{dT}{dR} = \text{constant} = -R^2q_x'' = -\frac{q_x}{4\pi}.$$

Note that, in this case, $q_x'' = q_x/4\pi R^2$ varies inversely with $R^2$. Separate the variables and integrate again, using the definition of $k_m$ from Eq. (2.13). The result is

*Sphere:*

$$q_x = \frac{4\pi R_1 R_2 k_m}{R_2 - R_1}(T_1 - T_2). \tag{2.18}$$

This is also quite different in form from the slab and cylinder cases.

For constant $k$, the integration yields the temperature distribution

*Sphere:*

$$\frac{T - T_1}{T_2 - T_1} = \frac{R - R_1}{R_2 - R_1}\frac{R_2}{R}. \tag{2.19}$$

These three cases are collected for convenience in Table 2.1. Typical temperature distributions for the cylinder and the sphere are shown in Fig. 2.5. For thick walls, if $T_1 > T_2$, these distributions are concave upward, as shown, because of the outward increase of heat transfer area $A_x$. As the wall thickness becomes small, the distributions become linear, like the slab of Fig. 2.4.

### 2.2.4 The Conduction Resistance

The results of Sections 2.2.1–2.2.3 for $q_x$ are algebraically similar to electric current relations: $I = \Delta E/R$, where $\Delta E$ is the voltage drop and $R$ the electric resistance. The conduction analogy is $q_x = \Delta T/R_k$, where $R_k$ is called the *conduction resistance* of the body. It is very useful, in the electric analogy approach of Section 2.4, to write one-dimensional conduction results in the form

$$q_x = \frac{1}{R_k}(T_1 - T_2), \tag{2.20}$$

where the values of $R_k$ for a slab, cylinder, and sphere are listed in Table 2.1 and will be used in Section 2.4.

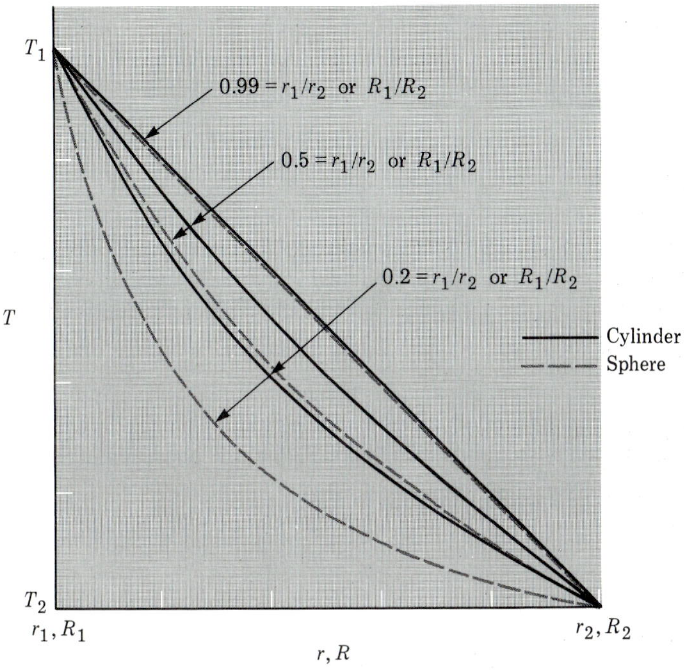


**Figure 2.5 Temperature distributions in a hollow cylinder or sphere for constant conductivity.**

## Example 2.2

An iron pipe, $D_1 = 3$ in. and $D_2 = 6$ in., contains steam such that its inner surface is at 340°F and its outer surface is at 220°F. Allowing for variable conductivity, compute $q_x$ per foot of pipe length.

*Solution* From Fig. 1.3 or Appendix C, the thermal conductivity of iron in Btu/(hr · ft · °F) is 40.7 at 220°F and 37.6 at 340°F. These differ by only 8%, so an arithmetic average should be quite adequate for evaluating the mean conductivity:

$$k_m \doteq \frac{(k_1 + k_2)}{2} = \frac{(37.6 + 40.7)}{2} = 39.15 \text{ Btu/(hr} \cdot \text{ft} \cdot \text{°F)}.$$

A more exact numerical or graphic integration of the $k(T)$ data using Eq. (2.13) would yield $k_m = 39.08$, or only 0.2% lower.

Then the heat flux for $L = 1$ ft is computed from the cylinder relation, Eq. (2.16):

$$q_x = \frac{2\pi L k_m}{\ln(r_2/r_1)}(T_1 - T_2) = \frac{2\pi(1\text{ ft})(39.15\text{ Btu/hr}\cdot\text{ft}\cdot{}^\circ\text{F})}{\ln(3.0\text{ in.}/1.5\text{ in.})}(340^\circ - 220^\circ)$$

$$= 42{,}600\text{ Btu/hr per foot of pipe length.} \quad [Ans.]$$

This is a typical application of variable conductivity to a conduction analysis. Unless the variation of $k$ across the slab is very large (50% or more), an arithmetic average of the endpoint values of $k$ is sufficient. One could also take $k_m$ to be the value of $k$ at the average temperature $T_m = (T_1 + T_2)/2$ with little error. In this case, $T_m = (340 + 220)/2$, or 280°F, at which point we estimate $k_m \doteq 39.1$ Btu/hr · ft · °F. ∎

## Example 2.3

An iron sphere has an outer diameter of 2 m, a wall thickness of 15 cm, and an outer temperature of 40°C. The heat transfer is $q_x$ = 250 kW outward. Accounting for variable $k(T)$, estimate the inner surface temperature.

*Solution* From the given data, $R_1 = 0.85$ m and $R_2 = 1.0$ m. We estimate $k(T)$ from Fig. 1.3 or Appendix C, using SI units. Substitute all available data into Eq. (2.18):

$$q_x = 250{,}000\text{ W} = \frac{4\pi R_1 R_2 k_m}{(R_2 - R_1)}(T_1 - T_2) = \frac{4\pi(0.85)(1.0)k_m}{(1.0 - 0.85)}(T_1 - 40^\circ\text{C}).$$

Solve for

$$T_1 = (40 + 3511/k_m)^\circ\text{C}. \tag{1}$$

Since $k_m$ depends on $T_1$, a bit of iteration is needed. Make an initial guess $T_1 \doteq 80$°C, from which $T_m \doteq (80 + 40)/2$, or 60°C. At 60°C, estimate $k_m \doteq 74.4$ W/(m · K). Equation (1) then allows an improved estimate for $T_1$:

$$T_1(\text{new}) = (40 + 3511/74.4) \doteq 87.2^\circ\text{C}.$$

Repeat once: $T_m(\text{new}) = (87.2 + 40)/2 = 63.6$°C, and $k_m(\text{new}) \doteq 74.1$ W/m · K, which is hardly changed from the first iteration. This should be sufficient convergence:

$$T_1(\text{final}) = (40 + 3511/74.1) = 87.4^\circ\text{C}. \quad [Ans.]$$

This is the last variable-conductivity problem that we will illustrate. From now on we will simply take $k_m$ to be an average value based on the expected temperature range in the given problem. ∎

## 2.3 Generalized One-Dimensional Conduction

The solutions given in Section 2.2 for slabs, cylinders, and spheres were found by reducing the heat conduction to its appropriate one-dimensional form. A similar approach holds for conduction through arbitrary one-dimensional bodies.

Figure 2.6 shows an idealized one-dimensional "slab" with arbitrary area distribution $A_x(x)$, insulated sides, and uniform end temperatures. For steady conduction, the heat transfer must be independent of position:

$$-\frac{dq_x}{dx} = \frac{d}{dx}\left(kA_x\frac{dT}{dx}\right) = 0,$$

or

$$k(T)A_x(x)\frac{dT}{dx} = -q_x = \text{constant}.$$

Separate the variables and integrate, using the definition of average conductivity $k_m$ from Eq. (2.13):

$$q_x = \frac{k_m(T_1 - T_2)}{\displaystyle\int_{x1}^{x2}\frac{dx}{A_x}}. \tag{2.21}$$

**Figure 2.6 Illustration of one-dimensional steady conduction through an insulated solid of variable cross section and conductivity.**

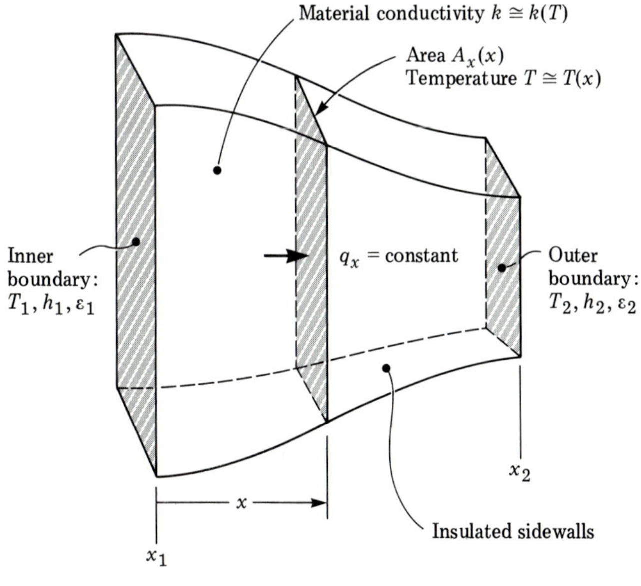

This is the general solution for heat transfer through a one-dimensional insulated body of area $A_x(x)$. For the particular case $A_x = 2\pi rL$, we obtain Eq. (2.16) for the cylinder. The case $A_x = 4\pi R^2$ yields the sphere solution, Eq. (2.18).

Equation (2.21) gives $q_x$ as a function of end conditions, which is usually all that is required. We may also compute the temperature distribution $T(x)$ by integrating $q_x$ only up to the particular point $x$:

$$\int_{T_1}^{T(x)} k(T)\, dT = -q_x \int_{x1}^{x} \frac{dx}{A_x}, \tag{2.22}$$

where $q_x$ is first evaluated from Eq. (2.21). The form of $T(x)$ depends on the area distribution and the particular property variation $k(T)$. Two examples are given for the slab in Fig. 2.4.

Finally, note from Eq. (2.21) that the conduction resistance of the body is given by

$$R_k = \frac{1}{k_m} \int_{x1}^{x2} \frac{dx}{A_x}.$$

Substitution of $A_x = 2\pi rL$ or $4\pi R^2$ leads to the cylinder and sphere results in Table 2.1.

## Example 2.4

The conical aluminum oxide plug in Fig. E2.4 has insulated side walls and a 200°C temperature difference. Estimate the steady conduction heat transfer in watts.

*Solution* Assuming one-dimensional conduction, we compute the inverse-area integral for use in Eq. (2.21). The local diameter is given by

$$D = D_1 + (D_2 - D_1)(x/L),$$

and thus the integral is

$$\int_{x1}^{x2} \frac{dx}{A_x} = \int_0^L \frac{dx}{\frac{\pi}{4}[D_1 + (D_2 - D_1)(x/L)]^2} = \frac{4L}{\pi(D_2 - D_1)}(D_1^{-1} - D_2^{-1}).$$

For the data in Fig. E2.4, the numerical value is

$$\frac{4(0.07\text{ m})}{\pi(0.05 - 0.03\text{ m})}(0.03\text{ m}^{-1} - 0.05\text{ m}^{-1}) = 59.42\text{ m}^{-1}.$$

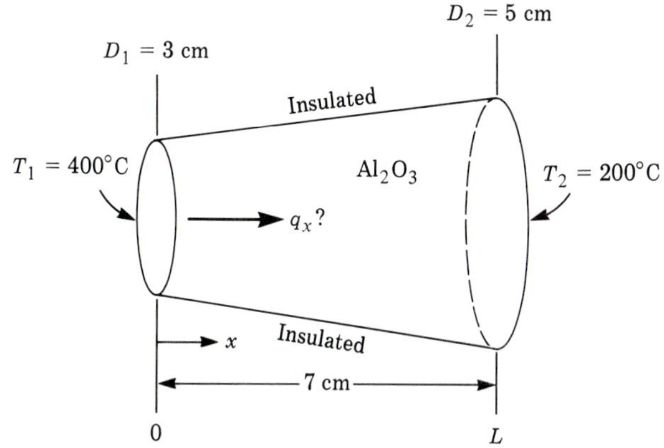


**Figure E2.4**

From Appendix D, $k$ for aluminum oxide is 22 W/(m · K) at 200°C and 13 W/(m · K) at 400°C. The average value is, approximately,

$$k_m \doteq \frac{22 + 13}{2} = 17.5 \text{ W/(m} \cdot \text{K)}.$$

Then, from Eq. (2.21), the heat transfer is

$$q_x = \frac{(17.5 \text{ W/m} \cdot \text{K})(400°\text{C} - 200°\text{C})}{59.42 \text{ m}^{-1}} = 59 \text{ W}. \quad [\textit{Ans.}]$$

Note that the units of "per °K" and "°C" cancel, since each degree increment is the same. That is, a change of 1K exactly equals a change of 1°C. ∎

## 2.4 Composite Walls: The Electric Analogy

Since, for one-dimensional steady heat flow, $q_x$ is the same for all cross sections along the flux path, the electric analogy mentioned in Section 2.2.4 is a very convenient scheme for multilayered slabs, cylinders, or spheres. Applications may involve either series or parallel flow or both.

All three of our basic one-dimensional heat transfer modes —

conduction, convection, and radiation — can be written as heat flux (current) equals a temperature difference (voltage drop) divided by a resistance. The resulting electric analogy can be applied to both series and parallel systems.

The conduction resistance $R_k = q_x/\Delta T$ was discussed in Eq. (2.20), and values were tabulated in Table 2.1.

Similarly, for convection of coefficient $h$ at the surface of a slab, Eq. (1.16) for the heat flux through the convection layer may be rewritten as

$$q_x = \frac{\Delta T}{R_c}, \qquad R_c = \frac{1}{hA_x} = \text{convection resistance.} \tag{2.23}$$

Here $\Delta T$ is the temperature change through the layer, $(T_\infty - T_w)$.

Finally, radiation, Eq. (1.14), can be written in a "temperature drop" form by using the algebraic identity

$$(T_1^4 - T_2^4) = C(T_1 - T_2), \qquad C = (T_1 + T_2)(T_1^2 + T_2^2).$$

Equation (1.14) can thus be rewritten as

$$q_x = \frac{\Delta T}{R_r}, \qquad R_r = \frac{1}{\sigma F_{12}(T_1 + T_2)(T_1^2 + T_2^2)A_x} = \text{radiation resistance,} \tag{2.24}$$

where $F_{12}$ is a "view factor" between bodies 1 and 2. Here $T_1$ would typically denote the radiating surface temperature and $T_2$ the environment.

### 2.4.1 Resistances in Series

For flow through series resistances, no matter what the geometry, the heat flux is given by

$$q_x = \frac{\Delta T}{\Sigma R}, \tag{2.25}$$

where $\Sigma R$ denotes the sum of all the resistances across the temperature difference $\Delta T$. Equation (2.25) applies either to the entire temperature change or to any part thereof.

Consider the three-layer cylinder in Fig. 2.7, with convection at both inner and outer surfaces. The inner and outer fluid temperatures are $T_i$ and $T_o$, respectively. Figure 2.7 also shows the thermal "circuit" or electric analogy. When applied to the total temperature drop, Eq. (2.25) becomes

$$q_x = \frac{T_i - T_o}{R_{ci} + R_{k1} + R_{k2} + R_{k3} + R_{co}}. \tag{2.26}$$

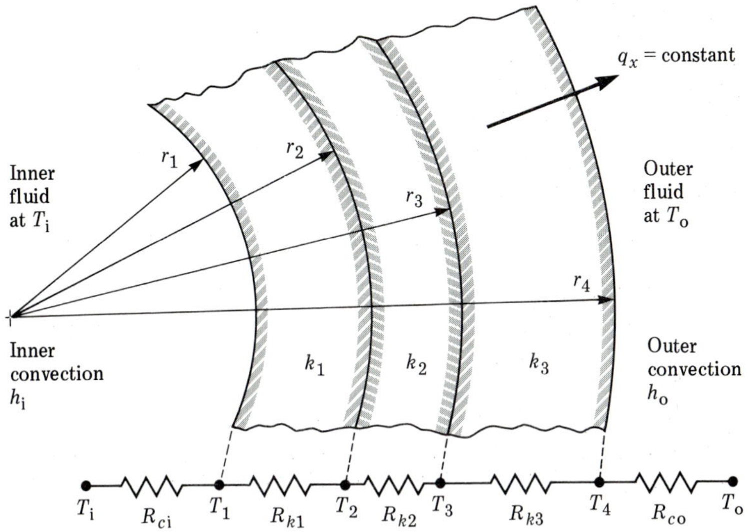


**Figure 2.7** A three-layer cylinder with convection at both boundaries. Solution by electric analogy.

The subscript $c$ means the convective resistance from Eq. (2.23).

$$R_{ci} = \frac{1}{h_i A_1}, \qquad A_1 = 2\pi r_1 L,$$

$$R_{co} = \frac{1}{h_o A_4}, \qquad A_4 = 2\pi r_4 L.$$

Note that, regardless of the geometry, $R_c = 1/hA$, where $A$ is the area of the surface at which the convection occurs.

The subscript $k$ denotes conductive resistance of a particular layer, from Table 2.1 for a cylinder:

$$R_{k1} = \frac{\ln(r_2/r_1)}{2\pi L k_1},$$

$$R_{k2} = \frac{\ln(r_3/r_2)}{2\pi L k_2},$$

$$R_{k3} = \frac{\ln(r_4/r_3)}{2\pi L k_3}.$$

If $k = k(T)$, use the mean values $k_{m1}$, $k_{m2}$, and $k_{m3}$ as computed from Eq. (2.13) for each layer.

### 2.4.2 Intermediate Temperatures

After $q_x$ is computed from Eq. (2.26), we can compute intermediate temperatures by applying Eq. (2.25) across a partial temperature change to the desired point. Suppose we wish to compute the interface temperature $T_3$ in Fig. 2.7. We would apply Eq. (2.25) from $T_i$ to $T_3$:

$$q_x = \frac{T_i - T_3}{R_{ci} + R_{k1} + R_{k2}}.$$

With $q_x$ already computed, we can solve this for $T_3$. Alternatively, we can take the shorter path from $T_o$ to $T_3$:

$$q_x = \frac{T_3 - T_o}{R_{k3} + R_{co}}.$$

These two relations will give identical results for $T_3$.

**Example 2.5**

A steam pipe has an inside diameter of 2.0 in. and consists of 0.22 in. of copper pipe covered by 1.0 in. of glass-fiber insulation and 0.008 in. of aluminum foil. The steam temperature is $T_i$ = 300°F, and the outside air is at $T_o$ = 80°F. Heat transfer coefficients are $h_i$ = 25 and $h_o$ = 12 Btu/hr · ft² · °F, respectively. If the pipe is 30 ft long, compute the heat loss in Btu/hr.

*Solution* This three-layer situation exactly fits Fig. 2.7, so Eq. (2.26) applies. For the given wall thicknesses, $r_1$ = 1.00 in., $r_2$ = 1.22 in., $r_3$ = 2.22 in., and $r_4$ = 2.228 in. From Appendices C and D, $k_1$ (copper) = 226, $k_2$ (glass fiber) = 0.022, and $k_3$ (aluminum) = 136 Btu/hr · ft · °F, respectively, assumed constant. Compute and sum the various resistances:

$$R_{ci} = \frac{1}{\left(25\dfrac{\text{Btu}}{\text{hr}\cdot\text{ft}^2\cdot{}^\circ\text{F}}\right)\left(2\pi\left(\dfrac{1.0}{12}\,\text{ft}\right)(30\text{ft})\right)} = 0.00255\,\frac{\text{hr}\cdot{}^\circ\text{F}}{\text{Btu}},$$

$$R_{k1} = \frac{\ln(1.22/1.0)}{2\pi(30\text{ft})(226\ \text{Btu/hr}\cdot\text{ft}\cdot{}^\circ\text{F})} = 0.000005\,\frac{\text{hr}\cdot{}^\circ\text{F}}{\text{Btu}},$$

$$R_{k2} = \frac{\ln(2.22/1.22)}{2\pi(30\ \text{ft})(0.022\ \text{Btu/hr}\cdot\text{ft}\cdot{}^\circ\text{F})} = 0.1444\,\frac{\text{hr}\cdot{}^\circ\text{F}}{\text{Btu}},$$

$$R_{k3} = \frac{\ln(2.228/2.22)}{2\pi(30\ \text{ft})(136\ \text{Btu/hr}\cdot\text{ft}\cdot{}^\circ\text{F})} = 0.0000001\,\frac{\text{hr}\cdot{}^\circ\text{F}}{\text{Btu}},$$

$$R_{co} = \frac{1}{\left(12\dfrac{\text{Btu}}{\text{hr}\cdot\text{ft}\cdot{}^\circ\text{F}}\right)\left(2\pi\dfrac{2.228}{12}\,\text{ft}\right)(30\ \text{ft})} = 0.00238\,\frac{\text{hr}\cdot{}^\circ\text{F}}{\text{Btu}}.$$

Thus the total resistance is $\Sigma R = 0.1493$ hr · °F/Btu, of which 97% is due to the insulation. The heat flux is given by

$$q_x = (T_i - T_o)/\Sigma R = \frac{300°\text{F} - 80°\text{F}}{0.1493\ \text{hr}\cdot°\text{F/Btu}} = 1470\,\frac{\text{Btu}}{\text{hr}} \quad [Ans.]$$ ■

*Additional Question for Extra Credit* Compute the inside insulation temperature, $T_2$, where it touches the copper pipe.

*Solution* Apply Eq. (2.25) from the steam out to $T_2$:

$$q_x = 1470 = \frac{T_i - T_2}{R_{ci} + R_{k1}} = \frac{300°\text{F} - T_2}{0.00255},$$

or

$$T_2 = 296°\text{F}. \quad [Ans.]$$

Alternatively, we could apply Eq. (2.25) from the air in to $T_2$:

$$q_x = 1470 = \frac{T_2 - T_o}{R_{k2} + R_{k3} + R_{co}} = \frac{T_2 - 80°\text{F}}{0.1467},$$

or

$$T_2 = 296°\text{F}. \quad [Ans.]$$ ■

---

Example 2.5 is a case in which a single resistance dominates: the insulation resistance in a system to be protected from heat loss. Another example is a gas/liquid heat exchanger, where the gas-side heat transfer coefficient gives the dominant resistance. In such cases, you should do your best to estimate the dominant resistance as closely as possible, since it controls heat output of the system. Meanwhile, the minor resistances need not be so accurate, since a large change in a non-dominating resistance only causes a small change in the results.

### 2.4.3 Other Geometries

Figure 2.7 might also represent a three-layer spherical system, in which case Eq. (2.25) would still apply with resistances appropriate to the spherical geometry. The convective resistance $R_{ci}$ would require inner surface area $A_1 = 4\pi r_1^2$, and $R_{co}$ would relate to outer area $A_4 = 4\pi r_4^2$. The conductive resistances would be obtained from Table 2.1

for a spherical layer:

$$R_{k1} = \frac{r_2 - r_1}{4\pi r_1 r_2 k_1},$$

$$R_{k2} = \frac{r_3 - r_2}{4\pi r_2 r_3 k_2},$$

$$R_{k3} = \frac{r_4 - r_3}{4\pi r_3 r_4 k_3}.$$

You should check these against the parameters labeled in Fig. 2.7 to verify the procedure.

Also, if one straightened out the curves, Fig. 2.7 could have been a three-layer slab. Equation (2.26) would still apply with constant $A_x$ and slab relations used from Table 2.1:

$$R_{ci} = 1/h_i A_x, \qquad R_{co} = 1/h_o A_x, \qquad R_{k1} = (x_2 - x_1)/A_x k_1,$$

$$R_{k2} = (x_3 - x_2)/A_x k_2, \qquad R_{k3} = (x_4 - x_3)/A_x k_3.$$

The point is that the generalized electric analogy $q_x = \Delta T/\Sigma R$ works for all geometries, as long as one uses the correct surface area for convection and the appropriate conductive resistances from Table 2.1 for the solid layers.

## Example 2.6

A furnace wall has an inner layer of silica brick, $k_1 = 1.07$ W/m · K, and an outer layer of masonry brick, $k_2 = 0.66$ W/m · K (see Fig. E2.6). Furnace gas conditions are $T_i = 320°C$, $h_i = 45$ W/m$^2$ · K, and outside air is $T_o = 25°C$, $h_o = 25$ W/m$^2$ · K. It is desired that the heat loss be 800 W/m$^2$ and that the interface temperature $T_2$ be 150°C. Compute $\Delta x_1$ and $\Delta x_2$ needed.

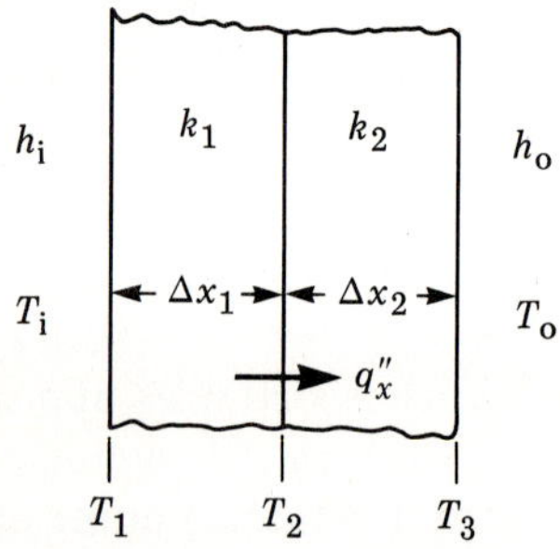


Figure E2.6

*Solution* With $T_2$ known, apply Eq. (2.25) in two stages. First, from the furnace gas to the interface,

$$q''_x = \frac{T_i - T_2}{1/h_i + \Delta x_1/k_1} = \frac{320 - 150}{1/45 + \Delta x_1/1.07} = 800 \text{ W},$$

or

$$\Delta x_1 = \frac{170 - 800/45}{800/1.07} = 0.204 \text{ m} = 20.4 \text{ cm.} \quad [Ans.]$$

Second, from the interface to the outside air,

$$q''_x = \frac{T_2 - T_o}{\Delta x_2/k_2 + 1/h_2} = \frac{150 - 25}{\Delta x_2/0.66 + 1/25} = 800 \text{ W},$$

or

$$\Delta x_2 = \frac{125 - 800/25}{800/0.66} = 0.077 \text{ m} = 7.7 \text{ cm.} \quad [Ans.]$$

Note that we never had to use the overall temperature drop or the total conductive resistance. ■

---

### 2.4.4 Contact Resistance

In making our series analyses of conduction through layered solids, we have assumed an idealized perfect thermal contact at the junction of two materials. This idealization is shown in Fig. 2.8(a). The real situation likely resembles Fig. 2.8(b), where there is an imperfect mating: Peaks and valleys of surface roughness do not fit perfectly, leaving small voids containing a gas, liquid, or vacuum. Such an imperfect bond has a nonzero *contact resistance,* which should be included in the series $\Sigma R$ for the heat flux computation.

Heat flow through an imperfect interface such as Fig. 2.8(b) is by conduction across contacting peaks and by conduction, convection, and radiation across the voids. Thus heat flux will result in a finite temperature change across the imperfect interface:

$$\Delta T_I = R_I q_x, \tag{2.27}$$

as shown in Fig. 2.8(b). The contact resistance $R_I$ depends on several interface characteristics: surface roughness, pressure holding the two surfaces together, the type of fluid in the void spaces, and the interface temperature. Resistance increases with roughness and decreases with

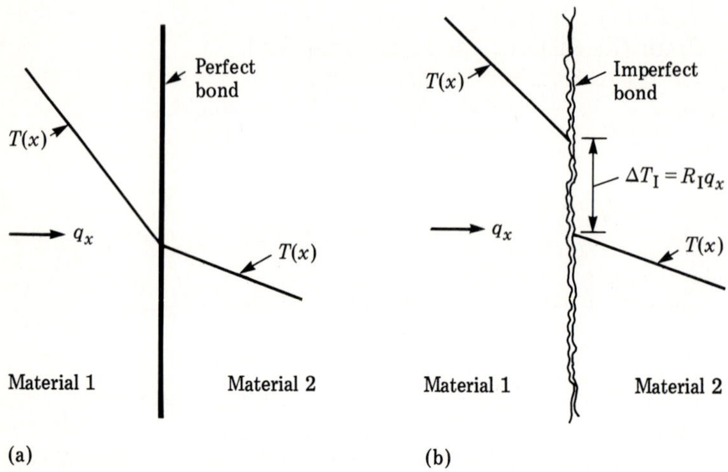


**Figure 2.8 Idealized and real conductive contact conditions: (a) idealized perfect bond, zero contact resistance, no interface temperature jump; (b) actual imperfect bond, finite contact resistance, $R_I$, finite interface temperature jump $\Delta T_I = R_I q_x$.**

pressure, but the data scatter is large, making it futile for us to prepare a table of reliable design resistance values.

To give a rough idea, we can cite the order of magnitude of measured contact resistances at moderate pressures, for (1) aluminum surfaces: $5 \times 10^{-5}\ m^2 \cdot °C/W$; (2) stainless steel surfaces: $3 \times 10^{-4}\ m^2 \cdot °C/W$; and (3) copper surfaces: $10^{-5}\ m^2 \cdot °C/W$, based on $A_x = 1\ m^2$. As a rule of thumb, the typical contact resistance is equivalent approximately to 5 mm of additional material thickness. Contact resistance may be reduced by use of special bonding greases or by insertion of a soft metal foil between the two surfaces. A bibliography of contact resistance literature is given by Moore et al. [1], and further details may be found in [2–4].

### 2.4.5 Resistances in Parallel

Although less common, it is possible for conduction systems to act in parallel, for either the solid components or the boundary conditions. A combined example is shown in Fig. 2.9, along with its electric analogy. The solid layer $\Delta x_1$ has two paths: material $k_1$ with area $A_1$ and material $k_2$ with area $A_2$. The electric analogy shows two

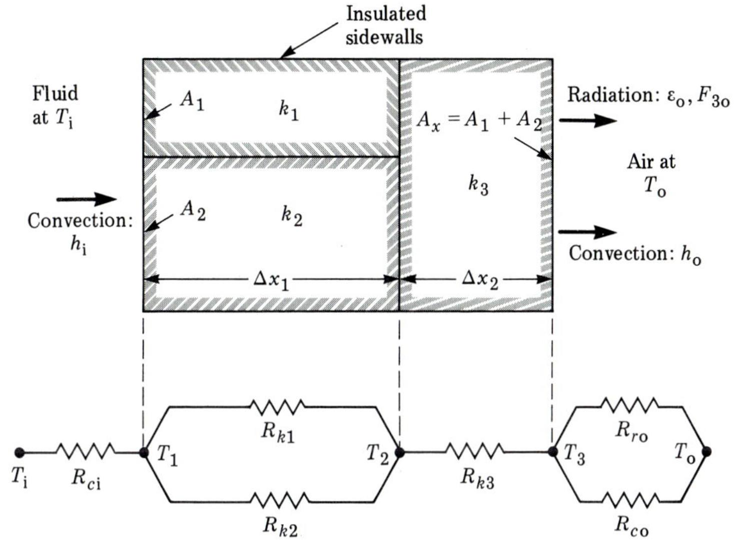


**Figure 2.9 An illustrative one-dimensional conduction problem with combined series and parallel heat flow paths and the equivalent electric analogy.**

resistances in parallel, $R_{k1}$ and $R_{k2}$, which together carry the total heat flux $q_x$. They could be combined into a single effective resistance $R_{12}$, which also follows the electric analogy:

$$\frac{1}{R_{12}} = \frac{1}{R_{k1}} + \frac{1}{R_{k2}}, \qquad \text{or} \qquad R_{12} = \frac{R_{k1}R_{k2}}{R_{k1} + R_{k2}}. \tag{2.28}$$

Note that the parallel resistances depend on their own individual conductivity and area:

$$R_{k1} = \Delta x_1/k_1A_1, \qquad R_{k2} = \Delta x_1/k_2A_2. \tag{2.29}$$

The effective resistance is a sort of mixture:

$$R_{12} = \Delta x_1/(k_1A_1 + k_2A_2). \tag{2.30}$$

A similar parallel system in Fig. 2.9 occurs at the right-hand boundary, which can both convect and radiate heat to an environment at temperature $T_o$. The convective resistance is $R_{co} = 1/h_oA_x$, where $A_x = A_1 + A_2$. From Eq. (2.24), the radiative resistance is

$$R_{ro} = [\sigma\varepsilon_0F_{3o}(T_3 + T_o)(T_3^2 + T_o^2)A_x]^{-1}, \tag{2.31}$$

where $F_{3o}$ is the configuration factor, assumed known. These boundary losses act in parallel and are therefore equivalent to a single effective boundary resistance, $R_b$:

$$R_b = \frac{R_{ro}R_{co}}{R_{ro} + R_{co}}$$

Once $R_{12}$ and $R_b$ have been computed, we are left with the effective series solution $q_x = (T_i - T_o)/(R_{ci} + R_{12} + R_{k3} + R_b)$. Since the radiative resistance $R_{ro}$ depends on the unknown boundary temperature $T_3$, it may be necessary to use iteration.

We will ask you to prove in Problem 2.42 that there is no "crosstalk" conduction between material $k_1$ and material $k_2$ in Fig. 2.9, so that the electric analogy is exact. For more complicated systems of multiple parallel elements, there can be finite cross-conduction between elements, so that the parallel circuit analogy is only approximate. An excellent discussion of such complex parallel systems is given in the text by Thomas [5].

### Example 2.7

A wall consists of wood, $k_1 = 0.11$ Btu/hr · ft · °F, and glass fiber, $k_2 = 0.022$, in parallel and a layer of brick, $k_3 = 0.29$, as shown in Fig. E2.7. The sidewalls are insulated by exactly the same patterns. Inside conditions are $T_i = 65$°F, $h_i = 3$ Btu/hr · ft$^2$ · °F, and outside conditions are $T_o = 10$°F, $h_o = 8$ Btu/hr · ft$^2$ · °F, and $\varepsilon_0 F_{3o} = 0.8$. Compute the heat loss per unit area, $q''_x$.

*Solution* This problem matches Fig. 2.9, with $\Delta x_1 = 6$ in. and $\Delta x_2 = 3$ in. For unit total area, $A_x = 1$ ft$^2$, $A_1 = 0.2$ ft$^2$ (wood), $A_2 = 0.8$ ft$^2$ (glass fiber). The wood and glass fiber are in parallel,

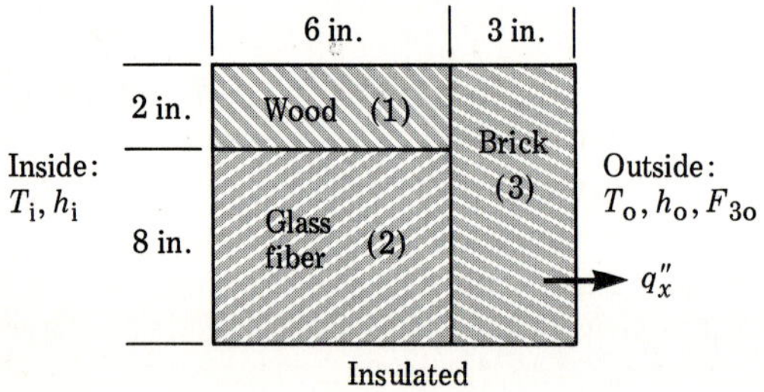


Figure E2.7

with effective resistance computed from Eq. (2.30):

$$R_{12} = \frac{6/12}{(0.11)(0.2) + (0.022)(0.8)} = 12.63 \, \frac{\text{hr} \cdot {}^\circ\text{F}}{\text{Btu}}.$$

The inside convective resistance is

$$R_{ci} = \frac{1}{h_i A_x} = \frac{1}{3(1)} = 0.33 \, \frac{\text{hr} \cdot {}^\circ\text{F}}{\text{Btu}}.$$

The brick has resistance

$$R_{k3} = \Delta x_2 / k_3 A_x = \frac{3/12}{0.29(1)} = 0.86 \, \frac{\text{hr} \cdot {}^\circ\text{F}}{\text{Btu}}.$$

Outside convection and radiation are in parallel:

$$R_{co} = \frac{1}{h_o A_x} = \frac{1}{8(1)} = 0.125 \text{ hr} \cdot {}^\circ\text{F/Btu}.$$

From Eq. (2.31), guessing that $T_3 \simeq 20^\circ\text{F} = 480^\circ\text{R}$, compute

$$R_{ro} = [1.712 \times 10^{-9}(0.8)(480 + 470)(480^2 + 470^2)(1.0)]^{-1} = 1.70 \, \frac{\text{hr} \cdot {}^\circ\text{F}}{\text{Btu}}.$$

Then the effective outer boundary resistance is

$$\frac{1}{R_b} = \frac{1}{R_{ro}} + \frac{1}{R_{co}} = \frac{1}{1.70} + \frac{1}{0.125}, \qquad \text{or} \qquad R_b = 0.12 \, \frac{\text{hr} \cdot {}^\circ\text{F}}{\text{Btu}}.$$

This reduces the problem to a series solution:

$$\begin{aligned} q_x &= (T_i - T_o)/(R_{ci} + R_{12} + R_{k3} + R_b) \\ &= (65 - 10)/(0.33 + 12.63 + 0.86 + 0.12) \\ &= 3.95 \text{ Btu/hr}. \end{aligned}$$

Since we took $A_x = 1.0 \text{ ft}^2$, then also

$$q''_x = 3.95 \, \frac{\text{Btu}}{\text{hr} \cdot \text{ft}^2}. \quad [\textit{Ans.}]$$

This problem was so crowded with computation that we omitted the intermediate units, which you should check as an exercise. ■

### 2.4.6 The Overall Heat Transfer Coefficient, *U*

In heat exchanger problems (Chapter 10) involving composite walls, it is useful to define an *overall heat transfer coefficient, U* ($\text{W/m}^2 \cdot \text{K}$), such that

$$q_x = UA\Delta T, \tag{2.32}$$

where $\Delta T$ is the overall temperature difference across the wall. Typically, $U$, which has the same units as $h$, will be a combination of convection, conduction, and fouling or scaling on the exchanger surfaces. The wall area $A$ should be carefully specified in curved-wall problems, since area varies with radius. For plane slabs, $A$ is constant.

For example, for the composite slab of Fig. E2.6, we have

$$q_x'' = \frac{T_i - T_o}{1/h_i + \Delta x_1/k_1 + \Delta x_2/k_2 + 1/h_o},$$

or

$$U = (1/h_i + \Delta x_1/k_1 + \Delta x_2/k_2 + 1/h_o)^{-1}.$$

The area $A$ is not ambiguous. However, for the layered cylinder of Fig. 2.7, one could base $U$ on any of the four radii:

$$q_x = U_jA_j(T_i - T_o), \quad \text{where} \quad U_1A_1 = U_2A_2 = U_3A_3 = U_4A_4.$$

Here the selection of area is at your discretion. Examples will be given in Chapter 10.

## Example 2.8

In the heat exchanger of Fig. E2.8, steam (fluid 1) at 100°C is being condensed by cooling water (fluid 2), which enters at $T_i$ = 10°C and a flow rate $\dot{m}_2$ = 900 kg/hr. If the wall area is 1.5 m$^2$ and the overall heat transfer coefficient is 65 W/m$^2$ · K, estimate (a) $q_x$, (b) $T_o$, and (c) the condensate rate $\dot{m}_1$ (kg/hr).

Figure E2.8

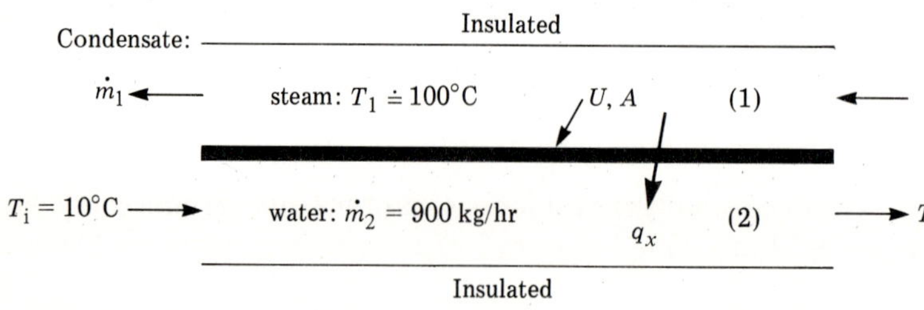

*Solution* The enthalpy of vaporization of steam at 100°C is 2.257 MJ/kg. From Appendix F for water, $c_{p2} = 4197$ J/kg · K. Assuming no losses to the environment, the enthalpy lost during condensation must equal the wall heat transfer and the enthalpy rise of the cooling water:

$$\dot{m}_1 h_{fg,1} = UA(T_1 - T_{2,\text{avg}}) = \dot{m}_2 c_{p2}(T_o - T_i). \tag{1}$$

Approximate the mean water temperature by an arithmetic average:

$$T_{2,\text{avg}} \doteq (T_i + T_o)/2.$$

(A better average, called the *log mean,* is derived in Section 10.3.) Introduce the given numerical values into Eq. (1):

$$\dot{m}_1(2{,}257{,}000) = (65)(1.5)\left(100 - \frac{10 + T_o}{2}\right) = \left(\frac{900}{3600}\right)(4197)(T_o - 10).$$

These are two equations in two unknowns. Note that the SI units require $\dot{m}_2$ in kg/s, not kg/hr. The solution is

$T_o = 18.0°\text{C}$, [*Ans. (b)*]

$\dot{m}_1 = 0.00372$ kg/s $= 13.4$ kg/hr. [*Ans. (c)*]

We may then compute $q_x$, which equals any of the three terms in Eq. (1). For example,

$$q_x = UA\Delta T = (65)(1.5)\left(100 - \frac{10 + 18}{2}\right) = 8400 \text{ W}. \quad [\textit{Ans. (a)}]$$

This problem illustrates a simple heat exchanger analysis. Later we will be able to estimate $U$ itself from the geometry and the properties of the flowing streams. ■

## †2.5 Critical Radius of Insulation

We learn in advertisements that adding insulation to a house wall always reduces heat loss. This is because $A_x$ is constant, so adding insulation always increases resistance. If, however, $A_x$ increases outward, as with a cylinder or sphere, adding insulation increases outer surface area and may actually decrease resistance.

†This section may be omitted without loss of continuity.

Consider the cylinder of radius $r_1$ in Fig. 2.10(a). It may be bare, or it may already contain some insulation. We wish to add more insulation, of conductivity $k_o$, out to a new radius $r_o$. What will happen to heat loss $q_x$?

Let $\Sigma R$ be the original cylinder resistance from all causes except outside convection. Then, after we add the insulation, the heat flux will be

$$q_x = \frac{T_i - T_o}{\Sigma R + \dfrac{\ln(r_o/r_1)}{2\pi L k_o} + \dfrac{1}{2\pi r_o L h_o}}. \tag{2.33}$$

Now, as we add more and more insulation, only $r_o$ varies on the right-hand side of Eq. (2.33). If we plot $q_x$ versus $r_o$ from this equation, the result is always as shown in Fig. 2.10(b): $q_x$ rises to a maximum and then drops. The maximum is found from Eq. (2.33) by setting $dq_x/dr_o = 0$, with the result that there is a unique *critical radius* of insulation:

$$h_o r_{crit}/k_o = 1.0, \qquad \text{or} \qquad r_{crit} = k_o/h_o. \tag{2.34}$$

The critical radius, where heat loss is maximum, depends only on insulation conductivity and outside convection coefficient, regardless of the original cylinder properties. The quantity $h_o r_{crit}/k_o$ is dimensionless and is called the *Biot number* of the cylinder, after J. B. Biot, a pioneer

**Figure 2.10** **Adding insulation to a cylinder may be counterproductive: (a) definition sketch; (b) regardless of original cylinder parameters, a plot of $q_x$ versus $r_o$ always shows a maximum at $r_{crit} = k_o/h_o$.**

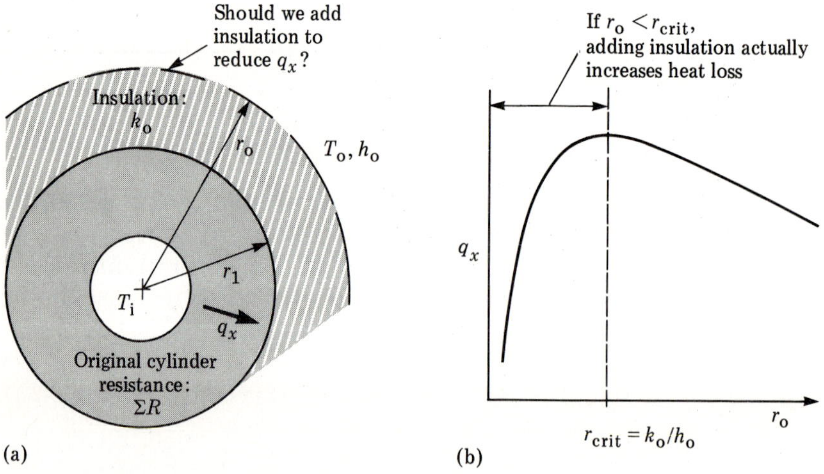

in heat transfer analysis along with Fourier. As we shall see in Chapters 3 and 4, the Biot number is the primary parameter affecting solid conduction with convection boundary conditions.

It turns out that $r_{crit}$ from Eq. (2.34) is rather small. Typical insulation has a conductivity $k_o \simeq 0.04$ W/m · K. From Table 1.1, the lowest value of $h_o$ is about 5 W/m$^2$ · K. Thus the highest practical value of $r_{crit}$ is $0.04/5 = 0.008$ m $= 8$ mm, corresponding to free convection of a gas. For liquids or forced convection, $r_{crit}$ may be about a millimeter. Steam pipes and other industrial fluid conduits are larger than $r_{crit}$ and should be insulated to reduce heat loss or gain. Electric wires may be smaller than $r_{crit}$, in which case they should be left bare to minimize the heat loss. A complication is that adding insulation somewhat decreases $h_o$ (Chapters 6 and 7). Also, as discussed by Sparrow and Prata [18], circumferential variations in $h_o$ can alter the critical radius concept considerably.

### Example 2.9

Modify Eq. (2.34) by adding blackbody radiation, with configuration factor $F_o$, to the convection at $r = r_o$.

*Solution* Add a radiation resistance $R_{ro}$ to the outside. For simplicity, assume that the insulation surface temperature approximately equals the environment temperature $T_o$. Then, from Eq. (2.31),

$$R_{ro} \simeq [\sigma F_o(T_o + T_o)(T_o^2 + T_o^2)A_o]^{-1} = \frac{1}{4\sigma F_o T_o^3(2\pi r_o L)}.$$

Introducing this into Eq. (2.32), we find, after some algebra, that the boundary resistance is

$$R_b = [2\pi r_o L(h_o + 4\sigma F_o T_o^3)]^{-1}.$$

This replaces the term $1/(2\pi r_o L h_o)$ in Eq. (2.33). The only effect is to replace $h_o$ by $(h_o + 4\sigma F_o T_o^3)$. The critical radius, including radiation, is thus

$$r_{crit} = \frac{k_o}{(h_o + 4\sigma F_o T_o^3)} \qquad [Ans.]$$

This effect makes $r_{crit}$ even smaller. For example, if $F_o = 0.8$ and $T_o = 100°C = 373K$, the extra term $4\sigma F_o T_o^3 = 9.4$ W/m$^2$ · K, far larger than the minimum $h_o$ of 5 W/m$^2$ · K. ■

## †2.6 Conduction with Heat Sources

The analysis of conduction in bodies undergoing internal heat generation from sources is a common problem in engineering. Electric currents generate resistance heating in solids. Atomic fission generates heat in nuclear fuel rods. Chemical reactions in a solid can either generate (exothermic) or absorb (endothermic) heat. Solid fuel combustion generates heat within the body. In one-dimensional conduction, the heat source causes a variable flux $q_x(x)$, unlike the previous analyses in this chapter.

Consider the body in Fig. 2.11 undergoing heat generation of amount $\dot{q}$ per unit volume of material. Assume one-dimensional conduction. In general, both $\dot{q}$ and $A_x$ can vary with $x$, and $k = k(T)$. The flow is steady, so $dE/dt$ is zero for any part of the body. Application of the first law, Eq. (2.1), to the differential slice $dx$ of material in Fig. 2.11 gives no net heat added:

$$q_x - (q_x + dq_x) + \dot{q}A_x dx = 0, \tag{2.35}$$

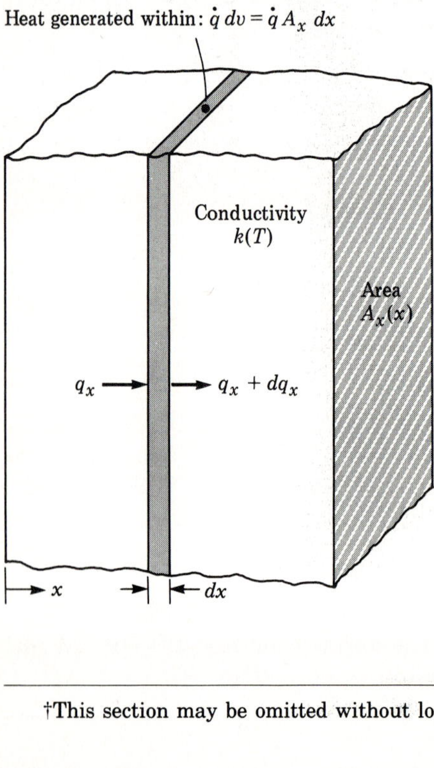


**Figure 2.11 One-dimensional steady heat conduction in a solid with internal heat generation $\dot{q}$ per unit volume.**

†This section may be omitted without loss of continuity.

where $A_x dx$ is the volume of the slice. Canceling $q_x$ and introducing Fourier's law, Eq. (1.4), we obtain

$$\dot{q}A_x dx = dq_x = d\left(-kA_x \frac{dT}{dx}\right).$$

Rearranging slightly, we have the basic second-order differential equation for one-dimensional conduction with heat sources:

$$\frac{1}{A_x}\frac{d}{dx}\left(kA_x \frac{dT}{dx}\right) = -\dot{q}. \tag{2.36}$$

Two boundary conditions are needed: for example, known temperatures at the left and right sides of the body in Fig. 2.11. In general, $k$, $A_x$, and $\dot{q}$ are variables, and the solution of Eq. (2.36) is therefore numerical (finite difference or finite element). The examples presented here assume constant $k$ and $\dot{q}$, leading to closed-form analytic solutions.

### 2.6.1 Heat Generation in a Slab

The simplest case is a slab, $A_x$ = constant; we also assume $k$ and $\dot{q}$ constant. Equation (2.36) reduces to

$$k\frac{d^2T}{dx^2} = -\dot{q} = \text{constant}. \tag{2.37}$$

For convenience, we adopt the coordinate system in Fig. 2.12(a), with $x = 0$ in the center and the slab boundaries at $x = \pm L$.

To isolate the source effect from, say, conduction caused by a hot inner and cold outer surface, let the boundaries be at the same temperature:

$$T = T_o \quad \text{at} \quad x = \pm L. \tag{2.38}$$

Equation (2.37) may be integrated twice to yield

$$T(x) = -\frac{\dot{q}}{2k}x^2 + C_1 x + C_2. \tag{2.39}$$

The constants $C_1$ and $C_2$ are determined by conditions (2.38), which give $C_1 = 0$ and $C_2 = T_o + \dot{q}L^2/2k$. The solution to this slab generation problem is thus

$$T(x) = T_o + \frac{\dot{q}L^2}{2k}(1 - x^2/L^2). \tag{2.40}$$

This distribution is parabolic; it is plotted in Fig. 2.12(a). The maximum temperature rise induced by the source is at the center of the slab

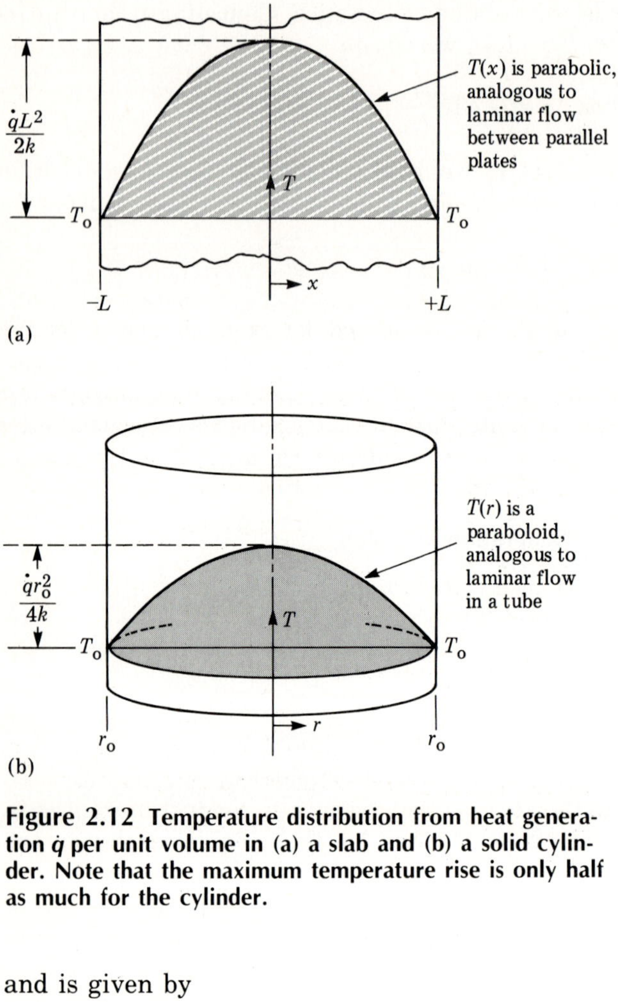


**Figure 2.12 Temperature distribution from heat generation $\dot{q}$ per unit volume in (a) a slab and (b) a solid cylinder. Note that the maximum temperature rise is only half as much for the cylinder.**

and is given by

$$\Delta T_{\text{max}} = \frac{\dot{q}L^2}{2k}. \tag{2.41}$$

Often we need to limit this rise to protect material integrity.

In Section 1.3.6 we expressed doubts about the so-called momentum–heat analogy, but this particular example fits perfectly: Eq. (2.40) is exactly analogous to laminar viscous flow between parallel plates $2L$ apart.† Temperature $T$ is analogous to fluid velocity $u$, $k$ is analogous

†See, for example, [6, pp. 323–324.]

to viscosity $\mu$, and $\dot{q}$ plays the role of pressure gradient. The boundary condition $T = T_o$ would be replaced by the no-slip condition $u = 0$ at both walls. In multidimensional or accelerating flows the analogy is, however, likely to fail.

One further remark about Eq. (2.40) should strengthen our confidence in the physics of heat conduction analysis. Figure 2.12(a) shows clearly that the slab is losing heat at both boundaries. Let us compute the total loss:

$$\begin{aligned} \left.\frac{dQ}{dt}\right|_{\text{removed}} &= -kA_x \left.\frac{dT}{dx}\right|_{x=+L} + kA_x \left.\frac{dT}{dx}\right|_{x=-L} \\ &= 2A_x L\dot{q} = \dot{q}\,(\text{slab volume}). \end{aligned} \tag{2.42}$$

But this exactly equals the total heat generated by the slab, thus satisfying the first law, Eq. (2.1), for the whole slab. This happy event is, in fact, guaranteed when we set up the differential balance, Eq. (2.37), correctly and use the proper boundary conditions, Eq. (2.38). That is, integration or differentiation of a correct physical relation still satisfies all the basic laws originally used.

### 2.6.2 Heat Generation in a Solid Cylinder

For a solid cylinder with uniform heat generation, let us adopt the coordinate system shown in Fig. 2.12(b). At any position $r$, the heat flows through area $A_x = 2\pi rL$, which we substitute into Eq. (2.36), assuming constant $k$ and $\dot{q}$. The term $2\pi L$ cancels, and we obtain the cylinder equation

$$\frac{1}{r}\frac{d}{dr}\left(r\frac{dT}{dr}\right) = -\frac{\dot{q}}{k} = \text{constant}. \tag{2.43}$$

We may separate the variables and integrate this equation twice, with the result

$$T = -\frac{\dot{q}r^2}{2k} + C_1 \ln r + C_2. \tag{2.44}$$

There appears to be only one boundary condition, uniform temperature at the outer surface:

$$T = T_o \quad \text{at} \quad r = r_o. \tag{2.45}$$

However, upon thinking it over, we realize that $C_1$ must be zero to avoid a logarithmic singularity at the origin.† With $C_1$ eliminated,

†Is this equivalent to the condition $dT/dr = 0$ at $r = 0$? Why?

Eq. (2.45) requires that $C_2 = T_o + \dot{q}r_o^2/4k$, so that the final solution may be written as

$$T(r) = T_o + \frac{\dot{q}r_o^2}{4k}(1 - r^2/r_o^2). \tag{2.46}$$

As shown in Fig. 2.12(b), this is a paraboloid distribution, with maximum source-induced temperature rise at the center:

$$\Delta T_{max} = \frac{\dot{q}r_o^2}{4k}. \tag{2.47}$$

This is only half the temperature rise of the slab, Eq. (2.41), because the cylinder has twice as much surface-area-to-volume ratio as a slab, so there is more heat dissipation.

The temperature profile, Eq. (2.46), is exactly analogous to the velocity profile for laminar flow in a circular tube [6, pp. 305–306]. In fact, heat generation in cylinders of any cross section can be computed simply by adapting the solution for laminar flow through that section.

### 2.6.3 Electric Resistance in Wires

A common case of heat generation is the passing of current through a wire, for which $\dot{q} = I^2R_e/v$, where $R_e$ is the electric resistance and $v$ is the volume of the wire. The wire resistance is a function of material resistivity $\gamma$:

$$R_e = \frac{\gamma L}{A}. \tag{2.48}$$

By inspection, the units of $\gamma$ are ohm · m (more likely stated as $\mu\Omega \cdot$ cm). Combining this with $\dot{q}$ and noting that $v = LA$, we have the following expression for wire heat generation:

$$\dot{q} = \frac{I^2\gamma}{A^2} \tag{2.49}$$

for a wire of any cross section $A$, independent of its length.

**Example 2.10**

A stainless steel wire ($k = 14$ W/m · K) of diameter 1 mm carries a current of 120 A. If $\gamma = 45\ \mu\Omega \cdot$ cm, what is the excess temperature at the center of the wire?

*Solution* The heat generation follows from Eq. (2.49):

$$\dot{q} = I^2\gamma/A^2 = \frac{(120\ \text{A})^2(45 \times 10^{-8}\Omega \cdot \text{m})}{[\pi(0.0005\ \text{m})^2]^2}$$
$$= 1.05 \times 10^{10}\ \text{W/m}^3.$$

Then the excess temperature for a cylinder is given by Eq. (2.47):

$$\Delta T_{\text{max}} = \frac{\dot{q}r_o^2}{4k} = \frac{(1.05 \times 10^{10}\ \text{W/m}^3)(0.0005\ \text{m})^2}{4(14\text{W/m} \cdot {}^\circ\text{C})}$$
$$= 47^\circ\text{C}. \quad [\textit{Ans.}]$$

■

## Example 2.11

As shown in Fig. E2.11, a heat-generating slab 3 in. thick is sandwiched between two 4-in. slabs immersed in a fluid. The center temperature is 700°F, and $k_1 = 5$ Btu/hr · ft · °F, $k_2 = 14$ Btu/hr · ft · °F, $h_\infty = 12$ Btu/hr · ft$^2$ · °F, $T_\infty = 60$°F. Compute $\dot{q}$.

*Solution* First relate the interface temperature $T_1$ to the heat generation from Eq. (2.41):

$$T_1 = T_{\text{max}} - \frac{\dot{q}L^2}{2k_2} = 700^\circ\text{F} - \frac{\dot{q}(1.5/12\ \text{ft})^2}{2(14\ \text{Btu/hr} \cdot \text{ft} \cdot {}^\circ\text{F})} \tag{1}$$
$$= 700 - \dot{q}/1792, \qquad \text{with } \dot{q} \text{ in Btu/hr} \cdot \text{ft}^3.$$

Figure E2.11

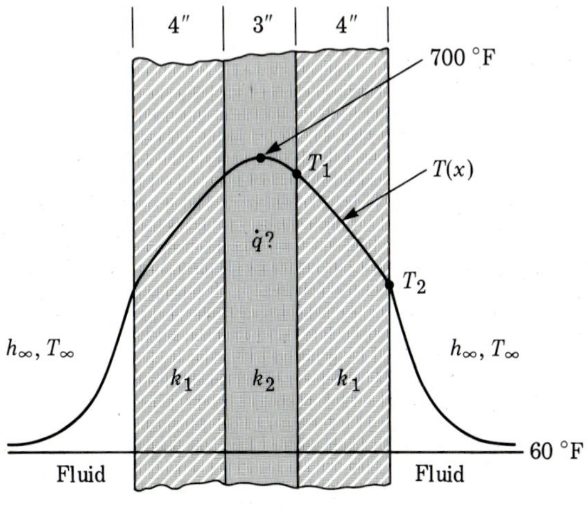

Owing to symmetry, the heat flux on either side is one-half of the total heat generated. Taking $A_x = 1\ \text{ft}^2$, we compute

$$q_x = q_x'' = \frac{1}{2}\dot{q}LA_x = \frac{1}{2}\dot{q}(3/12\ \text{ft})(1\ \text{ft}^2) = 0.125\dot{q}. \quad \textbf{(2)}$$

This is set equal to the series heat loss from $T_1$ to $T_\infty$:

$$q_x'' = \frac{T_1 - T_\infty}{\Delta x_1/k_1 + 1/h_\infty} = \frac{T_1 - 60}{(4/12)/5 + 1/12}. \quad \textbf{(3)}$$

Combining Eqs. (2) and (3) and substituting for $T_1$ from Eq. (1), we obtain

$$q_x'' = 0.125\dot{q} = \frac{700 - \dot{q}/1792 - 60}{0.0667 + 0.0833}.$$

Solving for $\dot{q}$, we get

$\dot{q} = 33{,}200\ \text{Btu/hr}\cdot\text{ft}^3$. [*Ans.*]

Because of the relatively high conductivity of the heat-generating center slab, its temperature rise is not large ($\Delta T_{max} = 20°\text{F}$). You may compute as an exercise that $T_1 = 682°\text{F}$ and $T_2 = 405°\text{F}$, so that most of the temperature rise occurs in the low-conductivity outer slabs and the poorly convecting outer thermal boundary layer. ■

---

The reader should note carefully that the electric analogy is used in Example 2.11 *only* in the non-heat-generating areas — see Eq. (3). The analogy *fails* in the generation area ($k_1$) because $q_x$ is not constant there.

## 2.7 Convection Enhancement: Fins

Fins are designed to increase heat transfer and thus are the opposite of insulation. In principle, a fin is an appendage attached to a base structure to increase its surface area and hence increase the flux. Fins were attached to the radiator surfaces of Fig. 1.1. Electronic equipment is finned to increase natural convection cooling. Heat exchanger surfaces are finned, especially on the gas side. The most common geometry is the finned tube, which may contain either inside or outside fins of various shapes. Some examples are shown in Fig. 2.13. The fins are fabricated in many ways: welded, extruded, embedded, wrapped on,

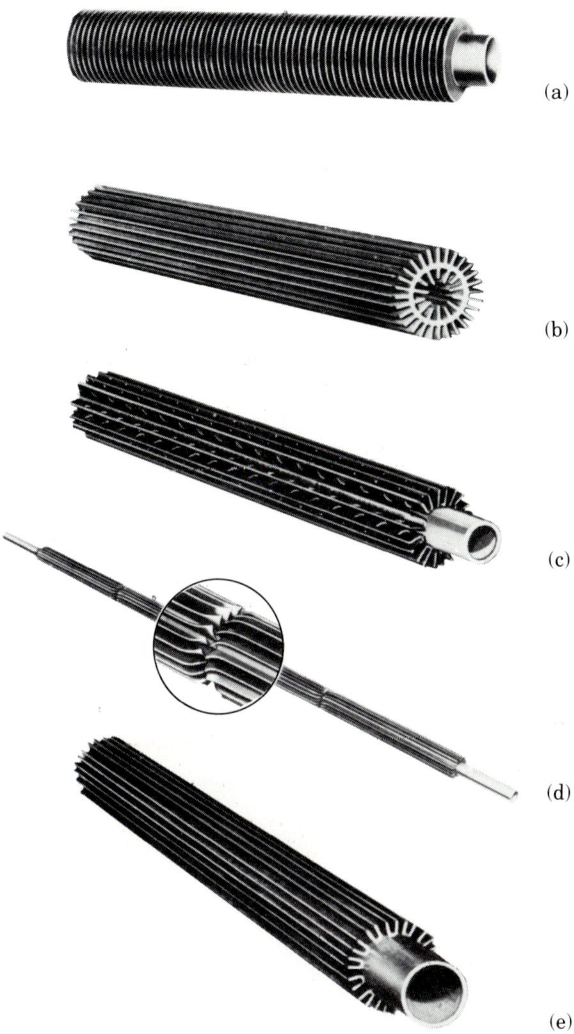


**Figure 2.13** **Finned tubes used to enhance convective heat transfer: (a) transverse fins, (b) internal and external fins, (c) perforated fins, (d) cut and twisted fins, (e) longitudinal fins. (Courtesy of Brown Fintube Company.)**

or machined from thick stock. Since they extend and expose more surface to convection (or, in vacuum applications, to radiation), they belong to a class of devices called "extended surfaces" about which whole books have been written [7]. Extended surfaces are in turn part of a more general class of designs called "heat transfer augmenters," which are the subject of an excellent review by Bergles [8] and a book by Bergles and Webb [15].

### 2.7.1 The One-Dimensional Fin Equation

The fin analysis that we present here is extremely idealized, as in Figure 2.14(a). The base temperature of the fin is assumed to be equal to the wall temperature of the main structure. The fin temperature is assumed to be one-dimensional, $T = T(x)$; it must drop with $x$ because of the convection losses. The external heat transfer coefficient

**Figure 2.14** Simplified fin analysis: (a) idealized one-dimensional distribution with fin base equal to wall temperature; (b) actual distribution with depressed temperature at fin base; (c) definition sketch for a fin.

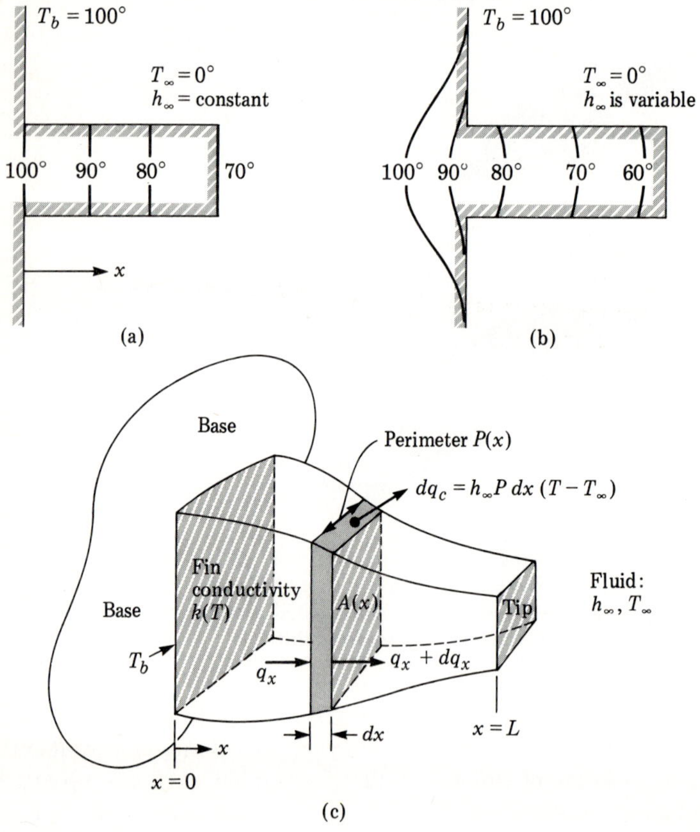

$h_\infty$ is assumed to be constant. The resulting analysis is simple but very useful.

The actual state of affairs resembles Fig. 2.14(b). Being an area of high conduction flux, the fin base is depressed below the wall temperature of the main structure, as was first noted by Sparrow and Hennecke [9]. Fin temperatures are two-dimensional, especially near the base, and the heat transfer coefficient varies along the fin surface and its tip. Later we may be able to analyze Fig. 2.14(b) using the multidimensional methods of Chapters 3 and 7, but the one-dimensional idealization is really quite adequate for design computations. In Fig. 2.14(b), the fin loses less heat, and the main wall loses more heat than in Fig. 2.14(a). For a short fin, the error can be large [10].

A "perfect" fin is so highly conductive that its tip temperature nearly equals its base value, $T_b$. Such a fin is truly a perfect "extended surface" and is said to have an efficiency of 100%. The two fins in Figs. 2.14(a, b) are both quite effective: Their tips, though cooler than 100°, are still much hotter than the fluid temperature of 0°. Fin (a) is about 85% efficient, and fin (b) about 72%. If a fin has less than about 60% efficiency, it is a poor design: Try again with a higher conductivity or shorter length or closer spacing or better shape. Now we will analyze fin efficiency and the fin Biot number that governs it. Figure 2.14(c) shows the one-dimensional approximation that we will adopt. Fin cross-section area $A$, perimeter $P$, and local temperature $T$ are all functions of $x$ in general. Fluid properties $T_\infty$ and $h_\infty$ are assumed to be known and constant. Base temperature $T_b$, at $x = 0$, is assumed to be known and equal to the main wall temperature. Tip temperature $T(L)$ is unknown and is part of the solution.

Write the first law, Eq. (2.1), for the differential slice $dx$ of the fin in Fig. 2.14(c):

$$q_x - (q_x + dq_x) - dq_c = 0, \tag{2.50}$$

where $q_x$ denotes conduction and $q_c$ is convection. Using Fourier's law and Newton's convection law, we obtain

$$dq_x = d\left(-kA\frac{dT}{dx}\right) = -dq_c = -h_\infty P dx(T - T_\infty) \tag{2.51}$$

Rearrange this relation and introduce the variable $\Theta = T - T_\infty$, noting that $d\Theta = dT$. The result is

$$\frac{d}{dx}\left(kA\frac{d\Theta}{dx}\right) - h_\infty P\Theta = 0, \qquad \Theta = T - T_\infty. \tag{2.52}$$

This is the general one-dimensional fin equation. It is a second-order ordinary differential equation that is nonlinear if $k = k(T)$. Even if $k$ is assumed constant, variation in $A(x)$ and $P(x)$ for, say, a tapered fin make it a formidable linear equation leading to complex solutions such as Bessel functions. Fortunately, solutions for almost every conceivable fin shape and boundary condition are given in [7]. For arbitrary $k$, $A$, and $P$, a finite-difference numerical solution is always possible, and a solution technique is outlined in Problem 3.69. Here we concentrate on the constant-area fin.

Two boundary conditions are needed for Eq. (2.52). One is the known base temperature at the wall.

*At $x = 0$:*

$$T = T_b, \qquad \Theta = \Theta_b = T_b - T_\infty. \tag{2.53}$$

The other is the tip condition, for which there are three common cases: (A) convection loss at the tip, (B) an insulated tip, and (C) a very long fin so that tip temperature $= T_\infty$. The mathematical statements of these three tip conditions are as follows. At $x = L$,

*Case A, tip convection:*

$$-k\frac{d\Theta}{dx} = h_\infty \Theta(L), \tag{2.54a}$$

*Case B, insulated tip:*

$$\frac{d\Theta}{dx} = 0, \tag{2.54b}$$

*Case C, very long fin:*

$$\Theta \to 0 \text{ as } x \to \infty. \tag{2.54c}$$

Case A occurs if the tip is exposed to the fluid. Case B would occur if the tip ended not in the fluid but against a low-conductivity wall. Case C simply means that the fin is too long and therefore very inefficient, since its outer portion is cold and convects away little heat.

### 2.7.2 Solution for a Constant-Area Fin

We will leave the variable-area fins for further reading in [7] or [12] and consider here only a uniform fin of constant $k$, $A$, and $P$. Equation (2.52) simplifies to

$$\frac{d^2\Theta}{dx^2} - \frac{h_\infty P}{kA}\Theta = 0. \tag{2.55}$$

The boundary conditions remain the same, Eqs. (2.53) and (2.54).

By inspection we see that the coefficient $(h_\infty P/kA)$ must have dimensions of inverse length squared. Let us define a characteristic length scale $\lambda$ using this parameter:

$$\lambda = \left(\frac{kA}{h_\infty P}\right)^{1/2}. \tag{2.56}$$

The general solution to Eq. (2.55) may be written in terms of either hyperbolic or exponential functions of $x/\lambda$:

$$\begin{aligned} \Theta = T - T_\infty &= C_1\cosh(L/\lambda - x/\lambda) + C_2\sinh(L/\lambda - x/\lambda) \\ &= C_3 e^{x/\lambda} + C_4 e^{-x/\lambda}. \end{aligned} \tag{2.57}$$

The first form is convenient to Cases A and B, while the second (exponential) form fits Case C.

*Case A, tip convection.* The hyperbolic form in Eq. (2.57) is appropriate, and the boundary conditions (2.53) and (2.54a) yield the following:

*At* $x = 0$:

$$\Theta_b = T_b - T_\infty = C_1\cosh(L/\lambda) + C_2\sinh(L/\lambda), \tag{2.58a}$$

*At* $x = L$:

$$kC_2/\lambda = h_\infty C_1. \tag{2.58b}$$

Solving for $C_1$ and $C_2$ and substituting back into our basic solution (2.57), we obtain the temperature distribution in a finite-length uniform fin with a convecting tip:

$$\frac{T - T_\infty}{T_b - T_\infty} = \frac{\cosh(L/\lambda - x/\lambda) + \mathrm{Bi}\sinh(L/\lambda - x/\lambda)}{\cosh(L/\lambda) + \mathrm{Bi}\sinh(L/\lambda)}, \tag{2.59}$$

where $\mathrm{Bi} = h_\infty\lambda/k = (h_\infty A/kP)^{1/2}$ = fin-tip Biot number.

*Case B, insulated tip.* Again the hyperbolic form of Eq. (2.57) is appropriate, and the base boundary condition (2.58a) is the same. The insulated-tip condition (2.54b) yields $C_2 = 0$. Thus the solution for a uniform fin with insulated tip is

$$\frac{T - T_\infty}{T_b - T_\infty} = \frac{\cosh(L/\lambda - x/\lambda)}{\cosh(L/\lambda)}. \tag{2.60}$$

Note that this is exactly equivalent to the convecting-tip solution (2.59) with $\mathrm{Bi} = 0$ (no tip convection).

*Case C, very long fin.* Here the exponential form of Eq. (2.57) is convenient, and in order for $\Theta$ to vanish as $x \to \infty$, $C_3$ must equal zero. Then the base condition (2.53) requires that $C_4 = (T_b - T_\infty)$. The final solution is

$$\frac{T - T_\infty}{T_b - T_\infty} = e^{-x/\lambda}. \tag{2.61}$$

You may show as an exercise that this is the limiting condition of Eq. (2.60) as $L$ becomes very large.

Figure 2.15(a) shows sample temperature profiles for these three cases. For a given length, the insulated tip (solid curves, Case B) cools down less than a typical convecting tip (dashed curves, Case A). The very long fin (dot-dash curve, Case C) is a single curve simulating an infinite length. We see that Cases A and B are beginning to approach

**Figure 2.15** Solutions for a constant-area fin: (a) temperature distributions, (b) efficiency. For Case B, use *L* instead of *L** in (b).

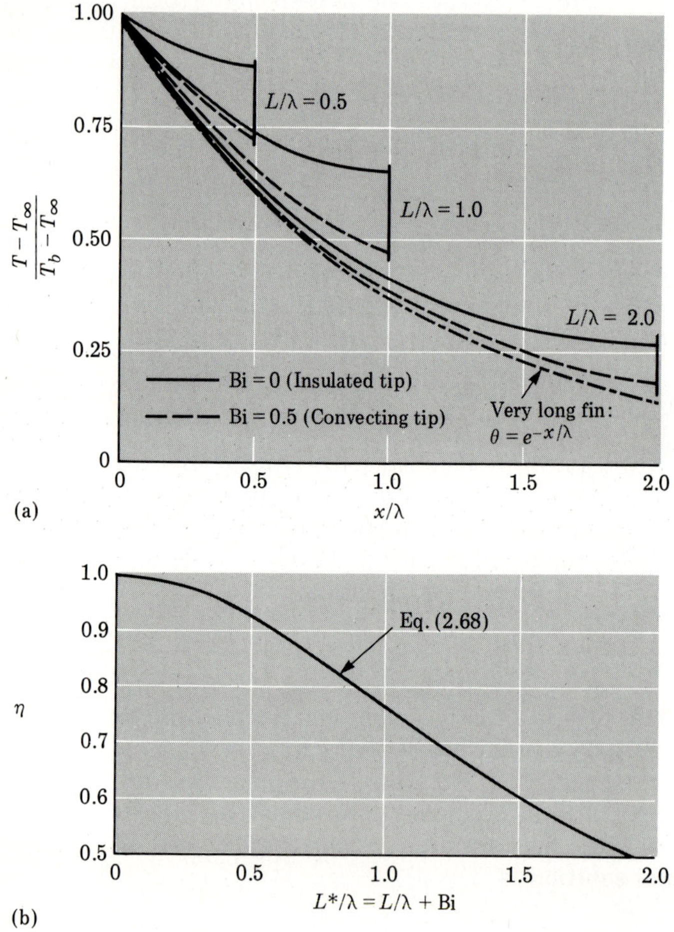

Case C for $L/\lambda \geq 2.0$, which means that the fin is too long to be efficient.

### 2.7.3 Heat Transfer and Fin Efficiency

An ideal fin would be perfectly conducting and would maintain an even surface temperature equal to its base temperature. Then its convection heat transfer would be a maximum:

$$q_{\text{ideal}} = h_\infty(T_b - T_\infty)A_{\text{wall}}. \tag{2.62}$$

An actual fin, however, cools off away from the base. Its heat transfer can be computed either by integrating the convection along the surface or by using Fourier's law at the base:

$$q_{\text{actual}} = \int_0^L h_\infty(T - T_\infty)dA_{\text{wall}} = -kA\left.\frac{dT}{dx}\right|_{x=0}. \tag{2.63}$$

The *efficiency* $\eta$ of a fin is defined as the ratio of the actual and the ideal:

$$\eta = \frac{q_{\text{actual}}}{q_{\text{ideal}}}, \tag{2.64}$$

and it is a number between zero and unity.

Equations (2.63) and (2.64) may be used to evaluate heat transfer and efficiency for each of the three cases.

*Case A, tip convection:*

$$\begin{aligned} q_{\text{actual}} &= (T_b - T_\infty)(h_\infty PkA)^{1/2}\frac{\sinh(L/\lambda) + \text{Bi}\cosh(L/\lambda)}{\cosh(L/\lambda) + \text{Bi}\sinh(L/\lambda)}, \\ \eta &= \frac{\sinh(L/\lambda) + \text{Bi}\cosh(L/\lambda)}{\cosh(L/\lambda) + \text{Bi}\sinh(L/\lambda)}\left(\frac{\lambda}{L + A/P}\right). \end{aligned} \tag{2.65}$$

*Case B, insulated tip:*

$$\begin{aligned} q_{\text{actual}} &= (T_b - T_\infty)(h_\infty PkA)^{1/2}\tanh(L/\lambda), \\ \eta &= \tanh(L/\lambda)/(L/\lambda). \end{aligned} \tag{2.66}$$

*Case C, very long fin:*

$$\begin{aligned} q_{\text{actual}} &= (T_b - T_\infty)(h_\infty PkA)^{1/2}, \\ \eta &= \lambda/L, \quad \text{if} \quad L \geq 3\lambda. \end{aligned} \tag{2.67}$$

As before, Case B is the limit of Case A if $\text{Bi} = 0$, and Case C is the limit of Case B if $L/\lambda$ is greater than about 3.0.

Case A is rather complicated algebraically. It was pointed out by Jakob [11] that the efficiency of Case A was very well approximated by using the Case B formula evaluated at an effective length $L^*$ equal to $L$ plus the "tip size" $A/P$:

$$\eta \doteq \frac{\tanh(L^*/\lambda)}{L^*/\lambda}, \qquad L^* = L + A/P. \tag{2.68}$$

This formula represents the efficiency of all three cases and is plotted in Figure 2.15(b). The efficiency is greater than 60% if

*Good Design:*

$$L^*/\lambda \leq 1.5. \tag{2.69}$$

I consider this the limit of a well-designed fin; adding additional length to the fin is only marginally effective.

As mentioned, additional solutions for a wide variety of other fin shapes are given in [7] and [12]. Figure 2.16 shows the efficiency of an important case, a cylindrical base with a radial fin of rectangular cross section, as computed by Gardner [12].

Formulas for temperature and heat transfer for Cases A, B, and C are collected in Table 2.2 for easy reference.

**Figure 2.16** Efficiency of a transverse circular fin of length $L$ and thickness $\delta$, after Gardner [12].

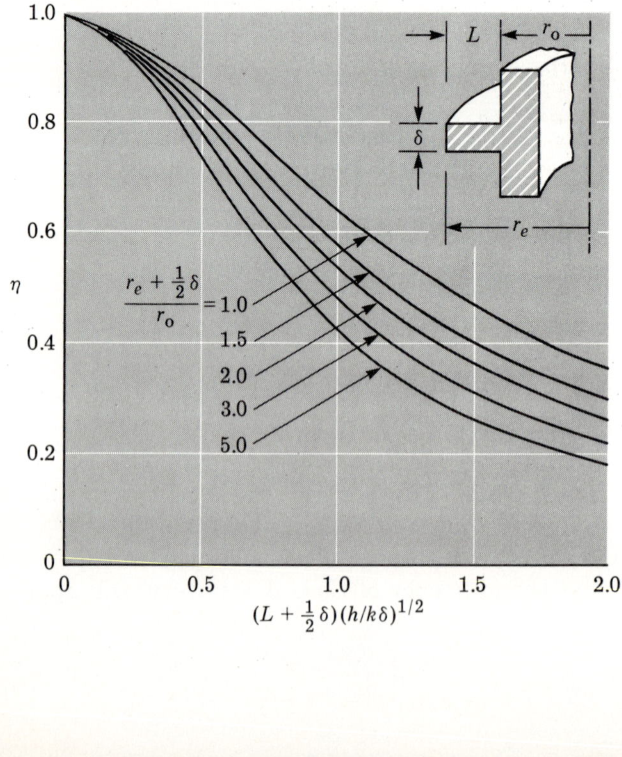

**Table 2.2** Temperature and heat transfer of uniform-area fins

| Case | Conditions | Temperature $(T - T_\infty)/(T_b - T_\infty)$ | Heat Transfer $q_b/[(T_b - T_\infty)(h_\infty PkA)^{1/2}]$ |
|---|---|---|---|
| A | Tip convection | $\dfrac{\cosh(L/\lambda - x/\lambda) + \mathrm{Bi}\sinh(L/\lambda - x/\lambda)}{\cosh(L/\lambda) + \mathrm{Bi}\sinh(L/\lambda)}$ | $\dfrac{\sinh(L/\lambda) + \mathrm{Bi}\cosh(L/\lambda)}{\cosh(L/\lambda) + \mathrm{Bi}\sinh(L/\lambda)}$ |
| B | Insulated tip | $\dfrac{\cosh(L/\lambda - x/\lambda)}{\cosh(L/\lambda)}$ | $\tanh(L/\lambda)$ |
| C | Very long fin: $L > 3\lambda$ | $e^{-x/\lambda}$ | 1.0 |

Notation: $\lambda = (kA/h_\infty P)^{1/2}$

$\mathrm{Bi} = h_\infty \lambda/k = (h_\infty A/kP)^{1/2}$

### Example 2.12

A rectangular fin has length $L = 10$ cm, width $W = 4$ cm, and thickness $t = 8$ mm, with a convecting tip. Its base temperature is 200°C, and fluid conditions are $h_\infty = 9\ \text{W/m}^2 \cdot °\text{C}$, $T_\infty = 20°\text{C}$. Compute the efficiency, the tip temperature, and the total heat transfer if the fin material is (a) aluminum, $k = 240\ \text{W/(m} \cdot °\text{C)}$, and (b) stainless steel, $k = 14\ \text{W/(m} \cdot °\text{C)}$.

*Solution* First compute $A = Wt = 0.00032\ \text{m}^2$, and $P = 2(W + t) = 0.096$ m. The solution then depends on fin conductivity.
a) Aluminum:

$$\lambda = \left(\frac{kA}{h_\infty P}\right)^{1/2} = \left(\frac{240(0.00032)}{9(0.096)}\right)^{1/2} = 0.298\ \text{m}.$$

Then $L/\lambda = 0.1/0.298 = 0.3354$ (<1.5, a good design). Also, $\text{Bi} = (h_\infty A/kP)^{1/2} = [9(0.00032)/240(0.096)]^{1/2} = 0.0112$, and $L^* = L + A/P = 0.1033$ m, $L^*/\lambda = 0.3466 = L/\lambda + \text{Bi}$. Now compute the exact efficiency from Eq. (2.65):

$$\eta = \frac{\sinh(0.3354) + 0.0112\cosh(0.3354)}{\cosh(0.3354) + 0.0112\sinh(0.3354)} \cdot \frac{1}{0.3466} = 0.962. \quad [\textit{Ans. (a)}]$$

Jakob's approximation (2.68) is identical to three decimal places:

$$\eta \doteq \frac{\tanh(0.3466)}{0.3466} = 0.962.$$

For the tip temperature, apply Eq. (2.59) at $x = L$:

$$T_L - T_\infty = (T_b - T_\infty)\frac{\cosh(0.0) + 0.0112\sinh(0.0)}{\cosh(0.3354) + 0.0112\sinh(0.3354)}$$

$$= (200° - 20°)\frac{1.0}{1.0606} = 170°\text{C},$$

or

$$T_L = 20 + 170 = 190°\text{C}. \quad [\textit{Ans. (a)}]$$

The ideal heat flux is computed from Eq. (2.62):

$$q_{\text{ideal}} = h_\infty(T_b - T_\infty)(PL + A) = (9)(200 - 20)[0.096(0.1) + 0.00032]$$
$$= 16.07\ \text{W}.$$

Then the actual flux is

$$q_{\text{fin}} = \eta q_{\text{ideal}} = 0.962 \cdot 16.07 = 15.5\ \text{W}. \quad [\textit{Ans. (a)}]$$

Compare this to the base heat transfer if there were no fin:

$$q_o = h_\infty(T_b - T_\infty)A = 9(200 - 20)0.00032 = 0.52 \text{ W}.$$

We see that the aluminum fin transfers about 30 times more heat than the base when unfinned.

b) For the stainless steel fin, $k = 14$ W/(m · °C), you can verify the following computations:

$$\lambda = 0.072 \text{ m}, \qquad L/\lambda = 1.389,$$
$$\text{Bi} = 0.0463, \qquad L^*/\lambda = 1.435.$$

By either Eq. (2.65) or Eq. (2.68), $\eta = 0.622$. [*Ans. (b)*]
By Eq. (2.59), $T_L = 101$°C. [*Ans. (b)*]
Since $q_{\text{ideal}} = 16.07$ W, $q_{\text{fin}} = 0.622 \cdot 16.06 = 10.0$ W. [*Ans. (b)*]
This is 19 times more heat transfer than the base wall — good, but not as good as the aluminum fin, which has smaller $L/\lambda$. ■

## Example 2.13

Consider the slab in Fig. E2.13 with fins of length $L$ on one side. The unfinned surface has area $\zeta A_x$, where $\zeta$ is a fraction. Derive a relation for heat flux $q_x$ with and without fins.

*Solution* Without the fins, the system consists of inner and outer convective resistance and slab conductive resistance:

$$q_x = \frac{T_i - T_o}{R_{ci} + R_k + R_{co}} = \frac{A_x(T_i - T_o)}{1/h_i + \Delta x/k + 1/h_o}. \quad [\textit{Ans.}]$$

With fins added, $R_{ci}$ and $R_k$ are the same, but we must derive a relation for the effective outer resistance $R'_{co}$. First write the heat

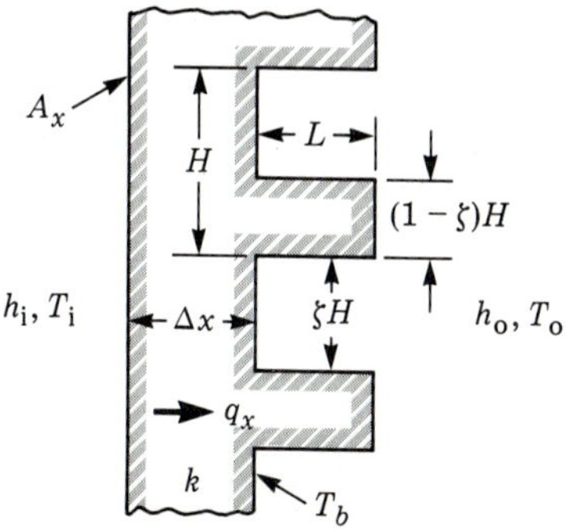


Figure E2.13

flux in terms of $T_o$ and the outer wall temperature $T_b$, assuming that the fin base is also at $T_b$:

$$q_x = h_o(T_b - T_o)(\zeta A_x + \eta A_{fin}), \qquad A_{fin} = 2LW + (1 - \zeta)A_x,$$

where W is the slab width into the paper. To eliminate W, note that

$$2LW = 2\frac{L}{H}HW = 2\frac{L}{H}A_x.$$

Then the effective resistance with fins is given by

$$q_x = \frac{T_b - T_o}{R'_{co}}, \quad \text{where} \quad \frac{1}{R'_{co}} = h_o A_x \left[\zeta + \eta\left(2\frac{L}{H} + 1 - \zeta\right)\right].$$

Since $R_{co} = 1/h_o A_x$, the term in brackets is the ratio of resistance without fins to resistance with fins. As an example, suppose $L = H$, $\eta = 0.8$, and $\zeta = 0.3$. Then we compute $R_{co}/R'_{co} = 0.3 + 0.8[2(1) + 1 - 0.3] = 2.46$. When combined with the slab and inner convection, the heat flux through the finned system will be

$$q_{x,fins} = \frac{(T_i - T_o)}{R_{ci} + R_k + R'_{co}}, \quad [Ans.]$$

where all resistances have been defined above.

The fins work well only if the outer resistance is dominant, $R_{co} >> R_{ci}, R_k$. Take the example above where $R_{co}/R'_{co} = 2.46$. If, in arbitrary units, $R_{ci} = R_k = 0.1$ and $R_{co} = 1$, then the use of fins will dramatically increase flux:

$$\frac{q_{x,fins}}{q_x} = \frac{0.1 + 0.1 + 1}{0.1 + 0.1 + 1/2.46} = 1.98,$$

or a 98% increase. If, however, $R_{ci} = R_k = 2$ and $R_{co} = 1$,

$$\frac{q_{x,fins}}{q_x} = \frac{2 + 2 + 1}{2 + 2 + 1/2.46} = 1.13,$$

or only a 13% increase. Sometimes fins are not cost-effective. ■

## Summary

This chapter begins by applying Fourier's law and the first law of thermodynamics to derive the general heat conduction equation, Eq. (2.6), and its boundary conditions, Eqs. (2.11). Applications are then restricted to a variety of steady one-dimensional conduction problems. Chapters 3 and 4 will consider multidimensional and unsteady problems.

If there are no internal heat sources, the one-dimensional conduction equation can be solved to obtain simple formulas for heat transfer and thermal resistance. These are summarized for a slab, a cylinder, and a sphere in Table 2.1.

When a body consists of layers of solids of different conductivities, an attractive approach is the electric analogy, in which layer thermal resistances and boundary resistances are all summed to compute the heat flux. The electric analogy is effective for both series and parallel resistance systems. The analogy fails if there are heat sources.

The chapter concludes with three special topics: critical radius of insulation, internal heat sources, and fins. The critical radius concept shows that when insulation is added to a small cylinder or sphere, it may actually increase the heat loss and thus be counterproductive.

When internal heat sources are added to the one-dimensional conduction problem, the basic differential equation becomes nonhomogeneous, Eq. (2.36). If the sources are uniformly distributed, the source-generated temperature profile is analogous to a fully developed laminar viscous flow profile in a duct, as in Fig. 2.12.

The chapter ends with a brief analysis of convection enhancement by fins or "extended surfaces." The basic fin conduction-convection equation, Eq. (2.52), can be solved for a variety of fin shapes and boundary conditions. The efficiency of the fin always drops off as the ratio $L/\lambda$ increases, where $L$ is fin length (Fig. 2.15b). Fins are most attractive for improving free convection to gases, but they have liquid and forced-convection applications also. In treating fin analysis this early, we have had to pull values of $h$ almost out of a hat, since we have not yet studied convection very thoroughly. In fact, fins have a strong effect on $h$ itself, so there is a possibility of optimizing the design by using the right spacing [13] or a staggered arrangement [14]. We will address these points in more detail in Section 7.3.7.

## References

1. C. J. Moore, Jr., H. A. Blum, and H. Atkins, "Studies Classification Bibliography for Thermal Contact Resistance Studies," ASME Paper 68-WA/HT-18, December 1968.
2. T. N. Veizirogen, "Correlation of Thermal Contact Conductance Experimental Results," *Progress in Astronautics and Aeronautics,* vol. 20, Academic Press, New York, 1967.

3. M. G. Cooper, B. B. Mikic, and M. M. Yovanovich, "Thermal Contact Conductances," *Int. J. Heat Mass Transfer,* vol. 12, 1969, pp. 279–300.
4. W. M. Rohsenow and J. P. Hartnett (Eds.), *Handbook of Heat Transfer,* sec. 3 by P. J. Schneider, McGraw-Hill, New York, 1973.
5. L. C. Thomas, *Fundamentals of Heat Transfer,* Prentice-Hall, Englewood Cliffs, N.J., 1980, pp. 52–56.
6. F. M. White, *Fluid Mechanics,* 2nd ed., McGraw-Hill, New York, 1986.
7. D. Q. Kern and A. D. Kraus, *Extended Surface Heat Transfer,* McGraw-Hill, New York, 1972.
8. A. E. Bergles, "Survey of Evaluation Techniques to Augment Convective Heat and Mass Transfer," *Progress in Heat and Mass Transfer,* vol. 1, 1969, pp. 331–424.
9. E. M. Sparrow and D. K. Hennecke, "Temperature Depression at the Base of a Fin," *J. Heat Transfer,* vol. 92, 1970, pp. 204–206.
10. N. V. Suryanarayana, "Two-Dimensional Effects on Heat Transfer Rates from an Array of Straight Fins," *J. Heat Transfer,* vol. 99, 1977, pp. 129–132.
11. M. Jakob, *Heat Transfer,* vol. 1, Wiley, New York, 1949.
12. K. A. Gardner, "Efficiency of Extended Surfaces," *Transactions ASME,* vol. 67, 1945, pp. 621–631.
13. W. Elenbaas, "Heat Dissipation of Parallel Plates by Free Convection," *Physica,* vol. 9, no. 1, 1942, pp. 1–28.
14. N. Sobel, F. Landis, and W. K. Mueller, "Natural Convection Heat Transfer in Short Vertical Channels Including the Effect of Stagger," *Proc. 3rd Intl. Heat Transfer Conference,* vol. 2, 1966, pp. 121–125.
15. A. E. Bergles and R. L. Webb, *Augmentation of Heat and Mass Transfer,* ASME Symposium Proceedings, New York, Dec. 2, 1970.
16. J. M. Chenoweth et al. (Eds.), "Advances in Enhanced Heat Transfer," 18th Natl. Heat Transfer Conference, ASME Symposium Vol. No. 100122, San Diego, 1979.
17. A. D. Kraus, *Extended Surface Thermal Systems,* Hemisphere, New York, 1982.
18. E. M. Sparrow and A. T. Prata, "Effect of Circumferential Conduction on the Critical Radius of Insulation for Forced Convection Crossflow", *J. Heat Transfer,* Nov. 1985, pp. 960–963.

## Review Questions

1. Explain what physical principles are represented by the general heat conduction equation, Eq. (2.6). Is it linear or nonlinear?
2. What are typical boundary conditions for heat conduction?

3. Explain the "one-dimensional" approximation for conduction.
4. Define the average conductivity of a conducting layer.
5. What is a thermal resistance? Name three types. Why does the concept fail when the layer contains heat sources?
6. Illustrate the electric analogy for a system in series.
7. How can the electric analogy be used to find interior temperatures in a series system?
8. Define contact resistance and the parameters that affect it.
9. Give an example of parallel resistances (a) for internal conduction, and (b) at a boundary.
10. Explain the concept of critical radius of insulation. Why does it happen to cylinders but not to slabs?
11. Explain and illustrate why conduction with internal heat sources is analogous to laminar flow in a duct.
12. List and criticize three basic assumptions in fin analysis.
13. List three tip boundary conditions for a fin.
14. Define the efficiency of a fin.
15. How long should a fin be for good efficiency?

## Problems

### Problem distribution by sections

**The Problem Assignments Are Organized as Follows:**

| Problems | Sections Covered | Topics Covered |
|---|---|---|
| 2.1–2.12 | 2.1 | The Heat Conduction Equation |
| 2.13–2.25 | 2.2 | Steady One-Dimensional Conduction |
| 2.26–2.29 | 2.3 | Conduction with Arbitrary Area |
| 2.30–2.49 | 2.4 | Composite Walls: The Electric Analogy |
| 2.50–2.54 | 2.5 | Critical Radius |
| 2.55–2.64 | 2.6 | Internal Heat Sources |
| 2.65–2.90 | 2.7 | Fins |

2.1 Explain what effects would cause either the heat conduction equation, Eq. (2.6), or its boundary conditions, Eqs. (2.11), to be nonlinear.

2.2 Given steady conduction with constant $k$ through the plate in Fig. P2.2, write out the appropriate differential equation and boundary conditions.

2.3 A sphere of radius $R_o$ is at temperature $T_i$ when suddenly the outside surface is subjected to convection conditions ($h_\infty$, $T_\infty$). Set up the appropriate differential equation and its conditions for this case. Sketch the final equilibrium temperature distribution, if any.

2.4 By analogy with Fig. 2.3, develop a boundary condition for radiation through a vacuum to surroundings at temperature $T_\infty$.

2.5 For the half-cylinder geometry of Fig. P2.5 and the parameters shown, set up the appropriate differential equation and boundary conditions for steady conduction.

2.6 Starting with a spherical shell of material between $R$ and $R + dR$, use a heat balance to derive the differential equation for unsteady radial

**Figure P2.2**

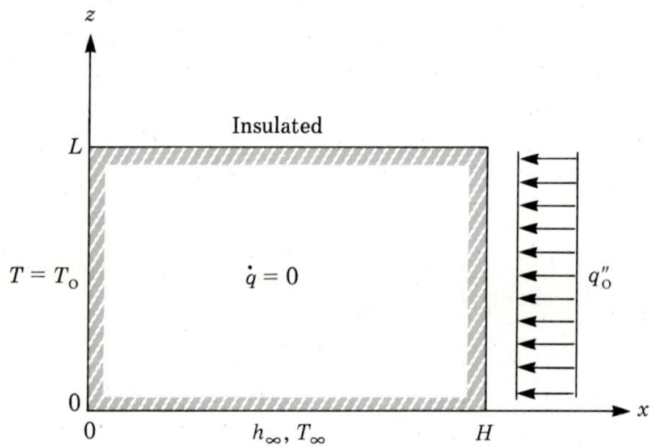


**Figure P2.5**

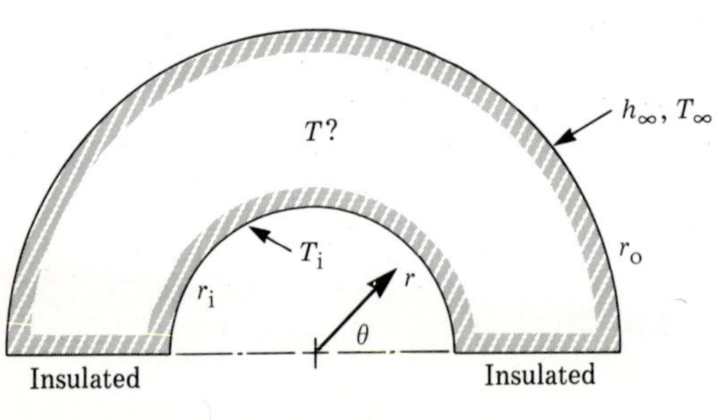

conduction in a sphere, including heat generation. Compare with the special case of Eq. (2.9).

2.7 Using the differential cylindrical-coordinate element of Fig. P2.7, use a heat balance to derive the heat conduction equation and compare with Eq. (2.8).

2.8 A solid cylinder of radius $r_o$ is initially at temperature $T_i$, with both ends ($z = 0, L$) insulated. Conductivity is constant, and there is no heat generation. The curved outer surface is suddenly subjected to convection ($h_\infty$, $T_\infty$). Set up the proper differential equation and conditions. What will be the final equilibrium temperature distribution?

2.9 The slab of thickness $H$ in Fig. P2.9 is at temperature $T_i$ when the electric heater and convection conditions are suddenly applied. Set up the proper differential equation and conditions. Sketch the temperature distributions for short, intermediate, and long times. What are $T(0)$ and $T(H)$ after a long time?

2.10 A slab of thickness 10 cm has conductivity 22 W/m · K, density 4500 kg/m$^3$, and specific heat 510 J/kg · K. It has an instantaneous temperature

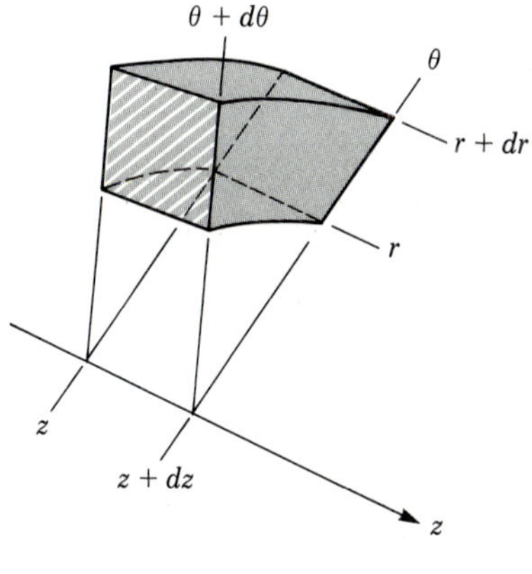


Figure P2.7

Figure P2.9

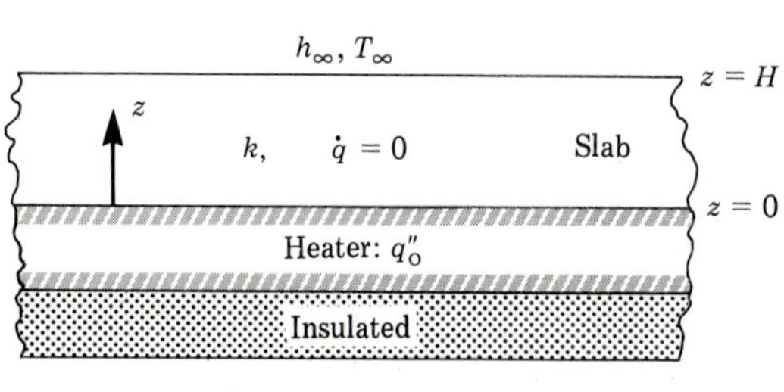

distribution

$T = 500 - 2500x + 6000x^2,$

with $T$ in °K and $x$ in meters, with $(x = 0)$ denoting the left surface. How much energy is being instantaneously received by the slab, in W/m$^2$? At what rate is the temperature at the center of the slab changing in °K/s?

2.11 Figure P2.11 shows three temperature distributions in a slab of constant conductivity and zero heat generation. Determine whether they represent heating, cooling, or steady-state conditions and explain your conclusions.

2.12 Develop expressions for heat flows $q''_r$, $q''_\theta$, and $q''_z$, which express Fourier's law in the cylindrical coordinates of Fig. 2.2(a).

2.13 Is the average conductivity concept, Eq. (2.13), valid for any one-dimensional approximation, such as spheres and thin hollow ellipsoids, or only for plane slabs?

2.14 A slab of aluminum oxide has inner temperature 300K and outer temperature 900K. Evaluate its average conductivity $k_m$ (a) as exactly as possible from the data of Appendix D, (b) by an arithmetic average of the endpoint conductivities, (c) at the average temperature of the boundaries.

2.15 If the slab of Problem 2.14 is 15 cm thick, evaluate (a) the heat flux $q''_x$ and (b) the temperature at mid-slab.

2.16 The slab in Fig. P2.16 is 1% carbon steel, subjected to steady heat flow. If $T_2 = 378$K and $T_3 = 361$K, determine the fluid temperatures $T_1$ and $T_4$.

2.17 Starting with the proper reduction of the heat conduction equation, Eq. (2.7), derive the temperature distributions, Eq. (2.17) for a cylinder and Eq. (2.19) for a sphere.

2.18 A plane slab is 12 cm thick and has its outer temperature at 0°C. With

**Figure P2.11**

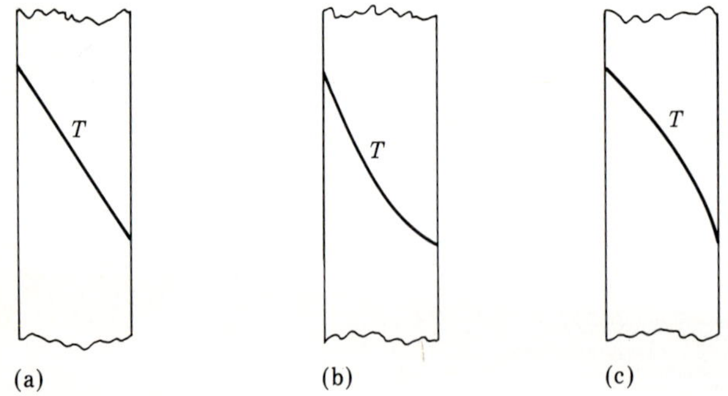

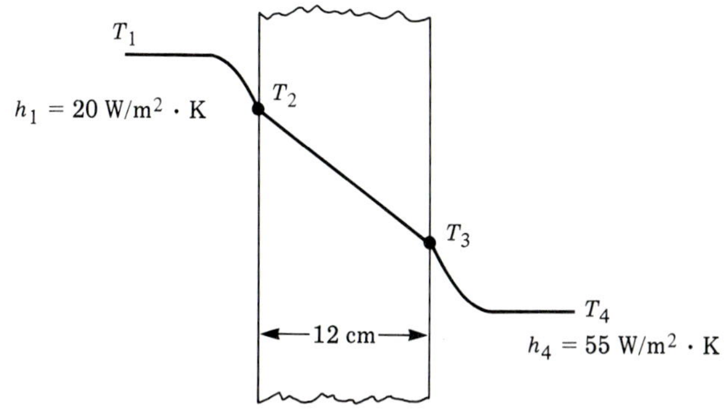


**Figure P2.16**

the inner temperature at 100°C, 200°C, and 300°C, the measured heat fluxes are 15, 38, and 72 kW/m$^2$, respectively. Does the slab conductivity increase or decrease with temperature? Using the given data, find an approximate curve-fit expression for $k(T)$.

2.19 Prove that, in the limit of very small shell thickness, the thermal resistances for the cylinder and sphere in Table 2.1 approach the slab result $\Delta x/kA_x$.

2.20 We may define the wind-chill factor in weather reporting as the calm-air temperature that causes the same heat flux as a given windy day. Suppose that human skin is 4 mm thick, with no heat generation, and maintained on its inside surface at 37°C. Using Fig. 1.8 as a guide, let $h_o$ be 15 W/m$^2$ · K on a calm day and 55 W/m$^2$ · K on a windy day. If air temperature $T_o$ is $-18$°C, what is the heat transfer in W/m$^2$ for (a) calm and (b) windy conditions? What is the wind-chill factor for these windy conditions?

2.21 A hollow cylinder has $r_1 = 1.5$ in., $r_2 = 2.0$ in., $L = 25$ ft, $T_1 = 70$°F, and $T_2 = 140$°F. It is surrounded by fluid with $T_o = 250$°F and $h_o = 125$ Btu/hr · ft$^2$ · °F. What is the average conductivity of the cylinder?

2.22 A water-heater tank is 100 cm in diameter and 180 cm high. It is covered with 5-cm-thick rock wool insulation. Metal wall resistance is negligible. If $T_i = 60$°C and $T_o = 15$°C, what is (a) the heat loss and (b) the monthly cost of this loss at 10¢/kwh?

2.23 A hollow iron cylinder has $d_i = 5$ cm, $d_o = 6$ cm, and $L = 18$ m. Water flows through the tube at an average velocity of 2 m/s, entering at 15°C and leaving at 50°C. If the inner wall is at $T_i = 75$°C, what is $T_o$ of the outer wall? If the outer surface is heated by air at $h_o = 450$ W/m$^2$ · K, what is the air temperature?

2.24 A hollow stainless steel sphere has $D_1 = 48$ cm, $D_2 = 60$ cm, and $T_2 =$

50°C. It is cooled by a fluid with $T_o$ = 15°C and $h_o$ = 145 W/m$^2$ · K. What is the inner temperature $T_1$?

2.25 A hollow sphere of uniform conductivity has steady inner, midpoint, and outer temperatures of 40°F, 75°F, and 100°F, respectively. If the outer diameter is 6 in., what is the inner diameter? Can the conductivity be estimated?

2.26 The slab in Fig. P2.26 has a linearly variable conductivity as shown. For steady conduction, compute and sketch the temperature distribution and compare with Fig. 2.4. What is $T$ at $x = L/2$?

2.27 A cone frustrum has $T_1$ = 200°F and $D_1$ = 4 in. at $x_1 = 0$; $T_2$ = 60°F, $D_2$ = 2 in. at $x_2 = L$ = 6 in. Assume one-dimensional heat flow with insulated curved sides. If $k$ = 40 Btu/hr · ft · °F = constant, evaluate (a) $q_x$ and (b) $T(L/2)$.

2.28 According to the two-dimensional shape-factor theory of Chapter 3, the hollow square solid with the conditions in Fig. P2.28 transmits a total heat flux of 13,700 W per meter of depth into the paper. Compare this result with a one-dimensional analysis for the same geometry.

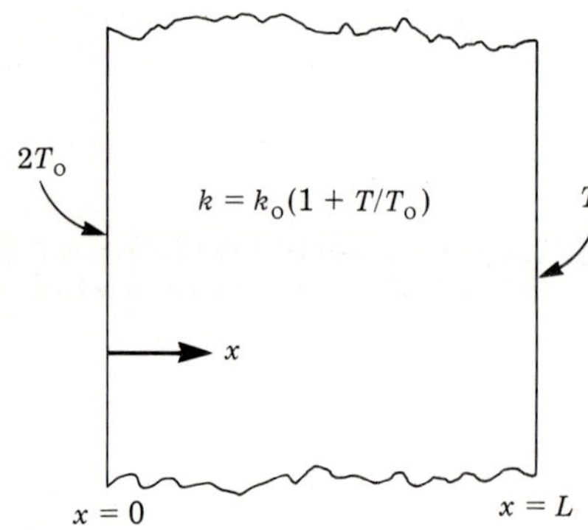


Figure P2.26

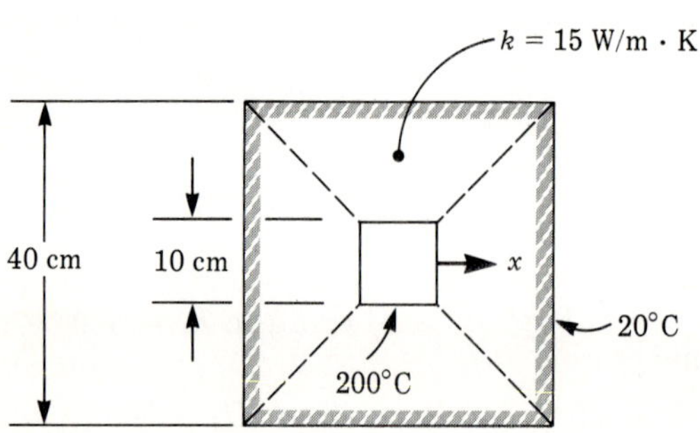


Figure P2.28

2.29 For the eccentric annular solid of Fig. P2.29, shape-factor theory from Chapter 3 predicts a heat flux of $q_x = 39.8k(T_1 - T_2)$ per unit depth. Compare this result to a one-dimensional approximation for this shape.

2.30 A house wall consists of 2 cm of plaster, 12 cm of glass-fiber insulation, and 4 cm of oak wood. Inner and outer heat transfer coefficients are 10 and 25 W/m$^2$ · K, and inner and outer temperatures are 20°C and −12°C, respectively. Compute $q_x''$.

2.31 A steam pipe is carbon steel with $d_1 = 1$ in. and $d_2 = 1.2$ in., covered by 1-in.-thick foam insulation ($k = 0.02$ Btu/hr · ft · °F). Steam conditions are $T_i = 250$°F and $h_i = 25$ Btu/hr · ft$^2$ · °F; outside air is at $T_o = 60$°F and $h_o = 5$ Btu/hr · ft$^2$ · °F. Compute (a) the heat loss per foot of length and (b) the temperature at the outside of the foam insulation.

2.32 A furnace is 3 m by 4 m by 5 m, with a composite wall of 15 cm silica brick, 5 cm glass fiber, and 1 cm carbon steel. Inner conditions are $T_i = 450$°C and $h_i = 15$ W/m$^2$ · K; outer conditions are $T_o = 20$°C and $h_o = 20$ W/m$^2$ · K. Compute (a) $q_x$ and (b) the temperature at the brick/glass fiber interface.

2.33 Suppose that the glass fiber in Problem 2.32 is limited to a maximum temperature of 320°C. How much extra silica brick should be added? What is the heat loss?

2.34 A cast-iron steam pipe has $d_1 = 4$ in. and $d_2 = 4.4$ in. Inner conditions are $T_i = 350$°F and $h_i = 40$ Btu/hr · ft$^2$ · °F. Outer conditions are $T_o = 30$°F and $h_o = 15$ Btu/hr · ft$^2$ · °F. Compute the heat loss if $L = 100$ ft. If 1 in. of insulation ($k = 0.022$ Btu/hr · ft · °F) is to be added, what is the heat loss for (a) inside and (b) outside installation?

2.35 Refrigerant flows at $T_i = -40$°C, $h_i = 11$ W/m$^2$ · K inside a copper pipe of inner diameter 3 cm and wall thickness 2 mm. Outside conditions are $T_o = 25$°C and $h_o = 18$ W/m$^2$ · K. What thickness of insulation

**Figure P2.29**

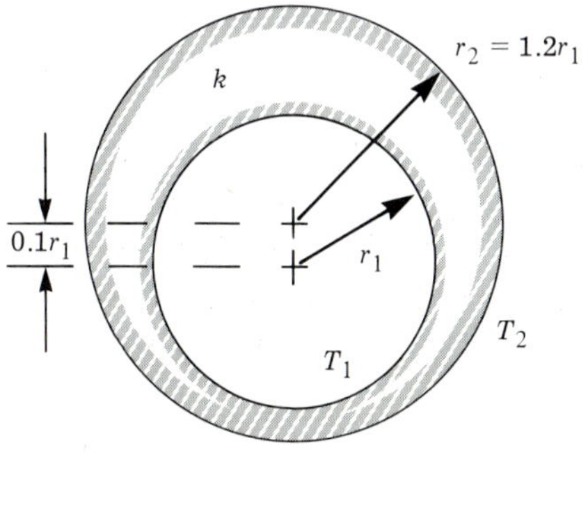

($k$ = 0.04 W/m · K) should be added outside to make the heat loss 25 W per meter of pipe length?

2.36 For the three-layer slab in steady conduction in Fig. P2.36, arrange the conductivities $k_1$, $k_2$, and $k_3$ in order of increasing magnitude and explain your choices.

2.37 An aluminum box is a 40-cm cube, 1 cm thick, filled with ice at $T_i$ = 0°C and $h_i$ = 12 W/m$^2$ · K. Outside conditions are $T_o$ = 35°C and $h_o$ = 8 W/m$^2$ · K. What thickness of foam insulation ($k$ = 0.04 W/m · K) is required to reduce the rate of ice melting to 200 g/hr?

2.38 A 2-ft-inner-diameter sphere is constructed of 1 in. of insulation sandwiched between two ¼-in. thicknesses of stainless steel. Inner and outer temperatures are $T_i$ = 10°F and $T_o$ = 70°F, with $h_i$ = $h_o$ = 15 Btu/hr · ft$^2$ · °F. What insulation conductivity is required to make the heat flux 250 Btu/hr?

2.39 Given the composite wall in Fig. P2.39, with $k_A$ = 18 W/m · K and $k_C$ = 40 W/m · K. If $T_3$ is measured to be 455°C, compute (a) $q''_x$, (b) $k_B$ and (c) $T_5$.

2.40 A composite house wall consists of 18 mm of gypsum plaster, 102 mm of loose vermiculite, and 25 mm of white pine wood. Inside conditions

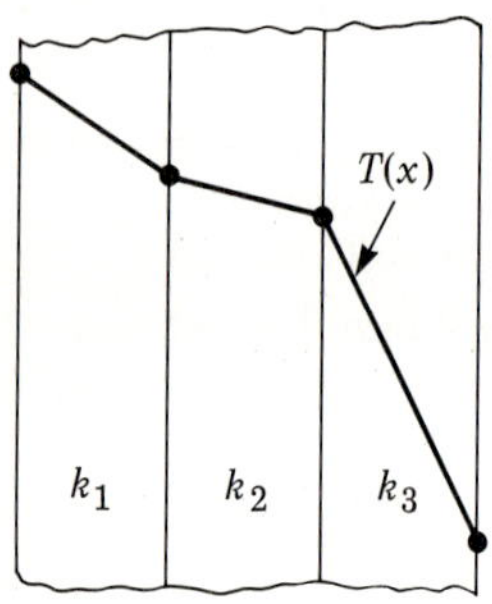


Figure P2.36

Figure P2.39

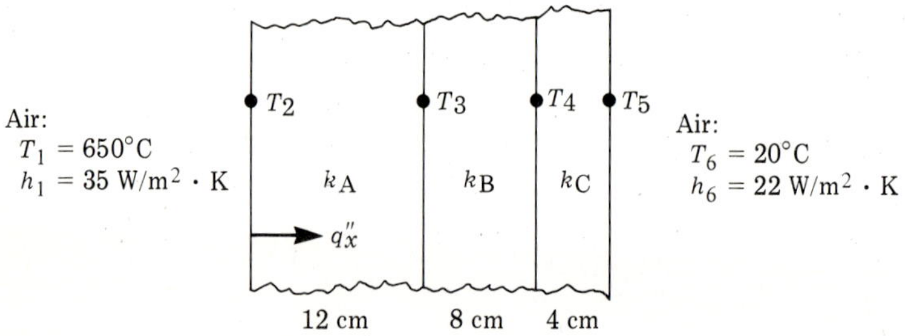

are $T_i = 20°\text{C}$ and $h_i = 25\ \text{W/m}^2 \cdot \text{K}$. Outside conditions are $h_o = 35\ \text{W/m}^2 \cdot \text{K}$ and an atmospheric temperature that varies diurnally:

$$T_o = -5°\text{C} - 12°\text{C}\cos(2\pi t/24),$$

with $t$ in hours. If the wall area is 90 m$^2$, find the total heat loss in joules over a 24-hr period, and the cost of heating per day at 10¢/kwh.

2.41 For the system of Fig. P2.41, take $k_1 = 5$, $k_2 = 8$, and $k_3 = 12\ \text{W/m} \cdot \text{K}$; $T_1 = 100°\text{C}$, $T_3 = 20°\text{C}$; $A_1 = 0.4\ \text{m}^2$, $A_2 = 0.6\ \text{m}^2$; $\Delta x_1 = 7$ cm, $\Delta x_2 = 5$ cm. Compute (a) $q_x$ and (b) $T_2$.

2.42 In Fig. P2.41, prove that for arbitrary properties $k_{1,2,3}$ there is no transverse cross-conduction between elements $k_1$ and $k_2$ if $T_1$ and $T_3$ are uniform.

2.43 A house has 3000 ft$^2$ of walls constructed of panels of the type shown in Fig. P2.43. The wood is pine, and the insulation has $k = 0.04$ Btu/hr · ft · °F. Inside conditions are $T_i = 68°\text{F}$ and $h_i = 8$ Btu/hr · ft$^2$ · °F; on the outside, $T_o = 0°$ and $h_o = 12$ Btu/hr · ft$^2$ · °F. Compute the total heat loss through these walls, neglecting radiation. If the furnace delivers 110,000 Btu/gal of fuel, how many gallons of fuel are burned per day for these conditions?

2.44 Repeat Problem 2.43 by adding in 300 ft$^2$ of single-pane window glass 3/16 in. thick. What do you conclude?

2.45 A picnic cooler is 16 in. by 12 in. by 10 in. high, made of ¾-in.-thick polystyrene. It contains ice at 32°F with $h_i = 9$ Btu/(hr · ft$^2$ · °F). Outside

**Figure P2.41**

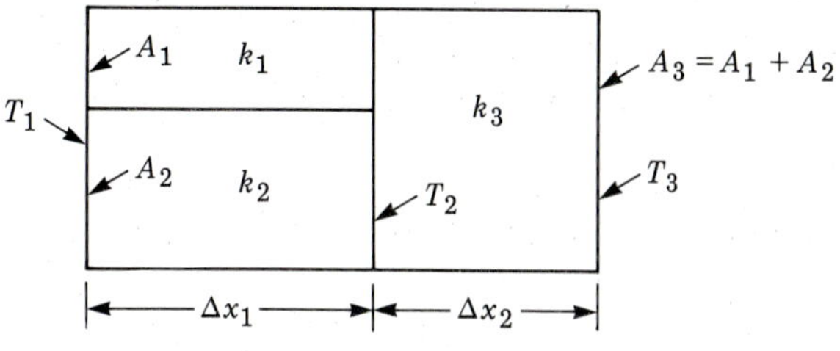


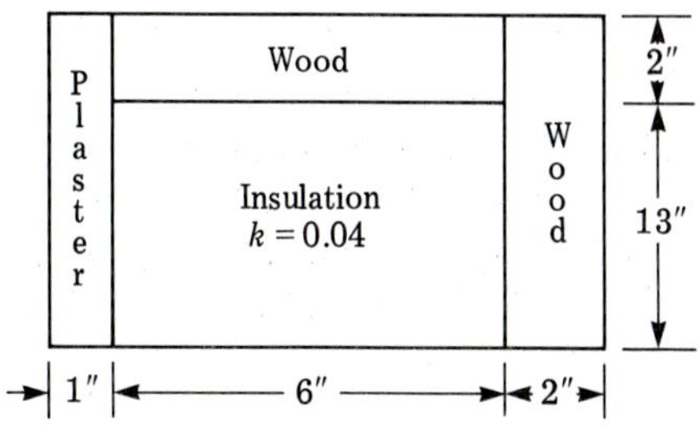


**Figure P2.43**

conditions are $T_o = 90°F$ and $h_o = 5$ Btu/(hr · ft$^2$ · °F). Estimate the rate of ice melting in lb$_m$/hr.

2.46 For the dual-parallel system of Fig. P2.46, under what conditions will there be conduction crosstalk between $k_1$ and $k_2$ and between $k_3$ and $k_4$? For what ratios of conductivity will the crosstalk be (a) zero and (b) very large? (For further details, see [5, pp. 53–55].)

2.47 For the system of Fig. P2.46, take the following data:

$$k_1 = 5, \quad k_2 = 8, \quad k_3 = 12, \quad k_4 = 3 \quad \text{W/m} \cdot \text{K};$$

$$A_1 = 0.4 \text{ m}^2, \quad A_2 = 0.6 \text{ m}^2;$$

$$T_1 = 200°\text{C}, \quad T_2 = 40°\text{C};$$

$$\Delta x_1 = 5 \text{ cm}, \quad \Delta x_2 = 7 \text{ cm}.$$

Evaluate the total heat transfer in watts for (a) two series systems in parallel, and (b) two parallel systems in series. Why are the two results different? Which is the better estimate?

2.48 A spherical container with an inner diameter of 40 cm consists of 4 cm of lead covered by 1 cm of 347 stainless steel and 3 cm of concrete. The container is filled with nuclear wastes that generate heat at a rate $\dot{q}$ (W/m$^3$). The sphere is placed in the deep ocean where $T_o = 8°C$ and $h_o = 115$ W/m$^2$ · K. If the lead temperature is to remain 100°C less than its melting point, what is the maximum allowable heat-generation rate?

2.49 For the system in Fig. P2.49, $T_i = 400°C$ and $h_i = 15$ W/m$^2$ · K. If radiation is negligible on the left side, compute the total heat flux.

2.50 An electrically heated wire with a diameter of 0.040 in. and a length of 5 ft develops a surface temperature of 400°F. Ambient air conditions are $T_o = 60°F$ and $h_o = 5$ Btu/hr · ft$^2$ · °F. Neglecting radiation, compute the heat loss (a) of the bare wire and (b) if the wire is covered with ⅛ in. of insulation ($k = 0.025$ Btu/hr · ft · °F). Can you explain these results?

2.51 What happens to the critical-radius concept when you add insulation to

**Figure P2.46**

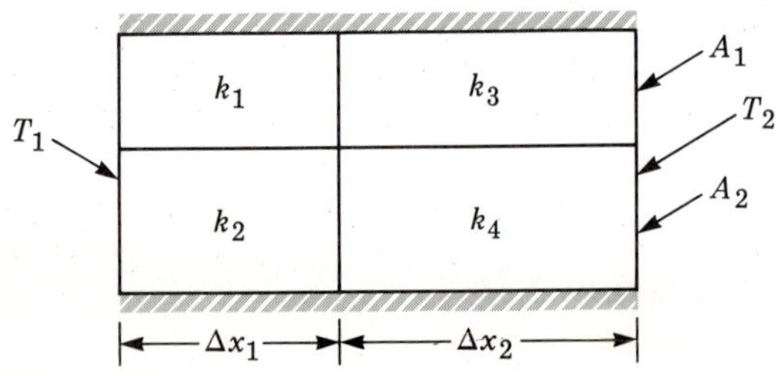

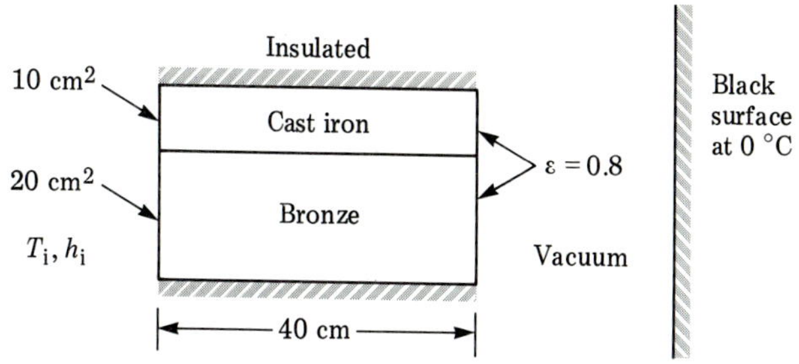


**Figure P2.49**

the *inside* of a cylinder? Let the system of Fig. P2.51 have initial resistance $\Sigma R$, not including inside convection conditions ($T_i$, $h_i$). Insulation of conductivity $k_i$ is added inside to reduce the radius to $r_i$. Evaluate $q_x$ as a function of $r_i$ and determine whether it has a maximum at a critical radius.

2.52 An aluminum wire has a diameter of 3 mm and a surface temperature of 150°C, with ambient conditions $T_o$ = 20°C and $h_o$ = 8 W/m$^2$ · K. If you want to add a small amount of insulation of conductivity $k$, what is the maximum allowable $k$ that will cause a reduction in heat flux? If, in fact, $k$ = 0.035 W/m · K, what is the critical radius and the maximum heat flux for that condition?

2.53 If you want to add insulation of conductivity $k$ to a sphere of radius $R_o$ with ambient conditions $T_o$ and $h_o$, derive an expression for the critical radius of insulation for the sphere. Does your result resemble Eq. (2.34)?

2.54 Evaluate the critical radius of insulation for the wire of Problem 2.50 when (a) radiation is neglected, and (b) blackbody radiation is included.

**Figure P2.51**

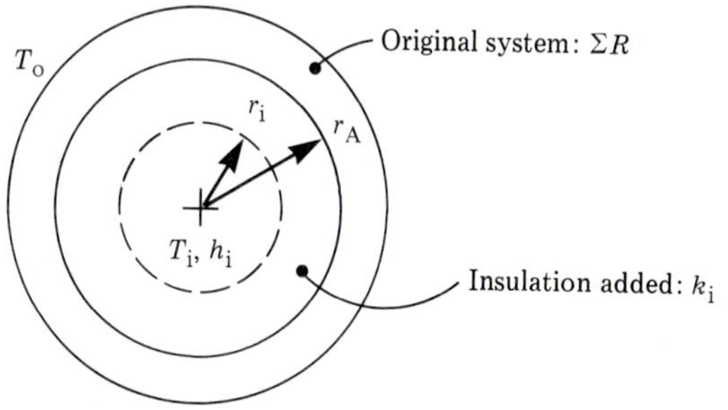

2.55 A slab 3 in. thick, with $k = 4.5$ Btu/hr · ft · °F, is subjected to uniform heat generation $\dot{q} = 70{,}000$ Btu/hr · ft$^3$. Ambient conditions on each side are $T_o = 60$°F, $h_o = 35$ Btu/hr · ft$^2$ · °F. Compute the maximum and surface temperatures in the slab for steady conditions.

2.56 A slab of thickness $L$ is subjected to steady uniform heat generation and unequal wall temperatures, as in Fig. P2.56. Find an expression for $T(x)$ using the coordinates shown.

2.57 Derive the temperature distribution $T(R)$ for a solid sphere of outer radius $R_o$, surface temperature $T_o$, constant $k$, and uniform heat generation $\dot{q}$. What is $\Delta T_{max}$?

2.58 A steel alloy wire ($k = 31$ W/m · K, $\gamma = 110$ $\mu\Omega$ · cm) carries a current of 6 A and has a diameter of 1 mm and a length of 50 cm. What is the total heat generated in watts? If ambient conditions are $T_\infty = 20$°C and $h_\infty = 18$ W/m$^2$ · K, what is the centerline temperature of the wire?

2.59 Would it reduce the center temperature in the wire of Problem 2.58 if it were wrapped with 2 mm of insulation ($k = 0.1$ W/m · K)? If your answer is yes, compute the new centerline temperature.

2.60 Using the parameters in Fig. P2.60, derive an analytic expression for the insulated-side temperature $T(x = 0)$ as a function of $\dot{q}$, $k$, $L$, $T_\infty$, and $h_\infty$. Does a Biot number appear?

2.61 A slab of $k = 15$ Btu/hr · ft · °F is 1 in. thick and subjected to uniform $\dot{q} = 500{,}000$ Btu/hr · ft$^3$. If the right-side temperature is $T_1 = 80$°F, what should the left-side temperature $T_o$ be so that all of the heat generated flows out of the right side only?

2.62 A long hollow cylinder of constant $k$, inner radius $r_i$, and outer radius $r_o$ is subjected to uniform heat generation $\dot{q}$. Wall temperature is $T_o$ at both inner and outer surfaces. Find an analytic expression for $T(r)$ in the cylinder. Is this analogous to laminar viscous flow through an annulus?

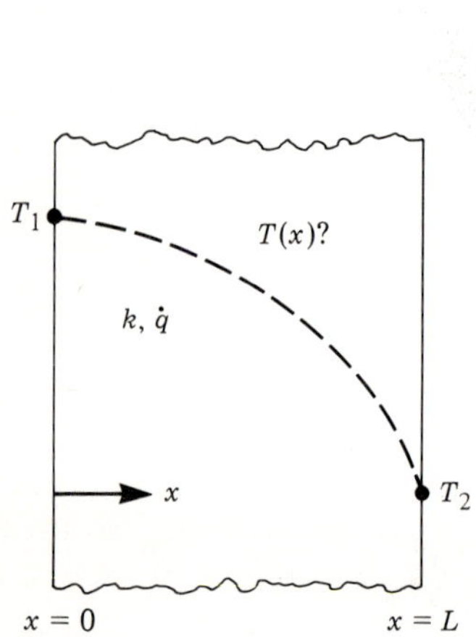


**Figure P2.56**

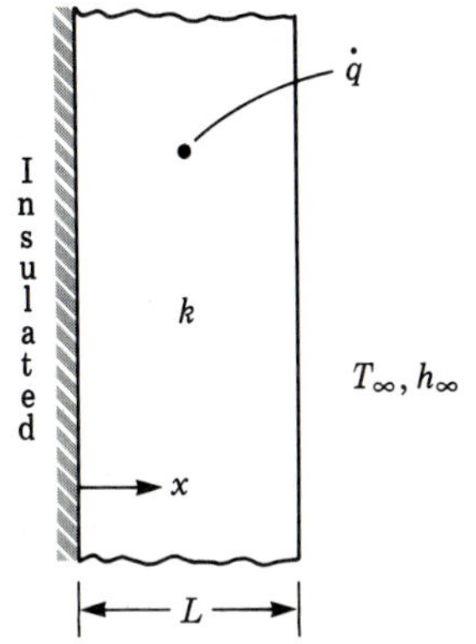


Figure P2.60

2.63 A stainless steel strip ($\lambda = 75\ \mu\Omega \cdot \text{cm}$) is 0.02 in. thick, ¼ in. wide, and 12 in. long. It carries a current of 40 A parallel to the long side. If ambient conditions are $T_\infty = 60°\text{F}$ and $h_\infty = 35\ \text{Btu/hr} \cdot \text{ft}^2 \cdot °\text{F}$, estimate the temperature in the center plane of the strip.

2.64 The 30-mm-radius fuel rod in Fig. P2.64 is to be clad with 2024T6 aluminum. The fuel generates heat uniformly and has conductivity $k_f = 55$ W/m · K. It is desired that the fuel generate maximum power but that fuel and cladding temperatures not exceed 1300°C and 500°C, respectively. Find appropriate values of (a) $r_o$, and (b) $\dot{q}$ to satisfy these conditions.

2.65 A rectangular fin of type 347 stainless steel is 1 mm thick, 8 mm wide, and 30 mm long. Base temperature is 240°C, and the surrounding air is at 20°C, with $h_\infty = 14\ \text{W/m}^2 \cdot \text{K}$. Estimate (a) the tip temperature, (b) the efficiency, and (c) the total heat transfer.

2.66 Repeat Problem 2.65 if the fin is 2024T6 aluminum.

Figure P2.64

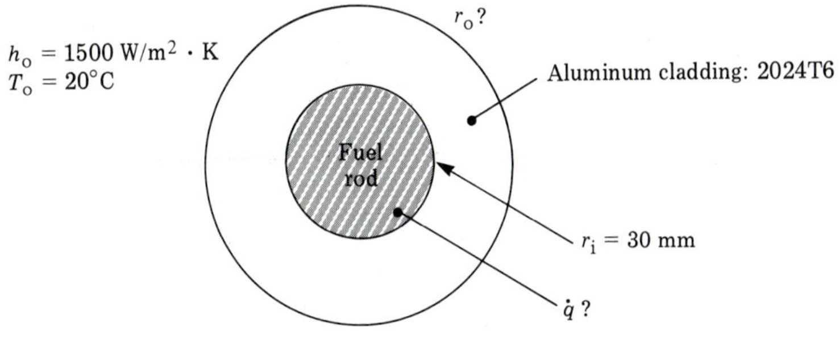

2.67 Consider a case D to be added to Table 2.2, in which the fin tip is held at a known temperature $T_L$. Derive expressions for the temperature distribution and the heat transfer rate.

2.68 Heat from an electronic device is to be dissipated by rectangular aluminum fins that are 0.06 in. thick and 0.85 in. wide, with their bases at 100°C and the ambient air at 55°C. Each fin should remove 1 W of heat transfer from the base. If $h_\infty = 9\ \mathrm{W/m^2 \cdot K}$, how long in inches should the fins be?

2.69 According to the criterion of Eq. (2.69), what is the lowest conductivity for which the fin of Problem 2.65 will be acceptably efficient?

2.70 The thermal conductivity of material X is to be measured by comparing two pin fins of diameter 5 mm and length 95 mm, each attached to a base at 150°C and exposed to identical convection environments. As shown in Fig. P2.70, the aluminum fin has a tip temperature of 125°C, while the fin of material X has $T_L = 90°\mathrm{C}$. If the tips are insulated, what is $k_X$? What are the possible inaccuracies of this measurement technique?

2.71 A bronze circular pin fin has $D = 0.07$ in. and $L = 0.5$ in. The tip is convecting, and $h_\infty = 20\ \mathrm{Btu/hr \cdot ft^2 \cdot °F}$. What is the fin efficiency? What thermal conductivity is needed for the efficiency to be 90%?

2.72 An aluminum rod 20 in. long and ½ in. in diameter is inserted through a hole in a furnace wall so that half the rod is inside and half outside. Neglect the furnace wall thickness. Furnace air conditions are $T_1 = 500°\mathrm{F}$ and $h_1 = 20\ \mathrm{Btu/hr \cdot ft^2 \cdot °F}$. Outside air conditions are $T_2 = 60°\mathrm{F}$ and $h_2 = 15\ \mathrm{Btu/hr \cdot ft^2 \cdot °F}$. Estimate the rod temperatures (a) at the midpoint (in the plane of the wall) and (b) at the outside tip in the room air.

2.73 Repeat the derivation of Eq. (2.52) to account for uniform heat generation $\dot{q}$ throughout the fin. Describe the mathematical character of the resulting equation.

2.74 The rod in Fig. P2.74 connects two walls and is surrounded by fluid.

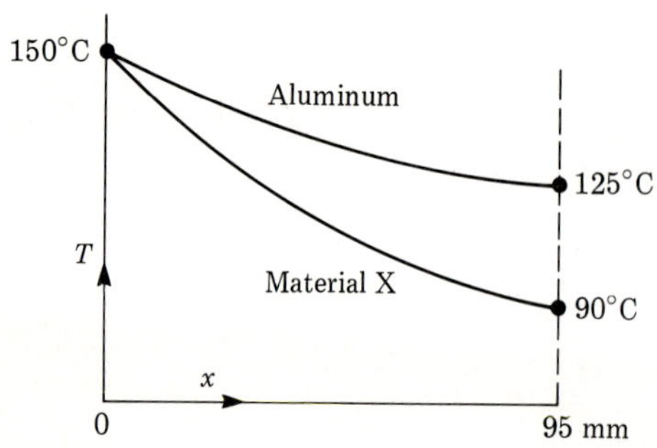


**Figure P2.70**

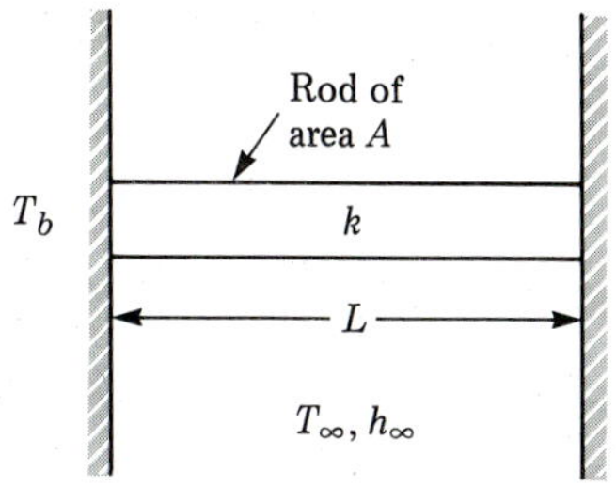


Figure P2.74

Explain how the geometry and boundary conditions could be adopted from fin analysis to compute the heat transfer from the rod.

2.75 Continuing our study of Fig. P2.74, suppose that the walls are at different temperatures, $T_{b1}$ and $T_{b2}$. Using fin theory, find an expression for the temperature distribution $T(x)$ in the rod.

2.76 In Problem 2.75, suppose that $T_{b1} = 200°C$ and $T_{b2} = 120°C$, with ambient air at $T_\infty = 20°C$ and $h_\infty = 15\ W/(m^2 \cdot K)$. If the rod is pure nickel of diameter 5 cm and length 45 cm, estimate the total heat loss from the rod.

2.77 A brass slab is 2 cm thick and has a gas on one side, $T_1 = 500°C$ and $h_1 = 15\ W/m^2 \cdot K$, and a liquid flow on the other side, $T_2 = 40°C$ and $h_2 = 2000\ W/m^2 \cdot K$. Compute the heat flux through this bare slab in $W/m^2$. It is desired to add brass fins, 1 cm long and 1 mm thick, spaced 1 cm apart on centers. Compute the percent increase in heat flux with the fins placed (a) on the gas side and (b) on the liquid side.

2.78 Consider a straight rectangular fin with an insulated tip, of length $L$, width $W$, and thickness $t \ll W$. Assume that both the width $W$ and the total fin volume $V = LWt$ are constant. Show that the optimum fin thickness,

$$t_{opt} = 0.993 h_\infty L^2/k,$$

maximizes the total heat transfer from the fin. Why is this solution inconsistent if the Biot number $(h_\infty L/k)$ is large?

2.79 Figure P2.79 shows a typical electronic cooling device made of stamped aluminum, with ten fins on a base that is bolted to a transistor. The base temperature is 70°C. Ambient conditions are $T_\infty = 30°C$ and $h_\infty = 5\ W/m^2 \cdot K$. Estimate the fin efficiency and the total heat dissipated.

2.80 According to the criterion of Eq. (2.69), what is the minimum thermal conductivity for which the ten-fin transistor cooler of Fig. P2.79 will have acceptable efficiency? What will be the total heat dissipated under these conditions?

2.81 A 3-cm-thick slab has $k = 30\ W/m \cdot K$ and generates heat at $\dot{q} = 5 \times$

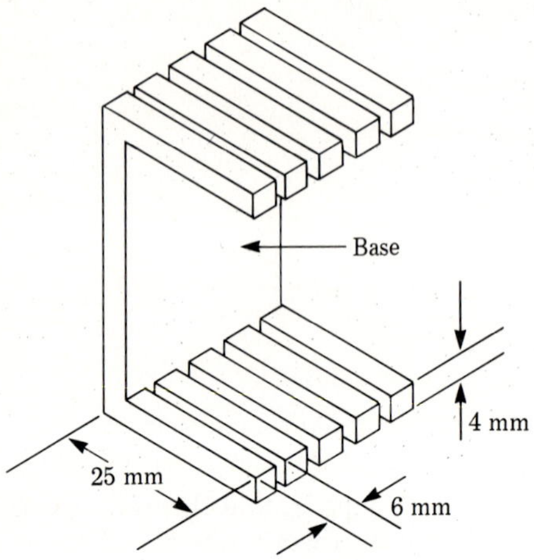


**Figure P2.79**

$10^6$ W/m$^3$. As shown in Fig. P2.81, it is sheathed on both sides by steel alloy sheets ($k$ = 52 W/m · K) 1 cm thick, fitted with fins 2 cm long and 1 mm thick, spaced 1 cm apart on centers. The outside air temperature is $T_\infty$ = 20°C. If the center-slab temperature should not exceed 400°C, what is the minimum allowable external heat transfer coefficient $h_\infty$?

2.82 An aluminum rectangular fin is 15 mm wide and 3 mm thick. Its base temperature is 180°C, and ambient conditions are $T_\infty$ = 30°C and $h_\infty$ = 18 W/m$^2$ · K. How long is the fin if its exposed tip temperature is 80°C? What is its efficiency? What is the heat transfer rate?

2.83 A stainless steel cylinder is 4 cm in diameter and 10 cm long with a uniform base temperature of 300°C. It is desired to add ten stainless steel radial fins (see Fig. 2.16), each 1 cm long and 1 mm thick. Neglecting radiation and end losses from the flat faces of the cylinder, estimate the

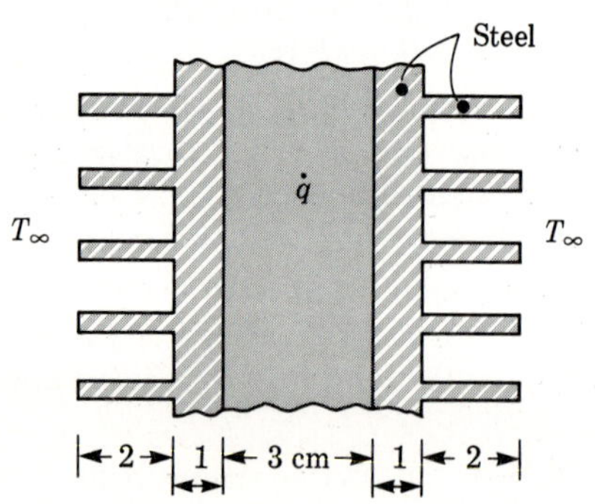


**Figure P2.81**

heat loss, for $T_\infty = 50°C$ and $h_\infty = 15\ W/m^2 \cdot K$, (a) without and (b) with the fins. Assume no change in base temperature.

2.84 Show that, for a constant-area fin with insulated or negligible tip heat transfer, the fin efficiency as defined by Eq. (2.64) is equal to the shaded area in Fig. P2.84 divided by the area of rectangle *abcd*.

2.85 A cast iron engine cylinder has an outer wall of temperature 300°C and diameter 10 cm. Its upper 8 cm are cooled by nine equally spaced circular fins of size $L = 3$ cm and $\delta = 4$ mm (see Fig. 2.16). Ambient air conditions are $T_\infty = 30°C$ and $h_\infty = 16\ W/m^2 \cdot K$. Estimate the total heat loss from this part of the cylinder.

2.86 A baseboard room heater delivers water at 65°C through a 2-cm-diameter copper tube, as in Fig. P2.86. The tube is finned by wrapping 1-mm-thick carbon-steel sheet in spirals so that the outer fin diameter is 6 cm. The room air is at 20°C with $h_\infty = 8\ W/m^2 \cdot K$. The system is to deliver 150 W per meter of length to the air. Estimate the number of spiral windings needed per meter.

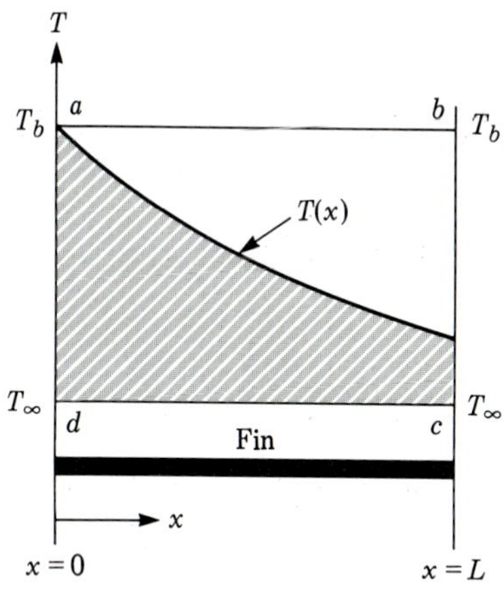


Figure P2.84

Figure P2.86

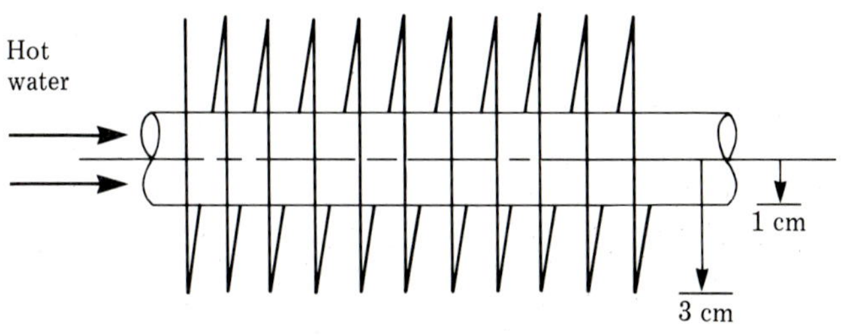

2.87 For fins with variable geometry $A(x)$ and $P(x)$, we may either look for a known exact solution in the standard references [7, 12, 17] or else attempt a numerical solution of Eq. (2.52). For a finite-difference model, divide the fin into $N$ equally spaced slices, as in Fig. P2.87. Derive a finite-difference form of Eq. (2.52) that relates $T_j$ at any node $j$ to its neighbors, $T_{j+1}$ and $T_{j-1}$. The formula will vary according to the particular approximations you make, but one suggested algorithm is

$$\Theta_{j-1}(A_j + A_{j-1}) - \Theta_j \times \left(A_{j-1} + 2A_j + A_{j+1} + \frac{2h\Delta x^2}{k} P_j\right) + \Theta_{j+1}(A_{j+1} + A_j) = 0.$$

This formula assumes that $h$, $k$, and $\Delta x$ are constant. Also suggest suitable approximations for (a) an insulated tip, and (b) a convecting tip.

2.88 Show that, for a fin of uniform area and perimeter, the algorithm in Problem 2.87 reduces to

$$\Theta_{j-1} - \left(2 + \frac{\Delta x^2}{\lambda^2}\right)\Theta_j + \Theta_{j+1} = 0.$$

Apply this formula to a uniform fin with an insulated tip and the following properties: $\Theta_b = 100°\text{C}$, $A = 1\ \text{cm}^2$, $P = 4$ cm, $k = 40$ W/m · K, $h = 10$ W/m$^2$ · K, and $L = 20$ cm. Take $\Delta x = 4$ cm ($N = 5$), compute the five nodal temperatures and the total heat transfer, and compare with the exact solution (Case B, Table 2.2).

2.89 Apply the variable-area formula of Problem 2.87 to a fin with linearly varying $A = A_o(1 - x/2L)$, $P = P_o(1 - x/2L)$, and $A_o = 1\ \text{cm}^2$, $P_o =$

**Figure P2.87**

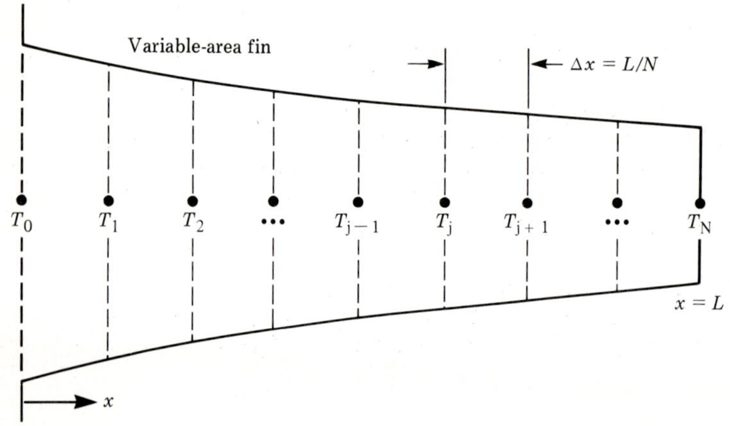

4 cm, $\Theta_b = 100°C$, $k = 40$ W/m · K, $h = 10$ W/m$^2$ · K, and $L = 20$ cm. Take $\Delta x = 4$ cm ($N = 5$) and compute the five nodal temperatures and the total heat transfer if the tip is insulated. Can you find the exact solution for this problem in [12]?

2.90 In Problem 2.87, you probably suggested the first-order insulated tip approximation $\Theta_N \simeq \Theta_{N-1}$. Do not reveal this result to your friends now attempting Problem 2.87. A second-order or three-point approximation would be

$$\Theta_N \simeq \frac{4}{3}\Theta_{N-1} - \frac{1}{3}\Theta_{N-2}.$$

The reasoning behind this formula is explained in Section 3.4.4. Use this formula to help solve the numerical problem posed in Problem 2.88. The nodal temperatures should be considerably more accurate.

# Multidimensional Steady Conduction

# 3

## 3.1 Introduction

In Chapter 2 we derived the general three-dimensional heat conduction equation, Eq. (2.6), and its boundary conditions, Eqs. (2.11). However, applications in Chapter 2 were limited to one-dimensional temperature distributions, such as slabs and hollow cylinders. These invariably led to neat algebraic formulas for heat transfer and temperature. Groups of one-dimensional elements could easily be hooked together by the electric analogy (Section 2.4). I think it is fair to say that Chapter 2 was quite straightforward.

Now we will look at multidimensional conduction in solids: steady here in Chapter 3 and unsteady in Chapter 4. We will find that multidimensional analyses are easy in principle but often lead to messy results. Exact solutions result in complex formulas (Section 3.2), and numerical solutions culminate in a bulky table of grid temperatures (Section 3.4). But the basic approach is relatively simple: Multidimensional conduction is a wonderfully mature topic directly related to many other fields of applied physics. The primary analytical technique is the method of Fourier series, used today in much the same way as Fourier proposed it in 1807. The basic numerical technique is a simple averaging process that you can easily remember even without a book.

One-dimensional conduction idealizations usually fail for one of two reasons: (1) the solid body is blocky or irregular in shape, or (2) the boundary conditions are irregular or spatially variable. As an example of the latter case, consider the hollow cylinder in Table 2.1. If the inner or outer boundary temperature $T_i$ or $T_o$ varies around the circumference or along the length of the cylinder, the resulting temperature distribution is definitely two- or three-dimensional, and the thermal resistance concept in Table 2.1 fails.

An example of a blocky or irregular shape is the thick hollow chimney shape in Fig. 3.1. Even if the boundary temperatures $T_i$ and $T_o$ are uniform, the shape itself causes two-dimensionality. Near the corners, the heat flux vectors take two greatly different directions, and the region influenced by these corners is a large portion of the chimney. No single variable $x$ can describe the flux paths for this shape. The exact isotherms, $T(x, y)$ = constant, are shown in Fig. 3.1(a) to be curves that crowd near the inner corners and shy away from the outer corners. The exact analysis is not known analytically to the author,† but a numerical solution can easily be obtained (Section

†An approximate shape factor is given in item 2 of Table 3.1.

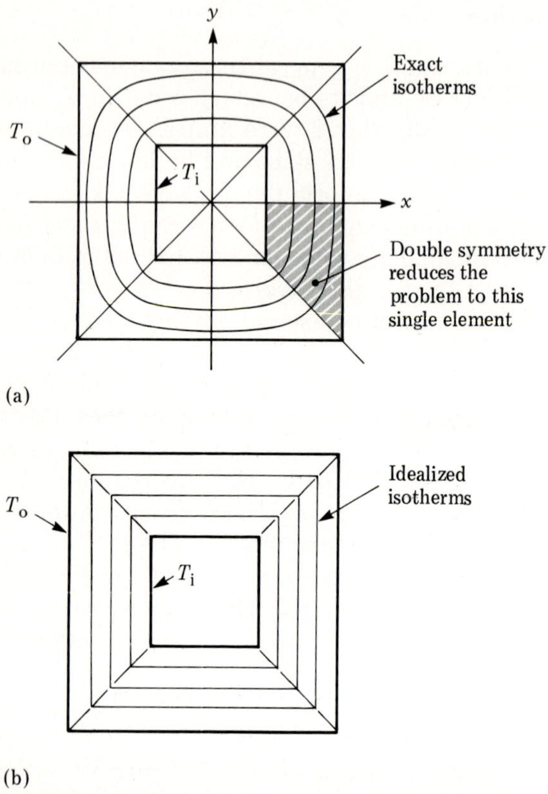


**Figure 3.1 Heat conduction through a hollow square chimney shape: (a) exact two-dimensional temperature distribution; (b) idealized one-dimensional approximation.**

3.4) for the shaded trapezoidal region in Fig. 3.1(a) to almost any accuracy. See Example 3.10 for a crude numerical model of a chimney.

The one-dimensional approximation to this chimney problem is shown in Fig. 3.1(b). The corner regions are not well modeled, so the results will be relatively crude, perhaps 30% in error for the temperatures and heat flux. If, however, the walls were very thin, the accuracy would be excellent.

A final remark is that Fig. 3.1(a), though exact mathematically, is not a good model for a real chimney. A real chimney is better represented by convection boundary conditions, which would result in boundary temperatures that vary both around the chimney and along its axis.

For further details or particular problems not treated in this chapter, you may consult several excellent advanced textbooks on multidimensional conduction analysis [1–6].

### 3.1.1 Steady Heat Conduction

The general heat conduction equation, including unsteady heat storage and variable conductivity, was derived and discussed as Eq. (2.6), repeated here:

$$\frac{\partial}{\partial x}\left(k\frac{\partial T}{\partial x}\right) + \frac{\partial}{\partial y}\left(k\frac{\partial T}{\partial y}\right) + \frac{\partial}{\partial z}\left(k\frac{\partial T}{\partial z}\right) + \dot{q} = \rho c_v \frac{\partial T}{\partial t}. \tag{3.1}$$

This equation looks formidable but is actually extremely well behaved. Even with variable properties ($k$, $\rho$, $c_v$, $\dot{q}$), it may be readily solved by the numerical (digital computer) techniques to be discussed in Section 3.4.

The present chapter considers only steady heat flux ($\partial/\partial t \equiv 0$) with constant $k$, for which Eq. (3.1) reduces to

$$\frac{\partial^2 T}{\partial x^2} + \frac{\partial^2 T}{\partial y^2} + \frac{\partial^2 T}{\partial z^2} = \nabla^2 T = -\frac{\dot{q}}{k}. \tag{3.2}$$

This is *Poisson's equation,* probably the second most heavily studied equation in applied physics. The nonhomogeneous term $(-\dot{q}/k)$ is called the *source function* for the solution $T(x, y, z)$, and solutions are known for a wide variety of geometries. Its two-dimensional form ($\partial/\partial z \equiv 0$) is exactly what one solves for the axial velocity $u(x, y)$ in fully developed laminar duct flow [8, pp. 119–128].

Finally, the bulk of our examples in this chapter are steady conduction with constant $k$ and no generation, $\dot{q} = 0$:

$$\frac{\partial^2 T}{\partial x^2} + \frac{\partial^2 T}{\partial y^2} + \frac{\partial^2 T}{\partial z^2} = 0. \tag{3.3}$$

This is *Laplace's equation,* probably the most thoroughly analyzed equation in applied physics. Its solutions are called "potential" or "harmonic" functions and occur in electromagnetic theory, inviscid fluid flow, low Reynolds number flows, and many other applications. Since it is linear, its solutions can be added together, which is the basis of the Fourier series method of Section 3.2. You will never find a more well-behaved equation than Laplace's [7].

### 3.1.2 Boundary Conditions

As discussed earlier in Section 2.1.4, the heat conduction equation, Eq. (3.2), is second-order in space and requires known conditions on either $T$ or its normal derivative at *every* boundary point. The four most common boundary conditions were shown in Fig. 2.3:

1. Known temperature:

$$T_b = T_o, \tag{3.4}$$

2. Insulated wall:

$$\left.\frac{\partial T}{\partial n}\right|_b = 0, \tag{3.5}$$

3. Known heat flux:

$$-k_b \left.\frac{\partial T}{\partial n}\right|_b = q''_o, \tag{3.6}$$

4. Convection environment:

$$-k_b \left.\frac{\partial T}{\partial n}\right|_b = h_\infty(T_b - T_\infty), \tag{3.7}$$

where $n$ denotes the outward normal coordinate. All four of these conditions are linear in temperature if the thermal conductivity is constant.

A fifth possible condition would be a radiation boundary:

$$-k_b \left.\frac{\partial T}{\partial n}\right|_b = \sigma F_{b\infty}(T_b^4 - T_\infty^4), \tag{3.8}$$

where $F_{b\infty}$ is a known radiation shape factor. This condition is nonlinear because of the term $T_b^4$, and hence elementary solutions to the heat conduction equation cannot be added to satisfy this requirement.

**Example 3.1**

Using a reference body temperature $T_o$ and a reference body length $L$, nondimensionalize the heat conduction equation, Eq. (3.2), and the four boundary conditions, Eqs. (3.5–3.8). Do any important dimensionless parameters arise?

*Solution* After much thought, guided by nearly two centuries of conduction analysis, we select the following definitions for dimensionless variables:

$$x^* = x/L; \quad y^* = y/L; \quad z^* = z/L; \quad n^* = n/L; \quad t^* = \alpha t/L^2;$$
$$\theta = (T - T_\infty)/(T_o - T_\infty).$$

Substitution of these variables into Eq. (3.2) gives the following nondimensional heat conduction equation:

$$\frac{\partial^2\theta}{\partial x^{*2}} + \frac{\partial^2\theta}{\partial y^{*2}} + \frac{\partial^2\theta}{\partial z^{*2}} + \frac{\dot{q}L^2}{k(T_o - T_\infty)} = \frac{\partial\theta}{\partial t^*}. \tag{3.9}$$

The only parameter is the dimensionless heat generation $\dot{q}L^2/k(T_o - T_\infty)$, whose form was predicted by Eq. (2.40) earlier.

The prescribed heat flux condition, Eq. (3.6), becomes

$$\left.\frac{\partial\theta}{\partial n^*}\right|_b = -\, q''_b L/k(T_o - T_\infty). \tag{3.10}$$

The parameter on the right is a sort of Biot number. The special case of an insulated boundary has no parameters:

$$\left.\frac{\partial\theta}{\partial n^*}\right|_b = 0. \tag{3.11}$$

The convection boundary condition introduces the more familiar Biot number:

$$\left.\frac{\partial\theta}{\partial n^*}\right|_b = -\text{Bi}\,\theta_b, \qquad \text{Bi} = h_\infty L/k. \tag{3.12}$$

Finally, the radiation condition, Eq. (3.8), is a mess if we use the variable $\theta$. Define instead a new dimensionless temperature:

$$T^* = T/T_\infty.$$

Then the radiation condition, in terms of $T^*$, becomes

$$\left.\frac{\partial T^*}{\partial n^*}\right|_b = -\frac{\sigma F_{b\infty} T_\infty^3 L}{k}(T_b^{*4} - 1). \tag{3.13}$$

The coefficient on the right is also a sort of Biot number. We conclude that nondimensionalizing a heat conduction problem creates (1) a heat generation parameter and (2) three different types of boundary Biot number due to (a) known flux, (b) convection, or (c) radiation. Note that the grouping in Eq. (3.13) was suggested by Example 2.9. ■

## Example 3.2

Set up the particular differential equation and boundary conditions for temperature in the rectangular body in Fig. E3.2. There is no heat generation, and gradients normal to the figure are negligible.

*Solution* The proper differential equation is found from Eq. (3.2)

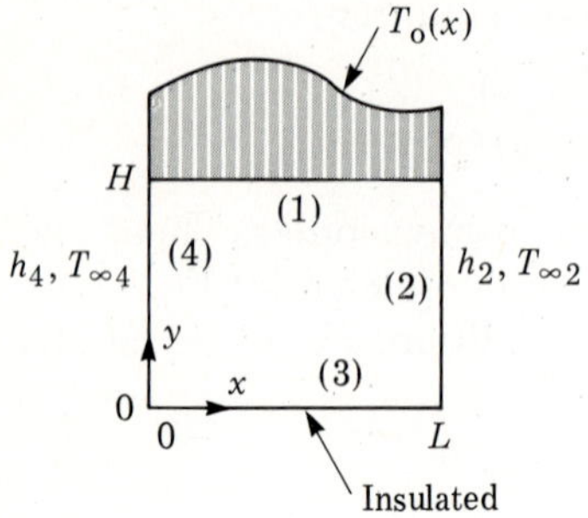


**Figure E3.2**

by setting $\dot{q} = 0$ and $\partial/\partial z = 0$:

$$\frac{\partial^2 T}{\partial x^2} + \frac{\partial^2 T}{\partial y^2} = 0,$$

which is Laplace's equation in two dimensions. The boundary conditions may be listed for each of the four sides:

*Side 1, along $y = H$:*

$$T(x, H) = T_o(x)$$

*Side 2, along $x = L$:*

$$\text{"}n\text{"} = x, \quad \text{hence} \; -k\frac{\partial T}{\partial x} = h_2(T - T_{\infty 2}),$$

*Side 3, along $y = 0$:*

$$\text{"}n\text{"} = -y, \quad \text{hence} \; \frac{\partial T}{\partial y} = 0,$$

*Side 4, along $x = 0$:*

$$\text{"}n\text{"} = -x, \quad \text{hence} \; k\frac{\partial T}{\partial x} = h_4(T - T_{\infty 4}).$$

This problem is well posed and will result in a unique solution $T(x, y)$ within the region bounded by the rectangle. Further, if $k$, $h_2$, and $h_4$ are all independent of temperature, the problem is linear, and it may be possible to treat the boundary conditions separately and add the results. ■

## Example 3.3

Consider a very long solid hot cylinder suddenly thrown into a cold bath such that the convection heat transfer coefficient is constant around its surface. Reduce the general heat conduction equation to apply to this problem, using cylindrical coordinates.

*Solution* The term *very long* is a code word meaning that end effects and axial variations $\partial/\partial z$ are negligible. A constant heat transfer coefficient implies that angular variations $\partial/\partial\theta$ are negligible. A sudden change from hot to cold environment implies a transient or unsteady condition, $\partial/\partial t \neq 0$. Internal heat generation was not mentioned, so we will neglect it. The general equation, Eq. (2.8), thus reduces to

$$\frac{1}{r}\frac{\partial}{\partial r}\left(rk\frac{\partial T}{\partial r}\right) = \rho c_v \frac{\partial T}{\partial t}. \quad [Ans.]$$

If $k$ is assumed constant, this further reduces to

$$\frac{\alpha}{r}\frac{\partial}{\partial r}\left(r\frac{\partial T}{\partial r}\right) = \frac{\partial T}{\partial t}.$$

This particular problem will be solved as an important application in Section 4.4.2. ∎

---

### 3.1.3 Orthogonality of Heat Flux Lines and Isotherms

A heat conduction problem is readily visualized by sketching or constructing the two families of constant temperature lines ("isotherms") and lines parallel to the heat flux vector ("flux lines"). From Fourier's law,

$$\mathbf{q}'' = -k\,\nabla T, \tag{3.14}$$

and the fact that the gradient vector acts in the direction of maximum change in a property, it follows that heat flux lines are everywhere normal to isotherms. This is illustrated in Fig. 3.2 and is true for both steady and unsteady conduction and for any boundary condition. The intersections of these lines form *curvilinear squares*.

This orthogonality condition enables one to construct an approximate graphical *flux plot* to solve a heat conduction problem. One takes pencil and paper (and a good eraser) and tries to fill the conducting region with a network of curvilinear squares. An example is shown in Fig. 3.3 for the chimney problem of Fig. 3.1. These flux lines and isotherms are reproduced, warts and all, from the author's own unskilled sketches. Even after considerable erasing and resketching, a perfect condition of "curvilinear squareness" has not been achieved. Never-

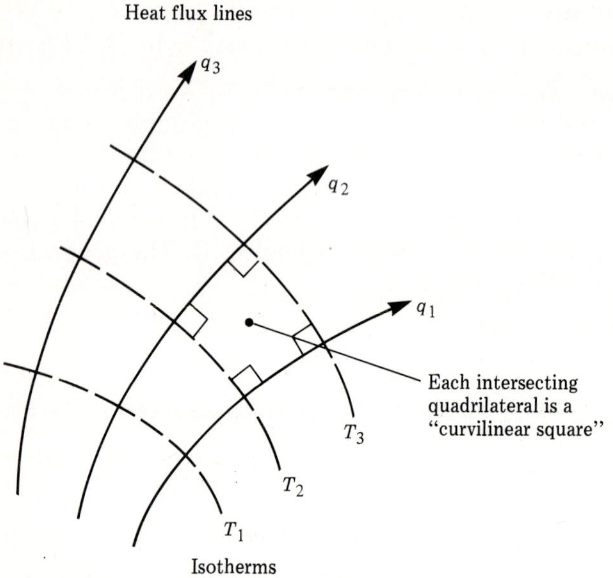


**Figure 3.2** **In heat conduction, for both steady and unsteady flow, the heat flux lines are everywhere normal to the isotherms.**

theless, the figure is approximately correct and gives good insight into the heat flow pattern. By estimating temperature gradients along the heat flux paths or "lanes" of squares, a crude estimate of total heat transfer can be made.

Some rules of thumb for constructing a flux plot are as follows:

1. The method applies only to regions with isothermal or insulated boundaries. It would be extremely difficult to develop a flux plot for convection or radiation conditions.
2. Sketch the outer boundaries, identifying the isothermal boundaries (the left and right sides in Fig. 3.3) and the insulated boundaries (the top and bottom in Fig. 3.3). If both the body and its boundary conditions are symmetric, the lines of symmetry *must be flux lines*. This is the case in Fig. 3.3.
3. Attempt to draw in a set of equally spaced isotherms that are normal to each insulated boundary and vary smoothly across the region.
4. Now add equally spaced flux lines that are normal to the isothermal

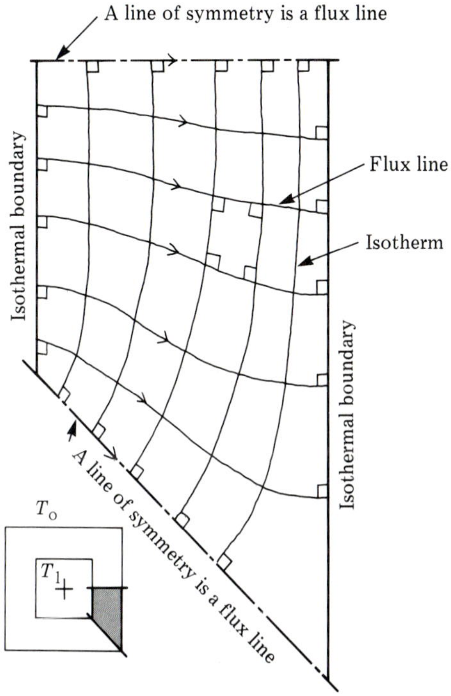


**Figure 3.3** **A hand-drawn flux plot for the chimney problem of Fig. 3.1. The inset shows the complete cross-section, of which only one-eighth is needed for the flux plot.**

boundaries and intersect the interior isotherms from part 3 above to form curvilinear squares.

5. If you are not satisfied with the resulting pattern, erase furiously and modify the interior lines to form better curvilinear squares.

This graphical method is explained in detail in Chapter 19 of [23]. It is not often used now except to give preliminary insight, since many accurate digital computer techniques are now available (see Section 3.4).

Occasionally, the boundary itself will contain a corner that is not at 90°. An example is the lower right corner in Fig. 3.3, which is a 45° angle. It is impossible to sketch a true curvilinear square in such a corner, but one does the best one can.

### 3.1.4 Superposition

If $k$, $\rho$, $c_v$, and $\dot{q}$ are independent of temperature, then the heat conduction equation is linear, and solutions can be added together to form more complex solutions. This is the basis for the Fourier series analysis to be given in the next section.

In a second type of superposition, solutions for various elementary boundary conditions can be added to give a more complex boundary situation. An example is shown in Fig. 3.4, where two parts of the boundary have different specifications of known temperature variation. We may split these up by letting the complete solution $T(x, y)$ be the sum of two simpler functions, $T_1(x, y)$ and $T_2(x, y)$. Then the sum of these two must satisfy each of the complex boundary conditions in Fig. 3.4(a):

*At* $y = H$:

$$T_1 + T_2 = f(x): \tag{3.15}$$

*At* $x = L$:

$$T_1 + T_2 = g(y); \tag{3.16}$$

*At* $y = 0$:

$$T_1 + T_2 = 0; \tag{3.17}$$

*At* $x = 0$:

$$T_1 + T_2 = 0. \tag{3.18}$$

Looking these conditions over, we see that it is very convenient to let $T_1$ take on the entire burden of $f(x)$ and let $T_2$ shoulder $g(y)$ alone. In other words, split up $T_1$ and $T_2$ into separate conditions, as follows:

*At* $y = H$:

$$T_1 = f(x), \qquad T_2 = 0; \tag{3.19}$$

**Figure 3.4 Solution of a complex boundary condition problem by splitting into two simpler problems, $T = T_1 + T_2$.**

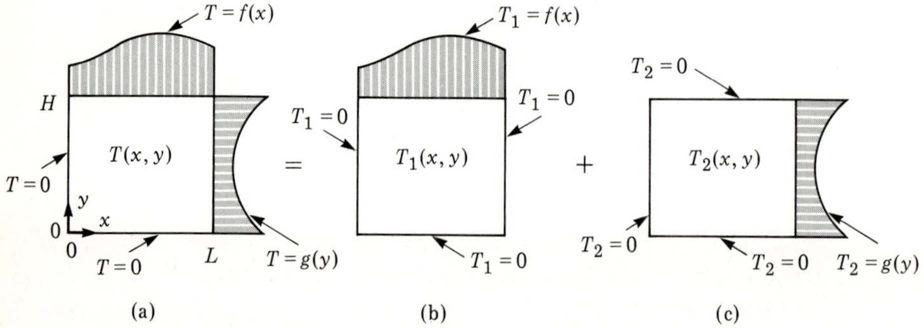

*At* $x = L$:

$$T_1 = 0, \qquad T_2 = g(y); \tag{3.20}$$

*At* $y = 0$:

$$T_1 = 0, \qquad T_2 = 0; \tag{3.21}$$

*At* $x = 0$:

$$T_1 = 0, \qquad T_2 = 0. \tag{3.22}$$

These two separate solutions are shown in Fig. 3.4(a, b). They are much easier to generate than a single function $T(x, y)$ satisfying Eqs. (3.15–3.18). Note that the boundary conditions themselves must be linear or the technique fails; for example, radiation conditions (Eq. 3.8) cannot be superimposed.

**Example 3.4**

A square region has different boundary temperatures on its four sides: 20°, 50°, 65°, and 95°. Show how to simplify this problem by superposition of four functions.

*Solution* Subtract 20° from every side, then take the remaining sides one at a time, with the results shown in Fig. E3.4. After some thought, we realize that $T_1 = 20°$ everywhere. Since the region is square, $T_2$, $T_3$, and $T_4$ are similar in form: $T_3$ is 1.5 times as large as $T_2$ and turned through 90 degrees, while $T_4$ is 2.5 times as large as $T_2$ and upside down. This particular superposition is an unequivocal success. ■

**Figure E3.4**

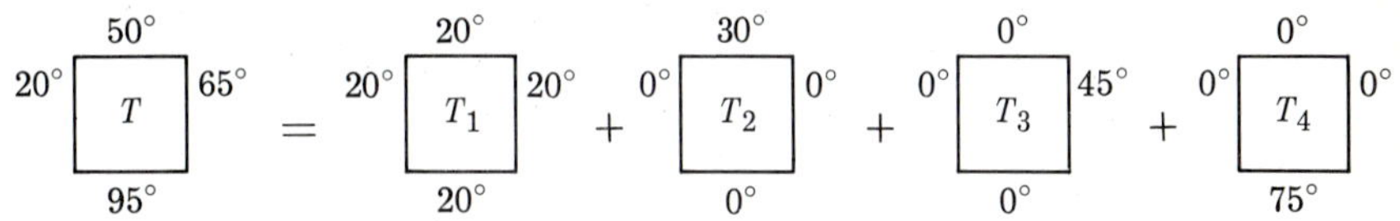


### 3.1.5 Variable Conductivity

If the thermal conductivity is variable, $k = k(T)$, the conduction equation (3.1) is nonlinear but still may be solved by the numerical techniques of Section 3.4. However, for steady conduction, $\partial T/\partial t = 0$, the equation is made linear by defining the following integral variable

$\Phi$ $(T)$:

$$\Phi = \int_{T_0}^{T} k(T)dT, \tag{3.23}$$

where $T_o$ is any convenient reference temperature. Note by its definition that, for example, $\partial\Phi/\partial x = k\ (\partial T/\partial x) = -q_x$. Then substitution of Eq. (3.23) into Eq. (3.1) for steady heat conduction yields a linear Poisson equation:

$$\nabla^2\Phi = \frac{\partial^2\Phi}{\partial x^2} + \frac{\partial^2\Phi}{\partial y^2} + \frac{\partial^2\Phi}{\partial z^2} = -\dot{q}, \tag{3.24}$$

similar to Eq. (3.2) for constant $k$. If the boundary conditions are of type 1, prescribed temperatures, then from Eq. (3.23) this is equivalent to prescribed values of $\Phi$; hence the problem is linear. However, both type 2, prescribed heat flux, and type 3, convection environment, conditions result in nonlinear functions of $\Phi$, so the attempt to achieve a linear problem fails for those cases. Type 4 or radiation boundary conditions (Eq. 3.8) are of course nonlinear for either the $T$ or $\Phi$ variable.

## 3.2 Exact Analysis of Classical Problems

If the geometry of the body is appropriate to the classic coordinate systems — rectangles, circles, spheres — a multitude of classical analytic solutions are known and treated in advanced texts [1–6]. Many of these use the technique of Fourier series superposition [7]. Here we consider one of these classic solutions for a rectangular shape. This serves several purposes: (1) to illustrate the Fourier series idea, (2) to introduce the concept of conduction shape factor, and (3) to provide exact values for comparison with the numerical methods of Section 3.4. Of course, it will also set up some tough homework problems.

### 3.2.1 Rectangle with Prescribed Boundary Temperatures

For these examples, we consider only two-dimensional $(x, y)$, steady conduction with constant $k$ and no heat generation. The general conduction equation, Eq. (3.1), reduces to the linear, two-dimensional Laplace equation:

$$\frac{\partial^2 T}{\partial x^2} + \frac{\partial^2 T}{\partial y^2} = 0. \tag{3.25}$$

Let us solve this equation on the rectangular region shown in Fig.

3.5(a), where three boundary temperatures are equal to $T_o$ and the fourth (along $y = H$) is an arbitrary function of $x$. We do not attack these four conditions at once. Instead, from our superposition examples in Section 3.1.4, we subtract out the uniform distribution $T_o$ (as in Fig. 3.5b). What remains is the simpler problem of Fig. 3.5(c), with temperature $f(x)$ on the upper surface and 0 on the other boundaries:

$$\text{At } y = H: \quad T(x, H) = f(x); \tag{3.26}$$

$$\text{At } x = 0,\ y = 0,\ x = L: \quad T = 0. \tag{3.27}$$

Let us now develop a Fourier series solution of this problem.

The basic technique is the method of *separation of variables*, outlined in any good engineering mathematics text, such as [7, 12]. The solution is assumed to be separable into a product of functions: $T(x, y) = X(x) \cdot Y(y)$. Substitution into Laplace's equation (3.25) yields

$$X''(x)Y(y) + X(x)Y''(y) = 0,$$

or **(3.28)**

$$\frac{X''}{X} = -\frac{Y''}{Y} = \text{constant} = -a^2.$$

The reason we obtain a constant in Eqs. (3.28) is that $X''/X$ is a function only of $x$, while $Y''/Y$ varies only with $y$. The two cannot be equated everywhere unless both are equal to the same constant.

**Figure 3.5** Analysis of steady conduction in a rectangular region with prescribed boundary temperatures. By subtracting out $T_o$ (b), the problem reduces to one nonzero boundary (c).

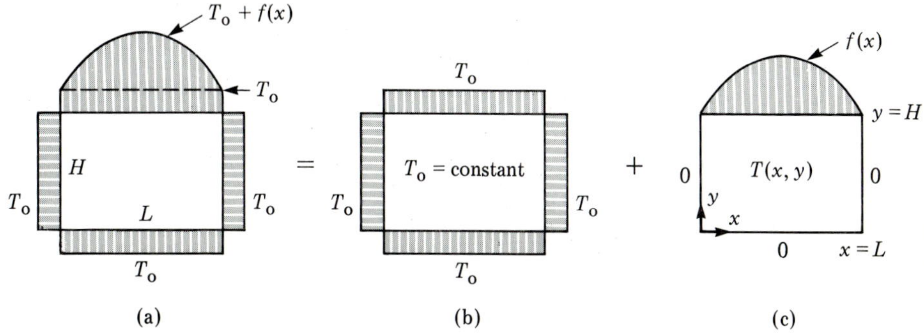

When rearranged, Eqs. (3.28) lead to separate solutions:

$$X'' + a^2X = 0, \quad \text{or} \quad X = C_1\sin(ax) + C_2\cos(ax),$$

$$Y'' - a^2Y = 0, \quad \text{or} \quad Y = C_3\sinh(ay) + C_4\cosh(ay).$$

We have thus found a trial solution to Laplace's equation:

$$T(x, y) = [C_1\sin(ax) + C_2\cos(ax)][C_3\sinh(ay) + C_4\cosh(ay)]. \tag{3.29}$$

We now attempt to satisfy the boundary conditions, Eq. (3.27):

*At $x = 0$:*

$T = 0 = [0 + C_2][C_3\sinh(ay) + C_4\cosh(ay)]$,

or $C_2 = 0$;

*At $y = 0$:*

$T = 0 = [C_1\sin(ax) + 0][0 + C_4]$,

or $C_4 = 0$.

With $C_2$ and $C_4$ eliminated, the trial solution reduces to

$$T(x, y) = C\sinh(ay)\sin(ax), \qquad C = C_1C_3.$$

Now try to satisfy the third condition of (3.27):

*At $x = L$:*

$T = 0 = C\sinh(ay)\sin(aL)$.

This can be true only if $\sin(aL) = 0$, or

$$a = n\pi/L, \qquad n = 1, 2, 3, \ldots. \tag{3.30}$$

Thus only special values of $a$ (*eigenvalues*) satisfy this particular problem. The trial solution now is

$$T(x, y) = C\sinh(n\pi y/L)\sin(n\pi x/L) \tag{3.31}$$

for any integer $n$. This allows a sine-wave shape only along the upper boundary ($y = H$), which is rather specialized but still interesting. Let us examine the case $n = 1$.

### 3.2.2 Solution for a Half-Sine-Wave Boundary Temperature

Suppose that the known temperature $f(x)$ along the upper surface in Fig. 3.5(c) has a half-sine-wave shape:

$$f(x) = \Delta T\sin(\pi x/L). \tag{3.32}$$

Set this equal to our trial solution (3.31) at $y = H$:

$$T(x, H) = C\sinh(n\pi H/L)\sin(n\pi x/L) = \Delta T\sin(\pi x/L).$$

This is satisfied if $n = 1$ and $C = \Delta T/\sinh(\pi H/L)$. The exact solution to this particular problem is, then,

$$T(x, y) = \Delta T \frac{\sinh(\pi y/L)}{\sinh(\pi H/L)} \sin(\pi x/L). \tag{3.33}$$

For any rectangle aspect ratio $(H/L)$, we may compute and plot the isotherms ($T$ = constant) and the heat flux lines (along the maximum temperature gradients). An example is shown in Fig. 3.6 for an aspect ratio $H/L = 1/3$. As expected, the isotherms are everywhere perpendicular to the flux lines except at the two lower corners, which are singularities.

Figure 3.6 shows that all heat enters the plate at the hot ($\Delta T > 0$) upper surface and leaves from the three lower surfaces. Fourier's law gives the heat flux at the upper surface:

$$q''_w|_{y=H} = -k\frac{\partial T}{\partial y}\bigg|_{y=H} = -k\frac{\pi}{L}\Delta T \coth(\pi H/L) \sin(\pi x/L).$$

The negative sign indicates heat transfer *into* the plate.

The total flux through the upper surface is given by the integral of $q''_w$ across the surface. If $W$ denotes the width of the plate into the paper, we obtain

$$\begin{aligned} q_w\bigg|_{y=H} &= -k\frac{\pi}{L}\Delta T \coth(\pi H/L)\int_0^L \sin(\pi x/L)\, W\, dx \\ &= -2kW \coth(\pi H/L)\Delta T. \end{aligned} \tag{3.34}$$

**Figure 3.6 Computed isotherms and heat flux lines from Eq. (3.33) for a half-sine-wave boundary temperature on a rectangular plate of aspect ratio $H/L$ = ⅓.**

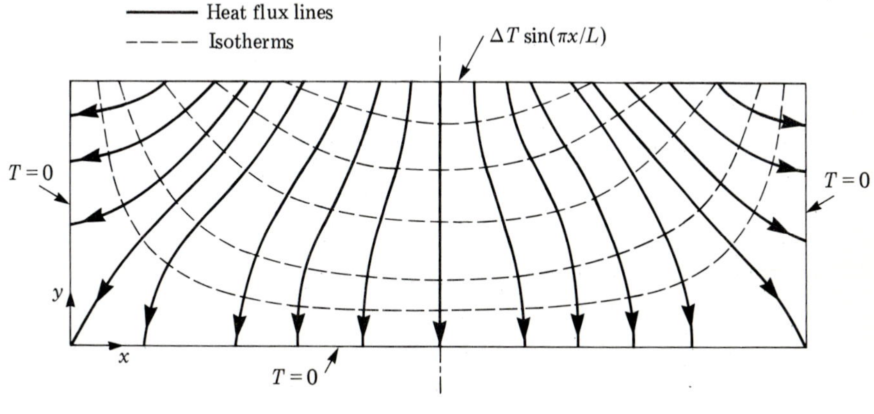

In a similar manner, we can compute the bottom heat flux:

$$q''_w\bigg|_{y=0} = -k\frac{\partial T}{\partial y}\bigg|_{y=0} = -k\frac{\pi}{L}\Delta T\,\mathrm{csch}(\pi H/L)\sin(\pi x/L),$$
$$q_w\bigg|_{y=0} = \int_0^L q''_w\, W\, dx = -2kW\,\mathrm{csch}(\pi H/L)\Delta T. \tag{3.35}$$

Finally, on the left and right sides we obtain

$$q''_w\bigg|_{x=L} = -q''_w\bigg|_{x=0} = k\frac{\pi}{L}\Delta T\frac{\sinh(\pi y/L)}{\sinh(\pi H/L)},$$
$$q_w\bigg|_{x=L} = -q_w\bigg|_{x=0} = kW\Delta T[\coth(\pi H/L) - \mathrm{csch}(\pi H/L)]. \tag{3.36}$$

The sum of the three heat fluxes out the bottom and two sides exactly equals the flux into the top surface, as expected for steady conduction. This first law requirement is guaranteed by our correct solution of the basic differential equation.

We may use this exact solution to illustrate how the rectangle approaches a one-dimensional "slab" as the aspect ratio $H/L$ becomes smaller. The four surface heat fluxes from Eqs. (3.34–3.36) are computed and shown in Fig. 3.7 for various aspect ratios. We see that end losses become negligible as the rectangle becomes slimmer. At $H/L = 1/10$, side fluxes total only 4.8%, with 95.2% of the heat flowing across the "slab" and out the bottom. At $H/L = 1/100$ (not shown), total end losses have dropped to less than 0.05%.

Another feature revealed by this exact solution is the "conduction shape factor," which is discussed in the next section. Still another benefit of these regular-geometry exact solutions is that they may be used as test cases for the numerical analyses of Section 3.4.

### 3.2.3 Solution for Uniform Surface Temperature

If the surface distribution $f(x)$ from Fig. 3.5(c) is not a sine wave, the Fourier series technique must be used. Our trial function from Eq. (3.31) is an exact solution for any value of $n$. Since Laplace's equation is linear, we may sum solutions of this type and generalize Eq. (3.31) to an infinite series:

$$T(x, y) = \sum_{n=1}^{\infty} C_n \sinh(n\pi y/L)\sin(n\pi x/L) \tag{3.37}$$

This satisfies Laplace's equation (3.25), is zero on both sides and the

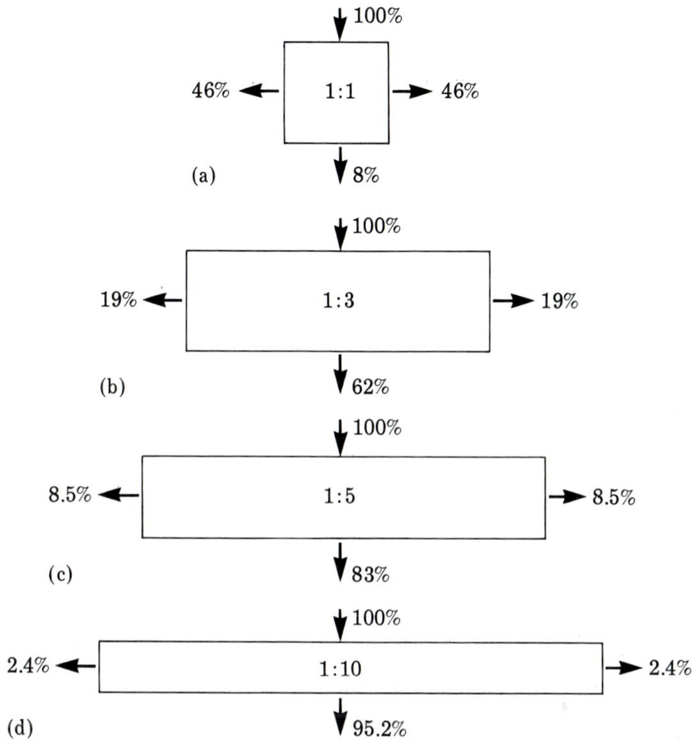


**Figure 3.7 Relative heat flux through each side of a rectangular plate with a half-sine-wave upper surface temperature, from Eqs. (3.34–3.36). For convenience the upper flux is denoted as 100%.**

bottom, and will be the correct solution if we can force an equality at the upper surface:

*At* $y = H$:

$$T(x, H) = f(x) = \sum_{n=1}^{\infty} C_n \sinh(n\pi H/L) \sin(n\pi x/L). \tag{3.38}$$

As shown in texts on applied mathematics [7], if $f(x)$ is piecewise continuous, the series expansion in Eq. (3.38) is uniformly valid. Fourier discovered that the coefficients $C_n$ can be evaluated by multiplying both sides of Eq. (3.38) by $\sin(m\pi x/L)$ and integrating over the plate length $L$:

$$\int_0^L f(x)\sin(m\pi x/L)dx$$
$$= \int_0^L \sum_{n=1}^{\infty} C_n \sinh(n\pi H/L)\sin(n\pi x/L)\sin(m\pi x/L)dx. \quad \textbf{(3.39)}$$

The integral of $[\sin(m\pi x/L)\sin(n\pi x/L)]$ is zero if $m \neq n$, and it equals $L/2$ if $m = n$. Because of this unusual property, the sine function is termed *orthogonal* [7], and Eq. (3.39) immediately isolates the $n$th coefficient of the series:

$$C_n = \frac{2}{L}\operatorname{csch}(n\pi H/L)\int_0^L f(x)\sin(n\pi x/L)dx. \quad \textbf{(3.40)}$$

Thus the desired *Fourier coefficients* $C_n$ are directly related to certain integrals over plate length of the boundary temperature function $f(x)$, after which the solution is given by Eq. (3.37). Equation (3.40) is valid for any piecewise continuous $f(x)$.

Let us take as an example the very common case of uniform temperature $T_o$ along the upper surface:

$$f(x) = T(x, H) = T_o = \text{constant}. \quad \textbf{(3.41)}$$

For this case, the Fourier coefficients in Eq. (3.40) are given by

$$\begin{aligned} C_n &= (4T_o/\pi n)\operatorname{csch}(n\pi H/L) \qquad \text{if } n = 1, 3, 5, \ldots \\ &= 0 \qquad \text{if } n = 2, 4, 6, \ldots. \end{aligned} \quad \textbf{(3.42)}$$

Substitution into Eq. (3.37) gives the exact solution in series form for a plate with a uniform upper-surface temperature, with $T = 0$ on the lower three sides:

$$T(x, y) = \frac{4T_o}{\pi}\sum_{n=1,3,5,\ldots}^{\infty} \frac{1}{n}\frac{\sinh(n\pi y/L)}{\sinh(n\pi H/L)}\sin(n\pi x/L). \quad \textbf{(3.43)}$$

This is one of many classic Fourier series solutions to steady-state heat conduction problems [1–6]. Other examples will be assigned as problem exercises.

The isotherms computed from Eq. (3.43) are shown in Fig. 3.8 for the particular case of a square plate, $H = L$. Note that the upper corners are points of discontinuity, where the temperature takes on all values between 0 and $T_o$. The Fourier series, Eq. (3.43), because of its oscillating nature [7], predicts an average temperature of $T_o/2$ at the upper corners.

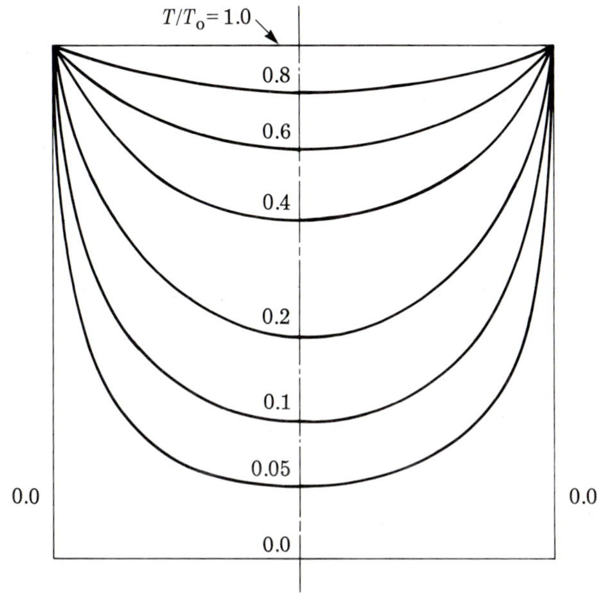


**Figure 3.8** **Isotherms in a square plate with uniform upper surface temperature, from Eq. (3.43). The upper corners are singular points where temperature and heat flux lines are indeterminate.**

A difficulty with Eq. (3.43) is that it cannot be used to predict the heat flux. This particular series has coefficients $C_n$ from Eq. (3.42), which decrease only as $1/n$; it is improper to differentiate such a series term by term [7], and such a process does not yield the correct value of $\partial T/\partial x$ or $\partial T/\partial y$.

## Example 3.5

The series in Eq. (3.43) has coefficients that decrease only as $1/n$. Does it converge reasonably quickly? If so, compute the temperature at the center $(L/2, L/2)$ of the square plate in Fig. 3.8 and interpret the result.

*Solution* Yes, the series (3.43) does converge in only a few terms as long as one is not too near the upper surface $y = H$. A hand calculator suffices for the calculation. For the given case $H = L$,

$y = x = L/2$, the series (3.43) gives

$$T(L/2, L/2) = (4T_o/\pi)\left[\frac{\sinh(\pi/2)}{\sinh(\pi)}\sin(\pi/2) + \frac{\sinh(3\pi/2)}{3\sinh(3\pi)}\sin(3\pi/2) + \frac{\sinh(5\pi/2)}{5\sinh(5\pi)}\sin(5\pi/2) + \cdots\right]$$

$$= (4T_o/\pi)(0.199268 - 0.002994 + 0.000078 - 0.000002 + \cdots)$$

$$= 0.25000\ T_o, \quad [Ans.]$$

correct to five decimal places. Actually, the temperature at the plate center is *exactly* $T_o/4$ because this point is equidistant from all four sides. If all four sides were at temperature $T_o$, the center would be at $T_o$. Thus *one* side, being equal to the other three, causes one-fourth of $T_o$ at the center. This argument does not apply to nonsquare rectangles or to any other point in the square plate. ■

## Example 3.6

Using the method of separation of variables as in Eq. (3.28), investigate steady conduction in plane polar coordinates $(r, \theta)$, with variable boundary temperature $f(\theta)$ at a given value of $r$. Is a Fourier series possible for this geometry? Neglect heat generation.

*Solution* The polar coordinate form of Laplace's equation is obtained by neglecting axial gradients $\partial/\partial z$ in Eq. (2.8):

$$\frac{\partial^2 T}{\partial r^2} + \frac{1}{r}\frac{\partial T}{\partial r} + \frac{1}{r^2}\frac{\partial^2 T}{\partial \theta^2} = 0. \qquad \textbf{(1)}$$

Assume that the solution can be "separated" into the product of functions of each variable: $T(r, \theta) = F(r)G(\theta)$. Substitution into Eq. (1) gives

$$r^2F''G + rF'G + FG'' = 0.$$

Divide by $FG$ and equate separate functions of each variable:

$$(r^2F'' + rF')/F = -G''/G. \qquad \textbf{(2)}$$

Since $r$ and $\theta$ are independent variables, the equality implied by Eq. (2) is possible only if both sides are equal to the same constant. If they equal a *negative* constant, the solution for $G(\theta)$ is an exponential function, which cannot be superimposed into a Fourier series. But it all works if we set the separation constant equal to a positive value, say $n^2$. Equation (2) becomes the two equations

$$r^2F'' + rF' - n^2F = 0, \tag{3}$$

$$G'' + n^2G = 0. \tag{4}$$

The solutions to these two are

$$F(r) = Ar^n + Br^{-n},$$

$$G(\theta) = C\cos(n\theta) + D\sin(n\theta).$$

The product $T = FG$ is a solution of Laplace's equation, and the sine and cosine can be summed into a Fourier series. The general form of such a series solution would be

$$T(r, \theta) = \sum_{n=1}^{\infty}(A_n r^n + B_n r^{-n})[C_n\cos(n\theta) + D_n\sin(n\theta)]. \tag{5}$$

Some of these constants and functions may drop out, depending on the particular type of circumferential boundary condition. See Section 6.5 of [2] for an example. ■

## 3.3 The Conduction Shape Factor Length

Exact analyses of the type illustrated in Section 3.2 have been published for many different steady conduction problems, as detailed in advanced texts [1–6]. Of primary interest to engineers is the total heat flux from the surface of the body. The example that we computed in Eq. (3.34) is characteristic of all steady conduction problems involving known boundary temperatures and a single important temperature difference. We may rewrite Eq. (3.34) in the form

$$|q_w| = k\,S\,\Delta T, \qquad S = 2\,W\coth(\pi H/L). \tag{3.44}$$

The quantity $S$ has the dimensions of length and is called the *conduction shape factor*. For the conditions stated above, $S$ is a function only of the system geometry.

We may verify the generality of the shape factor concept by dimensional analysis. If $q_w$ depends only on conductivity, temperature difference, body size, and geometry,

$$q_w = \text{fcn}(k, \Delta T, L, \text{geometry}), \tag{3.45}$$

then, since there are four fundamental dimensions involved — mass, length, time, and temperature — the only possible dimensionally consistent combination of these variables is

$$q_w = k\,\Delta T\,S, \qquad S = L\,\text{fcn(dimensionless geometry)}. \tag{3.46}$$

If, however, there are, for example, *convection* boundary conditions, an additional parameter such as Biot number must appear:

$$q_w\Big|_{\substack{\text{boundary}\\ \text{convection}}} = k\,\Delta T\,L\,\text{fcn}(h_\infty L/k, \text{dimensionless geometry}). \tag{3.47}$$

This usually means that a plot of the shape factor versus Biot number is needed rather than a single geometric formula.

A great many shape factors have been solved and tabulated, especially in [9–11]. Some of the known solutions are given in Table 3.1 for your interest and use. We stress again that the cases listed in Table 3.1 are not valid for convection boundary conditions.

**Table 3.1** Conduction shape factor $S = q_w/[k(T_1 - T_2)]$ for various geometries, after [9]

| Geometry | Shape Factor | Restrictions |
|---|---|---|
| 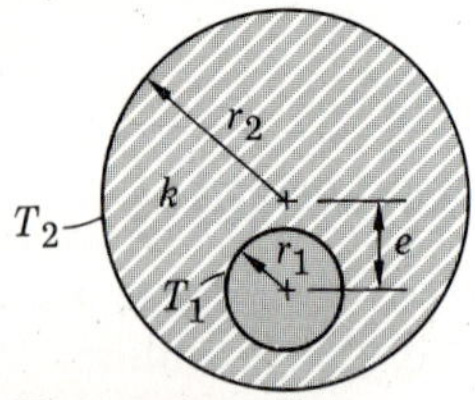 1. Eccentric cylinders of length $L$ | $\dfrac{2\pi L}{\cosh^{-1}\left(\dfrac{r_1^2 + r_2^2 - e^2}{2r_1r_2}\right)}$ | No axial heat flux |
| 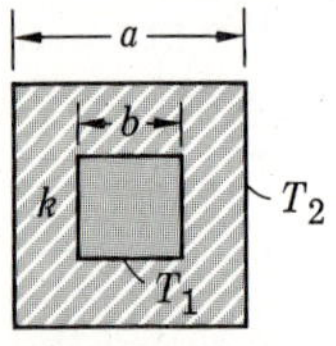 2. Square pipe of length $L$ | $\dfrac{2\pi L}{0.93\ln(0.948\,a/b)}$ | $a/b > 1.41$ |
| | $\dfrac{2\pi L}{0.785\ln(a/b)}$ | $a/b < 1.41$ |
| 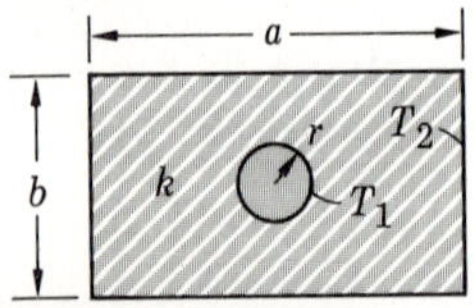 3. Circular pipe of length $L$ buried in a rectangular region | $\dfrac{2\pi L}{\ln\left[\dfrac{b}{r}(0.637 - 1.781\,e^{-2.9a/b})\right]}$ | $a \geqslant b$ |

**Table 3.1 (continued)**

| Geometry | Shape Factor | Restrictions |
|---|---|---|
| 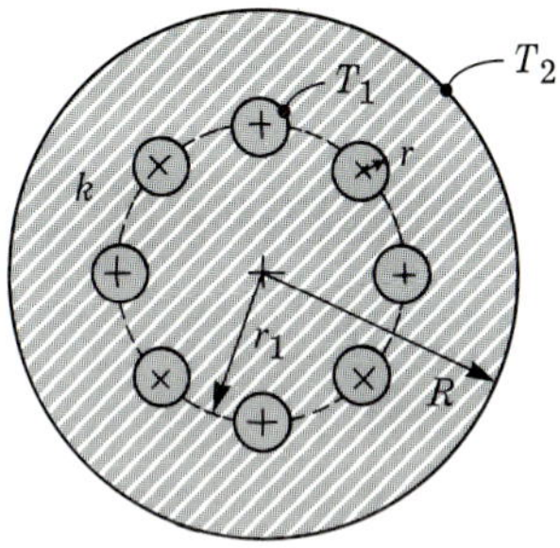 4. A row of $N$ equally spaced pipes of length $L$, buried in a circular region | $\dfrac{2\pi L}{\ln\left(\dfrac{R}{r_1}\right) - \dfrac{1}{N}\ln\left(\dfrac{Nr}{r_1}\right)}$ (for one pipe) | $r \ll r_1$ <br> $N \gg 1$ |
| 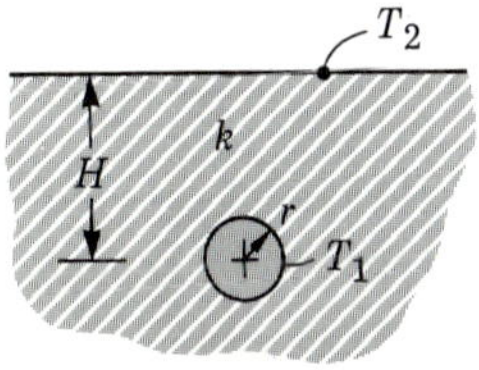 5. Pipe of length $L$ buried beneath the surface of a semi-infinite medium | $\dfrac{2\pi L}{\cosh^{-1}(H/r)}$ | $L \gg r$ |
| 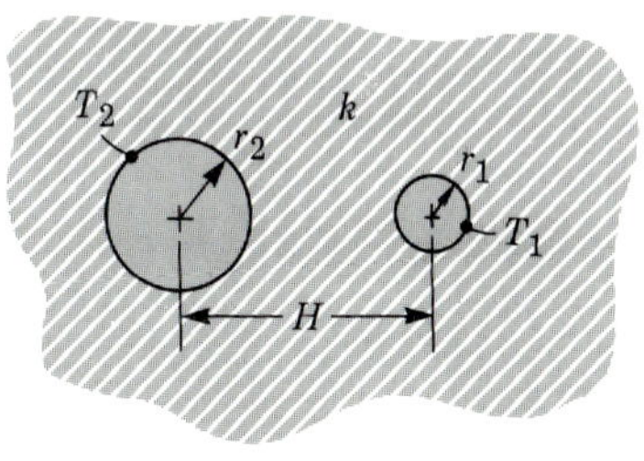 6. Two different-sized pipes of length $L$, buried in an infinite medium | $\dfrac{2\pi L}{\cosh^{-1}\left(\dfrac{H^2 - r_1^2 - r_2^2}{2r_1r_2}\right)}$ | $L \gg r, H$ |

**Table 3.1** (continued)

| Geometry | Shape Factor | Restrictions |
|---|---|---|
| 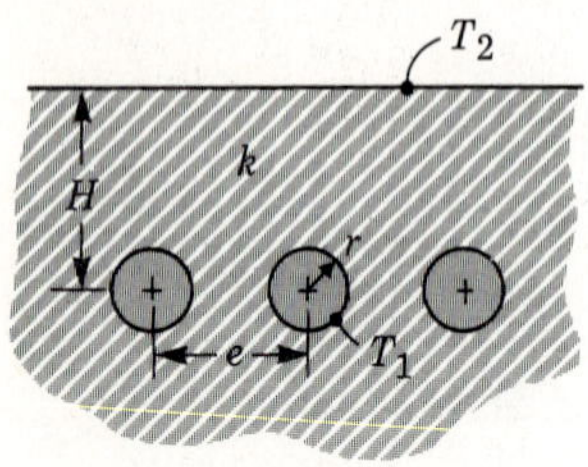<br>7. An infinite row of equally spaced pipes of length $L$, buried below the surface of a semi-infinite medium | $\dfrac{2\pi L}{\ln\left[\dfrac{e}{\pi r}\sinh\left(\dfrac{2\pi H}{e}\right)\right]}$ | $L >> r, H$<br>$e \geqslant 3r$ |
| 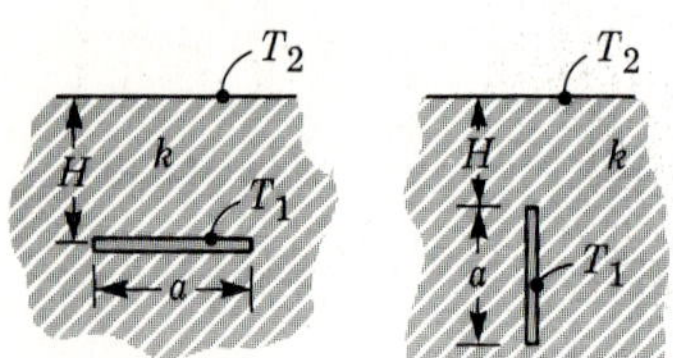<br>8. A thin strip of length $L$ buried beneath the surface of a semi-infinite medium | a) $2.94\ L\ (a/H)^{0.32}$<br>b) $2.38\ L\ (a/H)^{0.24}$ | $L >> a, H$ |
| 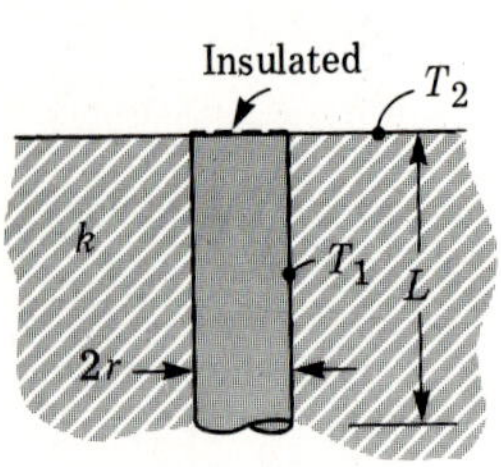<br>9. A pipe buried normal to and flush with the surface of a semi-infinite medium | $\dfrac{2\pi L}{\ln(2L/r)}$ | $L >>r$ |

**Table 3.1 (continued)**

| Geometry | Shape Factor | Restrictions |
|---|---|---|
| 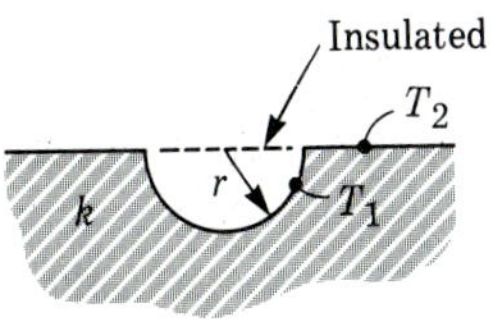<br>10. Buried hemisphere | $2\pi r$ | None |
| 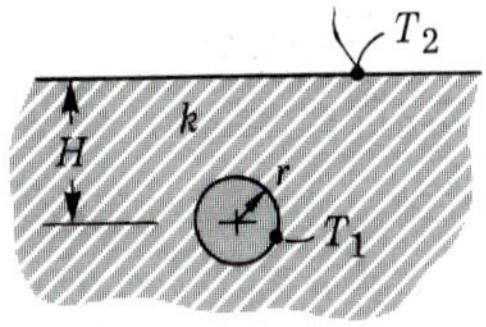<br>11. Buried sphere | $\dfrac{4\pi r}{1 - r/2H}$ | None |
| 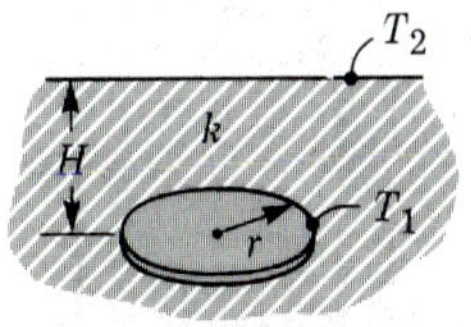<br>12. Disc buried parallel to the surface | a) $4r$<br>b) $8r$ | $H = 0$<br>$H >> r$ |
| 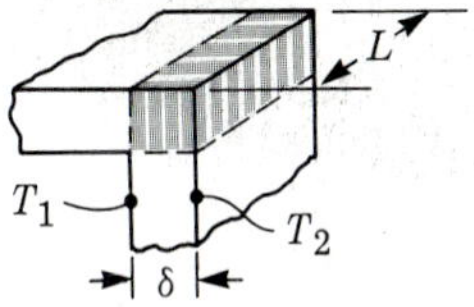<br>13. Two-dimensional corner or "edge" section | $0.54\ L$ | None |
| 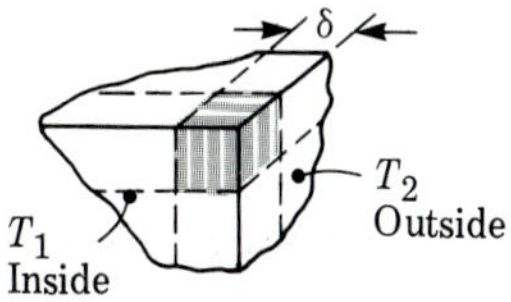<br>14. Three-dimensional corner formed by walls $\delta$ thick | $0.15\ \delta$ | None |

**Example 3.7**

The buildings of a certain university are heated by steam fed from a central boiler plant through 3 statute mi of 6-in.-diameter bare pipe buried 3 ft below the surface. The pipe wall temperature is 340°F, and the ground temperature is 50°F year round. If the ground is moist, $k = 1.2$ Btu/hr · ft · °F, compute the total heat lost to the ground in Btu/hr. If fuel oil delivers 110,000 Btu/gal and costs $1.00/gal, what is the annual cost of these heat losses?

*Solution* The geometry fits item 5 in Table 3.1. The total pipe length is 3 mi = 3(5280) = 15,840 ft. The shape factor is

$$S = \frac{2\pi L}{\cosh^{-1}(H/r)} = \frac{2\pi(15{,}840\text{ ft})}{\cosh^{-1}[3\text{ ft}/(\tfrac{3}{12}\text{ ft})]} = 31{,}300\text{ ft.}$$

If you don't have $\cosh^{-1}$ on your calculator, use the identity

$$\cosh^{-1}(x) = \ln[x + (x^2 - 1)^{1/2}].$$

The total heat lost to the ground is thus

$$\begin{aligned} q_w = k\,S\,\Delta T &= (1.2\text{ Btu/hr}\cdot\text{ft}\cdot{}^\circ\text{F})(31{,}300\text{ ft})(340^\circ - 50^\circ\text{F}) \\ &= 10{,}900{,}000\text{ Btu/hr.} \quad [Ans.] \end{aligned}$$

The total fuel wasted per year is

$$\begin{aligned} \text{Fuel} &= (10{,}900{,}000\text{ Btu/hr})/(110{,}000\text{ Btu/gal}) \\ &= (99\text{ gal/hr})(24\text{ hr/day})(365\text{ day/yr}) \\ &= 868{,}000\text{ gal/yr.} \end{aligned}$$

The total cost of this wasted fuel is

$$\text{Cost} = (868{,}000\text{ gal/yr})(\$1.00/\text{gal}) = \$868{,}000/\text{yr.} \quad [Ans.]$$

These buried pipes should be thoroughly insulated and waterproofed, or some alternative heating system should be used, such as heat plants in individual buildings. ■

**Example 3.8**

A spherical package containing radioactive wastes generates heat at 6 kW and is 1.5 m in diameter. It is to be buried 10 m deep in soil whose ambient temperature is 12°C and whose conductivity is 2.4 W/m · K. Estimate the surface temperature of the sphere under steady conditions.

*Solution* This case fits item 11 of Table 3.1, for which the shape factor is

$$S = \frac{4\pi r}{1 - r/2H} = \frac{4\pi(0.75 \text{ m})}{1 - 0.75 \text{ m}/2(10 \text{ m})} = 9.79 \text{ m}.$$

The steady heat flux is thus

$$q_w = 6000 \text{ W} = kS\,\Delta T = (2.4 \text{ W/m}\cdot\text{K})(9.79 \text{ m})(T_w - 12°\text{C}).$$

Solve for the sphere surface temperature:

$T_w = 267°\text{C}$. [*Ans.*] ■

## 3.4 Numerical Analysis

The exact analysis technique of Section 3.2 bears fruit only for regular geometries and well-behaved boundary conditions. The shape factors of Section 3.3 are valid only for the specific geometries shown and uniform wall temperature conditions. But there are many other practical heat conduction problems involving irregular geometries, complicated boundary conditions, or both. Such problems can be solved approximately by a technique called *numerical analysis,* in which the basic differential equation and boundary conditions are modeled by a set of simultaneous *algebraic* equations to predict temperatures at certain nodal points in the body.

The numerical analysis of steady heat conduction is an outstanding success. The algebraic model used is simple, accurate, well behaved, and stable — even for complex boundary conditions — and it can easily be adapted to a programmable calculator or a digital computer. After reading this section, you may agree that numerical conduction analysis can be a routine tool for any engineer to use in estimating heat flux and temperature distributions in arbitrary body shapes. Both the two- and three-dimensional formulations are easy to use.

In the engineering literature, one finds that applied mathematics texts [7, 12] have at least one chapter devoted to numerical analysis, and books on numerical methods [13, 14] devote a great deal of time to the heat conduction equation. There is also at least one heat transfer text with an extensive treatment of numerical methods [4]. Two basic techniques are widely used in partial differential equations such as the heat conduction equation: (1) the finite-difference method and (2) the finite-element method [19–22]. We will concentrate here on the finite-difference method.

### 3.4.1 Finite-Difference Formulation

A finite-difference formula simulates an ordinary or partial derivative by an algebraic approximation that is equivalent to the quotient for which limits are taken in calculus derivations. For example, if temperature $T$ is known at $x$ and at $x + \Delta x$, an algebraic approximation for the derivative $\partial T/\partial x$ is

$$\frac{\partial T}{\partial x} \simeq \frac{T(x + \Delta x) - T(x)}{\Delta x}. \tag{3.48}$$

This relation is exact in the limit as $\Delta x \to 0$, but in numerical analysis we let $\Delta x$ remain finite; hence $T(x + \Delta x) - T(x)$ is termed a *finite difference.* Equation (3.48) is equivalent to using the first term only in a Taylor series expansion:

$$T(x + \Delta x) = T(x) + \Delta x\frac{\partial T}{\partial x} + \frac{1}{2}\Delta x^2\frac{\partial^2 T}{\partial x^2} + \cdots. \tag{3.49}$$

Thus the error involved in using Eq. (3.48) is of order $\Delta x(\partial^2 T/\partial x^2)$, which is the first term neglected.

The second derivative may be approximated by taking differences of $\partial T/\partial x$ from Eq. (3.48):

$$\begin{aligned}\frac{\partial^2 T}{\partial x^2} &\simeq \frac{1}{\Delta x}\left[\frac{T(x + \Delta x) - T(x)}{\Delta x} - \frac{T(x) - T(x - \Delta x)}{\Delta x}\right] \\ &\simeq \frac{T(x + \Delta x) - 2T(x) + T(x - \Delta x)}{\Delta x^2}.\end{aligned} \tag{3.50}$$

From the Taylor series, we can find that the first term neglected in Eq. (3.50) is of order $\Delta x^2(\partial^4 T/\partial x^4)$.

The essence of the method is to apply these approximations to the basic differential equation(s) and boundary conditions. We divide the region where the solution is desired into a rectangular mesh of nodal points spaced $\Delta x$, $\Delta y$, $\Delta z$ apart. A sample mesh for a plane $(x, y)$ region is shown in Fig. 3.9. To avoid having to write $(x + \Delta x, y + \Delta y)$ and so on all the time, the custom is to denote nodal temperatures with integer subscripts $(m, n)$, increasing in the $x$ and $y$ directions, respectively. Referring to the nodal labels in Fig. 3.9, then, we may rewrite Eqs. (3.48) and (3.50) in the form

$$\begin{aligned}\frac{\partial T}{\partial x} &\simeq \frac{1}{\Delta x}[T_{m+1,n} - T_{m,n}], \\ \frac{\partial^2 T}{\partial x^2} &\simeq \frac{1}{\Delta x^2}[T_{m+1,n} - 2T_{m,n} + T_{m-1,n}].\end{aligned} \tag{3.51}$$

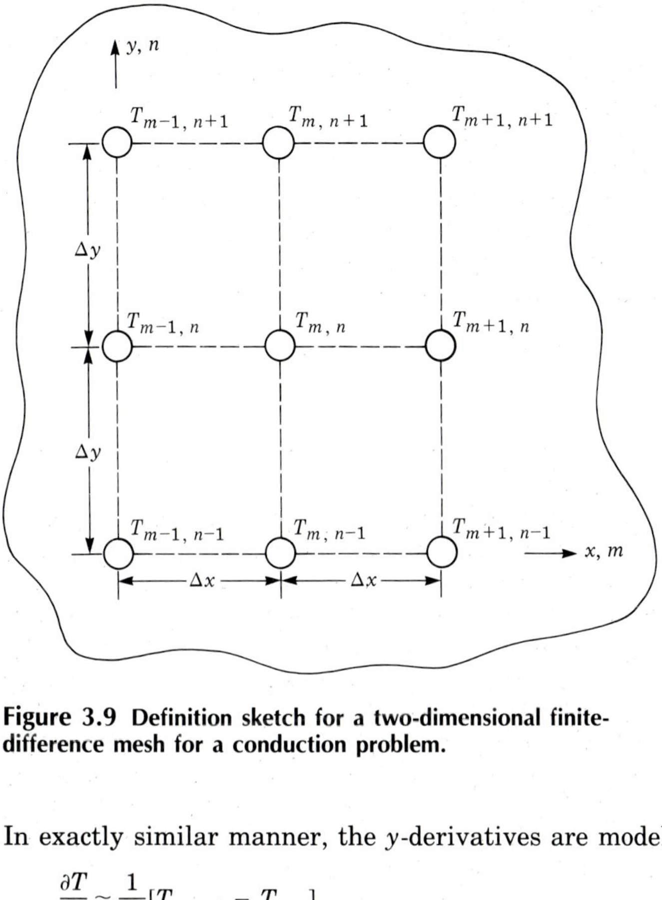


**Figure 3.9 Definition sketch for a two-dimensional finite-difference mesh for a conduction problem.**

In exactly similar manner, the $y$-derivatives are modeled as

$$\frac{\partial T}{\partial y} \simeq \frac{1}{\Delta y}[T_{m,n+1} - T_{m,n}],$$
$$\frac{\partial^2 T}{\partial y^2} \simeq \frac{1}{\Delta y^2}[T_{m,n+1} - 2T_{m,n} + T_{m,n-1}]. \quad \textbf{(3.52)}$$

Not only are these economical formulas, but they can now be programmed directly onto a digital computer, using subscript notation for the variable $T$.

The use of these formulas to simulate heat conduction can be developed in two ways: (1) directly substituting into the basic differential equation (3.12) or (2) using these formulas to make a heat balance of the region surrounding a given node. Let us use scheme 1 to derive the basic difference equation and scheme 2 to develop the modeled boundary conditions. There are, of course, whole books devoted to finite difference modeling [15, 16, 28, 30] to which you can refer for further application of these methods to physical problems.

Now let us model the differential equation of two-dimensional steady heat conduction with internal heat generation:

$$\frac{\partial^2 T}{\partial x^2} + \frac{\partial^2 T}{\partial y^2} = -\frac{\dot{q}}{k}. \tag{3.53}$$

Replacing the second derivatives in Eq. (3.53) by the algebraic formulas from Eqs. (3.51) and (3.52) and rearranging, we obtain

$$2(1 + \beta)T_{m,n} \doteq \beta T_{m,n+1} + T_{m+1,n} + \beta T_{m,n-1} + \beta T_{m-1,n} + \frac{\dot{q}}{k}\Delta x^2, \tag{3.54}$$

where $\beta = (\Delta x/\Delta y)^2$ is the mesh-size factor. The partial differential equation has been replaced by a linear algebraic equation relating the local nodal temperature $T_{m,n}$ to its nearest neighbors and the local value of $\dot{q}_{m,n}$. By applying Eq. (3.54) to each of the $N$ unknown nodal temperatures, we obtain $N$ simultaneous linear algebraic equations in $N$ unknowns. If $N$ is small, we may use a Gauss elimination or matrix inversion procedure [7]. But it is more likely in a practical engineering problem that $N$ is large (50 or more), in which case a solution is best achieved by an iterative procedure, which we explain in the next section.

By far the most common case encountered is a square mesh ($\beta = 1$) and zero heat generation, for which Eq. (3.54) becomes

$$T_{m,n} = \frac{1}{4}(T_{m,n+1} + T_{m+1,n} + T_{m,n-1} + T_{m-1,n}). \tag{3.55}$$

This is a remarkable formula, stating that an algebraic model for the two-dimensional Laplace equation (3.25) forces each nodal temperature to be the arithmetic average of its four nearest neighbors. It should be obvious that Eq. (3.55) is suitable for direct programming onto a digital computer by defining a subscripted variable $T(M, N)$.

A much more accurate model of Laplace's equation, which uses all eight nearest neighbors in Fig. 3.9, has been suggested by Milne [17], again for a square mesh, $\Delta x = \Delta y$:

$$\begin{aligned} T_{m,n} = \frac{1}{20}(&4T_{m,n+1} + T_{m+1,n+1} + 4T_{m+1,n} + T_{m+1,n-1} \\ &+ 4T_{m,n-1} + T_{m-1,n-1} + 4T_{m-1,n} + T_{m-1,n+1}). \end{aligned} \tag{3.56}$$

This is a weighted average, related to the familiar Simpson's quadrature rule [7], which for a given mesh size $\Delta x$ is anywhere from 10 to 100 times more accurate than the five-point formula, Eq. (3.55). In spite of its accuracy, however, this nine-point formula is not widely used.

The two formulas may be absorbed at a glance by drawing what Milne [17] calls a "stencil" sketch of the local nodes, as shown in Figure 3.10.

One reason that Eq. (3.56) is unpopular is that it is very difficult to formulate convection boundary condition models of equivalent accuracy. The nine-point formula is ideal, however, when the boundary temperatures are known (see Example 3.9).

### 3.4.2 Solution Procedure and Numerical Stability

Since the number of unknown nodes in a typical engineering conduction problem will be large (50 or more), *direct* solutions by matrix inversion or Gaussian elimination techniques [7] are uneconomical, except for one-dimensional problems such as fin conduction. Therefore we resort to *iterative* techniques that converge toward the correct nodal values.

There are many iteration techniques for simultaneous linear algebraic equations [15, 16, 18]. The simplest — but not the fastest converging — is the Gauss-Seidel iteration, in which Eq. (3.55) is applied successively to each and every node in the conduction region after initial guesses are made for the nodal values. We sweep the whole field of nodes over and over again with Eq. (3.55) until the nodal values change by only a negligible, small "cutoff" amount. The solution is then said to have "converged" to the correct steady-state temperature distribution.

The simplest boundary condition is that of known surface temperature — as in Fig. 3.4 or 3.5 — in which the boundary nodes are specified and held constant during the iteration, which takes place only over interior nodes. Suppose that the region is rectangular, with

| | | |
|---|---|---|
| 0 | 1 | 0 |
| 1 | −4 | 1 |
| 0 | 1 | 0 |

**Eq. (3.55) Five-Point Formula**

| | | |
|---|---|---|
| 1 | 4 | 1 |
| 4 | −20 | 4 |
| 1 | 4 | 1 |

**Eq. (3.56) Nine-Point Formula**

**Figure 3.10** Two finite-difference stencils that simulate Laplace's equation.

boundaries given by $n = 1$, $m = 1$, $n = n_f$, and $m = m_f$, where all temperatures are known and constant. Then a FORTRAN program to solve for the interior temperatures might look as follows:

```
    DIMENSION T(100,100), EPS(100,100)
C read in all boundary temperatures:
    READ (5,*) MF, NF
    READ (5,*) (T(I,1),T(I,NF),I=1,MF)
    READ (5,*) (T(1,J),T(MF,J),J = 1,NF)
C make some reasonable guesses of the interior temperatures:
      MLIM = MF - 1
      NLIM = NF - 1
      DO 8 M = 2,MLIM
      DO 8 N = 2,NLIM
    8 T(M,N) = 50.      (a rather crude guess)
C now sweep the interior with the Gauss-Seidel iteration:
    1 DO 6 M = 2,MLIM
      DO 6 N = 2,NLIM
      TZ = 0.25*(T(M,N+1)+T(M+1,N)+T(M,N-1)+T(M-1,N))
      EPS(M,N) = TZ - T(M,N)
    6 T(M,N) = TZ
C then test to see if the nodal changes are all small:
      DO 7 M = 2,MLIM
      DO 7 N = 2,NLIM
      IF(ABS(EPS(M,N)).GT.0.01) GO TO 1
    7 CONTINUE
C if you get this far, you have converged and can print out:
    WRITE(6,*)((T(M,N),M=1,MF),N=1,NF)
    END
```

This is a rather mediocre program and is not meant to be copied. It is intended only to show the simplicity of the finite-difference modeling of the problem. The actual iteration procedure takes only the three lines beginning with statement 1 in the program, while 80% of the program is merely involved with input, output, and testing of results.

How reliable is such a program? It is shown in advanced texts [18, p. 64] that for steady heat conduction the Gauss-Seidel process *always* converges uniformly to the proper solution of the modeled system of simultaneous equations. There is no instability whatever. For a large system of nodes, however, convergence can be rather slow, and one may wish to try some faster iteration schemes. Some faster techniques discussed in [16] and [18] are (1) line-by-line iteration with a tridiagonal back-substitution procedure, (2) the alternating-direction (ADI) method, (3) successive overrelaxation (SOR), and (4) "hopscotch" methods, none of which will be discussed here in this elementary text. Many efficient heat transfer programs are available to the public from the Computer Software Management and Information Center (COSMIC), 112 Barrow Hall, The University of Georgia, Athens, GA 30602.

How accurate are the finite-difference models? Milne [17] shows that the five-point formula, Eq. (3.55), will predict a nodal temperature that is in error by no more than

$$E = r^2 \Delta x^2 M_4 / 24, \qquad \textbf{(3.57)}$$

where $r$ is the radius of the smallest circle enclosing the region and $M_4$ is the maximum absolute value of the fourth partial derivatives of $T$ at the nodal point. This upper bound is difficult to estimate numerically [17, p. 219]; in practice, one might just decrease the mesh size to decrease the error.

### Example 3.9

Make a finite-difference analysis of the problem illustrated in Fig. 3.8 of a square plate with temperature $T_o = 100°C$ on the upper surface and $T = 0°C$ on the other three surfaces. Use $\Delta x = \Delta y = 0.25L$ and compare with the exact solution, Eq. (3.43). Compare the accuracy of the five-point and nine-point finite-difference formulas.

*Solution* For the given conditions, the finite difference mesh would be as sketched in Fig. E3.9. Since the problem is symmetric about the plate centerline ($x = L/2$), the nine interior mesh points in the figure yield only six unknowns, $T_1$–$T_6$, each of the three points $T_{1,3,5}$ having a mirror image. The exact temperatures can be computed by programming the Fourier series solution, Eq. (3.43), for a hand calculator. The exact computations are as follows, to three

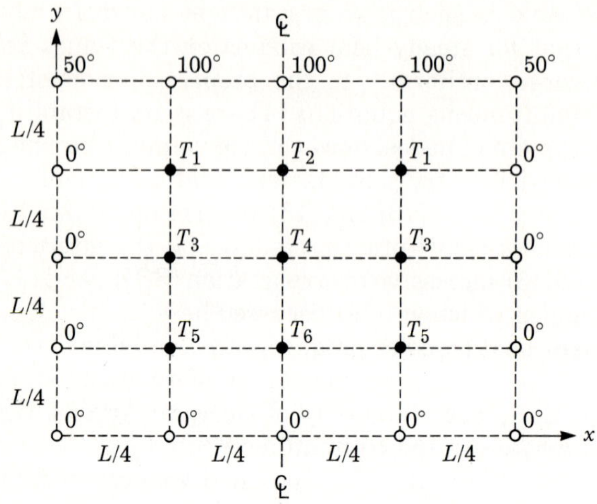


**Figure E3.9**

decimal places:

$$T_1 = 43.203°, \quad T_2 = 54.053°,$$
$$T_3 = 18.203°, \quad T_4 = 25.000°,$$
$$T_5 = 6.797°, \quad T_6 = 9.541°.$$

We will compare these with the finite-difference calculations.

First, the five-point formula, Eq. (3.55), produces the following formulas, proceeding clockwise from the top:

$$\begin{aligned}
T_1 &= \frac{1}{4}(100° + T_2 + T_3 + 0°),\\
T_2 &= \frac{1}{4}(100° + T_1 + T_4 + T_1),\\
T_3 &= \frac{1}{4}(T_1 + T_4 + T_5 + 0°),\\
T_4 &= \frac{1}{4}(T_2 + T_3 + T_6 + T_3),\\
T_5 &= \frac{1}{4}(T_3 + T_6 + 0° + 0°),\\
T_6 &= \frac{1}{4}(T_4 + T_5 + 0° + T_5).
\end{aligned} \tag{1}$$

As an initial guess, assume a linear distribution in $y$. Then evaluate

Eqs. (1) in order, repeating until the process converges. The results of this iterative process are shown in the following table.

| Nodal Value | Initial Guess | Iteration of Eqs. (1) | | | | | Percent Error |
|---|---|---|---|---|---|---|---|
| | | #1 | #3 | #5 | #10 | #15 | |
| $T_1$ | 75.0° | 56.250 | 46.338 | 43.736 | 42.885 | 42.857° | −0.80 |
| $T_2$ | 75.0° | 65.625 | 56.177 | 53.557 | 52.706 | 52.679° | −2.54 |
| $T_3$ | 50.0° | 32.813 | 22.266 | 19.269 | 18.777 | 18.750° | +3.01 |
| $T_4$ | 50.0° | 39.063 | 28.516 | 25.880 | 25.027 | 25.000° | 0.00 |
| $T_5$ | 25.0° | 14.453 | 8.905 | 7.582 | 7.157 | 7.143° | +5.09 |
| $T_6$ | 25.0° | 16.992 | 11.581 | 10.261 | 9.835 | 9.821° | +2.93 |

Thus in spite of the crude initial guesses, the computations converge in about 15 iterations. Do not confuse "convergence" with "exactness"; the converged temperatures are only approximate, with a maximum relative error of 5% and a maximum absolute error of 1.38°C. Note that the center temperature $T_4 = 25.000°$ is predicted exactly in this model.

Second, the nine-point formula, Eq. (3.56), leads to the following algebraic relations, again clockwise from the top:

$$
\begin{aligned}
T_1 &= \frac{1}{20}[4(100°) + 100° + 4T_2 + T_4 + 4T_3 + 0° + 0° + 50°],\\
T_2 &= \frac{1}{20}[4(100°) + 100° + 4T_1 + T_3 + 4T_4 + T_3 + 4T_1 + 100°],\\
T_3 &= \frac{1}{20}[4T_1 + T_2 + 4T_4 + T_6 + 4T_5 + 0° + 0° + 0°],\\
T_4 &= \frac{1}{20}[4T_2 + T_1 + 4T_3 + T_5 + 4T_6 + T_5 + 4T_3 + T_1],\\
T_5 &= \frac{1}{20}[4T_3 + T_4 + 4T_6 + 0° + 0° + 0° + 0° + 0°],\\
T_6 &= \frac{1}{20}[4T_4 + T_3 + 4T_5 + 0° + 0° + 0° + 4T_5 + T_3].
\end{aligned} \qquad (2)
$$

Note that we have taken the upper-corner temperatures in the example figure to be 50°C, or halfway between 0° on the side and 100° on the top. This is necessary to ensure the accuracy of the nine-point formula.

Again taking initial temperatures linear with $y$, we find that the iterations of Eqs. (2) lead to the following table.

| Nodal Value | Initial Guess | Iterations of Eqs. (2) | | | | | Percent Error |
|---|---|---|---|---|---|---|---|
| | | #1 | #3 | #5 | #10 | #15 | |
| $T_1$ | 75.0° | 55.000 | 45.916 | 43.761 | 43.217 | 43.206° | +0.01 |
| $T_2$ | 75.0° | 67.000 | 56.972 | 54.686 | 54.112 | 54.101° | +0.09 |
| $T_3$ | 50.0° | 30.600 | 20.773 | 18.709 | 18.192 | 18.182° | −0.12 |
| $T_4$ | 50.0° | 38.640 | 27.734 | 25.556 | 25.010 | 25.000° | 0.00 |
| $T_5$ | 25.0° | 13.052 | 8.026 | 7.044 | 6.798 | 6.793° | −0.05 |
| $T_6$ | 25.0° | 16.009 | 10.834 | 9.800 | 9.540 | 9.536° | −0.06 |

Again the computations converge in about 15 iterations, and accuracy is much better than with the five-point formula: maximum relative error of 0.12% (40 times better) and maximum absolute error of 0.05°C (30 times better). Again the center temperature, $T_4 = 25$°C, is predicted exactly. In spite of its demonstrated better accuracy, however, the nine-point formula is not widely used.

Additional accuracy can be achieved by reducing the mesh size. If we halve the mesh width to $\Delta x = \Delta y = L/8$, there will be 28 unknown nodal values, and the maximum errors of the five-point formula will be reduced from 1.4°C to 0.4°C and from 5.1% to 1.4%, or about a 4:1 reduction proportional to $\Delta x^2$. For the nine-point formula, the smaller mesh reduces errors from 0.05°C to 0.003°C and from 0.12% to 0.008%, or about a 16:1 reduction proportional to $\Delta x^4$. ■

---

### 3.4.3 Convection Boundary Condition Formulations

When temperatures are known at the boundary, as in Example 3.9, all boundary nodal values are specified, and the entire solution consists of satisfying Eq. (3.55) at every interior node. For convection boundary conditions, however, boundary temperatures will be unknown and will become part of the analysis. We must derive appropriate nodal formulas to supplement Eq. (3.55).

Consider, for example, in Fig. 3.11 a flat surface exposed to a convecting fluid ($h_\infty$, $T_\infty$). Boundary temperatures $T_{m,n}$ are unknown, and we need a numerical model for each boundary node. One modeling scheme is to write a heat balance for the shaded control volume in the figure, approximating conduction fluxes by finite differences. For

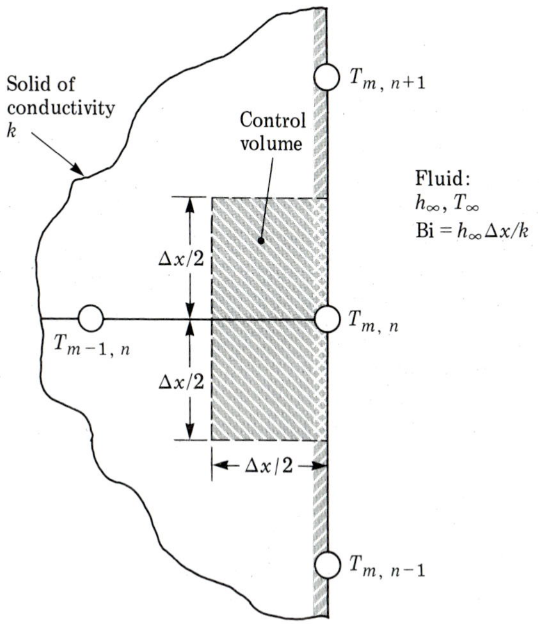


**Figure 3.11 Definition sketch for derivation of the numerical approximation to a convection boundary condition, Eq. (3.59).**

steady flow, we obtain

Heat conducted in = Heat convected out,

or

$$kb\Delta x\left(\frac{T_{m-1,n} - T_{m,n}}{\Delta x}\right) + kb\frac{\Delta x}{2}\left(\frac{T_{m,n+1} - T_{m,n}}{\Delta x}\right) + kb\frac{\Delta x}{2}\left(\frac{T_{m,n-1} - T_{m,n}}{\Delta x}\right) \simeq h_\infty b\Delta x(T_{m,n} - T_\infty). \tag{3.58}$$

where $b$ is the body width into the paper. Rearranging and canceling where possible, we obtain a convection approximation:

$$T_{m,n} \simeq \frac{1}{2 + \text{Bi}}\left[T_{m-1,n} + \frac{1}{2}(T_{m,n+1} + T_{m,n-1}) + \text{Bi}\, T_\infty\right], \tag{3.59}$$

where Bi $= h_\infty \Delta x/k$ is a *mesh-size Biot number* characteristic of finite-difference convection models. Equation (3.59) is written in a form that is suitable for sweeping along the boundary nodes in a Gauss-Seidel type of iteration. Interior points would meanwhile be swept with Eq. (3.55). The mesh-size Biot number, Bi, may have any (positive) magnitude without affecting the inherent stability of the iteration convergence.

Table 3.2 lists Eq. (3.59) and six other types of flux and geometry conditions commonly occurring in steady conduction problems. The double-subscript notation has been omitted for convenience. In all cases the modeled condition has been written in a form that is suitable for iteration. All of these models are comparable in accuracy to the five-point Laplacian model for interior nodes, Eq. (3.55). If used with the fancier nine-point interior model, Eq. (3.56), they would degrade its accuracy back to the level of the five-point formula. Nine-point equivalents to Table 3.2 are not widely used or known. All of the models in Table 3.2 result in stable, convergent iterations, regardless of the magnitude of the mesh Biot number, if any.

## Example 3.10

Make a finite-difference model of a square chimney similar to Fig. 3.1. Outside width is 1.5 m, inside width is 0.5 m, and brick conductivity is $k = 1.2$ W/m · K. Let $\Delta x = \Delta y = 0.25$ m and take full advantage of symmetry. Compute nodal temperatures and estimate the heat loss per meter of chimney depth into the paper, with convection shown in Fig. E3.10.

*Solution* Because of combined diagonal and central symmetry, we need analyze only one-eighth of the chimney, similar to the shaded region in Fig. 3.1(a). For the given mesh size, there are nine unknown nodes, shown as shaded circles in this figure: four at the outer boundary, three interior points, and two at the inner boundary. The two mesh-size Biot numbers are

$$\mathrm{Bi}_o = h_o \Delta x/k = (18\ \mathrm{W/m^2\cdot K})(0.25\ \mathrm{m})/(1.2\ \mathrm{W/m\cdot K}) = 3.75,$$

$$\mathrm{Bi}_i = h_i \Delta x/k = 5\mathrm{Bi}_o = 18.75.$$

We use several nodal points outside the one-eighth segment that are equal by symmetry to their images inside the solution zone, as shown in the figure. We set up the nodal algebraic relations in groups, as follows.

*One exterior corner from item 5 of Table 3.2:*

$$T_{11} = \frac{1}{1 + 3.75}\left[\frac{1}{2}(T_{21} + T_{21}) + 3.75(20)\right]. \qquad (1)$$

**Table 3.2 Numerical boundary condition approximations**

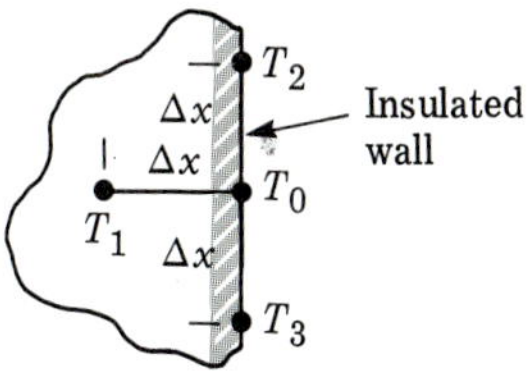


1. Flat, insulated wall

$$T_0 = \frac{1}{2}T_1 + \frac{1}{4}(T_2 + T_3)$$

---

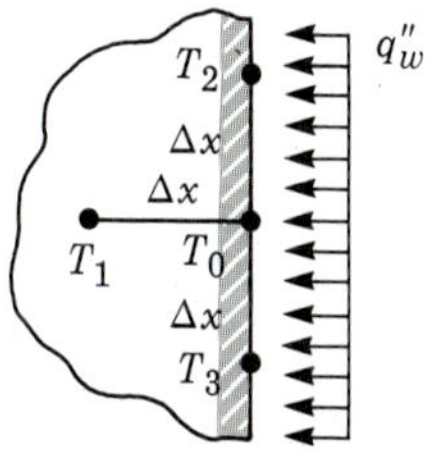


2. Flat wall with known heat flux $q''_w$

$$T_0 = \frac{1}{2}T_1 + \frac{1}{4}(T_2 + T_3) + \frac{q''_w\,\Delta x}{2k}$$

---

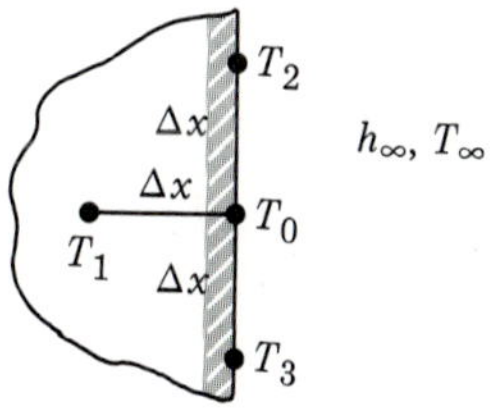


3. Flat wall in contact with a fluid

$$T_0 = \frac{1}{2 + \mathrm{Bi}}\left[T_1 + \frac{1}{2}(T_2 + T_3) + \mathrm{Bi}\,T_\infty\right]$$

$$\mathrm{Bi} = h_\infty \Delta x / k$$

---

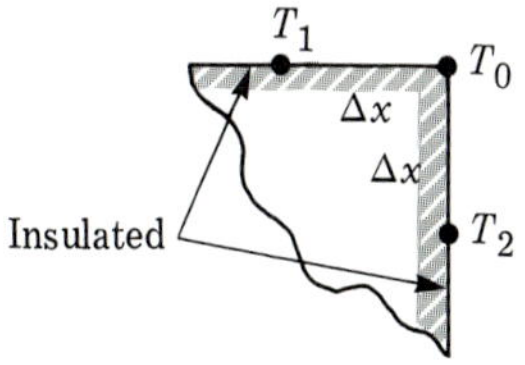


4. Exterior corner, both walls insulated

$$T_0 = \frac{1}{2}(T_1 + T_2)$$

**Table 3.2** (continued)

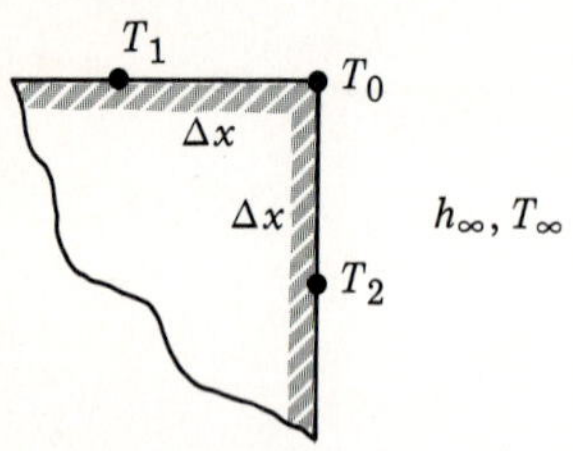


5. Exterior corner in contact with a fluid

$$T_0 = \frac{1}{1 + \text{Bi}}\left[\frac{1}{2}(T_1 + T_2) + \text{Bi}\, T_\infty\right]$$

$$\text{Bi} = h_\infty \Delta x / k$$

---

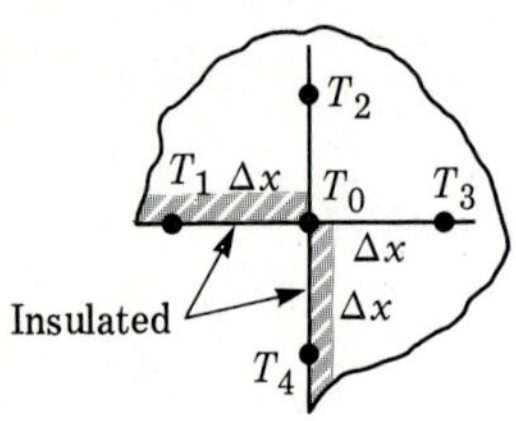


6. Interior corner, both walls insulated

$$T_0 = \frac{1}{3}(T_2 + T_3) + \frac{1}{6}(T_1 + T_4)$$

---

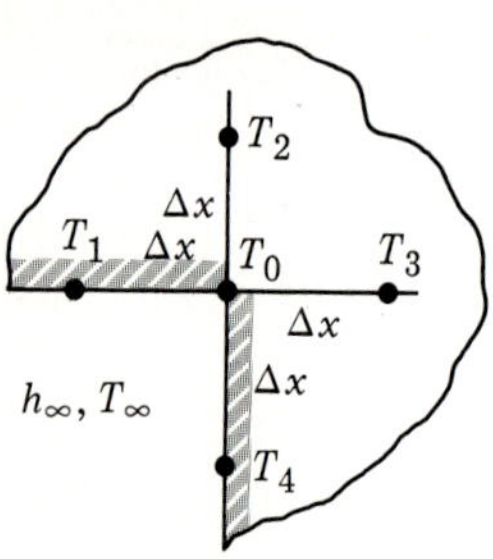


7. Interior corner in contact with a fluid

$$T_0 = \frac{1}{3 + \text{Bi}}\left[T_2 + T_3 + \frac{1}{2}(T_1 + T_4) + \text{Bi}\, T_\infty\right]$$

$$\text{Bi} = h_\infty \Delta x / k$$

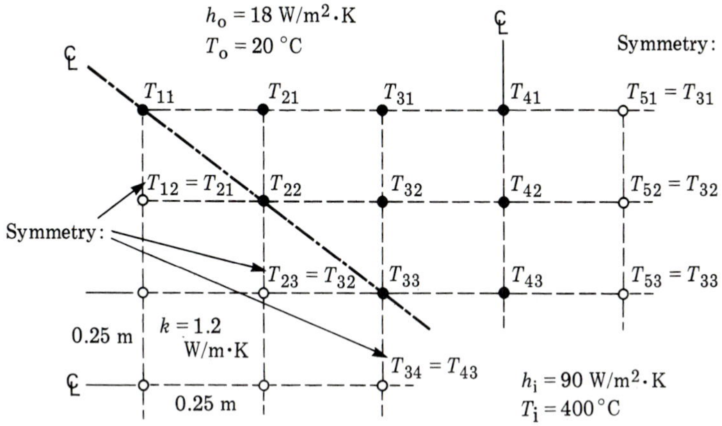


**Figure E3.10**

*Three flat-wall outside convection boundary nodes, item 3:*

$$T_{21} = \frac{1}{2 + 3.75}\left[T_{22} + \frac{1}{2}(T_{11} + T_{31}) + 3.75(20)\right], \tag{2}$$

$$T_{31} = \frac{1}{2 + 3.75}\left[T_{32} + \frac{1}{2}(T_{21} + T_{41}) + 3.75(20)\right], \tag{3}$$

$$T_{41} = \frac{1}{2 + 3.75}\left[T_{42} + \frac{1}{2}(T_{31} + T_{31}) + 3.75(20)\right]. \tag{4}$$

*Three interior points from Eq. (3.55):*

$$T_{22} = \frac{1}{4}(T_{21} + T_{21} + T_{32} + T_{32}), \tag{5}$$

$$T_{32} = \frac{1}{4}(T_{22} + T_{31} + T_{42} + T_{33}), \tag{6}$$

$$T_{42} = \frac{1}{4}(T_{32} + T_{41} + T_{32} + T_{43}). \tag{7}$$

*One interior corner from item 7 of Table 3.2:*

$$T_{33} = \frac{1}{3 + 18.75}\left[T_{32} + T_{32} + \frac{1}{2}(T_{43} + T_{43}) + 18.75(400)\right]. \tag{8}$$

*One flat-wall inside convection boundary node, item 3:*

$$T_{43} = \frac{1}{2 + 18.75}\left[T_{42} + \frac{1}{2}(T_{33} + T_{33}) + 18.75(400)\right]. \tag{9}$$

Starting with initial guesses of 30°C at the outer wall, 200°C for the interior nodes, and 350°C at the inner wall, we sweep through Eqs. (1–9) successively, repeating the iteration until the nodal values cease to change. Convergence to one-decimal-place accuracy occurs after seven iterations, and two-decimal-place convergence takes eleven sweeps. The converged values are as follows:

$$
\begin{aligned}
T_{11} &= 24.16°\text{C}, & T_{21} &= 39.75°\text{C}, & T_{31} &= 54.42°\text{C}, & T_{41} &= 58.41°\text{C},\\
& & T_{22} &= 114.28°\text{C}, & T_{32} &= 188.81°\text{C}, & T_{42} &= 206.44°\text{C},\\
& & & & T_{33} &= 380.11°\text{C}, & T_{43} &= 389.71°\text{C}.
\end{aligned}
$$

If we sketched in isotherms, they would crowd near the inside corner as in Fig. 3.1(a). Of course, it is fatuous to keep two decimal places: Because of the coarse mesh used, the accuracy of these computed temperatures is approximately $\pm 5°$C. Equation (3.57) for this case provides an upper bound for the error of $\pm 10.6°$C. Since no exact solution is known to this problem, one can only increase the accuracy of, for example, $T_{11} = 24 \pm 5°$C by decreasing the mesh size and setting up a new group of nodal equations.

The heat flux could be computed from the thermal conductivity and a numerical estimate of the temperature gradients in the chimney. However, numerical differentiation is avoided by using instead the average surface temperatures and the known convection coefficients. From the nodal values obtained, the average outer temperature is approximately (1/4)(24.16 + 39.75 + 54.42 + 58.41), or 44.2°C. Then the outer surface convection is, per unit depth,

$$
\begin{aligned}
q_{wo} &= h_o A_o \Delta T_o = (18\ \text{W/m}^2\cdot\text{K})[8(0.75\ \text{m})(1\ \text{m})](44.2° - 20°\text{C})\\
&= 2610\ \text{W}. \quad [\textit{Ans.}]
\end{aligned}
$$

Meanwhile, the average inner surface temperature is (1/2)(380.11 + 389.71) = 384.9°C, from which the inside convection can be estimated as

$$
\begin{aligned}
q_{wi} &= h_i A_i \Delta T_i = (90\ \text{W/m}^2\cdot\text{K})[8(0.25\ \text{m})(1\ \text{m})](400° - 384.9°\text{C})\\
&= 2720\ \text{W}. \quad [\textit{Ans.}]
\end{aligned}
$$

These two estimates should be identical, but they actually differ by 4% because of the coarse mesh size. A third estimate can be made from the average surface temperatures and the shape factor formula for a thick square pipe, item 2 in Table 3.1:

$a/b = 3.0$:

$$
S = \frac{2\pi(1\ \text{m})}{0.93\ \ln[0.948(3.0)]} = 6.46\ \text{m}.
$$

Then the conduction heat flux is given by

$$q_w = kS\,\Delta T_{\text{walls}} = (1.2\ \text{W/m}\cdot\text{K})(6.46\ \text{m})(384.9° - 44.2°\text{C})$$
$$= 2640\ \text{W}. \quad [Ans.]$$

All these estimates are comparable, so the actual heat flux might best be reported as $q_w = 2650 \pm 100$ W. ■

### 3.4.4 Conduction Heat Flux Estimation

In principle, once nodal temperatures are computed, the boundary heat flux can be calculated from Fourier's law, $q_w = -kA_w(\partial T/\partial n)_w$. However, this procedure requires a numerical approximation to the boundary derivative $(\partial T/\partial n)$. Numerical differentiation is a rather inaccurate process [13, 17].

Consider the problem of estimating wall heat flux from a number of equally spaced nodal temperatures, as in Fig. 3.12. If the wall is in contact with a fluid, it is much more accurate to estimate the average wall temperature and then use the convection formula: $q_w = h_\infty A_w(T_{w,\text{avg}} - T_\infty)$. This is because "averaging" is related to numerical integration, which is more accurate than numerical differentiation. If a convection estimate is not possible, however, as in Example 3.9, we must attempt a numerical derivative.

Referring to Fig. 3.12, we see that the simplest estimate of the

**Figure 3.12** Estimating the wall temperature gradient from a number of equally spaced interior nodal values. Recommended formulas are given as Eqs. (3.60)–(3.63).

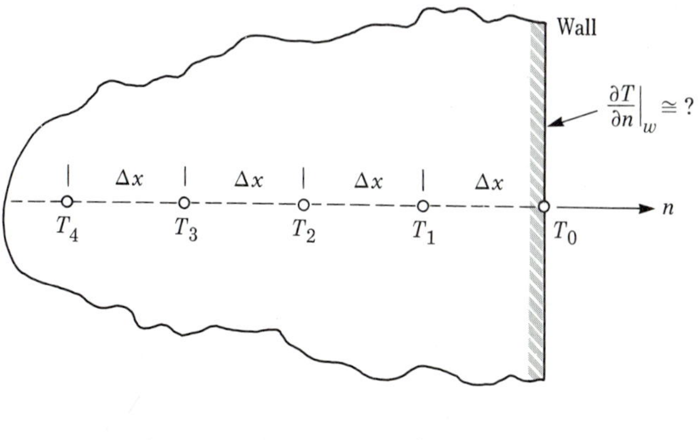

wall temperature gradient is a two-point formula:

$$(\partial T/\partial n)_w \simeq (T_0 - T_1)/\Delta x. \tag{3.60}$$

However, if more interior nodes are known, more accuracy can usually be obtained by utilizing multiple-point formulas [17]:

*Three points:*

$$(\partial T/\partial n)_w \simeq (3T_0 - 4T_1 + T_2)/2\Delta x; \tag{3.61}$$

*Four points:*

$$(\partial T/\partial n)_w \simeq (11T_0 - 18T_1 + 9T_2 - 2T_3)/6\Delta x; \tag{3.62}$$

*Five points:*

$$(\partial T/\partial n)_w \simeq (25T_0 - 48T_1 + 36T_2 - 16T_3 + 3T_4)/12\Delta x. \tag{3.63}$$

But we should remind ourselves that the nodal temperatures will be somewhat uncertain because of the unavoidable truncation error incurred in the finite-difference model. Thus no matter how powerful the numerical derivative formula, the computed gradients will have substantial uncertainty.

### Example 3.11

With reference to Fig. E3.10 and the numerical solution for Example 3.10, evaluate the heat flux $q''_w$ at the point (4, 1) on the outer chimney wall by two conduction formulas. Compare with a convection estimate at the same point.

*Solution* The nodal temperatures were computed and listed in Example 3.10. A two-point formula at (4, 1) gives the estimate

$$\begin{aligned} q''_{w41} &= -k(\partial T/\partial n)_{41} \simeq -k(T_{41} - T_{42})/\Delta x \\ &= -(1.2\ \mathrm{W/m\cdot K})(58.41^\circ - 206.44^\circ\mathrm{C})/(0.25\ \mathrm{m}) \\ &= 711\ \mathrm{W/m^2}. \quad [Ans.] \end{aligned}$$

Since one additional nodal value, $T_{43}$, is available along the normal to point (4, 1), try the three-point formula, Eq. (3.61):

$$\begin{aligned} q''_{w41} &\simeq -k(3T_{41} - 4T_{42} + T_{43})/2\Delta x \\ &= -(1.2\ \mathrm{W/m\cdot K})[3(58.41^\circ) - 4(206.44^\circ) + 389.71^\circ\mathrm{C}]/2(0.25\ \mathrm{m}) \\ &= 626\ \mathrm{W/m^2}. \quad [Ans.] \end{aligned}$$

The convection estimate at this same point gives

$$\begin{aligned} q''_{w41} &= h_o(T_{41} - T_\infty) = (18\ \mathrm{W/m^2\cdot K})(58.41^\circ - 20^\circ\mathrm{C}) \\ &= 691\ \mathrm{W/m^2}. \quad [Ans.] \end{aligned}$$

This last estimate is the most accurate. It appears that the more "powerful" three-point formula actually gives poorer results in this case because of the crude mesh used. ■

---

### 3.4.5 The Boundary Element Method

This text primarily stresses the finite-difference method and its implementation on a digital computer. Of comparable importance is the finite-element method, in which the computational region is divided into small *elements* over which the differential equation is approximately satisfied. We will not develop finite-element schemes here, but there are excellent treatments in [19–21] and special application to heat conduction in [4].

A new and striking alternative to the above two techniques is the *boundary element method,* in which the interior is bypassed and all computations take place on the boundaries. Instead of filling the region with nodes or small elements, there are only boundary nodes and boundary elements. This is accomplished by transforming the basic partial differential equation into an equivalent boundary integral equation. As explained in the text by Brebbia [24], the transformation usually involves multiple integration by parts, followed by an error-minimizing "weighted residual" computation. However, Laplace's equation may be handled directly by use of Green's Second Formula [7, p. 446], which for a two-dimensional region takes the form

$$\iint_A (u\,\nabla^2 v - v\nabla^2 u)\,dA = \int_C \left(u\frac{\partial v}{\partial n} - v\frac{\partial u}{\partial n}\right) dS, \qquad \textbf{(3.64)}$$

where $n$ is the outward normal to the boundary and $u$ and $v$ are any functions with continuous second derivatives.

Our specific interest is in plane, steady heat conduction, for which $\nabla^2 T(x, y) = 0$, with boundary conditions of either (a) known temperature or (b) known temperature gradient $\partial T/\partial n$. We therefore apply Green's formula, Eq. (3.64), to the specific functions $u = T$ and $v = \ln(r)$, which is a fundamental solution to Laplace's equation. Then Eq. (3.64) becomes

$$\iint_A [T\,\nabla^2(\ln r) - \ln(r)\,\nabla^2 T]\,dA = \int_C \left[T\frac{\partial}{\partial n}(\ln r) - \ln(r)\frac{\partial T}{\partial n}\right] dS. \qquad \textbf{(3.65)}$$

Now this relation is applied specifically to a point $P$ on the boundary, as sketched in Fig. 3.13. The area integral on the left-hand side is zero everywhere except at point $P$, where it is singular. Brebbia [24]

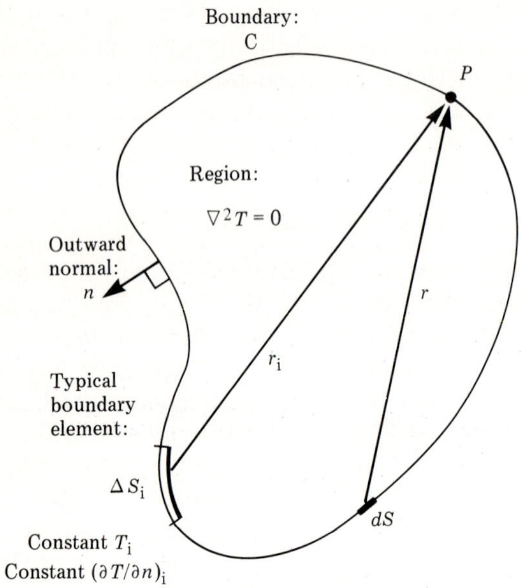


**Figure 3.13 Application of the boundary element method to plane, steady heat conduction.**

shows that the singular part of the integral yields the numerical value $\pi T_P$. Meanwhile the line integral on the right-hand side requires knowledge of the distribution of $T$ and $\partial T/\partial n$ along the boundary. It is here that the "boundary element" scheme arises.

As shown in Fig. 3.13, the boundary $C$ is divided into $N$ boundary elements of length $\Delta S_i$. Each element is assumed to have constant temperature $T_i$ and gradient $(\partial T/\partial n)_i$. The line integral then becomes the sum of $N$ elemental contributions, and Eq. (3.64) reduces to

$$\pi T_P = \sum_{i=1}^{N} T_i \int_{\Delta S_i} \frac{\partial}{\partial n}(\ln r_i)\, dS - \sum_{i=1}^{N} \left.\frac{\partial T}{\partial n}\right|_i \int_{\Delta S_i} \ln r_i\, dS. \tag{3.66}$$

The integrals involving $(\ln r_i)$ are evaluated for each point $P$ and element $i$ by using Gaussian quadrature formulas [24]. Finally, we let points $P$ be the centers of the boundary elements. Then the final form of Eq. (3.66) is

$$T_j = \sum_{i=1}^{N} T_i F_{ij} - \sum_{i=1}^{N} \left.\frac{\partial T}{\partial n}\right|_i G_{ij}, \quad j = 1, \cdots N, \tag{3.67}$$

where $F_{ij}$ and $G_{ij}$ represent the element integrals. Of the $2N$ element values $T_i$ and $(\partial T/\partial n)_i$ in these equations, $N$ of them are known from the boundary conditions of the problem. Thus Eq. (3.67) represents $N$ simultaneous algebraic equations for the $N$ unknown values of $T_i$ and $(\partial T/\partial n)_i$. Like its finite-difference counterpart, the boundary element method is extremely stable for heat conduction and converges rapidly.

Brebbia [24] gives a FORTRAN program for solving Eq. (3.67). At the author's institution, this program has been supplemented by some very friendly graphic interactive software developed by a senior mechanical engineering student. As a result, one can sketch an arbitrary shape on the screen, select nodes, enter boundary conditions, and solve Laplace's equation with a minimum of effort [25].

As an example, let us solve the problem posed by the one-eighth chimney shape in Fig. 3.3. We enter this quadrilateral shape and select ten equally spaced nodes on each side, as in Fig. 3.14. The ratio of outer to inner chimney side length is 2.0. We then specify insulated top and bottom boundaries ($\partial T/\partial n = 0$), temperature $T_1 = 1$ on the inner surface, and $T_o = 0$ on the outer surface. Brebbia's program is then solved for the unknown nodal temperatures on the top and bottom and the unknown temperature gradients on the left and right.

Examining the computed gradients in Fig. 3.14, we see that the heat transfer is highest at the inner corner and lowest at the outer corner, which was evident earlier from our flux plot in Fig. 3.3. The average inner gradient is 1.284, approximately twice as much as the average outer gradient of 0.639. Taking unit chimney depth into the paper and multiplying by 8 for a whole chimney, we compute a total heat transfer of

$$q = kA \left.\frac{\partial T}{\partial n}\right|_{\text{avg}} \doteq k(8)(1)(1.284) \doteq k(8)(2)(0.639) \doteq 10.25\,k.$$

In terms of the conduction shape factor idea of Section 3.3, this is equivalent to a shape factor

$$S \doteq 10.25 \qquad \text{(40 boundary elements)}.$$

Compare this with the correlation formula in Table 3.1, item 2a:

$$S_{\text{chimney}} \doteq \frac{2\pi(1.0)}{0.93 \ln[0.948(2.0)]} \doteq 10.56.$$

The difference is only 3%. If we double the number of nodes in Fig. 3.14, we compute only a slight change, to $S \doteq 10.24$. Therefore the discrepancy is probably in the formula of Table 3.1, which is a curve-fit correlation, not an exact solution to Laplace's equation.

Note from Figure 3.14 that the output of a boundary element

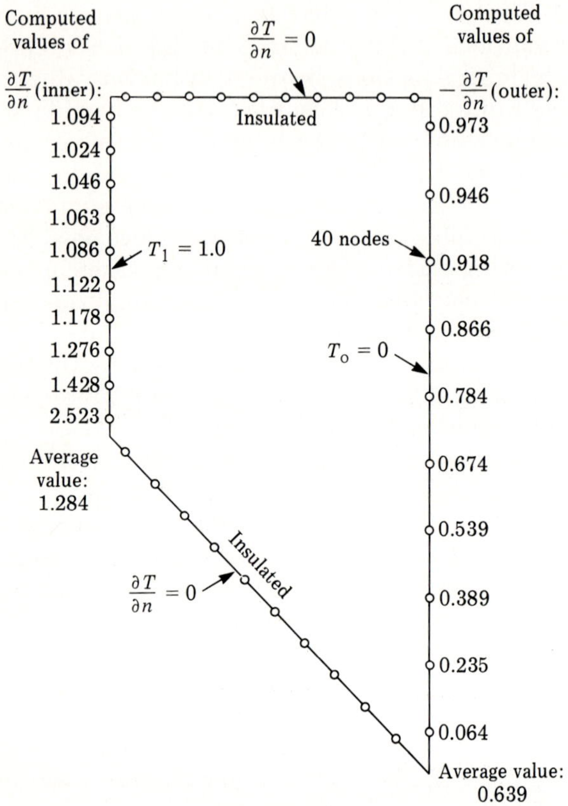


**Figure 3.14 Boundary element solution, using the computer program of [24], for the one-eighth chimney shape of Fig. 3.3. The computer output also includes nodal temperatures along the two insulated sides (not shown).**

program is entirely confined to *boundary* unknowns. Additional programming would be necessary to use the boundary element strengths to compute isotherms and heat flux lines in the body interior. We did not do this because the boundary heat transfer was our only concern.

Because it uses fewer nodes and is simpler to implement, the boundary element method will probably become the preferred method of the future. It has recently been applied [26] to more complex problems such as nonlinear heat conduction, transient wave motion, unsteady elastodynamics, laminar flow, and convective heat transfer.

## †3.5 Analog Methods

Since engineers now have ready access to programmable calculators and time-shared computers, the numerical methods of Section 3.4 constitute a fairly routine tool for solving almost any steady conduction problem. However, for historical and educational reasons, we should note here that several electrical and mechanical analogies exist for modeling a conduction problem.

The electric field $E(x, y)$ in a thin sheet of conducting solid or electrolyte, for example, satisfies the differential equation

$$\frac{\partial^2 E}{\partial x^2} + \frac{\partial^2 E}{\partial y^2} = 0,$$

which is Laplace's equation, exactly analogous to Eq. (3.26). Electric current $I(x, y)$ in the sheet is exactly analogous to conduction heat flux $q''$. Constant-temperature boundaries are simulated by specifying the potential $E$ along the boundary; an "insulated" boundary is electrically insulated. Potential lines (isotherms) and current lines (heat flux directions) can be traced out on the sheet with a pointer and a potentiometer. Such a device is called an *analog field plotter* and not long ago could be purchased commercially.

Unsteady conduction problems (Chapter 4) can be modeled electrically by an array of resistance and capacitance elements and can be subjected to a variety of complex transient boundary conditions. The electric response is very fast — a few seconds or less — and the results exactly model a conduction response that may take hours or days. Thus many transient analog cases can be run in a short time. A very large electric analog of this type was installed in 1937 at Columbia University, and it modeled many important problems, as discussed in the text by Jakob [23].

For steady two-dimensional conduction with uniform internal heat generation, Eq. (3.53), an exact analog is the deflection of a thin membrane, $\eta(x, y)$, which satisfies the equation

$$\frac{\partial^2 \eta}{\partial x^2} + \frac{\partial^2 \eta}{\partial y^2} = -P/\Upsilon = \text{constant}, \tag{3.68}$$

where $P$ is the pressure difference and $\Upsilon$ is the tension per unit length in the membrane. One stretches the membrane across a cutout of the

†This section may be omitted without loss of continuity.

shape of the desired conduction region, applies a pressure difference $P$, and measures deflection $\eta$, which is analogous to temperature $T$. The membrane analogy is still a popular device for demonstrating the solution to Poisson's differential equation, Eq. (3.68).

As we discussed in Section 3.1.3, there are also graphical hand-drawn construction techniques for producing a net of potential and flux lines to simulate the solution of Laplace's equation on an irregular region. At one time, all engineers were required to learn the pencil-and-paper method of "curvilinear squares" for quick solution of potential problems [23, Chapter 19].

## Summary

This chapter has generalized Fourier's law and the first law of thermodynamics into three-dimensional conditions. The basic heat conduction equation is derived in very general form as Eq. (3.1), and four different types of boundary condition are discussed. Cylindrical coordinates are introduced, and the concepts of superposition and variable thermal conductivity are discussed briefly.

Three approaches to conduction analysis in steady flow are introduced: (1) exact analysis by the method of Fourier series, (2) conduction shape factors reported and tabulated in engineering literature, and (3) numerical analysis. The exact analyses are shown to be limited to simple geometries and straightforward boundary conditions — the conduction analysis technique fails if there are nonlinearities such as a radiation boundary condition. The conduction shape factors are valid only for constant-temperature boundary conditions but are a very useful subset of known engineering solutions.

The numerical analyses are limited only by the ingenuity of the analyst and the availability of a large computer. Any type of steady conduction problem with any boundary conditions and/or nonlinearities can be solved for temperatures at selected nodes, with guaranteed convergence and no instability. However, numerical solutions are inherently approximate, and accuracy can be increased only by using finer and finer mesh sizes. It is noted here that certain electrical, mechanical, and graphical analogs are also applicable to conduction analysis.

## References

1. H. S. Carslaw and J. C. Jaeger, *Conduction of Heat in Solids,* 2nd ed., Oxford Press, London, 1959.

2. P. J. Schneider, *Conduction Heat Transfer,* Addison-Wesley, Reading, Mass., 1974.
3. V. S. Arpaci, *Conduction Heat Transfer,* Addison-Wesley, Reading, Mass., 1966.
4. G. E. Myers, *Analytical Methods in Conduction Heat Transfer,* McGraw-Hill, New York, 1971.
5. M. N. Ozisik, *Heat Conduction,* Wiley-Interscience, New York, 1980.
6. M. M. Yovanovich, *Advanced Heat Conduction,* Hemisphere, New York, 1981.
7. E. Kreyszig, *Advanced Engineering Mathematics,* 4th ed., Wiley, New York, 1979.
8. F. M. White, *Viscous Fluid Flow,* McGraw-Hill, New York, 1974.
9. E. Hahne and U. Grigull, "Formfaktor und Formweiderstand der stationärem mehrdimensionalen Wärmeleitung," *Int. J. Heat Mass Transfer,* vol. 18, 1975, pp. 751–767.
10. W. M. Rohsenow and J. P. Hartnett (Eds.), *Handbook of Heat Transfer,* McGraw-Hill, New York, 1973.
11. H. Y. Wong, *Heat Transfer for Engineers,* Longman, New York, 1977.
12. M. C. Potter, *Mathematical Methods in the Physical Sciences,* Prentice-Hall, Englewood Cliffs, N.J., 1978.
13. M. L. James, G. M. Smith, and J. C. Wolford, *Applied Numerical Methods for Digital Computation with FORTRAN and CSMP,* 2nd ed., Crowell, New York, 1977.
14. T. E. Shoup, *A Practical Guide to Computer Methods for Engineers,* Prentice-Hall, Englewood Cliffs, N.J., 1979.
15. G. D. Smith, *Numerical Solution of Partial Differential Equations — Finite Difference Methods,* 2nd ed., Clarendon Press, Oxford, 1978.
16. A. R. Mitchell and D. F. Griffiths, *The Finite Difference Method in Partial Differential Equations,* Wiley, New York, 1980.
17. W. E. Milne, *Numerical Solution of Differential Equations,* Wiley, New York, 1953.
18. S. V. Patankar, *Numerical Heat Transfer and Fluid Flow,* McGraw-Hill/Hemisphere, New York, 1980.
19. C. S. Desai, *Elementary Finite Element Method,* Prentice-Hall, Englewood Cliffs, N.J., 1979.
20. K. H. Huebner, *The Finite Element Method for Engineers,* Wiley, New York, 1975.
21. D. H. Norrie and G. DeVries, *An Introduction to Finite Element Analysis,* Academic Press, New York, 1978.
22. D. H. Norrie and G. DeVries, *Finite Element Bibliography,* Plenum Press, New York, 1976.

23. M. Jakob, *Heat Transfer,* vol. 1, Wiley, New York, 1949.
24. C. A. Brebbia, *The Boundary Element Method for Engineers,* Halsted Press, New York, 1978.
25. T. E. Archibald, "SHAPE — A Program for Interactive Graphic Input to Boundary Element Problems," Senior Project Report, Dept. of Mechanical Engineering, Univ. of Rhode Island, Kingston, RI, May 1984.
26. C. A. Brebbia (Ed.), *Topics in Boundary Element Research,* 2 vols., Springer-Verlag, New York, 1985.
27. S. Kakac and Y. Yener, *Heat Conduction,* 2nd ed., Hemisphere, New York, 1985.
28. Y. Jaluria and K. E. Torrance, *Computational Heat Transfer,* Hemisphere, New York, 1986.
29. U. Grigull and H. Sandner, *Heat Conduction,* Hemisphere, New York, 1984.
30. T. M. Shih, *Numerical Heat Transfer,* Hemisphere, New York, 1984.
31. J. P. Den Hartog, *Advanced Strength of Materials,* McGraw-Hill, New York, 1952.

## Review Questions

1. What basic laws of engineering are used to derive the so-called heat conduction equation?
2. What are the four different types of conduction boundary conditions?
3. Explain how superposition affects conduction solutions.
4. Name three common orthogonal coordinate systems.
5. Explain in words the Fourier series technique.
6. Can one construct a Fourier-type series using functions other than sines and cosines? Give an example.
7. Explain and justify the conduction shape factor.
8. Explain why there are no convection boundary examples among the 14 shape factors listed in Table 3.1.
9. Give the essence of the finite-difference method.
10. What is the advantage of the finite-difference method compared to exact Fourier-series type analysis?
11. What is a physical interpretation of the finite-difference analog for Laplace's equation?
12. Outline three different computation schemes for finding the solution to a set of finite-difference relations.
13. Explain the differences and relative merits of the finite-difference and boundary element methods.

# Problems

## Problem distribution by sections

**The Problem Assignments Are Organized as Follows:**

| Problems | Sections Covered | Topics Covered |
|---|---|---|
| 3.1 –3.14 | 3.1 | Heat Conduction Equation |
| 3.15–3.27 | 3.2 | Exact Analyses |
| 3.28–3.42 | 3.3 | Conduction Shape Factors |
| 3.43–3.71 | 3.4 | Numerical Analyses |
| 3.72–3.73 | 3.5 | Analog Methods |

3.1 The general heat conduction equation, Eq. (3.1), is of a type often described by mathematicians as "elliptic" in space and "parabolic" in time. Can you explain these terms?

3.2 Explain what would cause either the heat conduction equation, Eq. (3.1), or its boundary conditions, Eqs. (3.4–3.8), to be nonlinear.

3.3 For the square pipe with a concentric circular hole in Fig. P3.3, sketch the isotherms and flux lines by eye, without solving any equations. What coordinate system would be convenient for this problem?

3.4 For steady conduction with constant $k$ and no heat generation, write the basic equation and boundary conditions for the plate in Fig. P3.4.

3.5 For the plate of Fig. P3.5, variable $k$ and heat generation, write the basic differential equation and boundary conditions.

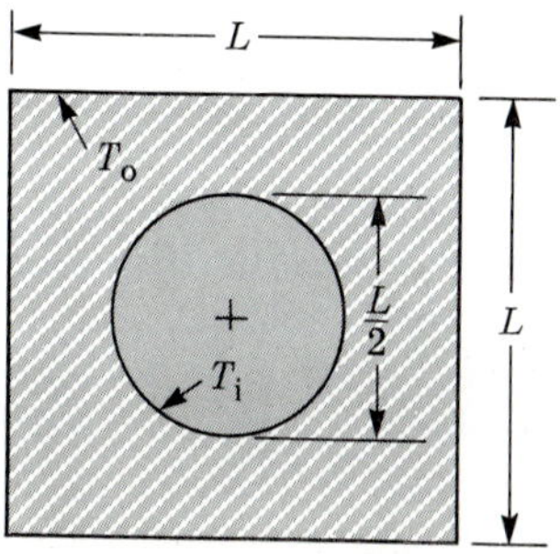


**Figure P3.3**

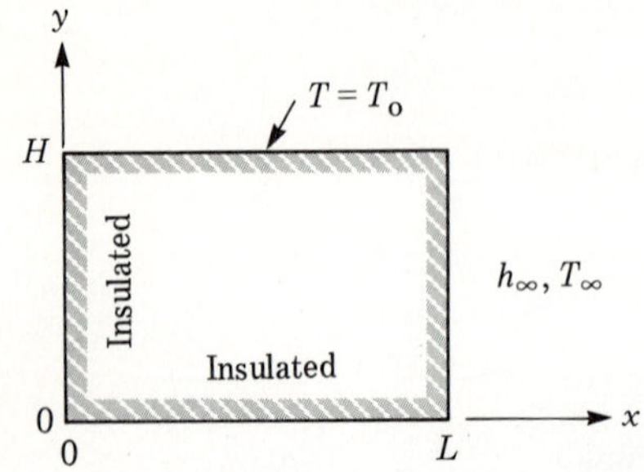


Figure P3.4

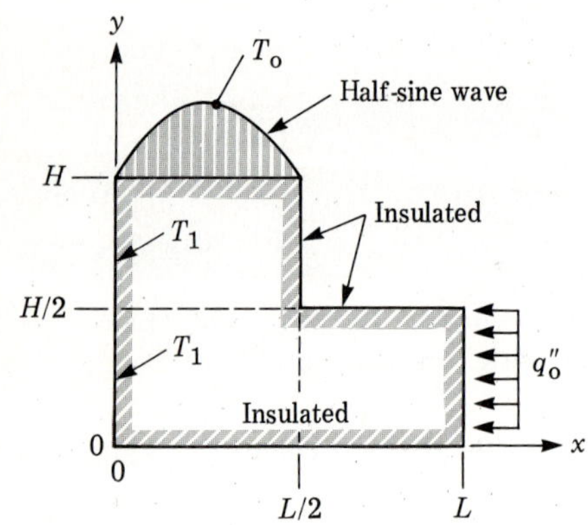


Figure P3.5

3.6 Indicate the differential equation and boundary conditions that should be solved for the exact temperature distribution in the stubby fin/wall configuration of Fig. 2.14(b).

3.7 When nondimensional variables were introduced in Example 3.1, if heat generation is zero the heat conduction equation, Eq. (3.9), contains no dimensionless parameters. Can you interpret this?

3.8 Consider a flux plot such as Fig. 3.3, bounded by two isotherms and two insulated surfaces. Suppose that there are $m$ flux lanes and $n$ approximately equal temperature increments. Show that the total heat transfer per unit depth through the body is given by

$$q \doteq k\frac{m}{n}(T_1 - T_o).$$

3.9 Figure P3.9 shows three instantaneous temperature distributions in a slab of constant conductivity and zero heat generation. Determine whether they represent heating, cooling, or steady-state conditions and justify your conclusions. Do not attack this problem if you have already solved Problem 2.11.

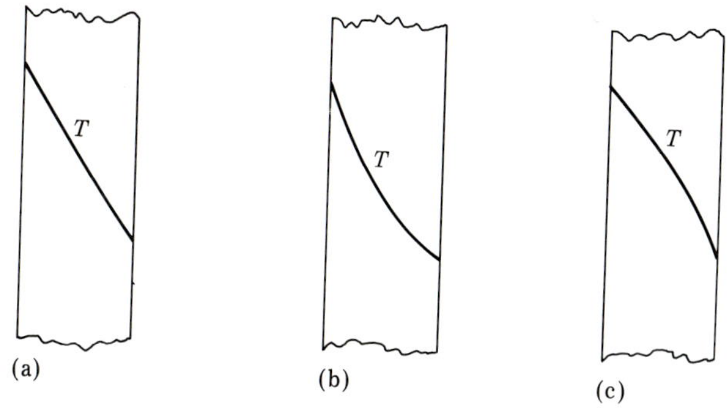


**Figure P3.9**

3.10 Using at least five temperature increments, sketch a flux plot for the hollow eccentric pipe of Fig. P3.10. Using the ideas in Problem 3.8, estimate the total heat transfer per unit depth if $k = 5.0$ W/(m · K).

3.11 Set up, for steady conduction with no heat generation, the basic equation and boundary conditions for the composite body of Fig. P3.11.

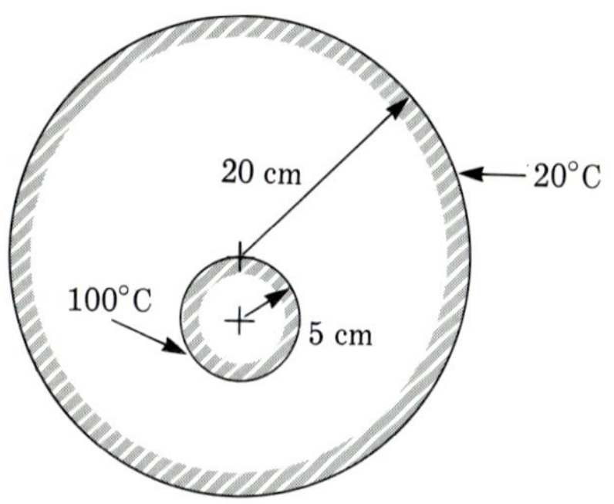


**Figure P3.10**

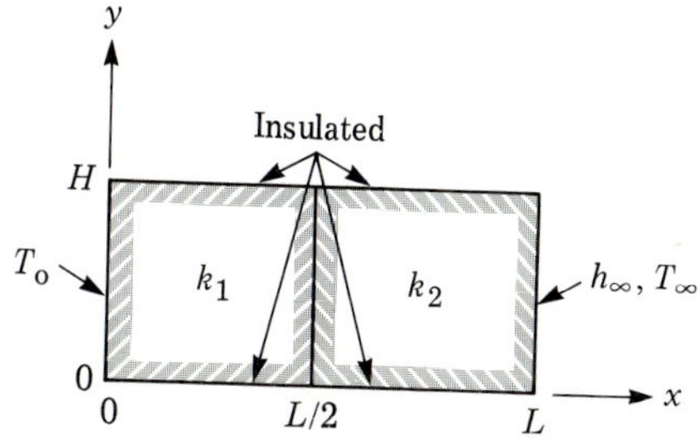


**Figure P3.11**

3.12 If the square solid in Fig. P3.12 satisfies Eq. (3.25), explain how superposition will simplify the problem. How many different types of solutions to Eq. (3.25) must be obtained? Using only the numerical result of Example 3.5, estimate the exact temperature at the center of this solid.

3.13 Return to the hollow solid in Fig. P2.28. Draw the body to scale and make an accurate flux plot, using at least five temperature increments. Using the ideas from Problem 3.8, estimate the total heat transfer per unit depth and compare to the supposedly accurate value of 13,700 W/m.

3.14 Sketch an accurate flux plot for the square solid in Fig. P3.14 using at least five temperature increments. Estimate, from Problem 3.8, the total heat transfer per unit depth.

3.15 Using only the exact analytical results of Example 3.9 for temperatures $T_{1-6}$ and the principle of superposition, compute the exact temperature in the square solid of Fig. P3.12 at $x = L/4$, $y = 3L/4$.

3.16 Using the known exact solution Eq. (3.33), write down, without further ado, the exact solution to the problem in Fig. P3.16.

3.17 Using the solution of Section 3.2.2 as a guideline, solve for the temperature distribution $T(x, y)$ in the problem of Fig. P3.17. Sketch the isotherms

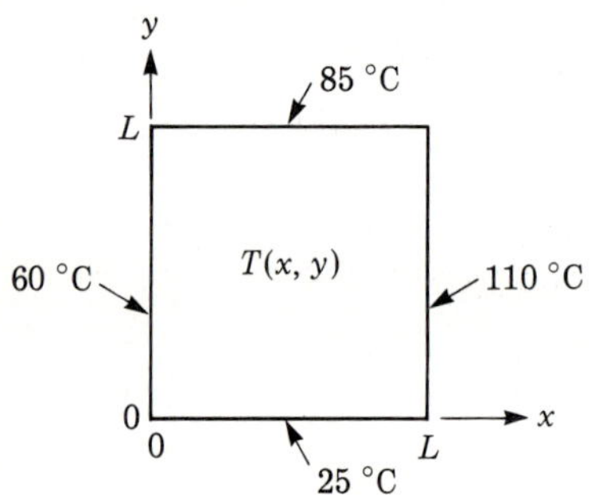


Figure P3.12

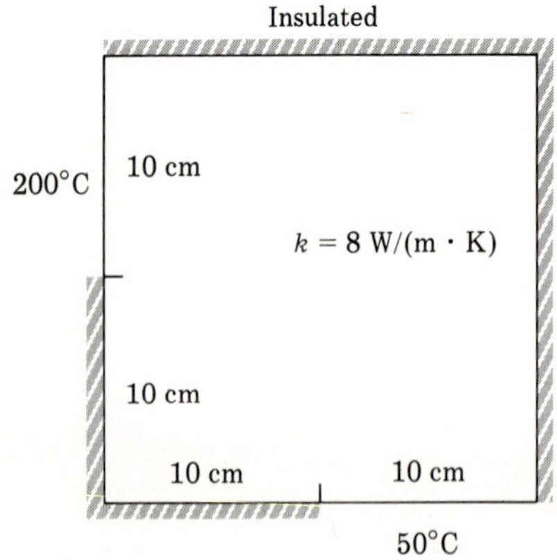


Figure P3.14

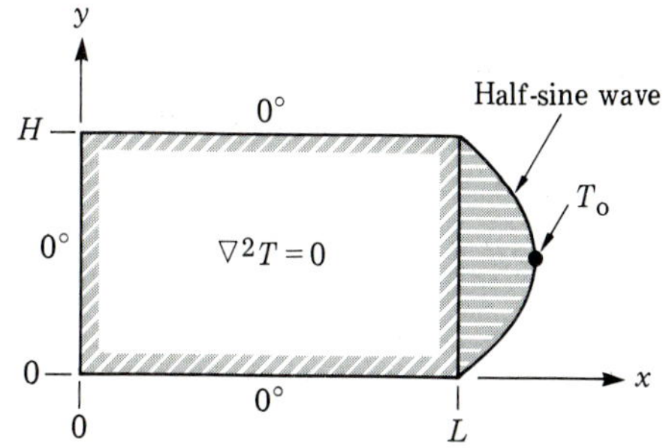


Figure P3.16

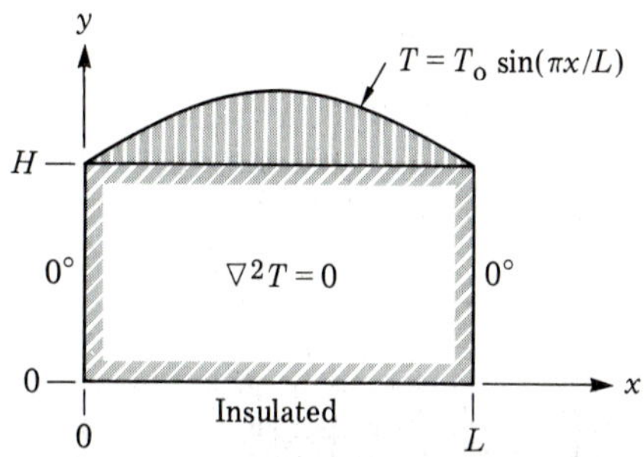


Figure P3.17

for $H/L = 1/3$ and compare with Fig. 3.6. Is the temperature constant along the bottom ($y = 0$)? For the same $T_o$ and $H = L/3$, compute the ratio of the heat flux entering the top surface ($y = H$) to that of Eq. (3.34).

3.18 Find the exact solution $T(x, y)$ to the problem shown in Fig. P3.18. Sketch the isotherms. Is there any net heat transfer through the top surface?

3.19 For the quarter-circular strip in Fig. P3.19, using the polar-coordinate

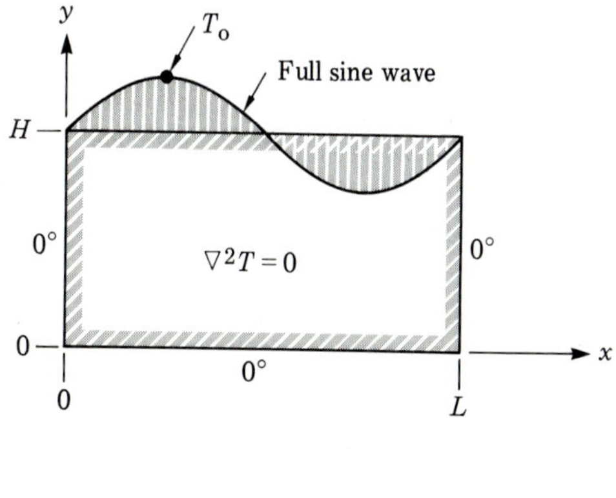


Figure P3.18

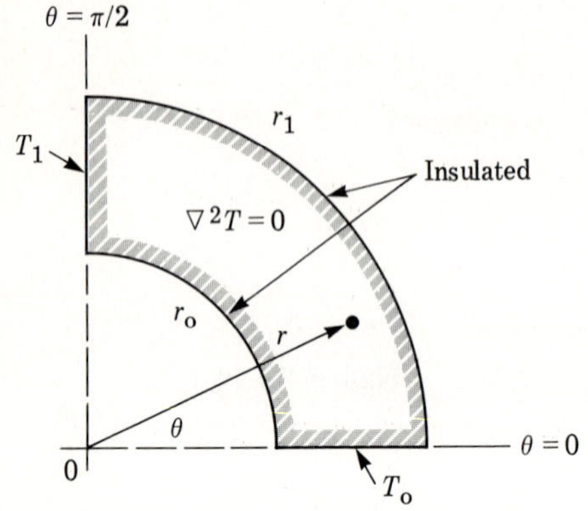


Figure P3.19

Laplace equation from Eq. (2.8), find the exact temperature distribution $T(r, \theta)$ and sketch the isotherms.

3.20 For the quarter-circular strip in Fig. P3.20, using the Laplace equation in the form of Eq. (2.8), find the exact temperature distribution $T(r, \theta)$ and sketch the isotherms.

3.21 For the quarter-circular plate of Fig. P3.21, using Eq. (2.8) for constant $k$ and zero heat generation, show that the exact solution is given by

$$\frac{T - T_o}{T_1 - T_o} = \frac{2}{\pi}\tan^{-1}\left[\frac{2r_1^2 r^2 \sin 2\theta}{r_1^4 - r^4}\right].$$

Plot the isotherms.

3.22 For the square plate solution given by Fig. 3.8, if $T_o = 150°C$, compute the exact temperature to two decimal places at the point ($x = 0.2L$, $y = 0.3L$). How many terms are needed?

Figure P3.20

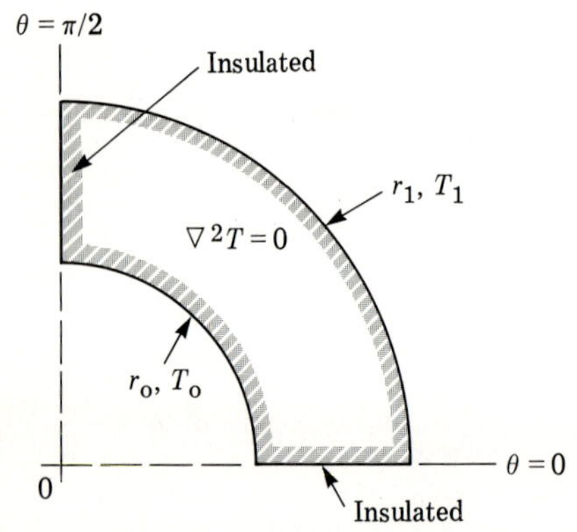

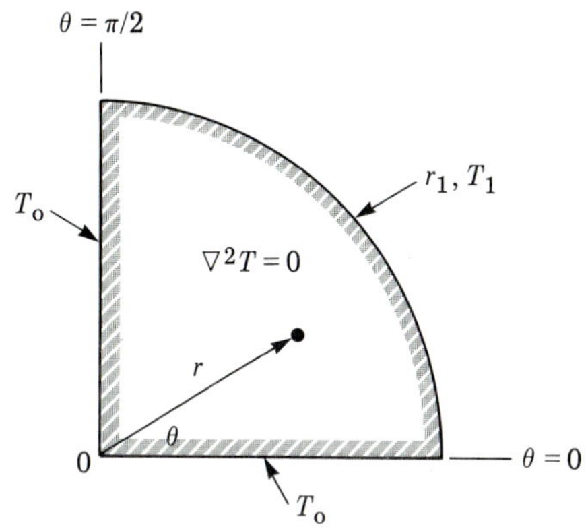


Figure P3.21

3.23 Using the analysis of Section 3.2.3 as a guide, find the exact Fourier series solution for the square plate with a linear upper temperature distribution as in Fig. P3.23. Compute the center temperature $T(L/2, L/2)$ and interpret.

3.24 With Section 3.2.3 as a guide, find the exact Fourier series solution for a square plate with a symmetric hat-shaped temperature distribution along the upper surface, as in Fig. P3.24. Compute $T(L/2, L/2)$.

3.25 For the semi-infinite strip of Fig. P3.25, use the Fourier series technique of Section 3.2.3 to find the exact temperature distribution in the strip.

3.26 Return to Problem 3.10 and use the shape factor concepts of Table 3.1 to estimate the total heat transfer per unit depth.

3.27 Return to Problem 2.28 and use the shape factor concepts of Table 3.1 to estimate the total heat transfer per unit depth.

3.28 A 1-in.-diameter copper hot-water pipe is buried down the center of a concrete ($k$ = 0.7 Btu/hr · ft · °F) slab 6 in. thick and 30 in. wide. The

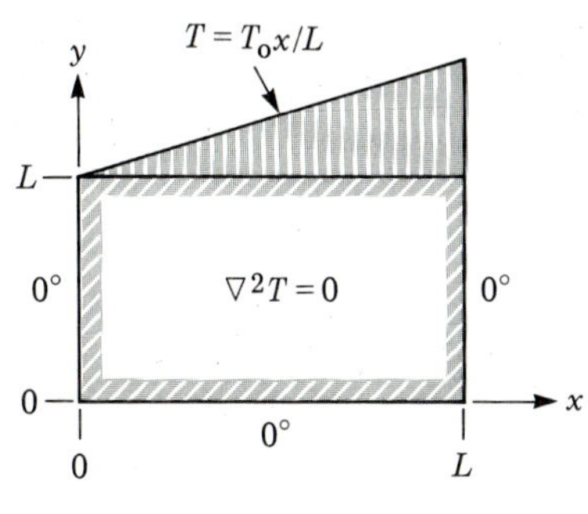


Figure P3.23

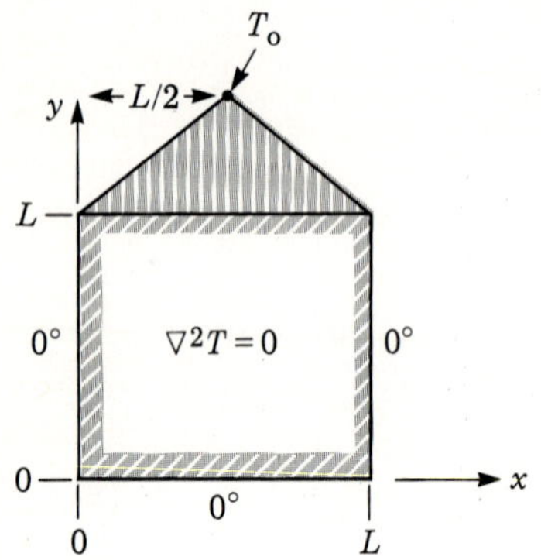


Figure P3.24

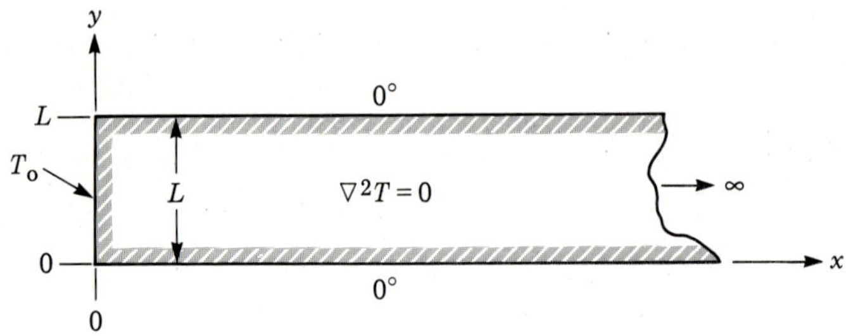


Figure P3.25

length of the slab and pipe is 50 ft. If the water is at 170°F and the outside of the slab is at 60°F, estimate the heat loss from the pipe in Btu/hr.

3.29 A hot-water pipe 2 cm in diameter and 20 m long is buried down the center of a concrete slab ($k = 1.2$ W/m · K) 20 m by 1 m by 15 cm. If the pipe surface is at 75°C and the outside of the slab is at 20°C, estimate the pipe's heat loss in watts and the annual cost of this loss for continuous operation if the water is heated electrically at 9¢/kwh.

3.30 A square chimney is 40 ft high and has an outside surface 3 ft by 3 ft at 60°F and an inside surface 2 ft by 2 ft at 500°F. The chimney brick has $k = 0.6$ Btu/hr · ft · °F. Estimate the heat loss in Btu/hr.

3.31 As shown in Fig. P3.31, a square hollow wrought iron duct, 50 m long, is encased in another square hollow magnesite brick duct. Using the shape factor concept and some inspiration from Section 2.4.1, estimate the total heat transfer rate and the temperature at the brick/iron interface.

3.32 Repeat Problem 3.30 by considering the chimney to consist of four constant-area plane walls plus four corner sections, using Table 3.1. Compare with Problem 3.30, for which the result was 203,000 Btu/hr.

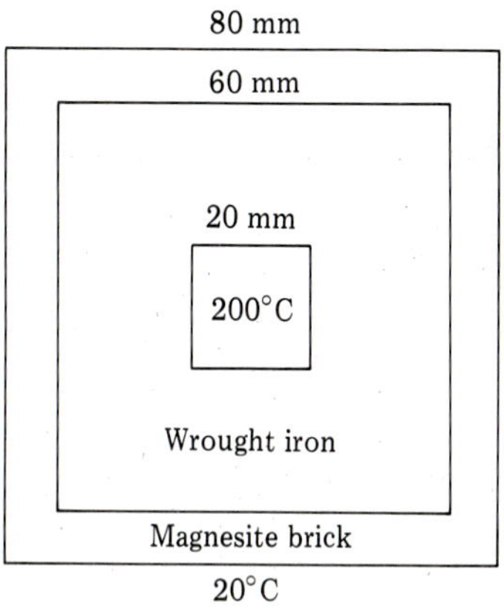


**Figure P3.31**

3.33 Repeat Problem 3.30 by taking the outside *air* temperature to be 60°F with $h_o$ = 5 Btu/hr · ft$^2$ · °F and letting the inside air temperature be 500°F with $h_i$ = 25 Btu/hr · ft$^2$ · °F. Assume that the chimney surface temperatures are constant but unknown and use Table 3.1 for the conduction flux.

3.34 A bare steam pipe at 225°C is 8 cm in diameter and 100 m long and is buried in earth ($k$ = 1.5 W/m · K) whose surface temperature is 10°C. Estimate how deep the pipe must be buried to keep the heat loss no more than 50 kW.

3.35 A 12-cm-diameter steam pipe at 200°C is deeply buried in earth ($k$ = 1.2 W/m · K). Parallel to it is buried a 3-cm-diameter water pipe at 15°C. Both pipes are 150 m long. How far apart must the pipe centerlines be placed to keep the total heat exchange below 40 kW?

3.36 A furnace is 4 ft by 3 ft by 2 ft in size and made of 6-in.-thick brick ($k$ = 0.8 Btu/hr · ft · °F). Inside and outside surface temperatures are at 800°F and 100°F, respectively. Estimate the total heat loss in Btu/hr.

3.37 A 150-cm-diameter sphere generates heat at 5 kW and is buried in earth ($k$ = 1.3 W/m · K) whose surface is at 15°C. How deep must the sphere be buried to keep its surface below 400°C?

3.38 Repeat Problem 3.37 assuming that the sphere generates 4 kW of heat. Please explain if your result seems peculiar.

3.39 A 30-cm-diameter steam pipe is sunk vertically 300 m into the earth ($k$ = 1.3 W/m · K) to recover oil. Pipe and earth surfaces are at 180°C and 15°C, respectively. Estimate the heat loss in watts.

3.40 Repeat Problem 3.39 and assume that the pipe is covered with 8 cm of insulation, $k_i$ = 0.05 W/m · K.

3.41 Liquid metal at 300°C flows through a 5-cm-diameter thin pipe that is buried eccentrically in a 12-cm diameter concrete cylinder, which itself

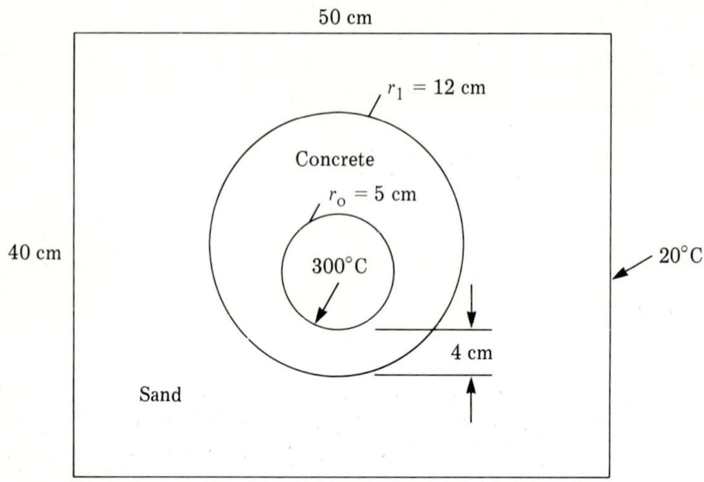


**Figure P3.41**

is buried symmetrically in a rectangular sand duct, as shown in Fig. P3.41. If the system is 60 m long, estimate the total heat loss from the liquid metal.

3.42 Make an approximate analysis of the heat transfer per unit length from the chimney of Example 3.10, using a conduction shape factor from Table 3.1 plus convection boundary conditions. Compare with the numerical results of Example 3.10.

3.43 Using the subscript notation of Fig. 3.9 for a square mesh, derive finite-difference approximations for (a) $\partial^3 T/\partial y^3$; (b) $\partial^4 T/\partial x^2 \partial y^2$; and (c) $\partial^4 T/\partial x^4$.

3.44 Using a hand calculator, apply Eqs. (3.48) and (3.50) to the approximate evaluation of $\partial T/\partial x$ and $\partial^2 T/\partial x^2$ for the function $T = \cos(x)$ at $x = 0.8$, for $\Delta x = 0.2$, 0.1, and 0.05. Do the numerical errors agree with the Taylor series truncation estimates? Are there any round-off errors?

3.45 Is the following expression, using the notation of Fig. 3.9, a valid simulation of Laplace's equation?

$$T_{m,n} = (1/4)(T_{m-1,n+1} + T_{m+1,n+1} + T_{m+1,n-1} + T_{m-1,n-1}).$$

What does its stencil look like? How should its numerical error compare with the familiar five-point formula, Eq. (3.55)?

3.46 Using a heat-balance approach in the spirit of Fig. 3.11, derive numerical boundary conditions for the interior corner in contact with a fluid, item 7 in Table 3.2, for (a) no heat generation (Table 3.2), and (b) uniform heat generation $\dot{q}$ in the solid.

3.47 To what extent can the principles of superposition and linearity apply to numerical solutions such as Example 3.10? That is, the stated conditions are $T_i = 400°C$ and $T_o = 20°C$. Can we solve the problem instead for $T_i = 380°C$ and $T_o = 0°C$ and then add 20°C to all nodal values? Can we go even further and solve the problem for $T_i = 1°$ and $T_o = 0°$ and afterward multiply each nodal value by 380 and then add 20°C? Couch your answers in general terms if you can.

3.48 Using the equally spaced polar coordinate mesh in Fig. P3.48, derive a finite-difference five-point approximation to the polar-coordinate Laplace equation, Eq. (2.8), with $\partial/\partial z = 0$. [*Note:* There are several slightly different numerical formulations for this equation, approximately comparable in accuracy.]

3.49 Using the rectangular mesh of Fig. 3.9 with $\Delta x = \Delta y$, modify the five-point Laplacian simulation, Eq. (3.55), to account for variable thermal conductivity $k_{m,n}$ at each node.

3.50 Using the mesh shown in Fig. P3.50, apply the nine-point formula, Eq. (3.56), to a square plate with a sine-wave upper temperature. Compute (a) the four interior nodes $T_{1-4}$ and (b) the heat flux from the upper surface, and compare with the exact solutions, Eqs. (3.32) and (3.34).

3.51 Derive the exterior insulated corner relation, item 4 of Table 3.2, using a heat-balance method similar to that in Figure 3.11.

3.52 Using the mesh indicated in Fig. P3.52 for steady heat conduction and the five-point formula, Eq. (3.55), solve for all unknown nodal temperatures.

3.53 For a square solid with the hat-shaped upper-surface temperature shown in Fig. P3.53, use the five-point formula, Eq. (3.55), to solve for all unknown nodal temperatures if $\Delta x = \Delta y = L/4$.

3.54 Using a square mesh $\Delta x = \Delta y = L/3$, make a numerical analysis of the square plate with a single convection boundary, as shown in Fig. P3.54. Compute all nodal temperatures to one decimal place.

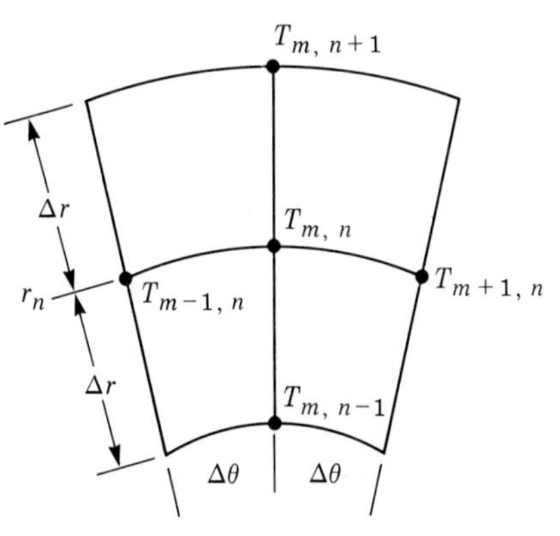


**Figure P3.48**

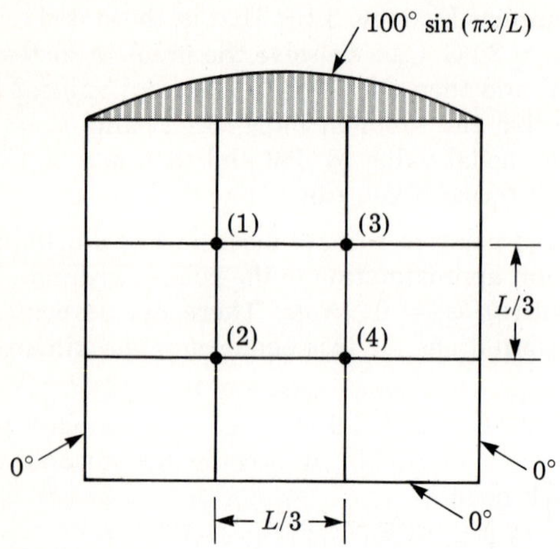


Figure P3.50

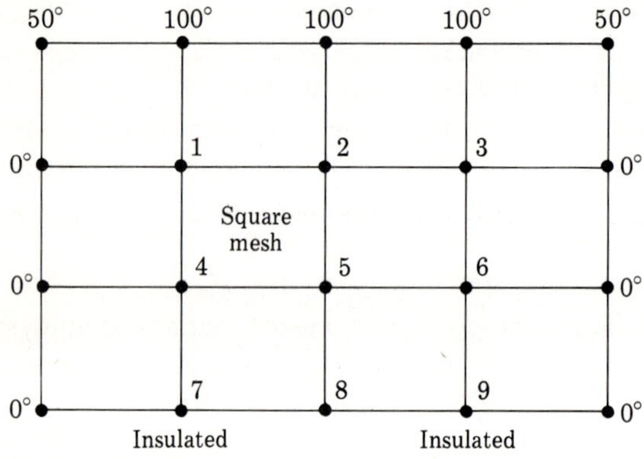


Figure P3.52

Figure P3.53

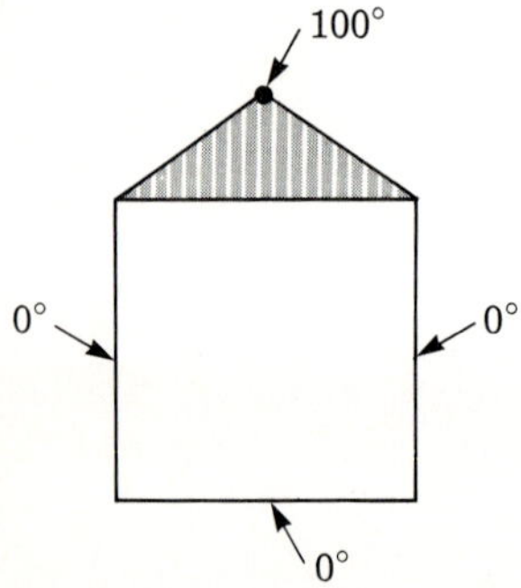

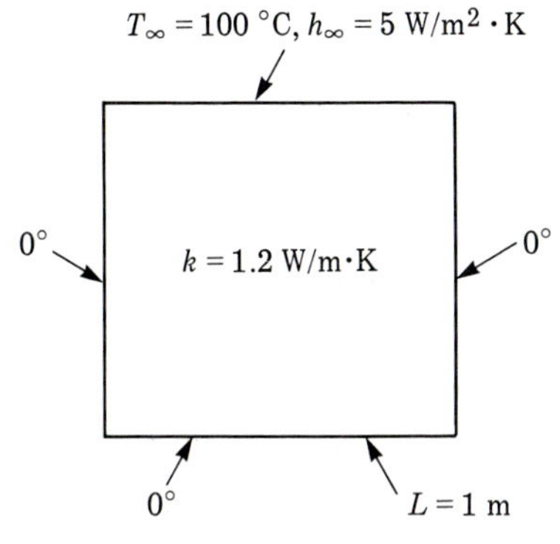


**Figure P3.54**

3.55 Square finite difference meshes are very cumbersome when simulating curved boundaries. In two dimensions, the general case is shown in Fig. P3.55, where the curved boundary shortens the left and upper legs of the mesh to be only $\zeta\Delta x$ and $\eta\Delta x$ long, respectively, where $\zeta$ and $\eta$ are fractions. Show that Laplace's equation (3.25) can be simulated for nodal point $T_0$ by

$$\left(\frac{1}{\zeta} + \frac{1}{\eta}\right)T_0 = T_2/(1 + \zeta) + T_3/(1 + \eta) + T_1/\eta(1 + \eta) + T_4/\zeta(1 + \zeta).$$

Note that these fractional lengths would require a great deal of detailed special input to a computer program.

3.56 Using the five-point formula, Eq. (3.55), solve for the temperatures at the four interior nodes in Fig. P3.56, assuming a square plate. The mesh size is $\Delta x = 30$ cm.

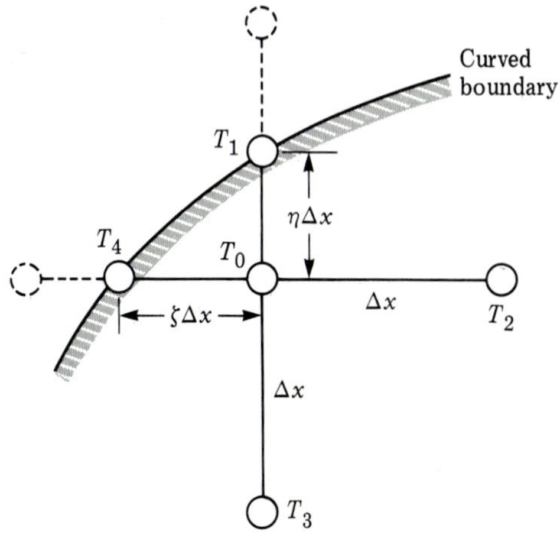


**Figure P3.55**

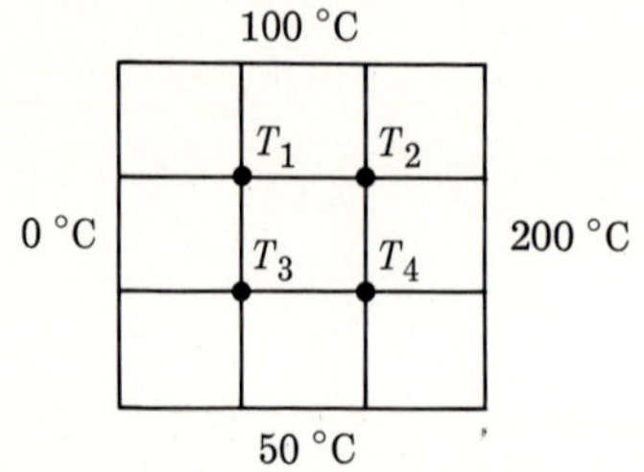


**Figure P3.56**

3.57 Repeat Problem 3.56 by assuming that the entire plate is subjected to a uniform heat generation $\dot{q} = 1600\ \text{W/m}^3$. Could we solve this problem by taking the results from Problem 3.56 and adding the nodal temperatures for heat generation with 0°C at all four boundaries? Take $k = 1.2\ \text{W/m}\cdot\text{K}$.

3.58 Using a heat-balance method, derive the insulated interior corner relation, item 6 of Table 3.2. Show that your result is a special case of item 7.

3.59 In Fig. P3.59, the mesh size is $\Delta x = \Delta y = 10$ cm and $k = 3.5\ \text{W/(m}\cdot\text{K)}$. The solid is subjected to uniform heat generation, $\dot{q} = 150{,}000\ \text{W/m}^3$. Find all unknown nodal temperatures. [*Hint:* You will need to derive new relations for an insulated boundary node.]

3.60 Program Problem 3.59 for a digital computer, using a smaller mesh size $\Delta x = \Delta y = 5$ cm. As output, show all nodal temperatures and an estimate of the total heat transfer out of the left and upper surfaces.

3.61 Using the five-point formula, Eq. (3.55), and Table 3.2, evaluate the nodal temperatures $T_{1\text{-}4}$ in Fig. P3.61, assuming a square plate.

3.62 Shown in Fig. P3.62 are nodal values of the function $T = (1000°\text{C})\cos(x)$,

**Figure P3.59**

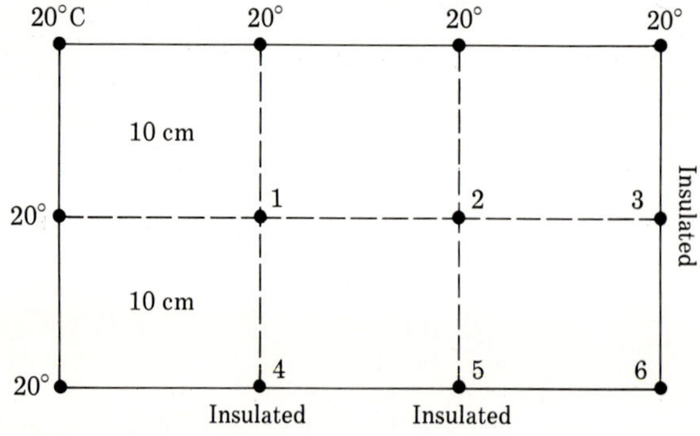

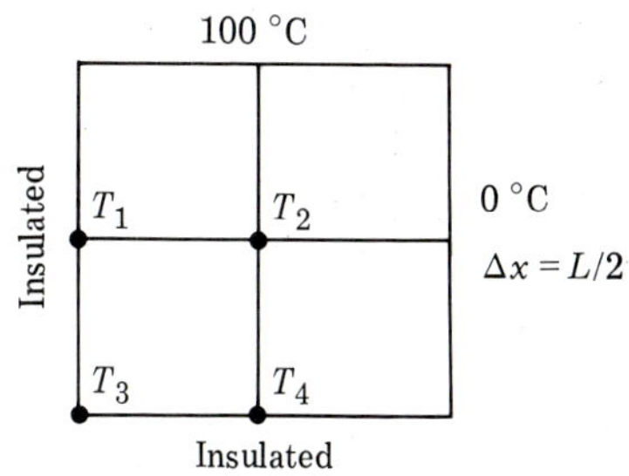


**Figure P3.61**

with $x$ in meters. The "wall" is at $x = 1.0$ m. Using the two- to five-point formulas for numerical differentiation, evaluate the wall temperature gradient in °C/m and compare with the exact slope. What error is incurred in each of the four formulas if the value of $T_1$ is in error by 1°?

3.63 Program Problem 3.50 for a digital computer with $\Delta x = L/5$ and print all nodal temperatures to two decimal places.

3.64 Program Problem 3.52 for a digital computer, using the five-point formula, Eq. (3.55), with $\Delta x = L/10$.

3.65 Program Problem 3.54 for a digital computer, using the five-point formula, Eq. (3.55), with $\Delta x = 0.2$ m.

3.66 Program Problem 3.56 for a digital computer, using the five-point formula, Eq. (3.55), with $\Delta x = 10$ cm.

3.67 Program Problem 3.57 for a digital computer with $\Delta x = 10$ cm.

3.68 Program Problem 3.61 for a digital computer with $\Delta x = L/5$.

3.69 Figure P3.69 shows an equally spaced one-dimensional finite-difference mesh for a variable-area fin similar to that in Fig. 2.14(c), for which the basic differential equation is Eq. (2.52). Let area $A_m$ and perimeter $P_m$

**Figure P3.62**

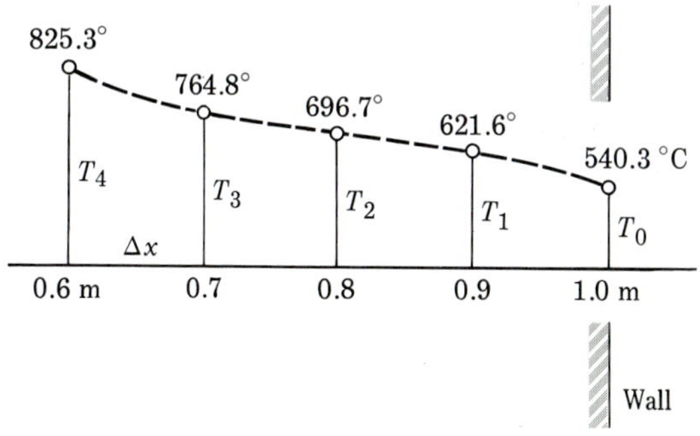

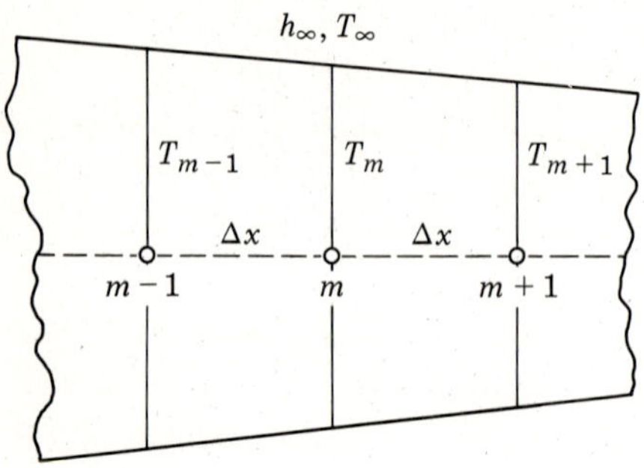


**Figure P3.69**

be different at each nodal point. Set up a general finite-difference formula for a given interior node $T_m$ and the boundary condition for a convection tip. Your results may differ slightly depending on your particular choices of, for example, forward or backward differences. I suggest the following result:

$$(2 + \beta_m)T_m \simeq T_{m-1} + T_{m+1} + \beta_m T_\infty, \qquad \beta_m = h_\infty P_m \Delta x^2/kA_m.$$

The suggested tip boundary condition is, at $m = N$, the last point:

$$(1 + \tfrac{1}{2}\beta_N + \text{Bi})T_N \simeq T_{N-1} + (\tfrac{1}{2}\beta_N + \text{Bi})T_\infty, \quad \text{Bi} = h_\infty \Delta x/k,$$

and of course the base condition ($m = 0$) is the known node $T_0$.

3.70 Using the finite-difference scheme of Problem 3.69, set up the solution of Example 2.12 for the stainless steel fin with $\Delta x = 2$ cm. Compute all nodal temperatures, the base heat flux, and the efficiency, and compare with Example 2.12.

3.71 In the spirit of Figs. 3.3 and 3.14, the Brebbia boundary element program [24] and Archibald software [25] were run for the case shown in Fig. P3.71. The resulting values of temperature gradient are shown in the figure. Does this case represent a hollow square solid as in item 2 of Table 3.1? Determine a shape factor for this case and compare your total heat transfer with a one-dimensional analysis from Eq. (2.21).

3.72 Figure P3.72 shows results obtained on an analog field plotter for a two-dimensional rectangular region. What heat transfer problem has been simulated by this solution? What boundary conditions have been simulated?

3.73 The differential equation for torsion of a straight solid shaft of elastic material is

$$\frac{\partial^2\Phi}{\partial x^2} + \frac{\partial^2\Phi}{\partial y^2} = -2G\theta, \tag{1}$$

where $\Phi$ is the torsional stress function, $G$ is the shear modulus, and $\theta$ is the angle of twist per unit shaft length [31]. The boundary condition

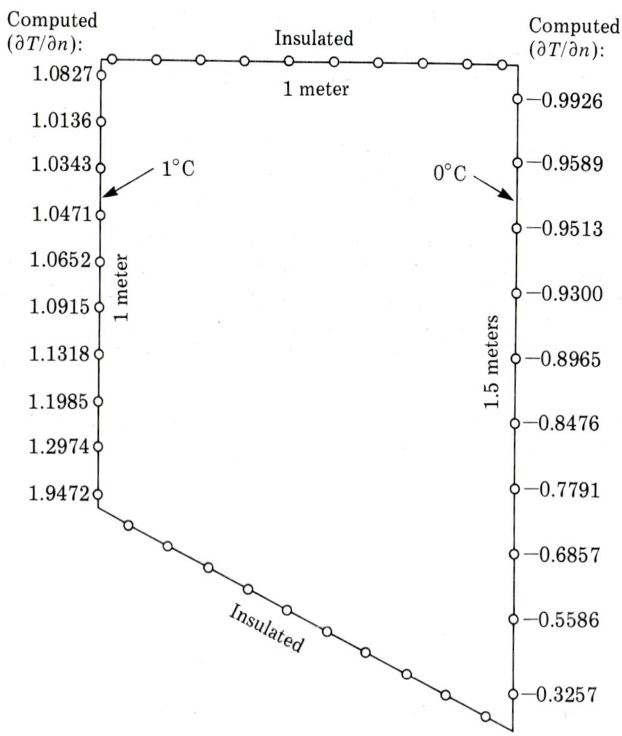


**Figure P3.71**

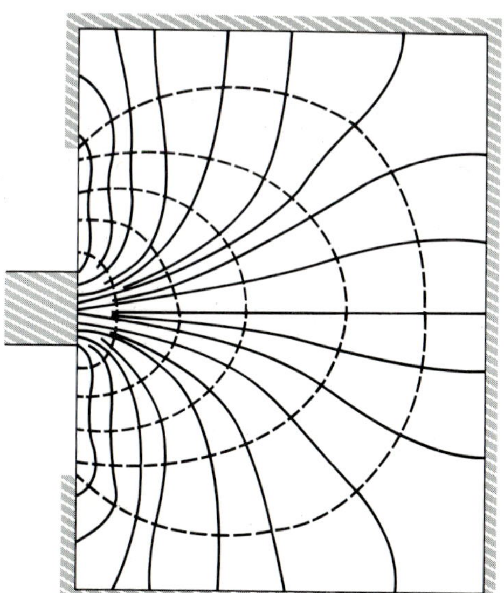

**Figure P3.72**

is $\Phi = 0$ at the outer surface of the shaft. The solution to Eq. (1) for an elliptical shaft of major axis $2a$ and minor axis $2b$ is [31]

$$\Phi = \frac{2G\theta a^2 b^2}{a^2 + b^2}(1 - x^2/a^2 - y^2/b^2). \tag{2}$$

Apply this solution, by analogy, to heat conduction in an elliptical solid with uniform heat generation $\dot{q}$, constant conductivity $k$, and uniform surface temperature $T_o$, as in Fig. P3.73. Determine (a) the maximum temperature in the solid and (b) the total heat loss, per unit depth, from the solid.

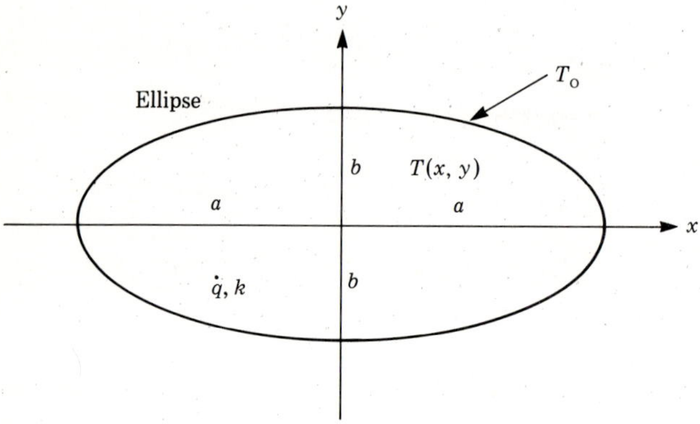


**Figure P3.73**

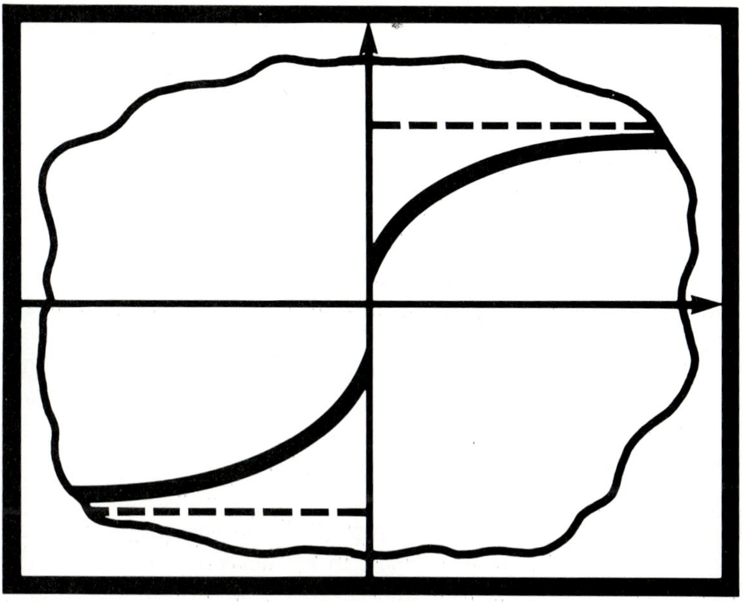

# Unsteady Heat Conduction

# 4

## 4.1 Introduction

In Chapter 3 we studied steady heat conduction for a variety of geometries, boundary conditions, and methods of solution. Steady problems are notable for their computational stability and the ease with which the analyst can visualize and sketch the approximate temperature and heat transfer patterns. In fact, for regular geometries and uniform-temperature boundary conditions, the solutions often lead to surprisingly simple algebraic formulas, such as the shape factors in Table 3.1.

In transient or unsteady conduction, temperature is a function of both time and space. A control volume energy balance of the body generally shows a net heat transfer to or from the body. The body energy thus rises or falls as heat is added or removed. As discussed in Section 1.3.5, the physical property that most strongly influences the rate of change of body temperature is the thermal diffusivity, $\alpha$. Unsteady conduction may result in rather complex analytical solutions (see Sections 4.3 and 4.4), and their numerical solutions (Section 4.6) may require certain stability limits. But, in general, transient conduction analysis rests on solid analytical foundations that a nonspecialist engineer can readily master and apply.

The starting point for a transient analysis is the general heat conduction relation, Eq. (3.1). Here we make two simplifying assumptions: (1) We neglect heat generation, $\dot{q}$, except for the special case of a "lumped" mass (Section 4.2); and (2) we neglect variations in $k$, $\rho$, and $c_p$. Thus we analyze the following form of the heat conduction equation:

$$\nabla^2 T = \frac{\partial^2 T}{\partial x^2} + \frac{\partial^2 T}{\partial y^2} + \frac{\partial^2 T}{\partial z^2} = \frac{1}{\alpha}\frac{\partial T}{\partial t}, \tag{4.1}$$

where $\alpha = k/\rho c_p$ is the thermal diffusivity, assumed constant. We will treat only a few examples here, but recall that there are at least six excellent textbooks ([1–6] in Chapter 3) that deal wholly with conduction analyses.

All unsteady problems require an *initial* condition that is a known temperature distribution at some time $t = t_0$:

$$T(x, y, z, t_0) = T_0(x, y, z). \tag{4.2}$$

For all our applications, we will take $T_0$ = constant or a uniform initial temperature distribution. We also take $t_0 = 0$.

At the boundaries of the body, temperature conditions might be any one of the four types discussed in Section 3.1.3 and Fig. 2.3. Most of our applications will be for a fluid convection environment so that, at the boundary $b$,

$$-k\frac{\partial T}{\partial n}\bigg|_b = h_\infty(T_b - T_\infty), \tag{4.3}$$

where $n$ is the outward normal coordinate (see Fig. 2.3). The environmental conditions ($h_\infty$, $T_\infty$) can be variable in practice but will be taken constant in our examples.

### 4.1.1 Dimensional Analysis

Before plunging into a series of conduction analyses, it is instructive to nondimensionalize Eqs. (4.1), (4.2), and (4.3) and think about the consequences. In fact, we did this in Example 3.1, selecting the following dimensionless variables†:

$$\begin{aligned} &x^* = x/L; \quad y^* = y/L; \quad z^* = z/L; \quad n^* = n/L; \quad t^* = \alpha t/L^2; \\ &\Theta = (T - T_\infty)/(T_0 - T_\infty), \end{aligned} \tag{4.4}$$

where $L$ is a characteristic body length and we have taken $T_0$ and $T_\infty$ to be constant.

Substitution of Eqs. (4.4) into Eqs. (4.1), (4.2), and (4.3) yields the following dimensionless equation and conditions:

$$\frac{\partial^2\Theta}{\partial x^{*2}} + \frac{\partial^2\Theta}{\partial y^{*2}} + \frac{\partial^2\Theta}{\partial z^{*2}} = \frac{\partial\Theta}{\partial t^*}, \tag{4.1a}$$

$$\Theta_0 = 1, \tag{4.2a}$$

$$\frac{\partial\Theta}{\partial n^*}\bigg|_b = -\mathrm{Bi}\,\Theta_b, \qquad \mathrm{Bi} = h_\infty L/k. \tag{4.3a}$$

We see that the basic equation (4.1a) and initial condition (4.2a) contain no parameters, and the convection boundary condition (4.3a) yields a single parameter, the Biot number. We conclude that, for this type of common transient conduction problem, solutions for the temperature at any point in systems of like geometry depend only on dimensionless time and the Biot number:

*At any $x^*$, $y^*$, $z^*$:*

$$\Theta = \mathrm{fcn}(\alpha t/L^2, \mathrm{Bi}, \text{geometry}). \tag{4.5}$$

Note that if the geometries are different, the functional relationships

---

†The variable $t^* = \alpha t/L^2$ is often called the *Fourier number,* Fo.

will be quite different: It makes no sense to compare cylinders to spheres to slabs in the same mathematical analysis.

Here is a special but important conclusion: If two bodies of like geometry have the same Biot number, they undergo exactly the same dimensionless temperature history $\Theta(t^*)$. If body 1 of size $L_1$ and diffusivity $\alpha_1$ cools down, say, in time $t_1$, body 2 will cool down such that $t_2^* = t_1^*$ or, if $\text{Bi}_1 = \text{Bi}_2$,

$$\frac{t_2}{t_1} = \left(\frac{L_2}{L_1}\right)^2 \left(\frac{\alpha_1}{\alpha_2}\right). \tag{4.6}$$

This is quite interesting: If diffusivities and Biot numbers are the same, transient conduction times in like-shaped bodies are proportional to body size squared (or body surface area). But we must be careful not to jump to unwarranted conclusions: For example, if a roast beef of size 20 cm cooks in 40 minutes, will not a roast beef twice as large ($L = 40$ cm) cook in $4 \times 40 = 160$ minutes? Not necessarily — the larger roast may cause different convection conditions in the oven and thus have a different Biot number, hence a different $\Theta(t^*)$ relation.

Probably the most important consequence of Eq. (4.5) is that we need not solve such transient problems over and over again. We find the dimensionless function in Eq. (4.5) for a given geometry and plot it up once and for all with Biot number as a parameter. Then transient convection problems of any body size or diffusivity can be picked off this single chart at a given Biot number. We will give several examples of such dimensionless charts in Section 4.3 and Appendix J.

### Example 4.1

When thrust into a cooling bath of convection coefficient 75 W/m$^2$ · K, a 12-cm-diameter hot cast-iron sphere cools down in 42 min. What convection coefficient is needed to give exactly similar conditions for a 30-cm-diameter stainless steel sphere? How long will the steel take to cool?

*Solution* For cast iron, take $k_1 = 52$ W/m · K, $\alpha_1 = 1.7 \times 10^{-5}$ m$^2$/s. For stainless steel, take $k_2 = 14$ W/m · K and $\alpha_2 = 3.9 \times 10^{-6}$ m$^2$/s. The cast iron is cooling at a boundary Biot number of

$$\text{Bi}_1 = h_{\infty 1}D_1/k_1 = (75 \text{ W/m}^2 \cdot \text{K})(0.12 \text{ m})/(52 \text{ W/m} \cdot \text{K}) = 0.173.$$

For exact similarity, the stainless steel must have the same Biot number:

$$\text{Bi}_2 = h_{\infty 2}D_2/k_2 = (h_{\infty 2})(0.30 \text{ m})/(14 \text{ W/m} \cdot \text{K}) = 0.173,$$

or

$h_{\infty 2} = 8.1 \text{ W/m}^2 \cdot \text{K}$. [*Ans.*]

This corresponds to a very weak free convection bath (see Table 1.1).

With Biot numbers matched, Eq. (4.6) is valid and the steel sphere cools down in time

$$t_2 = t_1(L_2/L_1)^2(\alpha_1/\alpha_2) = (42 \text{ min})(30/12)^2(1.7 \times 10^{-5}/3.9 \times 10^{-6})$$
$$= 1144 \text{ min} = 19.1 \text{ hr}. \quad [Ans.]$$

Note that we did not need to compute $t_1^* = t_2^* = 2.98$. ■

### 4.1.2 The Lumped-Mass Approximation at Small Biot Number

The boundary condition (4.3a) leads to the interesting simplification of the "lumped mass." Consider the two examples in Fig. 4.1 of a plate of width $L$ suddenly immersed in a convection environment ($h_\infty$, $T_\infty$) on both sides. In case (a) the boundary Biot number, $h_\infty L/k$, is small, Bi = 0.2, while in case (b) it is large, Bi = 20. In either case, the

**Figure 4.1 Illustration of transient temperature distributions for small and large boundary Biot numbers: (a) Bi = $h_\infty L/k$ = 0.2, nearly uniform temperature $T(t)$ — lumpable; (b) Bi = 20, strongly variable temperature $T(x, t)$ — not lumpable.**

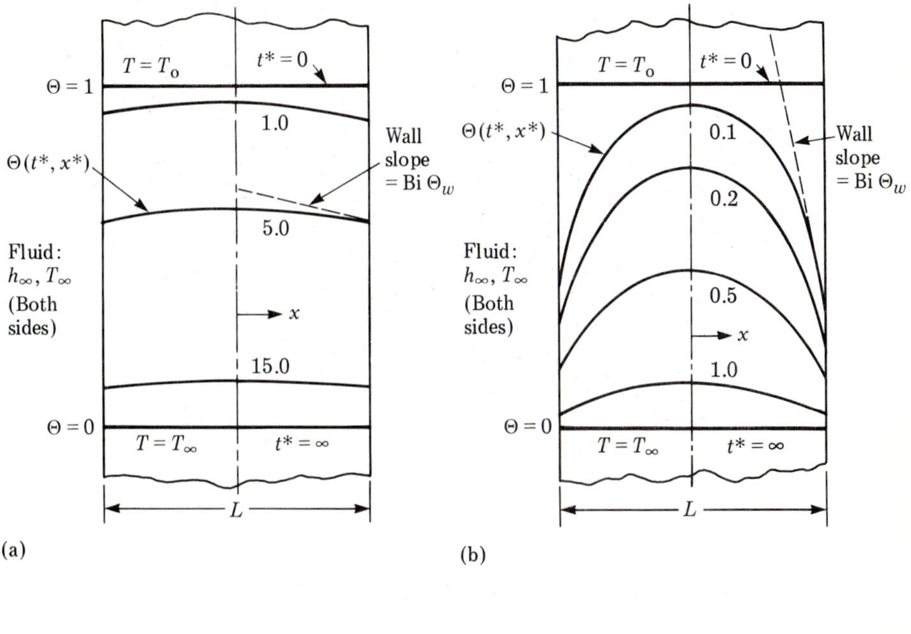

walls, being directly exposed to the fluid, cool faster than the center of the body. The instantaneous temperature profiles are rounded and concave downward, with maximum slope at the walls.

From Eq. (4.3a), the wall slope magnitude $|\partial\Theta/\partial n^*| = \text{Bi}\,\Theta_b$, and $\Theta_b$ is always less than or equal to unity. Thus the total temperature change from wall to center can be no greater than $\Delta\Theta = \Delta n^*\,\text{Bi}\Theta_b \leqslant \text{Bi}/2$. Thus, if Bi/2 is small, say less than 0.1, temperature variations across the plate will be less than 10% of $(T_o - T_\infty)$. This is the case in Fig. 4.1(a), and it is convenient to consider the plate "lumped" at a uniform temperature $T(t)$.

An alternative view of this concept is to compare the conductive resistance of the plate to the convective resistance at the boundary. If conduction resistance is small,

$$L/(kA_x) << 1/(h_\infty A_x),$$

or

$$h_\infty L/k << 1.$$

Again we see that the plate is "lumpable" if its Biot number is small. This is a great simplification, resulting in the algebraic "lumped" analysis of the next section.

## 4.2 Lumped Systems with Nearly Uniform Temperature

As we saw in Section 4.1.2, for small Biot number a body undergoing transient conduction can be considered a "lump" whose interior temperatures are nearly uniform. Put another way, the internal "conductive" resistance of a lump is much smaller than its external convective resistance. As shown in Section 4.1.2, we make no more than a $\pm 5\%$ error in estimating body temperature by lumping if its Biot number is less than approximately 0.1: If

$$\text{Bi} = h_\infty L/K \leqslant 0.1, \qquad T \simeq T(t) \text{ only.} \tag{4.7}$$

Here $L$ denotes the distance across which most significant temperature gradients occur: the plate thickness in Fig. 4.1.

If a body has an irregular shape with no unique "thickness," its characteristic length scale is taken to be its volume divided by its surface area, $L^* = v/A_w$. This length scale results in a somewhat less lenient Biot number restriction:

*Irregular body lumpability:*

$$\text{Bi}^* = h_\infty(v/A_w)/k \leqslant 0.05. \tag{4.8}$$

For a sphere, for example, $L^* = [(\pi/6)D^3]/(\pi D^2) = D/6$. Equation (4.8) predicts sphere lumpability if $h_\infty(D/6)/k \leq 0.05$, or $h_\infty D/k \leq 0.3$. Thus both the cast-iron and stainless steel spheres in Example 4.1 are lumpable for the given convection conditions.

Consider the lumped solid shown in Fig. 4.2. A simple energy balance shows that the energy convected into the surface must cause a rise in energy of the lump:

$$h_\infty A_w(T_\infty - T) = m\frac{de}{dt} \simeq \rho \upsilon\, c_\upsilon \frac{dT}{dt}, \tag{4.9}$$

where we have assumed the lump temperature to vary only with time. Recall that we made the same simple analysis in Example 1.6. Rearranging and noting that $dT_\infty/dt = 0$, we obtain a first-order linear differential equation:

$$\frac{d}{dt}(T - T_\infty) + \frac{h_\infty A_w}{\rho c_\upsilon \upsilon}(T - T_\infty) = 0, \tag{4.10}$$

subject to the initial condition

$$T(0) = T_o. \tag{4.11}$$

The solution is readily obtained as an exponential decay:

$$\frac{T - T_\infty}{T_o - T_\infty} = e^{-t/\tau_o}, \quad \tau_o = \frac{\rho c_\upsilon \upsilon}{h_\infty A_w}. \tag{4.12}$$

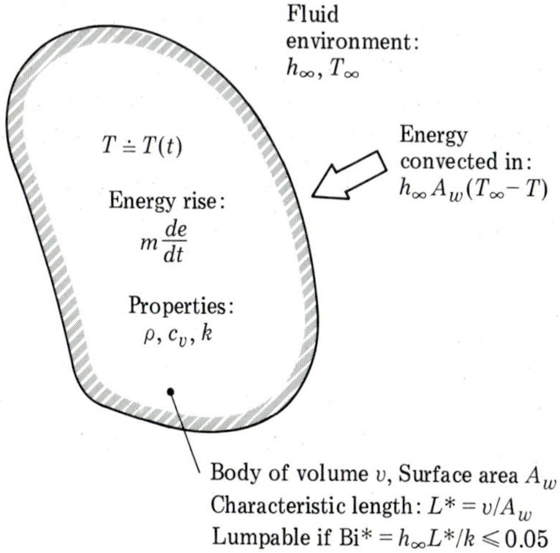


**Figure 4.2** Energy balance for a lumped solid subjected to fluid convection.

Thus, if external convection conditions remain constant, the temperature of a lump drops off exponentially with time. We may interpret $\tau_o$ as a characteristic decay time of the lump temperature, increasing with the lump heat capacity $\rho v c_v$ and decreasing with the convection rate $h_\infty A_w$.

The exponential argument $(t/\tau_o)$ can also be interpreted as the product of Biot number and Fourier number:

$$t/\tau_o = (h_\infty L^*/k)(\alpha t/L^{*2}) = \mathrm{Bi}^* t^*. \tag{4.13}$$

This is a common concept in the literature but rather misleading, because it introduces thermal conductivity into a problem for which it has a negligible effect.

## Example 4.2

Determine if the two spheres in Example 4.1 are lumpable and, if so, estimate the time for each sphere to cool from 200°C to 40°C if immersed in a fluid at $T_\infty = 20°\mathrm{C}$.

*Solution* Check the lump Biot numbers, which are the same for both. For a sphere, $L^* = D/6$, hence

$$\begin{aligned}\mathrm{Bi}_1^* &= h_{\infty 1}L_1^*/k_1 = (75\ \mathrm{W/m^2\cdot K})(0.12\ \mathrm{m}/6)/(52\ \mathrm{W/m\cdot K}) \\ &= 0.029 = \mathrm{Bi}_2^*.\end{aligned}$$

This is less than 0.05, hence both are lumpable by Eq. (4.8).

1. For cast iron take $\rho = 7270\ \mathrm{kg/m^3}$ and $c_v = 420\ \mathrm{J/kg\cdot K}$. Then

   $$\begin{aligned}\tau_o &= \rho c_v L^*/h_\infty = (7270\ \mathrm{kg/m^3})(420\ \mathrm{J/kg\cdot K}) \\ &\quad \cdot (0.12\ \mathrm{m}/6)/(75\ \mathrm{W/m^2\cdot K}) \\ &= 814\ \mathrm{s}.\end{aligned}$$

   Then, for $T_o = 200°\mathrm{C}$, $T = 40°\mathrm{C}$, and $T_\infty = 20°\mathrm{C}$, Eq. (4.12) predicts

   $$\frac{T - T_\infty}{T_o - T_\infty} = \frac{40 - 20}{200 - 20} = e^{-t/814\ \mathrm{s}},$$

   or

   $$t = 1790\ \mathrm{s} = 29.8\ \mathrm{min}. \quad [\mathit{Ans.\ (1)}]$$

2. For stainless steel, take $\rho = 7820\ \mathrm{kg/m^3}$ and $c_v = 460\ \mathrm{J/kg\cdot K}$.

Then

$\tau_o = (7820)(460)(0.3/6)/(8.1) = 22{,}200$ s.

The temperatures are the same for the stainless steel, so

$$\frac{40 - 20}{200 - 20} = e^{-t/22{,}200\ \mathrm{s}},$$

or

$t = 48{,}800$ s $= 813$ min. [*Ans. (2)*]

The stainless steel sphere, for the same Biot number, has a decay time 27.2 times longer than the cast iron, which is exactly the same ratio as in Example 4.1. ■

---

### 4.2.1 Transient or Cyclic Convection Conditions

The lumped analysis is also valid if the convection environment changes with time, as long as the Biot number remains small. Suppose, for example, the fluid temperature oscillates sinusoidally about a mean value:

$$T_\infty = T_m + \Delta T \cos(\omega t). \tag{4.14}$$

Assume that $h_\infty$ remains constant. Equation (4.10) does not apply because $T_\infty$ varies, but Eq. (4.9) does apply and may be rewritten as

$$\tau_0 \frac{dT}{dt} + T = T_\infty, \tag{4.15}$$

where $\tau_o$ is the decay time in Eq. (4.12). Substituting Eq. (4.14) into (4.15) and solving gives

$$T = T_m + \frac{\Delta T}{(1 + \omega^2\tau_o^2)^{1/2}} \cos(\omega t - \phi), \tag{4.16}$$

where $\phi = \tan^{-1}(\omega\tau_o)$. This is the "steady oscillation" that ensues after initial transients have damped out. We see that the lump also oscillates about $T_m$ but with a smaller amplitude than the fluid oscillation and a phase lag $\phi$.

One obvious advantage to the lumped approximation is that simple estimates such as Eq. (4.16) can be obtained to complex problems that would be extremely cumbersome to handle as multidimensional transient conduction analyses. You should keep in mind, however, that convection coefficients around a lumpy shape are probably nonuniform and quite difficult to estimate accurately. Thus the lumped analysis is subject to considerable uncertainty in practice.

**Example 4.3**

A 1-mm-diameter constantan thermocouple lead ($\rho = 8920\ \text{kg/m}^3$, $c_v = 410\ \text{J/kg}\cdot\text{K}$) is inserted in a gas stream whose temperature fluctuates as $T_\infty = 20°\text{C} + 10°\text{C}\cos(\omega t)$ with a period of 1 min. What will be the thermocouple response if the gas convection coefficient is $h_\infty = 350\ \text{W/m}^2\cdot\text{K}$?

*Solution* Assuming a spherical shape for the lead, we compute

$$\tau_o = \rho c L^*/h_\infty = (8920)(410)(0.001/6)/(350) = 1.74\ \text{s}.$$

The product $\omega\tau_o = (2\pi/60\ \text{s}^{-1})(1.74\ \text{s}) = 0.182$. From Eq. (4.16),

$$T = 20°\text{C} + \frac{10°\text{C}}{[1 + (0.182)^2]^{1/2}}\cos[\omega t - \tan^{-1}(0.182)]$$

$$= 20°\text{C} + 9.8°\text{C}\cos(\omega t - 10°). \quad [\textit{Ans.}]$$

The thermocouple follows the gas temperature rather closely. ■

## 4.3 Exact Solutions for a Semi-Infinite Slab

By a semi-infinite slab we mean a "large" solid with a single plane surface, with its other three surfaces being far enough away to ignore. An example is shown in Fig. 4.3. Suppose this slab is initially at uniform temperature $T_i$. Then, at $t = 0$, the (single) surface is suddenly exposed to a fluid at temperatures $T_o$ and heat transfer coefficient $h_o$. Compute the transient temperature distribution in the slab for (a) infinite $h_o$, and (b) finite $h_o$. The solutions are given here with three goals in mind: (1) They are useful exact solutions to a realistic problem; (2) they introduce some interesting and educational mathematical functions; and (3) they can be used as test cases for the numerical methods of Section 4.6.

### 4.3.1 Instantaneous Surface Temperature Change

If the fluid has a very large heat transfer coefficient $h_o \to\to \infty$, as might be approximated by condensation of a saturated vapor on the surface, the slab surface will immediately rise to $T_o$ and remain there, as is shown in Fig. 4.3. As time increases, the temperature profile $T(x, t)$ penetrates deeper and deeper into the slab.

Assuming constant $\alpha$ and no heat generation, we must solve the one-dimensional unsteady conduction equation:

$$\frac{\partial T}{\partial t} = \alpha \frac{\partial^2 T}{\partial x^2}, \tag{4.17}$$

subject to an initial condition,

$$T(x, 0) = T_i \tag{4.18}$$

and a surface boundary condition,

$$T(0, t) = T_o. \tag{4.19}$$

You may not have seen this type of problem before. What makes it different is its semi-infinite spatial coordinate. The temperature generated by the solution must vanish at large $x$. As $x \twoheadrightarrow \infty$,

$$T(x, t) \twoheadrightarrow T_i. \tag{4.20}$$

**Figure 4.3** **Instantaneous temperature distribution in a semi-infinite slab whose surface temperature is suddenly raised from $T_i$ to $T_o$.**

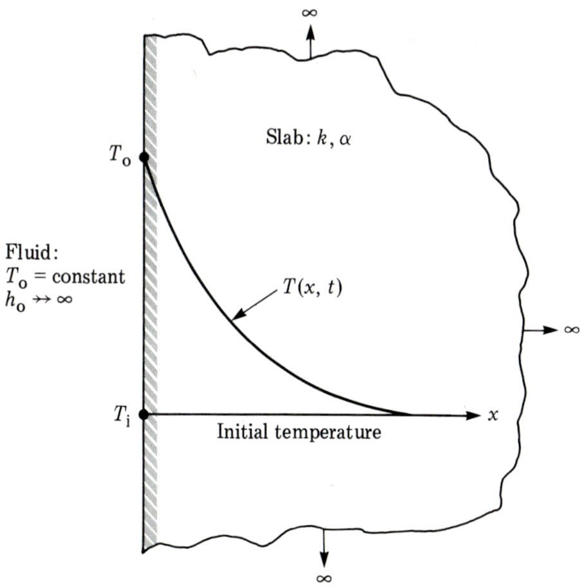

Thus the Fourier series technique of Section 3.2.3 does not work. Semi-infinite problems are typically solved either by Fourier integrals or by Laplace transforms (see, for example, [1, Chapter 11]). We will forgo these techniques here and find a solution by simply pointing out something interesting about semi-infinite coordinates.

A semi-infinite body has no length scale $L$. It has only the coordinate $x$ to indicate length. Therefore, length and time are not independent in this type of problem: The differential equation (4.17) must really be a function of only a *single* variable $\eta$, which is a combination of $x$, $t$, and the diffusivity $\alpha$. The only combination of these three that is dimensionless is the following variable:

$$\eta = \frac{x}{2\sqrt{(\alpha t)}}, \tag{4.21}$$

where we have introduced the factor of 2 merely for convenience. Similarly, an appropriate dimensionless temperature that fits the given conditions is

$$\Theta = \frac{T - T_i}{T_o - T_i}. \tag{4.22}$$

The correct solution to this problem is that $\Theta$ is a function only of the single "similarity" variable $\eta$; the complete rationale behind this thinking is given in advanced texts on *similarity theory,* such as [2].

If we assume that $\Theta = \Theta(\eta)$ only and substitute directly into Eq. (4.17), we obtain a simple ordinary differential equation:

$$\frac{d^2\Theta}{d\eta^2} + 2\eta\frac{d\Theta}{d\eta} = 0. \tag{4.23}$$

Conditions (4.18) and (4.19) are converted to

$$\Theta(\eta) \to\to 0, \qquad \text{as } \eta \to\to \infty,$$

and **(4.24)**

$$\Theta(0) = 1,$$

respectively. The general solution to Eq. (4.23) is

$$\Theta = C_1 + C_2 \int e^{-\eta^2} d\eta. \tag{4.25}$$

When the constants are made to satisfy Eqs. (4.24), we obtain $C_1 = 1.0$ and $C_2 = -2/\sqrt{\pi}$. The desired solution is thus

$$\Theta = \frac{T - T_i}{T_o - T_i} = 1 - \frac{2}{\sqrt{\pi}}\int_0^\eta e^{-z^2}\, dz = \mathrm{erfc}(\eta). \tag{4.26}$$

The function erfc is called the *complementary error function*. Since the integral is not known in closed form, values of erfc($\eta$) are tabulated in Appendix A and sketched in Fig. 4.4.

This solution can be used to estimate the "penetration depth" of a slab subjected to a sudden change in boundary temperature. Define $x = \delta$ as the interior point at which the temperature change is 1% of $(T_o - T_i)$. This is equivalent to $\Theta = 0.01 = \text{erfc}(\eta)$, or $\eta \doteq 1.82$ from Appendix A. Since $\eta(x = \delta) = \delta/[2\sqrt{\alpha t}]$, our estimates are

*Penetration depth:*

$$\delta = 3.64\sqrt{\alpha t};$$

*Penetration time:* **(4.27)**

$$t = 0.0755\,\delta^2/\alpha.$$

For a cast-iron slab, $\alpha = 1.7 \times 10^{-5}$ m$^2$/s, after 1 min surface heating will have penetrated to $\delta = 11.6$ cm. In contrast, the penetration depth in a brick slab, $\alpha = 4 \times 10^{-7}$ m$^2$/s, after 1 min would be only $\delta = 1.8$ cm.

**Figure 4.4** **Transient temperature profile for a semi-infinite slab with a suddenly changed wall temperature.**

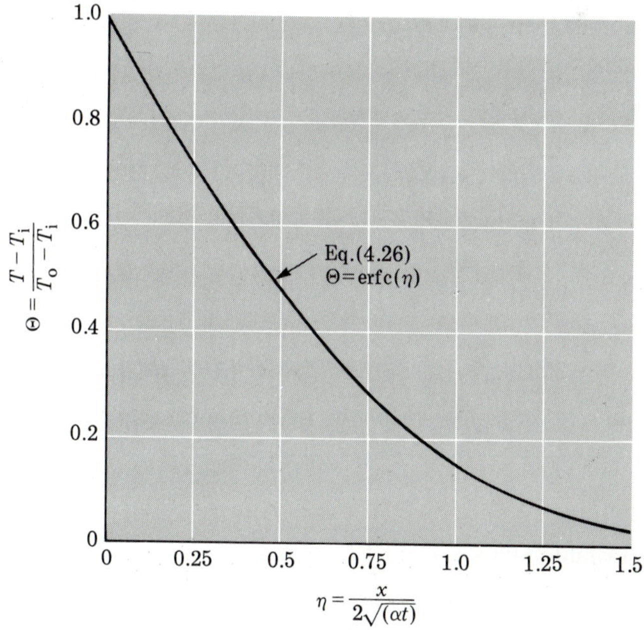

Note that Eq. (4.27) is analogous to the penetration depth of shear effects when the wall next to a viscous fluid is suddenly set into uniform motion (see, for example, [17, Eq. (3-114)]). The analogy is exact if $\alpha$ is replaced by the fluid kinematic viscosity, $\nu = \mu/\rho$.

### Example 4.4

A slab of wood, considered semi-infinite, has a uniform temperature of 20°C. Its surface temperature is suddenly changed and held at 100°C. (a) How long will it take the temperature 5 cm within the slab to reach 60°C? (b) What will be the temperature 1 cm deep after 1 hr?

*Solution* For wood, take $\alpha \doteq 1.2 \times 10^{-7}$ m²/s. An interior temperature of 60°C corresponds to

$$\Theta = \frac{T - T_i}{T_o - T_i} = \frac{60 - 20}{100 - 20} = 0.5.$$

From Fig. 4.4 or Appendix A, at $\Theta = 0.5$, $\eta = 0.48$. Thus

$$\frac{x}{2\sqrt{(\alpha t)}} = \frac{0.05}{2\sqrt{(1.2 \times 10^{-7})t}} = 0.48,$$

or

$t = 22{,}900$ s $= 6.4$ hr. [*Ans. (a)*]

After 1 hr at $x = 1$ cm, we know the parameter

$$\eta = \frac{x/2}{\sqrt{(\alpha t)}} = \frac{(0.01\ \text{m})/2}{\sqrt{(1.2 \times 10^{-7})(3600\ \text{s})}} = 0.24.$$

From Fig. 4.4 or Appendix A, read $\Theta = 0.734$. Thus, at this time and position,

$T = (100 - 20°\text{C})(0.734) + 20°\text{C} = 79°\text{C}$. [*Ans. (b)*]

Wood has a very slow response to conduction transients. ■

### 4.3.2 Instantaneous Surface Convection

A more realistic condition for the semi-infinite slab is the sudden application of surface convection at fluid conditions $T_o$ and finite $h_o$.

Equation (4.19) would be replaced by a convection condition:

*At* $x = 0$:

$$-k\frac{\partial T}{\partial x} = h_o(T_o - T). \tag{4.28}$$

This condition brings in an additional dimensionless parameter:

$$\lambda = (h_o/k)\sqrt{(\alpha t)} \tag{4.29}$$

The basic differential equation is still Eq. (4.23) and the initial condition $\Theta \to 0$ as $\eta \to \infty$ still applies. The solution is given in Section 10.15 of the text by Schneider [3]:

$$\Theta = \frac{T - T_i}{T_o - T_i} = \text{erfc}(\eta) - e^{\lambda(2\eta + \lambda)}\,\text{erfc}(\eta + \lambda), \tag{4.30}$$

where $\eta$ is the same variable from Eq. (4.21). This solution is plotted in Fig. 4.5. The upper curve ($\lambda = \infty$) in this figure is the sudden temperature solution, Eq. (4.26). For finite $h_o$, the surface temperature rises gradually from $T_i$ and approaches the fluid temperature $T_o$ at large times. Note that because $t$ appears in both $\eta$ and $\lambda$ this is a *nonsimilar* solution.

### 4.3.3 Instantaneous Surface Heat Flux

A final case is the semi-infinite slab suddenly subjected to a uniform constant heat flux $q''_o$ at the surface. The initial condition is $T = T_i$, and the surface condition is:

*At* $x = 0$:

$$-k\frac{\partial T}{\partial x} = q''_o. \tag{4.31}$$

The solution is given by Carslaw and Jaeger [4]:

$$T = T_i + (q''_o x/k)\left[\frac{1}{\eta\sqrt{\pi}}e^{-\eta^2} - \text{erfc}(\eta)\right]. \tag{4.32}$$

There is no final temperature $T_o$ in this case: The slab temperature continues to change as long as $q''_o$ is applied.

**Example 4.5**

A large carbon steel slab at 800°F is suddenly cooled at its surface by air with $T_o = 60$°F and $h_o = 5$ Btu/hr · ft$^2$ · °F. Compute (a)

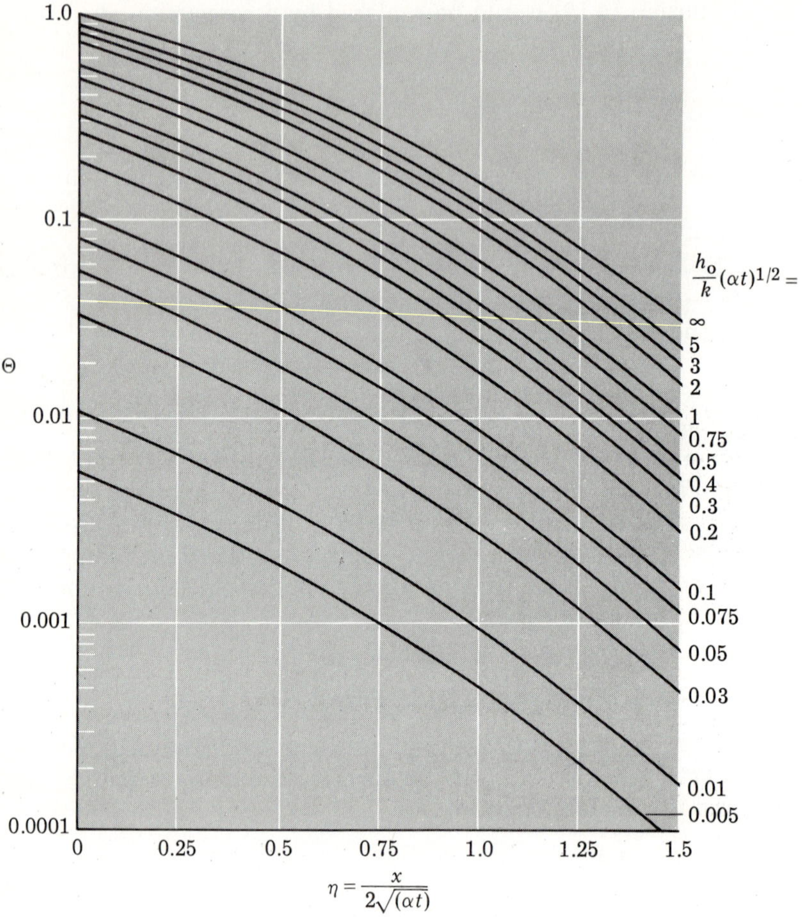


**Figure 4.5 Temperature distribution in a semi-infinite slab with sudden surface convection, from Eq. (4.30).**

the surface temperature at 5 hr of cooling, and (b) the interior depth at which the temperature has dropped 5% after 5 hr.

*Solution* For carbon steel take $k = 25$ Btu/hr · ft · °F and $\alpha = 1.3 \times 10^{-4}$ ft$^2$/s. At the surface, $\eta = 0$ and we compute

$$\lambda = (h_o/k)(\alpha t)^{1/2} = (5/25)\sqrt{1.3 \times 10^{-4}(5 \times 3600)} = 0.306.$$

From Fig. 4.5 at $\eta = 0$, read $\Theta \simeq 0.27$, or from Eq. (4.30) compute

$$\Theta = 1 - e^{(0.306)^2}\,\text{erfc}(0.306) = 0.2695.$$

Then the surface temperature at 5 hr is given by

$$\Theta = 0.27 = \frac{T - 800°\text{F}}{60 - 800}, \quad \text{or} \quad T = 600°\text{F}. \quad [Ans. (a)]$$

For (b), assume that a 5% drop means the condition $\Theta = 0.05$, with $\lambda = 0.306$ at $t = 5$ hr from (a). This defines a point on Fig. 4.5 from which we read $\eta \simeq 0.77$. We could also iterate Eq. (4.30) on a programmable calculator:

$$\Theta = 0.05 = \text{erfc}(\eta) - e^{0.306(2\eta + 0.306)}\,\text{erfc}(\eta + 0.306).$$

Solve for $\eta = 0.776 = x/\{2\sqrt{[1.3 \times 10^{-4}(5 \times 3600)]}\}$, or

$x = 2.37$ ft. $[Ans. (b)]$

Thus, after 5 hr, the transient conduction has begun to penetrate quite deep into the slab. Either the semi-infinite analysis is becoming invalid or this is quite a big slab. ■

## 4.4 Transient Conduction in Three Common Shapes

The semi-infinite geometry solutions of Section 4.3 are a special case corresponding to the early stages of unsteady conduction in a body of finite thickness. The complete solution for a finite-sized body of regular geometry can be accomplished by the method of Fourier series used in Section 3.2.3. We give here only the solutions themselves and refer you to advanced references [3, 4] for details of the method.

The three most common shapes for which exact transient solutions are known are the slab of finite thickness, the long cylinder, and the sphere. These geometries are defined in Fig. 4.6. The practical problem is one of suddenly applied convection at the surface. We assume that the bodies are at uniform initial temperature $T_i$ when, at $t = 0$, their surfaces are suddenly exposed to a uniform convection environment $h_o$ and $T_o$. We assume that temperature varies with time and only a single spatial coordinate: $x$ for the slab and $r$ for the cylinder and sphere, as in Fig. 4.6.

### 4.4.1 The Finite Slab

Consider the slab of thickness $2L$ in Fig. 4.6(a). We must solve the one-dimensional conduction equation,

$$\frac{\partial T}{\partial t} = \alpha \frac{\partial^2 T}{\partial x^2}, \tag{4.33}$$

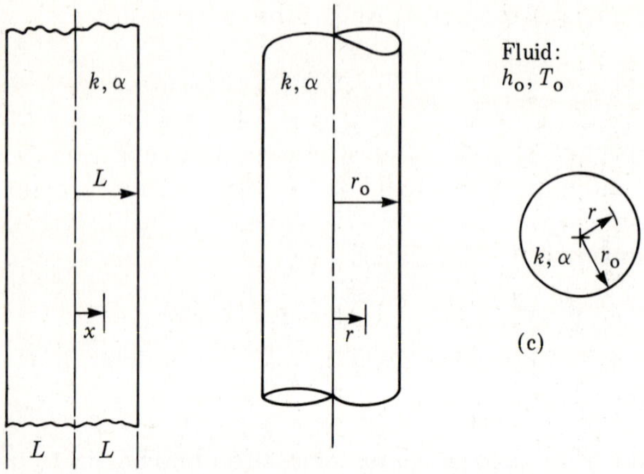


**Figure 4.6 Geometry and coordinate system for three classical transient conduction solutions: (a) finite slab, (b) cylinder, (c) sphere. These shapes have in common that only one spatial dimension (*x* or *r*) is required. All shapes have uniform initial temperature $T_i$.**

subject to an initial condition,

$$T(x, 0) = T_i \tag{4.34}$$

plus a uniform convection condition at both surfaces:

At $x = \pm L$:

$$-k\frac{\partial T}{\partial x} = \pm h_o(T - T_o). \tag{4.35}$$

The exact solution is given as a Fourier series in [3]:

$$\Theta = \frac{T - T_o}{T_i - T_o} = \sum_{i=1}^{\infty} C_i\, e^{-\beta_i^2 \alpha t/L^2} \cos(\beta_i x/L), \tag{4.36}$$

where

$$C_i = \frac{4 \sin\beta_i}{2\beta_i + \sin(2\beta_i)}. \tag{4.37}$$

The constants $\beta_i$ are the roots of the transcendental algebraic equation

$$\beta_i \tan(\beta_i) = \text{Bi} = h_o L/k. \tag{4.38}$$

Thus the solutions have the Biot number of the slab as a parameter. The roots of Eq. (4.38) are tabulated in Appendix A. Before discussing this solution, let us look at the other two cases.

### 4.4.2 The Cylinder

The suddenly immersed cylinder satisfies the equation

$$\frac{\partial T}{\partial t} = \frac{\alpha}{r}\frac{\partial}{\partial r}\left(r\frac{\partial T}{\partial r}\right), \tag{4.39}$$

subject to an initial condition,

$$T(r, 0) = T_{\mathrm{i}}, \tag{4.40}$$

and a convection condition at the surface:

At $r = r_0$:

$$-k\frac{\partial T}{\partial r} = h_{\mathrm{o}}(T - T_{\mathrm{o}}). \tag{4.41}$$

The solution is given as a Fourier series in [3]:

$$\Theta = \frac{T - T_{\mathrm{o}}}{T_{\mathrm{i}} - T_{\mathrm{o}}} = \sum_{i=1}^{\infty} C_i\, e^{-\beta_i^2 \alpha t/r_0^2} J_0(\beta_i r/r_0), \tag{4.42}$$

where

$$C_i = \frac{2}{\beta_i}\frac{J_1(\beta_i)}{J_0^2(\beta_i) + J_1^2(\beta_i)}. \tag{4.43}$$

The constants $\beta_i$ are the roots of the algebraic equation

$$\beta_i J_1(\beta_i)/J_0(\beta_i) = \mathrm{Bi} = h_{\mathrm{o}} r_{\mathrm{o}}/k. \tag{4.44}$$

The functions $J_0$ and $J_1$ are the Bessel functions of the first kind [1, Section 4.5], and their numerical values are tabulated in many places, such as [1, 5, 6]. We have tabulated $J_0$ and $J_1$ in Appendix A for a limited range of interest. The roots of Eq. (4.44) are tabulated in [3].

### 4.4.3 The Sphere

The suddenly immersed sphere satisfies the equation

$$\frac{\partial T}{\partial t} = \frac{\alpha}{r^2}\frac{\partial}{\partial r}\left(r^2\frac{\partial T}{\partial r}\right) \tag{4.45}$$

and is subject to the same conditions as the cylinder, Eqs. (4.40) and (4.41). The solution is given also in [3]:

$$\Theta = \frac{T - T_{\mathrm{o}}}{T_{\mathrm{i}} - T_{\mathrm{o}}} = \sum_{i=1}^{\infty} C_i \frac{r_0}{r\beta_i}\, e^{-\beta_i^2 \alpha t/r_0^2} \sin(\beta_i r/r_0), \tag{4.46}$$

where

$$C_i = \frac{4(\sin\beta_i - \beta_i \cos\beta_i)}{2\beta_i - \sin(2\beta_i)}. \tag{4.47}$$

This time the constants $\beta_i$ are the roots of the equation

$$1 - \beta_i \cot(\beta_i) = \mathrm{Bi} = h_o r_0/k. \tag{4.48}$$

These roots are also tabulated in [3].

### 4.4.4 Centerline Temperature for $t^* > 0.2$

It turns out that the summing of multiple terms in our three Fourier series solutions is necessary only for the very early stages of the transient — approximately the first 10–20% of the total cooling (heating) of the body. Heisler [7] points out that if the dimensionless time $t^* = \alpha t/L^2$ (or $\alpha t/r_0$) is greater than 0.2, a *single* term of the series is sufficient with an accuracy of 1% or better. This latter 80–90% of the cooling (heating) period is the region of greatest interest to the engineer. And the *point* of greatest interest is the center of the body ($x = 0$ or $r = 0$), which is the slowest to react to the surface convection. This means that the bulk of our practical computations are at the center for large times, for which the Fourier series solutions reduce to

$t^* > 0.2$, *center point:*

$$\Theta_c \simeq C_1 \, e^{-\beta_1^2 t^*}, \tag{4.49}$$

where $t^* = \alpha t/L^2$ for the slab and $\alpha t/r_0^2$ for the cylinder and sphere. Thus the center temperature difference decays exponentially, analogous to the lumped-mass approximation, Eq. (4.12), but with different constants.

Also, the temperatures at other points are related to the center temperatures by the following simple relations if $t^* > 0.2$:

*Slab:*

$$\Theta = \Theta_c \cos(\beta_1 x/L); \tag{4.50a}$$

*Cylinder:*

$$\Theta = \Theta_c J_0(\beta_1 r/r_0); \tag{4.50b}$$

*Sphere:*

$$\Theta = \Theta_c \frac{r_0}{r\beta_1} \sin(\beta_1 r/r_0). \tag{4.50c}$$

All points in the body thus decay at exactly the same rate as the center point if $t^* > 0.2$. Almost all practical problems associated with these three classical solutions simply require the evaluation of Eqs. (4.49) or (4.50). The necessary coefficients $\beta_1$ and $C_1$ are computed from Eqs. (4.37, 4.38), (4.43, 4.44), or (4.47, 4.48) and are tabulated versus Biot number in Table 4.1 and plotted in Fig. 4.7. Recall that the Biot number equals $h_o L/k$ for the slab, where $L$ is the half-thickness, and is $h_o r_o/k$ for both the cylinder and the sphere.

### 4.4.5 Total Transient Heat Flux

In addition to local temperatures, it is sometimes useful to compute the total heat lost (gained) by these three bodies during the transient cooling (heating) process. The result is expressed as a ratio to the total heat content:

$$\frac{Q(t)}{Q_o} = \frac{\rho\, c \int (T - T_i)\, dv}{\rho\, c\, (T_i - T_o)\, v_o} = \int (1 - \Theta)\frac{dv}{v_o}, \tag{4.51}$$

where $v_o$ is the total volume of the body. Substituting for $\Theta$ from Eqs. (4.50) and integrating, we obtain three results valid with 1% accuracy for $t^* \geqslant 0.2$:

*Slab:*

$$\frac{Q}{Q_o} = 1 - (\Theta_c/\beta_1)\sin(\beta_1); \tag{4.52a}$$

*Cylinder:*

$$\frac{Q}{Q_o} = 1 - (2\Theta_c/\beta_1)\, J_1(\beta_1); \tag{4.52b}$$

*Sphere:*

$$\frac{Q}{Q_o} = 1 - (3\Theta_c/\beta_1^3)[\sin(\beta_1) - \beta_1\cos(\beta_1)], \tag{4.52c}$$

where $\Theta_c$ is evaluated from Eq. (4.49) and of course the constants $\beta_1$ and $C_1$ are taken from Table 4.1 for the proper shape.

### 4.4.6 The Heisler/Gröber Charts

The above three classical solutions for conduction in a suddenly immersed solid, Eqs. (4.36), (4.42), and (4.46), were derived in 1947 by M. P. Heisler in a now famous paper [7]. The complexity of the three Fourier series and the need of engineers for numbers, coupled

**Table 4.1 Coefficients for the Heisler centerline formulas**

| | Slab | | Cylinder | | Sphere | |
|---|---|---|---|---|---|---|
| Bi† | $\beta_1$ | $C_1$ | $\beta_1$ | $C_1$ | $\beta_1$ | $C_1$ |
| 0.01 | 0.0998 | 1.0017 | 0.1412 | 1.0025 | 0.1730 | 1.0030 |
| 0.02 | 0.1410 | 1.0033 | 0.1995 | 1.0050 | 0.2445 | 1.0060 |
| 0.04 | 0.1987 | 1.0066 | 0.2814 | 1.0099 | 0.3450 | 1.0120 |
| 0.06 | 0.2425 | 1.0098 | 0.3438 | 1.0148 | 0.4217 | 1.0179 |
| 0.08 | 0.2791 | 1.0130 | 0.3960 | 1.0197 | 0.4860 | 1.0239 |
| 0.1 | 0.3111 | 1.0161 | 0.4417 | 1.0246 | 0.5423 | 1.0298 |
| 0.2 | 0.4328 | 1.0311 | 0.6170 | 1.0483 | 0.7593 | 1.0592 |
| 0.3 | 0.5218 | 1.0451 | 0.7465 | 1.0712 | 0.9208 | 1.0880 |
| 0.4 | 0.5932 | 1.0580 | 0.8516 | 1.0931 | 1.0528 | 1.1164 |
| 0.5 | 0.6533 | 1.0701 | 0.9408 | 1.1143 | 1.1656 | 1.1441 |
| 0.6 | 0.7051 | 1.0814 | 1.0185 | 1.1345 | 1.2644 | 1.1713 |
| 0.7 | 0.7506 | 1.0919 | 1.0873 | 1.1539 | 1.3525 | 1.1978 |
| 0.8 | 0.7910 | 1.1016 | 1.1490 | 1.1724 | 1.4320 | 1.2236 |
| 0.9 | 0.8274 | 1.1107 | 1.2048 | 1.1902 | 1.5044 | 1.2488 |
| 1.0 | 0.8603 | 1.1191 | 1.2558 | 1.2071 | 1.5708 | 1.2732 |
| 2.0 | 1.0769 | 1.1785 | 1.5995 | 1.3384 | 2.0288 | 1.4793 |
| 3.0 | 1.1925 | 1.2102 | 1.7887 | 1.4191 | 2.2889 | 1.6227 |
| 4.0 | 1.2646 | 1.2287 | 1.9081 | 1.4698 | 2.4556 | 1.7202 |
| 5.0 | 1.3138 | 1.2403 | 1.9898 | 1.5029 | 2.5704 | 1.7870 |
| 6.0 | 1.3496 | 1.2479 | 2.0490 | 1.5253 | 2.6537 | 1.8338 |
| 7.0 | 1.3766 | 1.2532 | 2.0937 | 1.5411 | 2.7165 | 1.8674 |
| 8.0 | 1.3978 | 1.2570 | 2.1286 | 1.5526 | 2.7654 | 1.8920 |
| 9.0 | 1.4149 | 1.2598 | 2.1566 | 1.5611 | 2.8044 | 1.9106 |
| 10.0 | 1.4289 | 1.2620 | 2.1795 | 1.5677 | 2.8363 | 1.9249 |
| 20.0 | 1.4961 | 1.2699 | 2.2881 | 1.5919 | 2.9857 | 1.9781 |
| 30.0 | 1.5202 | 1.2717 | 2.3261 | 1.5973 | 3.0372 | 1.9898 |
| 40.0 | 1.5325 | 1.2723 | 2.3455 | 1.5993 | 3.0632 | 1.9942 |
| 50.0 | 1.5400 | 1.2727 | 2.3572 | 1.6002 | 3.0788 | 1.9962 |
| 100.0 | 1.5552 | 1.2731 | 2.3809 | 1.6015 | 3.1102 | 1.9990 |
| $\infty$ | 1.5708 | 1.2732 | 2.4048 | 1.6021 | 3.1416 | 2.0000 |

† Bi = $h_o L/k$ for the slab and $h_o r_0/k$ for the cylinder and sphere (see Fig. 4.6).

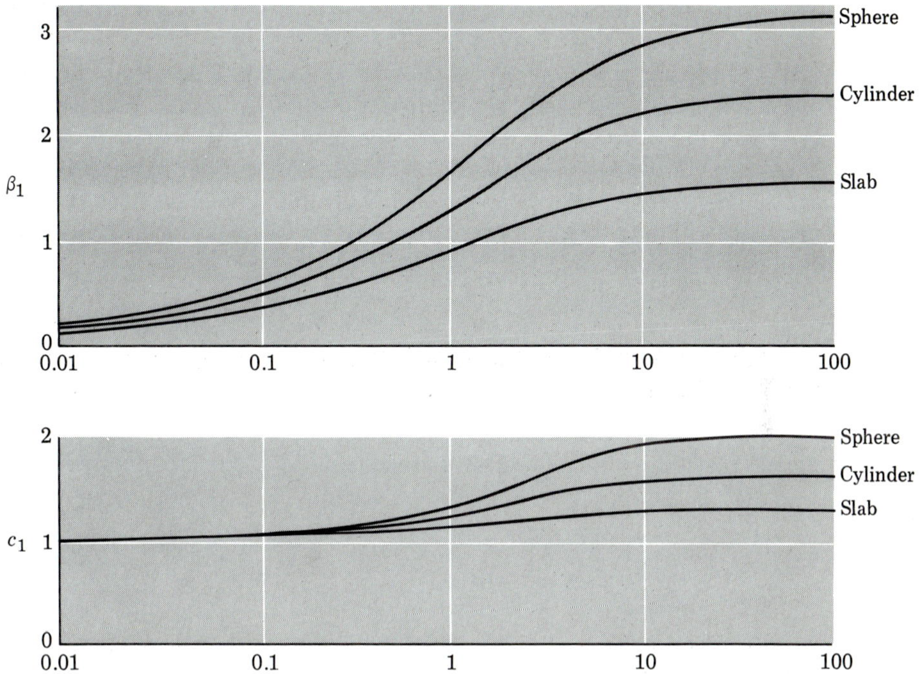


**Figure 4.7** Variation of the Heisler formula coefficients $\beta_1$ and $C_1$ with Biot number, for use in Eqs. (4.49)–(4.52).

with the lack of computer facilities at that time, led Heisler to construct a series of parametric graphs that are now called the Heisler charts. The text by Gröber et al. [8] added graphs of the total transient heat flux from Eqs. (4.52). Thus the Heisler/Gröber charts are a part of every text on heat transfer and are reproduced here in Appendix J. For each body shape — the slab, cylinder, and sphere — there are three charts:

1. a plot of center temperature $\Theta_c$ versus $t^*$ and inverse Biot number 1/Bi,
2. a plot of temperature at noncentral points $x$ or $r$ as compared to the center temperature, and
3. a plot of total heat transfer $Q/Q_o$ versus the parameter $\text{Bi}^2 t^*$.

All of the plots in Appendix J are limited to $t^* > 0.2$; that is, they are one-term approximations.

Examining the charts in Appendix J, you will see that they are difficult to read accurately, especially for small $t^*$. Minimum uncertainty is about $\pm 5\%$, which is adequate for preliminary design purposes. Additional accuracy, if necessary, is obtained by evaluating the Fourier series themselves. We will approach example problems here by both methods: the one-term series solution from Table 4.1 and the Heisler charts.

### Example 4.6

Modify Example 4.5 so that the slab has a finite thickness of 5 ft. Compute the center temperature and the surface temperature after 5 hours, using both Table 4.1 and the Heisler charts.

*Solution* For carbon steel, take $k = 25$ Btu/hr · ft · °F and $\alpha = 1.3 \times 10^{-4}$ ft²/s. We are given $h_o = 5$ Btu/hr · ft² · °F, $T_i = 800$°F, and $T_o = 60$°F. From Fig. 4.6(a), $L$ is the half-thickness of the slab, or $L = 2.5$ ft. Then compute

$$\begin{aligned}\text{Bi} &= h_o L/k = (5\ \text{Btu/hr}\cdot\text{ft}^2\cdot{}^\circ\text{F})(2.5\ \text{ft})/(25\ \text{Btu/hr}\cdot\text{ft}\cdot{}^\circ\text{F})\\ &= 0.5,\\ t^* &= \alpha t/L^2 = (1.3 \times 10^{-4}\ \text{ft}^2/\text{s})\,(5 \times 3600\ \text{s})/(2.5\ \text{ft})^2\\ &= 0.3744.\end{aligned}$$

We can solve the problem by two methods.

a) *The one-term Fourier series:*

From Table 4.1 at Bi = 0.5, read $C_1 = 1.0701$, $\beta_1 = 0.6533$ for a slab. Then, from Eq. (4.49), compute

$$\begin{aligned}\Theta_c &= (T_c - T_o)/(T_i - T_o) = 1.0701\, e^{-(0.6533)^2(0.3744)}\\ &= 0.9121.\end{aligned}$$

This should be quite accurate because $t^* > 0.2$. Solve

$$T_c = 60° + (800 - 60°)(0.9121) = 735°\text{F}. \quad [\textit{Ans.}]$$

For comparison, in Example 4.5, at $x = 2.5$ ft in the semi-infinite slab, the temperature at $t = 5$ hr was 767°F. The temperature at the surface is given by Eq. (4.50a) with $x = L$:

$$\Theta_s = \Theta_c \cos(\beta_1 L/L) = 0.9121 \cos(0.6533) = 0.7243,$$

whence

$T_s = 60° + (800 - 60°)(0.7243) = 596°\text{F}.$ [*Ans.*]

Again we may compare with Example 4.5, for which the computed surface temperature was 600°F. Thus this thick slab is reasonably "semi-infinite" up to $t = 5$ hours.

b) *The Heisler charts:*

For center temperature in a slab, the appropriate chart is Fig. J.1(a). Entering this chart at $t^* = 0.3744$ and $1/\text{Bi} = 2.0$, we read, approximately,

$$\Theta_c = \frac{T_c - 60}{800 - 60} \doteq 0.90 \pm 0.05.$$

It is in this area of small $t^*$ that the Heisler chart is hardest to read accurately. Solving for $T_c$, we have

$T_c = 60 + (740)(0.90 \pm 0.05) = 730 \pm 30°\text{F}.$ [*Ans.*]

The uncertainty is about 4% compared to the exact solution in part (a) above. In a similar manner, we compute the surface temperature at the same $t^*$ by entering Fig. J.1(b) at $1/\text{Bi} = 2.0$ and $x/l = 1.0$. We read, approximately,

$$\frac{T(x = L) - T_o}{T_c - T_o} = \frac{T_s - 60}{(730 \pm 30) - 60} \doteq 0.80 \pm 0.01,$$

or

$T_s = 60 + (670 \pm 30)\,(0.80 \pm 0.01) = 600 \pm 30°\text{F}.$ [*Ans.*]

Note that this solution depended upon having previously read Fig. J.1(a) to find $T_c$. We read this new chart quite accurately but still suffered a 5% uncertainty in $T_s$ because of the previous uncertainty in $T_c$. ■

## Example 4.7

For the cast iron sphere of Example 4.2, use both Table 4.1 and the Heisler charts to find the time for the center temperature to drop to 40°F and the surface temperature at that particular time.

*Solution* The sphere has a radius of 6 cm. For cast iron, take $k = 52$ W/m · K. We are given $h_o = 75$ W/m$^2$ · K. Then

$\text{Bi} = h_o r_o/k = (75\ \text{W/m}^2 \cdot \text{K})\,(0.06\ \text{m})/(52\ \text{W/m} \cdot \text{K}) = 0.0865.$

We are given $T_i = 200°F$, $T_o = 20°F$, and $T_c = 40°F$, so that

$$\Theta_c = \frac{40 - 20}{200 - 20} = 0.111.$$

This enables us to find the time by two methods, taking the thermal diffusivity of cast iron to be $1.7 \times 10^{-5}$ m$^2$/s.

a) *The one-term Fourier series:*

From Table 4.1 for a sphere at Bi = 0.0865, interpolate or use Eq. (4.47) or (4.48) to obtain

$$C_1 = 1.026, \qquad \beta_1 = 0.505.$$

Then, from Eq. (4.49),

$$\Theta_c = 0.111 = C_1 e^{-\beta_1^2 t^*} = 1.026 e^{-(0.505)^2 t^*},$$

from which we solve for $t^* = 8.71 = \alpha t/r_o^2$. Then solve for

$$t = 8.71 r_o^2/\alpha = (8.71)(0.06)^2/(1.7 \times 10^{-5})$$
$$= 1845 \text{ s} = 30.7 \text{ min.} \quad [Ans.]$$

Recall that the lumped analysis of Example 4.2 gave $t = 29.8$ min or 3% less. Meanwhile, with $\Theta_c$ known, the surface temperature follows from Eq. (4.50c):

$$\Theta_s = \Theta_c \sin(\beta_1)/\beta_1 = (0.1111)\sin(0.505)/0.505 = 0.1065,$$

or

$$T_s = 20° + 0.1065(180°) = 39.2°\text{C.} \quad [Ans.]$$

The center and surface temperatures differ by only 0.8°C, verifying that this sphere is approximately "lumpable."

b) *The Heisler charts:*

For center temperature, the appropriate chart is Fig. J.3(a), which we enter at $\Theta_c = 0.111$ and $1/\text{Bi} = 11.6$ to read

$$t^* \doteq 8.8 \pm 0.3 = (1.7 \times 10^{-5})t/(0.06)^2.$$

Solve for

$$t = 1860 \pm 60 \text{ s} = 31 \pm 1 \text{ min.} \quad [Ans.]$$

With $T_c$ known, enter Fig. J.3(b) at $1/\text{Bi} = 11.6$ and $r/r_o = 1.0$ to find

$$\frac{T_s - T_o}{T_c - T_o} = \frac{T_s - 20}{40 - 20} \doteq 0.95 \pm 0.02.$$

Solve for

$T_s = 39.0 \pm 0.4°C.$ [*Ans.*]

The Heisler charts are more accurate at large values of $t^*$. ■

---

In the above two examples, the Fourier series method works well because the Biot number is known and Table 4.1 can be entered immediately. If Bi is unknown, Table 4.1 cannot be used without guessing and iterating, so the Fourier approach becomes awkward. Meanwhile, the Heisler charts do not have this defect and can easily be read for an unknown Biot number. The following example illustrates.

### Example 4.8

A long brass cylinder of diameter 20 cm and temperature 20°C is to be heated by immersion in a fluid at 100°C. What heat transfer coefficient is required to heat the center of the cylinder to 80°C in 8 minutes?

*Solution* For brass, take $k = 110$ W/m · K and $\alpha = 3.4 \times 10^{-5}$ $m^2/s$. We are given sufficient information to compute dimensionless temperature and time:

$$\Theta_c = (T_c - T_o)/(T_i - T_o) = (80 - 100)/(20 - 100) = 0.25,$$

$$t^* = \alpha t/r_o^2 = (3.4 \times 10^{-5}\ \text{m}^2/\text{s})(8 \times 60\ \text{s})/(0.1\ \text{m})^2 = 1.632.$$

Now consider our two possible methods.

a) *The Heisler charts:*

The proper chart is Fig. J.2(a) for a cylinder. Enter at $\Theta_c = 0.25$ and $t^* = 1.632$ to read $1/\text{Bi} = 2.0 \pm 0.1$, or

$$\text{Bi} \doteq 0.5 \pm 0.025 = h_o r_o/k = h_o(0.1\ \text{m})/(110\ \text{W/m} \cdot \text{K}).$$

Solve for

$$h_o = (0.5 \pm 0.025)(110)/(0.1) = 550 \pm 30\ \text{W/m}^2\,\text{K}. \quad [\textit{Ans.}]$$

b) *The one-term Fourier series:*

The proper solution must satisfy Eq. (4.49) for a cylinder:

$$\Theta_c = 0.25 = C_1\, e^{-\beta_1^2(1.632)}, \qquad (1)$$

with both $C_1$ and $\beta_1$ dependent upon $\text{Bi} = h_o r_o/k$. However, $C_1$ is slowly varying. Assuming $C_1 \doteq 1.1$, Eq. (1) above gives the estimate $\beta_1 \doteq 0.95$, which occurs in Table 4.1 for the cylinder at Bi between 0.5 and 0.6. Read $\beta_1$ for these two Biot numbers and compare with values computed from Eq. (1):

| Guess Bi | Read $C_1$ | Read $\beta_1$ | $\beta_1$ from Eq. (1) | Difference |
|---|---|---|---|---|
| 0.5 | 1.1143 | 0.9408 | 0.9570 | + 0.0162 |
| 0.6 | 1.1345 | 1.0185 | 0.9627 | − 0.0558 |

Interpolate to

$$\text{Bi} \doteq 0.522 = h_o r_0/k = h_o(0.1\text{ m})/(110\text{ W/m}\cdot\text{K}),$$

or

$$h_o = 575\text{ W/m}^2\cdot\text{K}. \quad [\textit{Ans.}]$$

For this case of unknown Biot number, the Heisler chart is less accurate but clearly the handier of the two methods. ■

## 4.5 Multidimensional Solutions by the Product Method

The classic solutions for semi-infinite- and finite-thickness bodies in Sections 4.3 and 4.4 may be used to generate solutions for still more interesting shapes, using a very clever multiplicative superposition trick.

The use of this trick is illustrated by Fig. 4.8. We wish to solve the "sudden immersion" problem for a finite-length cylinder subjected to uniform convection conditions ($h_o$, $T_o$) on all sides and initial uniform temperature $T_i$. The differential equation to be solved is

$$\frac{1}{r}\frac{\partial}{\partial r}\left(r\frac{\partial \Theta}{\partial r}\right) + \frac{\partial^2 \Theta}{\partial x^2} = \frac{1}{\alpha}\frac{\partial \Theta}{\partial t}, \tag{4.53}$$

where $\Theta = (T - T_o)/(T_i - T_o)$. The initial condition is

$$\Theta(x, r, 0) = 1. \tag{4.54}$$

The convection boundary conditions are

*At top and bottom:*

$$-k\frac{\partial \Theta}{\partial x} = \pm h_o\Theta, \tag{4.55a}$$

*On the sides:*

$$-k\frac{\partial \Theta}{\partial r} = h_o\Theta. \tag{4.55b}$$

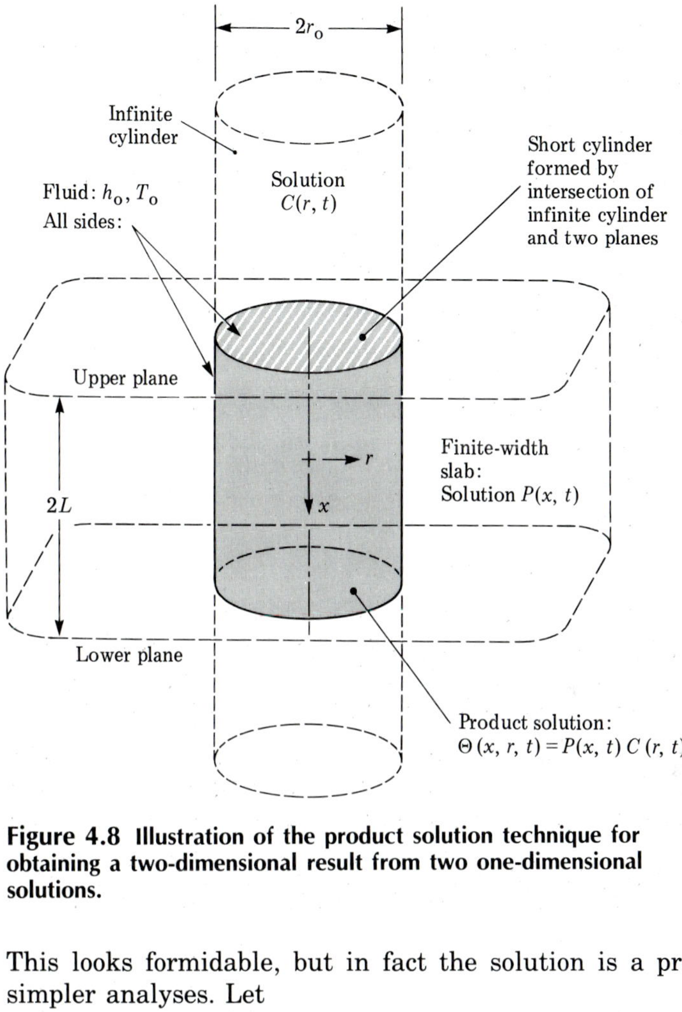


**Figure 4.8 Illustration of the product solution technique for obtaining a two-dimensional result from two one-dimensional solutions.**

This looks formidable, but in fact the solution is a product of two simpler analyses. Let

$$\Theta(x, r, t) = P(x, t)C(r, t). \tag{4.56}$$

Substitute Eq. (4.56) in Eqs. (4.53, 4.54, 4.55) and separate the variables. We obtain two independent problems for $P$ and $C$:

*Plate:*

$$\frac{\partial^2 P}{\partial x^2} = \frac{1}{\alpha}\frac{\partial P}{\partial t}$$

$$P(x, 0) = 1$$

$$-k\left.\frac{\partial P}{\partial x}\right|_{\pm L} = h_o P(\pm L, t),$$

*Cylinder:*

$$\frac{1}{r}\frac{\partial}{\partial r}\left(r\frac{\partial C}{\partial r}\right) = \frac{1}{\alpha}\frac{\partial C}{\partial t}$$

$$C(r, 0) = 1$$

$$-k\frac{\partial C}{\partial r}\bigg|_{r_o} = h_o C(r_o, t).$$

We see to our delight that this product trick splits the problem into two known solutions: (1) the "plate" or finite slab problem from Section 4.4.1 and (2) the "cylinder" problem from Section 4.4.2. Equation (4.56) is thus the exact solution to the sudden immersion of a short right circular cylinder.

Table 4.2 gives nine examples of the use of the product method to obtain solutions for a variety of sudden immersion geometries. These are all products of three basic solutions:

1. *Semi-infinite solid:*

$$S(x, t) = \frac{T(x, t) - T_o}{T_i - T_o} \qquad \text{from Eq. (4.30)}$$

2. *Finite-width plate:*

$$P(x, t) = \frac{T(x, t) - T_o}{T_i - T_o} \qquad \text{from Eq. (4.36)}$$

3. *Infinite cylinder:*

$$C(r, t) = \frac{T(r, t) - T_o}{T_i - T_o} \qquad \text{from Eq. (4.42)}$$

Note the mnemonic device of using the letter $S$ to mean semi-infinite, $P$ to mean plate, and $C$ for cylinder. Note also in Table 4.2 that the coordinates must be carefully defined: $x$ for function $S$ is from the surface inward, $x$ for the plate and $r$ for the cylinder must be measured from the centerlines of the equivalent plate and cylinder. Finally, note carefully that the function $S$ is defined as different from the semi-infinite solution $\Theta$ in Eq. (4.30), that is,

$$S = (T - T_o)/(T_i - T_o) = 1 - \Theta, \text{ since } \Theta = (T - T_i)/(T_o - T_i).$$

Further details and proofs of the applicability of the product method are given in Section 5.2 of the text by Arpaci [10]. The method works only if (1) *all* surfaces are subjected to the same uniform $(h_o, T_o)$ convection conditions and (2) all cutting planes and curved surfaces are orthogonal, that is, if the short cylinder in Fig. 4.8 had slanted upper and lower faces the solution (4.56) would not apply.

**Table 4.2 Multidimensional sudden immersion solutions as a product of one-dimensional results**

---

*0. Basic Solutions*

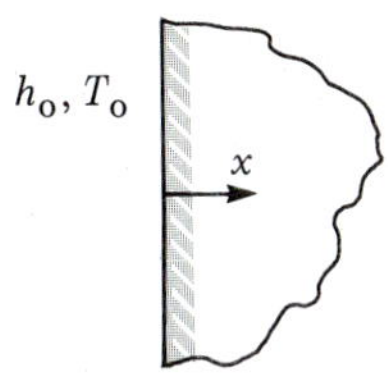


$S(x, t) = 1 - \Theta$
Eq. (4.30)

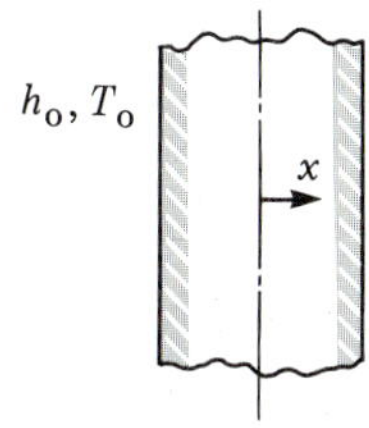


$P(x, t)$
Eq. (4.36)

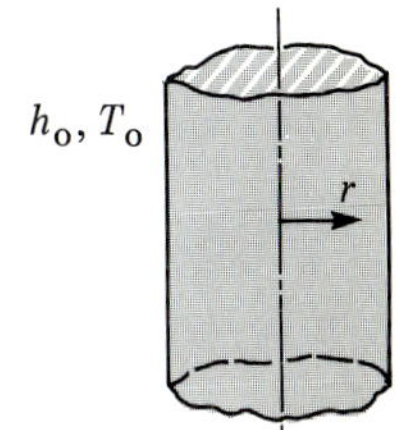


$C(r, t)$
Eq. (4.42)

---

*1. Two-Dimensional Corner Region*

$$\Theta = S(x, t)S(y, t)$$

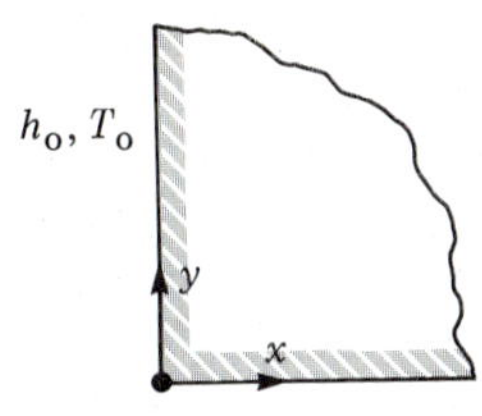


*2. Three-Dimensional Corner Region*

$$\Theta = S(x, t)S(y, t)S(z, t)$$

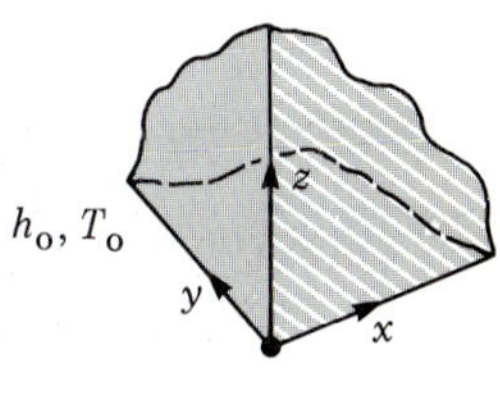


---

*3. Semi-Infinite Plate*

$$\Theta = P(x, t)S(y, t)$$

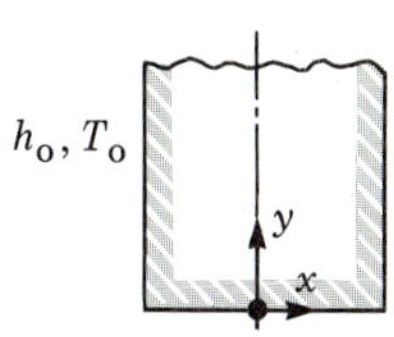


*4. Infinite Rectangular Bar (origin at centerline)*

$$\Theta = P(x, t)P(y, t)$$

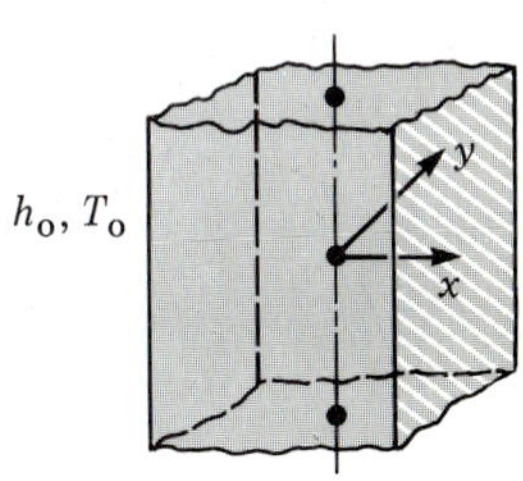

**Table 4.2** (*continued*)

5. *Semi-Infinite Rectangular Bar (origin in the center of the bottom face)*

$$\Theta = P(x, t)P(y, t)S(z, t)$$

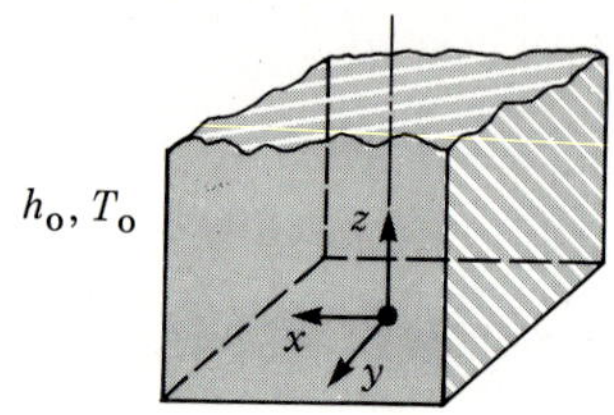


6. *Finite-Width Corner Region (origin in the center of the corner)*

$$\Theta = P(x, t)S(y, t)S(z, t)$$

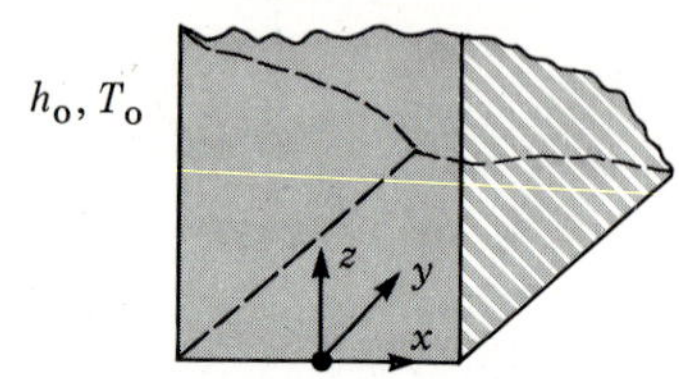


7. *Box Shape (origin at exact center of box)*

$$\Theta = P(x, t)P(y, t)P(z, t)$$

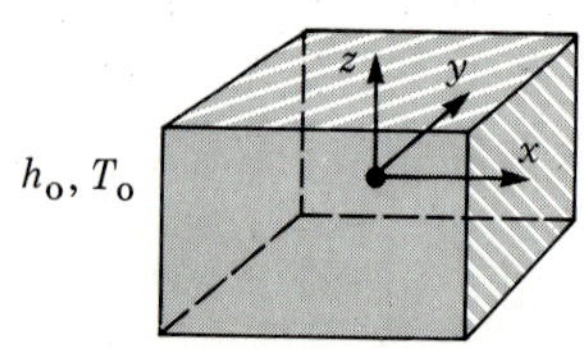


8. *Semi-Infinite Cylinder (origin at center of bottom face)*

$$\Theta = S(x, t)C(r, t)$$

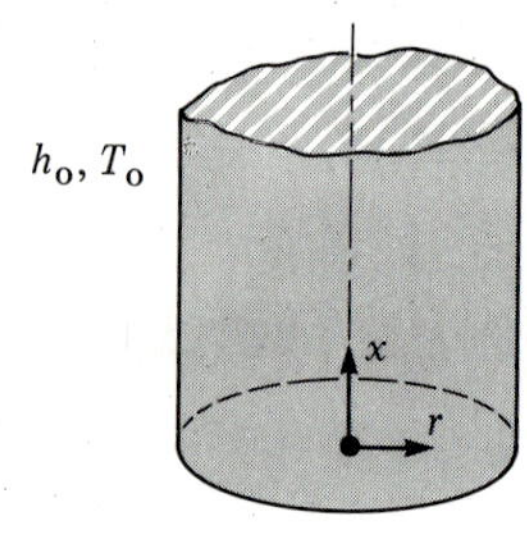


9. *Finite-Length Cylinder (origin at exact center)*

$$\Theta = P(x, t)C(r, t)$$

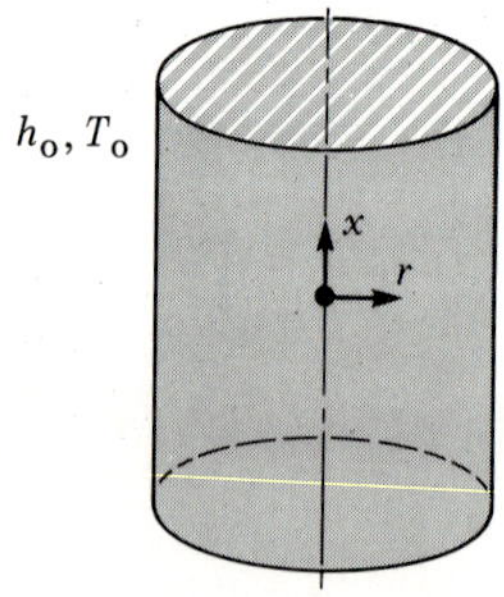

## Example 4.9

The short cylinder in Fig. E4.9 is initially at 40°C and then plunged into a fluid with $h_o = 300\ \text{W/m}^2 \cdot \text{K}$ and $T_o = 200$°C. The material is bronze, $k = 26\ \text{W/m} \cdot \text{K}$ and $\alpha = 8.6 \times 10^{-6}\ \text{m}^2/\text{s}$. Find the temperature at the center of the cylinder after 5 min.

*Solution* This configuration is item 9 of Table 4.2, also shown in Fig. 4.8. The center of the cylinder is the origin of coordinates, hence

$$\Theta(0, 0, t) = P(0, t)C(0, t) = \Theta_c(\text{slab})\, \Theta_c(\text{cylinder}), \tag{1}$$

where the $\Theta_c$ are computed from Eq. (4.49). There are two separate sets of Biot numbers, constants, and time scales:

*Slab:*

$$\text{Bi}_s = h_o L/k = (300\ \text{W/m}^2 \cdot \text{K})(0.08\ \text{m})/(26\ \text{W/m} \cdot \text{K}) = 0.923,$$

$$t_s^* = \alpha t/L^2 = (8.6 \times 10^{-6}\ \text{m}^2/\text{s})(300\ \text{s})/(0.08\ \text{m})^2 = 0.403.$$

From Table 4.1, interpolate $\beta_{1s} = 0.835$, $C_{1s} = 1.113$. From Eq. (4.49),

$$\Theta_c(\text{slab}) = 1.113\, e^{-(0.835)^2(0.403)} = 0.840.$$

This should be accurate because $t_s^*$ is greater than 0.2.

*Cylinder:*

$$\text{Bi}_c = h_o r_o/k = (300)(0.03)/26 = 0.346,$$

$$t_c^* = \alpha t/r_o^2 = (8.6 \times 10^{-6})(300)/(0.03)^2 = 2.867.$$

From Table 4.1 interpolate $\beta_{1c} = 0.797$, $C_{1c} = 1.081$.

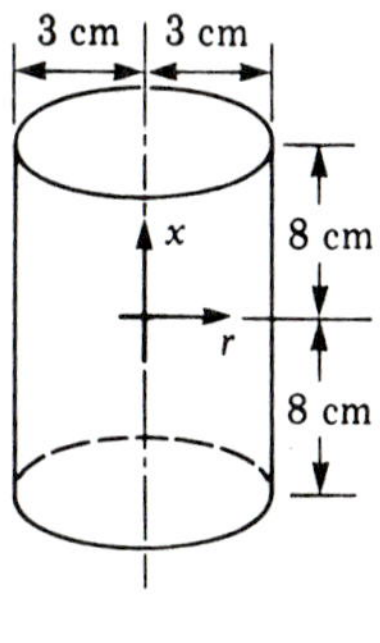


**Figure E4.9**

From Eq. (4.49),

$\Theta_c(\text{cyl}) = 1.081\, e^{-(0.797)^2(2.867)} = 0.175.$

This should be quite accurate because $t_c^* >> 0.2$. The product solution, Eq. (1) above, thus predicts

$\Theta(0, 0, 5 \text{ min}) = (0.840)(0.175) = 0.147 = (T - T_o)/(T_i - T_o),$

from which

$$T(0, 0, 5 \text{ min}) = (0.147)(40° - 200°) + 200°$$
$$= 176°\text{C}. \quad [Ans.]$$

Clearly most of the heat entered through the curved sides. ■

## Example 4.10

The long square glass rod ($k = 0.8$ W/m · K, $\alpha = 3.5 \times 10^{-7}$ m$^2$/s) in Fig. E4.10 is at 150°C when it is suddenly dipped into a fluid with $h_o = 120$ W/m$^2$ · K and $T_o = 20$°C. Estimate the temperature of the corner point $A$ of the rod after 4 min.

*Solution* This shape fits item 5 of Table 4.2. With the origin in the center of the bottom square face, the temperature at point $A$ is given by

$$\Theta_A = P(L, t)P(L, t)S(0, t) = P^2(L, t)S(0, t), \quad (1)$$

where $L = 1.5$ cm. The slab solution $P$ requires a Biot number:

$\text{Bi}_s = h_oL/k = (120 \text{ W/m}^2 \cdot \text{K})(0.015 \text{ m})/(0.8 \text{ W/m} \cdot \text{K}) = 2.25,$

$t_s^* = \alpha t/L^2 = (3.5 \times 10^{-7} \text{ m}^2/\text{s})(240 \text{ s})/(0.015 \text{ m})^2 = 0.373.$

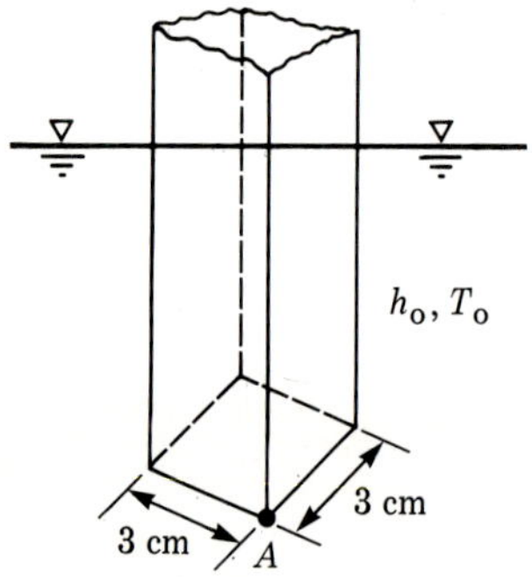


Figure E4.10

From Table 4.1 at $\mathrm{Bi}_s = 2.25$, interpolate $\beta_{1s} = 1.112$, $C_1 = 1.188$. Since $t^* > 0.2$, Eq. (4.49) applies:

$$\Theta_c(\text{slab}) = 1.188\, e^{-(1.112)^2(0.373)} = 0.749.$$

Then, from Eq. (4.50),

$$P(L, t) = \Theta_c \cos(\beta_1) = (0.749)\cos(1.112)$$
$$= 0.332.$$

The semi-infinite solid solution $S(0, t)$ requires the parameter $\lambda$ from Eq. (4.29):

$$\lambda = (h_o/k)(\alpha t)^{1/2} = (120/0.8)[(3.5 \times 10^{-7})(240)]^{1/2} = 1.375.$$

Then Eq. (4.30) predicts that, at $\eta = x/2(\alpha t)^{1/2} = 0$,

$$\Theta_{\text{slab}} = \operatorname{erfc}(0) - e^{(1.375)^2} \operatorname{erfc}(1.375) = 0.657.$$

Recall our warning that $\Theta_{\text{slab}}$ is defined opposite to $S$. Thus

$$S(0, t) = 1 - \Theta_{\text{slab}} = 1 - 0.657 = 0.343.$$

Then, from Eq. (1) above, the dimensionless temperature at point $A$ is

$$\Theta_A = P^2(L, t)S(0, t) = (0.332)^2(0.343) = 0.0378 = \frac{T_A - T_o}{T_i - T_o},$$

whence

$$T_A = (0.0378)(150° - 20°) + 20°$$
$$= 24.9°\text{C}. \quad [Ans.]$$

The corner cools off faster than any other point on the surface of the immersed rod. ■

---

### 4.5.1 Total Heat Transfer to a Composite Body

In a recent note, Langston [18] pointed out that superposition can also be applied to the total heat transfer of a suddenly immersed composite body formed by the intersection of planes and cylinders. The one-dimensional total heat flux $Q/Q_o$ for a slab and for a cylinder is computed from Eqs. (4.52a) and (4.52b), respectively.

Let $R = Q/Q_o$ for convenience. If the composite is formed by the intersection of body 1 and body 2, the total heat flux is

$$R = R_1 + R_2(1 - R_1). \tag{4.57}$$

In Fig. 4.8, for example, body 1 would be the slab and body 2 the infinite cylinder.

If the composite is formed by the intersection of three bodies, such as the box of item 7 in Table 4.2, Langston obtains

$$R = R_1 + R_2(1 - R_1) + R_3(1 - R_1)(1 - R_2), \qquad \textbf{(4.58)}$$

where $R_i = Q_i/Q_o$ is computed from Eqs. (4.52) for the given shape.

**Example 4.11**

Determine the total heat transfer for the cylinder of Example 4.9 after 5 min.

*Solution* Let body 1 be the slab and body 2 the infinite cylinder. For the slab, $\beta_{1s} = 0.835$ and $\Theta_c = 0.840$. From Eq. (4.52a),

$$R_1 = Q/Q_{os} = 1 - 0.840 \sin(0.835)/0.835 = 0.254.$$

Also, for the cylinder, from Ex. 4.9, $\beta_{1c} = 0.797$ and $\Theta_c = 0.175$. From Eq. (4.52b),

$$R_2 = Q/Q_{oc} = 1 - 2(0.175)\, J_1(0.797)/0.797 = 0.839.$$

Then, from Eq. (4.57) for the two-body intersection,

$$Q/Q_o\,(\text{composite}) = R_1 + R_2(1 - R_1)$$
$$= 0.254 + 0.839(1 - 0.254) = 0.88. \quad [Ans.]$$

The short cylinder has received 88% of the maximum possible heat from the fluid after 5 min. ■

## 4.6 Numerical Methods for Transient Conduction

In Sections 4.3 to 4.5 we found what appeared to be a wide variety of transient conduction solutions. Closer inspection reveals that they are actually limited to certain regular body shapes and uniform, constant wall temperature or convection conditions. Exact analysis becomes much more difficult for irregular shapes or wall temperatures and surface convection that vary with time. And of course there are essentially no exact analyses for radiation boundary conditions because the Fourier series superposition fails. However, all these problems can be handled with adequate accuracy by numerical finite-difference or finite-element methods. The steady conduction numerical models developed in Section 3.4 were found to be unconditionally stable no matter what mesh size was chosen. For a given mesh size, however, a transient numerical model can be *unstable* if the time step is taken too large. The stability criterion takes the form of a maximum allowable value of the *mesh Fourier number* $\sigma = \alpha(\Delta t)/(\Delta x)^2$, as we shall see.

### 4.6.1 One-Dimensional Explicit Finite-Difference Model

Suppose we wish to develop a finite-difference model for the one-dimensional transient conduction equation

$$\frac{\partial T}{\partial t} = \alpha \frac{\partial^2 T}{\partial x^2}, \tag{4.59}$$

subject to a variety of boundary conditions. Consider the one-dimensional mesh sketched in Fig. 4.9. We adopt the notation $T^j_m$, by analogy with Fig. 3.10, to denote the nodal temperature at a given position $m$ and at a given time $j$.

Now, recalling our derivative approximations from Eqs. (3.51), model the heat conduction equation (4.59) with a "forward" time difference and a "centered" spatial difference:

$$\frac{T^{j+1}_m - T^j_m}{\Delta t} \simeq \alpha \frac{T^j_{m+1} - 2T^j_m + T^j_{m-1}}{(\Delta x)^2}. \tag{4.60}$$

Rearrange and collect terms, solving for the "new" temperature

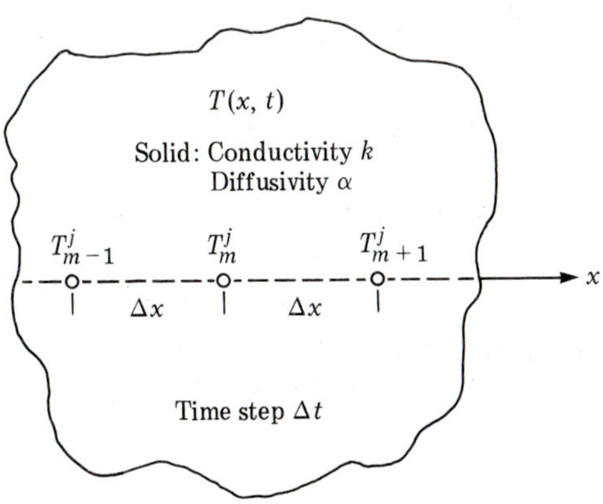


**Figure 4.9** Definition sketch for a finite-difference mesh to model a transient one-dimensional conduction analysis.

$T_m^{j+1}$ at node $m$:

$$T_m^{j+1} \simeq (1 - 2\sigma)\, T_m^j + \sigma(T_{m-1}^j + T_{m+1}^j), \tag{4.61}$$

where $\sigma = \alpha\Delta t/(\Delta x)^2$ is called the *mesh Fourier number.*

Equation (4.61) is the basic algebraic model of an interior node for the transient one-dimensional heat conduction equation. It relates the new nodal value at the next time step $j + 1$ to the old nodal value and its two old neighbors at time step $j$. Notice that the new value is immediately calculated; there is no iteration as happened in the steady conduction model of Section 3.4. Using known "old" temperature values, the first of which at $j = 0$ are the initial values, every new temperature is computed in sequence. This type of transient model is called an *explicit* method, meaning no iteration is necessary. The computation simply "marches" forward in time as long as boundary conditions continue to be known at a given time.

### 4.6.2 Stability of Interior Node Computations

The drawback to using the simple explicit model of Eq. (4.61) is that the time step $\Delta t$ is limited by a numerical stability condition. As proved for example, in [11], to avoid divergent oscillations of the nodal temperatures, the coefficients $\sigma$ and $(1 - 2\sigma)$ must both be positive. Since $\sigma$ is, by definition, positive, we require that $(1 - 2\sigma)$ be positive, or

$$\sigma = \alpha\,\Delta t/(\Delta x)^2 \leqslant 0.5. \tag{4.62}$$

Otherwise, the solution will be unstable. Physically, a value of $\sigma$ greater than 0.5 would cause the new nodal temperature $T_m^{j+1}$ to be less than the average of its neighbors, which violates the second law of thermodynamics.

Since we ordinarily choose mesh size $\Delta x$ first, the stability condition essentially limits the time step to $\Delta t \leqslant (\Delta x)^2/2\alpha$. If we stay below this limit, application of Eq. (4.61) to all interior nodes will give smooth and reasonably accurate results — not exact results, mind you, because we are using *finite* $\Delta t$ and $\Delta x$. If instead of known boundary temperatures there are convection, radiation, or insulation boundary conditions, then the next section develops *boundary stability* requirements that are more stringent than Eq. (4.62) and hence govern the time-step size.

### 4.6.3 Wall Convection Boundary Condition

An algebraic model simulating wall convection, comparable in accuracy to Eq. (4.61), can be derived with an energy balance on the control volume strip $\Delta x/2$ wide in Fig. 4.10. Let the wall node be $T_1^j$ and write the first law for the strip:

Heat convected in − Heat conducted out = Energy storage

or

$$h_o A(T_o^j - T_1^j) - kA\,\frac{T_1^j - T_2^j}{\Delta x} = \rho c\,A\,\frac{\Delta x}{2}\,\frac{T_1^{j+1} - T_1^j}{\Delta t}. \tag{4.63}$$

Rearrange and solve for the "new" nodal value at the wall:

$$T_1^{j+1} \approx 2\sigma(T_2^j + \mathrm{Bi}\,T_o^j) + (1 - 2\sigma - 2\sigma\mathrm{Bi})T_1^j, \tag{4.64}$$

where $\mathrm{Bi} = h_o\Delta x/k$ is the *mesh Biot number* (recall Eq. 3.59) and $\sigma = \alpha\Delta t/(\Delta x)^2$ as before. Equation (4.64) gives an explicit result for the new wall temperature $T_1^{j+1}$. Again all coefficients must be positive for stability, which requires $(1 - 2\sigma - 2\sigma\mathrm{Bi}) \geqslant 0$, or:

$$\sigma \leqslant \frac{1}{2(1 + \mathrm{Bi})}. \tag{4.65}$$

Since Bi is always positive, this is a stronger condition than Eq. (4.62) for interior points and further limits time step $\Delta t$.

Table 4.3 gives a list of additional boundary condition models, derived in a manner similar to Fig. 4.10 and Eq. (4.63). For completeness we include some two-dimensional boundary formulations also. Note that each type of condition results in a different stability requirement. When using these models in a transient conduction simulation, the strongest condition is the one that limits the size of the time step $\Delta t$.

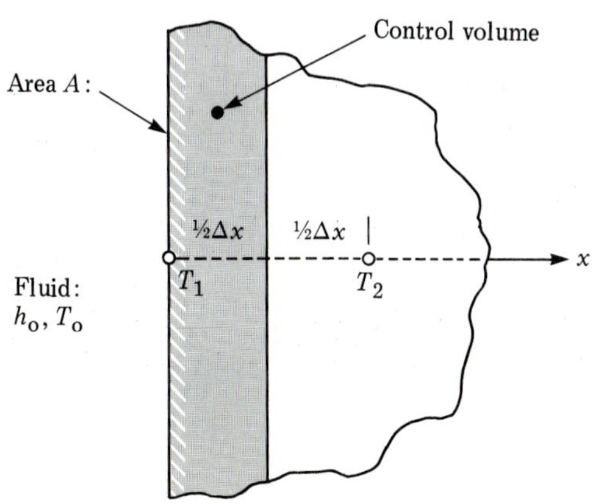


**Figure 4.10** Definition sketch for the derivation of the convection boundary model, Eq. (4.64).

**Table 4.3 Nodal equations for typical transient conduction boundary conditions**

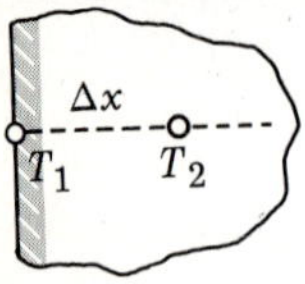


*1. One-dimensional geometry with convection boundary*

Explicit: $$T_1^{j+1} = 2\sigma(T_2^j + \text{Bi}T_o^j) + (1 - 2\sigma - 2\sigma\text{Bi})T_1^j$$

$$\sigma \leqslant \frac{1}{2(1 + \text{Bi})}$$

Implicit: $$T_1^{j+1} = \frac{2\sigma(T_2^{j+1} + \text{Bi}T_o^{j+1}) + T_1^j}{1 + 2\sigma + 2\sigma\text{Bi}}$$

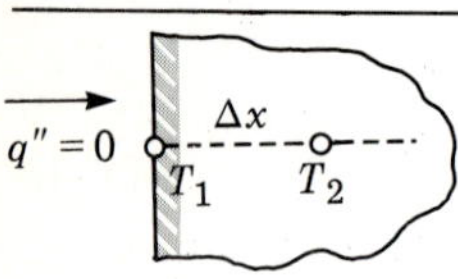


*2. One-dimensional geometry with insulated boundary*

Explicit: $$T_1^{j+1} = 2\sigma T_2^j + (1 - 2\sigma)\,T_1^j$$

$$\sigma \leqslant \frac{1}{2}$$

Implicit: $$T_1^{j+1} = \frac{2\sigma\, T_2^{j+1} + T_1^j}{1 + 2\sigma}$$

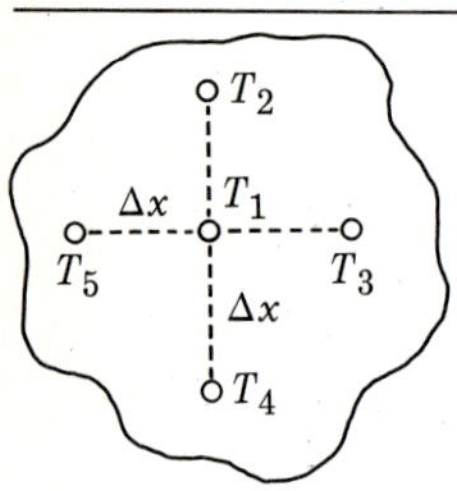


*3. Two-dimensional geometry, interior point*

Explicit: $$T_1^{j+1} = \sigma(T_2^j + T_3^j + T_4^j + T_5^j) + (1 - 4\sigma)T_1^j$$

$$\sigma \leqslant \frac{1}{4}$$

Implicit: $$T_1^{j+1} = \frac{\sigma(T_2^{j+1} + T_3^{j+1} + T_4^{j+1} + T_5^{j+1}) + T_1^j}{1 + 4\sigma}$$

*Notes:* 1. In all cases, $\sigma = \alpha\Delta t/(\Delta x)^2$ and $\text{Bi} = h_o\Delta x/k$. 2. The implicit formulas have no stability limits. 3. For an insulated two-dimensional boundary or corner, set $\text{Bi} = 0$ in items 4, 5, and 6.

**Table 4.3** *(continued)*

---

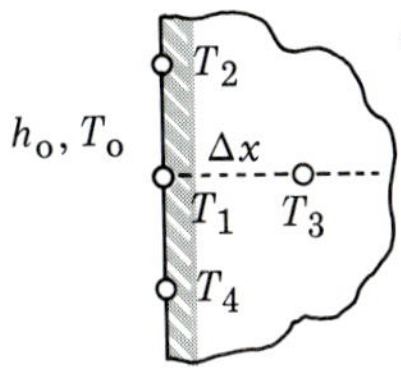


4. *Two-dimensional geometry, convection boundary node*

Explicit:
$$T_1^{j+1} = 2\sigma\left(T_3^j + \frac{1}{2}T_2^j + \frac{1}{2}T_4^j + \mathrm{Bi}\,T_o^j\right) + (1 - 4\sigma - 2\sigma\mathrm{Bi})\,T_1^j$$
$$\sigma \leqslant \frac{1}{2(2 + \mathrm{Bi})}$$

Implicit:
$$T_1^{j+1} = \frac{2\sigma\left(T_3^{j+1} + \frac{1}{2}T_2^{j+1} + \frac{1}{2}T_4^{j+1} + \mathrm{Bi}\,T_o^{j+1}\right) + T_1^j}{1 + 2\sigma(2 + \mathrm{Bi})}$$

---

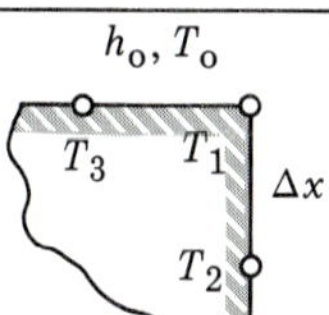


5. *Two-dimensional outside corner node with convection*

Explicit:
$$T_1^{j+1} = 2\sigma(T_2^j + T_3^j + 2\mathrm{Bi}\,T_o^j) + (1 - 4\sigma - 4\sigma\mathrm{Bi})T_o^j$$
$$\sigma \leqslant \frac{1}{4(1 + \mathrm{Bi})}$$

Implicit:
$$T_1^{j+1} = \frac{2\sigma(T_2^{j+1} + T_3^{j+1} + 2\mathrm{Bi}\,T_o^{j+1}) + T_1^j}{1 + 4\sigma(1 + \mathrm{Bi})}$$

---

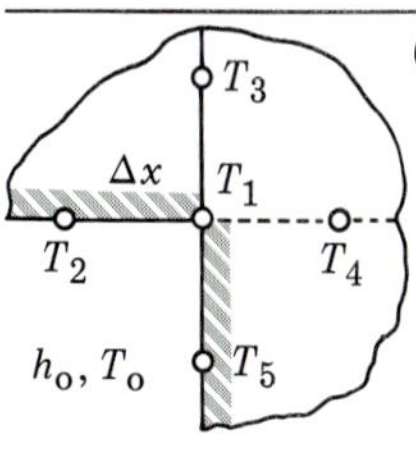


6. *Two-dimensional interior corner node with convection*

Explicit:
$$T_1^{j+1} = \frac{4}{3}\sigma\left(T_3^j + T_4^j + \frac{1}{2}T_2^j + \frac{1}{2}T_5^j + \mathrm{Bi}\,T_o^j\right) + \left(1 - 4\sigma - \frac{4}{3}\sigma\mathrm{Bi}\right)T_1^j$$
$$\sigma \leqslant \frac{3}{4(3 + \mathrm{Bi})}$$

Implicit:
$$T_1^{j+1} = \frac{\frac{4}{3}\sigma\left(T_3^{j+1} + T_4^{j+1} + \frac{1}{2}T_2^{j+1} + \frac{1}{2}T_5^{j+1} + \mathrm{Bi}\,T_o^{j+1}\right) + T_1^j}{1 + 4\sigma(1 + \mathrm{Bi}/3)}$$

## Example 4.12

Repeat Example 4.6 as a numerical simulation with $\Delta x = 0.5$ ft. Take advantage of symmetry. Compute nodal temperatures at $t = 5$ hr for (a) $t = 600$ s; (b) $t = 900$ s; and (c) $t = 1200$ s. Compare with the exact solution from Example 4.6 and interpret.

*Solution* Although the slab is 5 ft thick, we need only 6 nodes because of symmetry, as shown in Fig. E4.12(a). The centerline can be treated as an "insulated" boundary. The mesh size Biot number is $\text{Bi} = h_o\Delta x/k = (5.0)(0.5)/25 = 0.1$. There is one convection boundary, so the stability criterion is given by Eq. (4.65):

$$\Delta t_{\max} = \frac{\Delta x^2}{2(1 + \text{Bi})\alpha} = \frac{(0.5 \text{ ft})^2}{2(1 + 0.1)(0.00013 \text{ ft}^2/\text{s})} = 874 \text{ s}.$$

It follows that the requested step sizes will be (a) stable for 600 s; (b) gradually unstable for 900 s; and (c) very unstable for 1200 s. The nodal equations are of three types:

1. Node 1 is a convection boundary (Eq. 4.64):

$$T_1^{j+1} = 2\sigma[T_2^j + 0.1(60°)] + (1 - 2\sigma - 0.2\sigma)T_1^j. \tag{1}$$

2. Nodes 2 through 5 are interior points, Eq. (4.61):

$$T_m^{j+1} = (1 - 2\sigma)T_m^j + \sigma(T_{m-1}^j + T_{m+1}^j), \qquad m = 2, 3, 4, 5. \tag{2}$$

3. Node 6 is an insulated (centerline) boundary, item 2 of Table 4.3:

$$T_6^{j+1} = 2\sigma T_5^j + (1 - 2\sigma)T_6^j. \tag{3}$$

For $\Delta t = 600$, 900, and 1200 s, the three values of $\sigma$ are (a) 0.312, (b) 0.468, and (c) 0.624, respectively. The initial conditions at $j = 0$ are $T_{1,2,3,4,5,6} = 800$°F. We repeat Eqs. (1, 2, 3) over and

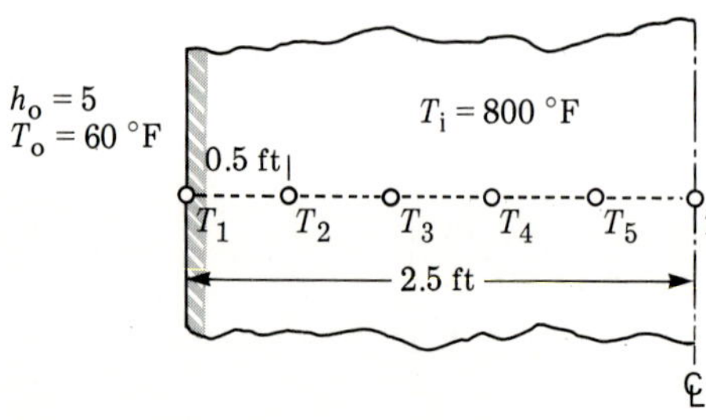


**Figure E4.12(a)**

over until we reach $j\Delta t = 18{,}000$ s, or five hours. The results for each time step are tabulated as follows.

| $\Delta t$ | $\sigma$ | $j$ for $t = 5$ hr | $T_1$ | $T_2$ | $T_3$ | $T_4$ | $T_5$ | $T_6$ |
|---|---|---|---|---|---|---|---|---|
| | | | **Computed Temperatures (°F) at $t = 5$ hr** | | | | | |
| 600 s | 0.312 | 30 | 596.9 | 645.8 | 684.4 | 712.1 | 728.8 | 734.4 |
| 900 s | 0.468 | 20 | 597.1 | 645.2 | 684.6 | 711.7 | 729.1 | 734.0 |
| 1200 s | 0.624 | 15 | −2262 | +3189 | −1512 | +2587 | −917 | +2297 |
| *Exact Solution, Example 4.6* | | | 597.1 | 645.8 | 684.2 | 711.7 | 728.3 | 733.8 |

We see that the results for $\Delta t = 600$ s are quite smooth and accurate, with a maximum error of 0.5°F. At $\Delta t = 900$ s, the results oscillate almost imperceptibly: The oscillations would continue as the time extended beyond $j = 20$. For $\Delta t = 1200$ s, the results are totally ridiculous and even at $j = 15$ have oscillated nearly off the map. In general, we must stay within the stability limits to avoid these oscillatory errors.

The oscillatory behavior of the two unstable solutions is shown in Fig. E.4.12(b) for the fluid boundary node $T_1$. In contrast, the $\Delta t = 600$ s solution (not shown) is a smooth curve dropping monotonically from 800°F at $t = 0$ to 597°F at $t = 5$ hr. ■

### 4.6.4 Schmidt's Graphical Method for $\sigma = 0.5$

An interesting interpretation of the explicit finite-difference model, Eq. (4.61), results in a graphical solution, now primarily of historical significance. The idea was first given by L. Binder in 1911 and then expanded and improved by E. Schmidt in 1924. The idea is to choose a time-step size so that $\sigma = 0.5$, or $\Delta t = (\Delta x)^2/2\alpha$. Then Eq. (4.61) becomes

$$T_m^{j+1} = \frac{1}{2}(T_{m-1}^j + T_{m+1}^j). \tag{4.66}$$

For this special time step, the nodal temperature at the new time $(t + \Delta t)$ is exactly equal to the arithmetic average of its two nearest neighbors at the old time $t$. Graphically, if one plots $T$ versus $x$, drawing a vertical line for each nodal position $m$, the new value $T_m^{j+1}$ is found by connecting $T_{m-1}^j$ and $T_{m+1}^j$ with a straight line and marking its intersection on the line denoting $T_m$. Thus, by connecting

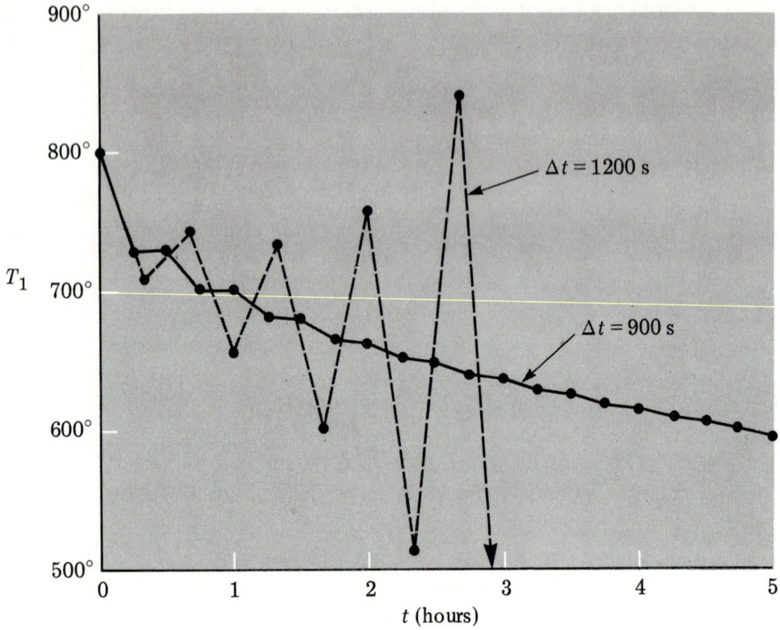


**Figure E4.12(b)**

every other nodal value with straight lines, one can mark off the entire distribution of temperature at the next time step. The idea can be adapted to convection boundaries and to cylindrical coordinates and other types of transient problem. Complete details are given in Chapter 19 of the text by Jakob [12].

The disadvantages of the Binder/Schmidt graphical method are (1) the selection $\sigma = 0.5$ is at the upper limit of numerical stability, where nodal values take on a series of annoying hops and pauses that are not very accurate and (2) the availability of computer terminals and programmable hand calculators makes graphical methods very unattractive. Forty years ago many such graphical aids were in wide use in engineering to avoid laborious computations.

### 4.6.5 An Implicit Finite-Difference Model

The explicit model of the previous sections is perfectly adequate and gives excellent results if stability limits are followed, as shown in Example 4.11. However, the stability criterion limits the time step to be no more than $\Delta t_{max} = (\Delta x)^2/2\alpha$ and much less than that if the mesh Biot number is large. Further, since $\Delta t \propto (\Delta x)^2$, if we seek more accuracy by, say, halving $\Delta x$, we must reduce $\Delta t$ by one-fourth. For a large, fine mesh, the number of computations may become excessive, to the point that an alternate, more stable scheme is desired.

One such scheme, with unlimited numerical stability, is the *implicit* finite-difference technique. This method models the heat conduction equation (4.59) in the same manner as the explicit technique, Eq. (4.60), with one subtle but profound difference: It evaluates the second derivative at the *new* rather than the old time. Thus our model (see Fig. 4.9) is

$$\frac{T_m^{j+1} - T_m^j}{\Delta t} \simeq \alpha \frac{T_{m+1}^{j+1} - 2T_m^{j+1} + T_{m-1}^{j+1}}{(\Delta x)^2}. \quad \textbf{(4.67)}$$

Rearrange and solve for the new nodal value at the center:

$$T_m^{j+1} = \frac{\sigma(T_{m+1}^{j+1} + T_{m-1}^{j+1}) + T_m^j}{1 + 2\sigma}, \quad \textbf{(4.68)}$$

where $\sigma = \alpha\Delta t/(\Delta x)^2$ as usual. This is the basic *implicit* model for an interior point in transient one-dimensional conduction. Notice that there are three unknowns: the three nodal values at the new time $(j + 1)$. Only $T_m^j$ is known. Thus, when applied to all nodes, Eqs. (4.68) are a set of simultaneous algebraic relations, each with three unknowns. While there are advanced numerical methods for solving such "tri-diagonal" algebraic equations (see, for example [16] and [18] in Chapter 3), the ordinary Gauss-Seidel iteration method also works well. At a given time $j$ where nodal values are known, one merely sweeps all nodes with Eq. (4.68) — or their boundary node equivalents — until the temperatures converge within some chosen deviation. The process is uniformly convergent to the new nodal values $T_m^{j+1}$. There is *no* instability: The time step $\Delta t$ may be chosen as large as you think prudent, and the inaccuracy of the model increases with increasing $\Delta t$. Generally speaking, the implicit model hops along at a much larger time step, for the same accuracy, than the explicit model. Even if $\Delta t \to 0$, neither model is "exact," the inaccuracy depending on the crudeness of the spatial mesh size $\Delta x$.

Implicit formulas that model various one- and two-dimensional boundary conditions are given in Table 4.3 along with their explicit cousins. None of these implicit boundary models has any limitation on the size of the time step.

This terminates our discussion of numerical solution of the transient heat conduction equation. There are, however, whole books that treat in detail the application of these and other more advanced numerical methods to heat transfer calculations. See, in particular, [13–16, 19].

### Example 4.13

Repeat Example 4.12 using the same mesh as in Fig. E4.12(a) but with an implicit method. Report results for $\Delta t = 600$, 900, 1200, 2000, and 3000 s.

*Solution* Again there are six nodes, and they satisfy three types of model relations. (The Biot number is again Bi = 0.1.)

1. Node 1 is a convection boundary from Table 4.3, item 1:

$$T_1^{j+1} = \frac{2\sigma[T_2^{j+1} + 0.1(60°)] + T_1^j}{1 + 2\sigma + 0.2\sigma}. \tag{1}$$

2. Nodes 2 through 5 are interior points, Eq. (4.68):

$$T_m^{j+1} = \frac{\sigma(T_{m+1}^{j+1} + T_{m-1}^{j+1}) + T_m^j}{1 + 2\sigma}, \qquad m = 2, 3, 4, 5. \tag{2}$$

3. Node 6 is an insulated boundary from Table 4.3, item 2:

$$T_6^{j+1} = \frac{2\sigma T_5^{j+1} + T_6^j}{1 + 2\sigma}. \tag{3}$$

Initial conditions at ($j = 0$) are $T_{1,2,3,4,5,6} = 800°$F. Select a time step $\Delta t$ so that $\sigma$ is known and, starting with any reasonable guess for the "new" nodal values (800°F will do), sweep Eqs. (1, 2, 3) until the six new values $T_m^{j+1}$ have converged. Then call these the "old" values and sweep again for convergence to the next time step. Stop when the time steps total 5 hr or 18,000 s as specified in Example 4.12. A digital computer is a necessity for this procedure, since approximately 20 iterations are needed in each time step to secure convergence. The final computed results at $t = 18{,}000$ s are listed below and compared with the exact solution:

| $\Delta t$ | $\sigma$ | $j$ for $t = 5$ hr | Computed temperatures (°F) at $t = 5$ hr: $T_1$ | $T_2$ | $T_3$ | $T_4$ | $T_5$ | $T_6$ |
|---|---|---|---|---|---|---|---|---|
| 600 s | 0.312 | 30 | 598.0 | 646.9 | 685.2 | 712.5 | 728.8 | 734.3 |
| 900 s | 0.468 | 20 | 598.3 | 647.2 | 685.4 | 712.6 | 728.8 | 734.2 |
| 1200 s | 0.624 | 15 | 598.6 | 647.5 | 685.6 | 712.7 | 728.8 | 734.2 |
| 2000 s | 1.040 | 9 | 599.4 | 648.3 | 686.1 | 712.9 | 728.8 | 734.0 |
| 3000 s | 1.560 | 6 | 600.5 | 649.3 | 686.8 | 713.2 | 728.7 | 733.8 |
| | *Exact Solution, Example 4.6* | | 597.1 | 645.8 | 684.2 | 711.7 | 728.3 | 733.8 |

All solutions are stable and surprisingly accurate. Even at $\Delta t$ = 3000 s the maximum error is only 3.5°F. At $\Delta t$ = 600 s the maximum error is 1.1°F and even at vanishingly small $\Delta t$ there would still be numerical error because of the crude mesh size $\Delta x$ = 0.5 ft. In this particular case the explicit method (Example 4.12) is faster and more accurate. The implicit method becomes faster if $\Delta x$ is reduced below 0.2 ft for this case. ■

## Summary

This chapter has dealt with a variety of methods for analyzing unsteady or transient conduction in solids. First we found by nondimensionalizing the basic equations that, if a time variable $t^* = \alpha t/L^2$ is used, only a single parameter Bi $= h_\infty L/k$ affects the solution behavior for a given geometry. We should check this Biot number immediately, for if it is much smaller than unity, the problem is "lumpable" and results in a simple first-order differential equation, Eq. (4.9). Some lumped analyses were given in Section 4.2 for both constant and time-varying convection environments.

In Section 4.3 we analyzed the semi-infinite slab, which simulates a body subjected to a sudden change for a short period of time. These mathematical solutions gave insight into transients in finite bodies, for which Fourier series solutions were obtained in Section 4.4. Again these solutions depend only on a single parameter, the Biot number. Section 4.5 shows how the classic one-dimensional transient solutions can be multiplied together to simulate sudden immersion of complex body shapes such as short cylinders or boxes. The example problems use both the Fourier series approach and the traditional Heisler/Gröber charts.

Finally, if the geometry or boundary conditions do not fit the classical cases, Section 4.6 shows that, no matter how complex the problem, the solution can always be obtained with adequate accuracy by numerical techniques. Both the explicit and implicit finite-difference methods are discussed and illustrated. Both are readily programmable by the average engineer.

## References

1. E. Kreyszig, *Advanced Engineering Mathematics*, 4th ed., Wiley, New York, 1979.

2. A. G. Hansen, *Similarity Analyses of Boundary Value Problems in Engineering*, Prentice-Hall, Englewood Cliffs, N.J., 1964.
3. P. J. Schneider, *Conduction Heat Transfer*, Addison-Wesley, Reading, Mass., 1955.
4. H. S. Carslaw and J. C. Jaeger, *Conduction of Heat in Solids,* 2nd ed., Oxford Univ. Press, London, 1959.
5. M. C. Potter, *Mathematical Methods in the Physical Sciences*, Prentice-Hall, Englewood Cliffs, N.J., 1978.
6. W. H. Beyer (Ed.), *Handbook of Mathematical Sciences*, 5th ed., CRC Press, Boca Raton, Fla., 1978.
7. M. P. Heisler, "Temperature Charts for Induction and Constant Temperature Heating," *ASME Transactions*, vol. 69, 1947, pp. 227–236.
8. H. Gröber, S. Erk, and U. Grigull, *Fundamentals of Heat Transfer,* McGraw-Hill, New York, 1961.
9. L. C. Thomas, *Fundamentals of Heat Transfer*, Prentice-Hall, Englewood Cliffs, N.J., 1980.
10. V. S. Arpaci, *Conduction Heat Transfer*, Addison-Wesley, Reading, Mass., 1966.
11. R. D. Richtmyer and K. W. Morton, *Difference Methods for Initial Value Problems,* Wiley-Interscience, New York, 1967.
12. M. Jakob, *Heat Transfer*, vol. 1, Wiley, New York, 1949.
13. G. M. Dusinberre, *Heat Transfer Calculations by Finite Differences,* International Textbook Co., Scranton, Pa., 1961.
14. H. Schenck, *FORTRAN Methods in Heat Flow*, Ronald Press, New York, 1963.
15. J. A. Adams and D. F. Rogers, *Computer-Aided Heat Transfer Analysis,* McGraw-Hill, New York, 1973.
16. S. V. Patankhar, *Numerical Heat Transfer and Fluid Flow*, Hemisphere Publishing, Washington, D.C., 1980.
17. F. M. White, *Viscous Fluid Flow*, McGraw-Hill, New York, 1974.
18. L. S. Langston, "Heat Transfer from Multidimensional Objects Using One-Dimensional Solutions for Heat Loss," *Int. J. Heat Mass Transfer*, vol. 25, no. 1, 1982, pp. 149–150.
19. Y. Jaluria and K. E. Torrance, *Computational Heat Transfer*, Hemisphere Publishing, New York, 1986.

## Review Questions

1. What dimensionless parameter basically determines the character of a transient conduction solution?

2. What is the value of dimensional analysis in classifying transient conduction problems and solutions?
3. If all dimensionless parameters remain the same and the body size is doubled, how does the conduction time change?
4. What is "lumpability"? When and why does it work?
5. What is the concept of "similarity" when applied to transient conduction in a semi-infinite slab?
6. Solutions are given here for sudden immersion of a finite slab, a long cylinder, and a sphere. Under what conditions can these solutions be used interchangeably?
7. What is a Bessel function? An error function?
8. What are the "Heisler/Gröber charts"? What are their advantages and disadvantages?
9. How does the "product method" of superposition work? How do you construct, for example, a three-dimensional corner using semi-infinite slab solutions?
10. Explain the distinction between an explicit and an implicit finite-difference numerical model.
11. What quantity in a finite-difference computation is the one that usually ends up limited by the stability requirement?
12. What is the "mesh Biot number" in a numerical model?
13. How does symmetry aid in forming a numerical model?

## Problems

### Problem Distribution by Sections

**The Problem Assignments Are Organized as Follows:**

| Problems | Sections Covered | Topics Covered |
|---|---|---|
| 4.1 –4.8 | 4.1 | General transient analysis |
| 4.9 –4.26 | 4.2 | Lumped analyses |
| 4.27–4.38 | 4.3 | Semi-infinite solutions |
| 4.39–4.53 | 4.4 | Slab, cylinder, sphere |
| 4.54–4.61 | 4.5 | Product solutions |
| 4.62–4.83 | 4.6 | Numerical solutions |
| 4.84–4.86 | All | Design projects |

4.1 Using the variables $r^* = r/r_0$, $z^* = z/r_0$, $t^* = \alpha t/r_0^2$, and $\Theta = (T - T_\infty)/(T_o - T_\infty)$ by analogy with Eqs. (4.4), nondimensionalize the cylindrical heat conduction equation (2.8) with $\dot{q} = 0$. How does the result compare with Eq. (4.1a)?

4.2 Nondimensionalize a wall radiation boundary condition, Eq. (3.8), letting $\Theta = T/T_\infty$, $n^* = n/L$. How does the result compare with Eq. (4.3a)? Does a Biot number appear?

4.3 A long 3-cm-diameter cylinder of nickel cools in 204 s when immersed in a fluid bath with $h_\infty = 450\ \text{W/m}^2 \cdot \text{K}$. What value of $h_\infty$ would cause "similar" conditions for a long 2-cm-diameter aluminum cylinder? How fast would the aluminum cool?

4.4 A 4-in.-diameter iron sphere at 400°F is immersed at $t = 0$ in a bath at $T_\infty = 50$°F and $h_\infty = 12\ \text{Btu/hr} \cdot \text{ft}^2 \cdot$°F. Its temperature is found to be 300°F after 320 s, 200°F after 790 s, and 100°F after 1800 s. Using only these data, estimate the time it would take this same sphere initially at 500°F to cool to 250°F if suddenly immersed in this same bath.

4.5 A plane slab 5 cm thick is heated with internal generation until it achieves a parabolic temperature distribution (see Fig. 2.12a) with both walls at 200°C and the center at 320°C. The heat generation then ceases and the walls are insulated. What will be the steady-state temperature distribution in the slab?

4.6 A slab 3 in. thick is conducting heat in steady state with one wall at 300°F and the other at 120°F (see Fig. 2.4 at constant $k$). The heat is removed, and both walls are insulated. What will be the steady-state temperature distribution in the slab? Sketch typical transient temperatures.

4.7 A slab of thickness $L$ is initially at temperature $T_i$ when the left wall is suddenly raised to temperature $T_1$ and the right wall to temperature $T_2$. Set up the differential equation and boundary and initial conditions and sketch the type of transient temperature profiles that occur. (This problem is solved in Article 36 of [4].)

4.8 A cylinder of inner radius $r_i$ and outer radius $r_o$ is in steady state with surface temperatures $T_i$ and $T_o$ when both surfaces are reduced to $T_a$ and held there. Set up the differential equation and boundary and initial conditions and sketch the type of transient temperature profiles that result.

4.9 We have seen in Section 4.2 that a small, blocky solid is "lumpable" if its Biot number is small when based on the characteristic body length $L^* = v/A$, where $v$ is body volume and $A$ is wetted area. How would you interpret lumpability if the body were squat but also rather "thin," as for example a short, hollow cylinder with $r_i = 1.0$ cm, $r_o = 1.1$ cm, and $L = 1.5$ cm? Would $L^*$ have a better interpretation in this case?

4.10 A cast iron block 1 by 2 by 3 cm is at 240°C when immersed in a fluid with $h_\infty = 80\ \text{W/m}^2 \cdot \text{K}$ and $T_\infty = 20$°C. Estimate the time required for the block to cool to 40°C.

4.11 Lead shot of diameter 0.105 in. is at 600°F when immersed in a quenching bath with $h_\infty = 25$ Btu/hr · ft$^2$ · °F and $T_\infty = 70$°F. How long should the shot be quenched to cool to 100°F?

4.12 Set up the differential equation for a lumpable body of area $A$, volume $\upsilon$, surface emissivity $\varepsilon$, and initial temperature $T_0$ suddenly exposed to a near-vacuum environment of temperature $T_\infty$. What are the conditions that ensure lumpability?

4.13 Carry out a numerical solution of Problem 4.12 for a carbon steel sphere of diameter 1 cm, $T_0 = 900$ K, $\varepsilon = 0.8$, and $T_\infty = 20$°C. Estimate the sphere temperature after 1 min of cooling.

4.14 A thermocouple lead approximates a sphere of constantan. When placed in hot air with $h = 300$ W/m$^2$ · K, it must have a "95%" reaction time of 10 s or less. What is the maximum allowable diameter of the lead?

4.15 A long, 1-cm-diameter copper rod immersed in a 25°C fluid stream cools from 100°C to 70°C in 35 s. Estimate the heat transfer coefficient. Is the rod lumpable?

4.16 An apple of diameter 8 cm approximates the physical properties of water at 20°C. It is initially at 18°C when the environment suddenly drops to $T_\infty = -10$°C and $h_\infty = 2$ W/m · K. How many hours will elapse before the apple begins to freeze?

4.17 How long will the cylinder of Problem 4.15 take to cool from 70°C to 30°C in the same fluid stream?

4.18 A cube of ice of side length $L$ is at 0°C when suddenly immersed in water of conditions ($h_\infty$, $T_\infty$). Set up and solve the equation to estimate the total time to melt the ice. Denote the latent heat of melting by $H_f$. To avoid free surface complications, assume the ice is fully submerged and that melt water is immediately removed by convection currents. Discuss the lumpability of an ice cube, which is an unusual case.

4.19 As a numerical example of Problem 4.18, estimate the time for a 2-in. ice cube at 32°F to melt completely if submerged in water at 60°F with $h_\infty = 40$ Btu/hr · ft$^2$ · °F. Take $H_f = 144$ Btu/lb$_m$.

4.20 A thermocouple lead approximates a 1-mm-diameter iron sphere. If placed in an air stream whose temperature oscillates from 90°C to 110°C with a period of 12 s, with $h_\infty = 85$ W/m$^2$ · K, what temperature oscillation will the thermocouple read? Comment on the maximum desirable lead diameter for this case.

4.21 A wrought iron cube is placed outside in winter, where air temperature varies sinusoidally each day from 20°F to 50°F. What is the cube size for which its phase lag with the air will be exactly ten degrees? What will be the temperature oscillation of the cube? Take $h = 3$ Btu/hr · ft$^2$ · °F.

4.22 A lumpable body is immersed in equilibrium with a fluid at ($h_\infty$, $T_{\infty 0}$). The fluid temperature suddenly begins to rise linearly, that is, $T_\infty =$

$T_{\infty 0} + \beta t$, with $h_\infty$ remaining constant. Solve for the resulting temperature distribution $T(t)$ of the lump, plot the distribution, and interpret.

4.23 As a numerical example of Problem 4.22, suppose that a 3-cm-diameter cast iron sphere is in equilibrium with a fluid at $h_\infty = 75\ \text{W/m}^2 \cdot \text{K}$ and $T_{\infty 0} = 20°\text{C}$. The fluid temperature suddenly begins to rise linearly at 10°C per minute. What will be the sphere temperature after 30 min?

4.24 As shown in Fig. P4.24, chrome steel is to be annealed by heating to 850°C in a 15-m-long convection oven and then cooling in air for an additional 50 m. The 2-cm-diameter bar stock moves continuously at axial velocity $V$. What velocity $V$ in cm/s is appropriate for this task? What will be the bar temperature at the end of the air run?

4.25 A lumpable body is in equilibrium with a fluid at conditions $(h_\infty, T_\infty)$ when it suddenly begins generating heat internally at a rate $\dot{Q}$ (watts). Derive an expression for the lump temperature $T(t)$, plot it, and interpret.

4.26 A recurrent question, indirectly related to Problem 4.25, is whether one saves fuel by turning back the home thermostat at night or whether instead the furnace "works too hard" to bring it back up in the morning. We can make a simple but quantitative analysis by assuming that a house is a "lump" (it isn't) of size 15 by 15 by 4 m, with all sides losing heat to outside conditions $h_\infty = 3\ \text{W/m}^2 \cdot \text{K}$ and $T_\infty = 0°\text{C}$. When running, the furnace provides 55 kW of heat to the interior. Analyze the following two scenarios: (a) the house is at 20°C when the thermostat is set back to 15°C at midnight and returned to a 20°C setting at 8:00 A.M., at which time the furnace comes on continuously to return the house to 20°C; or (b) there is no setback and the furnace runs intermittently all night to keep the house at 20°C. Compare the total fuel expended (in MJ of energy) from midnight to the time when the house again achieves 20°C in part (a). This problem is long but educational.

4.27 A large slab of aluminum at 300°C has its surface suddenly dropped to 30°C. How long will it take the temperature at a depth of 5 cm to drop to 150°C?

**Figure P4.24**

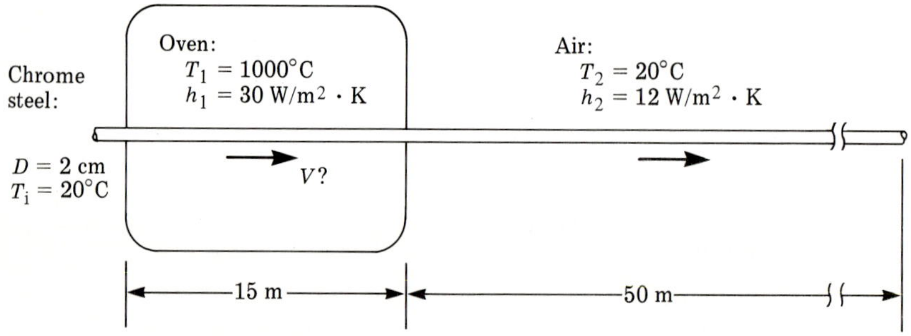

4.28 Show that, for the semi-infinite slab solution of Eq. (4.26) or Fig. 4.4, the surface heat flux at any instant $t$ is given by $q_o'' = k(T_o - T_i) \cdot (\pi\alpha t)^{-1/2}$. From this result find the total heat $Q_o''$ delivered to the slab in time $t$.

4.29 Soil, normally at 15°C, is subjected to a severe winter surface temperature of −13°C for four months. How deep should water pipes be buried to ensure that the soil surrounding them will not fall below 0°C? Take $\alpha = 2.5 \times 10^{-7}$ m²/s for the soil.

4.30 Two semi-infinite slabs of stainless steel, one at a uniform 0°F and the other at 200°F, are brought into perfect thermal contact at $t = 0$. Estimate the temperature after 5 min at a depth of 3 in. from the interface in each slab.

4.31 Clay earth, initially at 20°C, is suddenly subjected to an extremely cold environment. Compute the depth to which freezing temperatures will penetrate after 24 hours if (a) the soil surface is held at −20°C and (b) the air above the soil is at $T_o = -20$°C, $h_o = 15$ W/m² · K.

4.32 A semi-infinite nickel slab at 0°C is suddenly brought into perfect thermal contact with a similar nickel slab at 100°C. Estimate the temperature at a 5-cm depth in each slab after 5 min.

4.33 A semi-infinite slab of nickel at 60°F is exposed to a fluid with $T_o$ = 300°F, $h_o$ = 75 Btu/hr · ft² · °F. Compute the surface temperature after 30 min of immersion.

4.34 As shown in Fig. P4.34, a slab of yellow pine, initially at 20°C, is suddenly subjected to intense radiation heat flux of 7500 W/m². Neglecting re-radiation from the wood, estimate the time for the wood surface to reach 200°C. How would you modify your analysis to account for (blackbody) reradiation from the wood?

4.35 Soil at 15°C ($k$ = 1.2 W/m · K, $\alpha = 9 \times 10^{-7}$ m²/s) is suddenly subjected

**Figure P4.34**

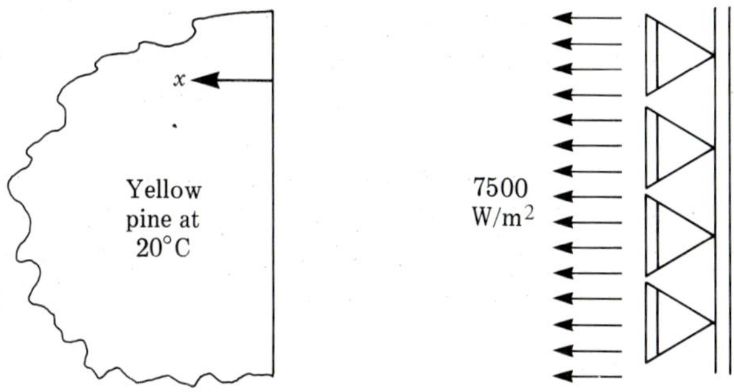

to a cold air spell at $T_o = -20°C$ and $h_o = 50$ W/m$^2$ · K. Estimate the depth where $T = 0°C$ after three days of cold.

4.36 A semi-infinite wood slab ($k = 0.17$ W/m · K, $\alpha = 1.2 \times 10^{-7}$ m$^2$/s) is subjected to a fire at $T_o = 1000°C$, $h_o = 150$ W/m$^2$ · K. How long will it take the surface to reach ignition temperature of 400°C?

4.37 The semi-infinite solid in Fig. P4.37, with a deep interior temperature of $T_0$, is subjected to a periodic temperature $T(x = 0) = T_o + \Delta T \cos(\Omega t)$. Show that, after initial transients have died down, points within the solid will have a steadily oscillating distribution

$$T(x, t) = T_o + \Delta T e^{-Bx} \cos(\Omega t - Bx), \quad B = (\Omega/2\alpha)^{1/2}.$$

Apply this to the practical problem of clay earth (Appendix D) subjected to diurnal variations, $\Omega = 2\pi/24$ hr$^{-1}$. At what depth in the soil will the amplitude of temperature oscillation have dropped to 5% of the surface variation?

4.38 A stainless steel slab at 20°C suddenly receives steady heat flux and after 8 min the temperature at 4-cm depth is 180°C. How much heat flux is being applied?

4.39 A wooden slab is 8 cm thick with one side insulated and the other side subjected to hot air at 300°C with $h_o = 85$ W/m$^2$ · K. If the initial slab temperature is 20°C, what are the temperatures at each slab surface after 3 hr?

4.40 A 2-in-thick cast-iron slab at 0°F is subjected on both sides to fluid with $T_o = 200°F$ and $h_o = 25$ Btu/hr · ft$^2$ · °F. Compute the time when the center reaches 5°F by (a) semi-infinite slab theory, and (b) finite slab theory.

4.41 An aluminum slab 3 cm thick at 20°C is suddenly immersed in a fluid

**Figure P4.37**

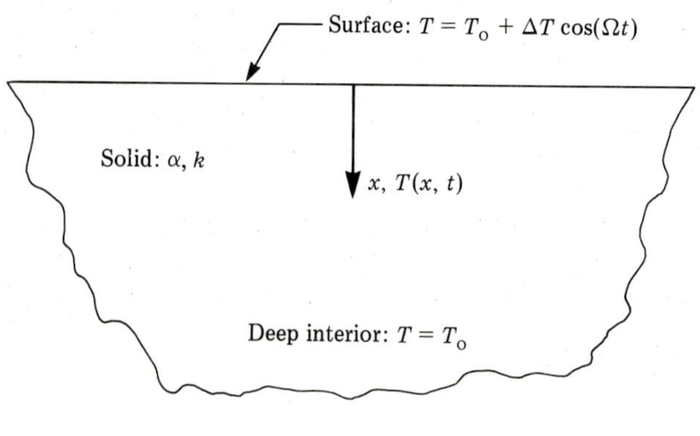

at 250°C with $h_o$ = 1500 W/m$^2$ · K. Estimate the center temperature and the surface temperature after 10 s.

4.42 A steak, considered to be a 35-mm-thick slab with the properties of water, is removed from a freezer at −18°C and thawed by exposure to air on both sides at $T$ = 20°C and $h$ = 15 W/m$^2$ · K. It is considered ready to cook when its center reaches 10°C. How long will this take?

4.43 A Nichrome slab is 20 mm thick and initially at 450°C when it is suddenly quenched in an oil bath at 20°C. After two minutes, the slab surfaces are observed to be at 50°C. Estimate the oil-bath heat transfer coefficient. What complicates this problem, especially when using the Heisler/Gröber charts?

4.44 A cast-iron slab 4 in. thick, when immersed in a fluid at $T_o$ = 100°F, cools from a uniform 400°F temperature to a center temperature of 300°F in 106 s. Estimate $h_o$ in Btu/hr · ft$^2$ · °F.

4.45 A 3-cm-diameter wooden rod at 20°C ($k$ = 0.17 W/m · K, $\alpha$ = 1.2 × 10$^{-7}$ m$^2$/s) is immersed in air at 400°C with $h_o$ = 12 W/m$^2$ · K. Estimate the time for the surface to reach ignition temperature of 250°C.

4.46 Inconel-X rods of diameter 4 cm are initially at 30°C. They are to be drawn through a 4-m-long furnace so that their center temperature is 400°C when leaving. Furnace conditions are $T$ = 600°C and $h$ = 150 W/m$^2$ · K. How fast should the rods be drawn through the furnace?

4.47 Estimate the time to cook a 5-lb$_m$ roast in a 325°F oven if initial temperature is 60°F and the desired center temperature is "medium," or 160°F. Assume a spherical shape with $h_o$ = 5 Btu/hr · ft$^2$ ·°F and roast physical properties equal to water. Compare with the estimate given in a cookbook. What is the surface temperature when cooking is complete?

4.48 Stainless steel 347 ball bearings initially at 20°C are to be heated to a center temperature of 550°C by passing on a conveyor belt through a convection oven whose air properties are $T$ = 600°C and $h$ = 225 W/m$^2$ · K. The minimum conveyor speed is 1 cm/s. What is the largest ball bearing diameter that can be treated by this process?

4.49 A cast-iron cylinder with 12-cm diameter is initially at 400°C and then cooled in air with $T_o$ = 20°C and $h_o$ = 70 W/m$^2$ · K. Estimate the center temperature after 30 min of cooling.

4.50 A 7-cm-diameter alloy steel cylinder ($k$ = 35 W/m · K and $\alpha$ = 1.2 × 10$^{-5}$ m$^2$/s) is initially at 700°F and is cooled in air with $T_o$ = 80°C and $h_o$ = 450 W/m$^2$ · K. At what time will the surface temperature reach 180°C?

4.51 A grapefruit, approximated as a 15-cm-diameter sphere with the properties of water, is at 10°C when the environment suddenly drops to −5°C with $h_o$ = 8 W/m$^2$ · K. How long will it be before (a) the surface and (b) the center reach 0°C?

4.52 Cookbooks suggest baking a potato at 450°F for 40 min. If the potato approximates a 3.5-in.-diameter sphere with the properties of water, and $h_o = 8$ Btu/hr · ft$^2$ · °F, what is the center temperature when the potato is cooked?

4.53 A carbon steel slab at 500°C is to be cooled to a center temperature of 200°C by immersion in a fluid with $T_o = 30$°C and $h_o = 120$ W/m$^2$ · K. What is the maximum thickness slab that can be handled if allowable cooling time is 20 min?

4.54 Repeat Problem 4.47, assuming that the roast is a short cylinder with length equal to its diameter.

4.55 A huge cube of aluminum at 200°C has one corner dipped in a fluid at 20°C with $h_o = 125$ W/m$^2$ · K. Estimate the corner temperature after 5 min of cooling.

4.56 Repeat Problem 4.47, assuming that the roast is a perfect cube.

4.57 Figure P4.57 shows the decay of center temperature of various bodies for a sudden temperature change, from the text by Schneider [3]. What Biot number do these curves represent? Using a one-term Fourier series, verify curve $A$ at a time $t^* = 0.6$. How are curves $B$ and $D$ derived from curve $A$? How is curve $E$ derived from curves $A$–$D$? Can curves $A$–$E$ be used to generate curve $F$?

4.58 One end of a 10-cm-diameter cast-iron cylinder at 20°C is dipped into fluid with $T_o = 200$°C and $h_o = 250$ W/m$^2$ · K. Compute the temperature of the bottom edge of this cylinder (assumed flat-ended) after 4 min.

4.59 The product-solution method of Table 4.2 does not work for truncated spheres cut by, for example, planes or cylinders. Thus sphere solutions were omitted from Table 4.2. Can you explain why the superposition scheme fails for spheres?

4.60 A long 1% carbon steel billet has a square 22-by-22-in. section and is initially at 20°C. It is heated in air at 700°C with $h_o = 85$ W/m$^2$ · K. Estimate when the center will reach 450°C.

4.61 Bricks ($k = 0.5$ W/m · K, $\alpha = 4 \times 10^{-7}$ m$^2$/s) are of size 5 by 10 by 20 cm. They are fired to 1200°C and then cooled in air with $T_o = 20$°C and $h_o = 20$ W/m$^2$ · K. Which point on the brick surface will cool the slowest? How long must we cool the bricks to ensure that all surface points reach 30°C or less?

4.62 Derive and verify the explicit formula for a two-dimensional interior mesh point given as item 3 in Table 4.3.

4.63 Derive and verify the explicit formula for a two-dimensional convection boundary point, given as item 4 of Table 4.3.

4.64 For constant $\alpha$, derive an explicit finite-difference formula, similar to Eq. (4.61), for unsteady cylindrical conduction from Eq. (2.8) with $T =$

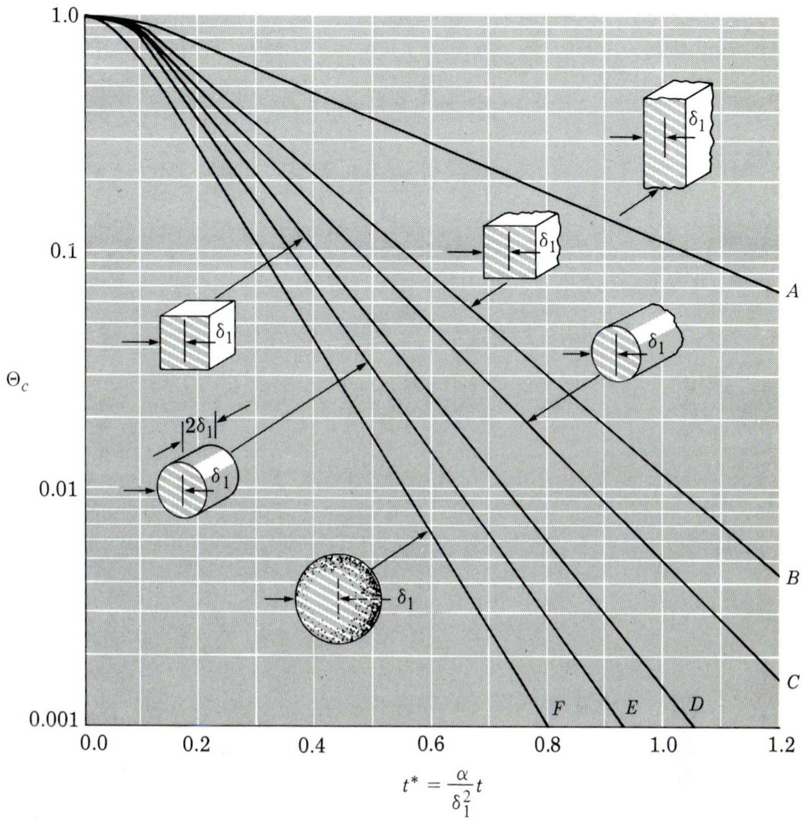


**Figure P4.57 Dimensionless center temperature in various bodies subjected to sudden surface temperature change. (Reprinted from [3, Fig. 10.8, p. 249] by permission of Addison-Wesley Publishing Co.)**

$T(r, t)$ only. Assume a uniform mesh size $\Delta r$. Does a mesh Biot number appear?

4.65 Extend Problem 4.64 to derive a formula for a two-dimensional polar coordinate interior point $T = T(r, \theta, t)$, assuming a uniform mesh size $\Delta r$ and $\Delta\theta$. Does a mesh Biot number appear?

4.66 Extend Problem 4.64 to derive a formula for an axisymmetric coordinate interior point $T = T(r, z, t)$, assuming a uniform mesh size $\Delta r = \Delta z$. Does a mesh Biot number appear?

4.67 A thick slab has surroundings of temperature $T_s$ that radiate to the slab surface, as shown in Fig. P4.67. For one-dimensional conduction, use the heat-balance approach to derive a finite-difference explicit formula for the surface nodal temperature $T_o^{j+1}$. How is the computation simplified if we approximate the surroundings by a radiation resistance as in Eq. (2.24)?

4.68 Solve Example 4.4 for the temperature at $x = 1$ cm and $t = 0.5$ hr by the explicit finite-difference formula, Eq. (4.61). Use a mesh size $\Delta x = 1$ cm with $\Delta t = 360$ s, with one boundary node and five interior nodes. Solve with a hand calculator. Compare with the exact temperature $T$(1 cm, 0.5 hr) computed from Eq. (4.26).

4.69 Solve Example 4.5 by the explicit method, Eq. (4.61), to find the surface temperature after 1 hr. Use a hand calculator with $\Delta x = 0.5$ ft, $\Delta t = 600$ s, and five interior nodes. Compare surface temperature at 1 hr with the exact result, Eq. (4.30).

4.70 A 1% carbon steel thick slab is at 20°C when it is suddenly exposed to a surface heat flux of 21,500 W/m$^2$. For $\Delta x = 5$ cm, use an explicit method to compute the temperature (a) at the surface and (b) at $x = 10$ cm, after 500 s of exposure. Compare your results with the analytical results from Section 4.3.3.

4.71 Solve Example 4.6 with an explicit numerical method, using $\Delta x = 10$ in. and $\Delta t = 2250$ s. Use a hand calculator if possible. What is the maximum allowable time step?

4.72 Solve Problem 4.39 by the explicit numerical method to find surface temperatures after 20 min, using $\Delta x = 4$ cm and $\Delta t = 300$ s. Use a hand calculator. What is $\Delta t_{max}$?

4.73 Solve Problem 4.41 by an explicit numerical method to find the center temperature after 1 s, using $\Delta x = 5$ mm and $\Delta t = 0.1$ s. What is the maximum allowable time step?

4.74 A cast-iron slab 4 in. thick ($\alpha = 1.83 \times 10^{-4}$ ft$^2$/s) is at 0°F when the

**Figure P4.67**

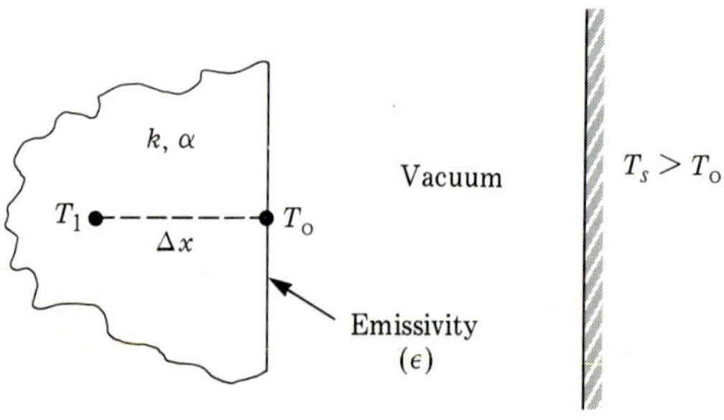

left boundary is suddenly raised to 100°F while the right boundary remains at 0°F. Using an explicit numerical method with $\Delta x = 1$ in. and $\Delta t = 15$ s, compute the interior nodal temperatures after 2 min. What will these nodal temperatures be after a very long time?

4.75 A long 3-cm-square aluminum bar ($\alpha = 9.8 \times 10^{-5}$ m$^2$/s) is at a uniform 0°C when the upper surface is suddenly raised to 100°C, as shown in Fig. P4.75. Using an explicit numerical method with $\Delta x = 1$ cm and $\Delta t = 0.25$ s, compute the four nodal temperatures after 2 s. Take advantage of symmetry if possible.

4.76 Modify Problem 4.75 to the case in which the three lower surfaces are insulated. Are additional nodes needed? Write out all nodal equations but do not solve.

4.77 Solve Problem 4.76 with a digital computer program and interpret.

4.78 Repeat Problem 4.68 by an implicit numerical method with $\Delta x = 1$ cm and $\Delta t = 1200$ s. Compute $T$ at $x = 1$ cm after 1 hr. How many interior nodes are needed to ensure the semi-infinite case?

4.79 Repeat Problem 4.69 by an implicit numerical method with $\Delta x = 0.5$ ft and $\Delta t = 1$ hr. Compute the surface temperature after 5 hr and compare with the result of Example 4.5.

4.80 Write a computer program to solve Problem 4.39 by an implicit numerical method to find the temperatures after 3 hr, using $\Delta x = 2$ cm and $\Delta t = 1200$ s. Compare with the exact results.

4.81 Using a computer program, solve Problem 4.40 by an implicit numerical method, with $\Delta x = 0.25$ in. and $\Delta t = 1$ min.

4.82 Using a computer program, solve Problem 4.74 by an implicit numerical

**Figure P4.75**

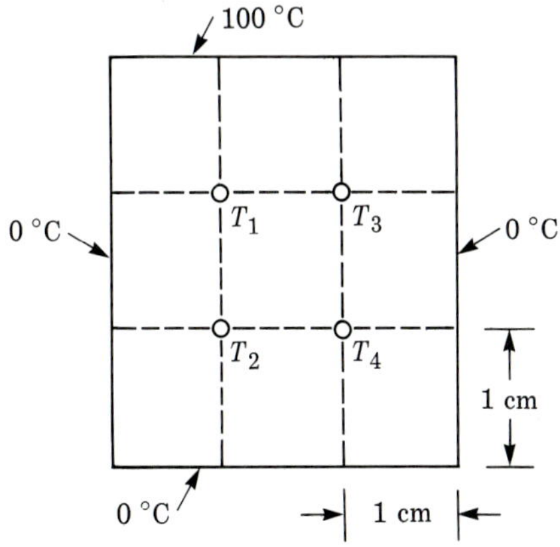

method, with $\Delta x = 1$ in. and $\Delta t = 30$ s. Run the solution out to $t = 30$ min to determine the near-steady-state condition.

4.83 Program and solve Problem 4.76 by a two-dimensional implicit numerical method, using $\Delta x = 1$ cm and $\Delta t = 1$ s. Print out all nodal temperatures up to $t = 10$ s.

## Short Design Projects

4.84 Figure P4.84 shows a proposed firewall for a safe, consisting of a slab of polystyrene sandwiched between thin metal sheets of negligible resistance. The slab is initially at 20°C, and interior convection conditions $T_i$, $h_i$ are shown. If an exterior fire should impose exterior conditions $T_o$, $h_o$ as shown, the interior metal sheet should rise to no more than 100°C after two hours. Using a numerical method with judicious choices for $\Delta x$ and $\Delta t$, find the minimum slab thickness required.

4.85 The rectangular bar in Fig. P4.85 is 10 cm by 5 cm and initially at 20°C. The upper surface is suddenly subjected to convection, and the three lower surfaces are insulated. It is desired that the temperature at point $A$ reach at least 65°C after 5 minutes. What materials in Appendices C and D are suitable for this requirement? What is the most inexpensive material for this task?

## Longer Design Project

4.86 Chrome steel bar stock of 5-cm-diameter is annealed to a uniform 400°C to remove residual stresses. It is then to be cooled to a surface temperature of 80°C by moving on a rack past a slot jet delivering air at 20°C, as shown in Fig. P4.86. The heat transfer coefficient is taken from Fig. 1.8: $\overline{h} = 8.0\ V_j^{0.8}$, with $\overline{h}$ in W/m² · K and $V_j$ in m/s. Economic analysis

**Figure P4.84**

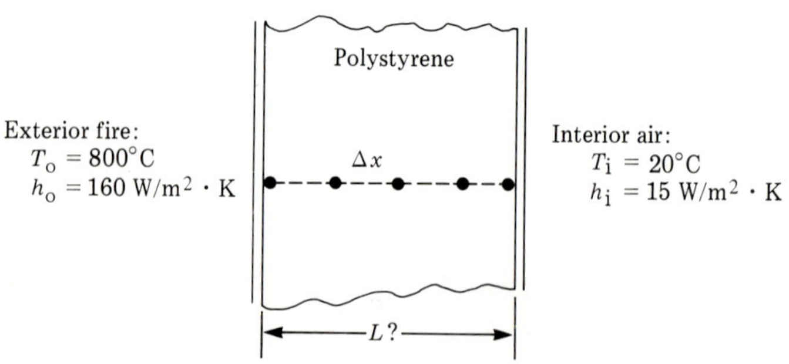

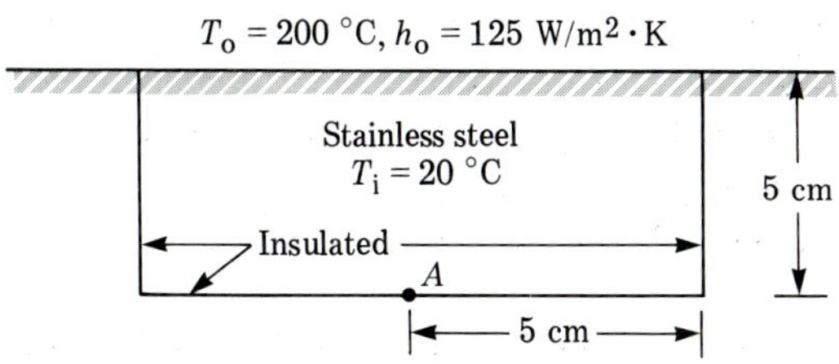


**Figure P4.85**

determines the annual cost of the system to be

$$C(\$) \doteq 45V_j^2L + 10{,}500V_rL,$$

with $V_j$ and $V_r$ in m/s and $L$ in meters. Design constraints are maximum slot length $L$ of 40 m, minimum rack speed $V_r$ of 1 cm/s, and maximum cooling time of 1 hr. Determine the minimum-cost system to meet these requirements.

**Figure P4.86**

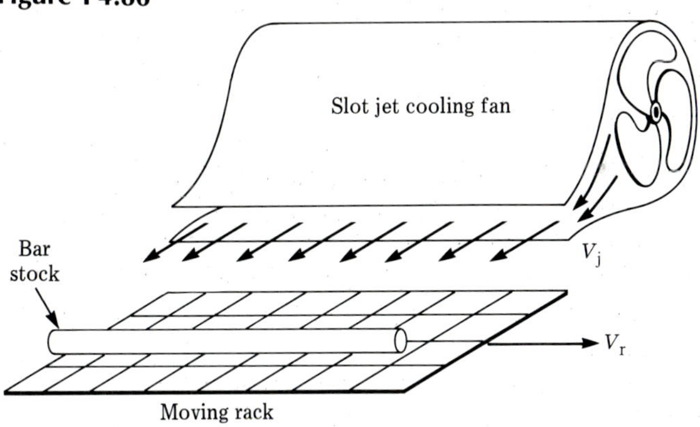

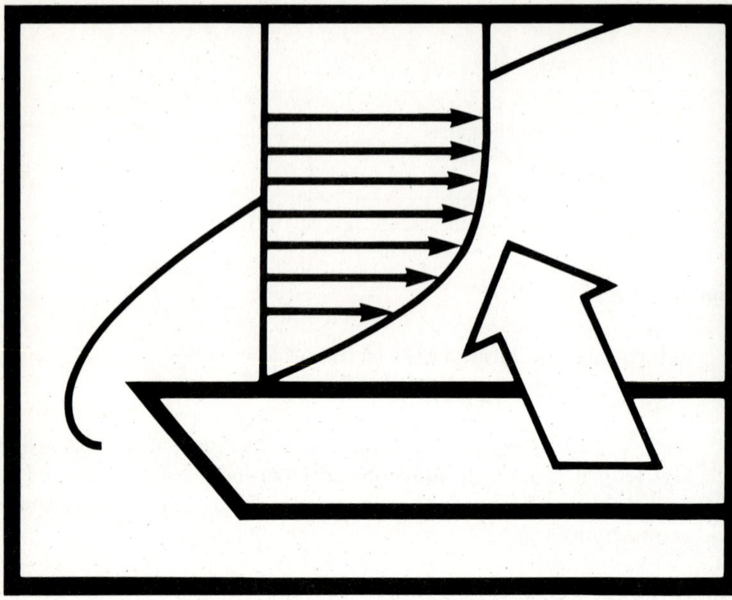

# Principles of Convection

# 5

## 5.1 Introduction

The previous three chapters have analyzed conduction problems, both steady and unsteady, in one-dimensional and multidimensional bodies. The next three chapters consider convection analysis without phase changes, which are taken up in Chapter 9.

Basically, *convection* is the study of conduction in a fluid as enhanced by its "convective transport," that is, its velocity with respect to a solid surface. It thus combines the energy equation, or first law of thermodynamics, with the continuity and momentum relations of fluid mechanics. We will need to review some of the principles of fluid mechanics as treated in undergraduate texts such as [1] and [2]. The velocity distributions of typical viscous flows — ducts and flat plates in particular — will be combined with fluid temperature distributions to predict the wall heat flux or heat transfer coefficient. Where theory is found to be inadequate, we will use dimensional analysis and experimental data to develop results.

In previous chapters we introduced convection only as a wall boundary condition for a conduction analysis. Usually this condition took the form of an assumed numerical value for the heat transfer coefficient $h$. In most cases the value was nominal or merely a guess or, at best, taken from Table 1.1. In practice, of course, $h$ cannot simply be guessed; it must be computed from a convection analysis and combined with the conduction and/or radiation modes also present in the problem. Estimating the actual value of $h$, or $q''_w$ when $h$ is not convenient, is the goal of the convection analyses to be presented in Chapters 5, 6, 7, and 9.

### 5.1.1 Types of Convection

It is customary to divide convection problems into two classes, depending on how the fluid motion arises. The two categories are forced and free convection.

In *forced convection,* the fluid has a nonzero streaming motion in the farfield away from the body surface, caused perhaps by a pump or fan or other driving force independent of the presence of the body. Two major examples are duct flows and bodies immersed in a uniform stream. The fluid streaming "forces" the convective effects near the body. Also in this class are bodies moving through a still fluid, since an observer on the body would see a streaming motion in the farfield.

In *free convection,* alternatively called natural convection, there is no farfield streaming. The fluid motion is due solely to local buoyancy differences caused by the presence of the hot or cold body surface. Most fluids near a hot wall, for example, will have their density decreased and an upward near-wall motion will be induced. Free convection velocities are relatively gentle and the resultant wall heat flux will generally be less than in forced motion. However, phase changes can greatly enhance free convection: Boiling causes highly agitated, rapidly rising bubbles, and condensation causes a dense, rapidly falling liquid film. Other effects may also appear: The melting of icebergs in seawater is enhanced by two additional mechanisms, bubble release into the water and dilution of the dense seawater by dissolving near-wall meltwater.

There is an overlap region, called *mixed convection,* where both buoyancy and forced motion effects may be important. Two examples are boiling near a surface immersed in a liquid stream and a very hot body immersed in a low-velocity gas stream (see Fig. 1.6c).

Chapter 6 treats forced convection of single-phase fluids, and Chap. 7 presents single-phase free and mixed convection analyses. Chapter 9 discusses the two-phase problems of boiling and condensation for both free and forced motions.

### 5.1.2 Internal Versus External Flow

A second way to classify convection flows is by their geometry: internal versus external. An *internal flow* is bounded on all sides by solid surfaces except possibly for an inlet and exit. The flow may be either forced or free. Examples are the flow in ducts, diffusers, nozzles, engine passages, turbomachines, or heat exchangers. Figure 5.1(a) shows a forced-convection internal flow. There is a nonzero stream velocity distribution $u(y)$ with an average value $V$, and there is a nonuniform temperature $T(y)$ with average value $T_m$. Evaluation of $V$ and $T_m$ is detailed in Chapter 6. Heat flows from the wall to be stored in the fluid as internal energy and transported down the duct. The appropriate temperature difference for specifying $h$ is $(T_w - T_m)$. Figure 5.1(b) shows a free-convection internal flow driven by differential buoyancy between a hot wall and a cold wall. There is typically an upward velocity profile near the hot wall and a nearly equal downward countercurrent near the cold wall. Heat flows across the fluid from the hot to the cold wall, and the appropriate temperature difference is $(T_1 - T_2)$.

In contrast, Figs. 5.1(c,d) show external forced- and free-convection flow, respectively. Conditions in the "freestream" are assumed known:

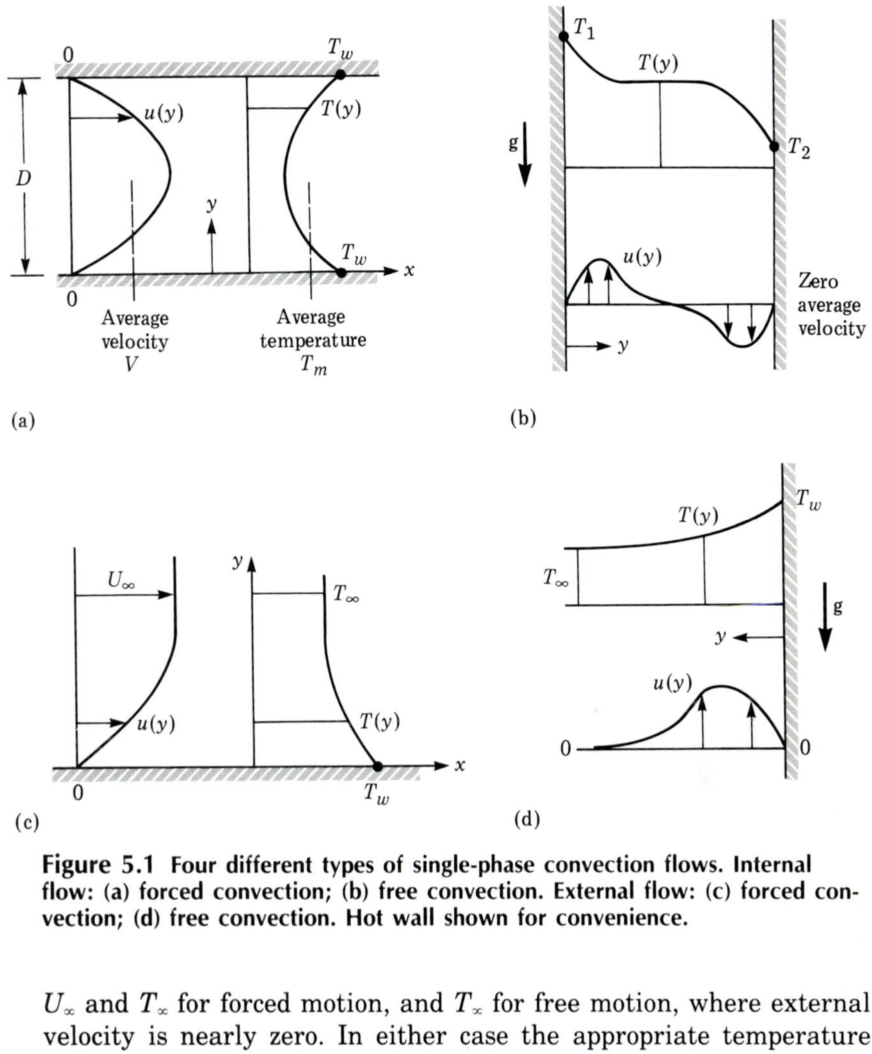


**Figure 5.1 Four different types of single-phase convection flows. Internal flow: (a) forced convection; (b) free convection. External flow: (c) forced convection; (d) free convection. Hot wall shown for convenience.**

$U_\infty$ and $T_\infty$ for forced motion, and $T_\infty$ for free motion, where external velocity is nearly zero. In either case the appropriate temperature difference is $(T_w - T_\infty)$, and there is no need to compute average velocity or temperature of the near-wall "boundary layer" profiles $u(y)$ and $T(y)$.

Generally speaking, internal flows are slightly easier to analyze for convection heat transfer, because the fluid is strictly bounded by the confining walls. However, internal flow requires computation of integrated average velocity and temperature, which is not necessary for external flow. As discussed, the two types of geometry lead to different definitions for the heat transfer coefficient, $h$:

*Internal flow:*

$$q''_w = h(T_w - T_m), \tag{5.1a}$$

*External flow:*

$$q''_w = h(T_w - T_\infty). \tag{5.1b}$$

Also, there are two different velocity scales: Average velocity $V$ for internal flow and freestream velocity $U_\infty$ for external flow. The reader is asked to accept these differences.

In all convection problems the goal is to find the temperature distribution in the fluid as a function of geometry, flow conditions, and fluid properties. Then the wall heat transfer is found by evaluating Fourier's law in the fluid at the wall. For example, in all four cases in Fig. 5.1, once $T(y)$ is known, we may compute

$$q''_w = -k_w \left.\frac{\partial T}{\partial y}\right|_{y=0}, \tag{5.2}$$

after which $h$ may be computed from Eqs. (5.1). Of course, $T(y)$ is different in each case in Fig. 5.1 because of flow and geometry differences. If $T(y)$ cannot be found by analysis or digital computer simulation, then $q''_w$ is measured experimentally.

## Example 5.1

In flow of helium past a flat cold surface, measurements of helium temperature near the wall, as shown in Fig. E5.1, may be fit by the empirical formula

$$T \doteq 300 + 25 \tanh(92\, y), \tag{1}$$

with $T$ in degrees K and $y$ in meters. Estimate (a) the wall heat transfer and (b) the heat transfer coefficient in SI units.

*Solution* This distribution fits case (c) in Fig. 5.1, except that the wall is cold. At $y = 0$, $T = 300$K, which we interpret as $T_w$. As $y$ gets large, $T$ approaches 325K, which we interpret as "freestream" temperature $T_\infty$.

From Eq. (1) the wall temperature gradient is

$$\left.\frac{\partial T}{\partial y}\right|_{y=0} = 25(92)\,\text{sech}^2(92\, y)_{y=0} = 2300\text{K/m}.$$

From Appendix G for helium at 300K, read $k_w \doteq 0.15$ W/(m · K).

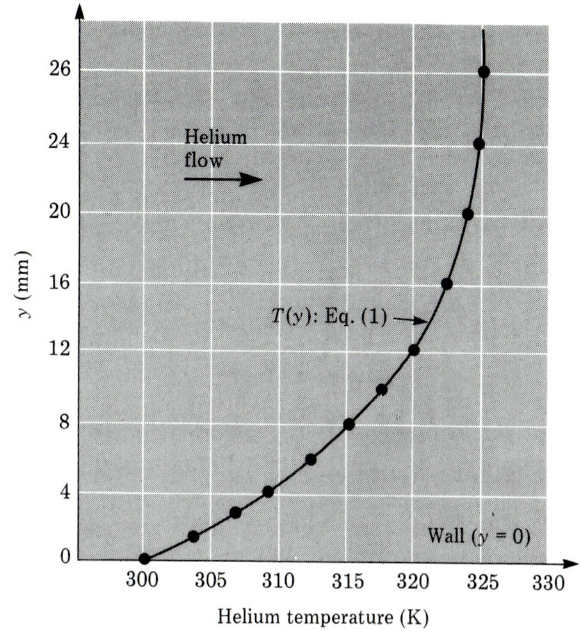


**Figure E5.1**

Then, from Eq. (5.2), the wall heat transfer is

$$q''_w = -k_w \frac{\partial T}{\partial y}\bigg|_{y=0} = -\,0.15\,\frac{\text{W}}{\text{m}\cdot\text{K}}\left(2300\,\frac{\text{K}}{\text{m}}\right).$$

$$= -345\ \text{W/m}^2 \quad (\textit{Ans. a})$$

The negative sign means that $q''_w$ is down, into the wall, since the helium stream is hotter than the wall. This is an *external* flow. We may compute the measured heat transfer coefficient from Eq. (5.1b):

$$q''_w = -345\ \text{W/m}^2 = h(T_w - T_\infty) = h(300 - 325\text{K})$$

or

$$h = \frac{345}{25} \doteq 13.8\ \text{W/(m}^2\cdot\text{K)}. \quad (\textit{Ans. b})$$

Uncertainty in heat transfer measurements is high (recall Section 1.8). After evaluating all sources of error in this (postulated) experiment, the researcher may actually report this result as $h = 14 \pm 3\ \text{W/(m}^2\cdot\text{K)}$.

*Remarks* The heat transfer coefficient is always positive. It is the negative temperature difference that causes $q''_w$ to be negative.

From Fig. E5.1 the thermal boundary layer thickness is about 24 mm. This seems small; actually, air and water would probably have even thinner layers for the same flow conditions because they have smaller kinematic viscosities.

Note that flow velocity did not enter directly into the computation. However, $T(y)$ depends implicitly on velocity, and hence its wall gradient will vary with flow conditions (see Chapter 6 for several examples).

Accurate fluid temperature measurements are difficult and delicate in such a thin layer. For details of the types of instrumentation used, see the monograph edited by Eckert and Goldstein [11]. ■

---

### 5.1.3 Laminar Versus Turbulent Flow

For a given geometry and temperature difference, the most important parameter that determines the character of a flow is the dimensionless Reynolds number. As discussed in fluid mechanics texts [1–4], flow at low Reynolds numbers is *laminar,* with smooth streamlines and shear and conduction effects owing entirely to the fluid's molecular viscosity and conductivity. Dye streaks and velocity measurements would be very smooth, as shown in Fig. 5.2(a).

At some finite Reynolds number, usually called the *transition point,* laminar flow is unstable and the flow begins to generate "bursts" of random turbulent fluctuations, as shown in Fig. 5.2(b). The dye streak becomes erratic and subject to large lateral fluctuations. At a slightly larger Reynolds number, the flow is *fully turbulent,* as in Fig. 5.2(c). The turbulent fluctuations are continuous in velocity, pressure, and temperature, and a dye streak rapidly expands to fill the entire duct. Turbulent flows undergo intense mixing because of the superimposed rapid fluctuations; hence both friction and heat transfer are generally much higher than in laminar flow. As expected from our earlier discussion, the Reynolds number is defined differently for internal (Fig. 5.1a) versus external (Fig. 5.1c) flow:

*Internal flow:*

$$\mathrm{Re} = \mathrm{Re}_D = VD/\nu$$

*External flow:*

$$\mathrm{Re} = \mathrm{Re}_x = U_\infty x/\nu$$

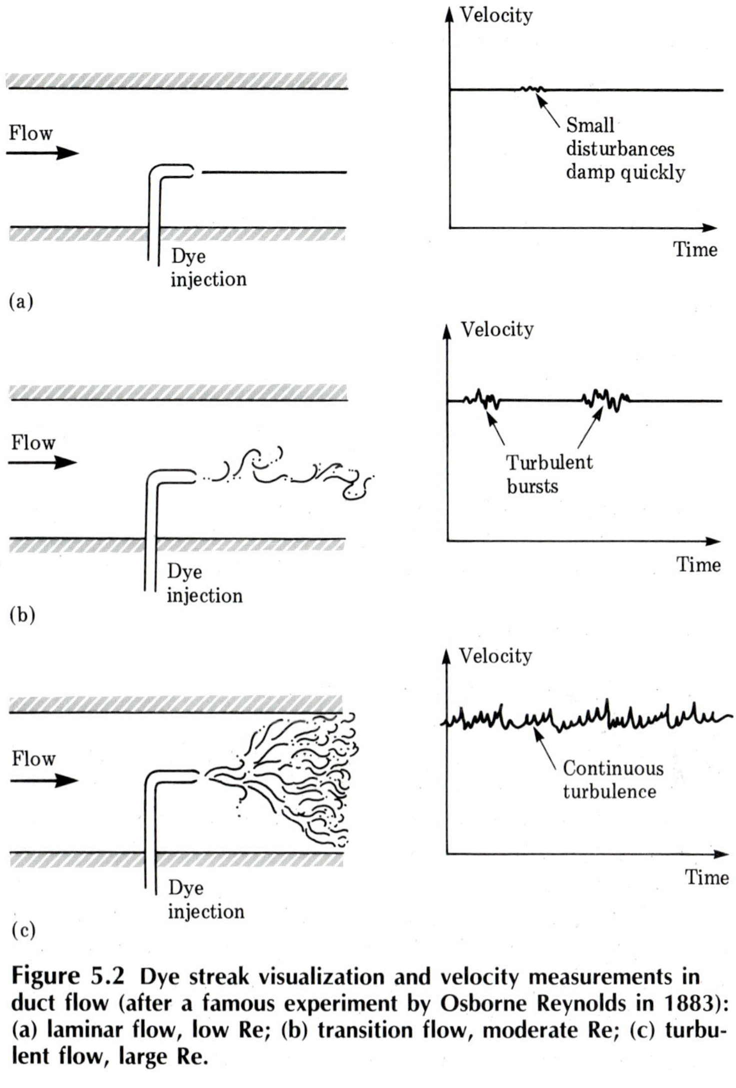


**Figure 5.2 Dye streak visualization and velocity measurements in duct flow (after a famous experiment by Osborne Reynolds in 1883): (a) laminar flow, low Re; (b) transition flow, moderate Re; (c) turbulent flow, large Re.**

where $\nu = \mu/\rho$ is the fluid kinematic viscosity. Again we ask the reader to be aware of these differences.

The Reynolds numbers at which transition flow and fully turbulent flow occur are dependent on several parameters: geometry, surface roughness, wall temperature, fluctuations in the freestream, and streamwise pressure gradient. For the common case of flow through a smooth circular pipe of diameter $D$, the accepted experimental values are

$$\begin{aligned} \mathrm{Re} &\simeq 2300 \quad \text{(transition)} \\ &\simeq 4000 \quad \text{(fully turbulent flow).} \end{aligned} \tag{5.3}$$

More details on the phenomenon of transition to turbulence may be found in advanced texts on viscous flow, such as [3] and [4].

**Example 5.2**

If the average velocity is 175 cm/s, what size of smooth circular tube will ensure that the flow is fully turbulent if the fluid is (a) air and (b) water at 1 atm and 20°C?

*Solution* The fluid kinematic viscosities are $1.57 \times 10^{-5}\ \mathrm{m^2/s}$ for air and $1.01 \times 10^{-6}\ \mathrm{m^2/s}$ for water for these conditions. From Eq. (5.3), fully turbulent flow occurs at about $VD/\nu = 4000$. Therefore, the pipe size should be $D \geqslant 4000\nu/V$. For the two given fluids, we predict:

*a) Air:*

$D \geqslant 4000(1.57 \times 10^{-5}\ \mathrm{m^2/s})/(1.75\ \mathrm{m/s}) = 0.036\ \mathrm{m}$, [*Ans. (a)*]

*b) Water:*

$D \geqslant 4000(1.01 \times 10^{-6}\ \mathrm{m^2/s})/(1.75\ \mathrm{m/s}) = 0.0023\ \mathrm{m}$. [*Ans. (b)*]

The required water pipe size is especially small: Most practical water flow problems result in fully turbulent flow. ■

### 5.1.4 Practical Convection Analysis

Although most of the convection theory in this text is for simple geometries such as circular tubes and flat plates, practical systems are often more complex. Consider the commercial fan-driven oil cooler shown in Fig. 5.3. The fan blows air over a set of finned tubes through which oil is pumped. Inside the tubes are so-called turbulator grooves, which help promote convection by swirling and agitating the flow.

Both air and oil flow, being driven by turbomachinery, are "forced" motions with a nonzero streaming velocity. Free convection is probably small but should be checked for the given temperature differences (see Chapter 7). There is little theory for external flow over tube bundles, even for the unfinned case. Experimental data are needed, especially for finned tubes. There is theory (see Chapter 6) for internal (oil) flow in smooth tubes, but not with turbulator grooves. More

**Figure 5.3** A practical convection apparatus: fan-driven oil cooler with air-flow up to 2.45 $m^3/s$ and oil flow up to 6300 $cm^3/s$. The oil flows through finned tubes containing "turbulator" convection enhancers. (Courtesy of Young Radiator Co.)

convection data are needed. Although it is possible, with the methods of Chapter 3, to analyze the conduction resistance through the grooves, tube walls, and fins, the geometry is complex and data may also be needed.

This example is intended to show that convection analysis relies heavily on empirical data for both internal and external flows, especially for practical devices such as that shown in Fig. 5.3. The theory, when successful, is based on the basic differential equations to be developed in Section 5.2. The experimental data should be organized with the aid of the dimensional analysis techniques to be discussed in Section 5.3. Because the theory is not as well developed as in conduction analysis (see Chapters 2–4), the field of convection abounds with empirical formulas to fit various flow conditions and geometries.

## 5.2 The Boundary Layer Equations

Most of the existing theory for convection heat transfer is based on the "boundary layer" equations for flow past a solid surface at high Reynolds numbers. These equations may be used for either internal or external flow and for laminar or turbulent flow. We limit our analyses in this text to two-dimensional, steady flow of an incompressible or nearly constant density fluid. A brief discussion of compressibility (very high speed) effects is given in Section 6.8. Further details on the analysis of three-dimensional, unsteady, or compressible flows are given in advanced texts [3–6]. Our treatment of turbulent-flow analysis is rather limited, and the reader is invited to consult [3–6] for more details. There is also a growing literature on digital computer modeling of viscous flow and heat transfer [7, 8, 12], which will not be treated here.

### 5.2.1 Boundary Layer Mass and Momentum Balances

A typical external boundary layer flow is shown in Fig. 5.4(a). The no-slip and no-temperature-jump conditions require that the fluid velocity be zero and the fluid temperature equal wall temperature at the body surface. This means that regions of high shear and heat flux will be developed in the flow near the surface: the "boundary layers." As shown in the figure, shear or velocity-gradient effects are confined to a velocity boundary layer of thickness $\delta$, much smaller than distance $x$ along the surface. Similarly, heat flux or temperature-gradient effects are confined to a thermal boundary layer of thickness $\Delta \ll x$. The relative values of these layer thicknesses depend on flow conditions and fluid properties (see Section 6.2). Detailed sketches of typical velocity and temperature distributions within a boundary layer are shown in Fig. 5.1(c).

The only requirement for these boundary layers to be thin, with both $\delta$ and $\Delta$ much smaller than $x$, is that the Reynolds number of the flow be large:

$$\mathrm{Re}_x = U_0 x/\nu \geqslant 1000, \tag{5.4}$$

where $U_0$ is the approach velocity of the oncoming stream. We have seen from Example 5.1 that condition (5.4) is generally met by flows of gases and light liquids such as water. Now let us develop some differential equations that relate the velocity and pressure distributions in a boundary layer.

Consider a differential element in the boundary layer as sketched in Fig. 5.4(b). Let the element width be $b$ as measured into the paper.

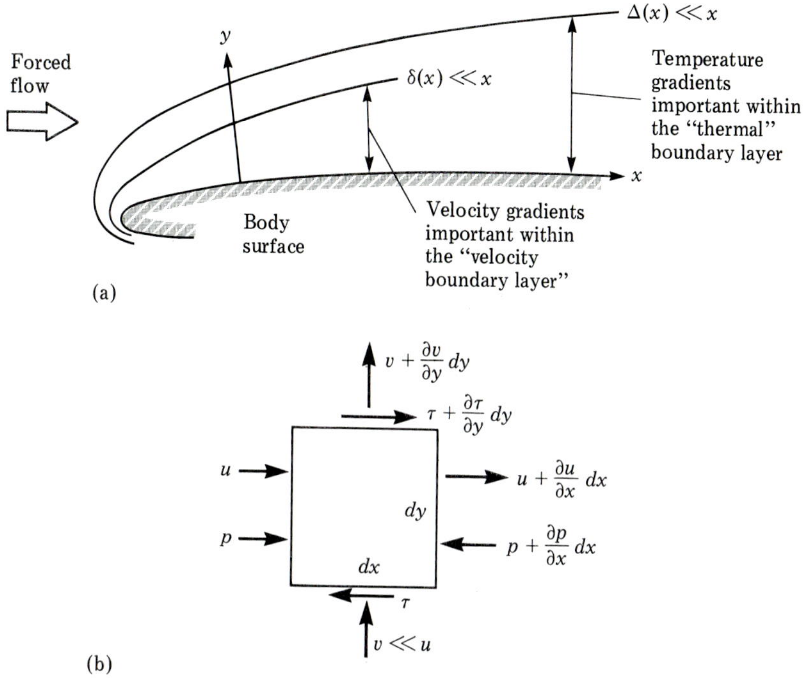


**Figure 5.4** Sketch of a two-dimensional boundary layer flow: (a) velocity and thermal boundary layer regions; (b) mass and momentum balance on a differential element within the boundary layer.

Assume that the fluid density $\rho$ is nearly constant. Then, since the flow is steady, the mass flow entering the element on the left and bottom sides must equal the mass flowing out at top and at right:

$$\rho u\, bdy + \rho v\, bdx = \rho\left(u + \frac{\partial u}{\partial x}dx\right)bdy + \rho\left(v + \frac{\partial v}{\partial y}dy\right)bdx. \quad \textbf{(5.5)}$$

After canceling out terms proportional to $u$ and $v$ and dividing by the element mass ($\rho bdxdy$), we obtain the basic *continuity* or conservation of mass relation for steady incompressible flow in two dimensions:

$$\frac{\partial u}{\partial x} + \frac{\partial v}{\partial y} = 0. \quad \textbf{(5.6)}$$

In a boundary layer, then, a mass balance requires that streamwise changes in the streamwise velocity $u$ must be accompanied by an equal and opposite transverse change in the transverse velocity $v$.

Equation (5.6) is valid for laminar or turbulent flow, with $u$ and $v$ being interpreted as time-averaged velocities in the latter case. Note that pressure, density, or viscosity do not enter directly into the balance implied by Eq. (5.6).

Now consider a balance of streamwise or $x$-directed forces on the element in Fig. 5.4(b). Since the flow is steady, the only acceleration of the element is its "convective" $x$-directed acceleration, $u(\partial u/\partial x) + v(\partial u/\partial y)$.† We may write Newton's second law for the forces on the element, with applied forces owing to pressure, shear stress, and gravity:

$$\sum(dF_x) = dm\left(u\frac{\partial u}{\partial x} + v\frac{\partial u}{\partial y}\right),$$

or

$$\begin{aligned} &pbdy - \tau bdx + \rho g_x bdxdy - \left(p + \frac{\partial p}{\partial x}dx\right)bdy \\ &+ \left(\tau + \frac{\partial \tau}{\partial y}dy\right)bdx = (\rho bdxdy)\left(u\frac{\partial u}{\partial x} + v\frac{\partial u}{\partial y}\right), \end{aligned} \tag{5.7}$$

where $g_x$ is the component of gravity in the $x$ direction. Canceling terms proportional to $p$ and $\tau$ and dividing by the element volume $(bdxdy)$, we obtain the basic differential equation for streamwise momentum in a boundary layer:

$$\rho\left(u\frac{\partial u}{\partial x} + v\frac{\partial u}{\partial y}\right) = -\frac{\partial p}{\partial x} + \rho g_x + \frac{\partial \tau}{\partial y}. \tag{5.8}$$

We have neglected the viscous normal stresses on the left and right faces, which are extremely small in a boundary layer.

If we were to write the force balance in the $y$ direction (normal to the surface) we would find that the pressure gradient normal to the wall is negligibly small [3, 4]:

$$\frac{\partial p}{\partial y} \ll \frac{\partial p}{\partial x}. \tag{5.9}$$

Thus, in boundary layer flow, $p \simeq p(x)$ only, and the pressure is assumed known from flow conditions. In external flow, for example, $p(x)$ is computed from the inviscid velocity distribution $U_\infty(x)$ outside the boundary layer:

*External flow:*

$$\frac{\partial p}{\partial x} \simeq -\rho U_\infty \frac{dU_\infty}{dx}. \tag{5.10}$$

†The reader should review this concept in [1] or [2].

Or, in internal (duct) flow, knowledge of inlet and exit pressures is usually sufficient to specify the gradient:

*Internal flow:*

$$\frac{\partial p}{\partial x} \simeq [p(L) - p(0)]/L, \tag{5.11}$$

where $L$ is the length of the duct.

With pressure $p(x)$ known, the unknowns in Eqs. (5.6) and (5.8) are the velocities $u$ and $v$ and the shear stress $\tau$. To complete the analysis, we need a relation between $\tau$ and velocity, which depends on whether the flow is laminar or turbulent.

### 5.2.2 Laminar Versus Turbulent Flow

In either laminar or turbulent flow, the shear stress is related to the fluid shear strain rate $d\gamma/dt$, which for two-dimensional flow is given by [1, 2]:

$$\frac{d\gamma}{dt} = \frac{\partial u}{\partial y} + \frac{\partial v}{\partial x}. \tag{5.12}$$

For boundary layer flow, the second term $(\partial v/\partial x)$ is neglected.

For laminar flow of a Newtonian fluid, shear stress is proportional to strain rate:

$$\tau_{\text{lam}} = \mu \frac{\partial u}{\partial y} = \rho \, \nu \frac{\partial u}{\partial y}, \tag{5.13}$$

where $\mu$ is the coefficient of viscosity. In turbulent flow, the rapid fluctuations cause an additional "turbulent" shear that can be correlated by an empirical function called the *eddy* viscosity, $\varepsilon_M$:

$$\tau = \tau_{\text{lam}} + \tau_{\text{turb}} = \rho(\nu + \varepsilon_M) \frac{\partial u}{\partial y}. \tag{5.14}$$

The eddy viscosity $\varepsilon_M >> \nu$ except very near the wall. It is *not* a true viscosity coefficient but rather varies with geometry and Reynolds number. A suggested formula is given in Section 6.4.

Substituting into the momentum relation, Eq. (5.8), we obtain a differential equation without shear stress as a variable:

$$u \frac{\partial u}{\partial x} + v \frac{\partial u}{\partial y} = -\frac{1}{\rho}\frac{\partial p}{\partial x} + g_x + \frac{\partial}{\partial y}\left[(\nu + \varepsilon_M)\frac{\partial u}{\partial y}\right], \tag{5.15}$$

to be solved simultaneously with Eq. (5.6) for $u$ and $v$. If the flow is laminar, set $\varepsilon_M$ equal to zero.

### 5.2.3 Boundary Conditions

Equations (5.6) and (5.15) are second order in $u$ and first order in $v$ and therefore require three boundary conditions, two on $u$ and one on $v$. These are the no-slip condition at the wall,

$$\text{At } y = 0:\quad u = v = 0, \tag{5.16}$$

and a smooth patching to the streamwise velocity outside the velocity boundary layer:

$$\text{As } y \to \infty:\quad u \to U_\infty(x) \quad \text{known.} \tag{5.17}$$

Assuming that the turbulent eddy viscosity $\varepsilon_M$ is a well-modeled smooth function of geometry and flow properties (see Section 6.4.1 for a suggested correlation), Eqs. (5.6, 5.15–5.17) are a well-posed set of partial differential equations and boundary conditions for which many analytical solutions or numerical computations are known, as detailed in [3–8]. In the present text we confine ourselves to two basic analyses: the flat plate boundary layer and fully developed pipe flow. We will consider both laminar and turbulent flow solutions.

### 5.2.4 The Boundary Layer Energy Equation

A third and final differential equation for the boundary layer is developed by applying the first law of thermodynamics,

$$\frac{dE}{dt} = \frac{dQ}{dt} + \frac{dW}{dt}, \tag{5.18}$$

to a fluid element, with $Q$ the heat added to the element and $W$ the work done on the element. Recall that the first law was also used to develop the basic conduction equation in Chapter 3.

Consider the element of size $b$ by $dx$ by $dy$ in Fig. 5.5. The total energy of this element is $dm(e + \frac{1}{2}V^2)$, where $dm = \rho b dx dy$. Its time rate of change is

$$\frac{dE}{dt} = \frac{\partial}{\partial x}\{\rho u(e + \tfrac{1}{2}V^2)\}b\,dx\,dy + \frac{\partial}{\partial y}\{\rho v(e + \tfrac{1}{2}V^2)\}b\,dx\,dy$$

for steady flow ($\partial/\partial t \equiv 0$). The total heat added to the element is due to heat generation $\dot{q}$ per unit volume (as in Fig. 2.11) plus the net heat conduction from Fig. 5.5(a):

$$\frac{dQ}{dt} = \dot{q}\, b\,dx\,dy - \frac{\partial q''}{\partial x}b\,dx\,dy - \frac{\partial q''}{\partial y}b\,dx\,dy.$$

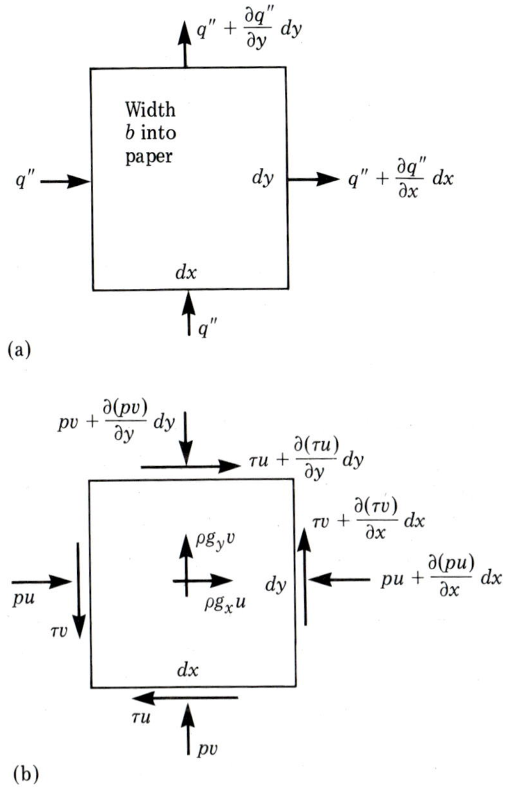


**Figure 5.5** Energy balances on a two-dimensional fluid element: (a) heat conduction terms; (b) pressure, shear, and gravity work terms.

Finally, the net work done on the element is due to the pressure, shear, and gravity terms shown in Fig. 5.5(b):

$$\frac{dW}{dt} = -\left[\frac{\partial(pu)}{\partial x} + \frac{\partial(pv)}{\partial y}\right] bdxdy + \left[\frac{\partial(\tau v)}{\partial x} + \frac{\partial(\tau u)}{\partial y}\right] bdxdy$$
$$+ (g_x u + g_y v)\,\rho bdxdy.$$

Combining these three terms into the first law, Eq. (5.18), and dividing by the element volume, we obtain a rather general form of the energy equation for fluid flow:

$$\frac{\partial}{\partial x}\{\rho u(e + \tfrac{1}{2}V^2)\} + \frac{\partial}{\partial y}\{\rho v(e + \tfrac{1}{2}V^2)\} = \dot{q} - \frac{\partial q''}{\partial x} - \frac{\partial q''}{\partial y}$$
$$+ \rho(g_x u + g_y v) - \frac{\partial(pu)}{\partial x} - \frac{\partial(pv)}{\partial y} + \frac{\partial(\tau v)}{\partial x} + \frac{\partial(\tau u)}{\partial y}, \tag{5.19}$$

where $V^2 = u^2 + v^2$. The only terms neglected are the viscous normal stresses [3, 4].

Equation (5.19) may be reduced to a boundary layer relation by (a) using the continuity and momentum relations to eliminate terms such as $(\partial p/\partial y)$ and $(\rho g_x)$, and (b) neglecting streamwise gradients, which are much smaller than transverse or cross-stream gradients in a boundary layer, for example, $(\partial q''/\partial x) \ll (\partial q''/\partial y)$. The result is the boundary layer energy equation:

$$\rho u \frac{\partial i}{\partial x} + \rho v \frac{\partial i}{\partial y} = \dot{q} - \frac{\partial q''}{\partial y} + u \frac{\partial p}{\partial x} + \tau \frac{\partial u}{\partial y}. \tag{5.20}$$

Here we have introduced the fluid enthalpy for convenience:

$$i = e + p/\rho \tag{5.21}$$

and, for the theories in Chapters 6 and 7, we will make the approximation that $di = c_p dT$, where $c_p$ is the specific heat at constant pressure.

In solving Eq. (5.20), the shear stress $\tau$ is taken from Eq. (5.14) and, in like manner, the heat flux $q''$ may be a combination of laminar and turbulent or "eddy" conduction terms:

$$q'' = -(k + \rho c_p \varepsilon_H) \frac{\partial T}{\partial y}, \tag{5.22}$$

where $\varepsilon_H$ is the "eddy" conductivity owing to the net effect of turbulent velocity and temperature fluctuations. A semi-empirical correlation for $\varepsilon_H$ is given in Section 6.4.

In most of our analysis here we neglect heat generation and pressure gradient effects in Eq. (5.20), so that the final desired form of the boundary layer energy equation is

$$\rho c_p \left( u \frac{\partial T}{\partial x} + v \frac{\partial T}{\partial y} \right) = \frac{\partial}{\partial y} \left[ (k + \rho c_p \varepsilon_H) \frac{\partial T}{\partial y} \right] + (\mu + \rho \varepsilon_M) \left( \frac{\partial u}{\partial y} \right)^2. \tag{5.23}$$

For laminar flow, set $\varepsilon_M = \varepsilon_H = 0$. The last term in Eq. (5.23) is called the *dissipation* term and, as we shall see, is rather small unless the fluid is very viscous or has a high kinetic energy relative to fluid enthalpy variations.

Equation (5.23) is second order in temperature and thus needs two boundary conditions. One is the no-temperature-jump condition at the wall:

$$\text{At } y = 0: \quad T = T_w(x) \quad \textit{known}. \tag{5.24}$$

The second is a smooth patching of temperature to the outer layer:

*As* $y \to \infty$:

$$T \to T_\infty(x) \qquad \textit{known.} \tag{5.25}$$

Again, if $\varepsilon_H$ is a smooth function, the thermal energy problem is well posed and many boundary layer solutions are known [3–7]. Note that solution for $T(x, y)$ depends on knowledge of the velocity fields $u(x, y)$ and $v(x, y)$ from Eqs. (5.6) and (5.15). Thus, in all convection problems, temperature is "coupled" to the velocity field. By inspection of Eq. (5.15), however, we see that velocity is *uncoupled* from temperature unless (a) the density variations produce significant buoyancy terms (free convection), or (b) the viscosity is a strong function of temperature. It is this uncoupling of velocity that enables us to study fluid mechanics without worrying about — or even mentioning, sometimes — the effect of temperature variations. The opposite, of course, is clearly not true: Convection *always* depends strongly on the velocity field.

### 5.2.5 Summary of the Basic Flow Equations

This section has been quite long but also quite necessary. We need to know the basic equations of flow and convection in order to solve for heat transfer relations in cases of simple geometry where accurate theoretical results are known. Let us summarize the basic boundary layer equations here for two-dimensional steady flow:

*Continuity:*

$$\frac{\partial u}{\partial x} + \frac{\partial y}{\partial y} = 0. \tag{5.6}$$

*Momentum:*

$$u\frac{\partial u}{\partial x} + v\frac{\partial u}{\partial y} = -\frac{1}{\rho}\frac{dp}{dx} + g_x + \frac{\partial}{\partial y}\left[(\nu + \varepsilon_M)\frac{\partial u}{\partial y}\right]. \tag{5.15}$$

*Energy:*

$$\rho c_p\left(u\frac{\partial T}{\partial x} + v\frac{\partial T}{\partial y}\right) = \frac{\partial}{\partial y}\left[(k + \rho c_p \varepsilon_H)\frac{\partial T}{\partial y}\right] + (\mu + \rho\varepsilon_M)\left(\frac{\partial u}{\partial y}\right)^2. \tag{5.23}$$

These are partial differential equations to be solved for $u(x, y)$, $v(x, y)$, and $T(x, y)$, subject to the following boundary conditions:

*At the wall,* $y = 0$: $\quad u = v = 0; \quad T = T_w$.

*Freestream,* $y \to \infty$: $\quad u \to U_\infty; \quad T \to T_\infty$.

For laminar flow, set the turbulent "eddy" terms $\varepsilon_M$ and $\varepsilon_H$ equal to zero in Eqs. (5.15) and (5.23).

Many different analyses and discussions of these equations are given in advanced texts on convection [3–6, 12–14]. Here we will solve only for forced motion in pipe flow and flat plate flow (Chapter 6) and for free convection past a vertical flat plate (Chapter 7).

### 5.2.6 The Compressible Navier-Stokes Equations

The equations developed in Section 5.2 are for incompressible boundary layer flow—that is, flow in the thin shear and thermal regions that occur in low-speed, high-Reynolds-number motion near a solid wall. Certain terms were negligible: (1) density variations, (2) pressure gradient normal to the wall, and (3) shear and conduction gradients parallel to the wall.

The full equations of motion, which account for all property variations, are treated in advanced texts [3–6, 12–14] and are not derived here. For completeness we include here the full equations for steady compressible laminar flow.

The continuity or mass-conservation equation is

$$\frac{\partial}{\partial x}(\rho u) + \frac{\partial}{\partial y}(\rho v) = 0. \tag{5.26}$$

The full $x$- and $y$-momentum equations are traditionally called the *Navier-Stokes equations*, after C. L. M. H. Navier (1823) and G. G. Stokes (1845), who first developed them:

$$\begin{aligned} \rho\left(u\frac{\partial u}{\partial x} + v\frac{\partial u}{\partial y}\right) &= -\frac{\partial p}{\partial x} + \rho g_x + \frac{\partial \tau_{xx}}{\partial x} + \frac{\partial \tau_{xy}}{\partial y}, \\ \rho\left(u\frac{\partial v}{\partial x} + v\frac{\partial v}{\partial y}\right) &= -\frac{\partial p}{\partial y} + \rho g_y + \frac{\partial \tau_{xy}}{\partial x} + \frac{\partial \tau_{yy}}{\partial y}, \end{aligned} \tag{5.27}$$

where

$$\begin{aligned} \tau_{xx} &= 2\mu\frac{\partial u}{\partial x} + \lambda\left(\frac{\partial u}{\partial x} + \frac{\partial v}{\partial y}\right), \\ \tau_{yy} &= 2\mu\frac{\partial v}{\partial y} + \lambda\left(\frac{\partial u}{\partial x} + \frac{\partial v}{\partial y}\right), \\ \tau_{xy} &= \mu\left(\frac{\partial v}{\partial x} + \frac{\partial u}{\partial y}\right). \end{aligned}$$

The complete two-dimensional energy equation is

$$\rho\left(u\frac{\partial i}{\partial x} + v\frac{\partial i}{\partial y}\right) = \dot{q} + \frac{\partial}{\partial x}\left(k\frac{\partial T}{\partial x}\right) + \frac{\partial}{\partial y}\left(k\frac{\partial T}{\partial y}\right) + u\frac{\partial p}{\partial x} + v\frac{\partial p}{\partial y} + \Phi, \tag{5.28}$$

where

$$\Phi = \tau_{xx}\frac{\partial u}{\partial x} + \tau_{yy}\frac{\partial v}{\partial y} + \tau_{xy}\left(\frac{\partial v}{\partial x} + \frac{\partial u}{\partial y}\right) + \lambda\left(\frac{\partial u}{\partial x} + \frac{\partial v}{\partial y}\right)^2.$$

The quantity $\lambda$, called the *second coefficient of viscosity*, is rarely important in fluid flows of engineering interest, hence its absence in elementary textbooks.

Note that the energy relation, Eq. (5.28), is the extension of the solid-body conduction equation, Eq. (2.6), to include fluid convection, pressure work, and viscous dissipation effects.

By analogy with Eqs. (5.16, 5.17, 5.24, 5.25) the boundary conditions for the full equations are (1) no-slip and no-temperature-jump at any solid surface and (2) known velocity, temperature, and pressure at any inlet or exit to the flow. In general, the Bernoulli relation (Eq. 5.10) is *not* used in a full Navier-Stokes analysis because viscous and heat conduction effects make it invalid.

### Example 5.3

For laminar flow, compare the dissipation and conduction terms in the energy relation, Eq. (5.23), assuming constant $k$ and $\mu$. Assume equal thermal and velocity boundary layer thicknesses, $\delta \simeq \Delta$. Show that dissipation is negligible unless the fluid is very viscous or has large kinetic energy.

*Solution* For these conditions the ratio of dissipation to conduction is

$$\lambda = \frac{\mu(\partial u/\partial y)^2}{k(\partial^2 T/\partial y^2)}.$$

For boundary layer flow, Fig. 5.1(c), $\partial u/\partial y$ is of order $U_\infty/\delta$ and $\partial^2 T/\partial y^2$ is of order $\Delta T/\Delta^2$, where $\Delta T$ is the temperature difference $(T_w - T_\infty)$ across the layer. Then, at least crudely, we may estimate the above ratio as

$$\lambda \simeq \frac{\mu(U_\infty/\delta)^2}{k\Delta T/\Delta^2} = (\mu c_p/k)\frac{U_\infty^2}{c_p\Delta T}(\Delta/\delta)^2.$$

The quantity $\mu c_p/k$ is the dimensionless *Prandtl number* Pr of the fluid. The parameter $U_\infty^2/c_p\Delta T$ is also dimensionless and is usually called the *Eckert number* [3].

To make some numerical comparisons compute $\lambda$ for (a) air,

(b) water, and (c) engine oil at 20°C, assuming $U_\infty = 10$ m/s and $\Delta T = 25$°C, and comment on the results. Assume $\Delta = \delta$.

| Fluid | Pr | $c_p$ (J/kg · K) | $\lambda$ |
|---|---|---|---|
| Air | 0.71 | 1010 | 0.0028 |
| Water | 7.0 | 4180 | 0.0067 |
| Engine oil | 1040 | 1880 | 2.21 |

Thus, for this typical example, dissipation is negligible for air and water but not for the highly viscous engine oil. The water result is quite general. However, air may move at much higher velocities than 10 m/s, in which case dissipation will become important; for example, at $U_\infty = 500$ m/s, $\lambda$(air) = 7.03. Also, the result for engine oil may be unrealistic because it requires a lot of power to move such a viscous fluid at 10 m/s. ■

## 5.3 Dimensional Analysis

We have developed the differential equations of convection in Section 5.2. But these are solvable only for simple geometries such as ducts and flat plates. They are boundary layer relations and thus not even valid in separated flow regions such as the rear surface of blunt bodies immersed in a freestream. They are difficult to solve for complex boundary conditions such as variable wall temperature, rough or grooved surfaces, or suction or blowing at the wall.

Thus theory helps to some extent, but convection is primarily an experimental science. You are probably aware from previous study in solid and/or fluid mechanics that such data become more general and more compact and useful when organized and presented in dimensionless form. Let us review the concepts of dimensional analysis here and present some of the most useful dimensionless convection parameters.

### 5.3.1 The Nusselt Number

The primary output of a convection experiment is the heat transfer coefficient $h$ or, alternatively, the heat flux $q''_w$. The traditional dimensionless form of $h$ is the *Nusselt number* Nu, which may be defined as the ratio of convection heat transfer to fluid conduction heat transfer under the same conditions. Consider a layer of fluid of width $L$ and temperature difference $(T_w - T_\infty)$. Assuming that the layer is moving so that convection occurs, the heat flux would be

$$q''_w = h(T_w - T_\infty).$$

If, however, the layer was stagnant and motionless, the heat flux would be entirely due to fluid conduction through the layer:

$$q''_w = k(T_w - T_\infty)/L.$$

We define the Nusselt number as the ratio of these two:

$$\text{Nu}_L = \frac{q''_w(\text{convection})}{q''_w(\text{conduction})} = hL/k.$$

A Nusselt number of order unity would indicate a sluggish motion little more effective than pure fluid conduction: for example, laminar flow in a long pipe. A large Nusselt number means very efficient convection: For example, turbulent pipe flow yields Nu of order 100 to 1000. The subscript on Nu denotes the length scale being used: $\text{Nu}_D$ is appropriate for pipe flow and $\text{Nu}_L$ or $\text{Nu}_x$ for a flat plate.

Note that Nu is identical in form to the Biot number used in conduction problems, such as Eq. (2.34) or (3.12). However, in Bi the quantity $k$ is the conductivity of the *solid,* while in Nu it is the fluid conductivity.

### 5.3.2 The Prandtl Number

In Example 5.3 we encountered a dimensionless parameter composed entirely of fluid properties, called the *Prandtl number:*

$$\text{Pr} = \mu c_p/k. \qquad \textbf{(5.29)}$$

This parameter, which is tabulated in Appendixes E, F, and G for various fluids, has a moderate to strong effect on fluid convection, especially for single-phase cases. It may be loosely interpreted as the ratio of viscous effects to conduction effects and has the following range of values:

| **Fluid** | **Pr** |
|---|---|
| Liquid metals | 0.004–0.03 |
| Gases | 0.7–1.0 |
| Water | 1.7–13.7 |
| Light organic liquids | 5–50 |
| Oils | 50–10,000 |
| Glycerin | 2000–100,000 |

The variation of Pr among fluids spans some seven orders of magnitude, but its effect on convection heat transfer is closer to three orders of magnitude — roughly, $\mathrm{Nu} \propto \mathrm{Pr}^{0.4}$, as we shall see in Chapter 6.

### 5.3.3 A General Dimensional Analysis

In a given convection problem, $h$ can vary with many system parameters: streaming velocity $U$, body shape, body size $L$, wall roughness height $\varepsilon$, wall and stream temperatures, fluid properties. If there are significant density changes across the boundary layer, the buoyant specific weight $g\Delta\rho$ may be important. Let us express this variation functionally:

$$h = f(U, L, \varepsilon, T_w, T_\infty, \rho_0, \mu, k, c_p, g\Delta\rho, \text{shape}). \tag{5.30}$$

This is a formidable list of twelve variables. They can all be expressed in terms of four *primary dimensions:* mass, length, time, and temperature. Then, by the principles of dimensional analysis,† we expect that Eq. (5.30) can be rewritten as a function of $12 - 4 = 8$ dimensionless variables. The particular eight variables obtained can differ according to the manner of grouping chosen. The reader should verify that the following functional list is a satisfactory grouping:

$$\frac{hL}{k} = f\left(\frac{\rho UL}{\mu}, \frac{\mu c_p}{k}, \frac{\varepsilon}{L}, \frac{T_w}{T_\infty}, \frac{g\Delta\rho\, L^3}{\nu^2\rho_0}, \frac{U^2}{c_p(T_w - T_\infty)}, \text{generic shape}\right)$$

or

$$\mathrm{Nu}_L = f\left(\mathrm{Re}_L, \mathrm{Pr}, \frac{\varepsilon}{L}, \frac{T_w}{T_\infty}, \mathrm{Gr}_L, \mathrm{Ec}, \text{generic shape}\right). \tag{5.31}$$

In this more efficient dimensionless list we recognize the Nusselt number $\mathrm{Nu}_L$, the Reynolds number $\mathrm{Re}_L$, the Prandtl number Pr, and the Eckert number Ec (from Example 5.3).

Of the three new parameters, the roughness ratio $\varepsilon/L$ is probably familiar from fluid mechanics, where it induces early transition to turbulence and changes the drag of cylinders [2, Fig. 5.3b] and greatly increases pipe friction in the Moody chart [2, Fig. 6.12]. Roughness generally increases convection heat transfer if the boundary layer is turbulent, but not as much as it increases friction.

The wall temperature ratio $T_w/T_\infty$ was not a factor in fluid mechanics but does have a weak effect on both laminar and turbulent convection.

†See, for example, [1, Chapter 7] or [2, Chapter 5].

This ratio essentially reflects the effect of variable fluid properties, especially viscosity and thermal conductivity, across a boundary layer.

The seventh parameter, $Gr_L = g\Delta\rho L^3/(\nu^2\rho_0)$ gives the effect of buoyancy and is called the *Grashof number*. It is the dominant parameter in free convection problems and will be thoroughly discussed in Chapter 7.

Finally, the generic shape is included in the list to remind us that body shape always has a strong effect in flow and convection problems. We cannot equate sphere data to cylinders or airfoils, and circular pipes give different results than annular or rectangular ducts. Also, the orientation or "angle of attack" of a given shape is a parameter.

### 5.3.4 Forced-Convection Correlations

Even though dimensional analysis reduced our "typical" convection problem in Eq. (5.31) to eight variables, eight is too many. For a given shape it would take at least $10^6$ experiments to establish the functional relationship. Thus we need to reduce the list in order to report some usable results. The following assumptions might be made:

1. The surface is "hydraulically smooth," or $\varepsilon/L \to 0$. If necessary, later we can add on a roughness correction.
2. Temperature differences are small, hence $T_w/T_\infty \to 1.0$. Later we can add on a correction for strongly variable physical properties such as occur in oil flows.
3. The flow is strongly forced so that we can neglect Grashof number or buoyancy effects.
4. Flow velocities are moderate and the fluid not too viscous, so we can neglect Eckert number or dissipation effects.

If these assumptions are valid, then Eq. (5.31) reduces to

$$Nu_L \doteq f(Re_L, Pr, \text{generic shape}). \tag{5.32}$$

This simplified correlation is the traditional type of forced-convection experiment reported in the literature. Since the data often fall on reasonably straight lines when plotted on log-log paper, it is customary to fit the functional relation to a power-law:

$$Nu_L \doteq C Re_L^n Pr^m, \tag{5.33}$$

where $m$ and $n$ are fractional exponents and the dimensionless constant $C$ depends on generic shape. The power-law relation is suggested by theory for both laminar and turbulent flows and generally fits the data well, at least over limited ranges of Reynolds and Prandtl numbers. In Chapter 6 we encounter some far more complex and presumably more accurate formulas to correlate forced-convection data.

### 5.3.5 Stanton Number and Friction Factor

The Nusselt number is not the only way to nondimensionalize the heat transfer coefficient. A widely used alternative parameter is called the *Stanton number*, St:

$$\mathrm{St} = h/(\rho U c_p). \tag{5.34}$$

Recall from the principles of dimensional analysis that such alternative parameters are new but not independent. The Stanton number is simply a regrouping of $\mathrm{Re}_L$, Pr, and $\mathrm{Nu}_L$:

$$\mathrm{St} \equiv \mathrm{Nu}_L/(\mathrm{Re}_L \mathrm{Pr}). \tag{5.35}$$

The length scale $L$ cancels out. Whereas the Nusselt number is of order unity or larger, the Stanton number is a small fraction of order 0.002. In fact, it is comparable in size to the dimensionless wall shear stress or *friction factor*, $C_f$, studied in fluid mechanics (see, for example, [2, Eq. 7.10]):

$$C_f = 2\tau_w/(\rho U^2). \tag{5.36}$$

More to the point, St is actually *related* to $C_f$, at least for certain geometries, notably smooth flat plates and pipes. When valid, the relationship takes the form

$$\mathrm{St} \doteq f(C_f, \mathrm{Pr}, \text{generic shape}) \tag{5.37}$$

and is an example of what is now called the *Reynolds analogy*, after Osborne Reynolds [9], whose studies of turbulent convection a century ago first suggested a proportionality between fluid heat flux and fluid friction. We take up the Reynolds analogy in Section 6.2.

### 5.3.6 Heat Flux Parameters

The heat transfer coefficient $h = q''/\Delta T$ is universally accepted among heat transfer workers. It is especially useful in single-phase forced convection, where $q''$ is nearly proportional to $\Delta T$ and hence $h$ will vary only with system parameters and not with $\Delta T$ itself. But there are three important convection problems in which $h$ varies with $\Delta T$: (1) free convection; (2) boiling; and (3) condensation. This makes the use of $h$ rather clumsy at times, especially when the heat flux is known and the temperature difference desired. One remedy would be to drop the use of $h$ entirely, and a recent monograph [10] attempted to sell this idea. But the widespread acceptance of $h$, together with its convenience as a boundary condition input to conduction problems (Chapters 2–4) makes the Nusselt number (and Biot number) impossible to abolish totally. There *are* alternative ways to nondimensionalize $q''$ without using $\Delta T$, however, and this text presents these alternatives in Chapters 7 and 9. For example, in free convection, a suitable parameter would be

$$\mathrm{Nu}^* = q''_w g\beta L^4/(k\nu^2), \tag{5.38}$$

where $\beta$ is the fluid coefficient of thermal expansion.

In condensation problems, a recommended parameter is

$$\mathrm{Nu}^{**} = \mu q''_w/(\rho^2 h_{fg} L^2 g), \tag{5.39}$$

where $h_{fg}$ is the latent heat of vaporization of the fluid.

In boiling problems, there is already a widely used parameter in the literature:

$$\mathrm{Nu}^{***} = (q''_w/\mu h_{fg})(\Upsilon/g\Delta\rho)^{1/2}, \tag{5.40}$$

where $\Upsilon$ is the fluid surface tension coefficient. All of these are formed without using $\Delta T$, the point being that traditional parameters are not always the most useful in every problem.

### 5.3.7 Nondimensionalizing the Basic Equations

In Section 5.3.3 we developed a list of dimensionless heat transfer parameters with a straightforward dimensional analysis. That is, we decided what variables affect heat transfer and then arranged them into dimensionless groups. This technique finds the parameters, such as Reynolds number and Eckert number, but does not reveal how important they are or when they can be neglected.

An alternative technique that both reveals the parameters *and* gives insight about their use is the nondimensionalization of the basic boundary layer equations. We learn a lot about Reynolds number and so on even if we are unable to solve the equations themselves. Let us illustrate with an example.

#### Example 5.4

Nondimensionalize the boundary layer momentum and energy equations for the case of laminar flow with constant $\mu$ and $k$ and negligible gravity and pressure gradient effects. Discuss the parameters that appear.

*Solution* For these conditions, the momentum relation, Eq. (5.15), and the energy relation, Eq. (5.23), reduce to

$$u\frac{\partial u}{\partial x} + v\frac{\partial u}{\partial y} = \nu\frac{\partial^2 u}{\partial y^2}, \tag{1}$$

$$\rho c_p\left(u\frac{\partial T}{\partial x} + v\frac{\partial T}{\partial y}\right) = k\frac{\partial^2 T}{\partial y^2} + \mu\left(\frac{\partial u}{\partial y}\right)^2. \tag{2}$$

Now assume that the flow has a characteristic velocity $U$ and length scale $L$ and use these to define dimensionless variables:

$x^* = x/L;\quad y^* = y/L;\quad u^* = u/U;\quad v^* = v/U;$

$T^* = T/\Delta T;\quad \Delta T = (T_w - T_\infty).$

Introduce these variables into Eqs. (1) and (2) and divide out all constants until the coefficient of the first term in each equation is unity. The following equations result:

$$u^*\frac{\partial u^*}{\partial x^*} + v^*\frac{\partial u^*}{\partial y^*} = \frac{1}{\mathrm{Re}_L}\frac{\partial^2 u^*}{\partial y^{*2}}, \tag{1a}$$

$$u^*\frac{\partial T^*}{\partial x^*} + v^*\frac{\partial T^*}{\partial y^*} = \frac{1}{\mathrm{Re}_L\mathrm{Pr}}\frac{\partial^2 T^*}{\partial y^{*2}} + \frac{\mathrm{Ec}}{\mathrm{Re}_L}\left(\frac{\partial u^*}{\partial y^*}\right)^2. \tag{2a}$$

The relevant dimensionless groups are immediately revealed as coefficients of certain terms. The Reynolds number and Prandtl number accompany the all-important shear and conduction terms, which can never be neglected in boundary layer analyses. Therefore, Reynolds number and Prandtl number are *always* important in forced-convection problems with negligible pressure gradient and gravity effects — a fact we proposed without proof as Eq. (5.32). The Eckert number accompanies the dissipation term, so that if the Eckert number is small, dissipation may be neglected — a fact we established by another method in Example 5.3. Thus this idea of nondimensionalizing the basic equations gives some very timely information about the use and significance of our traditional heat transfer parameters. ■

**Example 5.5**

Using the momentum and energy equations, determine the variation of velocity, pressure, and temperature outside the boundary layer by taking the limit of vanishing shear and conduction effects. Neglect gravity.

*Solution* Outside the boundary layer, $k$ and $\mu$ terms vanish, $y$-derivatives are negligibly small, $u \to U_\infty(x)$, $i \to i_\infty(x)$, and $v$ is negligibly small. Under these conditions the momentum equation (5.15) reduces to

$$\frac{d}{dx}(p_\infty) + \rho U_\infty \frac{d}{dx}(U_\infty) = 0.$$

Integrating this with respect to $x$, we obtain

$$p_\infty + \tfrac{1}{2}\rho U_\infty^2 = \text{constant}. \qquad \textbf{(1)}$$

Thus, outside the boundary layer, for incompressible flow, pressure and velocity are related, as expected, by the inviscid Bernoulli equation. See, for example, [2, Eq. (2.71)].

For these same conditions the energy equation (5.19) reduces to

$$\frac{d}{dx}[\rho U_\infty(e_\infty + \tfrac{1}{2}U_\infty^2) + p_\infty U_\infty] = 0.$$

Integrating this, we obtain

$$e_\infty + p_\infty/\rho + \tfrac{1}{2}U_\infty^2 = i_\infty + \tfrac{1}{2}U_\infty^2 = \text{constant}. \qquad \textbf{(2)}$$

This is the adiabatic steady-flow energy equation, as in [2, Eq. (9.22)], and holds even in compressible (variable-density) flow. Thus the outer flow is predicted to be adiabatic and inviscid. ∎

---

## Summary

This chapter has attempted to review and develop some basic principles relevant to convection analysis. Readers are advised to review their own studies in thermodynamics and fluid mechanics to assist in understanding these convection principles.

Convection analysis can be split in at least three ways: (1) forced versus free motion, (2) laminar versus turbulent flow, and (3) internal versus external flow. In each case different assumptions are made about the flow patterns and relevant parameters, and the heat transfer coefficient may have a slightly different definition.

We have derived and discussed the basic partial differential equations of continuity, momentum, and energy in a steady laminar or turbulent two-dimensional boundary layer. Boundary conditions are presented for both the wall and the outer stream. We will attempt to solve these equations for special cases in both forced flow (Chapter 6) and free convection (Chapter 7). We will have very little luck solving them for boiling problems, but a laminar flow condensation theory is possible (Chapter 9).

For many practical problems, such as the heat exchanger of Fig. 5.3, the theory is inadequate and experiments must be performed. The chapter presents some dimensional analysis techniques for heat transfer data and introduces, among others, the widely used parameters of Nusselt number, Stanton number, Reynolds number, Grashof number, Prandtl number, and Eckert number. Some alternate presentations are also discussed.

## References

1. R. W. Fox and A. T. McDonald, *Introduction to Fluid Mechanics,* 3rd ed., Wiley, New York, 1985.
2. F. M. White, *Fluid Mechanics*, 2nd ed., McGraw-Hill, New York, 1986.
3. H. Schlichting, *Boundary Layer Theory,* 7th ed., McGraw-Hill, New York, 1979.
4. F. M. White, *Viscous Fluid Flow,* McGraw-Hill, New York, 1974.
5. W. M. Kays and M. E. Crawford, *Convective Heat and Mass Transfer,* 2nd ed., McGraw-Hill, New York, 1980.
6. E. R. G. Eckert and R. M. Drake, Jr., *Analysis of Heat and Mass Transfer,* McGraw-Hill, New York, 1972.
7. C. Y. Chow, *An Introduction to Computational Fluid Mechanics,* Wiley, New York, 1979.
8. S. V. Patankar, *Numerical Heat Transfer and Fluid Flow,* Hemisphere Publishing, New York, 1980.
9. O. Reynolds, *Scientific Papers of Osborne Reynolds,* 2 vols., Cambridge Univ. Press, London, 1901.
10. E. F. Adiutori, *The New Heat Transfer,* Ventuno Press, Cincinnati, 1974.
11. E. R. G. Eckert and R. J. Goldstein (Eds.), *Measurements in Heat Transfer,* Hemisphere Publishing, New York, 1976.
12. A. Bejan, *Convection Heat Transfer,* John Wiley, New York, 1984.
13. V. S. Arpaci and P. S. Larsen, *Convection Heat Transfer,* Prentice-Hall, Englewood Cliffs, N.J., 1984.
14. L. C. Burmeister, *Convective Heat Transfer,* Wiley-Interscience, New York, 1982.

## Review Questions

1. Explain the differences between (a) forced and free convection, (b) laminar and turbulent flow, (c) internal and external flow.
2. Define and describe a boundary layer flow.
3. What could cause the velocity and thermal boundary layers to be of different thicknesses?
4. Under what conditions is boundary layer theory valid?
5. What is the no-slip condition? The no-temperature-jump condition?
6. Three boundary layer equations are developed here. What do they represent?
7. When is boundary layer viscous dissipation negligible?
8. Why is the pressure assumed known in boundary layer theory?
9. When is the velocity "uncoupled" from the temperature?
10. Define and interpret the Nusselt number.
11. Define and interpret the Prandtl number.
12. What is the primary parameter affecting free convection?
13. Explain the concept of "generic shape."
14. When might the Nusselt or Stanton number be inconvenient?
15. What is the Reynolds analogy?
16. Does wall roughness affect both shear and heat transfer? In both laminar and turbulent flow?
17. Compared to straightforward dimensional analysis, what is the advantage of nondimensionalizing the basic equations?
18. When do we expect a formula of the type $\mathrm{Nu} = C\mathrm{Re}^n\mathrm{Pr}^m$?

## Problems

### Problem distribution by sections

**The Problem Assignments Are Organized as Follows:**

| Problems | Sections Covered | Topics Covered |
|---|---|---|
| 5.1–5.7 | 5.1 | General discussion |
| 5.8–5.17 | 5.2 | Boundary layer equations |
| 5.18–5.30 | 5.3 | Dimensional analysis |

5.1 Consider the flow of hot combustion gases through the stator and rotor sections of a gas turbine. Discuss whether forced or free convection is important, if the flow is likely to be laminar or turbulent, if the geometry is internal or external or both, and whether boundary layers occur.

5.2 In Chapter 4 we solved many conduction problems involving immersion of a hot (cold) solid into a cold (hot) fluid bath. Discuss the flow patterns and types of convection that occur.

5.3 Consider the classic problem in elementary fluid mechanics of laminar, parabolic "Poiseuille" flow in a long pipe [2, pp. 305–306]. If, say, the fluid were hot and the pipe cold, would this be forced or free convection? Would there be boundary layer behavior? Would heat flux increase with velocity?

5.4 Figure 5.2 shows typical data for velocity versus time in laminar, transitional, and turbulent flow. Would the same sort of data occur for time plots of the fluid's (a) pressure, (b) temperature, (c) density, (d) enthalpy, (e) kinetic energy?

5.5 What is maximum flow rate in $cm^3/s$ to ensure laminar flow through a 3-cm-diameter tube if the fluid is (a) helium, (b) mercury, and (c) engine oil, at 20°C and 1 atm?

5.6 For flow at velocity $U_o$ past a flat plate of average roughness and stream disturbances, transition to turbulence occurs at about $\mathrm{Re}_x = 500{,}000$. If the plate is 7 in. long, what speed range in ft/s will ensure laminar flow for (a) water, (b) air, and (c) helium, at 20°C and 1 atm?

5.7 Repeat Problem 5.6 by assuming $U_o = 1$ m/s. Estimate the transition point, $x$, in meters for (a) air, (b) mercury, and (c) engine oil at 20°C and 1 atm.

5.8 Show that, in boundary layer flow (Fig. 5.1c), the local heat transfer coefficient $h_x$ can be computed from the relation

$$h_x = \frac{\left(-k\dfrac{\partial T}{\partial y}\right)_{y=0}}{T_w - T_\infty}$$

Do you expect the wall slope $(\partial T/\partial y)$ to vary with stream velocity?

5.9 In laminar boundary layer flow past a flat plate, it is known from theory (Section 6.2) that the local heat transfer coefficient $h_x \propto x^{-1/2}$, where $x$ is distance from the leading edge. Show by integration that the *average* coefficient $\bar{h}$ between 0 and $x$ equals exactly twice the value of $h_x$ at point $x$.

5.10 In laminar vertical free convection on a flat plate, it is known from theory (Section 7.2) that $h_x \propto x^{-1/4}$, where $x$ is distance from the leading edge. Show by integration that the average coefficient $\bar{h}$ between 0 and $x$ is one-third larger than the value of $h_x$ at point $x$.

5.11 Modify and rederive the continuity relation, Eq. (5.6), to account for variable density. Assume steady flow.

5.12 Repeat Problem 5.11 for the boundary layer momentum equation, Eq. (5.8). Discuss your surprising result.

5.13 Use the continuity and momentum relations and the boundary layer approximations to show the steps that reduce the energy relation, Eq. (5.19), to its simpler form, Eq. (5.20).

5.14 For flow of mercury past a flat surface that is 40 cm long and 20 cm wide, the local heat transfer coefficient is approximated by $h_x = 45{,}800x^{-0.2}$, with $h_x$ in W/(m$^2$ · K) and $x$ in meters. Estimate (a) the average heat transfer coefficient over the whole plate and (b) the total heat transferred to the plate if $(T_\infty - T_w) = 12$ K.

5.15 What are the dimensions of the turbulent eddy viscosity $\varepsilon_M$? Sketch what you think its variation would be with distance $y$ from the wall. Review your study of fluid mechanics and show how $\varepsilon_M$ is related to the turbulent fluctuations $u'(x, y, t)$ and $v'(x, y, t)$ in the boundary layer.

5.16 It is possible to solve the boundary layer equations, Eqs. (5.6, 5.15, 5.23), exactly for the case of laminar *Couette flow* between a moving and a fixed plate, as shown in Fig. P5.16. Assuming steady flow with constant $\rho$, $k$, $\mu$, and $c_p$, for the boundary conditions shown and zero pressure gradient, solve these equations for the velocity $u(y)$ and the temperature $T(y)$. Compute the shear stress and heat transfer rate at each plate surface.

5.17 Consider helium at 20°C and 1 atm, flowing in a tube such that $\Delta = \delta$. If the wall/fluid temperature difference is 65°C, what velocity range will ensure that dissipation is less than 10% of conduction effects?

5.18 A certain light oil has $\rho = 57$ lb$_m$/ft$^3$, $c_p = 0.43$ Btu/lb$_m$ · °F, $\nu = 0.00102$ ft$^2$/s, and $k = 0.133$ W/m · K. What is its Prandtl number?

5.19 In laminar duct flow, the average heat transfer coefficient $\bar{h}$ is a function of average velocity $V$, diameter $D$, distance $x$ from the entrance, and fluid properties $\rho$, $\mu$, $c_p$, and $k$. Using dimensional analysis, rewrite this relation in dimensionless form. Are there four parameters? In exact laminar theory (Section 6.3) only *three* parameters occur. Why the discrepancy?

**Figure P5.16**

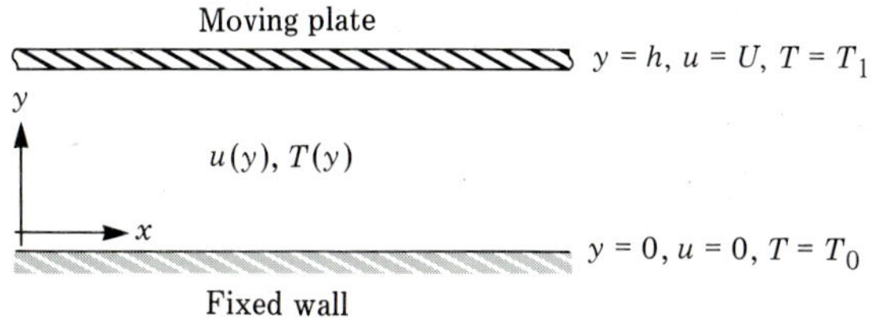

5.20 The Grashof number $\mathrm{Gr}_L = g\Delta\rho L^3/(\nu^2\rho_o)$ is a measure of fluid buoyancy and can vary greatly among fluids for the same temperature conditions. Let $L = 1$ m, $T_w = 60°\mathrm{C}$, and $T_\infty = 40°\mathrm{C}$. Taking $\Delta\rho = \rho_w - \rho_\infty$, compute $\mathrm{Gr}_L$ for (a) air, (b) water, and (c) engine oil, at 1 atm.

5.21 Is the concept of generic shape or "geometric similarity" clear to you? First, consider flow *along* versus flow *across* a very long cylinder. Are the two flows geometrically similar? Second, consider flow across a 2-cm-diameter, 30-cm-long cylinder versus flow across a 1-cm-diameter, 20-cm-long cylinder. Are the two flows geometrically similar? Discuss.

5.22 Using SI units, verify that the three alternate Nusselt numbers, Eqs. (5.38–5.40), are indeed dimensionless.

5.23 For flow through a long, 3-cm-diameter tube with wall temperature 25°C and average fluid temperature 15°C, the following heat transfer coefficients are reported for four velocities and three fluids at 1 atm:

| | **Heat Transfer Coefficient W/m² · K** | | | |
|---|---|---|---|---|
| **Fluid** | $V$ = 3 m/s | 6 m/s | 10 m/s | 20 m/s |
| Air | 17 | 30 | 46 | 80 |
| Water | 8000 | 14000 | 21000 | 37000 |
| Engine oil | 90 | 160 | 240 | 420 |

Convert these data to dimensionless form and fit it, perhaps using log-log graph paper, to a power-law relation similar to Eq. (5.33). Note the great disparity between air and water.

5.24 For the same pipe and temperature conditions of Problem 5.23, use the data of that problem to estimate $h$ and $q''_w$ for flow of helium at 1 atm and $V = 80$ m/s.

5.25 For laminar flow past a flat plate with gases or light liquids, the accepted forced convection correlation is $\mathrm{Nu}_x = 0.332\,\mathrm{Re}_x^{1/2}\,\mathrm{Pr}^{1/3}$. What percentage increase (or decrease) in $h_x$ is caused by a 20% increase in (a) $U_\infty$; (b) $k$; (c) $c_p$; (d) $\rho$; or (e) $\mu$?

5.26 Figure P5.26 shows a heated cylinder convection experiment, similar to Problem 1.34. The heated section has $D = 3$ cm, $L = 15$ cm, and $T_w = 25°\mathrm{C}$. The air has $T_\infty = 15°\mathrm{C}$ and moves in crossflow at speed $U$. The experimental data are as follows:

| $U$ (m/s): | 5 | 10 | 15 |
|---|---|---|---|
| $q$ (W): | 7.0 | 10.2 | 13.1 |

Fit these data to the correlation suggested by Eq. (5.33), assuming $m = 1/3$, and use your formula to estimate $\bar{h}$ in W/(m² · K) when $U = 18$ m/s.

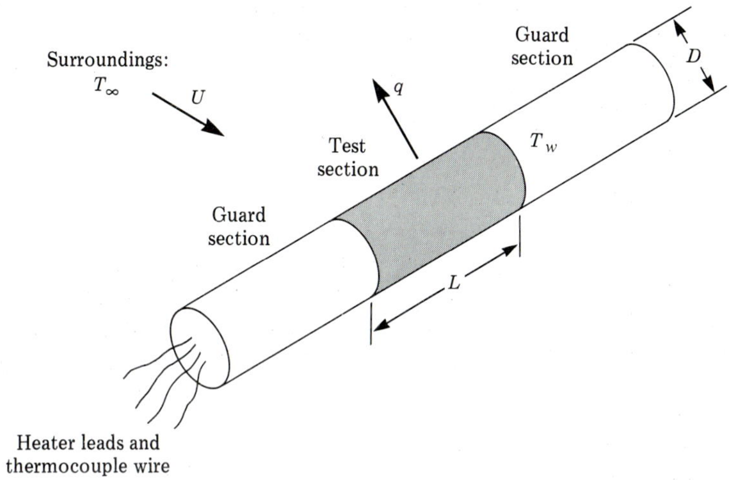


**Figure P5.26**

5.27 Using the dimensionless variables proposed in Example 5.4, rewrite the continuity equation (5.6) in dimensionless form and discuss the result.

5.28 What form does the inviscid pressure-velocity relation outside the boundary layer, Eq. (1) of Example 5.5, take if the density is variable (compressible flow of a gas)? What assumption is needed to integrate in closed form?

5.29 From Example 5.5, flow outside the boundary layer is nearly inviscid and adiabatic. Suppose fluid flows past a blunt body so that just outside the stagnation region the fluid temperature is 40°C. Further downstream the fluid velocity $U_\infty$ is 55 m/s. Estimate the fluid temperature at this point for (a) air, (b) water, and (c) helium, at 1 atm.

5.30 Electronic chips of length $L = 18$ mm and width $b = 16$ mm are mounted to a board and cooled by forced convection of air at 20°C and velocity $U = 15$ m/s. The experimental correlation of average Nusselt number is

$$\mathrm{Nu}_L = 0.093\ \mathrm{Re}_L^{0.7}\ \mathrm{Pr}^{0.4}.$$

For these conditions, estimate the chip temperature when it is dissipating 1 W of power. What air velocity is needed to maintain the chip at 40°C?

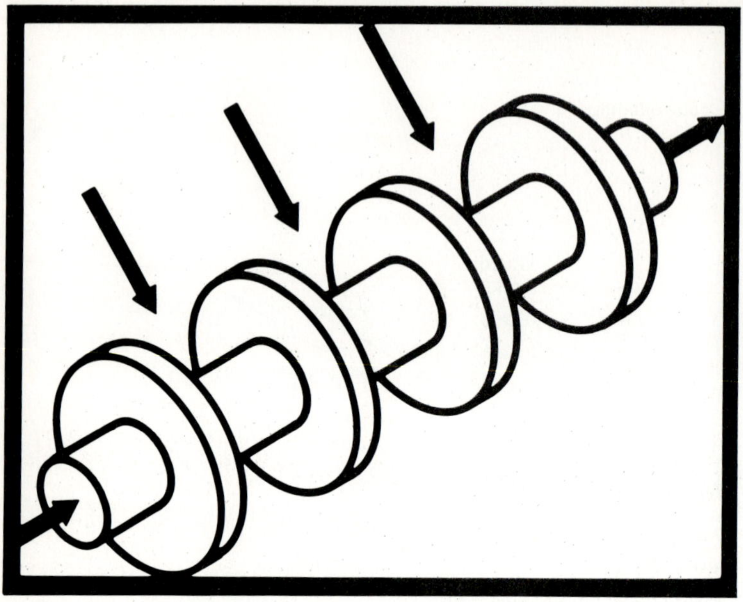

# Forced Convection

# 6

## 6.1 Introduction

In Chapter 5 we reviewed some basic concepts about fluid flow and convection, derived the boundary layer equations of mass, momentum, and energy, and discussed some important dimensionless parameters. This chapter now gives quantitative information for engineering analysis and experimental correlations of forced convection in various geometries. Chapter 7 then does the same for free convection.

As long as the boundary layer equations are valid, that is, no separated flow and not too small a Reynolds number, there are some excellent analytical methods, both laminar and turbulent, for predicting forced convection rates. Here we will limit ourselves primarily to flat plate or circular pipe flow, while the reader is referred to many other analytical cases in an advanced text on convection [1]. There are also specialized texts on particular problems such as laminar duct flow [2]. The availability of a large digital computer allows one to pursue numerical solutions of both laminar and turbulent, internal and external flows, with many successes now being reported [3]. Wherever possible, these analyses should be checked against experimental data, and the results should be presented in dimensionless form to broaden their generality.

### 6.1.1 The Boundary Layer Equations

Most of the existing analytical solutions for convection heat transfer are for boundary layer flows, where cross-stream gradients of velocity and temperature are much larger than streamwise gradients, and the streamwise velocity $u$ is much larger than the cross-stream velocity $v$. In Section 5.2, we used these approximations to derive the basic incompressible, two-dimensional, steady-flow boundary layer equations, which we collect and repeat here for convenience.

*Continuity:*

$$\frac{\partial u}{\partial x} + \frac{\partial v}{\partial y} = 0, \tag{6.1a}$$

*Momentum:*

$$u\frac{\partial u}{\partial x} + v\frac{\partial u}{\partial y} = -\frac{1}{\rho}\frac{dp}{dx} + \frac{\partial}{\partial y}\left[(\nu + \varepsilon_M)\frac{\partial u}{\partial y}\right], \tag{6.1b}$$

*Energy:*

$$u\frac{\partial T}{\partial x} + v\frac{\partial T}{\partial y} = \frac{\partial}{\partial y}\left[(\alpha + \varepsilon_H)\frac{\partial T}{\partial y}\right] + \frac{1}{c_p}(\nu + \varepsilon_M)\left(\frac{\partial u}{\partial y}\right)^2. \tag{6.1c}$$

We have neglected gravity, which is important in free convection (Chapter 7), and assumed constant $\rho$ and $c_p$ but allowed $\mu$ and $k$ to vary with $T$. In general, $x$ and $y$ are curvilinear coordinates, with $x$ along the wall and $y$ normal to the wall. Expressions for the turbulent eddy viscosity $\varepsilon_M$ and eddy conductivity $\varepsilon_H$ will be developed shortly.

The boundary conditions for Eqs. (6.1) are no-slip and no-temperature-jump at the wall, plus a smooth matching of velocity and temperature to the outer flow:

$$\begin{aligned} &\text{At } y = 0: \\ &u = v = 0, \qquad T = T_w(x), \\ &\text{As } y \to \infty: \\ &u \to U_\infty(x), \qquad T \to T_\infty(x). \end{aligned} \tag{6.2}$$

In addition, at some position $x_o$ along the boundary layer — usually at the leading edge — there must be known profiles of velocity and temperature:

$$\begin{aligned} &\text{At } x = x_o: \\ &u = u_o(y), \qquad T = T_o(y). \end{aligned} \tag{6.3}$$

Sometimes the wall temperature condition is replaced by a known surface heat flux, which specifies the wall temperature gradient of the fluid:

$$\begin{aligned} &\text{At } y = 0: \\ &\left(-k\frac{\partial T}{\partial y}\right)_{y=0} = q''_w(x) \end{aligned} \tag{6.4}$$

As mentioned, there are many exact and approximate solutions to these equations for particular cases, as treated in advanced texts [1, 4].

Note again that if $\nu$ and $\varepsilon_M$ are not functions of temperature (a common assumption), Eqs. (6.1a) and (6.1b) are totally *uncoupled* from temperature, Eq. (6.1c), and may be solved separately for the velocity field $u(x, y)$ and $v(x, y)$. The temperature field, however, is *coupled* to $u$ and $v$ through the convective and dissipation terms in Eq. (6.1c).

### 6.1.2 Momentum and Energy Integral Relations

Direct analytical solution of Eqs. (6.1) has been successful only for certain classical geometries and boundary conditions. An alternative approach is to satisfy Eqs. (6.1) in a thickness-averaged sense by using a control volume integral analysis. The integral technique may be used for either internal or external flows, under reasonably arbitrary conditions [1, 4, 13]. Here we consider only the special case of incompressible flow past a thin flat plate.

Consider steady viscous flow past a flat plate as sketched in Fig. 6.1. The flow may be either laminar or turbulent. The velocity profile $u(x, y)$ must drop to zero at the wall, and the temperature $T(x, y)$ must approach $T_w$. We have arbitrarily shown a hot wall and a thick thermal boundary layer (Pr $< 1.0$) in the sketch; the opposite may also apply. Note that as $y \to \delta(x)$, $u \to U_\infty$, and as $y \to \Delta(x)$, $T \to T_\infty$.

The oncoming freestream has conditions $U_\infty$, $T_\infty$, and $p_\infty$. Since the supposedly razor-sharp plate does not block the flow, the outer velocity $U_\infty$ is constant along the plate and thus, from Bernoulli's equation (5.10), the pressure is constant throughout the boundary layer.

By using the differential control volume $dx$ wide and $b$ deep in Fig. 6.1(b), we may derive very useful integral momentum and energy relations for this flow. As a preliminary, consider a mass balance of the large (energy) control volume [$\overline{123456}$]. The mass flux entering the left side [$\overline{123}$] is

$$\int_0^\Delta \rho\, ubdy,$$

**Figure 6.1 Sketch of boundary layer flow past a flat plate, showing (a) typical velocity and temperature profiles and (b) differential control volumes [1256] for momentum integral and [1346] for energy integral analysis.**

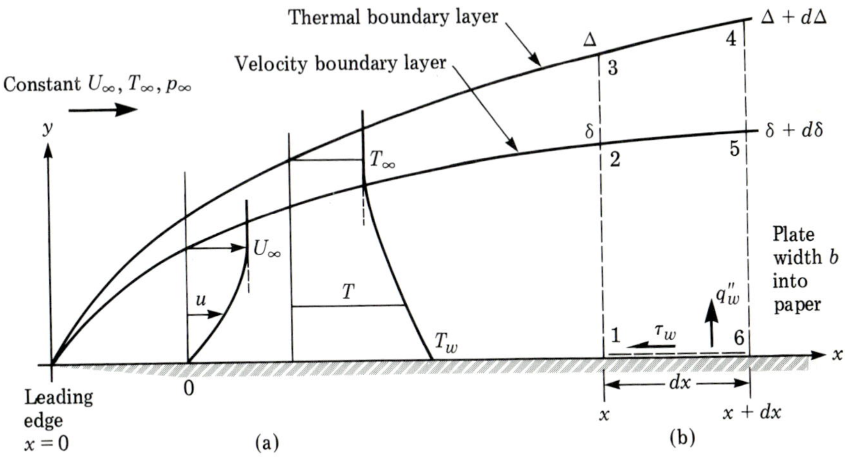

while the mass leaving the right side $[\overline{456}]$ differs by a differential amount:

$$\int_0^{\Delta} \rho u b dy + \frac{d}{dx}\left[\int_0^{\Delta} \rho u b dy\right] dx$$

Since net mass flux is zero in steady flow, the difference between these two must equal the mass entering the top $[\overline{34}]$:

$$d\dot{m}_{34} = \frac{d}{dx}\left[\int_0^{\Delta} \rho u b dy\right] dx.$$

Because of the solid plate, no mass enters the bottom side $[\overline{16}]$.

Now consider the $x$-momentum flux entering side $[\overline{12}]$:

$$\int_0^{\delta} \rho u^2 b dy.$$

The momentum flux leaving side $[\overline{56}]$ is slightly different:

$$\int_0^{\delta} \rho u^2 b dy + \frac{d}{dx}\left[\int_0^{\delta} \rho u^2 b dy\right] dx.$$

Finally, the $x$-momentum flux entering the top side $[\overline{25}]$ is

$$U_\infty d\dot{m}_{25} = U_\infty \frac{d}{dx}\left[\int_0^{\delta} \rho u b dy\right] dx.$$

There is no momentum flux entering the solid plate side $[\overline{16}]$. The difference between leaving and entering flux must equal the net applied force in the $x$-direction. Since the pressure is constant, the only force is due to the wall shear $\tau_w$ acting on the bottom side $[\overline{16}]$. Newton's law thus gives

$$-\tau_w\, b dx = \frac{d}{dx}\left[\int_0^{\delta} \rho u^2 b dy\right] dx - U_\infty \frac{d}{dx}\left[\int_0^{\delta} \rho u b dy\right] dx,$$

or

$$\tau_w = \frac{d}{dx}\left[\int_0^{\delta} \rho u(U_\infty - u) dy\right] = \left(\mu \frac{\partial u}{\partial y}\right)_{y=0}. \tag{6.5}$$

This is the *momentum integral equation* for a two-dimensional flat plate boundary layer. It is generalized to include pressure variations in advanced texts [1, p. 48] or [4, p. 306].

Now make an energy balance of control volume $[\overline{123456}]$, neglecting dissipation. The enthalpy entering side $[\overline{123}]$ is

$$\int_0^{\Delta} \rho u c_p T b dy,$$

and the enthalpy flux leaving the right side [$\overline{456}$] is slightly different:

$$\int_0^{\Delta} \rho u c_p T b dy + \frac{d}{dx}\left[\int_0^{\Delta} \rho u c_p T b dy\right] dx.$$

The enthalpy entering through the top side [$\overline{34}$] is

$$c_p T_\infty d\dot{m}_{34} = c_p T_\infty \frac{d}{dx}\left[\int_0^{\Delta} \rho u b dy\right] dx,$$

and of course there is no flux through the bottom surface [$\overline{16}$].

The difference between leaving and entering enthalpy flux must equal the heat transferred to the fluid at the wall, since there is no heat transferred at [$\overline{34}$], where $\partial T/\partial y \doteq 0$:

$$q''_w b dx = \frac{d}{dx}\left[\int_0^{\Delta} \rho u c_p T b dy\right] dx - c_p T_\infty \frac{d}{dx}\left[\int_0^{\Delta} \rho u b dy\right] dx,$$

or

$$q''_w = \frac{d}{dx}\left[\int_0^{\Delta} \rho c_p u (T - T_\infty) dy\right] = \left(-k\frac{\partial T}{\partial y}\right)_{y=0}. \tag{6.6}$$

This is the *energy integral relation* for a two-dimensional boundary layer. It assumes negligible dissipation but does not require flat plate flow, since pressure variations will not add any additional terms.

These two fundamental integral relations, (6.5) and (6.6), are very accurate if the exact $u$ and $T$ profiles are inserted. However, they are commonly solved *approximately* by assuming reasonable guesses for the profile shapes $u(x, y)$ and $T(x, y)$.

The integral relations are valid for either laminar or turbulent flow. We will apply them to the case of laminar flat plate flow in Section 6.2.3.

## 6.2 Laminar Flow Past a Flat Plate

For laminar flow past a flat plate as in Fig. 6.1, the boundary layer equations (6.1b) and (6.1c), assuming constant $\rho$, $\mu$, and $k$ and negligible dissipation, reduce to

*Momentum:*

$$u\frac{\partial u}{\partial x} + v\frac{\partial u}{\partial y} = \nu\frac{\partial^2 u}{\partial y^2}; \tag{6.7a}$$

*Energy:*

$$u \frac{\partial T}{\partial x} + v \frac{\partial T}{\partial y} = \alpha \frac{\partial^2 T}{\partial y^2}. \tag{6.7b}$$

There is no pressure gradient, that is, $U_\infty$ = constant.

Equation (6.7a) was first solved in 1908 by Ludwig Prandtl's student, H. Blasius. Equation (6.7b) was solved by another German worker, E. Pohlhausen, in 1921.

### 6.2.1 The Blasius Equation

Blasius reasoned that the velocity profiles along the plate would be *similar* in shape, that is, they would all look the same if $u$ were scaled by $U_\infty$ and $y$ by local boundary layer thickness:

$$u/U_\infty = fcn(y/\delta).$$

Earlier work by Lord Rayleigh had established that $\delta$ grew approximately as $x^{1/2}$. Thus Blasius, no doubt also aided by Prandtl's insight, defined the dimensionless *similarity variable*

$$\eta = y\,(U_\infty/\nu x)^{1/2}. \tag{6.8}$$

To eliminate the continuity equation (6.1a), he defined a dimensionless stream function $\psi(x, y)$:

$$\psi = (U_\infty \nu x)^{1/2} f(\eta). \tag{6.9}$$

Then, by definition of $\psi$, the velocity components may be computed by differentiation:

$$\begin{aligned} u &= \frac{\partial \psi}{\partial y} = U_\infty f'(\eta), \\ v &= -\frac{\partial \psi}{\partial x} = \tfrac{1}{2}(U_\infty \nu/x)^{1/2}(\eta f' - f). \end{aligned} \tag{6.10}$$

where the prime denotes differentiation with respect to $\eta$.

Blasius substituted $u$ and $v$ from Eqs. (6.10) into the momentum relation (6.7a). Some delightful cancellation occurs, and the result is a third-order nonlinear ordinary differential equation for the stream function $f$:

$$f''' + \tfrac{1}{2} f f'' = 0. \tag{6.11}$$

This is the *Blasius equation,* and its boundary conditions are found by combining Eqs. (6.2) and (6.10):

$$\begin{aligned} f(0) &= f'(0) = 0, \\ f'(\infty) &\longrightarrow 1.0. \end{aligned} \tag{6.12}$$

The Blasius equation has no known analytic solution, but it may readily be integrated at discrete points by a numerical technique such as the Runge-Kutta method. A typical numerical solution is shown in Table 6.1.

Since $u$ approaches $U_\infty$ only asymptotically, the concept of "thickness" is fuzzy. We define $\delta$ as the point where $u = 0.99U_\infty$, or $f' = 0.99$. From Table 6.1 this occurs at $\eta \simeq 5.0$. Thus the boundary layer thickness for laminar flat plate flow is accepted to be

$$\delta/x \simeq 5.0/\mathrm{Re}_x^{1/2}, \qquad \mathrm{Re}_x = U_\infty x/\nu. \tag{6.13}$$

**Table 6.1** The Blasius functions for laminar flat plate flow [13]

| $\eta$ | $f = \psi/(U_\infty \nu x)^{1/2}$ | $f' = u/U_\infty$ | $f'' = \tau(x/U_\infty^3 \rho\mu)^{1/2}$ |
|---|---|---|---|
| 0.0 | 0.0 | 0.0 | 0.3321 |
| 0.2 | 0.0066 | 0.0664 | 0.3320 |
| 0.4 | 0.0266 | 0.1328 | 0.3315 |
| 0.6 | 0.0597 | 0.1989 | 0.3301 |
| 0.8 | 0.1061 | 0.2647 | 0.3274 |
| 1.0 | 0.1656 | 0.3298 | 0.3230 |
| 1.2 | 0.2380 | 0.3938 | 0.3166 |
| 1.4 | 0.3230 | 0.4563 | 0.3079 |
| 1.6 | 0.4203 | 0.5168 | 0.2967 |
| 1.8 | 0.5295 | 0.5748 | 0.2829 |
| 2.0 | 0.6500 | 0.6298 | 0.2668 |
| 2.2 | 0.7812 | 0.6813 | 0.2484 |
| 2.4 | 0.9223 | 0.7290 | 0.2281 |
| 2.6 | 1.0725 | 0.7725 | 0.2065 |
| 2.8 | 1.2310 | 0.8115 | 0.1840 |
| 3.0 | 1.3968 | 0.8461 | 0.1614 |
| 3.2 | 1.5691 | 0.8761 | 0.1391 |
| 3.4 | 1.7470 | 0.9018 | 0.1179 |
| 3.6 | 1.9295 | 0.9233 | 0.0981 |
| 3.8 | 2.1160 | 0.9411 | 0.0801 |
| 4.0 | 2.3058 | 0.9555 | 0.0642 |
| 4.2 | 2.4981 | 0.9670 | 0.0505 |
| 4.4 | 2.6924 | 0.9759 | 0.0390 |
| 4.6 | 2.8883 | 0.9827 | 0.0295 |
| 4.8 | 3.0853 | 0.9878 | 0.0219 |
| 5.0 | 3.2833 | 0.9915 | 0.0159 |
| $\infty$ | $\infty$ | 1.0000 | 0.0 |

In like manner we may find the wall shear stress by differentiation:

$$\tau_w = \mu \left.\frac{\partial u}{\partial y}\right|_{y=0} = (U_\infty^3 \rho\mu/x)^{1/2} f''(0),$$

from which we see that shear increases as velocity to the 3/2 power. Noting that $f''(0) = 0.332$ from Table 6.1, we may rewrite the wall shear in dimensionless form, known as the *friction factor* or skin friction coefficient:

$$c_f = 2\tau_w/\rho U_\infty^2 = 0.664/\mathrm{Re}_x^{1/2}. \quad \textbf{(6.14)}$$

Equations (6.13) and (6.14), plus the profile shapes in Table 6.1, are the principal results of Blasius's theory.

E. Pohlhausen found the temperature profiles by assuming that the temperature ratio was a function of $\eta$ only:

$$\frac{T - T_w}{T_\infty - T_w} = \Theta(\eta). \quad \textbf{(6.15)}$$

Substituting into the energy equation (6.7b), with $u$ and $v$ given by Eqs. (6.10), results in the differential equation

$$\Theta'' + \tfrac{1}{2}\mathrm{Pr}\, f\Theta' = 0. \quad \textbf{(6.16)}$$

The boundary conditions (6.2) become

$$\Theta(0) = 0, \qquad \Theta(\infty) \longrightarrow 1.0. \quad \textbf{(6.17)}$$

This is a linear differential equation and has the solution

$$\Theta = \frac{\int_0^\eta [\exp(-\tfrac{1}{2}\mathrm{Pr} \int_0^\eta f\, d\eta)] d\eta}{\int_0^\infty [\exp(-\tfrac{1}{2}\mathrm{Pr} \int_0^\eta f\, d\eta)] d\eta}. \quad \textbf{(6.18)}$$

This solution is valid only for constant wall temperature.

Since $f(\eta)$ is only known numerically from Table 6.1, Equation (6.18) must also be evaluated numerically — a straightforward task for a microcomputer. Figure 6.2 shows temperature profiles $\Theta(\eta)$ from Eq. (6.18). We see that Prandtl number has a strong effect: Increasing Pr thins the thermal boundary layer and increases the wall heat transfer.

The heat transfer is computed from Fourier's law:

$$q_w'' = -k\left.\frac{\partial T}{\partial y}\right|_{y=0} = k(T_w - T_\infty)\,(U_\infty/\nu x)^{1/2}\,\Theta'(0).$$

Noting that the local heat transfer coefficient $h_x = q_w''/(T_w - T_\infty)$, we may rewrite this in the form of a local Nusselt number:

$$\mathrm{Nu}_x = h_x x/k = \mathrm{Re}_x^{1/2}\Theta'(0). \quad \textbf{(6.19)}$$

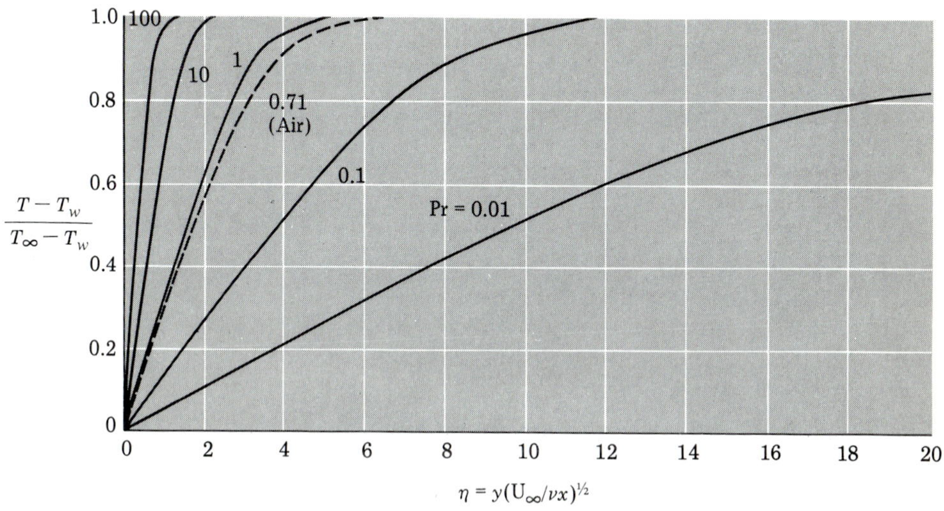


**Figure 6.2 Flat plate temperature profiles for various Prandtl numbers, computed from Eq. (6.18). The curve Pr = 1 is identical to the velocity profile $f'(\eta)$ from Table 6.1.**

The wall slope $\Theta'(0)$ equals the inverse of the denominator term in Eq. (6.18) and varies with Prandtl number. Table 6.2 tabulates some values obtained from a microcomputer. For Pr > 0.5 (which covers all but liquid metals), a popular and accurate correlation (±2%) is

$$\mathrm{Nu}_x \simeq 0.332\,\mathrm{Pr}^{1/3}\,\mathrm{Re}_x^{1/2} \tag{6.20}$$

Meanwhile, Churchill and Ozoe [48] proposed a curve fit that is accurate to ±1% over the entire range of Pr:

$$\mathrm{Nu}_x \simeq \frac{0.3387\,\mathrm{Pr}^{1/3}}{[1 + (0.0468/\mathrm{Pr})^{2/3}]^{1/4}}\,\mathrm{Re}_x^{1/2}. \tag{6.21}$$

These formulas are valid only for laminar boundary layer flow past a flat plate at constant wall temperature. They are the principal result of Pohlhausen's theory.

The thermal boundary layer thicknesses indicated in Fig. 6.2 may be approximated by the following correlation:

$$\Delta/\delta \simeq \mathrm{Pr}^{-0.37}. \tag{6.22}$$

When Pr = 1, $\Delta = \delta$ and, in fact, the shapes of the velocity and temperature profiles are identical. The case Pr = 1 means that viscous and conduction effects are equal, resulting in similar velocity and temperature profiles.

**Table 6.2** Flat plate laminar heat transfer parameter

| Pr | $\theta'(0) = \mathrm{Nu}_x/\mathrm{Re}_x^{1/2}$ |
|---|---|
| 0.01 | 0.0516 |
| 0.03 | 0.0844 |
| 0.1 | 0.1400 |
| 0.3 | 0.2148 |
| 0.71 (Air) | 0.2942 |
| 1.0 | 0.3321 |
| 3.0 | 0.4850 |
| 10. | 0.7281 |
| 30. | 1.052 |
| 100. | 1.572 |
| 300. | 2.267 |
| 1000. | 3.387 |

### 6.2.2 Average Drag and Heat Transfer

Equations (6.14) and (6.20) are for *local* friction $c_f(x)$ and local heat transfer $\mathrm{Nu}_x(x)$ on a flat plate. The *average* friction drag $\bar{c}_f$ and $\overline{\mathrm{Nu}}_L$ on a plate of length $L$ are defined by integration:

$$\bar{c}_f = \frac{\text{Drag}}{\frac{1}{2}\rho U_\infty^2 bL} = \frac{1}{L}\int_0^L c_f dx = 1.328\ \mathrm{Re}_L^{-1/2} = 2c_f(L), \tag{6.23}$$

$$\overline{\mathrm{Nu}}_L = \bar{h}L/k = \frac{1}{L}\int_0^L \mathrm{Nu}_x\, dx \doteq 0.664\ \mathrm{Pr}^{1/3}\mathrm{Re}_L^{1/2} = 2\mathrm{Nu}_x(L). \tag{6.24}$$

Thus, in laminar plate flow, where both $h_x$ and $\tau_w \propto x^{-1/2}$, the average values of $h_x$ and $\tau_w$ equal twice the local values at the trailing edge. In turbulent flow (Section 6.4.5) the averaging effect is much less dramatic.

### 6.2.3 Integral Solution for Laminar Flat Plate Flow

The usefulness of the integral technique is illustrated by assuming approximate shapes for flat plate $u$ and $T$ profiles. The technique was first introduced by von Kármán [12] in a pioneering paper on boundary layer analysis. In laminar flow, the profiles have a rounded, polynomial shape (in contrast to the logarithmic shape of turbulent profiles). We assume that the velocity profile is a polynomial:

$$u(x, y) \doteq U_\infty[A\eta + B\eta^2 + C\eta^3 + D\eta^4], \qquad \eta = y/\delta(x). \tag{6.25}$$

This satisfies the no-slip condition, $u(x, 0) = 0$. We need four realistic conditions to evaluate the constants $(A, B, C, D)$. One of these is found by evaluating Eq. (6.7a) at the wall:

*At $y = 0$:*

$$\frac{\partial^2 u}{\partial y^2} = 0. \tag{6.26}$$

The three remaining conditions ensure that $u$ merges smoothly into the outer layer:

*At $y = \delta$:*

$$u = U_\infty, \qquad \frac{\partial u}{\partial y} = \frac{\partial^2 u}{\partial y^2} = 0. \tag{6.27}$$

Substituting Eqs. (6.26) and (6.27) into (6.25), we find that $A = 2$, $B = 0$, $C = -2$, and $D = 1$. Then our realistic profile guess is

$$u/U_\infty \doteq 2\eta - 2\eta^3 + \eta^4. \tag{6.28}$$

This is not definitive: other guesses such as sine waves, hyperbolic tangents, or cubic polynomials could also be accurate shapes (see, for example, [13, Table 10.1]). Substitution of Eq. (6.28) into the integral relation (6.5) gives

$$\tau_w \doteq \frac{d}{dx}\left[\rho U_\infty^2 \frac{37}{315}\delta\right] \doteq \mu(2U_\infty/\delta),$$

or

$$\delta\, d\delta \doteq \frac{630}{37}\frac{\mu dx}{\rho U_\infty}.$$

Integrating, assuming that $\delta = 0$ at the leading edge ($x = 0$), we obtain

$$\delta^2 \doteq \frac{1260}{37}\frac{\mu x}{\rho U_\infty},$$

or

$$\frac{\delta}{x} \doteq \frac{5.84}{\mathrm{Re}_x^{1/2}}. \tag{6.29}$$

This predicts correctly that the laminar boundary layer thickness grows as $x^{1/2}$, but the constant 5.84 is about 17% high.

With $\tau_w \doteq 2\mu U_\infty/\delta$ and $\delta(x)$ known, we find that the friction factor is, approximately,

$$c_f = 2\tau_w/\rho U_\infty^2 \doteq \left(\frac{148/315}{\rho U_\infty x/\mu}\right)^{1/2} = \frac{0.685}{\mathrm{Re}_x^{1/2}}. \tag{6.30}$$

Thus, $\tau_w$ varies as $x^{-1/2}$, and the constant 0.685 is 3% high compared to the exact Blasius result, Eq. (6.14). Thus the simplicity and accuracy of the integral method has led to its wide use in analyzing both velocity and thermal boundary layers [1, 4, 13].

The heat transfer is estimated from a polynomial approximation to the temperature profile of Fig. 6.1(a), which satisfies

*At $y = 0$:*

$$T = T_w, \qquad \frac{\partial^2 T}{\partial y^2} = 0, \tag{6.31a}$$

*At $y = \Delta$:*

$$T = T_\infty, \qquad \frac{\partial T}{\partial y} = \frac{\partial^2 T}{\partial y^2} = 0. \tag{6.31b}$$

The polynomial that matches these conditions is similar to $u/U_\infty$:

$$\frac{T - T_w}{T_\infty - T_w} = 2\zeta - 2\zeta^3 + \zeta^4, \qquad \zeta = y/\Delta(x). \tag{6.32}$$

Substitution of $T$ from Eq. (6.32) and $u$ from Eq. (6.28) into the energy integral equation (6.6) gives the following:

$$q_w'' = \rho c_p U_\infty \frac{d}{dx}\left[(T_w - T_\infty)\frac{2\delta}{15}\left(r^2 - \frac{9r^4}{56} + \frac{r^5}{24}\right)\right] = \frac{2k(T_w - T_\infty)}{r\delta}, \tag{6.33}$$

where $r = \Delta/\delta$ is assumed less than unity in the integration, which is accurate for $\mathrm{Pr} \geqslant 0.7$ (gases and nonmetallic liquids). Given $\delta(x)$ from Eq. (6.29), this is a first-order ordinary differential equation for $r(x)$ that must be solved numerically if $T_w$ varies arbitrarily with $x$. However, if $(T_w - T_\infty)$ is constant, then $r$ is constant also, that is, the velocity and temperature profiles are *similar* in growth rate. Equation (6.33) reduces to

$$\left(r^3 - \frac{9r^5}{56} + \frac{r^6}{24}\right)\delta d\delta = \frac{15\,\alpha\,dx}{U_\infty}.$$

Introducing $\delta d\delta = 630\nu dx/(37U_\infty)$ from Eq. (6.29), we find simply that the constant ratio $r = \Delta/\delta$ is given by

$$r^3 - \frac{9r^5}{56} + \frac{r^6}{24} \doteq \frac{37}{42}\frac{\alpha}{\nu} = \frac{37}{42\,\text{Pr}} \tag{6.34}$$

for laminar flow past a flat plate with uniform wall temperature.

This approximate integral solution shows that $\Delta$ and $\delta$ grow in the same manner, as $x^{1/2}$, and their ratio is a function only of the Prandtl number.

From Eq. (6.33), the local heat transfer coefficient is $h_x = q''_w/(T_w - T_\infty) \doteq 2k/(r\delta)$. With $\delta(x)$ given by Eq. (6.29), the local Nusselt number thus becomes

$$\text{Nu}_x = h_x x/k \doteq 2x/r\delta = 0.342\,\text{Re}_x^{1/2} r^{-1}, \tag{6.35}$$

where $r(\text{Pr})$ is given approximately by Eq. (6.34) if $\text{Pr} \geqslant 0.7$. This is very close to the exact solution found numerically by E. Pohlhausen and discussed earlier:

$$\text{Nu}_x = 0.332\,\text{Re}_x^{1/2} f(\text{Pr}). \tag{6.36}$$

We may compare Pohlhausen's tabulated values of $f(\text{Pr})$ with values of $r^{-1}$ from Eq. (6.34) in the following table:

| Pr | 0.01 | 0.1 | 0.7 | 1.0 | 10.0 | 100.0 | 1000.0 |
|---|---|---|---|---|---|---|---|
| $f(\text{Pr})$ | 0.155 | 0.422 | 0.881 | 1.000 | 2.193 | 4.734 | 10.20 |
| $r^{-1}$ | — | — | 0.878 | 1.000 | 2.226 | 4.831 | 10.43 |

The agreement is good; no value is shown for $r^{-1}$ at low Pr because Eq. (6.34) is not valid. For $\text{Pr} \geqslant 0.7$, an excellent approximation to $f(\text{Pr})$ and $r^{-1}$ is $\text{Pr}^{1/3}$. Thus the accepted correlation for laminar flat plate flow with uniform wall temperature in the range $\text{Pr} \geqslant 0.7$ is

$$\text{Nu}_x \doteq 0.332\,\text{Re}_x^{1/2}\,\text{Pr}^{1/3}. \tag{6.37}$$

We take Eq. (6.37) as a design formula for laminar flow. When written out dimensionally, it takes the form

$$h_x = 0.332\,\rho^{1/2}\,U_\infty^{1/2}\,k^{2/3}\,c_p^{1/3}\,x^{-1/2}\,\mu^{-1/6}. \tag{6.38}$$

We see that $h_x$ increases with velocity as $U_\infty^{1/2}$, whereas from our previous result (6.14) the wall shear increased as $U_\infty^{3/2}$. Thus an attempt to increase forced convection by raising the flow rate will cause a dramatic increase in required pumping power. At some point the cost of pumping outweighs the gain in heat transfer efficiency. This is a common design difficulty in convection, and other examples will be cited.

Equation (6.37) may also be written in terms of the local Stanton number:

$$\mathrm{St}_x \equiv \frac{\mathrm{Nu}_x}{\mathrm{Re}_x \mathrm{Pr}} \cong 0.332\,\mathrm{Re}_x^{-1/2}\,\mathrm{Pr}^{-2/3}. \tag{6.39}$$

By comparing with the friction solution (6.14) we see that

$$\mathrm{St}_x \cong \frac{1}{2} c_f \mathrm{Pr}^{-2/3}. \tag{6.40}$$

This is the special case of the Reynolds analogy (5.37) for laminar flow past a flat plate. It is also used for turbulent flow (see, for example, [1, Eqs. 12–14]) but is less accurate. We will derive an alternate turbulent formula in Section 6.4.5.

### 6.2.4 Variable Properties—The Film Temperature

The preceding result (6.37) was based on an assumption of constant fluid properties ($\rho$, $\mu$, $c_p$, $k$), whereas in practice these quantities vary with temperature. Unless otherwise stated, it is recommended that, in convection formulas such as (6.37), fluid properties be evaluated at the *film temperature*

$$T_f = \frac{1}{2}(T_w + T_\infty). \tag{6.41}$$

This is a reasonable "boundary layer averaging" procedure and allows a simplified formula to achieve more generality.

A few empirical formulas in Chapters 6 and 7 will cite a different method of evaluating fluid properties, based on the particular correlation developed by that experimenter.

### 6.2.5 Transition to Turbulence

All of the formulas in this section are for *laminar* flow past a flat plate. But all laminar flows become unstable and undergo transition to turbulence if the Reynolds number is sufficiently high. For flat plate flows we will use the following transition estimate:

$$\mathrm{Re}_{x,\mathrm{tr}} \doteq 500{,}000. \tag{6.42}$$

This is strictly an average estimate. Transition Reynolds numbers may vary according to (1) the particular velocity and length scale used, (2) the geometry of the flow, and (3) additional effects such as wall roughness, freestream noise, and pressure gradient. For further details, see [4, Chap. 5] and [13, Chaps. 16 and 17].

## Example 6.1

A flat plate 120 cm long and 200 cm wide at 10°C is immersed in an air stream at 30°C and 1 atm with velocity 6 m/s. Estimate (a) $\tau_w$ at $x = 60$ cm; (b) $q''_w$ at $x = 60$ cm; (c) the total friction drag on one side of the plate; (d) the total heat flux on one side of the plate.

*Solution* At the film (average) temperature $T_f = \frac{1}{2}(10° + 30°) = 20°\text{C}$, the air properties are $\rho = 1.19\ \text{kg/m}^3$, Pr $= 0.71$, $k = 0.0251\ \text{W/m}\cdot\text{K}$, $\nu = 15 \times 10^{-6}\ \text{m}^2/\text{s}$, and $c_p = 1010\ \text{J/kg}\cdot\text{K}$. The overall Reynolds number is

$$\text{Re}_L = \frac{U_\infty L}{\nu} = \frac{(6\ \text{m/s})(1.2\ \text{m})}{15 \times 10^{-6}\ \text{m}^2/\text{s}} = 480{,}000.$$

Since this is less than our recommended $\text{Re}_{\text{tr}} = 500{,}000$, we assume that the flow is laminar over the entire plate. At $x = 60$ cm,

$$\text{Re}_x = \frac{(6)(0.6)}{15 \times 10^{-6}} = 240{,}000.$$

The local friction factor is given by Eq. (6.14):

$$c_f = \frac{0.664}{\text{Re}_x^{1/2}} = \frac{0.664}{240{,}000^{1/2}} = 0.00136,$$

whence

$$\tau_w\,(60\ \text{cm}) = c_f\left(\frac{1}{2}\rho U_\infty^2\right) = (0.00136)\left(\frac{1}{2}\right)(1.19)(6)^2$$
$$= 0.029\ \text{Pa}. \quad [\textit{Ans. (a)}]$$

The local Nusselt number at this point is, from Eq. (6.37), $\text{Nu}_x = 0.332\ \text{Re}_x^{1/2}\,\text{Pr}^{1/3} = 0.332(240{,}000)^{1/2}(0.71)^{1/3} = 145$, so that

$$h_x(60\ \text{cm}) = \frac{\text{Nu}_x k}{x} = \frac{(145)(0.0251)}{0.6} = 6.1\ \text{W/m}^2\cdot\text{K},$$

or

$$q''_w = h_x(T_w - T_\infty) = 6.1(10° - 30°) = -120\ \text{W/m}^2. \quad [\textit{Ans. (b)}]$$

The negative sign indicates heat flow fluid to the wall. From Eq. (6.23), the average friction drag coefficient is

$$\bar{c}_f = \frac{1.328}{\text{Re}_L^{1/2}} = \frac{1.328}{480{,}000^{1/2}} = 0.00192,$$

so that the drag force on one side of the plate, $A_w = bL$, is

$$F = \bar{c}_f\left(\frac{1}{2}\rho U_\infty^2\right)(bL) = (0.00192)\left(\frac{1}{2}\right)(1.19)(6)^2(1.2)(2.0)$$
$$= 0.10 \text{ N. } [Ans. (c)]$$

This is a very small force. The drag could be *much* larger if the plate were thick or shaped, thus creating pressure drag or "form drag" (see, for example, [14, Section 7.6]).

Finally, from Eq. (6.24) the average Nusselt number is

$$\overline{\text{Nu}}_L = 0.664\,\text{Re}_L^{1/2}\,\text{Pr}^{1/3} = 0.664(480{,}000)^{1/2}(0.71)^{1/3}$$
$$= 410$$

or

$$\bar{h} = \overline{\text{Nu}}_L k/L = 410(0.0251)/(1.2) = 8.6 \text{ W/m}^2\cdot\text{K}.$$

Then the total heat flux from one side of the plate is

$$q_w = \bar{h}(bL)(T_w - T_\infty) = (8.6)(1.2)(2.0)(10^\circ - 30^\circ)$$
$$= -410 \text{ W. } [Ans. (d)]$$

Again the negative sign indicates heat flux from fluid to wall. The local and average rates are rather modest, typical of gas flow at low velocity (Table 1.1). ■

## 6.3 Laminar Flow in Ducts

Flow through a uniform straight duct is a special case of internal flow for which numerous theories and experiments have been reported. The resulting formulas are similar to the flat plate results of Section 6.2, except that duct diameter $D$ is the length scale and average velocity $V$ is used, based on volume flow through the duct:

$$V = \frac{Q}{A} = \frac{1}{A}\int u\,dA. \tag{6.43}$$

Let us first outline some details of duct flow and heat transfer.

### 6.3.1 Flow in the Entrance Region

Figure 6.3 shows flow entering a uniform straight duct. The no-slip and no-temperature-jump conditions cause velocity and thermal boundary layers to grow, as in Fig. 6.3(b). Since the duct is of finite width, these boundary layers eventually meet in the center, after

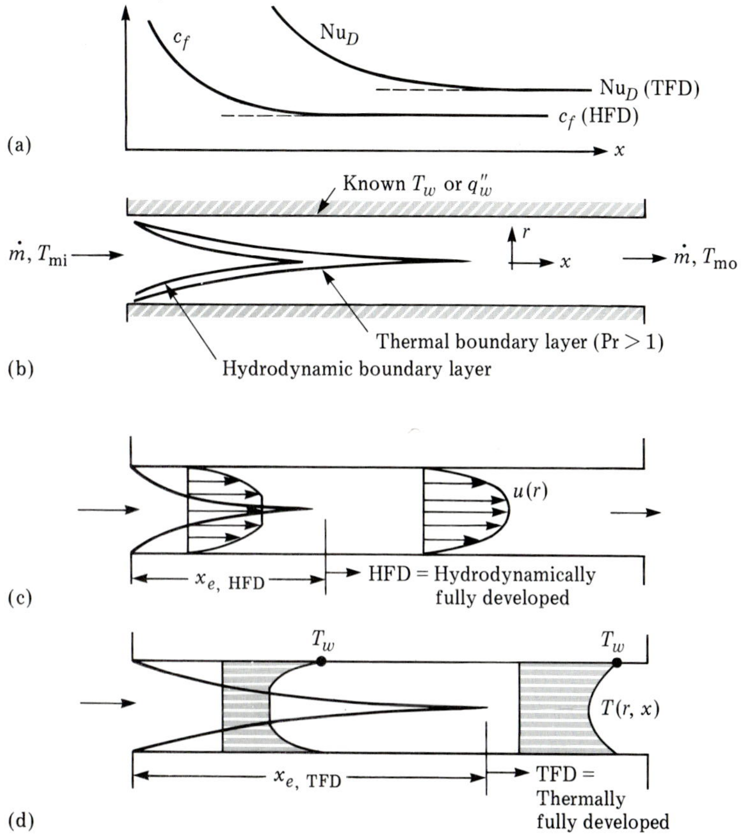


**Figure 6.3 Flow in the entrance region of a duct: (a) wall friction and Nusselt number distribution; (b) boundary layer development; (c) developing velocity profiles; (d) developing temperature profiles.**

which the duct is entirely filled with boundary layer. If $\mathrm{Pr} \neq 1$, the hydrodynamic and thermal layers grow at different rates.

In the entrance region, the velocity and temperature profiles change in shape, as in Figs. 6.3(c, d). Wall gradients are high, hence friction and heat transfer are higher than average, just as in the flat plate boundary layer. After the boundary layers meet, the wall friction and heat transfer coefficient level off and approach constant values in the so-called fully developed region, as sketched in Fig. 6.3(a). The distance required to achieve fully developed conditions is called the *entrance length*, $x_e$. We distinguish between hydrodynamically fully developed flow (HFD) and thermally fully developed flow (TFD); the two have different entrance lengths if $\mathrm{Pr} \neq 1$, as illustrated in Fig. 6.3(b).

In the fully developed region, the velocity profile is independent of axial distance $x$, $u = u(r)$, and the transverse velocity $v = 0$. The fully developed temperature profile $T(r, x)$ has a "shape" that is independent of $x$, but its magnitude may vary with $x$ because of the wall heat transfer.

It is beyond the scope of this text to make a complete analysis of flow in the entrance region. We will concentrate on finding suitable correlations of $c_f$ and $\mathrm{Nu}_D$ as functions of $\mathrm{Re}_D$, Pr, $x/D$, and duct shape. Further analytical and experimental details may be found in [1] and [2].

### 6.3.2 Average Velocity and Temperature

Consider a straight duct of length $L$, cross-sectional area $A$, cross-section perimeter $P$, and wall area $A_w = PL$. If the duct is noncircular, we define its *hydraulic diameter:*

$$D_h = 4A/P. \tag{6.44}$$

This is a useful length scale for correlating noncircular duct data (see, for example, [14, Section 6.6]). For a circular tube, $D_h = D$ exactly. For noncircular ducts, we define $\mathrm{Nu}_{Dh} = hD_h/k$ and $\mathrm{Re}_{Dh} = \rho V D_h/\mu$.

In steady flow, the mass flow in the duct is constant:

$$\dot{m} = \rho V A = GA = \text{constant}. \tag{6.45}$$

If density is nearly constant, we define $V$ by Eq. (6.43). If density varies significantly down the duct, it is convenient to use the "mass velocity,"

$$G = \frac{\dot{m}}{A} = \overline{\rho V} = \frac{1}{A}\int \rho u dA, \tag{6.46}$$

which remains constant. The appropriate Reynolds number for variable density is $\mathrm{Re}_D = \mathrm{G}D/\mu$.

Meanwhile, the concept of average or *mixed-mean* fluid temperature arises from the steady-flow energy equation (1.2), which states that the heat transferred from the wall must equal the increase in fluid enthalpy:

$$q_w = \dot{m}c_p\,(T_{\mathrm{mo}} - T_{\mathrm{mi}}) \tag{6.47}$$

where subscripts o and i denote the outlet and inlet of the duct, respectively. The mixed-mean temperature, $T_m$, is such that $\dot{m}c_pT_m$ equals the integrated enthalpy flux over the cross section:

$$T_m = \frac{1}{\dot{m}c_p}\int \rho u c_p T\, dA. \tag{6.48}$$

Since $T_m$ is averaged over the cross section, it follows that it is a function of $x$ only. We will illustrate the evaluation of $T_m$ and $V$ for both laminar and turbulent fully developed flow.

### 6.3.3 Mean Temperature Distribution

If through theory or experiment we can establish the heat transfer rate $q_w''(x)$ or the coefficient $h(x)$ in a duct, we can immediately calculate the distribution of $T_m$ or $T_w$ along the duct by writing the steady-flow energy equation for a control volume $dx$ wide:

$$q_w''\,P dx = \dot{m}c_p\, dT_m = h(T_w - T_m)P dx. \tag{6.49}$$

We may readily integrate these from $x = 0$ to any $x$ if conditions on $T_w$ or $q_w''$ are known. The two most common and practical conditions are (1) constant wall heat flux $q_w''$, and (2) constant wall temperature $T_w$.

For the first condition, constant $q_w''$, equate the first and second parts of Eq. (6.49) and integrate, with the result:

$$T_m = T_{mi} + (q_w''P/\dot{m}c_p)x. \tag{6.50}$$

The fluid mean temperature varies linearly throughout the entire tube, as sketched in Fig. 6.4(a). With $T_m$ known, the wall temperature follows from the first and third parts of Eq. (6.49):

$$T_w = T_m + \frac{q_w''}{h(x)}. \tag{6.51}$$

As shown also in Fig. 6.4(a), in the developing region, where $h$ is large (see Fig. 6.3a), $T_w$ changes rapidly with $x$. In the TFD region, where $h$ is constant, $T_w$ varies linearly with the same slope as $T_m$ and differs from $T_m$ by the constant amount $q_w''/h_{TFD}$.

For the second boundary condition, $T_w$ = constant, equate the second and third parts of Eq. (6.49), separate the variables, and integrate. The result is

$$T_m = T_w + (T_{mi} - T_w)\exp\left[-(P/\dot{m}c_p)\int_0^x h\, dx\right], \tag{6.52}$$

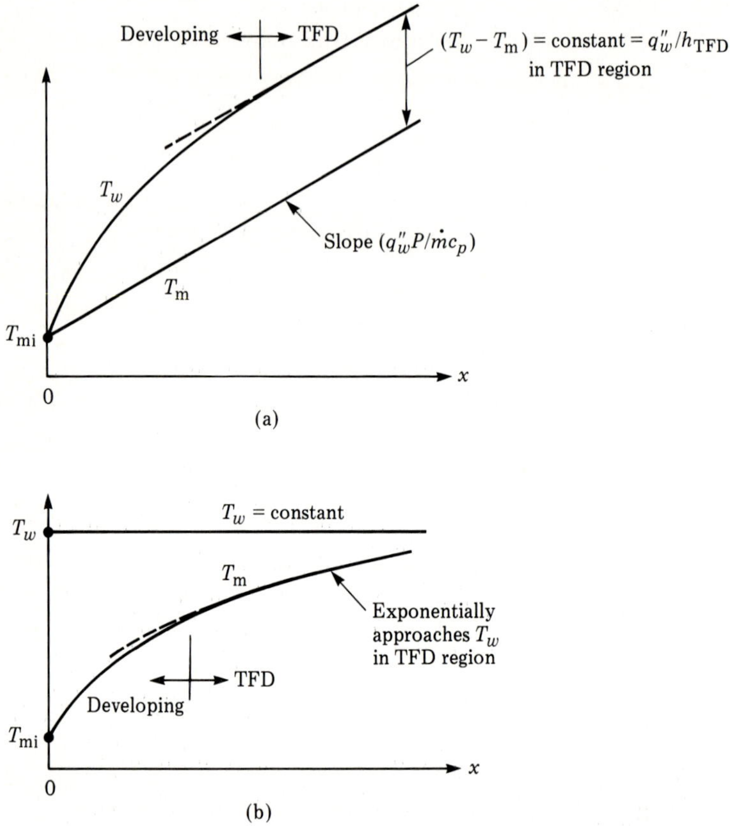


**Figure 6.4 Mean fluid temperature and wall temperature variation along a duct: (a) constant wall heat flux; (b) constant wall temperature. Hot wall shown for convenience.**

where again $T_{mi}$ is the initial fluid temperature at $x = 0$. This distribution is shown in Fig. 6.4(b). In the TFD region, where $h = h_{TFD}$ = constant, this approaches a pure exponential that asymptotically approaches $T_w$ at large $x$. Note, however, that $h_{TFD}$ has a different value for constant $q''_w$ versus constant $T_w$ conditions.

Equations (6.50), (6.51), and (6.52) apply to laminar or turbulent flow, either developing or fully developed. It is necessary only to provide accurate estimates of the heat transfer coefficient $h(x)$ for the given wall conditions.

### 6.3.4 The Log-Mean Temperature Difference

In the case of constant wall temperature, Fig. 6.4(b), the driving temperature difference $(T_w - T_m)$ varies down the duct. Hence, to apply Newton's formula $q_w = hA\Delta T$ to the entire duct, we need to define a suitable *average* temperature difference:

$$\begin{aligned} q_w &= \bar{h}PL(T_w - T_m)_{avg} = \dot{m}c_p(T_{mo} - T_{mi}) \\ &= \dot{m}c_p[(T_{mo} - T_w) - (T_{mi} - T_w)], \end{aligned} \tag{6.53}$$

where we have added and subtracted $T_w$ in the last step to make a point. The appropriate average value of $(T_w - T_m)$ is found by applying Eq. (6.52) across the entire duct from 0 to $L$:

$$(T_{mo} - T_w) = (T_{mi} - T_w)e^{-\bar{h}PL/\dot{m}c_p}, \qquad \bar{h} = \frac{1}{L}\int_0^L h\,dx. \tag{6.54}$$

Solving for $\dot{m}c_p$ in Eq. (6.54) and substituting into Eq. (6.53) give the following result:

$$(T_w - T_m)_{avg} = \Delta T_{LM} = \frac{\Delta T_o - \Delta T_i}{\ln(\Delta T_o/\Delta T_i)}, \tag{6.55}$$

where $\Delta T_o = T_w - T_{mo}$ and $\Delta T_i = T_w - T_{mi}$. The desired average is thus the *log-mean temperature difference* (LMTD), which will be shown in Chapter 10 to be valid under very general conditions in heat exchangers of various geometries. In the limit as $\Delta T_o \to \Delta T_i$, the LMTD is identical to the arithmetic average $\frac{1}{2}(\Delta T_o + \Delta T_i)$. The error in this arithmetic approximation is less than 1% if $\Delta T_o$ differs from $\Delta T_i$ by no more than 40%.

Incidentally, Eq. (6.54) is very handy for computing $T_{mo}$ when $T_w$ and $T_{mi}$ and $\bar{h}$ are known, without bothering to compute $\Delta T_{LM}$ or $q_w$, as we shall see in the following example. This equation is, however, limited to the case of constant wall temperature.

**Example 6.2**

Freon-12 at 0°F and 5 atm enters a 0.25-in.-diameter tube 4 ft long at a flow rate of 200 $\text{lb}_m$/hr. The tube wall temperature is 60°F and $\bar{h} = 370$ Btu/hr · ft$^2$ · °F. Estimate (a) the fluid exit temperature; (b) the LMTD; and (c) $q_w$ in Btu/hr.

*Solution* For Freon-12 take $c_p = 0.22$ Btu/$\text{lb}_m$ · °F. With $\bar{h}$ known, we can use Eq. (6.54) to compute $T_{mo}$:

$$T_{mo} - 60° = (0° - 60°)\exp\left[-\frac{(370)(0.25\pi/12)(4.0)}{(200)(0.22)}\right] = -6.6°F,$$

or

$$T_{mo} = 60 - 6.6 = 53.4°F. \quad [Ans. (a)]$$

Then, from Eq. (6.55), with $T_{mo}$ known,

$$\text{LMTD} = \Delta T_{LM} = \frac{(60 - 53.4) - (60 - 0)}{\ln[(60 - 53.4)/(60 - 0)]} = 24.2°F. \quad [Ans. (b)]$$

Finally, $q_w$ may be computed either from the wall flux or from the fluid enthalpy change:

$$q_w = \bar{h}PL\,\Delta T_{LM} = (370)(0.25\pi/12)(4)(24.2)$$
$$= 2350 \text{ Btu/hr},$$

or

$$q_w = \dot{m}c_p(T_{mo} - T_{mi}) = (200)(0.22)(53.4) - 0)$$
$$= 2350 \text{ Btu/hr}. \quad [Ans. (c)]$$

We will check later, in Problem 6.80, to see if $\bar{h}$ was accurate. ∎

---

### 6.3.5 Fully Developed Laminar Tube Flow

In the HFD region of a circular tube, $v = 0$ and the equation of continuity (6.1a) requires that $\partial u/\partial x = 0$. Hence $u = u(r)$ only, which means that the convective acceleration in the $x$-direction is also zero. The $x$-momentum equation (6.1b), written in cylindrical coordinates, reduces to

$$\frac{\mu}{r}\frac{d}{dr}\left(r\frac{du}{dr}\right) = \frac{dp}{dx} = \text{constant}, \quad \textbf{(6.56)}$$

for laminar flow with constant viscosity. Note that density does not appear because the acceleration is zero. The boundary conditions are no-slip at the wall ($r = r_o$) and radial symmetry at the centerline ($r = 0$):

$$(u)_{r=r_o} = 0; \qquad \left(\frac{du}{dr}\right)_{r=0} = 0. \quad \textbf{(6.57)}$$

This is a classic problem in elementary fluid mechanics (see, for example,

[14, Section 6.4]), and its solution is the fully developed Poiseuille paraboloid profile:

$$u_{\text{HFD}} = u_{\max}\left(1 - \frac{r^2}{r_o^2}\right), \tag{6.58}$$

where

$$u_{\max} = \left(-\frac{dp}{dx}\right)\left(\frac{r_o^2}{4\mu}\right).$$

Inserting $u(r)$ from Eq. (6.58) into Eq. (6.43), we find that

$$V = \tfrac{1}{2}u_{\max} = \left(-\frac{dp}{dx}\right)\left(\frac{r_o^2}{8\mu}\right). \tag{6.59}$$

Then, by differentiation, we find that

$$\tau_w = -\mu\left(\frac{du}{dr}\right)_{r=r_o} = \frac{2\mu u_{\max}}{r_o},$$

or, in dimensionless form,

$$c_f = \frac{2\tau_w}{\rho V^2} = \frac{16\mu}{\rho V D} = \frac{16}{\text{Re}_D}. \tag{6.60}$$

These are the basic hydrodynamic results for laminar HFD tube flow. Note that the pressure gradient is a negative constant, that is, the HFD pressure drops linearly in the flow direction. By comparison of Eqs. (6.59) and (6.60) we find that $(dp/dx) = -2\tau_w/r_o$, or the pressure drop over a length $L$ of HFD flow is

$$\Delta p = 4\frac{L}{D_h}\left(\frac{1}{2}\rho V^2\right)c_f. \tag{6.61}$$

This relation also holds for turbulent flow, with a different formula for $c_f$. Note that $c_f$ is related to average velocity $V$, not to maximum or centerline velocity.

To compute the laminar TFD temperature or Nusselt number, we solve the boundary layer energy equation (6.1c), which for cylindrical coordinates $(r, x)$ is

$$u\frac{\partial T}{\partial x} = \frac{\alpha}{r}\frac{\partial}{\partial r}\left(r\frac{\partial T}{\partial r}\right), \tag{6.62}$$

where $\alpha = k/\rho c_p$ is assumed constant and dissipation is neglected. The boundary conditions are no-temperature-jump at the wall and radial symmetry at the centerline:

$$(T)_{r=r_0} = T_w; \qquad \left(\frac{\partial T}{\partial r}\right)_{r=0} = 0. \tag{6.63}$$

The HFD velocity $u(r)$ is given by Eq. (6.58). Equation (6.62) can be solved for $T(r, x)$ for either constant $q''_w$ or constant $T_w$.

For *constant wall heat flux,* Fig. 6.4(a) shows that all temperatures in the TFD region rise linearly with $x$ at the rate $\partial T/\partial x = q''_w P/\dot{m}c_p$ = constant. Equation (6.62) may be integrated twice, using conditions (6.63) to yield the profile:

$$T = T_w - \frac{Vr_0^2}{8\alpha}\left(\frac{\partial T}{\partial x}\right)\left(3 - \frac{4r^2}{r_0^2} + \frac{r^4}{r_0^4}\right). \tag{6.64}$$

The entire profile changes linearly with $x$ at the rate $(q''_w P/\dot{m}c_p)$.

Inserting $T$ from Eq. (6.64) into Eq. (6.48) yields the mixed-mean temperature for TFD laminar flow with constant $q''_w$:

$$T_{\mathrm{m}} = T_w - \frac{11\,V\,r_0^2}{48\,\alpha}\frac{\partial T}{\partial x}. \tag{6.65}$$

Meanwhile, the heat flux is found by differentiation of Eq. (6.64)

$$q''_w = -k\left.\frac{\partial T}{\partial r}\right|_{r=r_0} = \frac{1}{2}\frac{k}{\alpha}Vr_0\frac{\partial T}{\partial x}. \tag{6.66}$$

Eliminating $(\partial T/\partial x)$ between Eqs. (6.65) and (6.66) gives

$$q''_w = \frac{48}{11}\frac{k}{D}(T_w - T_{\mathrm{m}}),$$

or, since $h = q''_w/(T_w - T_{\mathrm{m}})$ for duct flow,

$$\mathrm{Nu}_{\mathrm{TFD}} = \frac{hD}{k} = \frac{48}{11} = 4.36, \qquad \textit{constant heat flux.} \tag{6.67}$$

Laminar tube flow with uniform $q''_w$ should approach this constant Nusselt number in the fully developed region.

The companion analysis for *constant wall temperature* is much more difficult, since $(\partial T/\partial x)$ is not constant. The details are given in [1, 2, 4], and the final result is

$$\mathrm{Nu}_{\mathrm{TFD}} = 3.66, \qquad \textit{constant wall temperature.} \tag{6.68}$$

These are not large numbers. Recall from Section 5.3.1 that the Nusselt number is the ratio of convection to fluid conduction. A value Nu $\doteq$ 4 indicates that laminar tube flow is a rather ineffective way to enhance the heat transfer rate.

### 6.3.6 Laminar TFD Flow in Noncircular Ducts

The fully developed laminar flow equations (6.56, 6.62) can be readily solved for any shape of cross section. The monograph by Shah and London [2] discusses an astounding number of such solutions, for almost every imaginable duct shape. A partial list of their solutions is given in Table 6.3.

**Table 6.3** **Friction factor and Nusselt number for fully developed flow in ducts [2]**

| Duct Shape | | $Nu_H$† | $Nu_T$† | $c_f Re$‡ |
|---|---|---|---|---|
| 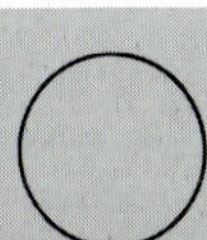 | | 4.36 | 3.66 | 16.00 |
| 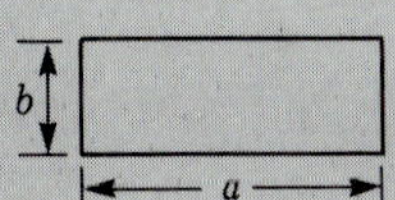 | $a/b =$ 1 | 3.61 | 2.98 | 14.23 |
| | 2 | 4.12 | 3.39 | 15.55 |
| | 3 | 4.79 | 3.96 | 17.09 |
| | 4 | 5.33 | 4.44 | 18.23 |
| | 6 | 6.05 | 5.14 | 19.70 |
| | 8 | 6.49 | 5.60 | 20.58 |
| | $\infty$ | 8.24 | 7.54 | 24.00 |
| 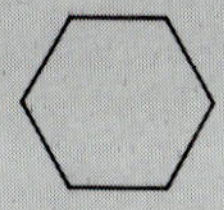 | | 4.00 | 3.34 | 15.05 |
| 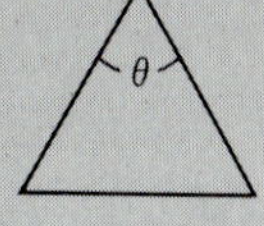 | $\theta =$ 10° | 2.45 | 1.61 | 12.47 |
| | 30° | 2.91 | 2.26 | 13.07 |
| | 60° | 3.11 | 2.47 | 13.33 |
| | 90° | 2.98 | 2.34 | 13.15 |
| | 120° | 2.68 | 2.00 | 12.74 |
| 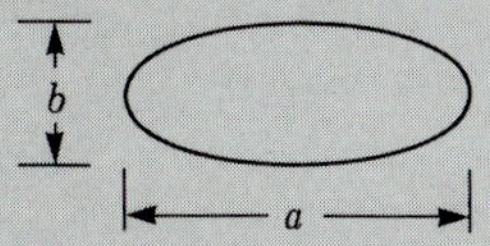 | $a/b =$ 1 | 4.36 | 3.66 | 16.00 |
| | 2 | 4.56 | 3.74 | 16.82 |
| | 4 | 4.88 | 3.79 | 18.24 |
| | 8 | 5.09 | 3.72 | 19.15 |
| | 16 | 5.18 | 3.65 | 19.54 |

†$Nu = hD_h/k$

‡$Re = VD_h/\nu, \quad D_h = 4A/P$

In all cases, the values in Table 6.3 are based on the duct hydraulic diameter $D_h = 4A/P$. The Nusselt numbers correspond to constant $q''_w$ and $T_w$, respectively:

$$\mathrm{Nu}_H = hD_h/k, \qquad \textit{constant wall heat flux,} \tag{6.69a}$$

$$\mathrm{Nu}_T = hD_h/k, \qquad \textit{constant wall temperature.} \tag{6.69b}$$

The friction factor and Reynolds number are defined as

$$c_f\mathrm{Re} = (2\bar{\tau}_w/\rho V^2)(VD_h/\nu). \tag{6.70}$$

The tabulated Nusselt numbers range from 2 to 8, only slightly better than pure conduction. The relative merit of these shapes is obscured by the large variation of $D_h$ among the various ducts. Actually, the circular section is the most efficient in the sense that, for a given wall area, it delivers maximum heat flux per unit friction loss.

## Example 6.3

Engine oil at 60°C flows at 0.5 kg/s in a duct with $T_w = 20°\mathrm{C} =$ constant. Assuming fully developed flow, estimate (a) $q''_w$ and (b) $\Delta p/L$, for a 3-cm-diameter tube and also for a $3 \times 1$ rectangular duct of equal wall area. Compare.

*Solution* For engine oil at $T_f = (20 + 60)/2 = 40°\mathrm{C}$, $\rho = 876$ kg/m$^3$, $k = 0.144$ W/m · K, and $\nu = 2.4 \times 10^{-4}$ m$^2$/s.

For the circular tube, $D = D_h = 3$ cm, and average velocity is

$$V = \dot{m}/\rho A = (0.5\ \mathrm{kg/s})/[876\ \mathrm{kg/m^3}\left(\frac{\pi}{4}\right)(0.03\ \mathrm{m})^2] = 0.8074\ \mathrm{m/s}.$$

Then the Reynolds number of the tube is

$$\mathrm{Re}_D = VD/\nu = (0.8074)(0.03)/(2.4 \times 10^{-4}) = 101.$$

The flow is definitely laminar, and from Table 6.3, $\mathrm{Nu}_T = 3.66$. The heat transfer coefficient is

$$h = \frac{\mathrm{Nu}_T k}{D} = \frac{(3.66)(0.144)}{(0.03)} = 17.6\ \mathrm{W/m^2 \cdot K}.$$

The heat flux is thus

$$q''_w = h(T_w - T_m) = 17.6(20 - 60) = -702\ \mathrm{W/m^2}. \quad [\textit{Ans. (a)}]$$

The friction factor is $c_f = 16/\mathrm{Re} = 16/101 = 0.159$, and from Eq. (6.61),

$$\Delta p/L = (4/D)\left(\frac{1}{2}\rho V^2\right)c_f = (4/0.03)\left[\frac{1}{2}(876)(0.8074)^2(0.159)\right]$$

$$= 6040\ \mathrm{Pa/m}. \quad [Ans.\ (b)]$$

A rectangular $3 \times 1$ duct with the same wall area should have the same perimeter: $2(b + 3b) = \pi D = \pi(0.03)$, or $b = 1.18$ cm. Its hydraulic diameter would be $4A/P = 4b(3b)/2(b + 3b) = 1.767$ cm. Its average velocity is $\dot{m}/\rho(3b^2) = V = 1.371$ m/s. Finally, its Reynolds number is

$$VD_h/\nu = \frac{(1.371)(0.01767)}{2.4 \times 10^{-4}} = 101.$$

From Table 6.3, $c_f\mathrm{Re} = 17.09$; hence $c_f = 17.09/101 = 0.169$. Also, $\mathrm{Nu}_T = 3.96$, so

$$h = \frac{(3.96)(0.144)}{0.01767} = 32.2\ \mathrm{W/m^2 \cdot K}.$$

The heat transfer is

$$q''_w\,(\text{rectangle}) = 32.2(20 - 60) = -1290\ \mathrm{W/m^2}. \quad [Ans.\ (a)]$$

The pressure drop is

$$\Delta p/L(\text{rectangle}) = (4/0.01767)\left[\frac{1}{2}(876)(1.371)^2\right](0.169)$$

$$= 31{,}500\ \mathrm{Pa/m}. \quad [Ans.\ (b)]$$

Being smaller for the same wall area, the rectangle has more heat flux but much more pressure drop. For the tube, the ratio

$$\frac{q''_w}{\Delta p/L} = \frac{702}{6040} = 0.116\ \mathrm{W/Pa \cdot m},$$

whereas for the $3 \times 1$ rectangle the ratio is

$$\frac{q''_w}{\Delta p/L} = \frac{1290}{31{,}500} = 0.041\ \mathrm{W/Pa \cdot m},$$

or 65% less heat flux per unit pumping power. As stated, for a given perimeter, the circular tube is the most efficient cross section in this sense. ■

### 6.3.7 Laminar Entrance Region Correlations

As mentioned, the analysis of flow in a duct entrance is beyond the scope of this book, but excellent treatments are contained in [1] and [2]. Both the pressure drop and wall heat flux can be computed with good accuracy for a wide variety of duct shapes. Here we report results only for a circular tube.

The overall pressure drop $p(0) - p(x)$ reflects not only the increased shear owing to the thin entry boundary layers but also the momentum change as the flow develops from a "slug" or uniform profile into a paraboloid shape. We define an "apparent" friction factor associated with total $\Delta p$:

$$c_{f,\,\text{app}} = \frac{p(0) - p(x)}{1/2\,\rho\,V^2}\frac{D}{4x}. \tag{6.71}$$

Although dimensional analysis would predict $c_{f,\,\text{app}} = f(x/D, \text{Re}_D)$, the theory shows that only a single parameter $(x/D)/\text{Re}_D$ is needed. The computed variation of entrance friction is shown in Fig. 6.5. As $(x/D)/\text{Re}_D$ becomes large ($>1$), $c_{f,\,\text{app}}$ approaches the fully developed value $c_f = 16/\text{Re}_D$. The hydrodynamic entry length is defined as the point where local $c_f$ is within 2% of $c_{f,\,\text{HFD}}$. The accepted estimate is

$$x_{e,\,\text{HFD}} \doteq 0.05\,\text{Re}_D D. \tag{6.72}$$

Thus, large laminar Reynolds numbers yield very long entry lengths — up to 100 diameters if $\text{Re}_D = 2000$, at which point transition to turbulence is usually encountered.

An excellent curve-fit expression for $c_{f,\,\text{app}}\text{Re}_D$ in Fig. 6.5 has been suggested by Shah [18]:

$$c_{f,\,\text{app}}\text{Re}_D \doteq \frac{3.44}{\zeta^{1/2}} + \frac{0.31/\zeta + 16 - 3.44/\zeta^{1/2}}{1 + 0.00021/\zeta^2}, \tag{6.73}$$

where $\zeta = (x/D)/\text{Re}_D$. The error is no more than 3%.

Also plotted in Fig. 6.5 (from tabulated values in [2]) are the mean Nusselt numbers for constant wall heat flux, $\overline{\text{Nu}}_H$, and for constant wall temperature, $\overline{\text{Nu}}_T$, versus a parameter called the Graetz number, after L. Graetz (1883):

$$\lambda = \text{Graetz number} = \frac{x/D}{\text{Pe}}, \qquad \text{Pe} = \text{Re}_D\text{Pr} \tag{6.74}$$

The product $(\text{Re}_D\text{Pr})$ is called the Péclet number, after E. Péclet (1860). A curve-fit for $\overline{\text{Nu}}_T$ as $\text{Pr} \to \infty$ was given by Hausen [19]:

$$\overline{\text{Nu}}_T(\text{Pr} \to \infty) \doteq 3.66 + \frac{0.0668/\lambda}{1 + 0.04/\lambda^{2/3}}. \tag{6.75}$$

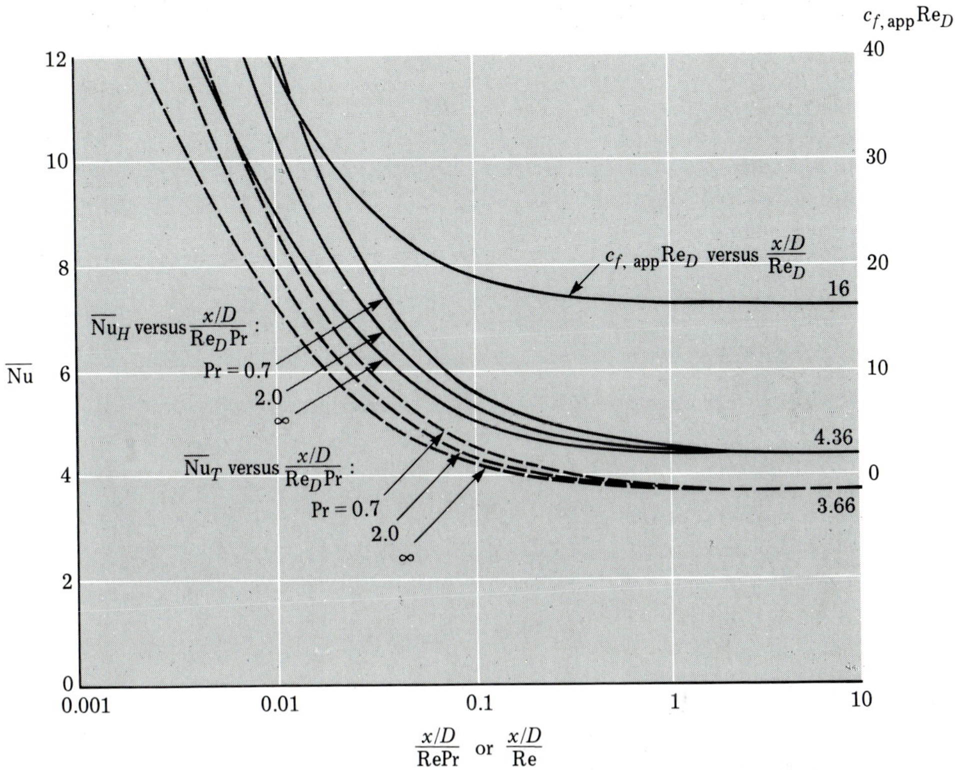


**Figure 6.5** **Laminar friction and Nusselt number in the entry of a circular tube (plotted from tabulated values in [2]).**

This represents a flow with a very long thermal entry length and a very short hydrodynamic entry length, such as occurs, for example, with a heavy oil flow.

The thermal entry length may be defined as the point where local Nu has dropped to within 2% of $\mathrm{Nu}_{\mathrm{TFD}}$. This occurs approximately at $\lambda = 0.05$ [1, p. 111]. Thus

$$x_{\mathrm{e,\,TFD}} \doteq 0.05\,\mathrm{Re}_D\mathrm{Pr}D, \tag{6.76}$$

similar to Eq. (6.72). This length is short for liquid metals (Pr << 1) but quite long for oils (Pr >> 1). In fact, most practical problems involving oil convection in ducts are for thermally developing flow conditions.

A very popular experimental correlation by Sieder and Tate [20] accounts for the strong variation $\mu(T)$ of oils and is valid in the developing entry range

$$\overline{\mathrm{Nu}}_T \doteq 1.86\, \lambda^{-1/3} (\mu_m/\mu_w)^{0.14}, \tag{6.77}$$

for $\lambda = (x/D)/(\mathrm{Re}_D \mathrm{Pr}) \leq 0.125$. The data correlated cover the ranges $0.5 < \mathrm{Pr} < 17{,}000$ and $0.0044 < \mu_m/\mu_w < 9.8$. The viscosity correction is important when $(T_w - T_m)$ is large.

## Example 6.4

Transformer oil at 20°C enters a 1-cm-diameter, 1-m-long tube at 0.25 kg/s. The wall temperature is 0°C. Estimate (a) the pressure drop and (b) outlet temperature $T_{mo}$.

*Solution* For transformer oil at $T_f = (20° + 0°)/2 = 10°\mathrm{C}$, $\rho = 885\ \mathrm{kg/m^3}$, $c_p = 1650\ \mathrm{J/kg \cdot K}$, $\nu = 3.8 \times 10^{-5}\ \mathrm{m^2/s}$, $\mathrm{Pr} = 495$, and $k = 0.11\ \mathrm{W/m \cdot K}$. Compute the average velocity:

$$V = \dot{m}/\rho A = \frac{0.25\ \mathrm{kg/s}}{(885\ \mathrm{kg/m^3})\frac{\pi}{4}(0.01\ \mathrm{m})^2} = 3.6\ \mathrm{m/s}.$$

Then

$$\mathrm{Re}_D = \frac{VD}{\nu} = \frac{(3.6)(0.01)}{3.8 \times 10^{-5}} = 947.$$

The flow is laminar ($\mathrm{Re}_D < 2000$), and the friction parameter is

$$\zeta = \frac{L}{(D\mathrm{Re}_D)} = \frac{1.0}{0.01(947)} = 0.106.$$

From Eq. (6.73),

$$c_{f,\,\mathrm{app}}\mathrm{Re}_D \doteq \frac{3.44}{(0.106)^{1/2}} + \frac{0.31/0.106 + 16 - 3.44/(0.106)^{1/2}}{1 + 0.00021/(0.106)^2} = 18.8.$$

Then $c_{f,\,\mathrm{app}} = 18.8/947 = 0.0198$. Then, from Eq. (6.71),

$$p(0) - p(L) = c_{f,\,\mathrm{app}}(4L/D)\left(\frac{1}{2}\rho V^2\right) = (0.0198)(400)\left(\frac{1}{2}\right)(885)(3.6)^2$$

$$= 45{,}400\ \mathrm{Pa}. \quad [\mathit{Ans.\ (a)}]$$

The Graetz number is $\lambda = (L/D)/\mathrm{Pe} = (100)/[947(495)] = 0.000213$. The flow is barely beginning to develop thermally. Since $\mathrm{Pr} \rightarrow$

$\infty$, use Eq. (6.75):

$$\overline{\text{Nu}}_T \doteq 3.66 + \frac{0.0668/0.000213}{1 + 0.04/(0.000213)^{2/3}} = 29.3.$$

(Alternatively, the Sieder-Tate relation (6.77) gives $\overline{\text{Nu}}_T = 27.2$.) Then

$$\bar{h} = \overline{\text{Nu}}_T k/D = \frac{(29.3)(0.11)}{0.01} = 323 \text{ W/m}^2 \cdot \text{K}.$$

The exit temperature difference is computed from Eq. (6.54):

$$T_{\text{mo}} - 0° = (20° - 0°) \exp\left[-\frac{(323)\pi(0.01)(1.0)}{(0.25)(1650)}\right] = 19.51°\text{C},$$

or

$T_{\text{mo}} = 19.5°\text{C}$. [*Ans. (b)*]

The oil bulk temperature drop is only 0.5°C while the pressure drop is nearly half an atmosphere! If this is meant to be an oil cooler, some design changes are in order. ■

## 6.4 Turbulent Boundary Layer Convection

The previous sections were restricted to laminar flows, which invariably become unstable at a finite Reynolds number. The flow then undergoes transition to turbulence, which is a random, chaotic, unsteady motion with time-mean values that can be correlated with reasonable accuracy.

Figure 6.6 is a sketch of the transition process in flat plate flow. The initial leading edge flow is laminar and stable. At about $\text{Re}_x \doteq 10^5$, certain wavelengths of possible disturbances become unstable and grow, but the flow does not immediately become turbulent. There is a finite transition length where the disturbances amplify, become three-dimensional, and form streamwise vortices and small "bursts" of turbulence. The flow then becomes fully turbulent at what we will call the *transition Reynolds number,* which for flat plate flow is about $\text{Re}_{x,\,\text{tr}} \doteq 5 \times 10^5$ on the average. Entirely new turbulent formulas are then needed for $\text{Re}_x > \text{Re}_{x,\,\text{tr}}$. The theory of turbulence is beyond the scope of this text (see [1, 4, 7, 13]). However, before leaping directly into turbulence correlations, we will present some introductory turbulence modeling concepts.

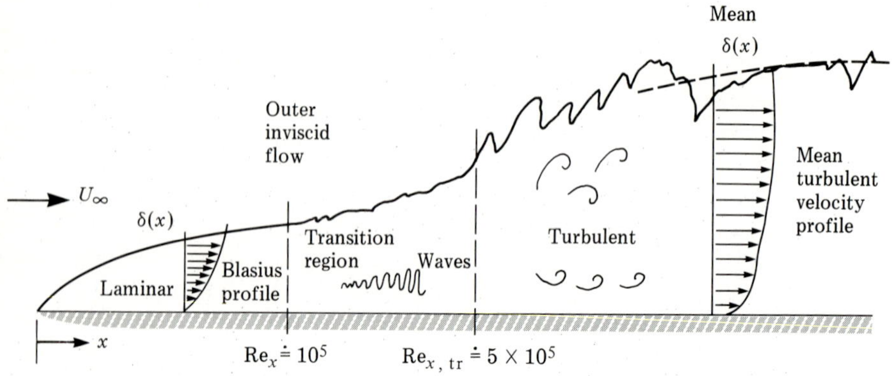


**Figure 6.6 Illustration of the transition process in viscous flow past a sharp flat plate. The boundary layer thickness is exaggerated.**

### 6.4.1 The Turbulent Eddy Viscosity

In laminar flow, we set $\varepsilon_M = \varepsilon_H = 0$ and solve Eqs. (6.1) for the velocities $u$ and $v$, the temperature $T$, and the wall heat transfer coefficient $h_x(x) = q''_x(x)/(T_w - T_\infty)$. The profile shapes $u$ and $T$ are typically well rounded and parabolic in appearance, similar to Fig. 5.1(c). The analysis of laminar flat plate flow is given in Section 6.2.

In turbulent flow, we need formulations for the empirical eddy coefficients $\varepsilon_M(x, y)$ and $\varepsilon_H(x, y)$. The expressions we use here were developed from dimensional and physical arguments put forth by Ludwig Prandtl and Theodore von Kármán and their students in a series of pioneering papers in the 1930s.

Prandtl and von Kármán reasoned that, except for the outermost part of a turbulent boundary layer, eddy viscosity must be dominated by wall effects: wall shear stress $\tau_w$, distance from the wall $y$, fluid density $\rho$, and fluid viscosity $\mu$. Thus, dimensionally,

$$\varepsilon_M = f(\tau_w, y, \rho, \mu).$$

The reader should show from dimensional analysis that Eq. (6.5) is equivalent to the following dimensionless form:

$$\varepsilon_M/\nu = f(y^+), \qquad y^+ = yv^*/\nu, \tag{6.78}$$

where $v^* = (\tau_w/\rho)^{1/2}$ is called the *wall friction velocity*. The variable $y^+$ is a sort of wall-related Reynolds number.

In modeling the empirical function $f(y^+)$ in Eq. (6.78) Prandtl and von Kármán further reasoned that, physically, $\varepsilon_M$ should vanish rapidly near the wall, $\varepsilon_M \propto y^4$, since the wall must strongly damp all turbulent fluctuations. Farther from the wall, $\varepsilon_M$ must approach a linear variation, $\varepsilon_M \propto y$. Data for $\varepsilon_M$ fit these hypotheses, and various empirical formulas have been proposed for the variation $f(y^+)$. Here we adopt the most popular formula, from van Driest [5]:

$$\varepsilon_M/\nu \doteq \tfrac{1}{2}[1 + 4\kappa^2 y^{+2}(1 - e^{-y^+/A})^2]^{1/2} - \tfrac{1}{2}, \tag{6.79}$$

where $\kappa \doteq 0.41$ is called von Kármán's constant and $A \doteq 25$ for smooth wall flat plate or pipe flow. For further formulations of eddy viscosity, see, for example, [4, pp. 476–477]. The constant $A$ is called van Driest's damping factor and may vary with external pressure gradient, wall roughness, and wall suction, as shown in [6].

Examination of Eq. (6.79) reveals three different regions characterizing a near-wall turbulent boundary layer:

1. $y^+ \leqslant 5$: Viscous sublayer, $\varepsilon_M \doteq \nu\kappa^2 y^{+4}/A^2 << \nu$;
2. $5 < y^+ < 40$: Buffer layer, $\varepsilon_M$ comparable to $\nu$;
3. $y^+ \geqslant 40$: Logarithmic layer, $\varepsilon_M \doteq \kappa\, y^+\, \nu >> \nu$.

The significance of the term *logarithmic layer* will become evident in the next section.

### 6.4.2 The Turbulent Law-of-the-Wall Velocity Profile

Using our eddy viscosity formulation from Eq. (6.79), we may develop an accurate expression for the velocity profile $u(y)$ near the wall of a turbulent shear flow. In flat plate flow (zero streamwise pressure gradient), the shear stress in the boundary layer is nearly constant near the wall:

$$\tau(x, y) = (\mu + \rho\varepsilon_M)\frac{\partial u}{\partial y} \doteq \tau_w(x).$$

Using Eq. (6.78) as a guide, we may rewrite this in dimensionless form:

$$\frac{\partial u^+}{\partial y^+} = \frac{1}{1 + \varepsilon_M/\nu}, \qquad u^+ = u/v^*. \tag{6.80}$$

The quantity $u^+$ is called the *law-of-the-wall velocity variable.* We may integrate Eq. (6.80) with respect to $y^+$ at constant $x$, assuming no-slip at the wall, $u^+ = 0$ at $y^+ = 0$:

$$u^+ = \int_0^{y+} \frac{dy^+}{1 + \varepsilon_M/\nu}. \tag{6.81}$$

We use the van Driest expression (6.79) for eddy viscosity. The integral is not known in closed form but may readily be evaluated numerically by, for example, a Runge-Kutta procedure. The results are shown as the solid line in Fig. 6.7, which compares very well with flat plate and pipe-flow data.

Two limiting cases of the turbulent velocity profile are plotted as dashed lines:

*Viscous sublayer, $y^+ \leq 5$:*

$$u^+ \doteq y^+, \tag{6.82a}$$

*Logarithmic layer, $y^+ \geq 40$:*

$$u^+ \doteq \frac{1}{\kappa}\ln(y^+) + B, \tag{6.82b}$$

where $B \doteq 5.0$. The buffer layer, $5 < y^+ < 40$, forms a smooth merging from the sublayer to the log-layer. We see that the log-layer is a good

**Figure 6.7 Semilogarithmic plot of the turbulent velocity law-of-the-wall. The wall is assumed smooth and impermeable.**

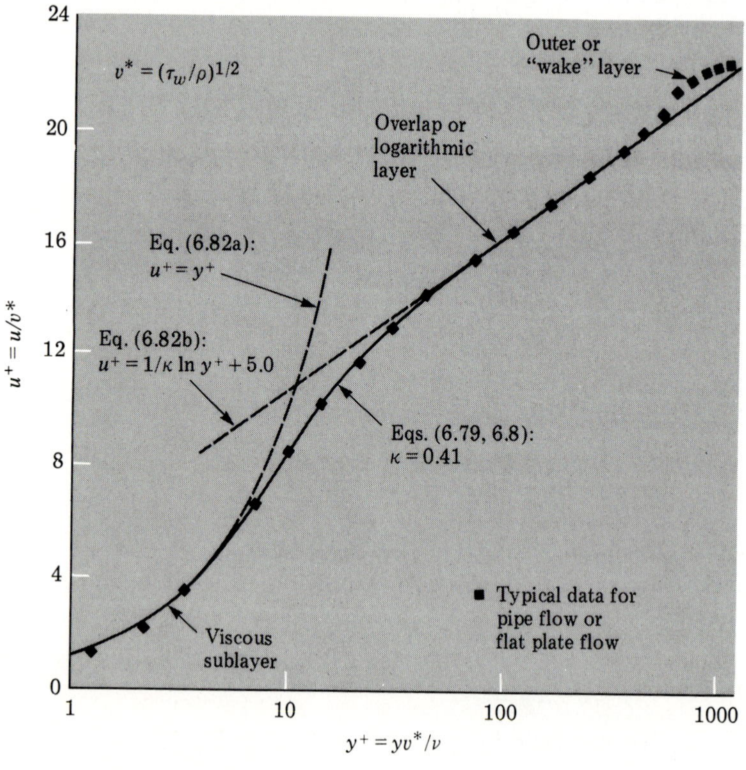

approximation over a large portion of the turbulent boundary layer. As shown in Fig. 6.7, the velocity data tend to rise above the log-law in the outer or "wake" layer, where the velocity matches the stream velocity $U_\infty$. We will ignore this wake effect here.

Finally, we remark that the law-of-the-wall shown in Fig. 6.7 is sensitive to wall roughness and wall suction effects, as shown in [1, 4, 6, 7]. We will use the log-law, Eq. (6.82b), to estimate friction and heat transfer coefficients in a turbulent smooth-wall flow in Section 6.4.5.

## Example 6.5

Air at 20°C and 1 atm flows past a smooth flat plate at 18 m/s. The flow is turbulent, and at $x = 1.5$ m the measured wall shear stress is 0.65 Pa. Using Fig. 6.7, estimate, at this position, (a) $v^*$; (b) the sublayer thickness $y_s$; and (c) the velocity $u$ at $y = 1$ cm.

*Solution* For air take $\rho = 1.19\ \text{kg/m}^3$ and $\nu = 15.7 \times 10^{-6}\ \text{m}^2/\text{s}$. Check $\text{Re}_x = U_\infty x/\nu = (18\ \text{m/s})(1.5\ \text{m})/(15.7 \times 10^{-6}\ \text{m}^2/\text{s}) = 1.72 \times 10^6$. This is greater than the accepted transition Reynolds number $\text{Re}_{\text{tr}} \doteq 500{,}000$ and verifies the assumption of turbulent flow. The friction velocity follows by definition from Eq. (6.78)

$$v^* = (\tau_w/\rho)^{1/2} = (0.65\ \text{N/m}^2/1.19\ \text{kg/m}^3)^{1/2} = 0.74\ \text{m/s}. \quad [\textit{Ans. (a)}]$$

The sublayer thickness is defined by Eq. (6.82a):

$$y_s^+ = y_s v^*/\nu = 5 = y_s(0.74\ \text{m/s})/(15.7 \times 10^{-6}\ \text{m}^2/\text{s}),$$

or

$$y_s = 0.000106\ \text{m} = 0.106\ \text{mm}. \quad [\textit{Ans. (b)}]$$

The sublayer is typically quite thin and cannot be seen on a linear plot of $u$ versus $y$, which is why the semilogarithmic plot of Fig. 6.7 is employed.

To estimate $u$ at $y = 1$ cm, compute

$$y^+ = yv^*/\nu = (0.01)(0.74)/(15.7 \times 10^{-6}) = 471.$$

From Fig. 6.7, this is in the logarithmic region and may be approaching the wake layer, which we ignore. The velocity may be estimated from the log-law, Eq. (6.82b):

$$\frac{u}{v^*} = u/0.74 \doteq \frac{1}{0.41}\ln(471) + 5.0 = 20.01,$$

or

$u = (0.74)(20.01) = 14.8$ m/s. [*Ans. (c)*]

Although quite near the wall, this velocity is 82% of the stream velocity of 18 m/s, which is typical of turbulent flows. ■

---

### 6.4.3 Turbulent Eddy Conductivity

The analysis of turbulent-flow temperature profiles depends on knowledge of the eddy conductivity $\varepsilon_H(x, y)$ in Eq. (6.1c). A crucial simplification of this problem was given over a century ago by Osborne Reynolds [8], who observed that turbulent fluctuations cause a mean transport of both heat and momentum and hence the two processes must be similar. Indeed, measurements over a wide range of gases, liquids, and liquid metals indicate a strong similarity between turbulent heat flux and turbulent shear. Reynolds thus postulated that the two eddy coefficients are equal:

$$\varepsilon_M = \varepsilon_H. \tag{6.83}$$

This is the fundamental form of what is now called the *Reynolds analogy,* a special application of which was mentioned before, Eq. (5.37). By analogy with the molecular Prandtl number $\text{Pr} = \nu/\alpha$, the ratio of eddy coefficients is called the *turbulent Prandtl number:*

$$\text{Pr}_t = \varepsilon_M/\varepsilon_H. \tag{6.84}$$

The Reynolds analogy thus suggests that $\text{Pr}_t = 1.0$, which measurements indicate is quite reasonable. Figure 6.8 shows the data of Blackwell [9] for air flow at various pressure gradients. We see that $\text{Pr}_t$ is approximately unity across the entire boundary layer, except very near the wall, $y^+ < 10$. The increase of $\text{Pr}_t$ in the sublayer, implying that turbulent momentum transport is stronger near the wall, is not important numerically because both $\varepsilon_M$ and $\varepsilon_H$ are very small in this region. Note that the effect of pressure gradient on $\text{Pr}_t$ is slight and can be neglected. Other effects — wall roughness, blowing and suction, and molecular Prandtl number — are also found experimentally to be small, so that we can recommend for analysis purposes that $\text{Pr}_t$ be taken as a constant:

$$\text{Pr}_t \doteq 0.9, \quad \text{or} \quad \varepsilon_H \doteq \varepsilon_M/0.9. \tag{6.85}$$

Equation (6.85), together with van Driest's eddy viscosity formula,

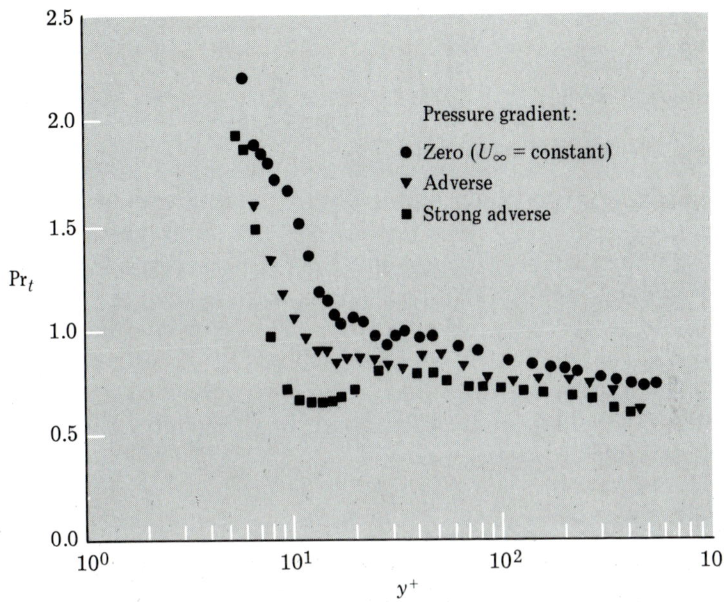


**Figure 6.8 Experimental variation of turbulent Prandtl number across the boundary layer. Data for airflow by Blackwell [9].**

Eq. (6.79), is an accurate algebraic model for the variation of $\varepsilon_H(x, y)$ in a smooth-wall turbulent boundary layer.

### 6.4.4 The Temperature Law-of-the-Wall

With the eddy conductivity known from Eqs. (6.85) and (6.79), we can compute turbulent temperature profiles from Eq. (6.1c). An important special case occurs for modest pressure gradients, such as flat plate or pipe flow, for which the heat flux is approximately constant in the inner and overlap layers. For this case, Eq. (5.22) becomes, at constant $x$,

$$q'' \doteq q''_w = -(k + \rho c_p \varepsilon_H)\frac{dT}{dy} = -\rho c_p(\alpha + \varepsilon_H)\frac{dT}{dy}.$$

We may separate the variables and integrate to obtain

$$\int_{T_w}^{T} dT = T - T_w = -\frac{q''_w}{\rho c_p}\int_0^y \frac{dy}{\alpha + \varepsilon_H}.$$

Introducing the wall-related variable $y^+ = yv^*/\nu$ and substituting $\alpha$

$= \nu/\text{Pr}$ and $\varepsilon_H = \varepsilon_M/\text{Pr}_t$, we obtain the dimensionless law-of-the-wall temperature distribution:

$$T^+ = (T_w - T)\rho c_p v^*/q''_w = \int_0^{y^+} \frac{dy^+}{1/\text{Pr} + \varepsilon_M/\nu \text{Pr}_t}. \tag{6.86}$$

Equation (6.86) is the exact thermal analogy of the velocity law-of-the-wall, Eq. (6.81). Assuming $\text{Pr}_t = 0.9$ and taking $\varepsilon_M/\nu$ from van Driest's formula (6.79) we may carry out the integration numerically for various values of Pr and plot the temperature profiles $T^+(y^+, \text{Pr})$ in Fig. 6.9. There is a family of curves, with molecular Prandtl number as a parameter. The curve for Pr = 1.0 is very similar to the velocity curve $u^+(y^+)$ in Fig. 6.7.

The profiles in Fig. 6.9 have a linear sublayer very near the wall, where $T^+ = \text{Pr}\ y^+$, and then bend over through a "buffer layer" to

**Figure 6.9** **The temperature law-of-the-wall, computed from Eqs. (6.79) and (6.86).**

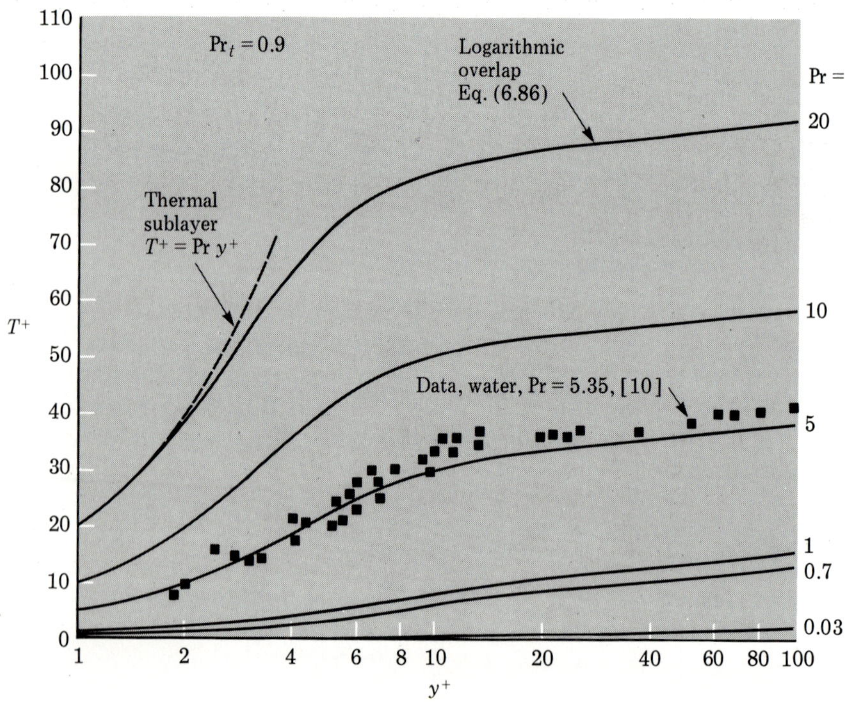

a logarithmic line of the form

*$y^+$ 40:*

$$T^+ = \frac{\mathrm{Pr}_t}{\kappa} \ln(y^+) + A(\mathrm{Pr}), \qquad \textbf{(6.87)}$$

where the intercept parameter $A(\mathrm{Pr})$ increases with Prandtl number, as shown in Fig. 6.9.

The theory of Eq. (6.86) is in good agreement with data shown for water from [10] and also with other data (not shown) for air with $\mathrm{Pr} = 0.7$ [9, 10], a water-glycol solution with $\mathrm{Pr} = 14.3$ [11], transformer oil with $\mathrm{Pr} = 64$ [10], and even mercury with $\mathrm{Pr} = 0.026$ [10]. Thus we may use this law-of-the-wall correlation with confidence for predicting turbulent convection coefficients.

An accurate curve-fit to the computed values of $A(\mathrm{Pr})$ from Eq. (6.87) is as follows:

*$0.7 \leq Pr \leq 10^5$:*

$$A(\mathrm{Pr}) \doteq 12.7\, \mathrm{Pr}^{2/3} - 7.7. \qquad \textbf{(6.88)}$$

We will use this curve-fit in the theories of Sections 6.

**Example 6.6**

Continuing Example 6.5, estimate the wall heat transfer if the wall temperature is 23°C and the air temperature is 17°C at $y$ = 1 cm.

*Solution* For air take $c_p = 1012$ J/kg · °C and $\mathrm{Pr} = 0.71$. At $y$ = 1 cm, $y^+ = 471$ from Example 6.5. From Eq. (6.88) the intercept constant $A$ is

$$A \doteq 12.7(0.71)^{1/3} - 7.7 = 2.41.$$

Since $y^+ = 471$ is in the log-law region of Fig 6.9, we use Eq. (6.87):

$$T^+ \doteq \frac{0.9}{0.41} \ln(471) + 2.41 = 15.92 = (T_w - T)\rho c_p v^*/q''_w,$$

or

$$q''_w = (23 - 17)(1.19)(1012)(0.74)/15.92 = 335 \text{ W/m}^2. \quad [Ans.]$$

The uncertainty in this estimate is ±20%. ■

### 6.4.5 Solution for Turbulent Flat Plate Flow

Transition from laminar to turbulent flow along a flat plate occurs at approximately $\mathrm{Re}_{\mathrm{tr}} \doteq 500{,}000$. The turbulent friction and heat transfer may be analyzed in two ways: (1) use of turbulent profile shapes $u(y)$ and $T(y)$ in the integral relations (6.5) and (6.6), or (2) substitution of $\varepsilon_M$ and $\varepsilon_H$ from Eqs. (6.79) and (6.85) into the boundary layer relations (6.1) and subsequent numerical solution on a digital computer [6, 7]. We will not consider method 2 here.

The classic integral solution for friction, method 1, assumes a one-seventh power-law profile, $u \propto y^{1/7}$, and a wall shear correlation from pipe-flow data, $\tau_w \propto \delta^{-1/4}$. Substitution into Eq. (6.5) and integration (not shown here) leads to formulas first given by Prandtl in 1921 for moderate Reynolds numbers (see, for example, [1, p. 174]):

$$\begin{aligned} &5 \times 10^5 < Re_x < 10^7: \\ c_f &\doteq 0.0592\,\mathrm{Re}_x^{-1/5}, \\ \bar{c}_f &\doteq 0.074\,\mathrm{Re}_L^{-1/5}, \\ \delta/x &\doteq 0.37\,\mathrm{Re}_x^{-1/5}. \end{aligned} \tag{6.89}$$

A more accurate formula uses the log-law, Eq. (6.82b), to represent both $u(y)$ and $\tau_w$ in the integral relation (6.5). The result of this analysis [4, Section 6.6.2] is valid for a smooth, flat plate at any turbulent Reynolds number:

$$\begin{aligned} c_f &\doteq 0.455/\ln^2(0.06\,\mathrm{Re}_x), \\ \bar{c}_f &\doteq 0.523/\ln^2(0.06\,\mathrm{Re}_L). \end{aligned} \tag{6.90}$$

With $c_f$ known, application of the Reynolds analogy, Eq. (6.40), gives a heat transfer formula adequate for Pr near unity:

$$\begin{aligned} &0.7 \leqslant Pr \leqslant 3: \\ \mathrm{St}_x &\doteq 0.0296\,\mathrm{Re}_x^{-1/5}\,\mathrm{Pr}^{-2/3}, \qquad 5 \times 10^5 < \mathrm{Re}_x < 10^7 \\ \overline{\mathrm{St}} &\doteq 0.037\,\mathrm{Re}_L^{-1/5}\,\mathrm{Pr}^{-2/3}. \end{aligned} \tag{6.91}$$

These are traditional formulas utilizing Prandtl's power-law expressions from Eqs. (6.89). They may also be written in Nusselt number form, since Nu = StPrRe:

$$\begin{aligned} &0.7 \leqslant Pr \leqslant 3: \\ \mathrm{Nu}_x &\doteq 0.0296\,\mathrm{Re}_x^{4/5}\,\mathrm{Pr}^{1/3}, \\ \overline{\mathrm{Nu}}_L &\doteq 0.037\,\mathrm{Re}_L^{4/5}\,\mathrm{Pr}^{1/3}. \end{aligned} \tag{6.92}$$

An alternative, more accurate, and more general formula can be found from the temperature law-of-the-wall, Eq. (6.87), as shown in

[4] and [15]. Let $\Pr_t = 1.0$ for convenience and evaluate the velocity and temperature log-laws at the outer edge of the boundary layer, $u = U_\infty$, $y = \delta$, $T = T_\infty$, $y = \Delta$:

$$\frac{U_\infty}{v^*} \doteq \frac{1}{\kappa}\ln\left(\frac{\delta v^*}{\nu}\right) + B,$$

$$(T_w - T_\infty)\rho c_p v^*/q''_w \doteq \frac{1}{\kappa}\ln\left(\frac{\Delta v^*}{\nu}\right) + A.$$

Subtracting these two and rearranging yields

$$\mathrm{St}_x = \frac{c_f/2}{1 + [A - B - (1/\kappa)\ln(\Delta/\delta)](c_f/2)^{1/2}}.$$

The term $\ln(\Delta/\delta)/\kappa$ is much smaller than $(A - B)$ for all Prandtl numbers and is neglected. Introducing $A(\Pr)$ from Eq. (6.88), we obtain the desired result for local Stanton number in turbulent flow past a flat plate:

$$\mathrm{St}_x = \frac{c_f/2}{1 + 12.7(\Pr^{2/3} - 1)(c_f/2)^{1/2}}, \tag{6.93}$$

where $c_f$ is computed from Eq. (6.90). This formula is accurate for any turbulent Reynolds number and for any Prandtl number greater than about 0.5. It is not accurate for liquid metals $(0.004 \leqslant \Pr \leqslant 0.03)$ because the temperature log-law is invalid there (see Fig. 6.9).

Integration over the length of the plate reveals that the average turbulent heat transfer is about 15% higher than the trailing edge local value:

$$\overline{\mathrm{St}} \doteq 1.15\,\mathrm{St}_{x=L}; \qquad \overline{\mathrm{Nu}}_L \doteq 1.15\,\mathrm{Nu}_L. \tag{6.94}$$

Equations (6.93) and (6.94) are believed to be fundamentally sound and agree well with experiments for gases and light liquids. There are few data for comparison at high Prandtl numbers or for $\mathrm{Re}_x$ greater than $10^8$.

### 6.4.6 Combined Laminar/Turbulent Plate Flow

A typical flat plate flow will be laminar up to a transition Reynolds number $\mathrm{Re}_{tr} \doteq 500{,}000$ and turbulent thereafter. If $\mathrm{Re}_L$ is only slightly greater than $\mathrm{Re}_{tr}$, the laminar contribution is important and we need a combined laminar/turbulent formula. The overall heat transfer is a sum of two contributions:

$$\bar{h} = \frac{k}{L}\left(\int_0^{x_{tr}} \mathrm{Nu}_{lam}\frac{dx}{x} + \int_{x_{tr}}^{L} \mathrm{Nu}_{turb}\frac{dx}{x}\right),$$

or

$$\overline{\mathrm{Nu}}_L = \frac{\bar{h}L}{k} = \overline{\mathrm{Nu}}_{\mathrm{tr,\ lam}} + (\overline{\mathrm{Nu}}_L - \overline{\mathrm{Nu}}_{\mathrm{tr}})_{\mathrm{turb}}. \tag{6.95}$$

This is more accurate than simply assuming fully turbulent flow as in Eq. (6.92), but it requires evaluation of three Nusselt numbers, one at $\mathrm{Re}_L$ and two at transition $\mathrm{Re}_{\mathrm{tr}}$.

For Prandtl number near unity and $\mathrm{Re}_L \leqslant 10^7$, use of the power-law expression (6.95) leads to a simple combined formula:

$$\frac{\bar{h}L}{k} \doteq [0.664\ \mathrm{Re}_{\mathrm{tr}}^{1/2} + 0.037(\mathrm{Re}_L^{4/5} - \mathrm{Re}_{\mathrm{tr}}^{4/5})]\ \mathrm{Pr}^{1/3}. \tag{6.96}$$

For $\mathrm{Re}_{\mathrm{tr}} = 500{,}000$, the formula reduces to

$$\frac{\bar{h}L}{k} \doteq (0.037\ \mathrm{Re}_L^{4/5} - 870)\ \mathrm{Pr}^{1/3}. \tag{6.97}$$

Figure 6.10 shows $\overline{\mathrm{Nu}}_L$ versus $\mathrm{Re}_L$ for various Pr as computed from Eqs. (6.94) and (6.95). Equation (6.96), which is not shown, is accurate

**Figure 6.10 Heat transfer for flow past a smooth flat plate assuming $\mathrm{Re}_{\mathrm{tr}}$ = 500,000.**

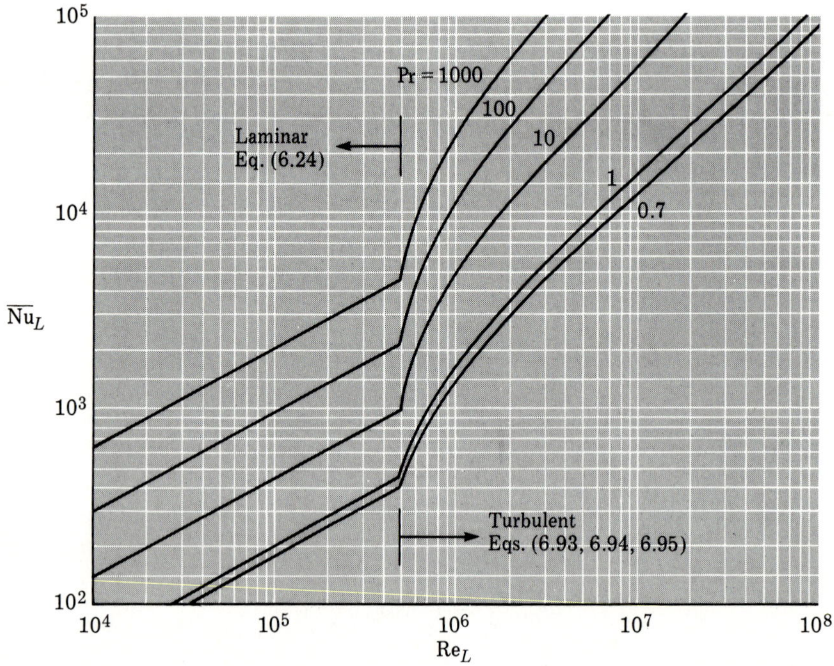

at Pr $\doteq$ 1 but predicts low from 10%–50% at higher Prandtl and Reynolds numbers.

Note from Eqs. (6.89) and (6.91) that, in turbulent flat plate flow $\tau_w \propto U_\infty^{1.8}$ while $h_x \propto U_\infty^{0.8}$. Thus, just as in laminar plate flow, pumping power increases much faster than convective heat transfer as velocity increases, creating an optimization problem for the designer.

## Example 6.7

Repeat Example 6.1 if the fluid is water at 30°C.

*Solution* For water at $T_f$ = 20°C, $\rho$ = 998 kg/m$^3$, Pr = 7.0, $k$ = 0.597 W/m · K, $\nu$ = 1.01 × 10$^{-6}$ m$^2$/s, and $c_p$ = 4180 J/kg · K. Then

$$\text{Re}_L = \frac{U_\infty L}{\nu} = \frac{(6)(1.2)}{1.01 \times 10^{-6}} = 7.13 \times 10^6,$$

and the flow is turbulent over 93% of its surface if $\text{Re}_{\text{tr}}$ = 500,000.

At $x$ = 60 cm, $\text{Re}_x = 3.56 \times 10^6$, which is turbulent also. From Eq. (6.93),

$$c_f = \frac{0.455}{\ln^2[0.06(3.56 \times 10^6)]} = 0.00302,$$

so

$$\tau_w(60\text{ cm}) = c_f(\tfrac{1}{2})\rho U_\infty^2 = (0.00302)(\tfrac{1}{2})(998)(6)^2 = 54\text{ Pa}. \quad [Ans.\ (a)]$$

This is 1900 times more than the airflow of Example 6.1.

Using $c_f/2$ = 0.00151 at $x$ = 60 cm, compute the local Stanton number from Eq. (6.93).

$$\text{St}_x = \frac{0.00151}{1 + 12.7(7.0^{2/3} - 1)(0.00151)^{1/2}} = 0.000653,$$

from which

$$h_x = \text{St}_x \rho c_p U_\infty = (0.000653)(998)(4180)(6) = 16{,}300\text{ W/m}^2.$$

and thus

$$q_w''(60\text{ cm}) = 16{,}300(10° - 30°) = -326{,}000\text{ W/m}^2. \quad [Ans.\ (b)]$$

This is 2700 times more than the air flow of Example 6.1.

The average friction drag from Eq. (6.90) is

$$\bar{c}_f = \frac{0.523}{\ln^2[0.06(7.13 \times 10^6)]} = 0.00311,$$

whence

$$F = \bar{c}_f \frac{1}{2} \rho U_\infty^2 bL = (0.00311)\left(\frac{1}{2}\right)(998)(6)^2(2.0)(1.2)$$
$$= 134 \text{ N.} \quad [\textit{Ans. (c)}]$$

This is 1300 times more than the airflow of Example 6.1 and could be still larger if there were roughness or form-drag effects.

Finally, to evaluate $\bar{h}$ from Eq. (6.95), we need three average Nusselt numbers. With $\mathrm{Re}_L = 7.13 \times 10^6$, compute $c_f(L) = 0.00271$ from Eq. (6.89) and $\mathrm{St}_L = 0.000604$ from Eq. (6.93), whence $\mathrm{Nu}_L = \mathrm{St}_L \mathrm{Re}_L \mathrm{Pr} = 30{,}100$ and $\overline{\mathrm{Nu}}_L = 1.15\ \mathrm{Nu}_L = 34{,}600$.

Then, at $\mathrm{Re}_{tr} = 500{,}000$, $\overline{\mathrm{Nu}}_{tr}(\mathrm{lam}) = 0.664(5 \times 10^5)^{1/2}\,(7.0)^{1/3} = 900$ from Eq. (6.24). Similarly, assuming turbulent flow, $c_f(\mathrm{tr}) = 0.00428$ from Eq. (6.90) and $\mathrm{St}_{tr} = 0.000835$ from Eq. (6.93), whence

$$\overline{\mathrm{Nu}}_{tr,\ turb} = 1.15(0.000835)(5 \times 10^5)(7.0) = 3400.$$

Thus, after this hard work, Eq. (6.95) predicts

$$\overline{\mathrm{Nu}}_L = 900 + 34{,}600 - 3400 = 32{,}100,$$

so that

$$\bar{h} = \overline{\mathrm{Nu}}_L k/L = \frac{(32{,}100)(0.597)}{1.2} = 16{,}000 \text{ W/m}^2 \cdot \text{K}.$$

Finally,

$$q_w = \bar{h}(bL)(T_w - T_\infty) = (16{,}000)(2.0)(1.2)(10^\circ - 30^\circ)$$
$$= -768{,}000 \text{ W.} \quad [\textit{Ans. (d)}]$$

This is 1830 times more than the airflow value in Example 6.1. These impressive increases in shear and heat flux are due to (1) the high density and conductivity of water and (2) the fact that the flow is turbulent instead of laminar.

The reader should verify that Eq. (6.97) predicts $\overline{\mathrm{Nu}}_L = 19{,}800$ and $q_w = -474{,}000$ W, or about 38% lower than our solution. ■

---

### 6.4.7 Plate Flow with Unheated Starting Length

All of the previous flat plate heat transfer formulas in this section have been for uniform wall temperature. But there are many practical problems involving nonuniform wall temperature, such as (1) partially heated plates or (2) uniform wall heat flux conditions because of electrical heating. All cases of variable $T_w(x)$ may be generalized from the solution to the fundamental problem of a sudden discontinuity in wall temperature.

Figure 6.11 sketches this fundamental case: a flat plate with "unheated starting length," that is, wall temperature equal to $T_\infty$ for $0 \leq x \leq \xi$, followed by a uniform value $T_w \neq T_\infty$ for $x \geq \xi$. The velocity boundary layer $\delta(x)$, being uncoupled from the temperature as previously discussed, grows from the leading edge ($x = 0$). The thermal boundary layer $\Delta(x)$ grows from the point of initial heating, $x = \xi$. The analysis, either for laminar or turbulent flow, is performed by the integral method of Section 6.2.3 and is given as a problem at the end of the chapter. Details of the solution are given in [1] and [4]. The basic result for laminar flow is the local Nusselt number for $x > \xi$:

$$\mathrm{Nu}_x = \frac{0.332\,\mathrm{Re}_x^{1/2}\,\mathrm{Pr}^{1/3}}{[1 - (\xi/x)^{3/4}]^{1/3}}. \tag{6.98}$$

Note that when $\xi = 0$ this reduces to Eq. (6.37).

The turbulent-flow solution is very similar, with slightly different

**Figure 6.11** **Sketch of flat plate flow with an unheated starting length.**

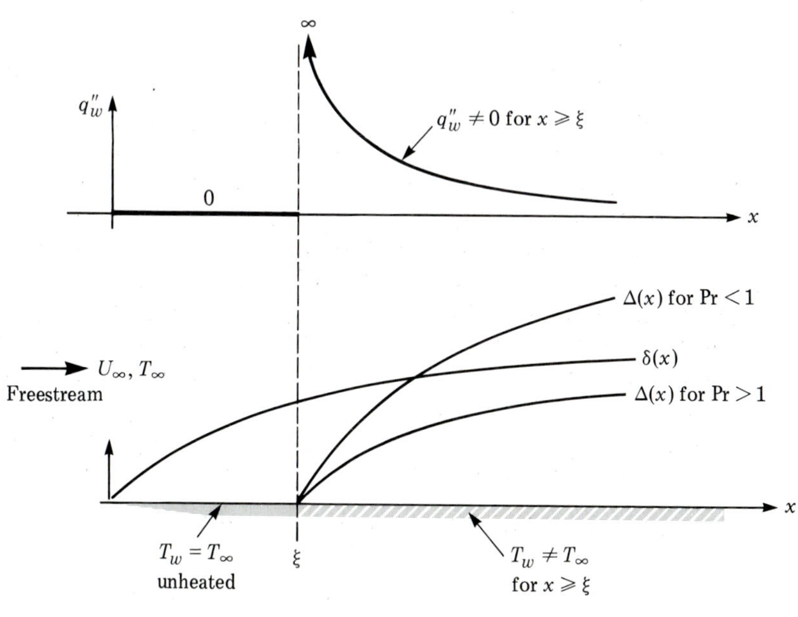

exponents on the starting length correction:

$$\mathrm{Nu}_x = \frac{(\mathrm{Nu}_x)_{\xi=0}}{[1 - (\xi/x)^{9/10}]^{1/9}}, \tag{6.99}$$

where $\mathrm{Nu}_x(\xi = 0)$ is predicted from Eq. (6.93) or from the simple power-law approximation, Eq. (6.92). No closed-form expression for the average value $\overline{\mathrm{Nu}}_L$ is known to the writer for either laminar or turbulent flow. It is suggested that Eqs. (6.98) and (6.99) be integrated numerically.

### Example 6.8

Repeat part (b) of Example 6.7 if the first 30 cm of the plate are unheated.

*Solution* If $\xi = 30$ cm, $\mathrm{Re}_\xi = 1.8 \times 10^6$, so the flow over the heated downstream section is definitely turbulent. From part (b) of Example 6.7, at $x = 60$ cm, we found $\mathrm{St}_x(\xi = 0) = 0.000653$, so that

$$(\mathrm{Nu}_x)_{\xi=0} = \mathrm{St}_x\mathrm{Re}_x\mathrm{Pr} = (0.000653)(3.56 \times 10^6)(7.0) = 16{,}300.$$

With the unheated length, from Eq. (6.99),

$$\mathrm{Nu}_x(60 \text{ cm}) = \frac{16{,}300}{[1 - (30/60)^{9/10}]^{1/9}} = 17{,}750.$$

Then

$$h_x = \mathrm{Nu}_x k/x = (17{,}750)(0.597)/(0.6) = 17{,}660 \text{ W/m}^2 \cdot \text{K},$$

and

$$q''_w(60 \text{ cm}) = 17{,}660(10° - 30°) = -353{,}000 \text{ W/m}^2. \quad [\textit{Ans.}]$$

This is 9% more than the result in Example 6.7(b). However, the total heat transferred is less because of the unheated length. ■

#### 6.4.8 Plate Flow with Uniform Wall Heat Flux

The unheated starting length solution of Section 6.4.7 can be developed through a superposition integral into the local Nusselt number for uniform wall heat flux, $q''_w$ = constant. Details of the solution are given in [1]. The basic results are

*Laminar flow:*

$$\mathrm{Nu}_x = 0.453\, \mathrm{Re}_x^{1/2}\, \mathrm{Pr}^{1/3}, \tag{6.100a}$$

*Turbulent flow:*

$$\mathrm{Nu}_x \doteq 1.04(\mathrm{Nu}_x)_{T_w=\text{constant}}. \qquad \textbf{(6.100b)}$$

The laminar Nusselt number is 36% higher than the laminar constant wall temperature result, Eq. (6.37), but the turbulent flux is only 4% higher.

Recently an elegant turbulent-flow analysis for this constant wall flux case was given by Thomas [16], resulting in an explicit formula for the local Stanton number:

$$\mathrm{St}_x = \frac{(c_f/2)^{1/2}}{2.16\ \ln[\mathrm{Re}_x(c_f/2)^{1/2}] + 12.7\ \mathrm{Pr}^{2/3} - 13.8}.$$

Since $q''_w$ is a known constant, the output of this type of problem is the wall temperature distribution:

$$T_w(x) = T_\infty + \frac{q''_w x}{k\ \mathrm{Nu}_x}, \qquad \textbf{(6.101)}$$

and, of course, by definition $\mathrm{Nu}_x = \mathrm{St}_x\mathrm{Re}_x\mathrm{Pr}$.

## Example 6.9

Repeat Example 6.1 to find $T_w(x)$ when the heat flux equals the average value of $(-175\ \mathrm{W/m^2})$ from part (d). Compute $T_w$ at (a) $x = 60$ cm, and (b) $x = 120$ cm. (c) At what value of $x$ does $T_w = 10°\mathrm{C}$ as in Example 6.1?

*Solution* This flow is laminar so Eq. (6.100a) applies:

$$\mathrm{Nu}_x = 0.453\ \mathrm{Re}_x^{1/2}\ \mathrm{Pr}^{1/3} = 0.453(6x/15 \times 10^{-6})^{1/2}(0.71)^{1/3}$$
$$= 255.6x^{1/2}, \qquad x \text{ in meters.}$$

Substituting this into Eq. (6.101) gives

$$T_w = T_{oo} + q''_w x/k\mathrm{Nu}_x = 30° - 175x/[0.0251(255.6x^{1/2})],$$

or

$$T_w = 30 - 27.28^{1/2}, \qquad x \text{ in meters}$$

At $x = 60$ cm,

$$T_w = 30 - 27.28(0.6)^{1/2} = 8.87°\mathrm{C}. \quad [\textit{Ans. (a)}]$$

At $x = 120$ cm,

$$T_w = 30 - 27.28(1.2)^{1/2} = 0.12°\mathrm{C}. \quad [\textit{Ans. (b)}]$$

If

$$T_w = 10° = 30 - 27.28x^{1/2},$$

then

$$x = 0.54 \text{ m.} \quad [Ans. (c)]$$

We see that the wall temperature equals the stream temperature of 30°C at the leading edge and drops parabolically to 0.12°C at the trailing edge. The average wall temperature is

$$\overline{T_w} = \frac{1}{1.2}\int_0^{1.2} (30 - 27.28x^{1/2})dx = 10.1°\text{C}.$$ ■

---

## 6.5 Turbulent Flow in Ducts

Laminar flow in a tube undergoes transition at about $\text{Re}_D \doteq 2000$, and the flow becomes fully turbulent for $\text{Re}_D > 4000$. These are low Reynolds numbers for flow of gases and light liquids, which are generally turbulent in practical cases.

Experiments show that the log-law correlations for $u^+(y^+)$ from Eq. (6.82b) and for $T^+(y^+, \text{Pr})$ from Eq. (6.87) are valid in tube flow also, where $y$ is interpreted as distance inward from the tube wall, $y = r_o - r$. Thus, for fully developed flow, one may readily evaluate the wall shear and heat transfer coefficient by averaging the log-laws across the tube cross section. The details of the analysis are given in [4, Chapter 6]. The coefficients are defined in the same manner as for laminar tube flow:

$$c_f = \frac{2\tau_w}{\rho V^2} = 2\left(\frac{v^*}{V}\right)^2, \qquad \text{Nu} = \frac{hD}{k}, \qquad h = \frac{q''_w}{T_w - T_m}. \tag{6.102}$$

### 6.5.1 Fully Developed Tube Flow

Averaging of the velocity log-law (6.82b) gives the following friction law for HFD smooth-wall turbulent flow in a tube:

$$(4c_f)^{-1/2} \doteq 2.0 \log_{10}\left[\frac{\text{Re}_D(4c_f)^{1/2}}{2.51}\right]. \tag{6.103}$$

This formula was derived by L. Prandtl in 1933.

Recall from elementary fluid mechanics, for example, [14], that turbulent tube friction is strongly influenced by wall roughness. Data for rough pipes were correlated by Colebrook [22] into the following curve-fit relation:

$$(4c_f)^{-1/2} \doteq -2.0 \log_{10}\left[\frac{\varepsilon/D}{3.7} + \frac{2.51}{\mathrm{Re}_D(4c_f)^{1/2}}\right], \tag{6.104}$$

where $\varepsilon$ is the average wall roughness height. This popular formula is the basis of the celebrated Moody friction factor chart [23], reproduced in Fig 6.12. It is valid for any Reynolds number in HFD flow and includes Eq. (6.103) as a special case as $\varepsilon \to 0$. Unfortunately, because of its implicit form, iteration is required to compute $c_f$ from Eq. (6.104). To circumvent this problem, S. E. Haaland† recently suggested a very accurate alternative explicit formula:

$$(4c_f)^{-1/2} \doteq -1.8 \log_{10}\left[\left(\frac{\varepsilon}{3.7D}\right)^{1.11} + \frac{6.9}{\mathrm{Re}_D}\right]. \tag{6.105}$$

The error compared to Eq. (6.104) is less than 1.5%. Meanwhile, the literature abounds with other friction factor formulas that have less accuracy or a more limited range of Reynolds number or roughness ratio. See, for example, [1, Chapter 11].

Averaging of the temperature log-law from Eq. (6.87) results in the following formula for heat transfer in TFD turbulent smooth-wall tube flow:

$$\mathrm{St}_{\mathrm{TFD}} \doteq \frac{c_f/2}{1 + 12.7(\mathrm{Pr}^{2/3} - 1)(c_f/2)^{1/2}}. \tag{6.106}$$

This is identical in form to the flat plate formula given earlier in Eq. (6.92), except that here $c_f$ denotes the smooth-wall *tube* friction factor from Eq. (6.103). Equation (6.106) was first derived by Petukhov and Kirillov in 1960 [21]. If St is known, the Nusselt number is given by definition as $\mathrm{Nu}_D = \mathrm{St}\mathrm{Re}_D\mathrm{Pr}$. Strictly speaking, Eq. (6.106) is for $\mathrm{Nu}_T$, but for $\mathrm{Pr} \geqslant 0.7$ the constant heat flux value $\mathrm{Nu}_H$ is only about 1% higher [24]. This is in contrast to laminar tube flow, Fig. 6.5, where $\mathrm{Nu}_H$ is about 20% greater than $\mathrm{Nu}_T$.

Nusselt numbers computed from Eqs. (6.103) and (6.106) are shown in Fig. 6.13 for $0.7 \leqslant \mathrm{Pr} \leqslant 1000$. The curves are nearly straight lines, implying a power-law relation. They are often approximated by the

†S. E. Haaland, "Simple and Explicit Formulas for the Friction Factor in Turbulent Pipe Flow." *J. Fluids Engineering.* March, 1983.

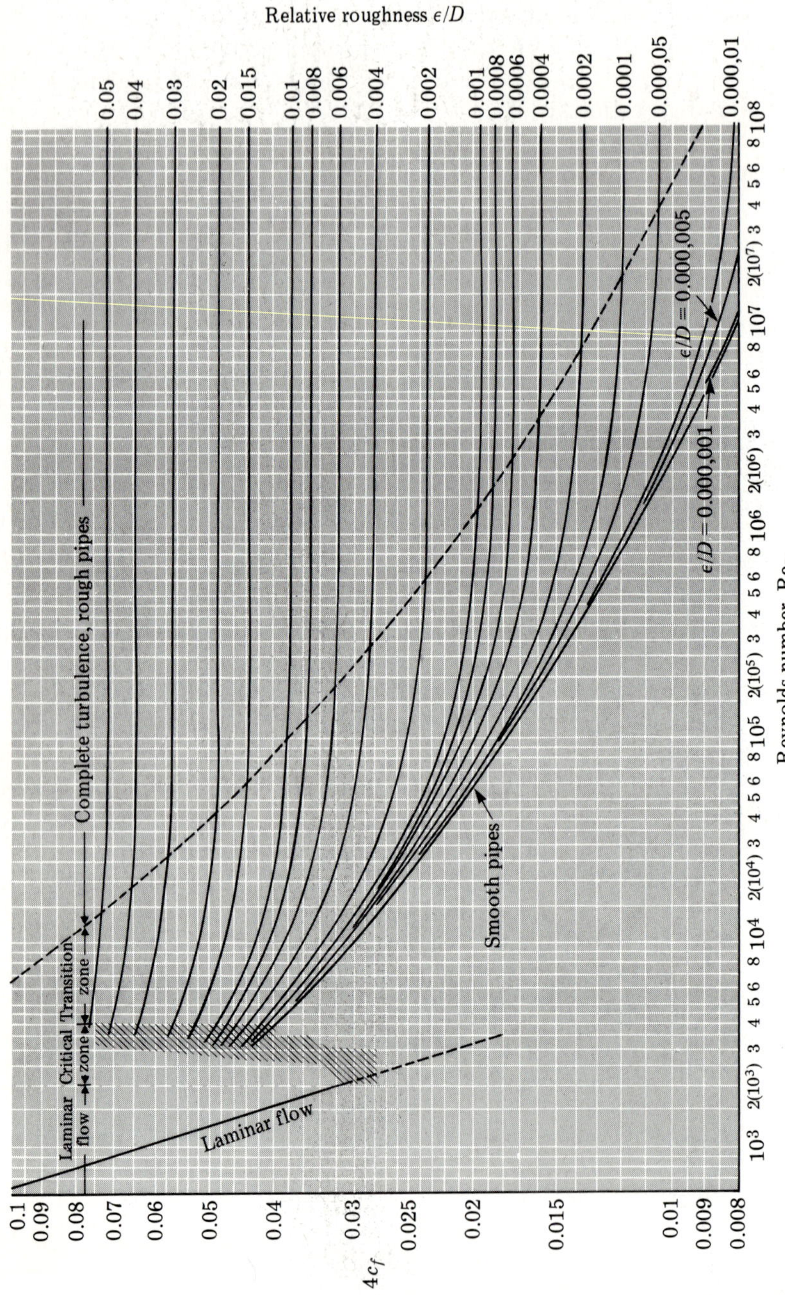


**Figure 6.12** The Moody chart for fully developed turbulent pipe friction, Eq. (6.104) (from [23]).

popular *Dittus-Boelter* formula for turbulent TFD smooth-wall flow, as modified in [20]:

$$\mathrm{Nu}_D \doteq 0.027\,\mathrm{Re}_D^{0.8}\,\mathrm{Pr}^{1/3}\,(\mu_\mathrm{m}/\mu_w)^{0.14}. \qquad \textbf{(6.107)}$$

This formula is accurate for Pr $\doteq$ 1 but predicts up to 50% low at high Prandtl numbers.

Equation (6.106) fails for liquid metals, Pr $\leqslant$ 0.03, because the log-law (6.87) is not accurate. A suggested correlation for TFD liquid metal flow is given in [24]:

**Figure 6.13** **Nusselt numbers for fully developed turbulent flow in smooth tubes.**

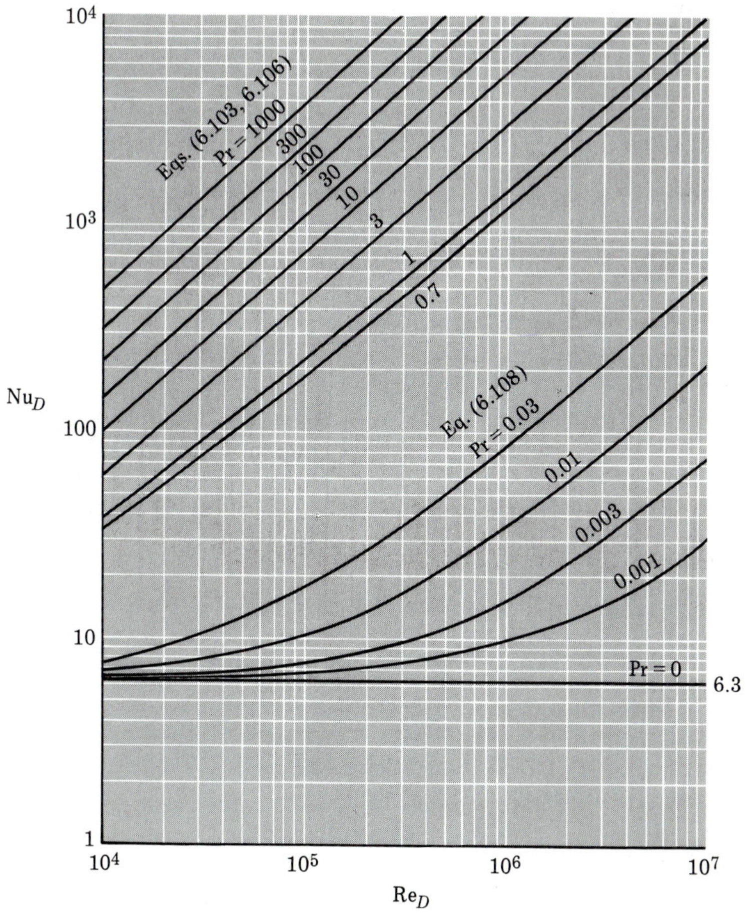

*Pr* $\leqslant$ *0.03:* **(6.108)**

$$\mathrm{Nu}_D \doteq 6.3 + 0.0167\,\mathrm{Re}_D^{0.85}\,\mathrm{Pr}^{0.93}.$$

This expression is also plotted in Fig. 6.13.

### 6.5.2 Thermally Developing Flow

The thermal entry region in turbulent flow is generally shorter than in laminar flow. For any Reynolds or Prandtl number, the local Nusselt number in the entrance tends to approach $\mathrm{Nu}_{\mathrm{TFD}}$ (Fig. 6.13) in 20–40 diameters, as shown in Fig. 6.14 from a comprehensive analysis by Sleicher and Tribus [25] as modified by Kays [1]. For $x/D \geqslant 10$, the mean Nusselt number may be approximated by

$$\overline{\mathrm{Nu}}_x/\mathrm{Nu}_{\mathrm{TFD}} \doteq 1 + \frac{C}{x/D}, \tag{6.109}$$

where $C \doteq 9$, 2, and 0.7 for Pr = 0.01, 0.7, and 10, respectively. The

**Figure 6.14 Local Nusselt number in the entrance of a tube for turbulent flow with smooth walls. Calculations from Sleicher and Tribus [25] as modified and improved by Kays [1].**

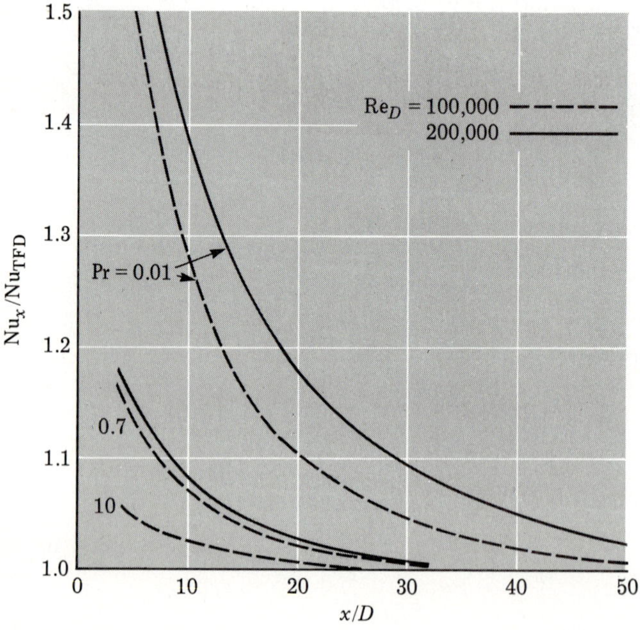

turbulent-flow thermal entry length decreases with Prandtl number, which is opposite to the laminar flow result of Eq. (6.76).

Finally, wall roughness effects were investigated by Dipprey and Sabersky [17], who found that roughness can increase heat transfer by a factor of up to 2.5. It is tempting to estimate St for rough walls by simply introducing $c_f$ for rough walls into Eq. (6.106); indeed, some texts recommend this, but it is not accurate. Based on an idea of W. Nunner in 1956, Norris [26] presents a simple correlation:

$$\frac{\mathrm{Nu}}{\mathrm{Nu}_{\mathrm{smooth}}} \doteq \left(\frac{c_f}{c_{f,\,\mathrm{smooth}}}\right)^n, \qquad \textbf{(6.110)}$$

for $0.7 \leqslant \mathrm{Pr} \leqslant 6$, where $n \doteq 0.68\ \mathrm{Pr}^{0.215}$. Equation (6.110) is to be used up to $c_f/c_{f,\,\mathrm{smooth}} = 3.0$, after which there is no further increase in Nu. Recall that roughness has little or no effect on laminar flow heat transfer.

## Example 6.10

Repeat Example 6.4 if the fluid is ammonia ($NH_3$).

*Solution* For ammonia at $T_f = 10°\mathrm{C}$, $\rho = 626\ \mathrm{kg/m^3}$, $\mathrm{Pr} = 2.04$, $c_p = 4714\ \mathrm{J/kg\cdot K}$, $\nu = 3.7 \times 10^{-7}\ \mathrm{m^2/s}$, and $k = 0.531\ \mathrm{W/m\cdot K}$. The average velocity and Reynolds number are

$$V = \dot{m}/\rho A = \frac{0.25}{626(\pi/4)(0.01)^2} = 5.08\ \mathrm{m/s},$$

$$\mathrm{Re}_D = VD/\nu = \frac{(5.08)(0.01)}{3.7 \times 10^{-7}} = 137{,}000.$$

The flow is definitely turbulent. Use Eq. (6.105) for $c_f$:

$$(4c_f)^{-1/2} \doteq -1.8 \log_{10}[0.0 + 6.9/137{,}000] = 7.74,$$

or $c_f = 0.00418$. Then the pressure drop is, by definition,

$$\Delta p = c_f(4L/D)\left(\frac{1}{2}\rho V^2\right) = 0.00418(400)\left(\frac{1}{2}\right)(626)(5.08)^2$$

$$= 13{,}500\ \mathrm{Pa}. \quad [Ans.\ (a)]$$

Since $L/D = 100$, the development region may be neglected. With $c_f/2 = 0.00209$, Eq. (6.106) predicts

$$\mathrm{St}_{\mathrm{TFD}} = \frac{0.00209}{1 + 12.7(2.04^{2/3} - 1)(0.00209)^{1/2}} = 0.00154,$$

whence

$$\overline{\mathrm{Nu}}_D = \mathrm{St}\mathrm{Re}_D\mathrm{Pr} = (0.00154)(137{,}000)(2.04) = 432,$$

$$\bar{h} = \overline{\mathrm{Nu}}_D k/D = \frac{(432)(0.531)}{0.01} = 22{,}900\ \mathrm{W/m^2\cdot K}.$$

The exit temperature is computed from Eq. (6.54):

$$T_{mo} - 0° = (20° - 0°)\exp\left[-\frac{(22{,}900)\pi(0.01)(1.0)}{(0.25)(4714)}\right] = 10.86°,$$

or

$$T_{mo} = 10.9°\mathrm{C}. \quad [Ans.\ (b)]$$

This is 19 times more temperature change than the transformer oil in Example 6.4 achieved with 70% less pressure drop. This is typical of the convection enhancement of turbulent flow. ■

---

### 6.5.3 Convection Enhancement

The theories and correlations presented so far have been for unadorned, uncomplicated geometries: straight ducts and flat plates. In the past 20 years, many methods have been developed to "enhance" or augment convection rates in such flows. One example just seen is the roughening of the wall, Eq. (6.110), which can double the heat flux for the same flow rate.

There have been two ASME symposiums devoted to convection enhancement papers, in 1970 [27] and 1979 [28]. The work prior to 1969 has been reviewed in a major survey paper by Bergles [29], and subsequent studies are the subject of a recent collection of papers [30]. Among the enhancement methods in use today, for both single-phase and two-phase flows, are the following.

1. Surface roughness, either distributed, isolated, or machined
2. Surface modification by coating with chemicals or metals
3. Addition of chemicals or particles to the fluid
4. Fins and ribs, either wall-attached or displaced
5. Twisted tapes, helical corrugations, and other swirl and turbulence promoters
6. Mechanical vibration or acoustic resonance
7. Application of electrostatic fields

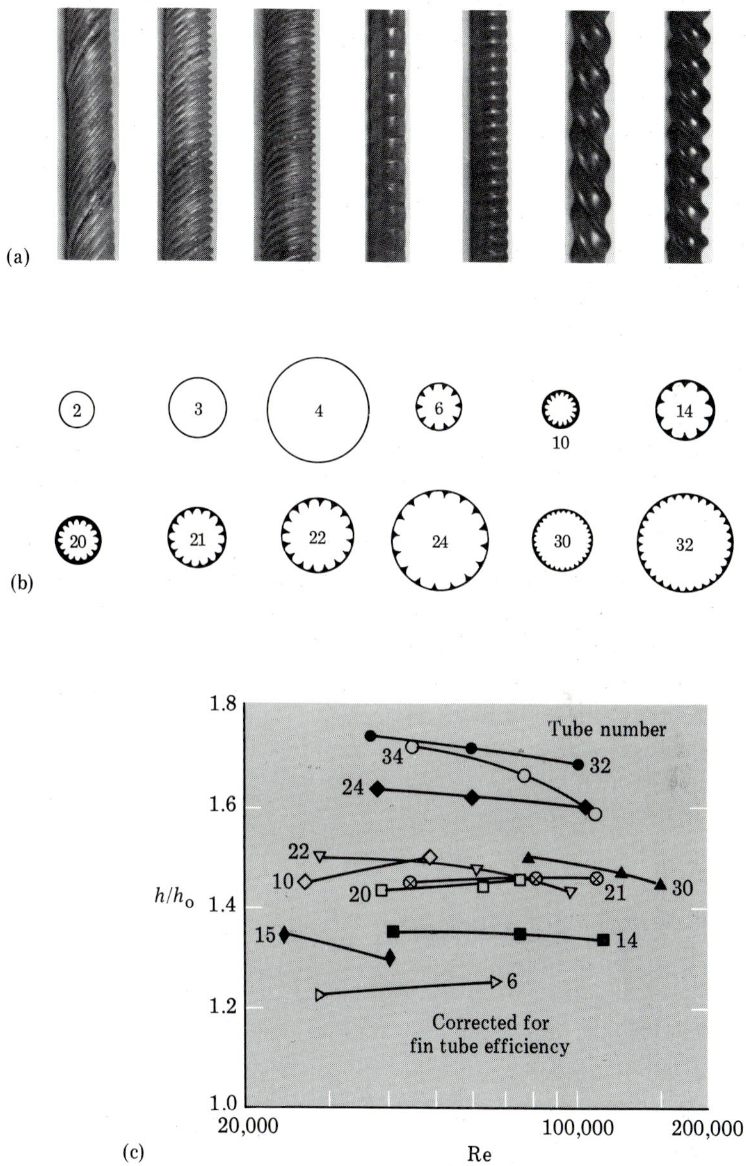


**Figure 6.15** Convection enhancement in tubes: (a) corrugated tubes studied by Marto et al. [31]; (b) internally finned tubes studied by Carnavos [32]; (c) heat transfer ratio for the tubes in (b), compared to a smooth tube at the same pumping power [32].

These and many other related ideas are discussed in [27–30], along with bibliographies containing hundreds of references.

We saw some fin designs in Fig. 2.13. Other fin and rib designs are shown in Fig. 6.15. Such tubes both increase surface area and promote turbulence and swirling motion. They are commonly compared to a smooth tube at the same flow rate, diameter, temperature difference, and pumping power, after correcting for the increased area and efficiency of the attached fins. The performance criterion could then be either (1) the surface area ratio, which should be small for the enhanced tube or (2) the heat flux ratio, which should be large.

Figure 6.15(a) shows the corrugated tubes tested by Marto et al. [31]. These tubes are costly, but all of them show dramatically decreased area ratios, so that the tubes can be from 10% to 50% shorter for a given heat load and pumping power. Figure 6.15(b) shows internally finned tubes tested by Carnavos [32], and Fig. 6.15(c) gives the test results. We see that all 11 fin designs gave increased heat flux, ranging from 22% to 73%, for the same heat load and pumping power as a smooth tube.

Enhancement designs are now used in all areas of convection: forced and free single-phase flows, boiling, and condensation, as well as mass transfer applications. For forced single-phase convection as studied in this chapter, wall roughness, fins, corrugations, twisted tapes, flow interruptors, and acoustic excitation all increase heat flux and have promising application. Application of electrostatic fields also increases heat flux but requires a costly voltage generator and dangerously high voltages.

## 6.6 Forced Convection in Separated Flows

In Sections 6.2–6.5 we were able to develop some theoretical relations for laminar and turbulent flow in ducts and past flat plates. Unfortunately, our ability to theorize declines rapidly if the boundary layer about the body shape suffers *flow separation*. Boundary layer theory [1, 2, 4, 13] works well up to the point of separation but not beyond. There is presently no accurate general analysis available for predicting surface pressure, friction, or heat flux in the separated flow region of either internal or external flows. Thus, the greater the degree of flow separation, the poorer our ability to predict overall drag and heat transfer in the flow.

An example is the blunt-body flow sketched in Fig. 6.16. The boundary layer on the front (windward) side of the body is "attached"

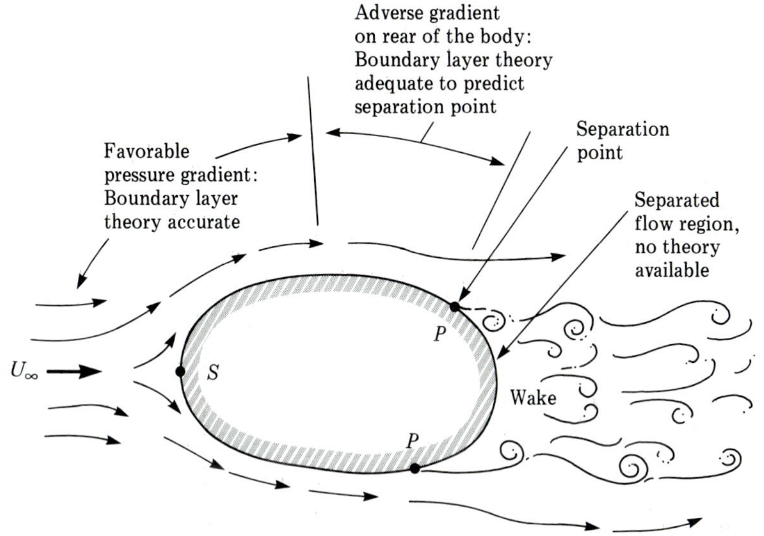


**Figure 6.16** Sketch of the flow pattern of a blunt body immersed in a uniform stream.

and does not separate. Pressure is maximum at the stagnation point, $S$, and drops along the surface toward the shoulder of the body. Pressure decreasing downstream is termed a *favorable gradient,* where separation cannot occur. Downstream of the shoulder the surface pressure increases: an unfavorable or *adverse gradient* that promotes flow separation. At the separation point, $P$, the boundary layer flow breaks away from the surface and forms a broad, pulsating wake that persists far downstream of the body. Except for digital computer simulations at low Reynolds numbers ($\mathrm{Re}_L < 10^4$), as in [3], there is no accurate theory for the wake region on the rear (lee) side of the body. However, boundary layer theory is adequate for estimating $c_f$ and $\mathrm{St}_x$ from the stagnation point back to the separation point. In particular, Ambrok [33] has developed a simple integral-relation estimate that is discussed in detail in the text by Kays and Crawford [1, p. 219]. Here we confine our attention to prediction of heat flux at the stagnation point.

### 6.6.1 Heat Transfer at the Stagnation Point

In any flow past a blunt body (Fig. 6.16) the stagnation point $S$ will be a region of high heat flux. The flow is almost always laminar at this point, so an exact boundary layer solution is possible. The theoretical result for local heat transfer coefficient $h_S$ at point $S$ is given by [1, p. 140] or [4, p. 183]:

$$h_S \doteq Ck\left(\frac{U_\infty}{\nu R}\right)^{1/2} \mathrm{Pr}^{0.4}, \tag{6.111}$$

for $0.1 \leqslant \mathrm{Pr} \leqslant 10$, where $U_\infty$ is the stream velocity approaching the body (see Fig. 6.16) and $R$ is the body radius of curvature at point $S$. The constant $C = 0.806$ for a two-dimensional body (such as a cylinder) and equals 0.933 for an axisymmetric body (such as a sphere). Experimental data for $h_S$ [34] are apt to be about 10% higher than predicted by Eq. (6.111) because of turbulence and noisiness in the oncoming stream. Note that $h_S$ increases as body radius $R$ decreases.

**Example 6.11**

In airflow past a circular cylinder at $\mathrm{Re}_D = 140{,}000$, Giedt [34] has measured a Nusselt number at the stagnation point of $h_S D/k = 410$. Compare this with theory, Eq. (6.111).

*Solution* For air take $\mathrm{Pr} = 0.7$. Using $D = 2R$ for a cylinder, we may rewrite Eq. (6.111) in the Nusselt number form

$$\frac{h_S D}{k} = 2C\left(\frac{U_\infty D}{2\nu}\right)^{1/2} \mathrm{Pr}^{0.4}, \qquad C = 0.806 \text{ (two-dimensional).}$$

Introducing the given Reynolds number, $U_\infty D/\nu = 140{,}000$, we have

$$h_S D/k = 2(0.806)(140{,}000/2)^{1/2}(0.7)^{0.4}$$
$$= 370. \quad [Ans.]$$

The fact that the experimental value of 410 is 10% higher is attributed to turbulence in the freestream (Section 6.6.3). ■

### 6.6.2 Crossflow Past a Circular Cylinder

Probably the most common convection geometry in external flow is a circular cylinder immersed in a crossflow, as in Fig. 6.17. There may be a single cylinder or a bank of cylinders. In either case, there is a broad separation zone, and boundary layer theory is not adequate to predict $\bar{h}$.

Figure 6.17 shows experimental data by Giedt [34] for local $h_x =$

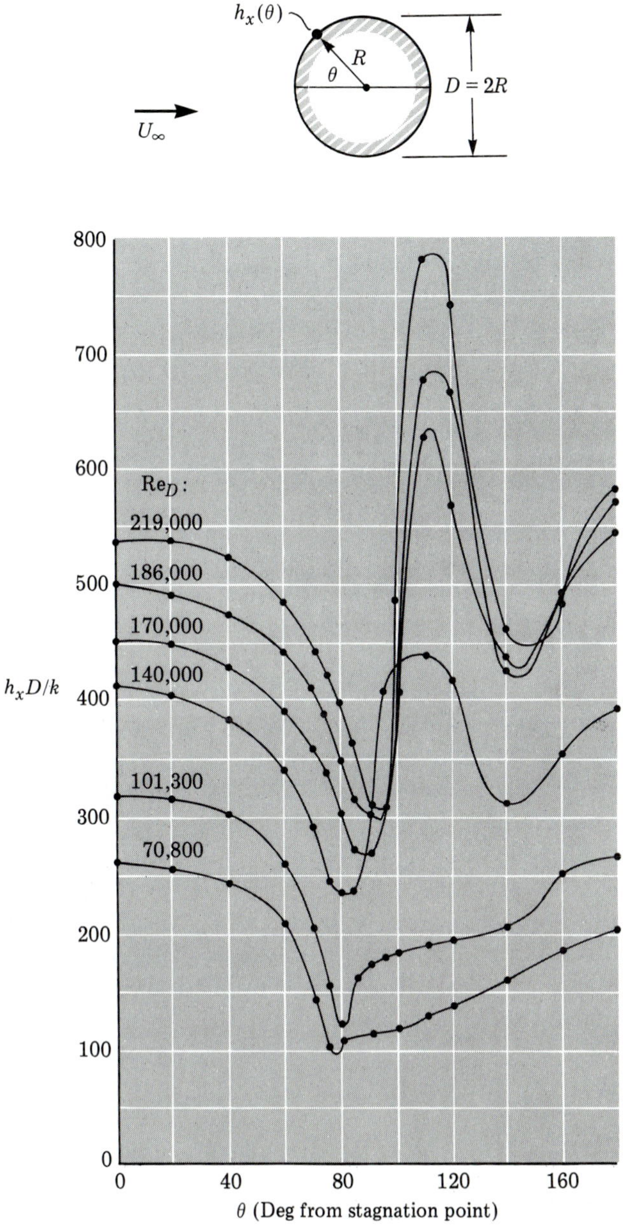


**Figure 6.17** Measurements of local Nusselt number in crossflow of air (Pr = 0.7) past a nearly isothermal circular cylinder (from [34]).

$q''_w/(T_w - T_\infty)$ around the periphery of a single immersed cylinder, for nearly constant $T_w$. For $\mathrm{Re}_D = 70{,}800$ and $101{,}300$, the boundary layer is laminar and separates at $\theta \doteq 80°$, after which $h_x$ increases moderately in the separation zone.

For $\mathrm{Re}_D = 140{,}000$ to $219{,}000$, there are *two* minima in $h_x$. The first occurs at about $\theta = 80°–90°$, where the laminar boundary layer undergoes transition to turbulence. Then the turbulent heat flux rises to a peak at about $\theta = 115°$, after which it drops to a second minimum at the turbulent separation point, $\theta \doteq 140°$. Finally, there is a moderate increase in $h_x$ in the turbulent-flow separation zone, $\theta > 140°$.

For engineering design, instead of $h_x(\theta)$, we need the average coefficient $\bar{h}$, relating total heat transfer to total cylinder surface area:

$$q_w = \bar{h}\,\pi DL\,(T_w - T_\infty). \tag{6.112}$$

The mean Nusselt number and drag coefficient are defined by

$$\overline{\mathrm{Nu}}_D = \frac{\bar{h}D}{k} = f(\mathrm{Re}_D, \mathrm{Pr}),$$

$$C_D = \frac{\mathrm{Drag}}{\frac{1}{2}\rho U_\infty^2 DL} = f(\mathrm{Re}_D). \tag{6.113}$$

The drag coefficient does not depend on Prandtl number.

There have been many experiments with circular cylinders (see [35, pp. 269–274] for details) and subsequently many empirical curve-fit formulas for $\overline{\mathrm{Nu}}_D$. A popular and relatively simple formula is due to Whitaker [38]:

$$\overline{\mathrm{Nu}}_D \doteq (0.4\,\mathrm{Re}_D^{1/2} + 0.06\,\mathrm{Re}_D^{2/3})\mathrm{Pr}^{0.4}(\mu_\infty/\mu_w)^{1/4}, \tag{6.114}$$

where all fluid properties are evaluated at $T_\infty$, not $T_f$. The correlation is valid for $0.7 \leqslant \mathrm{Pr} \leqslant 300$ and $10 \leqslant \mathrm{Re}_D \leqslant 10^5$. A complicated but more comprehensive formula was given by Churchill and Bernstein [36], who correlated all available data:

$$\overline{\mathrm{Nu}}_D \doteq 0.3 + \frac{0.62\,\mathrm{Re}_D^{1/2}\mathrm{Pr}^{1/3}}{[1 + (0.4/\mathrm{Pr})^{2/3}]^{1/4}}\left[1 + \left(\frac{\mathrm{Re}_D}{282{,}000}\right)^{5/8}\right]^{4/5}, \tag{6.115}$$

valid for $\mathrm{Re}_D\mathrm{Pr} \geqslant 0.2$ with an uncertainty of $\pm 30\%$. This relation is plotted in Fig. 6.18, showing a strong effect of both $\mathrm{Re}_D$ and Pr on the heat transfer. Also shown are the cylinder drag data of Wieselberger [37]; the drag drops sharply at about $\mathrm{Re}_D \doteq 300{,}000$, where the boundary layer becomes turbulent and the separation point moves farther back.

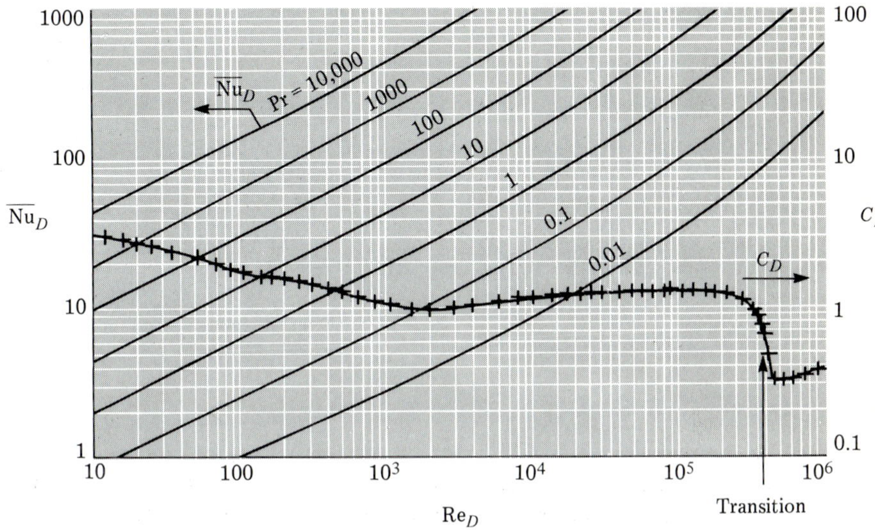


**Figure 6.18** **Average Nusselt number (Eq. 6.115) and drag coefficient [37] in crossflow past a smooth circular cylinder. The uncertainty in $\overline{\text{Nu}}_D$ is ±30%.**

There is no single formula for $C_D(\text{Re}_D)$, but for $1 \leqslant \text{Re}_D \leqslant 10^4$, [4] suggests

$$C_D(\text{cylinder}) \doteq 1 + \frac{10}{\text{Re}_D^{2/3}}. \tag{6.116}$$

Note that $\overline{\text{Nu}}_D$ does *not* change significantly at the transition point.

## Example 6.12

Air at 1 atm and 20°C flows at 8 m/s over a 3-cm-diameter, 4-m-long cylinder with $T_w$ = 100°C. Estimate (a) the total drag, and (b) the total heat transfer.

*Solution* For air at $T_f = (20° + 100°)/2 = 60°\text{C}$, $\rho = 1.045$ kg/m$^3$, $k = 0.028$ W/m · K, $\nu = 1.9 \times 10^{-5}$ m$^2$/s, and Pr = 0.71. Compute

$$\text{Re}_D = U_\infty D/\nu = \frac{(8.0)(0.03)}{1.9 \times 10^{-5}} = 12{,}600.$$

From Fig. 6.18 at $\text{Re}_D = 12{,}600$ read $C_D \doteq 1.1$. The drag is

$F = C_D 1/2 \rho U_\infty^2 DL = (1.1)(1/2)(1.045)(8)^2(0.03)(4) = 4.4\ \text{N}.$ [*Ans. (a)*]

For the heat transfer, from Eq. (6.115) or Fig. 6.18,

$$\overline{\text{Nu}}_D \doteq 0.3 + \frac{0.62(12{,}600)^{1/2}(0.71)^{1/3}}{[1 + (0.4/0.71)^{2/3}]^{1/4}}\left[1 + \left(\frac{12{,}600}{282{,}000}\right)^{5/8}\right]^{4/5}$$
$$= 61.0.$$

Then $\bar{h} = \overline{\text{Nu}}_D k/D = \dfrac{(61.0)(0.028)}{0.03} = 56.9\ \text{W/m}^2 \cdot \text{K}.$

Finally, the total heat lost by the cylinder is

$$q_w = \bar{h}\,\pi DL\,(T_w - T_\infty) = (56.9)\pi(0.03)(4)(100 - 20)$$
$$= 1700\ \text{W}. \quad [\textit{Ans. (b)}]$$

As mentioned, the uncertainty in this estimate is $\pm 30\%$. ■

---

### 6.6.3 Parametric Effects on Cylinder Crossflow Convection

Section 6.6.2 discusses the classical problem of forced convection in two-dimensional flow past a smooth constant-temperature cylinder whose axis is normal to the stream. The correlated result, Fig. 6.18, is commonly used for design estimates and has only two parameters, Reynolds number and Prandtl number.

Figure 6.18, while widely accepted, neglects other parameters that have a significant effect on cylinder convection in a real design situation. Let us list five of these parametric effects:

1. freestream turbulence
2. cylinder yaw
3. root and tip effects
4. multiple rows of cylinders (Section 6.7)
5. buoyancy effects (Section 7.3.4)

Here we discuss the first three of these effects, both as a warning of the limitations of Fig. 6.18 and also because they are intrinsically interesting.

Random fluctuations or *turbulence* in the stream approaching a cylinder tends to increase the local heat transfer coefficients, especially near the front of the body. Figure 6.19 shows data for cylinder crossflow from the monograph by Zukauskas and Ziugzda [49]. The turbulence

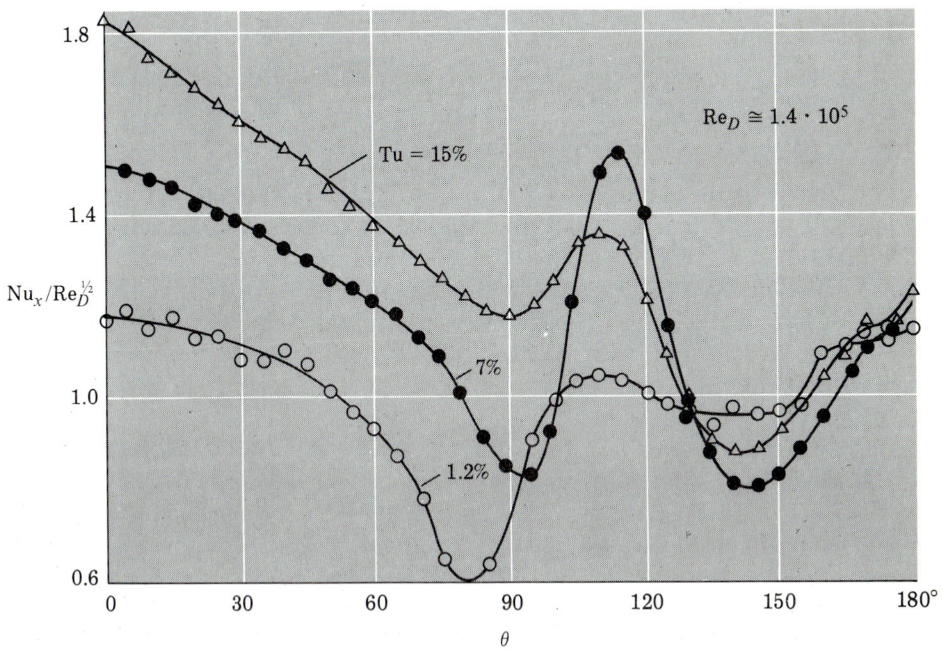


**Figure 6.19** Effect of freestream turbulence Tu on local heat transfer in cylinder crossflow. Contrast this with the low-turbulence data in Fig. 6.17 at the same Reynolds number. (Reprinted from [49] by permission of Hemisphere Publishing Corp.)

intensity Tu is defined as the rms fluctuation velocity expressed as a percent of stream velocity. The effect is strong, and the increase also varies with Reynolds number.

The curve for Tu = 1.2% is comparable to Giedt's data from Fig. 6.17 at the same Reynolds number. The increase in heat transfer with Tu is probably due to two different causes: (1) earlier transition to turbulence on the rear of the body and (2) formation of "Goertler" longitudinal vortices on the front of the body [Ref. 13, p. 526]. The effect is not limited to cylinders. McCormick et al. [50] report increases in $Nu_x$ up to 90% for a rectangular body when their wind tunnel turbulence rose to 40%. They report an astounding 300% increase in $Nu_x$ when the same body was placed outside in natural winds of 40% gustiness. There is no theory yet for stream turbulence effects.

Body yaw is our second effect. If the cylinder axis is yawed from the freestream at an angle $\phi \neq 90°$, the laminar equations of motion are unchanged in terms of the flow component normal to the cylinder axis. This is termed the *independence principle* by Schlichting [13, p. 250]. Thus the overall Nusselt number should be well correlated by the normal Reynolds number

$$\mathrm{Re}_N = U_\infty \sin\phi\, D/\nu. \tag{6.117}$$

However, the data of Groehn [51] in Fig. 6.20 show that this is true only up to a point. We see that transition to turbulence occurs earlier as the yaw angle $\phi$ decreases. The laminar heat transfer also increases slowly as $\phi$ decreases. Groehn attributes both effects to increased turbulence in the wake of the yawed cylinder. Note also that our basic unyawed formula, Eq. (6.115), falls about 15% low in the Reynolds number range $2 \times 10^4$ to $2 \times 10^5$. This is an artifact in the formula that has been reported by several writers.

Finally, consider tip and root effects. Equation (6.115) is for a very long "two-dimensional" cylinder, but naturally it is tempting to use it for a short cylinder. Consider crossflow past a short circular pin fin, typical of heat exchanger designs, as in Fig. 6.21(a). Experiments

**Figure 6.20** **Effect of yaw angle on mean cylinder heat transfer in crossflow, from Groehn [51]. (Reprinted by permission of Hemisphere Publishing Corp.)**

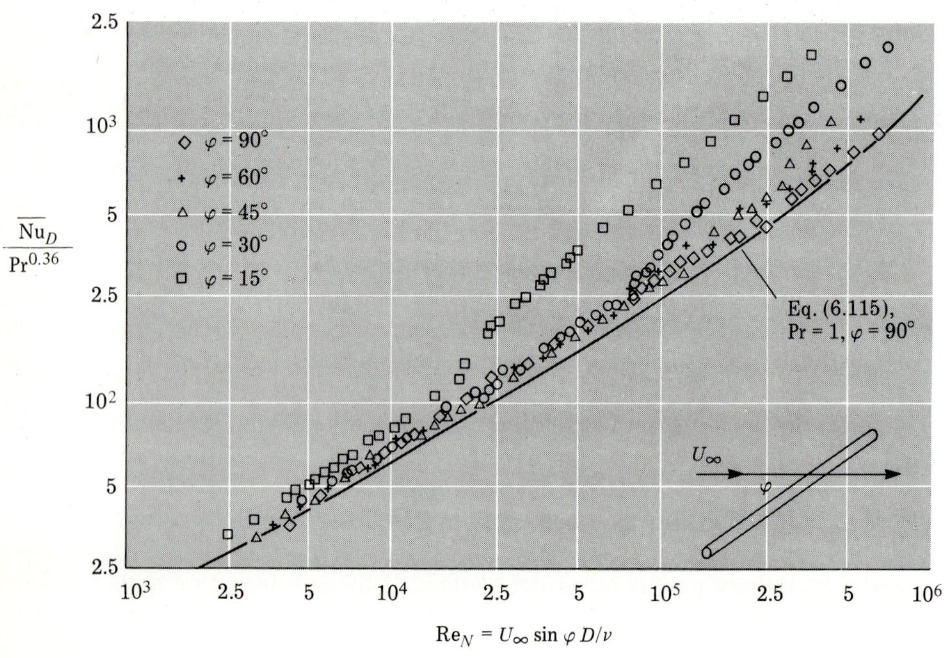

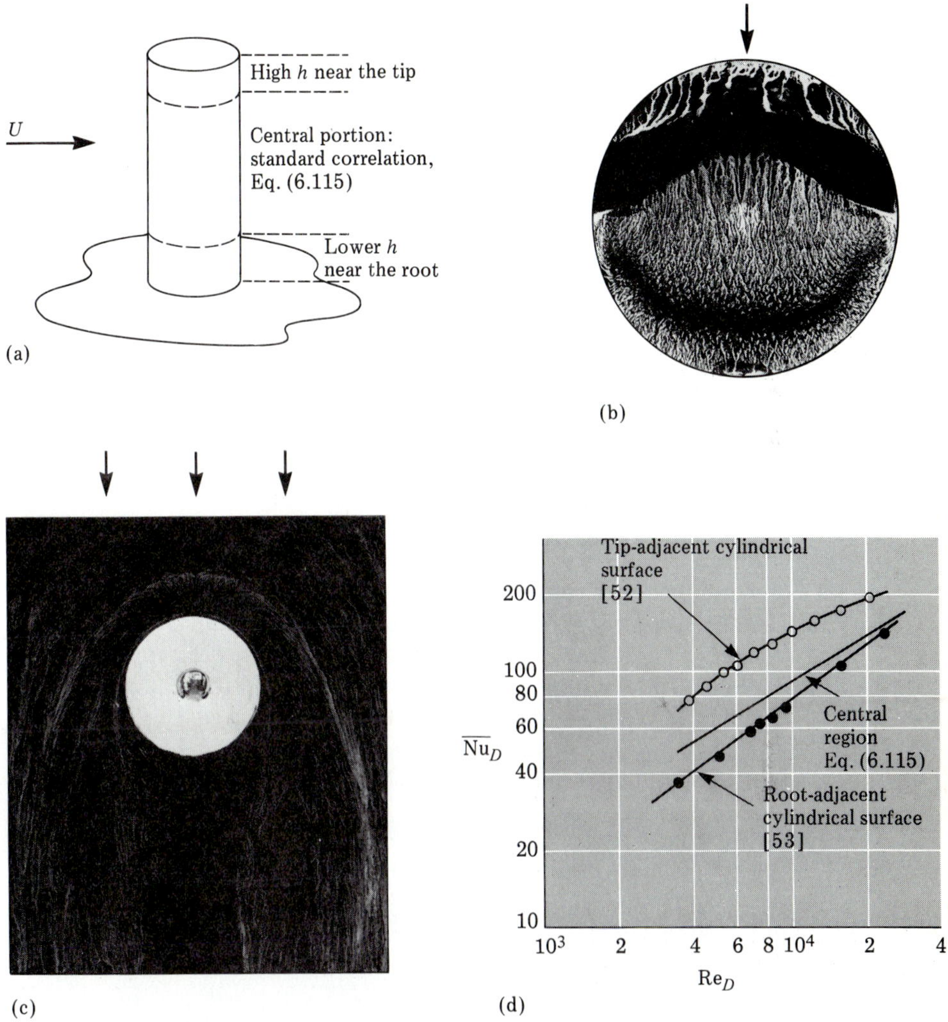


**Figure 6.21** **Tip and root effects on cylinder crossflow: (a) basic pin fin geometry; (b) tip flow visualization (from [52] by permission of ASME TRANSACTIONS); (c) root flow visualization (from [53] by permission of Pergamon Press); (d) heat transfer data.**

by Sparrow and co-workers [52, 53] show that, compared to Eq. (6.115), the heat transfer is high at the tip and lower at the root, at least for $\mathrm{Re}_D$ between 3000 and 20,000. Their data are shown in Fig. 6.20(d). The central region is well correlated by Eq. (6.115).

Figure 6.21(b) shows flow visualization, by erosion of a lampblack oil–paint mixture, of the flow over the tip. The flow separates at the broad black upstream band and reattaches at the narrow black smile-shaped downstream region. Between these two regions the surface flow is actually upstream.

Similar lampblack flow visualization of the root region is shown in Fig. 6.11(c), revealing a complex horseshoe vortex pattern on the flat floor near the root. This vortex extends upward from the floor about one cylinder radius and decreases the local heat transfer.

### 6.6.4 Crossflow Past Noncircular Cylinders

There is a limited amount of data for mean heat transfer in crossflow past noncircular cylinders at moderate Reynolds numbers. The data are correlated in the simple power-law form

$$\overline{\mathrm{Nu}} = \bar{h}D_e/k \doteq B(U_\infty D_e/\nu)^n \tag{6.118}$$

where $D_e$ = (Perimeter)/$\pi$ is the equivalent circular diameter of the cross-section. Table 6.4 shows values of $B$ and $n$ for various cross sections and orientations, from the text by Jakob [41]. Jakob arranged the table in order of increasing $n$, which masks certain discrepancies. For example, rows 1 and 3 do not predict the same $\overline{\mathrm{Nu}}$ in the range of Re = 5000–7500 where they should overlap. And there is a startling

**Table 6.4** **Experimental constants in Eq. (6.118) for crossflow past noncircular cylinders (from [41] by permission of John Wiley & Sons, Inc.)**

| Flow Direction and Profile | $U_\infty D_e/\nu$ From | To | $n$ | $B$ |
|---|---|---|---|---|
| → ◇ | 5,000 | 100,000 | 0.588 | 0.222 |
| → ⬭ | 2,500 | 15,000 | 0.612 | 0.224 |
| → ◇ | 2,500 | 7,500 | 0.624 | 0.261 |
| → ⬡ | 5,000 | 100,000 | 0.638 | 0.138 |
| → ⬡ | 5,000 | 19,500 | 0.638 | 0.144 |
| → □ | 5,000 | 100,000 | 0.675 | 0.092 |
| → □ | 2,500 | 8,000 | 0.699 | 0.160 |
| → \| | 4,000 | 15,000 | 0.731 | 0.205 |
| → ⬡ | 19,500 | 100,000 | 0.782 | 0.035 |
| → ⬯ | 3,000 | 15,000 | 0.804 | 0.085 |

Note: $D_e = \frac{1}{\pi}$(Perimeter) = Equivalent circular diameter.

change in $n$ from 0.638 to 0.782 for the hexagonal section at Re = 19500, presumably because of sudden transition to turbulence. No Prandtl number correction is listed, but presumably a factor $\mathrm{Pr}^{0.36}$ can be added to Eq. (6.118) to estimate this effect.

### 6.6.5 Flow Past a Sphere

Many experiments have been performed on drag and convection in uniform stream flow past a sphere, and many limited-range formulas have been proposed. A general correlation for a wide range of data was proposed by Whitaker [38]:

$$\overline{\mathrm{Nu}}_D \doteq 2 + (0.4\mathrm{Re}_D^{1/2} + 0.06\mathrm{Re}_D^{2/3})\mathrm{Pr}^{0.4}(\mu_\infty/\mu_w)^{1/4}, \tag{6.119}$$

where all fluid properties are evaluated at the freestream temperature, $T_\infty$, except $\mu_w$. The correlation is valid for $3.5 \leqslant \mathrm{Re}_D \leqslant 8 \times 10^4$ and $0.7 \leqslant \mathrm{Pr} \leqslant 380$.

The drag of a sphere is discussed in detail in the text by Clift et al. [39], who give accurate but limited formulas. A reasonably general curve-fit was given by [4]:

*$0 \leqslant Re_D \leqslant 10^5$:*

$$C_D \doteq 0.4 + \frac{24}{\mathrm{Re}_D} + \frac{6}{1 + \mathrm{Re}_D^{1/2}} = \frac{\mathrm{Drag}}{\frac{1}{2}\rho U_\infty^2\left(\frac{\pi}{4}\right)D^2}. \tag{6.120}$$

Like the cylinder, Fig. 6.18, the smooth-walled sphere experiences a sharp drop in drag coefficient at about $\mathrm{Re}_D = 2 \times 10^5$, where the boundary layer undergoes transition to turbulence.

For liquid metals, $\mathrm{Pr} \leqslant 0.03$, Witte [40] suggests the following heat transfer correlation for $3.6 \times 10^4 \leqslant \mathrm{Re}_D \leqslant 1.5 \times 10^5$:

$$\overline{\mathrm{Nu}}_D \doteq 2 + 0.386(\mathrm{Re}_D\mathrm{Pr})^{1/2}. \tag{6.121}$$

As a first approximation, these correlations may also be used for flow past liquid droplets. The effects of droplet deformation and flow when exposed to a uniform stream are discussed in [39].

#### Example 6.13

A 3-cm-diameter copper sphere at 90°C is cooled by being dropped into a 2-m-deep tank of water at 10°C. Assuming that the sphere immediately reaches terminal velocity in the water, estimate its center temperature when it hits bottom.

*Solution* For copper, take $\rho = 8930$ kg/m$^3$, $k_c = 400$ W/m$^2 \cdot$ K, and $\alpha_c = 1.17 \times 10^{-4}$ m$^2$/s. For water at $T_\infty = 10°$C, $\rho = 999$ kg/m$^3$, $k = 0.58$ W/m $\cdot$ K, $\mu = 1.31 \times 10^{-3}$ kg/m $\cdot$ s, and Pr = 9.4. For water at $T_w = 90°$C, $\mu_w = 3.19 \times 10^{-4}$ kg/m $\cdot$ s. First find the terminal velocity, which occurs when sphere drag equals buoyant weight:

$$C_D \frac{1}{2} \rho U_\infty^2 \frac{\pi}{4} D^2 = \frac{\pi}{6} D^3(\rho_s - \rho)g,$$

or

$$C_D U_\infty^2 = \frac{4}{3} D\left(\frac{\rho_s}{\rho} - 1\right) g = \frac{4}{3}(0.03)\left(\frac{8930}{999} - 1\right)(9.81) \qquad (1)$$
$$= 3.12, \qquad \text{with } U_\infty \text{ in m/s.}$$

The sphere Reynolds number is $\text{Re}_D = \rho U_\infty D/\mu_\infty$, or

$$\text{Re}_D = 22{,}900\ U_\infty, \qquad \text{with } U_\infty \text{ in m/s.} \qquad (2)$$

Iterating Eqs. (1) and (2) plus the drag correlation (6.120) and the initial guess $C_D \doteq 0.4$, we converge to the value

$$U_\infty = 2.71 \text{ m/s}, \qquad \text{Re}_D = 62{,}000.$$

Then, from Eq. (6.119), the mean Nusselt number is

$$\overline{\text{Nu}}_D \doteq 2 + [0.4(62{,}000)^{1/2} + 0.06(62{,}000)^{2/3}](9.4)^{0.4}\left(\frac{1.31}{0.319}\right)^{1/4}$$
$$= 677,$$

whence

$$\bar{h} = \frac{\text{Nu}_D k_\infty}{D} = \frac{677(0.58)}{0.03}$$
$$= 13{,}100 \text{ W/m}^2 \cdot \text{K}.$$

This completes the convection calculation. Now we are ready for a Chapter 4 transient conduction analysis. The sphere Biot number is

$$\text{Bi} = \frac{\bar{h} r_o}{k_c} = \frac{13{,}100(0.015)}{400} = 0.49.$$

This is much greater than 0.1 so the sphere is not lumpable. Instead, from Table 4.1 read $\beta_1 = 1.16$ and $C_1 = 1.14$ for a sphere. The sphere falls to the bottom in time $t = L/U_\infty = (2.0 \text{ m})/(2.71 \text{ m/s}) = 0.74$ s. Then the dimensionless conduction time is

$$t^* = \frac{\alpha_c t}{r_o^2} = \frac{(1.17 \times 10^{-4})(0.74)}{0.015^2} = 0.384.$$

Finally, from Eq. (4.49) for $t^* > 0.2$, compute

$$\theta_c = \frac{T_c - T_\infty}{T_i - T_\infty} = \frac{T_c - 10}{90 - 10} = 1.14\, e^{-(1.16)^2(0.384)} = 0.679,$$

or

$$T_{\text{center}} = 80(0.679) + 10 = 64^\circ\text{C}. \quad [\textit{Ans.}]$$

The uncertainty in this result is $\pm 10^\circ$C. ■

---

Additional data and correlation formulas for external forced convection past other shapes, such as square and hexagonal cylinders, are given in the text by Jakob [41, Section 26-7].

## †6.7 Crossflow over Tube Banks

A popular design geometry for a shell-and-tube heat exchanger is a (large) number of parallel rows of tubes or rods. Some examples are shown in Fig. 6.22. We use the term *tube bank* to denote crossflow past the tubes and the term *tube bundle* for axial flow along the tubes (Fig. 6.22c). Both are external flows, but the tube bundle, considered as a whole including the shell, is analogous to a duct flow with a complex cross section [44].

Tube banks may be arranged in many ways, but the two most widely used geometries are the in-line (Fig. 6.22a) and the staggered (Fig. 6.22b) arrangements. We need to know both the mean heat flux delivered to the bank plus the pressure drop across the bank. The staggered arrangement produces higher heat flux but also higher pressure drop. Thus the in-line geometry is often used to minimize the power required to force the outer stream past the tube bank. The difference in performance between the two geometries is not great.

### 6.7.1 Heat Transfer to Tube Banks

By dimensional analysis, the overall mean Nusselt number of a tube bank in crossflow should vary as follows:

$$\overline{\text{Nu}}_D = f\left(\text{Re}_D, \text{Pr}, \frac{S_L}{D}, \frac{S_T}{D}, N, \text{arrangement}\right), \qquad \textbf{(6.122)}$$

---

†This section may be omitted without loss of continuity.

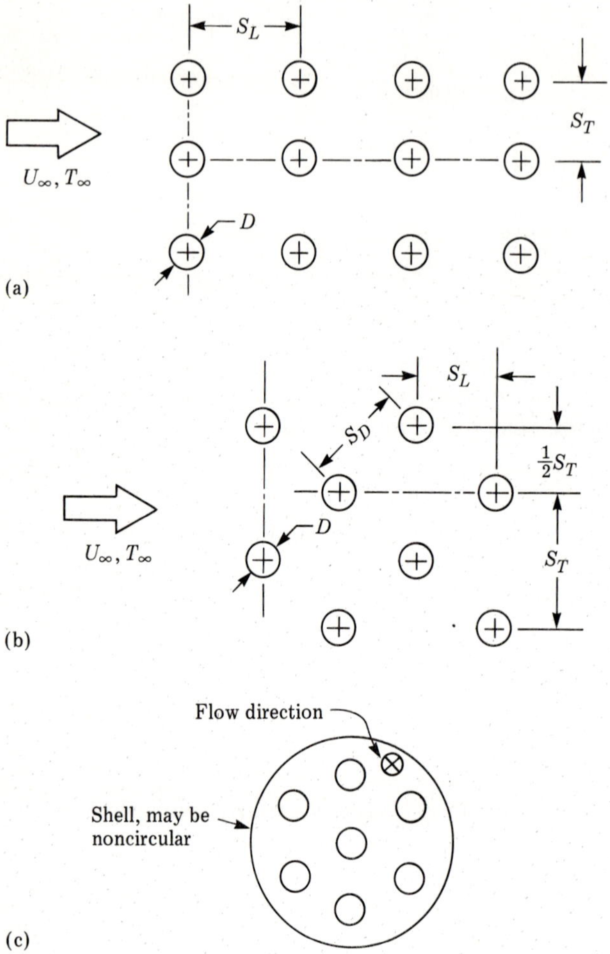


**Figure 6.22 Multiple-tube geometries: (a) crossflow past an in-line tube bank; (b) crossflow past a staggered tube bank; (c) internal axial flow along a tube bundle.**

where $S_L$ is the longitudinal tube spacing, $S_T$ the transverse spacing (see Fig. 6.22), and $N$ the number of rows of tubes (counting parallel to the freestream).

Much of the variability owing to spacing ratios can be eliminated by basing $\mathrm{Re}_D$ on the maximum average velocity between either the transverse or the diagonal openings between tubes. Thus we define, for a tube bank,

$$\mathrm{Re}_D = \frac{\rho U_{\max} D}{\mu}, \tag{6.123}$$

where, for the in-line arrangement,

$$U_{\max} = \frac{U_\infty S_T}{S_T - D}. \tag{6.124}$$

For the staggered geometry, $U_{\max}$ is given by Eq. (6.124) for large diagonal spacing $S_D$ and alternatively by

$$U_{\max} = \frac{U_\infty S_T}{2(S_D - D)}, \tag{6.125}$$

if $S_D < (S_T + D)/2$. With $\mathrm{Re}_D$ defined by Eq. (6.123), forced-convection data past tube banks is reasonably well correlated by a power-law formula in the review article by Zukauskas [42]:

$$(\overline{\mathrm{Nu}}_D)_{N\geq 10} \doteq C\, \mathrm{Re}_D^n\, \mathrm{Pr}_\infty^{0.36} \left(\frac{\mathrm{Pr}_\infty}{\mathrm{Pr}_w}\right)^{1/4}, \tag{6.126}$$

where all fluid properties except $\mathrm{Pr}_w$ are evaluated at the freestream temperature $T_\infty$. Zukauskas gives the following best-fit values for the constants $C$ and $n$:

1. *Staggered tube bank:*

   $100 < \mathrm{Re}_D < 2 \times 10^5$:

   $$\begin{aligned} n &= 0.6 \\ C &= 0.35(S_T/S_L)^{0.2} \qquad \text{for } S_T < 2S_L \\ &= 0.4 \qquad\qquad\qquad\ \text{for } S_T > 2S_L, \end{aligned}$$

   $\mathrm{Re}_D > 2 \times 10^5$:

   $n = 0.84, \qquad C = 0.022.$

2. *In-line tube banks:*

   $100 < \mathrm{Re}_D < 2 \times 10^5$:

   $n = 0.63, \qquad C = 0.27,$

   $\mathrm{Re}_D > 2 \times 10^5$:

   $n = 0.84, \qquad C = 0.021.$

The above correlations are for a deep bank with ten or more rows, and comparison of these formulas with Fig. 6.18 for a single isolated cylinder indicates that a deep bank has from 60% to 80% more heat flux. Thus, for a short bank with less than ten rows, Eq. (6.126) should be modified by a factor that decreases with the number of rows. A curve-fit to the data of [43] is as follows:

$N < 10$:

$$\frac{\mathrm{Nu}_{(N)}}{\mathrm{Nu}_{(N\to\infty)}} \doteq \left(\frac{N}{10}\right)^{0.18} \tag{6.127}$$

### 6.7.2 Pressure Drop across Tube Banks

The total pressure drop across a tube bank has been correlated in the review article by Jakob [45] in the form

$$\Delta p = f\rho\, U_{\max}^2\, N\, (\mu_w/\mu_\infty)^{0.14}, \tag{6.128}$$

where

$$f = \left[C_1 + \frac{C_2}{(S_T/D - 1)^n}\right] \mathrm{Re}_D^{-\mathrm{m}}$$

Jakob gives the following best-fit values for the constants:

1. *Staggered tube bank:*

   $C_1 = 0.5, \quad C_2 = 0.236, \quad n = 1.08, \quad \mathrm{m} = 0.16.$

2. *In-line tube bank:*

   $C_1 = 0.088, \quad C_2 = 0.16 S_L/D, \quad n = 0.43 + 1.13D/S_T, \quad \mathrm{m} = 0.15.$

The uncertainty is $\pm 30\%$. As mentioned, the staggered arrangement causes higher pressure drop at a given flow rate.

The concept of LMTD (Section 6.3.3) is valid for temperature change through a tube bank. If the crossflow enters the bank at $T_{\infty 1}$ and leaves at $T_{\infty 2}$, an overall heat balance gives

$$q_w = \dot{m} c_p (T_{\infty 2} - T_{\infty 1}) = \bar{h}\, M\, \pi D L\, \Delta T_{LM}, \tag{6.129}$$

where

$$\Delta T_{LM} = \frac{(T_{w1} - T_{\infty 1}) - (T_{w2} - T_{\infty 2})}{\ln[(T_{w1} - T_{\infty 1})/(T_{w2} - T_{\infty 2})]}$$

and where $\dot{m}$ is the total mass flow and $M$ the total number of tubes. If $T_w$ is constant, Eq. (6.54) also holds.

A graphical correlation for finding the friction factor $f$ in Eq. (6.128) is given in the review article by Zukauskas [42].

#### Example 6.14

Air at 20°C flows at 15 m/s over an in-line tube bank 6 rows deep, with $D = 0.04$ m, $L = 1.5$ m, and $T_w = 100$°C. Estimate: (a) $\bar{h}$, (b) $T_{\infty 2}$, and (c) $\Delta p$ across the bank, if $S_L = 6$ cm and $S_T = 7$ cm.

*Solution* For air at 20°C take $\rho = 1.19$ kg/m$^3$, $c_p = 1012$ J/kg · K, $k = 0.0251$ W/m · K, $\nu = 1.5 \times 10^{-5}$ m$^2$/s, and $\mathrm{Pr}_\infty \doteq \mathrm{Pr}_w$

= 0.71. From Eq. (6.124) for in-line tubes,

$$U_{max} = \frac{(15)(7)}{7 - 4} = 35 \text{ m/s}.$$

Then

$$\text{Re}_D = \frac{U_{max}D}{\nu} = \frac{(35)(0.04)}{1.5 \times 10^{-5}} = 93{,}300,$$

which is less than $2 \times 10^5$. From Zukauskas's correlation (6.126) for in-line tubes take $n = 0.63$ and $C = 0.27$. Then

$$(\overline{\text{Nu}}_D)_{N \geqslant 10} \doteq 0.27(93{,}300)^{0.63}(0.71)^{0.36}(0.71/0.71)^{1/4} = 323.$$

Since we have only 6 rows of tubes, correct this with Eq. (6.127):

$$(\overline{\text{Nu}}_D)_{N=6} \doteq (323)(6/10)^{0.18} = 294.$$

Then

$$\bar{h} = \frac{\overline{\text{Nu}}_D k}{D} = \frac{(294)(0.0251)}{0.04} = 185 \text{ W/m}^2 \cdot \text{K}. \quad [\textit{Ans. (a)}]$$

The mass flux over one column of tubes $S_T$ wide is

$$\dot{m} = \rho U_\infty S_T L = (1.19)(15)(0.07)(1.5) = 1.87 \text{ kg/s}.$$

This air flows over 6 tubes with total wall area

$$A_w = 6\pi DL = 6\pi(0.04)(1.5) = 1.13 \text{ m}^2.$$

Then, from Eq. (6.54),

$$(T_{\infty 2} - 100) = (20 - 100)e^{-\bar{h}A_w/\dot{m}c_p} = -80 \exp\left[\frac{-185(1.13)}{1.87(1012)}\right] = -72°,$$

or

$$T_{\infty 2} = 100 - 72 = 28°\text{C}. \quad [\textit{Ans. (b)}]$$

For the pressure drop, Eq. (6.128) applies for an in-line arrangement, with $S_L/D = 1.5$, $S_T/D = 1.75$, $C_2 = 0.16(1.5) = 0.24$, and $n = 0.43 + 1.13/1.5 = 1.183$. This gives

$$f = \left[0.088 + \frac{0.24}{(0.75)^{1.183}}\right](93{,}300)^{-0.15} = 0.0764.$$

For $T_w = 100°\text{C}$ and $T_\infty = 20°\text{C}$, $\mu_w/\mu_\infty = 1.24$ for air. Equation (6.128) predicts

$$\Delta p = (0.0764)(1.19)(35)^2(6)(1.24)^{0.14} = 690 \text{ Pa}. \quad [\textit{Ans. (c)}]$$

This is a modest pressure drop, accompanied by a modest temperature rise of the air. The pumping power per row of 6 tubes is

$$\dot{m}\Delta p/\rho = \frac{(1.87)(690)}{1.19} \doteq 1.1 \text{ kW}.$$

■

## †6.8 Forced Convection at Very High Speed

So far all the analyses and correlations in this chapter have been for "low-speed" flow, that is, when fluid compressibility and viscous dissipation are negligible. We saw in Example 5.3 that low speed means a very small value of the Eckert number $U_\infty^2/c_p\Delta T$. It follows that, for a given value of $c_p\Delta T$, there is a "high" speed of order $(c_p\Delta T)^{1/2}$ for which dissipation is not negligible. Let us see what changes this condition makes in a convection analysis.

When dissipation is important in boundary layer flow, mechanical energy is converted by shear stresses into thermal energy, increasing the boundary layer temperature above the freestream value $T_\infty$. This occurs even if there is no wall heat transfer ("adiabatic wall"). The parameter governing this temperature increase is a close cousin of the Eckert number, called the Mach number, after Ernst Mach (1887):

$$\text{Ma}_\infty = U_\infty/a_\infty, \tag{6.130}$$

where $a_\infty$ is the speed of sound of the freestream fluid. If $\text{Ma}_\infty$ is of the order of 0.3 or greater, dissipation and compressibility may be important. Being nearly incompressible, liquids cannot move at such velocities unless several hundred atmospheres of pressure drop are applied. Gases, on the other hand, need only a factor-of-two pressure change to achieve supersonic speeds. Therefore, "high-speed" flow analysis really concerns gasdynamics, most likely airflow.‡

For simplicity in gasdynamic computations, we commonly make the ideal-gas assumption with constant specific heat:

$$p = \rho RT; \quad \gamma = \frac{c_p}{c_v}; \quad c_p = \frac{\gamma R}{(\gamma - 1)}; \quad a = (\gamma RT)^{1/2}, \tag{6.131}$$

where $R$ is the gas constant. For air, $R \doteq 287\ \text{m}^2/(\text{s}^2 \cdot \text{K})$ and $\gamma \doteq 1.4$.

†This section may be omitted without loss of continuity.

‡The reader should review compressible fluid mechanics, for example [14, Chapter 9].

Figure 6.23 shows typical boundary layer temperature profiles for "low" and "high" speeds. At low speed ($\mathrm{Ma}_\infty \ll 1$) the adiabatic wall temperature $T_{aw} \doteq T_\infty$, but at high speed it is much greater. From Example 5.5, if the fast-moving freestream is slowed adiabatically to zero velocity, its enthalpy will rise to the "stagnation enthalpy" $i_o = i_\infty + U_\infty^2/2$. Assuming constant $c_p$, this is equivalent to a *stagnation temperature*

$$\begin{aligned} T_o &= T_\infty + \frac{U_\infty^2}{2c_p} \\ &= T_\infty\left(1 + \frac{\gamma - 1}{2}\mathrm{Ma}_\infty^2\right), \end{aligned} \tag{6.132}$$

where the latter form follows by substituting for $c_p$ from Eq. (6.131). This is an "ideal" conversion of kinetic energy into thermal energy. For gas flow, the adiabatic wall temperature in Fig. 6.23 is comparable to $T_o$ but slightly less, because of irreversible dissipation effects. The exact theory of boundary layer dissipation [1, 4] gives the result

$$T_{aw} = T_\infty\left(1 + r\frac{\gamma - 1}{2}\mathrm{Ma}_\infty^2\right), \tag{6.133}$$

**Figure 6.23 Boundary layer temperature profiles at high and low free-stream velocity.**

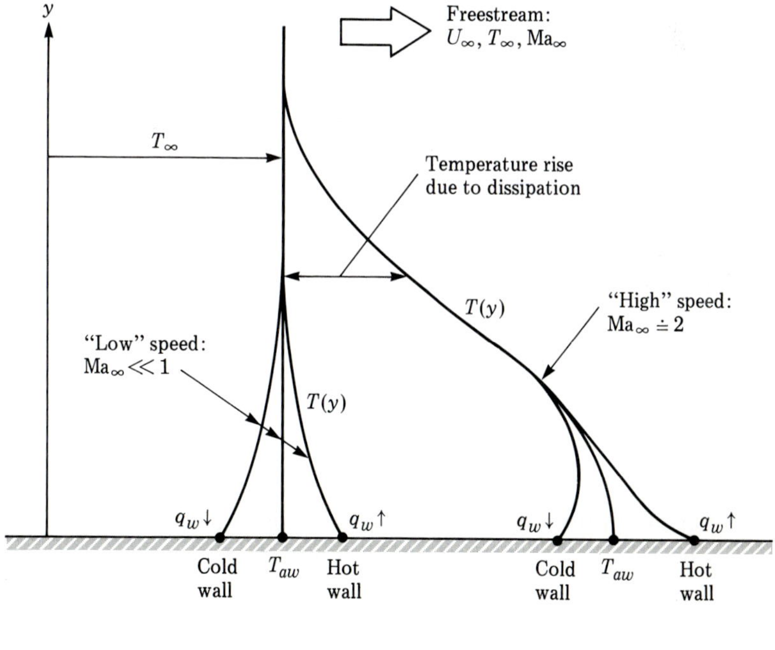

where $r$ is the dimensionless *recovery factor,* which is a function only of Prandtl number and of whether the flow is laminar or turbulent:

*Laminar flow:*

$$r \doteq \mathrm{Pr}^{1/2}, \tag{6.134a}$$

*Turbulent flow:*

$$r \doteq \mathrm{Pr}^{1/3}. \tag{6.134b}$$

Thus, for air, $\mathrm{Pr} \doteq 0.71$, $r_{\mathrm{lam}} \doteq 0.84$ and $r_{\mathrm{turb}} = 0.89$.

Refer again to Fig. 6.23. By definition, if $T_w = T_{aw}$, $q_w = 0$. Therefore, regardless of the Mach number, a "hot" wall is one in which $T_w > T_{aw}$, with heat transfer to the stream, and, conversely, a "cold" wall has $T_w < T_{aw}$, as shown. The appropriate definition of heat transfer coefficient is thus

$$q''_w = h_x(T_w - T_{aw}). \tag{6.135}$$

At low speed, this reduces to $q''_w = h_x(T_w - T_\infty)$, as in Eq. (5.1b).

For high-speed flat plate flow, numerous theories and experiments [1, 4, 13] show that, with $h_x$ defined by Eq. (6.135), the low-speed formulas are still valid if all fluid properties are evaluated at a reference temperature $T^*$. There are many correlations for $T^*$ (see, for example, [4, Table 7.3]). Here we suggest the popular formula from Eckert [46]:

$$T^* = T_\infty(0.5 + 0.039\,\mathrm{Ma}_\infty^2) + 0.5\,T_w. \tag{6.136}$$

At low speed ($\mathrm{Ma}_\infty << 1$), $T^*$ equals the film temperature $(T_w + T_\infty)/2$.

For laminar flow, we modify Eqs. (6.14) and (6.20):

$$\begin{aligned} c_f^* &= \frac{2\tau_w}{(\rho^* U_\infty^2)} \doteq \frac{0.664}{(\rho^* U_\infty x/\mu^*)^{1/2}}, \\ \mathrm{Nu}_x^* &= \frac{h_x x}{k^*} \doteq \frac{1}{2} c_f^* \, \mathrm{Pr}^{*1/3}\left(\frac{\rho^* U_\infty x}{\mu^*}\right). \end{aligned} \tag{6.137}$$

Similarly, for turbulent flow we modify Eqs. (6.90) and (6.93):

$$\begin{aligned} c_f^* &= \frac{2\tau_w}{(\rho^* U_\infty^2)} \doteq \frac{0.455}{\ln^2(0.06\,\rho^* U_\infty x/\mu^*)}, \\ \mathrm{St}_x^* &= \frac{h_x}{\rho^* U_\infty c_p^*} \doteq \frac{c_f^*/2}{1 + 12.7(\mathrm{Pr}^{*2/3} - 1)(c_f^*/2)^{1/2}}. \end{aligned} \tag{6.138}$$

These formulas correlate flat plate data within $\pm 20\%$ over a wide range of laminar and turbulent Reynolds numbers and Mach numbers, for both hot and cold walls. Because of the algebraic complexity of the formulas, it is necessary to find the average friction drag and overall heat flux by integrating the local $\tau_w(x)$ and $h_x(x)$ distributions numerically or graphically.

### Example 6.15

Repeat Example 6.1(b) if the air velocity is 600 m/s.

*Solution* First compute the Reynolds number at $x = 60$ cm:

$$\mathrm{Re}_x = \frac{U_\infty x}{\nu_\infty} = \frac{(600)(0.6)}{(15 \times 10^{-6})} = 2.4 \times 10^7.$$

Thus the local flow is turbulent, requiring formulas (6.138). The freestream speed of sound is $a_\infty = (\gamma R T_\infty)^{1/2}$, or

$$a_\infty = [1.4(287\ \mathrm{m^2/s^2 \cdot K})(273 + 30\mathrm{K})]^{1/2} = 349\ \mathrm{m/s},$$

whence the freestream Mach number is $\mathrm{Ma}_\infty = 600/349 = 1.72$. From Eq. (6.136), the reference temperature is

$$T^* = (303)[0.5 + 0.039(1.72)^2] + 0.5(283) = 328\mathrm{K}.$$

(*Note:* This computation *requires* °K and cannot be done in °C.) Meanwhile, the adiabatic wall temperature is, from Eq. (6.133),

$$T_{aw} = (303)\left[1 + (0.71)^{1/3}\left(\frac{1.4 - 1}{2}\right)(1.72)^2\right] = 463\mathrm{K}.$$

Whereas before the wall was cold, $T_w - T_\infty = 10° - 30° = -20°\mathrm{C}$, now it is extremely cold, $T_w - T_{aw} = 283 - 463 = -180\mathrm{K}$. Now evaluate the reference properties: for air at 1 atm and $T^* = 328\mathrm{K}$, $\rho^* = 1.077\ \mathrm{kg/m^3}$, $k^* = 0.0275\ \mathrm{W/m \cdot K}$, $\nu^* = 1.8 \times 10^{-5}\ \mathrm{m^2/s}$, $c_p^* = 1016\ \mathrm{J/kg \cdot K}$, and $\mathrm{Pr}^* = 0.71$.

From Eq. (6.138), compute

$$c_f^* \doteq \frac{0.455}{\ln^2[0.06(600)(0.6)/(1.8 \times 10^{-5})]} = 0.00232,$$

from which

$$\mathrm{St}_x^* \doteq \frac{(0.00232/2)}{1 + 12.7[(0.71)^{2/3} - 1](0.00232/2)^{1/2}}$$
$$= 0.00127.$$

Then the local heat transfer coefficient is

$$h_x = \mathrm{St}_x^* \rho^* U_\infty c_p^* = (0.00127)(1.077)(600)(1016) = 836\ \mathrm{W/m^2 \cdot K}.$$

Finally, the local heat flux is

$$(q''_w)_{x=60\text{cm}} = h_x(T_w - T_{aw}) = 836(283 - 463)$$
$$= -151{,}000 \text{ W/m}^2. \quad [Ans.]$$

This is much larger than the low-speed airflow of Example 6.1(b) but not as large as the low-speed water flow of Example 6.7(b).

■

---

## Summary

This chapter has presented theory and experiment for various forced-convection problems, mostly in the simplified form Nu = *f*(Re, Pr, shape). First, the boundary layer momentum and energy equations from Chapter 5 are modified into integral relations. Solutions are then given for laminar friction and heat transfer in flat plate flow and in circular duct flow.

In a similar manner, turbulent eddy viscosity and conductivity correlations are developed and applied to turbulent flow in ducts and along a flat plate. These classic formulas are then modified to account for entrance effects, an unheated starting length, wall heat flux conditions, noncircular cross sections, and wall roughness. Brief mention is made of convection-enhancement devices such as turbulators and finned tubes.

When geometries other than the flat plate and straight duct are encountered, the theory becomes weak and, instead, experiments are correlated over a range of Reynolds and Prandtl numbers. Engineering formulas are given here for flow about cylinders, spheres, and tube banks.

The chapter ends with a brief discussion of very high speed (compressible, dissipating) gas flows. The concepts of recovery factor and adiabatic wall temperature are introduced, and high-speed heat flux is shown to be proportional to $(T_w - T_{aw})$. Formulas are given for high-speed laminar and turbulent flow past a flat plate.

It cannot be overemphasized that convection analysis is heavily dependent on experiment, so that the common engineering formulas may have uncertainties of $\pm 30\%$. For details about heat transfer experimentation, consult the text by Eckert [47].

Since convection estimates involve dozens of formulas that are hard to remember and find, the most important correlations are summarized in Table 6.5.

Table 6.5 Summary of forced convection correlations

| Geometry, Conditions | Correlation |
|---|---|
| 1. Flat Plate Flow | |
| A. Laminar ($Pr > 0.6$, $Re_x < 5 \times 10^5$) | $\delta/x = 5.0/Re_x^{1/2}$ |
| | $c_f = 0.664/Re_x^{1/2}$ |
| | $\bar{c}_f = 1.328/Re_L^{1/2}$ |
| | $Nu_x = 0.332\ Re_x^{1/2}\ Pr^{1/3}$ |
| | $\overline{Nu}_L = 0.664\ Re_L^{1/2}\ Pr^{1/3}$ |
| B. Turbulent ($Pr > 0.6$, $Re_x > 5 \times 10^5$) | $\delta/x \doteq 0.37/Re_x^{1/5}$ |
| | $c_f = 0.455/\ln^2(0.06\ Re_x)$ |
| | $\bar{c}_f = 0.523/\ln^2(0.06\ Re_L)$ |
| | $Nu_x \doteq 0.0296\ Re_x^{4/5}\ Pr^{1/3}$ |
| | $\overline{Nu}_L \doteq 0.037\ Re_L^{4/5}\ Pr^{1/3}$ |
| 2. Duct Flow, Constant Wall Temperature | |
| A. Laminar: | |
| Circular, TFD, HFD | $c_f = 16/Re_D$ |
| | $Nu_D = 3.66$ |
| Noncircular, TFD, HFD | Table 6.3 |
| Entrance region, circular | Fig. 6.9 |
| B. Turbulent: | |
| Circular, TFD, HFD | $c_f$: Eq. (6.105) |
| | St: Eq. (6.106) |
| Entrance region | Fig. 6.14 |
| Liquid metals | $Nu_D$: Eq. (6.108) |
| 3. Cylinder in Crossflow | Fig. 6.18 |
| 4. Sphere in Uniform Stream | $\overline{Nu}_D$: Eq. (6.119) |

## References


1. W. M. Kays and M. E. Crawford, *Convective Heat and Mass Transfer,* McGraw-Hill, New York, 1980.
2. R. K. Shah and A. L. London, *Laminar Flow Forced Convection in Ducts,* Academic Press, New York, 1979.
3. S. V. Patankar, *Numerical Heat Transfer and Fluid Flow,* McGraw-Hill, New York, 1980.
4. F. M. White, *Viscous Fluid Flow,* McGraw-Hill, New York, 1974.

5. E. R. van Driest, "On Turbulent Flow Near a Wall," *J. Aero. Sciences,* vol. 23, 1956, pp. 1007–1011.

6. T. Cebeci and P. Bradshaw, *Momentum Transfer in Boundary Layers,* McGraw-Hill, New York, 1977.

7. P. Bradshaw, T. Cebeci, and J. H. Whitelaw, *Engineering Calculation Methods for Turbulent Flow,* Academic Press, New York, 1981.

8. O. Reynolds, "On the Extent and Action of the Heating Surface for Steam Boilers," *Proc. Manchester Lit. Phil. Soc.,* vol. 14, 1874, pp. 7–12.

9. B. F. Blackwell, "The Turbulent Boundary Layer on a Porous Plate: An Experimental Study of the Heat Transfer Behavior with Adverse Pressure Gradients," Ph.D. thesis, Stanford Univ., 1973.

10. A. Slanciauskas, A. Pedisius, and A. Zukauskas, "Universal Profiles of the Temperature and Turbulent Prandtl Number in the Boundary Layer on a Plate in a Stream of Fluid," *Heat Transfer — Soviet Research,* vol. 5, no. 6, 1973, pp. 63–73.

11. R. A. Gowen and J. W. Smith, "The Effect of the Prandtl Number on Temperature Profiles for Heat Transfer in Turbulent Pipe Flow," *Chem. Eng. Science,* vol. 22, 1967, pp. 1701–1711.

12. T. von Kármán, "Über laminare und turbulente Reibung," *ZAMM,* vol. 1, 1921, pp. 233–252. English translation as N.A.C.A. Tech. Memo. 1092, 1946.

13. H. Schlichting, *Boundary Layer Theory,* 7th ed., McGraw-Hill, New York, 1979.

14. F. M. White, *Fluid Mechanics,* 2nd ed., McGraw-Hill, New York, 1986.

15. B. S. Petukhov and V. V. Kirillov, "Heat Exchange for Turbulent Flow of Liquid in Tubes," *Soviet Teploenergetika,* vol. 5, 1958, pp. 63–68.

16. L. C. Thomas and M. M. Al-Sharif, "An Integral Analysis for Heat Transfer in Turbulent Incompressible Boundary Flow," *J. Heat Transfer,* vol. 103, Nov. 1981, pp. 772–777.

17. D. F. Dipprey and D. H. Sabersky, "Heat and Momentum Transfer in Smooth and Rough Tubes at Various Prandtl Numbers," *Int. J. Heat & Mass Transfer,* vol. 6, 1963, pp. 329–353.

18. R. K. Shah, "A Correlation for Laminar Hydrodynamic Entry Length Solutions for Circular and Non-Circular Ducts," *J. Fluids Engineering,* vol. 100, 1978, pp. 177–179.

19. H. Hausen, "Darstellung des Wärmeuberganges in Rohren durch verallgemeinerte Potenzbeziehungen," *Z. VDI Beih. Verfahrenstech,* no. 4, 1943, p. 91.

20. E. N. Sieder and C. E. Tate, "Heat Transfer and Pressure Drop of Liquids in Tubes," *Ind. Eng. Chem.,* vol. 28, 1963, p. 1429.

21. B. S. Petukhov, "Heat Transfer and Friction in Turbulent Pipe Flow with

Variable Physical Properties," in *Advances in Heat Transfer,* J. P. Hartnett and T. F. Irvine, Jr., Eds., vol. 6, Academic Press, New York, 1970, pp. 504–564.

22. C. F. Colebrook, "Turbulent Flow in Pipes, with Particular Reference to the Transition Between the Smooth and Rough Pipe Laws," *J. Inst. Civ. Eng. London,* vol. 11, 1938/39, pp. 133–156.
23. L. F. Moody, "Friction Factors for Pipe Flow," *Transactions ASME,* vol. 66, 1944, pp. 671–684.
24. C. A. Sleicher and M. W. Rouse, "A Convenient Correlation for Heat Transfer to Constant and Variable-Property Fluids in Turbulent Pipe Flow," *Intl. J. Heat & Mass Transfer,* vol. 18, 1975, pp. 677–683.
25. C. A. Sleicher and M. Tribus, "Heat Transfer in a Pipe with Turbulent Flow and Arbitrary Wall Temperature Distribution," *Heat Transfer & Fluid Mechanics Institute,* 1956, pp. 59–78.
26. R. H. Norris, "Some Simple Approximate Heat Transfer Correlations for Turbulent Flow in Ducts with Rough Surfaces," pp. 16–26 of [27].
27. A. E. Bergles and R. L. Webb, Eds., *Augmentation of Convective Heat and Mass Transfer,* Proceedings of ASME Symposium, Dec. 2, 1970, New York.
28. J. M. Chenoweth et al., *Advances in Enhanced Heat Transfer,* ASME Symposium Vol. No. I00122, Aug. 1979, San Diego.
29. A. E. Bergles, "Survey and Evaluation of Techniques to Augment Convective Heat Transfer," *Progress in Heat and Mass Transfer,* vol. 1, 1969, pp. 331–424.
30. C. L. Tien, V. P. Carey, and J. K. Ferrell, Eds., *Heat Transfer 1986,* vol. 6, "Heat Transfer Augmentation," Hemisphere Publishing, New York, 1986, pp. 2831–2976.
31. P. J. Marto et al., "An Experimental Comparison of Enhanced Heat Transfer Condenser Tubing," pp. 1–9 of [28].
32. T. C. Carnavos, "Heat Transfer Performance of Internally Finned Tubes in Turbulent Flow," pp. 61–67 of [28].
33. G. S. Ambrok, "Approximate Solutions of Equations for the Thermal Boundary Layer with Variations in the Boundary Layer Structure," *Soviet Phys. — Tech. Phys.,* vol. 2, no. 9, 1957, pp. 1979–1986.
34. W. H. Giedt, "Investigation of Variation of Point Unit Heat Transfer Coefficient Around a Cylinder Normal to an Air Stream," *ASME Transactions,* vol. 71, 1949, pp. 375–381.
35. M. N. Ozisik, *Basic Heat Transfer,* McGraw-Hill, New York, 1977.
36. S. W. Churchill and M. Bernstein, "A Correlating Equation for Forced Convection from Gases and Liquids to a Circular Cylinder in Crossflow," *J. Heat Transfer,* vol. 99, 1977, pp. 300–306.

37. C. Wieselberger, "Der Luftwiderstand von Kugeln," *Z. Flugtechnik Motorluftschiffahrt,* vol. 5, 1914, pp. 140–144.

38. S. Whitaker, "Forced Convection Heat Transfer Correlations for Flow in Pipes, Past Flat Plates, Single Cylinders, Single Spheres, and Flow in Packed Beds and Tube Bundles," *AIChE J.,* vol. 18, 1972, pp. 361–371.

39. R. Clift, J. R. Grace, and M. E. Weber, *Bubbles, Drops, and Particles,* Academic Press, New York, 1978.

40. L. C. Witte, "An Experimental Study of Forced-Convection Heat Transfer from a Sphere to Liquid Sodium," *J. Heat Transfer,* vol. 90, 1968, pp. 9–12.

41. M. Jakob, *Heat Transfer,* vol. 1, Wiley, New York, 1949.

42. A. Zukauskas, "Heat Transfer from Tubes in Crossflow," in *Advances in Heat Transfer,* J. P. Hartnett and T. F. Irvine, Jr., Eds., vol. 8, 1972, pp. 93–160.

43. W. M. Kays and A. L. London, *Compact Heat Exchangers,* 2nd ed., McGraw-Hill, New York, 1964.

44. S. C. Yao and P. A. Pfund, Eds., *Fluid Flow and Heat Transfer Over Rod or Tube Bundles,* ASME Symposium Vol. G00157, 1979.

45. O. C. Jones, Jr., "An Improvement in the Calculation of Turbulent Friction in Rectangular Ducts," *J. Fluids Engrg.,* June 1976, pp. 173–181.

46. E. R. G. Eckert, "Engineering Relations for Friction and Heat Transfer to Surfaces in High Velocity Flow," *J. Aero. Sciences,* vol. 22, 1955, pp. 585–587 (see also *ASME Transactions,* vol. 78, 1956, pp. 1273–1284).

47. E. R. G. Eckert and R. J. Goldstein, Eds., *Measurement in Heat Transfer,* McGraw-Hill Hemisphere, New York, 1976.

48. S. W. Churchill and H. Ozoe, "Correlations for Laminar Forced Convection in Flow over an Isothermal Flat Plate and in Developing and Fully Developed Flow in an Isothermal Tube," *J. Heat Transfer,* vol. 95, Feb. 1973, pp. 78–84.

49. A. Zukauskas and J. Ziukzda, *Heat Transfer in a Cylinder in Crossflow,* Hemisphere Publishing, New York, 1985.

50. D. C. McCormick, F. L. Test, and R. C. Lessmann, "The Effect of Freestream Turbulence on Heat Transfer from a Rectangular Prism," *J. Heat Transfer,* vol. 106, May 1984, pp. 268–275.

51. H. Groehn, "Integral and Local Heat Transfer of a Yawed Single Circular Cylinder up to Supercritical Reynolds Numbers," *Heat Transfer 1986,* vol. 3, Hemisphere Publishing, New York, 1986, pp. 1013–1018.

52. E. M. Sparrow and F. Samie, "Measured Heat Transfer Coefficients at and Adjacent to the Tip of a Wall-Attached Cylinder in Crossflow — Application to Fins," *J. Heat Transfer,* vol. 103, Nov. 1981, pp. 778–784.

53. E. M. Sparrow, T. J. Stahl, and P. Traub, "Heat Transfer Adjacent to the Attached End of a Cylinder in Crossflow," *Int. J. Heat & Mass Transfer,* vol. 27, no. 2, 1984, pp. 233–242.

54. S. W. Churchill, "A Comprehensive Correlating Equation for Forced Convection from Flat Plates," *AIChE Journal,* vol. 22, 1976, pp. 264–268.

## Review Questions

1. What are the boundary layer approximations?
2. What are eddy transport coefficients? Are they the same for turbulent shear and heat transport?
3. Define the turbulent Prandtl number. How large is it?
4. Define the law-of-the-wall. Is it the same for velocity and temperature? Does it hold for laminar and turbulent flow?
5. What is the difference between the boundary layer integral and differential relations? Are they independent? Can they both be solved exactly?
6. Give two forms of the Reynolds analogy. Does the analogy hold for both laminar and turbulent flow?
7. What is the viscous sublayer? The thermal sublayer?
8. What are the two most common wall heat transfer boundary conditions? Which of the two gives a higher Nusselt number?
9. What is the effect of roughness on wall friction and heat flux?
10. Define HFD and TFD flow in ducts.
11. Define the log-mean temperature difference.
12. Which gives higher heat flux, laminar or turbulent duct flow?
13. What is the hydraulic diameter? Does this concept work for both laminar and turbulent noncircular duct flows?
14. Describe some convection enhancement devices.
15. What is the chief difficulty in applying boundary layer theory to such problems as crossflow past a cylinder?
16. Is there a relation between drag coefficient and mean Nusselt number for flow past an immersed body?
17. Define a tube bank and a tube bundle.
18. Does a tube bank produce more or less average heat flux per unit area than a single cylinder in crossflow? Why?
19. Define the recovery factor for flat plate flow.
20. Discuss the adiabatic wall temperature concept for high-speed flow. Why not use $(T_w - T_\infty)$ to define $h$ at high speeds?

## Problems

### Problem distribution by sections

The Problem Assignments are Organized as Follows:

| Problems | Sections Covered | Topics Covered |
|---|---|---|
| 6.1–6.7 | 6.1 | Boundary layer equations |
| 6.8–6.19 | 6.2.1, 6.2.2 | The Blasius solution |
| 6.20–6.26 | 6.2.3 | Flat plate integral solutions |
| 6.27–6.34 | 6.3.5 | Laminar TFD tube flow |
| 6.35–6.38 | 6.3.6 | Laminar noncircular ducts |
| 6.39–6.46 | 6.3.7 | Developing laminar flow |
| 6.47–6.51 | 6.4.1–4 | Eddy viscosity and conductivity |
| 6.52–6.64 | 6.4.5–6 | Turbulent flat plate flow |
| 6.65–6.67 | 6.4.7 | Unheated starting length |
| 6.68–6.70 | 6.4.8 | Uniform wall heat flux |
| 6.71–6.86 | 6.5 | Turbulent duct flow |
| 6.87–6.95 | 6.6.2 | Crossflow past a cylinder |
| 6.96–6.100 | 6.6.5 | Flow past a sphere |
| 6.101–6.103 | 6.7 | Tube banks |
| 6.104 | 6.7 | Tube bundles |
| 6.105–6.106 | 6.8 | High-speed flow |
| 6.107–6.109 | All | Design problems |

6.1 Write the boundary layer momentum and energy equations for laminar steady flow with all fluid properties constant except temperature and velocity. Are any exact solutions known?

6.2 For the equations of Problem 6.1, show that, for Pr = 1, a possible exact solution is $T = T(u)$ only. This is one form of the Reynolds analogy. Assume $dp/dx = 0$.

6.3 Carry Problem 6.2 a step further by showing that the postulated exact solution has the form $T = a + bu + cu^2$. Evaluate $(a, b, c)$ in terms of $(T_w, T_\infty, U_\infty, c_p)$. This is called Crocco's relation [4, p. 580].

6.4 When deriving Eq. (6.5), von Kármán [12] wrote it in the form $\tau_w/\rho U_\infty^2 = d\theta/dx$. What are the dimensions of $\theta$? What might it represent? (See [4, p. 244] if you give up.)

6.5 By analogy with Problem 6.4, one could write the energy integral relation

(6.6) in the form $\text{St}_x = d\theta/dx$. Define $\theta$ as an integral and give its dimensions. What might it represent? (See [4, p. 250] if you give up.)

6.6 Consider laminar flow between two infinite parallel plates induced by moving the upper plate at velocity $U$, as in Fig. P6.6. This is called Couette flow. Assuming constant ($\rho$, $c_p$, $k$, $\mu$, $p$) and $v = 0$, use the boundary layer equations (6.1) to solve for the velocity and temperature distributions $u(y)$ and $T(y)$. Do not neglect dissipation. Also find expressions for $\tau_w$ and $q''_w$ at both plates.

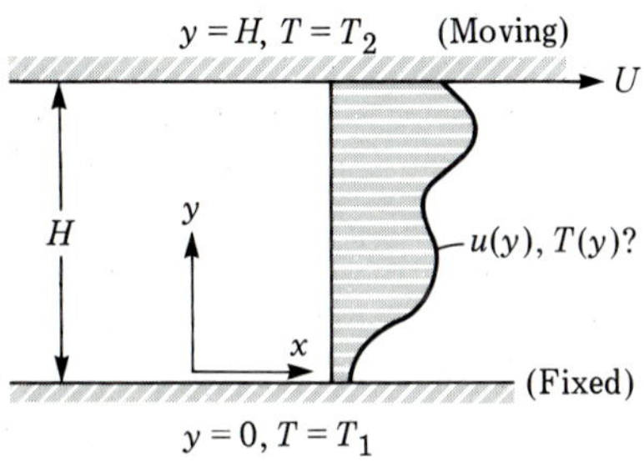


**Figure P6.6**

6.7 For flow of air at 20°C and 1 atm past a flat surface, the velocity profiles at two adjacent sections are measured and approximated as follows:

$x = 0.50$ m: $\quad u \doteq 7.2 \tanh(357y);$

$x = 0.52$ m: $\quad u \doteq 7.2 \tanh(350y),$

with $y$ in meters and $u$ in m/s. Use the momentum integral relation (6.5) to estimate the wall shear stress in pascals. Compare your integral estimate with $\mu(\partial u/\partial y)_{y=0}$.

6.8 Verify by direct substitution that the use of the similarity variables $\eta$ and $f(\eta)$ from Eqs. (6.8) and (6.9) reduces the momentum equation (6.7a) to the Blasius equation (6.11).

6.9 Using any numerical method (Runge-Kutta, predictor-corrector, etc.) and the hindsight knowledge that $f''(0) = 0.33206$, integrate the Blasius equation (6.11) with conditions (6.12) and compare your results with Table 6.1.

6.10 Following up on Problem 6.9, take $f(\eta)$ from your numerical solution and insert into Eq. (6.18) for any Prandtl number from Table 6.2. Compare the resulting $\Theta(\eta)$ with Fig. 6.2 and the heat transfer parameter $\Theta'(0)$ with Table 6.2.

6.11 Consider flow of air at 20°C and 1 atm past a flat plate at a stream velocity of 7.2 m/s. Assuming laminar flow, use the Blasius solution to evaluate the wall shear stress at $x = 0.50$ m and 0.52 m and compare your results with direct differentiation of the data correlations in Problem 6.7.

6.12 An array of 400 computer chips, each 7 by 7 mm and insulated on the bottom side, is to be cooled by flow of air at 20°C and 1 atm, as in Fig. P6.12. Each chip must dissipate at least 50 mW of heat and may not exceed 65°C. Assuming constant wall temperature conditions, estimate the minimum $U_0$ in m/s.

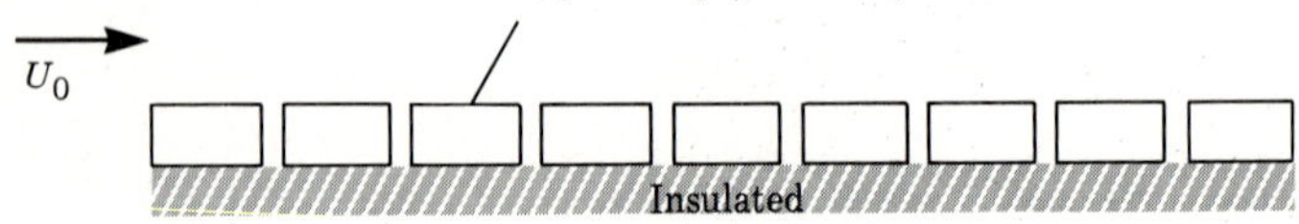


**Figure P6.12**

6.13 A flat plate 1 ft wide and 2 ft long at 100°F is immersed in an air stream at 0°F. What should be the air velocity to remove 1000 Btu/hr from both sides of the plate combined?

6.14 Repeat Problem 6.13 if the fluid is ammonia. Explain your peculiar and perhaps uncertain result.

6.15 Repeat Problem 6.13 if the fluid is ammonia and the desired heat removal is 30,000 Btu/hr. Why is ammonia more effective than air for this purpose?

6.16 An air heater consists of a square honeycomb made of thin flat plates, as in Fig. P6.16. The air enters at 20 m/s and 20°C and the plate temperature is 160°C. Using flat plate theory and neglecting boundary layer displacement effects, find the exit temperature $T_o$ if $H = 1$ cm and $L = 2$ cm.

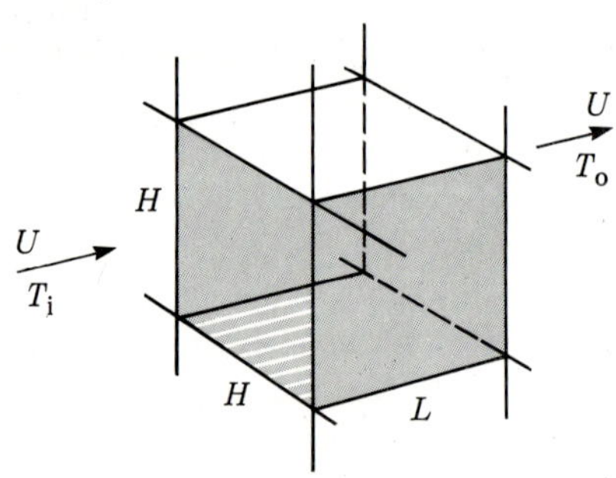


**Figure P6.16**

6.17 Modify Problem 6.16 to find the velocity $U$ that will cause the exit temperature to be 45°C.

6.18 For a liquid metal (Pr $\ll$ 1) the thermal boundary layer $\Delta$ is much thicker than $\delta$ in Fig. 6.1, and we may approximate the velocity profile as $u \doteq U_\infty$, $v \doteq 0$ to solve the energy equation (6.7b). Show that the energy equation and its boundary conditions then become analogous to semi-infinite conduction with a sudden change in wall temperature (Section

4.3.1) and the solution for local Nusselt number is $\mathrm{Nu}_x \doteq 0.564(\mathrm{Re}_x\mathrm{Pr})^{1/2}$. Compare this result with the values in Table 6.2 and with Eq. (6.21).

6.19 Engine oil at 80°C is to be cooled by strip coolers which approximate a flat plate 25 cm by 25 cm with both sides exposed to the oil flow. Each strip should remove 400 W of heat from the oil. If the oil velocity is 1 m/s, what should be the wall temperature of the strips? Try to account for the unknown film temperature effects by iteration.

6.20 Make an integral analysis for laminar flat plate flow assuming the velocity profile to have a cubic shape, $u = a + by + cy^2 + dy^3$. Dropping the condition $(\partial^2 u/\partial y^2) = 0$ at $y = \delta$, find the appropriate dimensionless profile similar to Eq. (6.28) and solve the integral momentum relation (6.5) for boundary layer thickness and friction factor. Compare your results with the Blasius solution.

6.21 Repeat Problem 6.20, but use only a quadratic velocity profile shape $u = a + by + cy^2$. You will have to drop the additional boundary condition $(\partial^2 u/\partial y^2) = 0$ at $y = 0$ because there are not enough free constants ($a$, $b$, $c$) available.

6.22 Repeat Problem 6.20 with the profile $u/U_\infty \doteq \sin(\pi\eta/2)$.

6.23 Make an integral energy analysis similar to that which resulted in Eq. (6.35), using the profile approximations $u/U_\infty \doteq 2\eta - \eta^2$ and $(T - T_w)/(T_\infty - T_w) \doteq 2\zeta - \zeta^2$.

6.24 Repeat Problem 6.23 with $u/U_\infty \doteq \sin(\pi\eta/2)$ and $(T - T_w)/(T_\infty - T_w) \doteq \sin(\pi\zeta/2)$.

6.25 For the air flow of Problem 6.7 the temperature profiles are also measured at the two sections, approximately:

$$x = 0.50\ \text{m}: \quad T \doteq 15 + 5\tanh(400y),$$

$$x = 0.52\ \text{m}: \quad T \doteq 15 + 5\tanh(392y),$$

with $T$ in °C and $y$ in meters. Use the energy integral relation (6.6) to estimate $q''_w$ between these two sections and compare with Fourier's law and the Pohlhausen solution, Eq. (6.20).

6.26 An engineer has proposed that the profile approximation $u/U \doteq 1 - e^{-ay}$, where $a$ = constant, can be used to solve Eq. (6.5) for $c_f$ since it satisfies $u = 0$ at $y = 0$ and $u \to U_\infty$ as $y \to \infty$. However, the resulting $c_f$ is grossly inaccurate. What is the reason for the discrepancy? (The trouble is at the wall.)

6.27 Water flowing through a 3-cm-diameter tube is to be heated from 10°C to 40°C by resistance heating that generates wall heat flux of $4 \times 10^4$ W/m$^2$ uniformly in the tube. The water volume flow is 1.5 m$^3$/hr. Estimate (a) the length of tube required and (b) the local heat transfer coefficient at the tube midlength, where $T_w = 42$°C.

6.28 Water at 10°C and a flow rate of 2.7 m$^3$/hr enters a tube of diameter 4 cm and length 3.5 m. The tube wall is at 70°C, and the average heat transfer coefficient is 5060 W/(m$^2$ · K). Estimate (a) the outlet mean water temperature and (b) the log mean temperature difference.

6.29 A fluid at 20°C with $c_p$ = 2.4 kJ/kg · K enters at 3 kg/s into a 6-cm-diameter duct whose wall temperature is 120°C. If $\bar{h} = [4 + 8(D/x)^{1/3}]$ W/m$^2$ · K, which approximates laminar flow conditions, estimate the distance $x$ downstream where the average fluid temperature reaches 30°C, and interpret. What is the log mean temperature difference?

6.30 Repeat Problem 6.29 if the duct instead contributes a uniform wall heat flux of 410 W/m$^2$, again simulating laminar flow.

6.31 Oil at 20°C flows at 0.8 kg/s through a very long tube of area 20 cm$^2$ and wall temperature 30°C. For constant properties $\rho$ = 850 kg/m$^3$, $c_p$ = 2100 J/kg · K, $k$ = 0.14 W/m · K, Pr = 510, and $\nu = 4 \times 10^{-5}$ m$^2$/s, estimate (a) $h_{\mathrm{TFD}}$; (b) $\Delta T_{\mathrm{fluid}}$ per meter of length; and (c) $\Delta p$ per meter of length.

6.32 In laminar HFD and TFD duct flow, to what power of $V$ are $\Delta p$ and $h$ proportional? What does this say about optimizing pumping power for a given heat flux?

6.33 Show with a control volume analysis that, for steady HFD flow in a circular duct, $dp/dx = -2\tau_w/r_o$ for either laminar or turbulent flow.

6.34 Air at 20°C and 1 atm flows through a circular duct 1 cm in diameter and 2 m long with $V$ = 2.2 m/s and $T_w$ = 5°C. Assuming HFD and TFD flow, compute the pressure drop and the exit temperature.

6.35 Repeat Problem 6.31 if the duct has a square cross section.

6.36 Repeat Problem 6.31 if the duct section is an equilateral triangle.

6.37 Transformer oil flows through a very long 2-cm-by-4-cm rectangular duct at 0.4 kg/s. If the oil temperature is 40°C and the wall temperature 0°C, estimate (a) the pressure drop, and (b) the oil temperature change per meter of duct length. (See Example 6.4 for physical properties.)

6.38 A thin sheet metal square duct, 16 mm on a side and 3 m in length, contains a layer of ice 3 mm thick, as shown in Fig. P6.38. To melt the ice, water at 20°C enters the duct at 0.07 m$^3$/hr. The latent heat of ice is 334 kJ/kg, and its density is 900 kg/m$^3$. Assuming fully developed laminar flow, estimate the time required to melt the ice completely.

6.39 Repeat Problem 6.34 if the duct is only 20 cm long, so that the flow is not fully developed.

6.40 In Problem 6.38 the duct is actually too short to be thermally fully developed. Repeat the analysis by accounting for flow development effects.

6.41 Helium at 300°C and 1 atm enters a tube 4 cm in diameter amd 1 m long. If the flow rate is 0.0005 kg/s and $T_w$ = 100°C, estimate (a) $\Delta p$, (b) exit helium temperature, and (c) $q_w$.

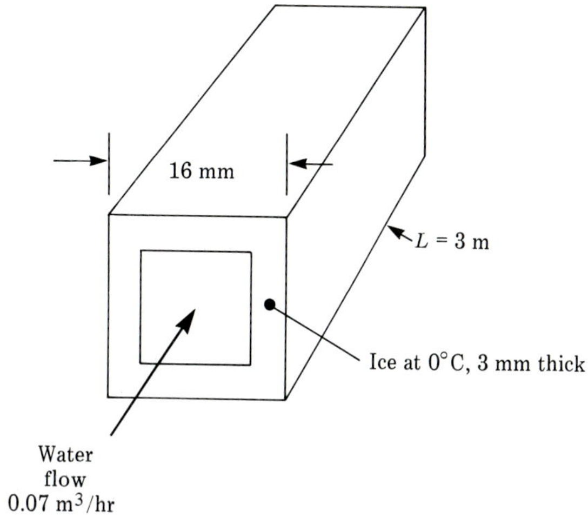


Figure P6.38

6.42 Engine oil at 100°F is to be cooled by flowing through a 3-in.-diameter tube with $T_w$ = 0°F. If the flow rate is 20 $lb_m$/s, estimate the tube length required to cool the oil down to 80°F. Interpret your perhaps unexpected result.

6.43 Instead of analyzing Problem 6.16 as flat plate flow, repeat the problem by assuming that the square passages represent laminar developing flow in an extremely short square duct.

6.44 Carbon dioxide at 0°C and 1 atm is to be heated to 100°C by passing through a 3-cm-diameter tube 80 cm long with $T_w$ = 200°C. Estimate the required laminar mass flux in kg/hr.

6.45 Show that, in laminar flow, the length of tube required to achieve a given fluid temperature change at constant $T_w$ is approximately proportional to the fluid average velocity. Try to prove this for both developed and developing flows.

6.46 Oil flowing in a buried Arctic pipeline is heated to lower its viscosity and then loses heat to the cold soil. Let the pipe be 1.5 m in diameter with its centerline buried 3 m below the surface of clay earth soil at −20°C. The pipeline is 200 km long, and the oil enters at 40°C and 375 kg/s. Let the oil properties be approximated by unused engine oil and assume laminar developing flow. Estimate the exit temperature of the oil. (*Hint:* Use Item 5 of Table 3.1 to estimate the soil heat loss.)

6.47 Using van Driest's eddy viscosity relation (6.79), carry out the numerical integration of Eq. (6.81) on a digital computer, using any method of quadrature. Compare with Fig. 6.7.

6.48 According to van Driest's correlation (6.79), at what point in a turbulent boundary layer does the eddy viscosity equal 2% of the molecular viscosity?

How does this compare with the definition of viscous sublayer from Eq. (6.82a)?

6.49 Freon-12 at 20°C flows in a 6-cm-diameter tube so that the centerline velocity is 12 m/s. Is the flow turbulent? If so, use the log-law (6.82b) to estimate (a) $\tau_w$ and (b) the velocity at $r = 2$ cm.

6.50 Using any numerical quadrature method you wish, carry out the integral in Eq. (6.86) on a digital computer for some value of Pr $\neq 1$ and compare with Fig. 6.9 and Eqs. (6.87, 6.88).

6.51 Ethyl alcohol at 20°C flows past a slightly hot smooth surface. At $y =$ 3.6 mm the measured temperature difference $(T_w - T) = 5°\text{C}$, and the velocity is $u = 1.8$ m/s. Using the log-laws (6.82b) and (6.87), estimate the friction velocity and wall heat flux.

6.52 There are no exact formulas for turbulent boundary layers, but Ludwig Prandtl suggested that the velocity and temperature profiles in flat plate flow could be approximated by

$$u/U \doteq (y/\delta)^{1/7}, \tag{1}$$

$$(T - T_\infty)/(T_w - T_\infty) \doteq 1 - (y/\Delta)^{1/7}, \tag{2}$$

and that $\Delta \doteq \delta$. Further, experimental data gave the following wall shear stress variation with boundary layer thickness:

$$\tau_w \doteq 0.0228\, \rho\, U_\infty^2\, (U_\infty\, \delta/\nu)^{-1/4}. \tag{3}$$

Substitute Eqs. (1–3) into the boundary layer momentum and energy integral relations (6.5) and (6.6) and derive formulas for $\delta(x)$, $c_f(x)$, and $\text{Nu}_x(x)$. Compare your results with Eqs. (6.89) and (6.92).

6.53 Water at 20°C flows at 5 m/s past a flat plate 80 cm long and 60 cm wide with $T_w = 50°\text{C}$. Estimate (a) $\tau_w$ and $q''_w$ at the trailing edge and (b) the drag and heat flux on one side.

6.54 A flat, smooth 30-m-by-30-m roof is heated to 15°C in the sun. Estimate the heat loss if exposed to a horizontal breeze of air at 8 m/s and 0°C.

6.55 Consider a motorcycle engine to have a cooling fin 2 cm high and 12 cm long at 200°C. If the fin moves at 90 km/hr parallel to 20°C still air, what is the heat loss?

6.56 An aircraft wing, idealized as a flat plate 4 ft long, moves at 250 mi/hr in air at 11 $\text{lb}_f/\text{in.}^2$ and 25°F. Estimate its average heat transfer coefficient in $\text{Btu/hr} \cdot \text{ft}^2 \cdot °\text{F}$.

6.57 A flat plate solar collector is flush-mounted in a roof as in Fig. P6.57. Its surface is at 20°C, and cold air is blowing over the roof as shown. Estimate the convective heat loss from the plate when it is mounted at the leading edge, $H = 0$. What downstream mounting distance $H$ would reduce this heat loss by 40%?

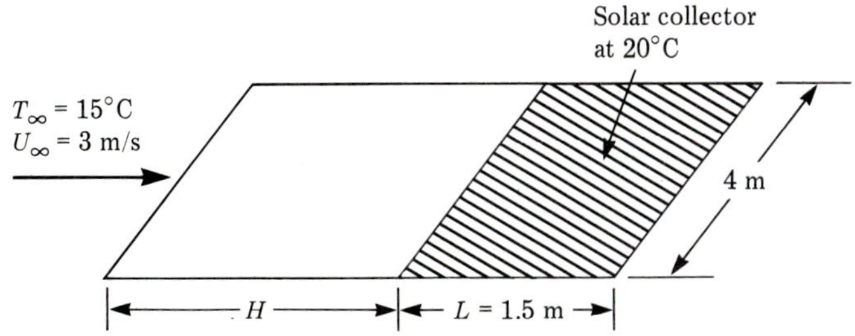


**Figure P6.57**

6.58 Ethylene glycol at 100°C flows at 4 m/s past a thin copper plate 1 m long and 1 m wide. It is cooled by water at 10°C flowing at 1.5 m/s on the bottom of the plate in the opposite direction. Estimate the total heat transfer through the plate. (Note that the wall temperature is unknown in this case.)

6.59 There were international conferences in 1977 and again in 1980 to discuss towing icebergs to arid lands as a water supply. Suppose an iceberg is 1 km by 1 km and is towed at 1.8 km/hr. If the water temperature is 8°C, estimate the average melting rate of the flat bottom during towing in m/day. The latent heat of melting of ice is 334 kJ/kg and its density is 900 kg/m$^3$.

6.60 A stainless steel 304 plate is initially at 400°C when laid on an insulated surface and cooled by air flow as in Fig. P6.60. Estimate the time required for the bottom surface of the plate to cool to 100°C.

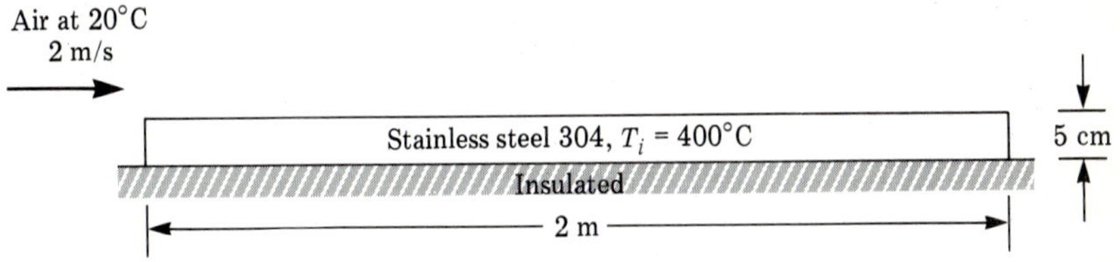


**Figure P6.60**

6.61 A brass plate, 1 cm by 40 cm by 60 cm, is initially at 90°C when suddenly immersed in a 10°C water flow at 1.5 m/s parallel to its long side. If both sides of the plate are exposed to the water, estimate the time required for the center plane of the plate to cool to 20°C.

6.62 Equation (6.95) is a reasonable approximation to combined laminar-turbulent flat plate flow, but actual data will not show the sudden change in slope indicated in Fig. 6.10. Churchill [54] suggests the following correlation formula, which varies smoothly through the transition region:

$$\overline{\mathrm{Nu}}_L \doteq 0.45 + 0.6774Z^{1/2}\left(1 + \frac{(Z/12500)^{0.6}}{[1 + (Z_t/Z)^{3.5}]^{0.4}}\right)^{1/2},$$

where $Z = \mathrm{Re}_L\mathrm{Pr}^{2/3}\,[1 + (0.0468/\mathrm{Pr})^{2/3}]^{-1/2}$.

The quantity $Z_t$ is a *transition parameter* whose value is $10^5$–$10^7$ and is fit to the data. Evaluate Churchill's expression for air (Pr = 0.71) in the Reynolds number range $10^4$–$10^7$ and plot $\overline{\mathrm{Nu}}_L$ versus $\mathrm{Re}_L$ on 3-by-3 log-log paper for values of $Z_t$ = (a) $10^5$; (b) $3 \times 10^5$; and (c) $10^6$. Comment on the utility and effectiveness of this approximation.

6.63 A thin copper plate separates a flow of 1 atm air at 500°C and 15 m/s from a parallel concurrent flow of water at 80°C. Assuming that $\mathrm{Re}_t$ = 500,000, what is the minimum water stream velocity that will keep the water from boiling on the plate surface?

6.64 Repeat Problem 6.13 if the heat to be dissipated is 15,000 Btu/hr. Assume that $\mathrm{Re}_t = 5 \times 10^5$.

6.65 Air at 20°C and 1 atm flows past a 60-cm-long flat plate at 8 m/s. The last 20 cm of plate are heated to 40°C while the first 40 cm are unheated. Estimate $h_x$ at $x$ = 60 cm.

6.66 Repeat Problem 6.65 if the fluid is water.

6.67 Air at 60°F and 1 atm flows at 30 ft/s past a flat plate 10 in. long whose last 5 in. are heated to 100°F. The first 5 in. are unheated. Plot $h_x$ versus $x$ for the entire plate and estimate $q_w$ per foot of plate width.

6.68 Air at 20°C and 1 atm flows at 6 m/s past an 80-cm-long flat plate heated electrically at a uniform 120 W/m$^2$. Estimate the wall temperature at the trailing edge of the plate.

6.69 Repeat Problem 6.68 for water. Interpret your result.

6.70 An electrically heated flat plate is 1 ft long and 2 ft wide and dissipates 600 Btu/hr from both sides combined when immersed in a 22-ft/s flow of air at 60°F and 1 atm. Estimate the plate wall temperature at $x$ = 6 in.

6.71 Air at 20°C and 1 atm flows turbulently in a 4-in.-diameter smooth-walled tube. Measured velocities at three points are

$r/r_o$ = 0.4, 0.6, 0.8

$u$ (ft/s) = 28.4, 27.0, 24.8

Estimate (a) $\tau_w$, and (b) the pressure drop per 100-ft length.

6.72 For fully developed turbulent flow in a smooth tube, assuming that the log-law (6.82b) is valid across the entire tube, integrate to show that the average velocity takes the form

$$V = u_{\max} - 3v^*/2\kappa.$$

Show further that, if $\kappa = 0.41$, this expression may be rewritten in the form

$$u_{max}/V = 1 + 2.59c_f^{1/2}.$$

6.73 Water at 20°C flows in a 6-cm-diameter smooth tube so that the centerline velocity is 12 m/s. Is the flow turbulent? If so, estimate the wall shear stress and the average velocity (a) from the expressions developed in Problem 6.72 and (b) from the Moody chart (Fig. 6.12).

6.74 Water is to be heated from 10°C to 30°C by passing through a 4-cm-diameter, 350-cm-long tube at 1.2 kg/s. Estimate the required constant wall temperature (smooth wall) and the pressure drop.

6.75 Water at 60°F enters a 1-in.-diameter smooth tube 10 ft long. If the mass flux is 2 $lb_m$/s and $T_w = 200$°F, estimate the pressure drop and exit water temperature.

6.76 For the conditions of Problem 6.75, what mass flux in turbulent flow will cause the pressure drop to be 1 $lb_f/in.^2$? What exit water temperature will result?

6.77 Air at 45°C and 1 atm flows through a 12-cm-square duct 20 m long whose wall temperature is 10°C. If the flow rate is 50 $m^3$/hr, estimate the exit air temperature and the heat loss.

6.78 A smooth tube is insulated on the outside and contains a central rod of ice at 0°C, carefully mounted so as not to break. Water at 20°C flows through the annular region at 0.25 kg/s. If ice has a latent heat of 334 kJ/kg and density of 900 kg/$m^3$, estimate the time for the ice rod to melt completely. The geometry is shown in Fig. P6.78.

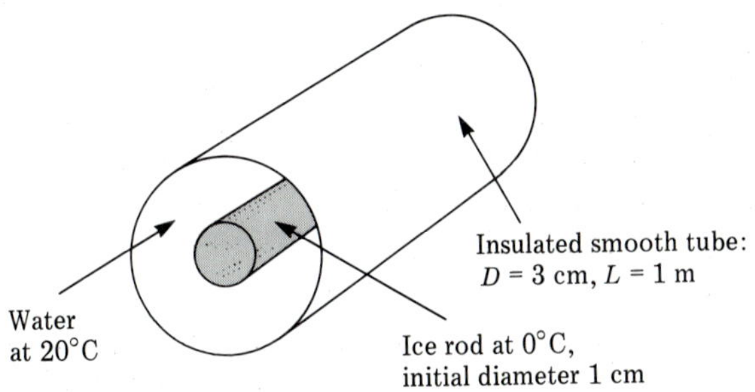


**Figure P6.78**

6.79 Modify Problem 6.38 by increasing the water flow to 0.7 $m^3$/hr, for which the flow is turbulent.

6.80 Determine whether the assumed value $\bar{h} = 370$ Btu/hr · ft$^2$ · °F for Freon-12 flow in Example 6.2 was realistic or just wildly made up by the writer.

6.81 Water is to be heated from 10°C to 30°C by passing at 5 kg/s through a smooth 5-cm-diameter tube whose wall temperature is 80°C. Estimate the required tube length.

6.82 Repeat Problem 6.81 if the fluid is mercury. What is the main reason the required length is so much shorter than for water flow?

6.83 Water at 10°C enters a 4-cm-diameter smooth tube 5 m long at 1 kg/s. The tube is electrically heated with uniform $q''_w$. If the wall temperature is to be nowhere greater than 95°C to avoid boiling, estimate the maximum allowable $q''_w$ and the corresponding exit water temperature, electric power, and $\Delta p$.

6.84 Liquid ammonia is to be cooled from 50°C to 20°C by passing through a smooth tube of diameter 3 cm and length 4 m. If $T_w$ = 0°C, what is the maximum allowable mass flux to achieve this cooling? What is the corresponding pressure drop? What is the minimum mass flux to achieve this cooling?

6.85 Size and effectiveness are important in design. Suppose we wish to cool ethyl alcohol at 40°C and 4 kg/s by passing it through tubes whose wall temperature is 0°C and whose length is 1 m. Would it be better to use one 3-cm-diameter tube or split the flow into two 2-cm-diameter tubes? Which configuration cools better? Which has the higher pressure drop?

6.86 In Problem 6.81 it is desired to enhance the convection by roughening the wall to reduce the required length to 2.5 m. Estimate the required wall roughness in millimeters.

6.87 Air at 200°C and 1 atm flows at 12 m/s across a 4-cm-diameter cylinder with $T_w$ = 20°C. Estimate the heat transfer per meter length of cylinder. What air velocity would cause this heat flux to be 2500 W/m?

6.88 In Problem 6.87, let the air velocity be 12 m/s, but let the cylinder size vary. What diameter would cause the heat transfer to be 2500 W/m?

6.89 Following up on Problem 6.88, how does diameter affect heat transfer to a cylinder in crossflow? If $\mathrm{Re}_D \doteq 10^4$ and $\mathrm{Pr} \doteq 0.7$ and only the *diameter* changes, find approximate exponents $m$ and $n$ such that $\bar{h} \propto D^m$ and $q_w \propto D^n$. Make some interpretive comments.

6.90 Water at 60°F flows at 8 ft/s across a 1-in.-diameter cylinder with $T_w$ = 100°F. Estimate, per foot of tube length, (a) the heat flux, and (b) the drag.

6.91 Arctic wind at 9 m/s and −40°C blows over an exposed 45-cm-diameter pipeline containing hot oil. If the pipe surface temperature is 35°C, estimate the heat loss per km of pipe.

6.92 It is desired to heat water flowing in a 3-cm-diameter tube with $T_w$ = 30°C by blowing air at 200°C and 1 atm across the tube. The desired heat flux is 1300 W per meter of tube length. What should be the air velocity?

6.93 A good marathon runner goes 26 miles in 2.5 hr. It is desired to estimate the total heat loss during the run, assuming skin temperature of 92°F and air at 60°F. As a model, assume the runner is a cylinder of length 70 in. and volume 4700 in.$^3$ Compute the total heat loss in calories (1 Btu = 252 cal) and compare to (a) typical daily food intake; and (b) total body enthalpy relative to a 60°F datum. Neglect sweat losses.

6.94 To help cool an electronic device whose base is at 120°C, air flows across a 1% carbon steel pin fin mounted to the device as in Fig. P6.94. Estimate the total heat loss from the fin in watts.

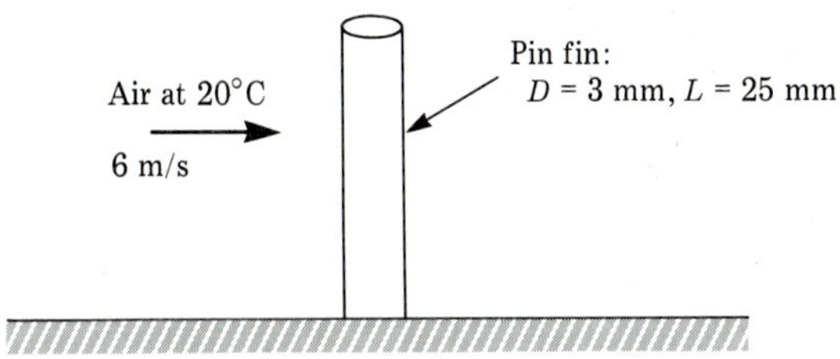


**Figure P6.94**

6.95 The Ocean Thermal Energy Conversion (OTEC) system proposes a vertical pipe carrying cold deep water up to the surface to use in a low-temperature energy cycle. Suppose, as in Fig. P6.95, the pipe is bare aluminum, with $D_i$ = 1 m and $D_0$ = 1.15 m. The internal flow is at 4°C with $Q_i$ = 1.3 m$^3$/s. There is an external crossflow current at 16°C with $U_o$ = 60 cm/s. Estimate (a) the current drag on the pipe per meter of length, (b) the heat transfer through the pipe per meter of length, and (c) the temperature rise of the internal flow per meter of length. Is insulation needed for the pipe?

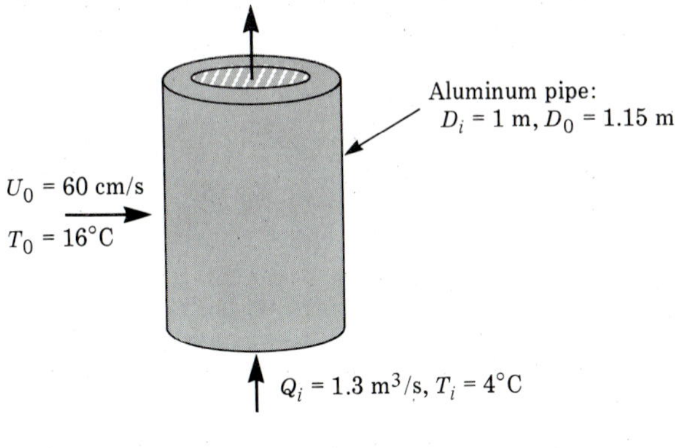


**Figure P6.95**

6.96 Mercury at 20°C and 1 atm flows at 10 m/s past a 14-cm-diameter sphere with $T_w$ = 120°C. Estimate the heat transfer and the drag force. What practical engineering difficulty arises?

6.97 Prove that, as the Reynolds number approaches zero in flow past a sphere (pure conduction into the fluid), the Nusselt number approaches the exact value of 2.0.

6.98 Buckshot may be made by dropping molten lead spheres from a tower through still air into a tank of water. The spheres solidify in air and then cool further in the water. Let the spheres be 2 mm in diameter at the temperature of molten lead, 601 K. Let the air be at 20°C and 1 atm. First estimate whether it is adequate to use the terminal velocity of the spheres as the fall velocity. Then estimate the needed tower height for solidification if the latent heat of lead is 25 kJ/kg.

6.99 A $\frac{1}{4}$-in.-diameter steel sphere at 210°F is dropped through 10 ft of 40°F air and then through 6 ft of 40°F water. Assuming free fall through air and thermal velocity in water, estimate the temperature drop of the sphere in each fluid.

6.100 Suppose Isaac Newton heats a 5-cm-diameter steel sphere to 400°C, runs outside and holds it up in a 20°C wind, and finds that it cools to 200°F in 10 min. Estimate the wind speed.

6.101 Air at 300°C and 1 atm flows at 6 m/s toward an in-line bank of 10 rows of 20 tubes each, with $D$ = 2.5 cm, $S_L$ = 6 cm, $S_T$ = 5 cm, $L$ = 1.5 m, and $T_w$ = 30°C. Estimate (a) total heat flux to the 200 tubes, (b) exit air temperature, and (c) pressure drop across the bank.

6.102 Repeat Problem 6.101 for a staggered tube bank.

6.103 It is desired to heat air at 1 atm flowing in a duct 21 by 48 in. from 60°F to 75°F by passing it over a 6-row-by-6-column tube bank with $D$ = 1 in. and $L$ = 2 ft. The bank is an in-line array with $S_T = S_L$ = 3 in. If the tubes are heated by condensing steam inside, estimate the desired steam temperature and pressure. Neglect tube conduction resistance.

6.104 A nuclear fuel reactor consists of a 2-m-diameter shell packed with a bundle of 2500 axial rods of 2-cm diameter arranged in an equilateral triangular array. Liquid sodium at a bulk temperature of 300°C flows axially along the rods at a mass rate of 1500 kg/s. If the rods' surface temperature is 600°C, estimate (a) the heat transfer coefficient, and (b) total $q_w$ per meter of bundle length.

6.105 A flat plate 40 cm long and 60 cm wide is tested in a wind tunnel where Ma = 2.5, $p$ = 35 kPa, and $T$ = 0°C. Estimate (a) $T_{aw}$, and (b) the total $q_w$ needed to cool both sides of the plate to 200°C.

6.106 A supersonic transport flies at 1450 mi/hr at an altitude of 40,000 ft, where $T$ = −69.7°F and $p$ = 2.72 lb$_f$/in.$^2$ Estimate the adiabatic wall temperature. If a certain region 2 ft from the leading edge of the wing is at $T_w$ = 150°F, estimate the heat flux $q''_w$ at this point in Btu/hr · ft$^2$.

## Design Problems

6.107 Water at 20°C enters a 2-cm-diameter, 1-m long smooth tube. It is to be heated by passing air at 1 atm, 400°C, and 12 m/s across the outside of the tube. Assume a thin tube with negligible conduction resistance. Make a careful analysis to estimate the minimum mass flow rate in kg/s for which the water will not begin to boil at the inside tube surface. What will be the exit water temperature for this condition? What power is needed to pump the water for this case?

6.108 Let us upgrade Problem 6.91 into a design problem. Let the pipeline be 100 km long, and approximate the oil as unused engine oil. Assume that the wind and air temperature are steady. The oil can be heated up to 80°C at the entrance to reduce its viscosity. Pumping power is available up to 500 kW at an annual cost of \$3500 per kW. The pipe may be insulated ($k_i = 0.05\ \text{W/m} \cdot \text{K}$) at an annual maintenance and replacement cost of \$6000 per cm thickness. The oil flow rate is 160 kg/s (100,000 barrels per day), and the annual cost of heating it is \$50000 for each degree above 20°C. Try to optimize this system to achieve a near-minimum annual operating cost.

6.109 Let us generalize Problem 6.59 into a minimum-cost design problem. Let the icebergs be 100 m thick initially and square horizontally, with side lengths up to 2 km. They may be towed at any speed up to 5 m/s, for a distance of 7000 km through water of average temperature 8°C. Let the total drag be the sum of friction drag on the bottom and sides plus a pressure drag coefficient of 1.1 based on the submerged frontal area of the iceberg. Towing costs are \$18,000 per day fixed expenses plus 60¢ per kilowatt-hour of towing power needed. Assume that whatever is left of the iceberg can be completely converted to fresh water at the destination. Try to find an optimum initial iceberg size and optimum towing speed to minimize the cost per cubic meter of freshwater delivered.

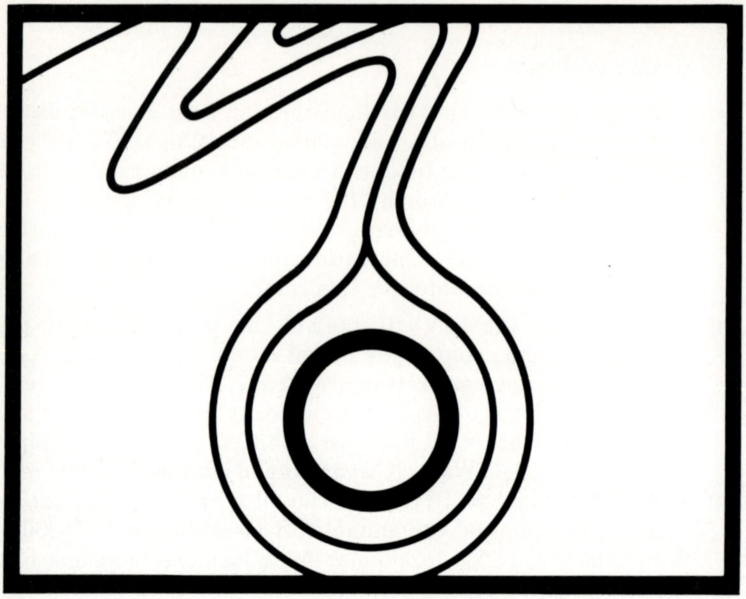

# Free Convection

# 7

## 7.1 Introduction

Free convection is a convective motion arising from the heat flux process itself rather than from an external freestream or forced motion. The addition or subtraction of heat from the otherwise still fluid sets up density differences, which, in a body force field, result in local streaming. The density differences may be caused by temperature gradients, as in a hot surface next to a cold fluid, or by fluid mixture concentration gradients, as in the melting of freshwater icebergs in the saline ocean. The body force field may be gravity, a centripetal acceleration field as in rotating machinery, or a Coriolis acceleration as in geophysical problems. Pressure gradients are almost always a negligible effect.

We will confine ourselves here to analysis only of gravity or "buoyancy" forces and density differences caused only by temperature gradients. These arise naturally in almost any earthbound heat transfer problem. Thus free convection is often called *natural* or *buoyant* convection. In any given problem we should assess the relative importance of free versus forced convection. If both are important, the flow is termed *mixed* convection. (See Fig. 1.6.)

Some illustrative examples of pure free convection are sketched in Fig. 7.1. Parts (a) and (b) show *free boundary flows,* which form buoyant plumes that rise in the heavier ambient fluid without interacting with solid surfaces. Free boundary flows are common in atmospheric and oceanic problems but will not be treated here; details may be found in [1, Chapter 4]. Figure 7.1(c,d) show a free convection induced by a heated (cooled) vertical plate surface. The boundary layer grows up from the bottom of the heated plate (Fig. 7.1c) and down from the top of the cooled plate (Fig. 7.1d); the flow patterns are antisymmetric images of each other. If the plate is long enough, transition from laminar flow to turbulence will occur in a manner similar to forced convection past a plate. Note that, unlike convection, the thermal and velocity boundary layers in Fig. 7.1(c,d) are approximately the same thickness, with the velocity layer being somewhat thicker for $\mathrm{Pr} >> 1$.

Free convection-induced velocities are relatively small, rarely more than 1 m/s. Thus free-convection heat flux is generally smaller than forced convection and forms a lower bound for the convection capability of a given geometry. This does not make free convection unimportant at all. For reasons of economy or scale, many practical engineering problems are concerned solely with free convection: for example, cooling of electronic devices or domestic baseboard and steam heating. Even in forced-convection designs, free convection provides an estimate of

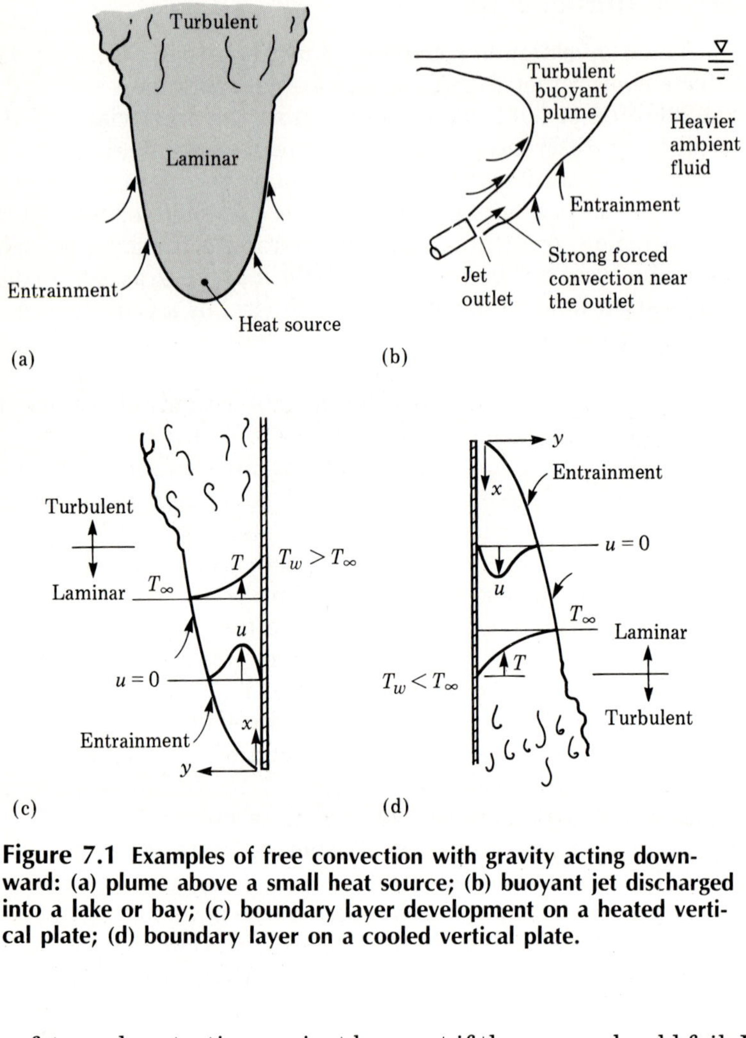


**Figure 7.1** **Examples of free convection with gravity acting downward: (a) plume above a small heat source; (b) buoyant jet discharged into a lake or bay; (c) boundary layer development on a heated vertical plate; (d) boundary layer on a cooled vertical plate.**

safety and protection against burnout if the power should fail. Finally, in geophysical problems, free convection is nearly the whole story: atmospheric and oceanic motions are dominated by massive natural convection effects arising from spatial and temporal variations of solar heating.

An interesting aspect of free convection is the *entrainment* of fluid from the nearly still ambient region into the boundary layer. This is shown in all four cases in Fig. 7.1, and generally the boundary layer mass flux increases with distance along the flow in all free-convection problems.

Obviously this brief chapter will not give the full story about free convection. There is at least one recent textbook devoted entirely to free convection [1], and the undergraduate heat transfer text by Gebhart [2] has an unusually rich discussion of the subject. Also of interest are the major review articles by Ede [3] and by Gebhart [4].

### 7.1.1 Characteristic Free-Convection Velocity

Since there is no "freestream" in pure free convection, there is no external velocity scale $U_\infty$, and we must look elsewhere to characterize the boundary layer streaming speed. An approximate energy balance reveals the result. With reference to Fig. 7.1(c), as the fluid moves up along the plate it loses potential energy and gains kinetic energy. The two do not quite balance because of wall friction losses. The average density difference in the boundary layer is approximately $\frac{1}{2}(\rho_\infty - \rho_w)$. Denote the average velocity at position $x$ by $u_{avg}$ and note that $u \doteq 0$ at the leading edge section where ambient (still) conditions exist. Then, by neglecting friction and equating potential energy loss to kinetic energy gain at position $x$, we estimate

$$\frac{1}{2}(\rho_\infty - \rho_w)\, g\, x \doteq \frac{1}{2}\rho_\infty u_{avg}^2,$$

or

$$u_{avg} \doteq [(1 - \rho_w/\rho_\infty)\, g\, x]^{1/2}. \tag{7.1}$$

Thus the boundary layer velocities increase up the plate at a rate proportional to $x^{1/2}$, and magnitudes across the layer are everywhere somewhat less than Eq. (7.1) because of wall friction losses, especially for high Prandtl number fluids. The form of the estimate (7.1) is quite correct, however.

### 7.1.2 The Grashof Number

Continuing the argument that led to Eq. (7.1), as we move up the plate in Fig. 7.1(c), the wall friction and heat flux should be functions of a "local Reynolds number" given by $u_{avg}\, x/\nu$. Introducing $u_{avg}$ from Eq. (7.1) and squaring the result, we discover the appropriate dimensionless local free-convection parameter:

$$\mathrm{Gr}_x = (u_{avg} x/\nu)^2 = g\left(1 - \frac{\rho_w}{\rho_\infty}\right)\left(\frac{x^3}{\nu^2}\right). \tag{7.2}$$

We recognize this as the *Grashof number*, which was found by the dimensional analysis technique in Eq. (5.31). In a typical free-convection problem (Fig. 7.1c), the local Nusselt number depends only on Grashof and Prandtl numbers and the shape of the body:

$$Nu_x = h_x x/k = f(Gr_x, Pr, \text{shape}). \tag{7.3}$$

We will establish the functional relationship for several body shapes in the remainder of this chapter.

In free convection, then, Grashof number takes the place of Reynolds-number-squared in forced convection. Indeed, if a streaming motion is superimposed on a buoyant heat transfer situation, we may assess its relative importance by comparing Grashof number to $(U_\infty x/\nu)^2$:

*Dominant free convection:*

$Gr_x >> Re_x^2$,

*Mixed convection:*

$Gr_x \simeq Re_x^2$,

*Dominant forced convection:*

$Gr_x << Re_x^2$.

We shall return to this discussion in Section 7.5.

### 7.1.3 The Coefficient of Thermal Expansion

The definition of $Gr_x$ in Eq. (7.2) is suitable for geophysical problems but inconvenient for engineering analysis, where temperature rather than density differences are more directly useful. Since pressure gradients are negligible in free convection, the boundary layer of a pure (uniform composition) substance will have a density profile $\rho \simeq \rho(T)$ only. Figure 7.2 shows the variation of density with temperature at constant pressure for several common fluids. The curves are nearly linear, at least over a moderate temperature range. The negative slope of the density curve is defined as the *coefficient of thermal expansion*, $\beta$:

$$\beta = -\frac{1}{\rho}\left(\frac{\partial \rho}{\partial T}\right)_p. \tag{7.4}$$

The most nonlinear curve in Fig. 7.2 is for water, which exhibits a density maximum at $T = 4°C$ and an anomalous region, $0 \leqslant T < 4°C$, where $\beta$ is actually negative. (The anomaly persists in saline water up to a concentration of 25 ppt and then vanishes.)

All common, nearly ideal gases follow the same curve in Fig. 7.2 and have the same expansion coefficient:

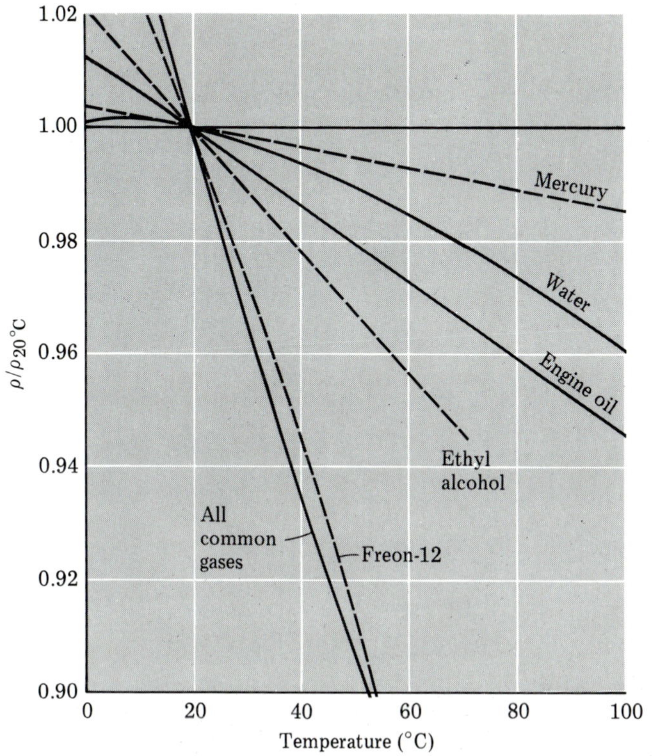


**Figure 7.2** Density versus temperature at constant pressure for several common fluids.

*Ideal gas:*

$$\beta \equiv 1/T. \tag{7.5}$$

Low-boiling-point refrigerants such as Freon-12 and liquid ammonia have a strong density variation nearly comparable to ideal gases.

Assuming a constant or average value of $\beta$, we can replace the density parameter in Eq. (7.2) by

$$(1 - \rho_w/\rho_\infty) \simeq \beta(T_w - T_\infty). \tag{7.6}$$

Then the Grashof number for engineering heat transfer analysis is more appropriately defined as

$$\mathrm{Gr}_x = g\beta(T_w - T_\infty)x^3/\nu^2. \tag{7.7}$$

We will use this definition for the remainder of the chapter.

The grouping $(g\beta/\nu^2)$ forms a free-convection fluid property team

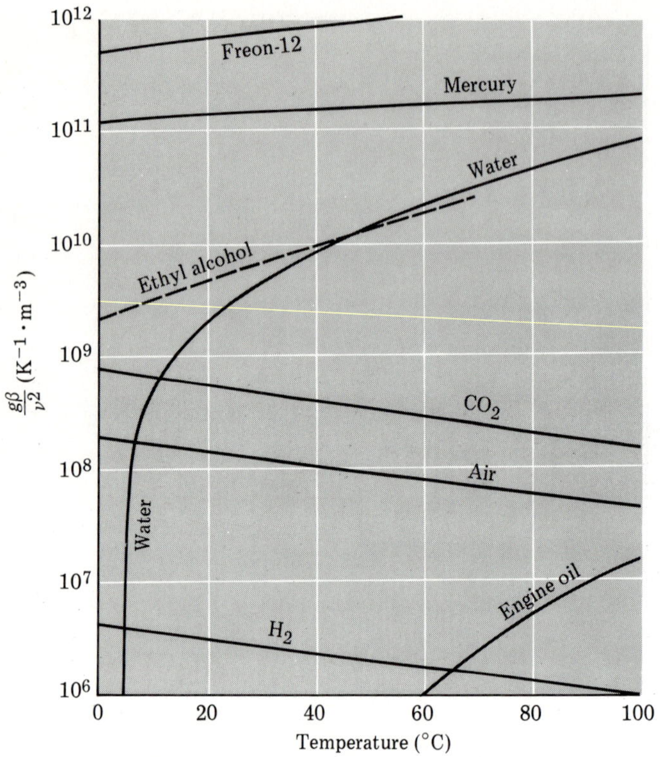


**Figure 7.3 Values of the Grashof number fluid parameter ($g\beta/\nu^2$) for various fluids, with $g$ = 9.81 m/s$^2$.**

that can be tabulated or plotted versus temperature for any fluid, as shown in Fig. 7.3. Its magnitude (in units of $K^{-1}$ $m^{-3}$) is a measure of the potential of the fluid for generating significant free-convection effects. We see in Fig. 7.3 that Freon-12 and mercury have very high potential, water is highly nonlinear, and engine oil has very low potential. Gases do not form a single curve here because of their differing kinematic viscosities. Hydrogen is low because of its high viscosity and low density.

## Example 7.1

For free convection on a vertical plate (Fig. 7.1c), transition to turbulence typically takes place at about $Gr_x = 10^9$. If $T_w = 40°C$

and $T_\infty = 20°C$, estimate at what position $x$ this occurs for (a) air, (b) water, and (c) Freon-12.

*Solution* To simulate average thermal expansion we should evaluate properties at the film temperature $T_f = \frac{1}{2}(T_w + T_\infty) = 30°C$.

a) For air at 30°C, $g\beta/\nu^2 \doteq 1.2 \times 10^8\ K^{-1}\ m^{-3}$. Then, from Eq. (7.7), set

$$Gr_x = (1.2 \times 10^8)(40 - 20)x^3 = 10^9$$

and solve for

$x = 0.75$ m. [*Ans. (a)*]

b) For water at 30°C, $g\beta/\nu^2 \doteq 4.3 \times 10^9$, whence

$$Gr_x = (4.3 \times 10^9)(40 - 20)\, x^3 = 10^9,$$

or

$x = 0.23$ m. [*Ans. (b)*]

c) For Freon-12 at 30°C, $g\beta/\nu^2 \doteq 7.4 \times 10^{11}$, so that

$$Gr_x = (7.4 \times 10^{11})(40 - 20)x^3 = 10^9,$$

or:

$x = 0.041$ m. [*Ans. (c)*]

Thus for common "low-viscosity" fluids, only moderate distances are required for a free-convection boundary layer to become turbulent. ■

---

### 7.1.4 The Free-Convection Momentum Equation

The boundary layer equations for continuity, momentum, and energy (6.1) hold also for free-convection flow, with two modifications: (1) dissipation is negligible (Example 7.2), and (2) gravity is a very important force term. Let us look at the desired form of the gravity (buoyancy) term.

The boundary layer momentum equation (6.1b) is, for two-dimensional steady flow,

$$\rho\left(u\frac{\partial u}{\partial x} + v\frac{\partial u}{\partial y}\right) = -\frac{dp}{dx} + \rho g_x + \frac{\partial \tau}{\partial y}, \tag{7.8}$$

where $g_x$ is the component of gravity in the direction of flow and

$\tau = \mu(\partial u/\partial y)$ for laminar flow and $\tau = (\mu + \rho\varepsilon_M)(\partial u/\partial y)$ for turbulent flow.

In free convection, outside the boundary layer ($y \to \infty$), $u$, $v$, and $\tau$ approach zero, so Eq. (7.8) becomes

$$0 = -\frac{dp}{dx} + \rho_\infty g_x,$$

or

$$-\frac{dp}{dx} + \rho g_x = (\rho - \rho_\infty)g_x. \tag{7.9}$$

Thus the combination of pressure gradient and particle weight component in Eq. (7.8) is equivalent to a differential weight. If we now assume a nearly constant coefficient of thermal expansion, $\beta$, Eq. (7.9) becomes

$$(\rho - \rho_\infty)g_x \doteq \rho\beta(T_\infty - T)\, g_x,$$

which we substitute into Eq. (7.8) to obtain the desired form of the free-convection momentum equation:

$$\rho\left(u\frac{\partial u}{\partial x} + v\frac{\partial u}{\partial y}\right) = \rho\beta(T_\infty - T)\, g_x + \frac{\partial \tau}{\partial y}. \tag{7.10}$$

Thus the momentum equation contains temperature $T$ explicitly, so that velocity and temperature are *coupled* and we cannot solve for them separately as we did for forced convection in Chapter 6. Since the variation in density across the boundary layer is only a few percent (Fig. 7.2), it is acceptable to let $\rho$ be constant in Eq. (7.10) and to use the incompressible form of the continuity equation (6.1a).† Although coupling somewhat complicates things mathematically, several exact solutions are known for free-convection boundary layers [1].

## Example 7.2

Show that dissipation is negligible in free-convection boundary layers, even on an atmospheric scale.

*Solution* From Examples 5.3 and 5.4 we saw that dissipation is negligible if the Eckert number $\text{Ec} = U^2/c_p\Delta T << 1$. For free

†The idea of letting density be constant everywhere except in the differential buoyancy term is called the *Boussinesq approximation* [1, p. 19].

convection, there is no freestream velocity, $U$, so we use the boundary layer maximum-velocity estimate $U = [g\beta(\Delta T)L]^{1/2}$ from Eq. (7.1) to establish the inequality

$$\mathrm{Ec} = \frac{U^2}{c_p\Delta T} = \frac{g\beta(\Delta T)L}{c_p\Delta \mathrm{T}} << 1,$$

or:

$$L << \frac{c_p}{g\beta}.$$

Dissipation is thus negligible in free convection if the vertical length scale $L$ is much smaller than a "convection length scale" $c_p/g\beta$. For air, water, and mercury at 20°C, the scale $c_p/g\beta$ equals 30, 2030, and 78 km, respectively. Typical free-convection heights, even in the atmosphere, are small compared to these values, so we neglect dissipation. ■

---

## 7.2 Free Convection on a Vertical Plate

A simple but extremely important geometry to illustrate free-convection effects is the vertical flat plate, either heated (Fig. 7.1c) or cooled (Fig. 7.1d). In either case the buoyancy term $\rho\beta(T_\infty - T)g_x$ is positive if $x$ is in the direction of flow, and the mathematical solution is the same.

Taking the heated plate (Fig. 7.1c) for convenience, the two-dimensional steady-flow boundary layer equations are

*Continuity:*

$$\frac{\partial u}{\partial x} + \frac{\partial v}{\partial y} = 0, \qquad \textbf{(7.11a)}$$

*Momentum:*

$$u\frac{\partial u}{\partial x} + v\frac{\partial u}{\partial y} = g\beta(T - T_\infty) + \frac{1}{\rho}\frac{\partial \tau}{\partial y}, \qquad \textbf{(7.11b)}$$

*Energy:*

$$u\frac{\partial T}{\partial x} + v\frac{\partial T}{\partial y} = -\frac{1}{\rho c_p}\frac{\partial q''}{\partial y}, \qquad \textbf{(7.11c)}$$

where $\tau = (\mu + \rho\varepsilon_M)(\partial u/\partial y)$ and $q'' = -(k + \rho c_p\varepsilon_H)(\partial T/\partial y)$.

The boundary conditions are no-slip at the wall and a still ambient fluid,

$$u(0) = v(0) = u(\infty) = 0, \qquad T(\infty) = T_\infty, \tag{7.12}$$

plus either known temperature or known heat flux at the wall,

$$T(0) = T_w \quad \text{or} \quad \frac{\partial T}{\partial y}(0) = -\frac{1}{k}q''_w. \tag{7.13}$$

The problem is well posed and has been solved for both laminar- and turbulent-flow conditions in the literature [6, 17].

### 7.2.1 Integral Solution for Laminar Flow

The system of equations (7.11–7.13) has been solved exactly for laminar flow with constant ($\nu$, $\alpha$, $\beta$), using a similarity transformation suggested by E. Pohlhausen and reported in a paper by Schmidt and Beckmann [5]. Their theory was later extended with a wide variety of digital computer numerical solutions by Ostrach [6]. The wall temperature is assumed constant, and details of the solution are given in advanced texts such as [1, pp. 31–53] or [7, pp. 377–380].

To get some insight into vertical plate convection, let us consider an approximate but very effective *integral* solution, similar to the forced-convection method outlined in Section 6.2.3. The boundary layer momentum integral relation is similar to Eq. (6.5) with no freestream ($U_\infty = 0$) and a buoyancy term added:

$$\tau_w = \int_0^\delta \rho g\beta(T - T_\infty)\,dy - \frac{d}{dx}\left[\int_0^\delta \rho u^2\,dy\right]. \tag{7.14}$$

The boundary layer energy integral relation is identical to Eq. (6.6), since buoyancy makes no direct change to energy:

$$q''_w = -k\left.\frac{\partial T}{\partial y}\right|_{y=0} = \frac{d}{dx}\left[\int_0^\delta \rho c_p u(T - T_\infty)dy\right]. \tag{7.15}$$

These are exact relations but are to be solved *approximately* by assuming reasonable shapes for $u(x, y)$ and $T(x, y)$. For simplicity, we assume that Pr is not too far from unity, so that the boundary layer thicknesses are nearly equal, $\delta \simeq \Delta$.

We approximate the velocity profiles sketched in Fig. 7.1(c,d) by a fourth-order polynomial:

$$u \doteq a\eta + b\eta^2 + c\eta^3 + d\eta^4, \qquad \eta = y/\delta. \tag{7.16}$$

The constants are found by matching the following four conditions:

$$u(1) = \frac{\partial u}{\partial \eta}(1) = \frac{\partial^2 u}{\partial \eta^2}(1) = 0, \qquad \textbf{(7.17)}$$
$$\frac{\partial^2 u}{\partial \eta^2}(0) = -\delta^2 g\beta(T_w - T_\infty)/\nu.$$

The last of these conditions follows by evaluating the differential momentum equation (7.11b) at the wall. Substitution of conditions (7.17) into the profile (7.16) yields the constants $(a, b, c, d)$, with the result

$$u/u_0 \doteq \eta - 3\eta^2 + 3\eta^3 - \eta^4, \qquad u_0 = u_0(x). \qquad \textbf{(7.18)}$$

For the temperature profile, we adopt a simple approximation suggested by Eckert [8]:

$$\frac{T - T_\infty}{T_w - T_\infty} \doteq (1 - \eta)^2. \qquad \textbf{(7.19)}$$

This satisfies $T(0) = T_w$, $T(\eta = 1) = T_\infty$, and $(\partial T/\partial \eta)(1) = 0$.

The profile shapes (7.18) and (7.19) are shown in Fig. 7.4 and are compared with the exact shapes computed for Pr = 1 by Ostrach [6]. The temperature profile is in very good agreement but the approximate velocity profile decays too rapidly in the outer stream. Of course, we normally do not know the exact shape or else we would not be using an approximate integral method. Figure 7.4 is shown only to illustrate the degree of approximation involved.

Assuming that $T_w$ and $T_\infty$ are constant, the only unknowns in Eqs. (7.18) and (7.19) are the thickness $\delta$ and velocity $u_0$. Following Eckert, we make the assumption of a power-law [8]:

$$\delta(x) = C_1 x^m, \qquad \textbf{(7.20)}$$
$$u_0(x) = C_2 x^n.$$

By differentiating our approximations (7.18) and (7.19) we find that the wall friction and heat flux are given by

$$\tau_w \doteq \mu u_0/\delta \qquad q''_w \doteq 2k(T_w - T_\infty)/\delta. \qquad \textbf{(7.21)}$$

Thus the problem boils down to finding $\delta(x)$ and $u_0(x)$ from the integral relations. Substituting the approximations (7.18–7.20) into Eqs. (7.14) and (7.15) and carrying out the integrations, we obtain

*Momentum:*

$$\frac{m + 2n}{252} C_1 C_2^2 x^{m+2n-1} = \frac{-\nu C_2 x^{n-m}}{C_1} + \frac{1}{3} g\beta(T_w - T_\infty)C_1 x^m,$$

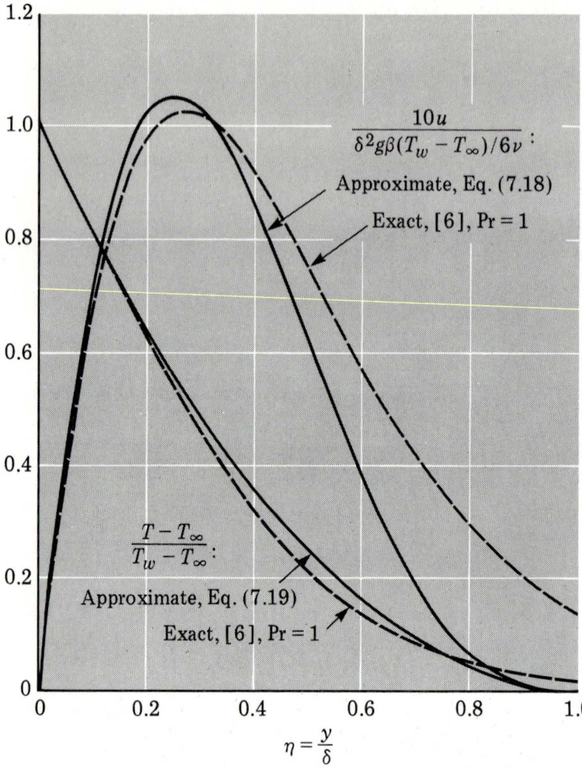


**Figure 7.4** **Approximate and exact velocity and temperature profiles for free convection over a vertical plate. For purposes of this comparison, $\delta$ is computed from Eq. (7.23).**

*Energy:*

$$\frac{m+n}{42} C_1 C_2 x^{m+n-1} = \frac{2\alpha}{C_1} x^{-m}.$$

Equating exponents, we find that both equations are satisfied if $m = \frac{1}{4}$ and $n = \frac{1}{2}$. We may then solve simultaneously for $C_1$ and $C_2$. The final results are

$$C_1 = \left[\frac{(336)\,(\mathrm{Pr} + 5/9)}{\mathrm{Pr}^2}\right]^{1/4} \left[\frac{g\beta(T_w - T_\infty)}{\nu^2}\right]^{-1/4},$$

$$C_2 = 112[336(\mathrm{Pr} + 5/9)]^{-1/2}\,[g\beta(T_w - T_\infty)]^{1/2}. \tag{7.22}$$

The boundary layer thickness is thus $\delta = C_1 x^{1/4}$, or

$$\frac{\delta}{x}\,\mathrm{Gr}_x^{1/4} \doteq \left[\frac{336\,(\mathrm{Pr} + 5/9)}{\mathrm{Pr}^2}\right]^{1/4}. \tag{7.23}$$

Note that the local Grashof number $\mathrm{Gr}_x = g\beta(T_w - T_\infty)x^3/\nu^2$ arises quite naturally in this solution.

Similarly, the characteristic velocity $u_0 = C_2 x^{1/2}$, or

$$\frac{u_0}{[g\beta(T_w - T_\infty)x]^{1/2}} \doteq \frac{112}{[336(\mathrm{Pr} + 5/9)]^{1/2}}. \tag{7.24}$$

Note that the velocity scale $[g\beta(T_w - T_\infty)x]^{1/2}$ also arises naturally in the solution, as suggested by Eq. (7.1).

From Eq. (7.21), $h_x = q''_w/(T_w - T_\infty) = 2k/\delta$; hence with $\delta$ from Eq. (7.23) we obtain the local Nusselt number:

$$\mathrm{Nu}_x = \frac{h_x x}{k} \doteq \frac{2\,\mathrm{Pr}^{1/2}}{[336(\mathrm{Pr} + 5/9)]^{1/4}}\,\mathrm{Gr}_x^{1/4}. \tag{7.25}$$

For a given Prandtl number, then, $\mathrm{Nu}_x$ varies as $\mathrm{Gr}_x^{1/4}$ or, equivalently, $h_x \propto x^{-1/4}$. The mean Nusselt number is easily found by integration:

$$\bar{h} = \frac{1}{L}\int_0^L h_x\,dx = \frac{4}{3}(h_x)_{x=L},$$

or

$$\overline{\mathrm{Nu}}_L = \frac{\bar{h}L}{k} \doteq \frac{(8/3)\,\mathrm{Pr}^{1/2}}{[336(\mathrm{Pr} + 5/9)]^{1/4}}\,\mathrm{Gr}_L^{1/4}. \tag{7.26}$$

Note the interesting fact that as $\mathrm{Pr} \to 0$, $\mathrm{Nu} \propto \mathrm{Pr}^{1/2}$, while as $\mathrm{Pr} \to \infty$, $\mathrm{Nu} \propto \mathrm{Pr}^{1/4}$, as verified by exact theory. All of the above results are valid for laminar flow in free convection past a vertical plate with constant wall temperature.

### 7.2.2 Exact Laminar Flow Computations

The laminar flow integral theory of the previous section leads to an expression $\mathrm{Nu}_x = f(\mathrm{Pr})\mathrm{Gr}_x^{1/4}$, which has the same form as the exact digital computer results of Ostrach [6]. In fact, the integral theory, Eq. (7.25), predicts $\mathrm{Nu}_x$ to within $\pm 6\%$ for all Prandtl numbers in the laminar flow range $10^4 \leq \mathrm{Gr}_x \leq 10^9$. Equation (7.25) is thus recommended as an engineering formula.

Ostrach [6] tabulated exact results for certain specific Prandtl numbers. There is no exact formula for his tabulations, but LeFevre [9], using asymptotic results at Pr = 0 and Pr = ∞ as a guide, suggested the following interpolation formula:

$$\frac{\mathrm{Nu}_x}{\mathrm{Gr}_x^{1/4}} \doteq \frac{\frac{3}{4}\mathrm{Pr}^{1/2}}{(2.435 + 4.884\,\mathrm{Pr}^{1/2} + 4.953\,\mathrm{Pr})^{1/4}}, \tag{7.27}$$

which fits Ostrach's results to within ±0.5% over the entire Prandtl number range. The same formula may be used for $\overline{\mathrm{Nu}}_L/\mathrm{Gr}_L^{1/4}$ by dropping the coefficient ($\frac{3}{4}$). Ostrach's paper also shows the exact temperature and velocity profile shapes for laminar flow and indicates that $\delta > \Delta$ if Pr > 1. The velocity profiles are shown in Fig. P7.16.

### Example 7.3

A flat plate 10 cm wide and 18 cm high at 250°C is placed vertically in air at 1 atm and 20°C. Estimate the total heat loss from the two sides of the plate.

*Solution* For air at $T_f = (250 + 20)/2 = 135°\mathrm{C}$, $\rho = 0.85$ kg/m³, $k = 0.0328$ W/m · K, Pr = 0.71, and $g\beta/\nu^2 = 3.3 \times 10^7\ \mathrm{K^{-1}\,m^{-3}}$. The overall Grashof number is thus

$$\mathrm{Gr}_L = (g\beta/\nu^2)(T_w - T_\infty)L^3 = (3.3 \times 10^7)(250 - 20)(0.18)^3 = 4.4 \times 10^7.$$

This is considerably less than $10^9$ so we assume laminar flow. From Eq. (7.27) at Pr = 0.71 we compute

$$\frac{\overline{\mathrm{Nu}}_L}{\mathrm{Gr}_L^{1/4}} = \frac{(0.71)^{1/2}}{[2.435 + 4.884(0.71)^{1/2} + 4.953(0.71)]^{1/4}} = 0.473,$$

or

$$\overline{\mathrm{Nu}}_L = (0.473)(4.4 \times 10^7)^{1/4} = 38.5.$$

Then

$$\bar{h} = \frac{\overline{\mathrm{Nu}}_L k}{L} = \frac{(38.5)(0.0328)}{0.18} = 7.02\ \mathrm{W/m^2 \cdot K}.$$

Finally, the total heat flux from both sides of the plate is

$$q_w = \bar{h}\,2bL\,(T_w - T_\infty) = (7.02)(2)(0.1)(0.18)(250 - 20) = 58\ \mathrm{W}. \quad [Ans.]$$

The integral theory (7.26) predicts $q_w = 61$ W, or 5% higher. Air does not remove heat by free convection very effectively except for large-size bodies. ∎

---

### 7.2.3 The Rayleigh Number

For laminar vertical plate convection as $\mathrm{Pr} \to \infty$, the Prandtl and Grashof number effects are the same on Nu: a fourth root. In fact, from Ostrach's computer results [6] for $\mathrm{Pr} > 7$,

$$\mathrm{Nu}_x \doteq 0.503\, \mathrm{Gr}_x^{1/4}\, \mathrm{Pr}^{1/4} \tag{7.28}$$

with an error of less than 10%. This suggests that the product of Grashof and Prandtl numbers is a useful parameter, at least over a moderate range of Prandtl numbers. The parameter is called the *Rayleigh number*:

$$\mathrm{Ra}_x = \mathrm{Gr}_x \mathrm{Pr} = \frac{\rho^2 c_p g \beta (T_w - T_\infty) x^3}{k\mu}. \tag{7.29}$$

The Rayleigh number arises in a fundamental way in the theory of geophysical flows, because it is the parameter that governs the overturning of an unstable (density increasing upward) ocean or atmosphere. Here it is more or less a computational coincidence, but a popular one because of its simplicity compared to more exact correlations such as Eq. (7.27). For example, near $\mathrm{Pr} = 1$ (gases and light liquids), we can use the approximation

$$\overline{\mathrm{Nu}}_L \doteq 0.535\, \mathrm{Ra}_L^{1/4}. \tag{7.30}$$

The accuracy for $0.5 \leqslant \mathrm{Pr} \leqslant 2$ is about $\pm 6\%$. It is a very popular formula, especially in the older literature predating the wide use of handheld calculators. Even today it is common to propose a Rayleigh number type correlation, $\mathrm{Nu} = f(\mathrm{Ra}, \mathrm{Pr})$, rather than the traditional $\mathrm{Nu} = f(\mathrm{Gr}, \mathrm{Pr})$. For some examples, see Eq. (7.37) or (7.40).

### 7.2.4 Constant Heat Flux Correlation

Previous results in this section, such as Eq. (7.27), are for laminar free convection with constant wall temperature. An alternate boundary condition is *constant wall heat flux*, $q''_w$, generated, for example, by electric heating of the surface. This problem was solved exactly by Sparrow and Gregg for various Prandtl numbers. Since temperature difference $(T_w - T_\infty)$ is unknown a priori and varies with $x$, an alternate Grashof number is needed that does not contain $(T_w - T_\infty)$. Sparrow and Gregg suggest the *modified Grashof number* [10]:

$$\mathrm{Gr}_x^* = \mathrm{Gr}_x \mathrm{Nu}_x = \frac{g\beta q_w'' x^4}{(k\nu^2)}, \tag{7.31}$$

which we mentioned earlier as Eq. (5.38). The final correlation for $h_x$ is then of the form $\mathrm{Nu}_x = f(\mathrm{Gr}_x^*, \mathrm{Pr})$, and [1] and [10] give digital computer results for various Pr. In the spirit of LeFevre's interpolation (7.27) these may be curve-fit into a similar formula accurate to $\pm 0.5\%$:

$$\frac{\mathrm{Nu}_x}{\mathrm{Gr}_x^{*\,1/5}} \doteq \frac{\mathrm{Pr}^{2/5}}{(3.91 + 9.32\mathrm{Pr}^{1/2} + 9.95\mathrm{Pr})^{1/5}}. \tag{7.32}$$

With $\mathrm{Nu}_x$ known, the local temperature difference is found from the defining relation $(T_w - T_\infty) = q_w'' x/(k\mathrm{Nu}_x)$. Note the power of $\frac{1}{5}$ on $\mathrm{Gr}^*$, which is analogous to the power of $\frac{1}{4}$ on Gr in Eq. (7.27). For a given Gr (or $\mathrm{Gr}^*$) and Pr, values of $\mathrm{Nu}_x$ computed from Eq. (7.32) are about 15% higher than the comparable constant-wall-temperature values from Eq. (7.27).

## Example 7.4

Modify the conditions of Example 7.3 to assume that total heat loss is 100 W and the plate has a uniform surface heat flux. Estimate the variation $T_w(x)$ and maximum $T_w$.

*Solution* This is nearly twice the heat loss of Example 7.3 so the wall temperatures could be upwards of 400°C or more. Guess $T_f \doteq 200$°C, so that $k = 0.0370$ W/m · K, $g\beta/\nu^2 = 1.64 \times 10^7\ \mathrm{K}^{-1}\ \mathrm{m}^{-3}$, and $\mathrm{Pr} = 0.71$. The local modified Grashof number would be

$$\mathrm{Gr}_x^* = \frac{q_w''(g\beta/\nu^2)x^4}{k} = \frac{(2780)(1.64 \times 10^7)x^4}{0.037}$$
$$= (1.23 \times 10^{12})x^4 \qquad (x \text{ in meters}),$$

where

$$q_w'' = \frac{q_w}{2bL} = \frac{100}{2(0.1)(0.18)} = 2780\ \mathrm{W/m^2}.$$

Then, from Eq. (7.32),

$$\frac{\text{Nu}_x}{\text{Gr}_x^{*1/5}} \doteq \frac{(0.71)^{2/5}}{[3.91 + 9.32(0.71)^{1/2} + 9.95(0.71)]^{1/5}}$$
$$= 0.485,$$

or

$$\text{Nu}_x = 0.485(1.23 \times 10^{12}x^4)^{1/5} = 126.9x^{0.8},$$

again of course with $x$ in meters. Then the local temperature difference is

$$T_w - T_\infty = T_w - 20 = \frac{q''_w x}{(k\text{Nu}_x)} = \frac{2780x}{0.037(126.9x^{0.8})},$$

or

$$T_w(x) = 20°\text{C} + 591.4x^{0.2}(°\text{C}). \quad [Ans.] \tag{1}$$

The maximum occurs at the trailing edge, $x = 0.18$ m:

$$T_w(\text{max}) = 20 + 591.4(0.18)^{0.2} = 440°\text{C}. \quad [Ans.] \tag{2}$$

An array of values may be computed as follows:

| $x$(cm): | 0 | 3 | 6 | 9 | 12 | 15 | 18 |
|---|---|---|---|---|---|---|---|
| $T_w$(°C): | 20 | 313 | 357 | 385 | 407 | 425 | 440 |

The average wall temperature is found by integrating Eq. (1):

$$T_w(\text{avg}) = \frac{1}{0.18}\int_0^{0.18} T_w(x)dx = 370°\text{C}.$$

Then a new estimate for the film temperature is $T_f = (370 + 20)/2 = 195°\text{C}$. This is not enough of a change from our initial guess of 200°C to warrant a second iteration. ■

### 7.2.5 Transition to Turbulence

If a vertical plate is long enough, the laminar free-convection boundary layer will undergo transition to turbulence. The process begins at a relatively low Grashof number of about $10^7$, where a laminar (viscous) instability occurs that amplifies any disturbances to the boundary layer. The amplified waves grow along the plate until they are large enough to trigger "bursts" of turbulence, which then

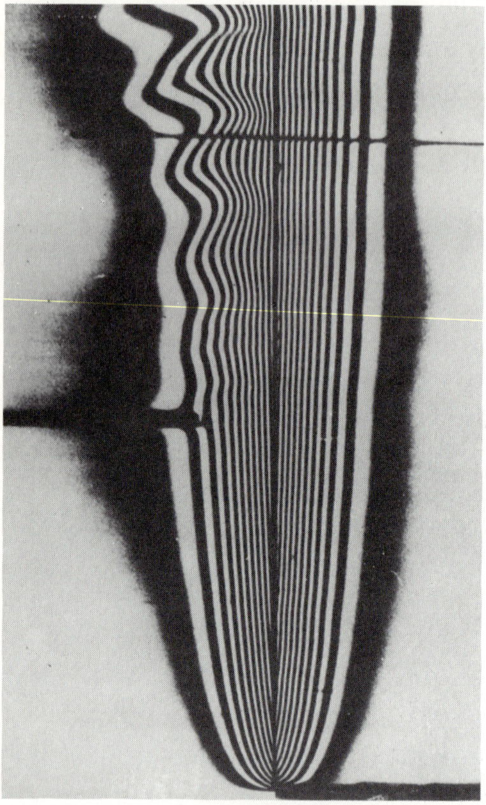

**Figure 7.5** Interferogram of an amplified disturbance in a laminar free-convection boundary layer on a heated vertical plate. The disturbance is introduced at $Gr_x^* \doteq 4.3 \times 10^9$. (From Polymeropoulos and Gebhart [12].)

cascade downstream into fully turbulent flow at $Gr_x = 10^9$ or larger. The theory of free convection instability is discussed in a review article by Gebhart [11] and flow visualization experiments are reported by Polymeropoulos and Gebhart [12].

Figure 7.5 shows an interferogram from [12] of free-convection airflow past a heated vertical plate. The dark lines are equivalent to the isotherms of the flow. A periodic disturbance has been introduced on the left side of the plate at about $Gr_x^* \doteq 4.3 \times 10^9$ (equivalent to $Gr_x \doteq 1.1 \times 10^8$). The disturbance is clearly seen to amplify downstream, hence the laminar boundary layer is unstable. Such amplified waves occur naturally, even on a very smooth plate and very still ambient fluid, at about $Gr_x^* \doteq 2 \times 10^{11}$ ($Gr_x \doteq 1.7 \times 10^9$), so that turbulence always occurs regardless of one's care in maintaining a disturbance-free environment.

Experiments by Vliet and Liu [13] with water on a heated plate at constant $q''_w$ indicate that natural transition to turbulence occurs in the range $10^{12} \leqslant \mathrm{Gr}_x^*\mathrm{Pr} \leqslant 10^{14}$, which corresponds approximately to $10^9 \leqslant \mathrm{Gr}_x \leqslant 4 \times 10^{10}$. Since it is difficult to predict the degree of disturbance in a typical environment for an engineering application, it is generally recommended that the transition Grashof number be taken as the minimum value of $\mathrm{Gr}_x(\mathrm{crit}) \doteq 10^9$. We will use this assumption in the problems and examples in this text.

### 7.2.6 Turbulent Free Convection on a Vertical Plate

As in the laminar case, turbulent free convection may be analyzed either by an integral method or by computer solution of the (turbulent) boundary layer equations. Eckert and Jackson [15] published a very effective integral theory using the following profile and wall shear approximations:

$$\begin{aligned} \frac{u}{u_0} &\doteq \eta^{1/7}(1-\eta)^4, \qquad \eta = y/\delta, \\ \frac{T-T_\infty}{T_w - T_\infty} &\doteq 1 - \eta^{1/7}, \\ \tau_w &\doteq 0.0228\, \rho\, u_0^2\, (\nu/u_0\delta)^{1/4}, \end{aligned} \tag{7.33}$$

plus the (forced-convection) flat plate Reynolds analogy $\mathrm{St}_x \doteq c_f/2\mathrm{Pr}^{2/3}$. The wall shear formula is based on a correlation for pipe-flow wall shear proposed by Prandtl's student H. Blasius in 1911 [16, p. 597].

By substituting Eqs. (7.33) into the boundary layer integral relations (7.14) and (7.15), Eckert and Jackson were able to develop a closed-form expression for local heat flux in turbulent free convection past a vertical plate:

$$\mathrm{Nu}_x \doteq 0.0295 \left( \frac{\mathrm{Pr}^{7/6}\,\mathrm{Gr}_x}{1 + 0.494\mathrm{Pr}^{2/3}} \right)^{2/5}, \tag{7.34}$$

valid with reasonable accuracy for $\mathrm{Gr}_x \geqslant 10^9$. This implies that $h_x \propto x^{1/5}$, so that integration yields $\overline{\mathrm{Nu}}_L = \frac{5}{6}\,\mathrm{Nu}_x\ (x = L)$.

There also have been several digital computer solutions of the turbulent boundary layer equations (7.11), utilizing various eddy viscosity and eddy conductivity assumptions. The reader is especially referred to the computations of Noto and Matsumoto [17], who assumed

$Pr_t = 1.0$ and an eddy viscosity correlation similar to the van Driest formula, Eq. (6.79). Their computed velocity profiles show a maximum at about $y/\delta \doteq 0.035$, far closer to the wall than the laminar profiles in Fig. 7.4. The Nusselt number computations are in agreement with Eq. (7.34) and also with the turbulent convection data of [13], [18], and [19].

### 7.2.7 A General Correlation Formula

We have discussed laminar and turbulent flow formulas for vertical plate free convection. Churchill and Chu [20] suggest a single interpolation formula that correlates all constant-wall-temperature data in the range $0.1 \leqslant \mathrm{Ra}_L \leqslant 10^{12}$:

$$\overline{\mathrm{Nu}}_L^{1/2} \doteq 0.825 + \frac{0.387\,\mathrm{Ra}_L^{1/6}}{[1 + (0.492/\mathrm{Pr})^{9/16}]^{8/27}} \tag{7.35}$$

where $\mathrm{Ra}_L = \mathrm{Gr}_L\mathrm{Pr}$ is the plate length Rayleigh number. This is a valuable formula since it holds even in the low range $\mathrm{Gr} < 10^4$ where boundary layer theory is invalid. However, it is generally not quite as accurate as Eq. (7.27) or (7.34) in range of applicability. Equation (7.35) is plotted in Fig. 7.6 for a range of $\mathrm{Gr}_L$ and Pr. Although it shows no change in slope between laminar and turbulent flow, it does agree reasonably ($\pm 30\%$) with experimental data.

**Figure 7.6** Mean Nusselt number for free convection on a vertical flat plate, from the correlation (7.35) suggested by Churchill and Chu [20].

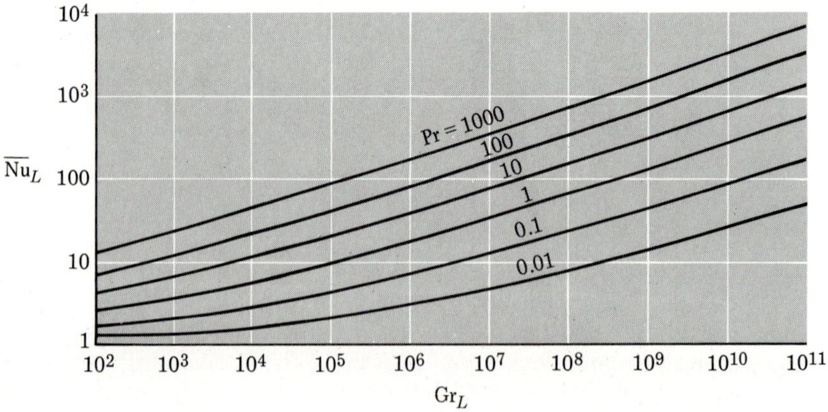

Churchill and Chu [20] also suggest that Eq. (7.35) can be used for constant heat flux conditions if $\overline{\mathrm{Nu}}_L$ and $\mathrm{Ra}_L$ are based on the difference between $T_w$ at $x = L/2$ and $T_\infty$. This is based on a suggestion of Sparrow and Gregg [10]. However, the value of $T_w$ at $x = L/2$ is a priori unknown so iteration is required for this use of the correlation.

**Example 7.5**

Repeat Example 7.3, assuming the fluid is mercury.

*Solution* For mercury at $T_f = 135°\mathrm{C}$, $k = 9.85\ \mathrm{W/m \cdot K}$, $\mathrm{Pr} = 0.0161$, and $g\beta/\nu^2 = 2.31 \times 10^{11}$. Then the Grashof number is

$$\begin{aligned}\mathrm{Gr}_L &= (g\beta/\nu^2)(T_w - T_\infty)L^3 = (2.31 \times 10^{11})(250 - 20)(0.18)^3 \\ &= 3.1 \times 10^{11}.\end{aligned}$$

This is greater than $10^9$, so we conclude the flow is turbulent. From the Eckert-Jackson correlation (7.34), compute

$$\overline{\mathrm{Nu}}_L \doteq \frac{5}{6}(0.0295)\left[\frac{(0.0161)^{7/6}(3.1 \times 10^{11})}{1 + 0.494(0.0161)^{2/3}}\right]^{2/5} = 140.$$

By comparison, the Churchill-Chu correlation (7.35) predicts $\overline{\mathrm{Nu}}_L \doteq 91$, or 35% lower. Both are within the scatter of the data. Continuing with the first prediction, we have

$$\bar{h} = \frac{\overline{Nu}_L k}{L} = \frac{(140)(9.85)}{0.18} = 7640\ \mathrm{W/m^2 \cdot K}.$$

The total heat transfer from both sides of the plate is

$$\begin{aligned}q_w &= \bar{h}\, 2bL\,(T_w - T_\infty) = (7640)(2)(0.1)(0.18)(250 - 20) \\ &= 63{,}000\ \mathrm{W}. \quad [Ans.]\end{aligned}$$

This is 1090 times more than the solution for air ($q_w = 58$ W) in Example 7.3. The difference is due primarily to the high conductivity and buoyancy of mercury. ■

## 7.3 External Free Convection on Other Geometries

The only theory we have presented is for free convection past a vertical plate. Let us briefly review some empirical correlations for other geometries in free-convection flow.

### 7.3.1 The Vertical Cylinder

Consider a free-convection boundary layer forming on the outside of a vertical cylinder of length $L$ and diameter $D$. If $D$ is large, the geometry simulates a vertical plate; if $D$ is small, the transverse curvature increases the heat transfer somewhat. An interesting non-similar laminar flow perturbation solution by Minkowycz and Sparrow [21] suggests a correction factor for average Nusselt number:

$$\overline{\mathrm{Nu}}_{L,\,\mathrm{cyl}} \doteq \overline{\mathrm{Nu}}_{L,\,\mathrm{plate}}(1 + 1.3\,\zeta^{0.9}), \qquad \textbf{(7.36)}$$

where $\zeta = (L/D)\mathrm{Ra}_L^{-1/4}$ and $\overline{\mathrm{Nu}}_{L,\,\mathrm{plate}}$ is taken from Eq. (7.35). The correction factor for air is less than 6% if $\zeta < 0.024$. Digital computer results for various Prandtl numbers are reported by Cebeci [22].

### 7.3.2 Inclined Plates

Consider a plate inclined at an angle $\theta$ from the vertical. Boundary layer theory, Eq. (7.10), predicts that the vertical plate relations of the previous section should be valid if $g$ is replaced by the component $g \cos\theta$ along the inclined plate, whether the convecting surface is on the bottom or the top of the plate. This is verified by the experiments of Vliet [23] for laminar flow, $\mathrm{Gr}_x^*\mathrm{Pr} < 10^{10}$, but turbulent flow seems to correlate better with $g$ used instead of $g \cos\theta$. Vliet measured plate inclinations up to $\theta = 60°$, and Fujii and Imura [24] reported nearly the same results for $\theta$ as high as 88°.

Both papers [23, 24] indicate that transition to turbulence begins much earlier as the plate is inclined. Vliet's data for transition may be tabulated as follows:

| $\theta$: | 0° | 15° | 30° | 45° | 60° | 75° |
|---|---|---|---|---|---|---|
| $\mathrm{Gr}_{\mathrm{tr}}^*$: | $6 \times 10^{12}$ | $3 \times 10^{11}$ | $2 \times 10^{10}$ | $10^9$ | $7 \times 10^7$ | $4 \times 10^6$ |

The early transition is especially noticeable when the heated plate faces upward.

### 7.3.3 Horizontal Plates

For a horizontal plate, the induced convection currents are more or less normal to the plate and surprisingly effective. The following correlations are recommended from [24]:

*Upper surface heated or lower surface cooled:*

$$\overline{\mathrm{Nu}}_L \doteq 0.54\ \mathrm{Ra}_L^{1/4} \quad \text{for} \quad 10^4 \leqslant \mathrm{Ra}_L \leqslant 10^7 \tag{7.37}$$

$$\doteq 0.15\ \mathrm{Ra}_L^{1/3} \quad \text{for} \quad 10^7 \leqslant \mathrm{Ra}_L \leqslant 10^{11}, \tag{7.38}$$

*Upper surface cooled or lower surface heated:*

$$\overline{\mathrm{Nu}}_L \doteq 0.27\ \mathrm{Ra}_L^{1/4} \quad \text{for} \quad 10^5 \leqslant \mathrm{Ra}_L \leqslant 10^{11}. \tag{7.39}$$

Note that Eq. (7.37) is identical to Eq. (7.30) for a *vertical* plate of the same size! However, Eq. (7.39) instead predicts only half as much heat flux, because the plate tends to block the induced convection currents.

The characteristic length $L$ equals the average of the length and width of a rectangular plate, it equals $0.9D$ for a circular plate, and it equals the area divided by the perimeter [25] for a plate of irregular shape.

### 7.3.4 Horizontal Cylinders

For a horizontal cylinder in free convection, the local streamwise $g_x = g \sin \phi$, where $\phi$ is the angle measured from the bottom of the cylinder (Fig. 7.7). Figure 7.7 shows the free-convection isotherms near a horizontal cylinder as measured by Kennard [26]. Note that there is almost no boundary layer separation, because pressure gradients are nearly negligible in free convection. It follows that boundary layer theory should be effective for this problem. Whereas on a vertical plate the integrated gravity effect is $\int g\, dx = gL$, on a horizontal cylinder the equivalent effect is $\int g_x dx = \int_0^\pi g \sin\phi\, (\frac{1}{2}D d\phi) = gD$. This implies that heat flux data on a horizontal cylinder should be closely approximated by vertical plate correlations if the plate length is replaced by cylinder diameter. This is indeed the case.

A general correlation for mean Nusselt number in free convection over a horizontal cylinder is given by Churchill and Chu [27], valid over a large range of data:

$$\overline{\mathrm{Nu}}_D^{1/2} \doteq 0.60 + \frac{0.387\ \mathrm{Ra}_D^{1/6}}{[1 + (0.559/\mathrm{Pr})^{9/16}]^{8/27}}, \tag{7.40}$$

for $10^{-5} \leqslant \mathrm{Ra}_D \leqslant 10^{12}$. Note the similarity with Eq. (7.35): Except for liquid metals ($\mathrm{Pr} < 0.04$), the two correlations differ by only a few percent for $\mathrm{Ra} \geqslant 10^5$.

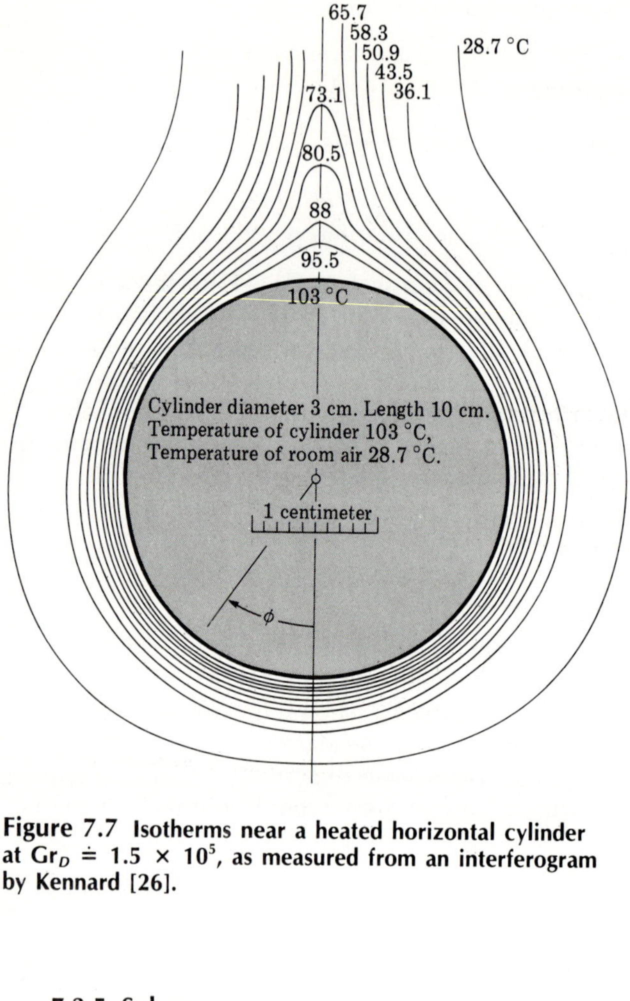


**Figure 7.7** Isotherms near a heated horizontal cylinder at $Gr_D \doteq 1.5 \times 10^5$, as measured from an interferogram by Kennard [26].

### 7.3.5 Spheres

Free convection to spheres in air has been measured by Yuge [28] in the laminar range $1 \leq Ra_D \leq 10^5$. The recommended correlation is

$$\overline{Nu}_D \doteq 2 + 0.43\, Ra_D^{1/4}. \tag{7.41}$$

This formula is probably inaccurate unless Pr is near unity. Recall that the lower limit of 2.0 was predicted for pure conduction near a sphere in problem 6.97.

### 7.3.6 Blocks and Other Complex Shapes

The preceding correlations are for classic shapes with a single distinctive length scale: wide plates, long cylinders, and spheres. Free convection on more complex three-dimensional shapes has not been thoroughly studied. One needs a rule of thumb to estimate the heat transfer from such shapes.

In a classic review article, King [29] recommended a power-law correlation of the form

$$\overline{\mathrm{Nu}}_L \doteq 0.55\, \mathrm{Ra}_L^{1/4}, \tag{7.42}$$

where $L$ is an average characteristic dimension for the shape. The problem, of course, is to determine $L$ for a complex blocky body. King recommended the scale

$$L = (1/L_h + 1/L_v)^{-1},$$

where $L_h$ and $L_v$ are the horizontal and vertical length scales of the body, respectively. More recently, Lienhard [49] proposed that $L$ should be the average length of travel of a particle in the free convection boundary layer. For a sphere, for example, King's formula gives $L = D/2$, while Lienhard's idea suggests that $L = \pi D/2$.

In a recent experiment with a short cylinder, Sparrow and Ansari [50] showed that, compared to actual data, King's idea predicts the heat transfer to be 50% high and Lienhard's suggestion is 20% low, on the average. Therefore neither length scale is very accurate for a blocky body. Lacking any other information, Lienhard's suggestion could be followed.

A more laborious but also more accurate prediction is given in a series of papers by Hollands and co-workers [51]. Their method involves the computation of laminar and turbulent conduction layers about the body. The resulting prediction has an accuracy of about 3% when compared to data for bi-spheres and for cubes at various orientations.

### 7.3.7 Optimum Spacing of Vertical Plate Fins

An array of vertical plates, as shown in Fig. 7.8, is common in the cooling of electronic equipment. Typically, the plates are hot, and cool air convects up between them. Natural convection will suffice if the desired heat flux is less then 1 W/cm$^2$, as discussed in review articles on electronic cooling by Chu [52] and by Oktay et al. [53].

If the plates in Fig. 7.8 are widely spaced or essentially isolated,

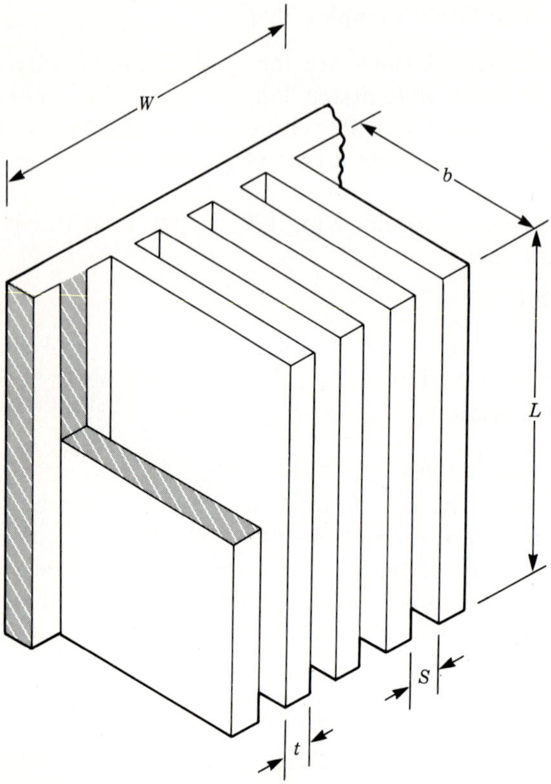


**Figure 7.8** Geometry and dimensions of a rectangular vertical fin array (after Bar-Cohen [55]).

heat transfer rates are high but the total fin area is low. If the plates are closely packed, total fin area is high but heat flux is low because of the boundary layer interaction between plates. Elenbaas [30] was the first to show that there is an optimum intermediate spacing that gives maximum total heat transfer for a given base size $WL$. His analysis was recently modified by Bar-Cohen and Rohsenow [54] to give better agreement with experiment. Their improved analysis is expressed in terms of a parameter $P$ defined by

$$P = (g\beta\Delta T/\nu^2)(\mathrm{Pr}/L) = \mathrm{Ra}_L/L^4,$$

where $\Delta T$ is the temperature difference between fin and fluid. For isothermal plates of thickness $t$ and spacing $S$, Bar-Cohen and Rohsenow show that optimum spacing occurs when

$$S + 1.5t - 0.0025P^{1.5}S^7 = 0. \tag{7.43}$$

Once the spacing is known, the heat transfer coefficient is given by

$$\overline{\mathrm{Nu}}_S = \bar{h}S/k = [576/(P^2S^8) + 2.873/(P^{1/2}S^2)]^{-1/2}. \tag{7.44}$$

The total heat transfer from the plate array is then

$$q = \bar{h}(2nLb\,\Delta T),$$

where $n = W/(S + t)$ is the total number of plate fins. If the fin thickness $t$ is negligible, Eqs. (7.43) and (7.44) yield the simple result

$$S = 2.714/P^{1/4} = 2.714L/\mathrm{Ra}_L^{1/4},$$

$$\overline{\mathrm{Nu}}_S = 1.31.$$

The optimum fin spacing thus depends on the Rayleigh number based on fin length $L$ parallel to the flow.

The analysis above depends upon the fins being isothermal or very efficient in the sense of Fig. 2.15(b) in our fin analysis of Chapter 2. Bar-Cohen [55] studied the case of fins whose temperature varies with distance from the base. His solutions for optimum fin length $b$ and spacing $S$ depend upon the fin thickness and its thermal conductivity.

## Example 7.6

A bare 3-in.-diameter horizontal steam pipe is 200 ft long and has a surface temperature of 340°F. Estimate the total heat loss by free convection to air at 60°F.

*Solution* For air at $T_f = (340 + 60)/2 = 200°\mathrm{F}$, $\mathrm{Pr} = 0.71$, $k = 0.0174$ Btu/hr · ft · °F, and $g\beta/\nu^2 = 8.54 \times 10^5\ °\mathrm{F}^{-1}\ \mathrm{ft}^{-3}$. The Rayleigh number based on pipe diameter is

$$\mathrm{Ra}_D = (g\beta/\nu^2)(T_w - T_\infty)D^3\mathrm{Pr} = (8.54 \times 10^5)(340 - 60)(3/12)^3(0.71)$$
$$= 2.65 \times 10^6.$$

Then, from Eq. (7.40) for horizontal cylinders,

$$\overline{\mathrm{Nu}}_D^{1/2} = 0.60 + \frac{0.387\,(2.65 \times 10^6)^{1/6}}{[1 + (0.559/0.71)^{9/16}]^{8/27}} = 4.38,$$

or $\overline{\mathrm{Nu}}_D = 19.2$. Then

$$\bar{h} = \frac{\overline{\mathrm{Nu}}_D k}{D} = \frac{(19.2)(0.0174)}{3/12} = 1.34 \text{ Btu/hr} \cdot \text{ft}^2 \cdot {}^\circ\text{F}.$$

The total heat loss is

$$q_w = \bar{h}\pi DL\,(T_w - T_\infty) = (1.34)\pi(3/12)(200)(340 - 60)$$
$$= 59{,}000 \text{ Btu/hr} \quad (17 \text{ kW}). \quad [Ans.]$$

The cost of generating this much heat, at 1988 prices, is from $20 to $45 per day, depending on the fuel used. It would be well worth it to insulate this pipe. ■

## †7.4 Free Convection Inside Enclosures

The previous two sections dealt with external free-convection flows. A topic of recent interest is the internal flow in an enclosure, induced by opposing hot and cold walls. Typical applications are cooling of electronic equipment, multipane windows, air gaps in house walls, fire detection equipment, stoves, and flat plate solar collectors. The subject was reviewed by Ostrach in 1972 [31] and by Catton in 1978 [32]. While external free convection is primarily a boundary layer type problem even for complex shapes (as in Fig. 7.7), enclosure flows are more complicated especially in the core region, away from the walls. At high Rayleigh numbers, the core region can develop multicellular two- and three-dimensional motions coupled with both steady and irregular traveling waves.

We confine ourselves here to the classic rectangular inclined enclosure in Fig. 7.9. The tilt angle $\phi$ can vary from 0° (horizontal with heated surface on the bottom) to 90° (vertical heated and cooled surfaces) to 180° (horizontal with heated surfaces on the top), or anywhere in between. The top, bottom, and side walls are unheated — there is only one heated and one cooled wall of area $Wd$ — and they may be either insulated or conducting. The Nusselt number and Rayleigh number are customarily based on the gap width $L$ between the hot and cold walls. The aspect ratio of most importance in Fig. 7.9(b) is the vertical ratio $A = d/L$; the sidewall ratio $W/L$ is less important and less studied.

†This section may be omitted without loss of continuity.

### 7.4.1 Vertical Box Enclosure

Consider the vertical enclosure ($\phi = 90°$) in Fig. 7.9(b). If convection were negligible, the heat flux would be by pure conduction in the fluid, $\bar{q}''_w = k(T_H - T_C)/L$, or:

$$\mathrm{Nu}_L(\text{conduction}) = \bar{h}L/k = 1, \tag{7.45}$$

since $\bar{h} = \bar{q}''_w/(T_H - T_C)$ by definition. This is the lower bound on heat

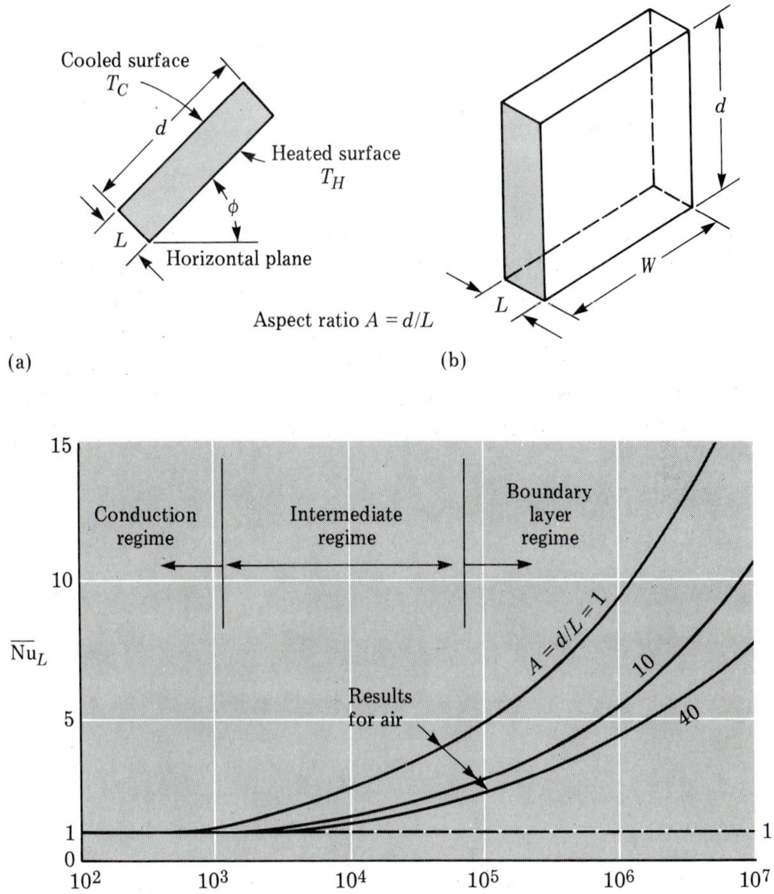


**Figure 7.9** Free convection in rectangular enclosures, modified from [32]: (a) inclined rectangular enclosure; (b) basic dimensions; (c) measured Nusselt numbers for a vertical enclosure, $\phi = 90°$.

flux and persists up to $\text{Ra}_L = [g\beta(T_H - T_C)L^3/\nu^2]\text{Pr} = 10^3$. This does not mean there is no motion: There is a gentle circulation up the hot wall, across the top, down the cold wall, and back across the bottom wall. But the motion is weak and does not contribute significantly to heat flux.

For $\text{Ra}_L > 10^3$, the circulating motion becomes stronger and contributes to total heat flux, and it is especially effective for short boxes, $A \doteq 1$. Some experimental data for air (reported in [32]) are shown in Fig. 7.9(c). In the intermediate range $10^3 < \text{Ra}_L < 10^5$, the circulation continues to fill the entire box, but for $\text{Ra}_L > 10^5$, boundary layers develop on the hot and cold walls and the central core is relatively motionless. For $\text{Ra}_L \doteq 10^6$, secondary and tertiary multiple-cell motions begin to appear in the core but there is no sharp increase in the heat flux. For $\text{Ra}_L = 10^7$, instability sets in as amplified traveling waves on both walls. Turbulent flow ensues at a *vertical* Rayleigh number $\text{Ra}_d = \text{Ra}_L A^3 \doteq 10^9$, again with no sharp change in the Nusselt number.

For aspect ratios near unity, Catton [32] recommends correlation formulas proposed by Berkovsky and Polevikov in a symposium volume on buoyant convection [33]:

*$2 < A < 10$:*

$$\overline{\text{Nu}}_L \doteq 0.22\,\zeta^{0.28} A^{-1/4}, \tag{7.46a}$$

*$1 < A < 2$:*

$$\overline{\text{Nu}}_L \doteq 0.18\,\zeta^{0.29}. \tag{7.46b}$$

where $\zeta = \text{Ra}_L\text{Pr}/(0.2 + \text{Pr})$. The formulas are valid for $\text{Pr} < 10^5$ and $\zeta > 10^3$. There are few experimental data for low aspect ratios, $A < 1$, but theoretical computer simulations by Catton, Ayyaswamy, and Clever [34] show that $\overline{\text{Nu}}_L$ drops off again, so that the highest attainable Nusselt numbers are for $A = 1$–$2$.

Since an enclosure has finite boundaries, it is an attractive candidate for numerical analysis on a digital computer, and successful simulations have been reported for Grashof numbers as high as $10^7$. An example of a finite-difference method is given in [34], and [35] reports a Galerkin method.

The two upper curves in Fig. 7.9(c) are equivalent to Eqs. (7.46). Note that the enclosure produces somewhat smaller Nusselt numbers than the isolated vertical plate in Fig. 7.7, a consequence of the confining walls reducing the wall flows. The enclosure does, however, produce an "effective conductivity," $k_{\text{eff}} = \overline{\text{Nu}}_L k$, considerably greater than unity.

### 7.4.2 Tilted Box Enclosure

A vertical box ($\phi = 90°$) always induces a buoyant flow, no matter how small the Rayleigh number. A substantially tilted box ($\phi < 80°$ or $\phi > 100°$) does not cause flow until a critical Rayleigh number, $Ra_L(\text{crit}) = 1708/\cos\phi$, is reached, as shown by Hart [36]. Hart's result is for large aspect ratios, $A > 10$. At smaller $A$, the critical $Ra_L$ is even greater (see, for example, Fig. 12 of [32]).

If we confine our attention to large $A > 10$, a tilted box (Fig. 7.9a) when heated from below ($\phi < 75°$) will exhibit pure conduction for $Ra_L \cos\phi < 1708$, above which there is a sharp "onset" of convection effects, with $\overline{Nu}_L > 1$. The convection is somewhat dependent upon $\phi$ in the range $1708 < Ra_L \cos\phi < 3 \times 10^4$ and then dependent only upon $Ra_L \cos\phi$ above $3 \times 10^4$. The experimental correlation proposed by Hollands and co-workers [37] at $A = 44$ may be used with reliability for all $A \geqslant 10$ and $0 \leqslant \phi \leqslant 75°$:

$$\overline{Nu}_L \doteq 1 + 1.44\{1 - 1708/Ra_e\}^{\dagger}[1 - 1708(\sin 1.8\phi)^{1.6}/Ra_e] + \{(Ra_e/5830)^{1/3} - 1\}^{\dagger}, \qquad Ra_e = Ra_L \cos\phi, \tag{7.47}$$

where the † on the term in braces means the term should be set equal to zero if it is negative. The predictions of Eq. (7.47) are slightly higher than the curve marked $A = 10$ in Fig. 7.9(c). Hollands's data extend only to $Ra_L \cos\phi = 10^6$ but the correlation should be valid at least up to $10^7$.

Free-convection results have also been reported for other enclosure geometries. Flow between concentric cylinders and spheres has been studied by Raithby and Hollands [38], Ostrach reviewed work on horizontal cylinders [31], and Catton reviewed *convection suppression* by use of honeycomb structures [32]. A very interesting study of free convection induced by the warm skin of the human body is contained in [40].

#### Example 7.7

A flat plate solar collector has $L = 8$ cm, $W = 1$ m, and $d = 1.6$ m and is tilted at 40° to the horizontal. The inner wall is at 70°C and the outer wall at 10°C, and the enclosure is filled with air at 1 atm. Estimate the heat loss.

*Solution* For air at $T_f = (70 + 10)/2 = 40°\text{C}$, $g\beta/\nu^2 = 1.01 \times 10^8\ \text{K}^{-1}\,\text{m}^{-3}$, $k = 0.0265$ W/m · K, and $\text{Pr} = 0.71$. The aspect ratio is $d/L = 20$, which is large enough to assume that Hollands's correlation (7.47) is valid. The gap Rayleigh number is

$$\mathrm{Ra}_L = (g\beta/\nu^2)(T_H - T_C)L^3\mathrm{Pr} = (1.01 \times 10^8)(70 - 10)(0.08)^3(0.71) = 2.20 \times 10^6$$

and

$$\mathrm{Ra}_e = \mathrm{Ra}_L \cos\phi = (2.20 \times 10^6)\cos(40°) = 1.69 \times 10^6.$$

Equation (7.47) then predicts

$$\overline{\mathrm{Nu}}_L \doteq 1 + 1.44(1 - 1708/\mathrm{Ra}_e)[1 - 1708(\sin 72°)^{1.6}/\mathrm{Ra}_e] + [(\mathrm{Ra}_e/5830)^{1/3} - 1] = 1 + 1.437 + 5.615 = 8.05.$$

Thus

$$\bar{h} = \frac{\overline{\mathrm{Nu}}_L k}{L} = \frac{(8.05)(0.0265)}{0.08} = 2.67\ \mathrm{W/m^2 \cdot K},$$

and

$$q_w = \bar{h}\,Wd(T_H - T_C) = 2.67(1)(1.6)(70 - 10) = 256\ \mathrm{W}. \quad [Ans.]$$

This is about 16% of the maximum clear-day solar insolation of 1600 W that is received by such a 1.6-m$^2$ collector at mid-latitudes. The loss might be even greater if strong winds flow over the outside of the collector [39]. The reader may verify as an exercise that the loss is increased by 17% if the gap size $L$ is halved to 4 cm. ■

## †7.5 Combined Free and Forced Convection

The term *forced convection* really means that we are neglecting free convection, because buoyancy effects always occur when fluid temperature differences exist in a gravity field. We saw in Section 7.1.2 that free-convection effects are negligible when $\mathrm{Gr}_L \ll \mathrm{Re}_L^2$. At the other extreme, free convection dominates when $\mathrm{Gr}_L \gg \mathrm{Re}_L^2$, which is the assumption underlying the earlier sections of this chapter.

The present section considers the mixed-convection region, where $\mathrm{Gr}_L$ is comparable to $\mathrm{Re}_L^2$ and both free- and forced-convection effects are important. The mixed-convection region is in the range 0.01 <

†This section may be omitted without loss of continuity.

$Gr_L/Re_L^2 < 100$, and this condition occurs in practice for both internal and external flows.

We may distinguish at least three separate cases of combined free and forced convection, as described, for example, in [56]:

1. Aiding or *assisting* flow, in which the buoyant motion is in the same direction as the forced motion. Examples are upward forced flow past a hot vertical plate and downward forced flow past a cold vertical plate.
2. *Opposing* flow, in which the buoyant motion is opposite to the forced motion. An example is upward forced flow past a cold vertical plate.
3. *Transverse* flow, in which the buoyant motion is at right angles to the forced motion. Examples are horizontal forced flow past a hot or cold cylinder or sphere.

In all these cases the forced and free motions interact but are not strictly additive or subtractive. For laminar flow, based on both theoretical and experimental evidence, Churchill [57] suggests the following correlation:

$$Nu_{overall} \doteq (Nu_{forced}^n \pm Nu_{free}^n)^{1/n}, \quad \textbf{(7.48)}$$

where $Nu_{free} = f(Gr_L, Pr)$ and $Nu_{forced} = f(Re_L, Pr)$ are the appropriate correlations for the particular geometry for purely free and forced motions, respectively. For assisting or opposing flow, Churchill recommends $n = 3$, with the positive and negative signs denoting assisting and opposing flows, respectively. Figure 7.10(a) shows the results of applying Eq. (7.48) to mixed convection on a vertical plate. The straight dash/dot lines for pure forced and pure free convection are computed from Eqs. (6.21) and (7.25), respectively. Note that the heat transfer drops off sharply in opposing flow as the opposing buoyancy becomes strong enough to cause flow separation inside the boundary layer [58]. Figure 7.10(a) is in good agreement with both theory and experiment for vertical plate flows [41, 43].

For laminar transverse mixed convection, Churchill [56] recommends that values of $n = 3.5$–$4.0$ be fit to the specific geometry, such as horizontal flow past cylinders [43] or spheres [28].

Turbulent mixed convection is more complex than simply using turbulent formulas in Eq. (7.48) because both assisting and opposing flows strongly affect the transition point. Some example computations for turbulent vertical plate flow are given by Armaly et al. [59].

Internal combined flows may be still more complex. Figure 7.10(b) shows a flow-regime map constructed by Metais and Eckert [44] for

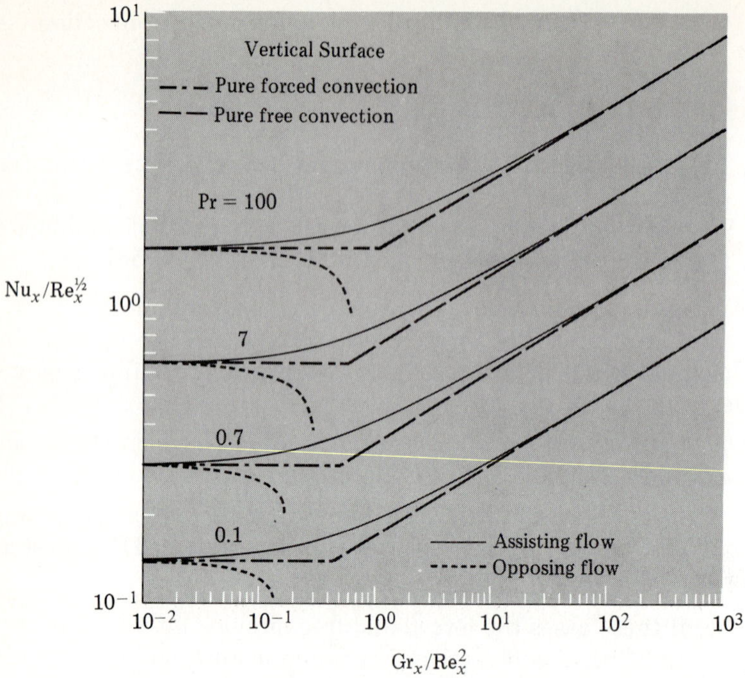


(a)

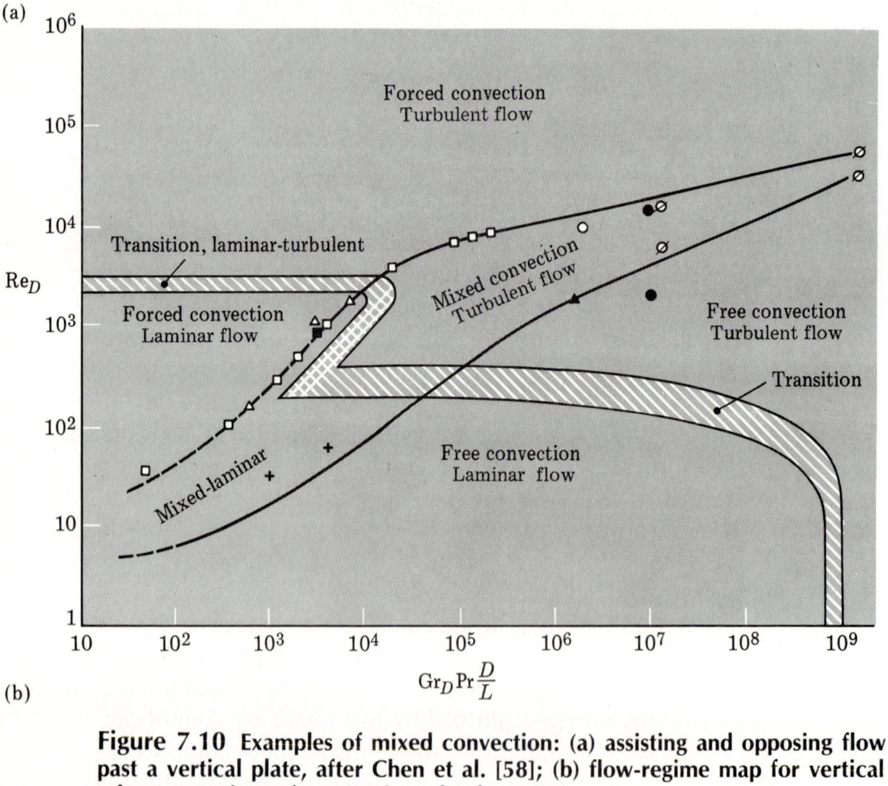


(b)

**Figure 7.10** Examples of mixed convection: (a) assisting and opposing flow past a vertical plate, after Chen et al. [58]; (b) flow-regime map for vertical tube convection, after Metais and Eckert [44].

both aiding and opposing flow in a vertical tube. There are nine different combinations of hydrodynamic (laminar, transitional, turbulent) and convective (forced, mixed, free) conditions. Note that the transition Reynolds number is reduced by buoyancy. The overall heat transfer is difficult to predict. For further details, consult [45–47].

## Summary

Free convection is a widely occurring heat transfer mechanism in both natural and human-made systems. Its effects should always be investigated when designing for temperature or species concentration differences in a fluid. The engineer will have a constant need to understand free convection in tackling heat transfer problems.

The chapter introduction shows that induced velocities are proportional to $(g\Delta\rho L)^{1/2}$, where $L$ is the vertical length scale and $\Delta\rho$ is the expected density difference. This leads to the fundamental parameter of Grashof number, $\mathrm{Gr}_L$, which is analogous to the Reynolds number squared in forced convection. For flows induced by thermal expansions, a buoyant force term $\rho g\beta(T - T_w)$ is shown to drive the differential momentum equation.

A detailed treatment is given for free convection on a vertical flat plate, which exhibits induced phenomena analogous to forced motion: creeping flow, laminar boundary layers, transition regions, and turbulent boundary layers. Extensive correlations are given for plates, cylinders, spheres, and other geometries in external free convection.

Consideration of flow within tilted enclosures gives rise to the concept of critical Rayleigh number required for a free convection flow to ensue. Correlations are given for box shapes.

Finally, there is a brief discussion of "mixed" flows, where free and forced convection are of comparable importance. These flows are typically in the region where $\mathrm{Gr}_L \doteq \mathrm{Re}_L^2$.

## References


1. Y. Jaluria, *Natural Convection Heat and Mass Transfer*, Pergamon Press, New York, 1980.
2. B. Gebhart, *Heat Transfer*, 2nd ed., McGraw-Hill, New York, 1970.
3. A. J. Ede, "Advances in Free Convection," in *Advances in Heat Transfer*, vol. 4, Academic Press, 1967, pp. 1–64.

4. B. Gebhart, "Buoyancy Induced Fluid Motions Characteristic of Applications in Technology," *J. Fluids Engineering*, vol. 101, March 1979, pp. 5–28.
5. E. Schmidt and W. Beckmann, "Das Temperatur und Geschwindigkeitsfeld vor einer Wärme abgebenden senkrechter Platte bei natürlicher Konvektion," *Tech. Mech. u. Thermodynamik*, bd. 1, no. 10, Oct. 1930, pp. 341–349, and bd. 1, no. 11, Nov. 1930, pp. 391–406.
6. S. Ostrach, "An Analysis of Laminar Free Convection Flow and Heat Transfer About a Flat Plate Parallel to the Direction of the Generating Body Force," NACA Report 1111, 1953 (see also NACA Tech. Note 2863, 1952).
7. F. M. White, *Viscous Fluid Flow*, McGraw-Hill, New York, 1974.
8. E. R. G. Eckert, *Introduction to the Transfer of Heat and Mass*, McGraw-Hill, New York, 1950.
9. E. J. LeFevre, "Laminar Free Convection from a Vertical Plane Surface," *Proc. 9th Intl. Congress Appl. Mechanics*, vol. 4, 1956, p. 168.
10. E. M. Sparrow and J. L. Gregg, "Laminar Free Convection from a Vertical Plate with Uniform Surface Heat Flux," *ASME Transactions*, vol. 78, Feb. 1956, pp. 435–440.
11. B. Gebhart, "External Natural Convection Flow," *Applied Mechanics Reviews*, vol. 22, July 1969, pp. 691–701.
12. C. E. Polymeropoulos and B. Gebhart, "Incipient Instability in Free Convection Laminar Boundary Layers," *J. Fluid Mechanics*, vol. 30, part 2, 1967, pp. 225–239.
13. G. C. Vliet and C. K. Liu, "An Experimental Study of Turbulent Natural Convection Boundary Layers," *J. Heat Transfer*, vol. 91, Nov. 1969, pp. 517–531.
14. B. Gebhart, "Natural Convection Flows and Stability," *Advances in Heat Transfer*, vol. 9, 1973, pp. 273–348.
15. E. R. G. Eckert and T. W. Jackson, "Analysis of Turbulent Free Convection Boundary Layer on a Flat Plate," NACA Report 1015, 1951 (see also NACA Tech. Note 2207, 1950).
16. H. Schlichting, *Boundary Layer Theory*, 7th ed., McGraw-Hill, New York, 1979.
17. K. Noto and R. Matsumoto, "Turbulent Heat Transfer by Natural Convection Along an Isothermal Vertical Flat Surface," *J. Heat Transfer*, vol. 97, Nov. 1975, pp. 621–624.
18. R. Cheesewright, "Turbulent Natural Convection from a Vertical Plane Surface," *J. Heat Transfer*, vol. 90, Feb. 1968, pp. 1–8.
19. R. D. Flack and C. L. Witt, "Velocity Measurements in Two Natural Convection Air Flows Using a Laser Velocimeter," *J. Heat Transfer*, vol. 101, May 1979, pp. 256–260.

20. S. W. Churchill and H. H. S. Chu, "Correlating Equations for Laminar and Turbulent Free Convection from a Vertical Plate," *Int. J. Heat Mass Transfer*, vol. 18, 1975, pp. 1323–1329.

21. W. J. Minkowycz and E. M. Sparrow, "Local Nonsimilar Solutions for Natural Convection on a Vertical Cylinder," *J. Heat Transfer*, vol. 96, May 1974, pp. 178–183.

22. T. Cebeci, "Laminar Free-Convective Heat Transfer from the Outer Surface of a Vertical Slender Circular Cylinder," *Fifth Int. Heat Transfer Conf.*, vol. 3, NC 1.4, 1974, pp. 15–19.

23. G. C. Vliet, "Natural Convection Local Heat Transfer on Constant-Heat-Flux Inclined Surfaces," *J. Heat Transfer*, vol. 91, Nov. 1969, pp. 511–516.

24. T. Fujii and H. Imura, "Natural Convection Heat Transfer from a Plate with Arbitrary Inclination," *Int. J. Heat Mass Transfer*, vol. 15, 1972, pp. 755–767.

25. R. J. Goldstein, E. M. Sparrow, and D. C. Jones, "Natural Convection Mass Transfer Adjacent to Horizontal Plates," *Int. J. Heat Mass Transfer*, vol. 16, 1973, pp. 1025–1035.

26. R. B. Kennard, "An Optical Method for Measuring Temperature Distribution and Convective Heat Transfer," *Bur. Stds. J. Research*, vol. 8, no. 6, June 1932, pp. 787–805.

27. S. W. Churchill and H. H. S. Chu, "Correlating Equations for Laminar and Turbulent Free Convection from a Horizontal Cylinder," *Int. J. Heat Mass Transfer*, vol. 18, 1975, pp. 1049–1053.

28. T. Yuge, "Experiments on Heat Transfer from Spheres Including Combined Natural and Forced Convection," *J. Heat Transfer*, vol. 82, Aug. 1960, pp. 214–220.

29. W. J. King, "The Basic Laws and Data of Heat Transmission," *Mechanical Engineering*, vol. 54, 1932, pp. 347–353.

30. W. Elenbaas, "Heat Dissipation of Parallel Plates by Free Convection," *Physica*, vol. 9, no. 1, 1942, pp. 1–28.

31. S. Ostrach, "Natural Convection in Enclosures," *Advances in Heat Transfer*, vol. 8, 1972, pp. 161–227.

32. I. Catton, "Natural Convection in Enclosures," *Proc. 6th Intl. Heat Transfer Conference*, Toronto, vol. 6, 1978, pp. 13–31.

33. D. B. Spalding and H. Afgan (Eds.), *Heat Transfer and Turbulent Buoyant Convection*, 2 vols., Hemisphere Publishing, Washington, D.C., 1977.

34. I. Catton, P. S. Ayyaswamy, and R. M. Clever, "Natural Convection Flow in a Finite, Rectangular Slot Arbitrarily Oriented with Respect to the Gravity Vector," *Int. J. Heat Mass Transfer*, vol. 17, 1974, pp. 173–184.

35. C. D. Mallinson and G. de vahl Davis, "Three-Dimensional Natural Con-

vection in a Box: A Numerical Study," *J. Fluid Mechanics*, vol. 83, part 1, 1977, pp. 1–31.

36. J. Hart, "Stability of the Flow in a Differentially Heated Inclined Box," *J. Fluid Mechanics*, vol. 47, 1971, pp. 547–576.
37. K. G. T. Hollands, T. E. Unny, G. D. Raithby, and L. Konicek, "Free Convective Heat Transfer Across Inclined Air Layers," *J. Heat Transfer*, vol. 98, May 1976, pp. 189–193.
38. G. D. Raithby and K. G. T. Hollands, "A General Method of Obtaining Approximate Solutions to Laminar and Turbulent Free Convection Problems," *Advances in Heat Transfer*, vol. 11, Academic Press, 1975, pp. 265–315.
39. F. Test, R. C. Lessmann, and A. Johary, "Heat Transfer During Wind Flow over Rectangular Bodies in the Natural Environment," *J. Heat Transfer*, vol. 103, May 1981, pp. 262–267.
40. H. E. Lewis et al., "Aerodynamics of the Human Microenvironment," *The Lancet*, 28 June 1969, pp. 1273–1277.
41. J. R. Lloyd and E. M. Sparrow, "Combined Forced and Free Convection Flow on Vertical Surfaces," *Int. J. Heat Mass Transfer*, vol. 13, 1970, pp. 434–438.
42. J. R. Kliegel, "Laminar Free and Forced Convection Heat Transfer from a Vertical Flat Plate," Ph.D. Thesis, Univ. of California, Berkeley, 1959.
43. P. H. Oosthuizen and S. Madan, "Combined Convective Heat Transfer from Horizontal Cylinders in Air," *J. Heat Transfer*, vol. 92, Feb. 1970, pp. 194–196.
44. B. Metais and E. R. G. Eckert, "Forced, Mixed, and Free Convection Regimes," *J. Heat Transfer*, vol. 86, May 1964, pp. 295–296.
45. C. K. Brown and W. H. Gauvin, "Combined Free and Forced Convection: I, Aiding Flow; II, Opposing Flow," *Canadian J. of Chemical Engineering*, Dec. 1965, pp. 306–318.
46. V. T. Morgan, "The Overall Convective Heat Transfer from Smooth Circular Cylinders," *Advances in Heat Transfer*, vol. 11, 1975, pp. 199–264.
47. *Heat Transfer 1978: Proceedings of the 6th Intl. Heat Transfer Conference*, Hemisphere Publishing, Washington, D.C., 1979.
48. K. E. Torrance and I. Catton (Eds.), "Natural Convection in Enclosures," ASME Symposium Vol. No. G00168, July 27–30, 1980.
49. J. H. Lienhard, "On the Commonality of Equations for Natural Convection from Immersed Bodies," *Int. J. Heat Mass Transfer*, vol. 16, 1973, p. 2121.
50. E. M. Sparrow and M. A. Ansari, "A Refutation of King's Rule for Multidimensional External Natural Convection," *Int. J. Heat Mass Transfer*, vol. 26, no. 9, 1983, pp. 1357–1364.
51. M. J. Chamberlain, K. G. T. Hollands, and G. D. Raithby, "Experiments

and Theory on Natural Convection Heat Transfer from Bodies of Complex Shape," *J. Heat Transfer,* vol. 107, Aug. 1985, pp. 624–629.

52. R. C. Chu, "Heat Transfer in Electronic Systems," *Heat Transfer 1986,* vol. 1, Hemisphere Publishing, New York, 1986, pp. 293–305.
53. S. Oktay, R. Hannemann, and A. Bar-Cohen, "High Heat from a Small Package," *Mechanical Engineering,* March 1986, pp. 36–42.
54. A. Bar-Cohen and W. M. Rohsenow, "Thermally Optimum Spacing of Vertical, Natural Convection Cooled Parallel Plates," *J. Heat Transfer,* vol. 106, Feb. 1984, pp. 116–123.
55. A. Bar-Cohen, "Fin Thickness for an Optimized Natural Convection Array of Rectangular Fins," *J. Heat Transfer,* vol. 101, Aug. 1979, pp. 564–566.
56. S. W. Churchill, "Combined Free and Forced Convection Around Immersed Bodies," Section 2.5.9 in *Heat Exchanger Design Handbook,* Hemisphere Publishing, New York, 1986.
57. S. W. Churchill, "A Comprehensive Correlating Equation for Laminar Assisting Forced and Free Convection," *AIChE Journal,* vol. 23, 1977, pp. 10–16.
58. T. S. Chen, B. F. Armaly, and N. Ramachandran, "Correlations for Laminar Mixed Convection Flows on Vertical, Inclined, and Horizontal Flat Plates," *J. Heat Transfer,* vol. 108, Nov. 1986, pp. 835–840.
59. B. F. Armaly, N. Ramachandran, and T. S. Chen, "Prediction of Turbulent Mixed Convection Along a Vertical Plate," *Heat Transfer 1986,* vol. 3, Hemisphere Publishing, New York, 1986, pp. 1445–1450.
60. R. Anderson and A. Bejan, "Natural Convection on Both Sides of a Vertical Wall Separating Fluids at Different Temperatures," *J. Heat Transfer,* vol. 102, 1980, pp. 630–635.

## Review Questions

1. What is the characteristic velocity scale for free convection?
2. Define and interpret the Grashof number. Is there more than one definition? Does it equal a ratio of physical effects?
3. What should the Nusselt number depend upon in free convection?
4. Define and interpret the Rayleigh number.
5. Can free convection develop in a gravity-free environment such as the inside of an orbiting space capsule?
6. What is "coupling"? Are free-convection flows *always* coupled and forced-convection flows *never* coupled? Explain.
7. What is a reasonably quantitative criterion for transition to turbulence in free convection?

8. Is there an analogy between (free) Grashof number and (forced) Reynolds number?
9. Is the Reynolds analogy valid for free and forced convection?
10. How is the Grashof number modified to correlate free convection with constant heat flux conditions?
11. What is the approximate criterion for "mixed" free and forced convection to occur?
12. Cite some differences between internal and external free convection.
13. How is a moderately inclined flat plate analyzed?
14. What is the "critical Rayleigh number" concept for tilted enclosures?
15. Why is there very little boundary layer separation in free convection?
16. At what Grashof number does free convection in a box enclosure become significantly stronger than a pure conduction condition?

## Problems

### Problem distribution by sections

**The Problem Assignments are Organized as Follows:**

| Problems | Section | Topics Covered |
|---|---|---|
| 7.1 –7.11 | 7.1 | General concepts |
| 7.12–7.34 | 7.2 | Vertical plates |
| 7.35–7.53 | 7.3 | Other external flows |
| 7.54–7.58 | 7.4 | Box enclosures |
| 7.59–7.63 | 7.5 | Mixed convection |
| 7.64–7.65 | All | Special Projects |

7.1 Prove that the thermal expansion coefficient of an ideal gas is exactly equal to the inverse absolute temperature.

7.2 Add data for liquid ammonia ($NH_3$) to Figs. 7.2 and 7.3 and comment on its potential to undergo free-convection effects.

7.3 From Fig. 7.2, water reaches its maximum density at about 4°C. Suppose that the vertical surface in Fig. 7.1(c) is submerged in water with $T_w$ = 8°C and $T_\infty$ = 2°C. Could the concept of thermal expansion coefficient be used in this case? Sketch how the velocity and temperature profiles might be modified in this case.

7.4 A 3-m-high vertical plate is at 40°C and is immersed in a fluid at 1 atm and 0°C. Estimate the Grashof number $Gr_L$ by both Eqs. (7.2) and (7.7) for (a) air, (b) water, (c) mercury.

7.5 A horizontal cylinder with a diameter of 3 in. and a surface temperature of 300°C is immersed in steam at 400°C and 1 atm. Estimate the Grashof number $Gr_D$. Suppose there is also forced crossflow past the cylinder. For what crossflow velocity would free and forced convection effects be comparable in magnitude?

7.6 Consider the special case of the free-convection vertical-plate momentum and energy equations (7.11b, c) for which the convective terms (the left-hand sides) are negligible. Solve for the velocity and temperature distributions assuming a finite boundary layer thickness $\delta$. What complication makes it impossible to solve for the numerical value of the Nusselt number at any position $x$?

7.7 Using dimensional analysis only, form a dimensionless group from the quantities heat flux $q''_w$, thermal expansion $\beta$, body length $L$, conductivity $k$, kinematic viscosity $\nu$, and gravity $g$, (a) without using $g$ and (b) using $g$. Comment on the results.

7.8 Nondimensionalize the vertical plate momentum equation (7.11b) for laminar flow with constant ($\rho$, $\beta$, $\mu$), using the dimensionless variables $x^* = x/L$, $y^* = y/L$, $T^* = (T - T_\infty)/(T_w - T_\infty)$, $u^* = u/u_0$, and $v/u_0$, where $u_0 = [g\beta(T_w - T_\infty)L]^{1/2}$. Comment on any dimensionless parameters that result.

7.9 Nondimensionalize the boundary layer energy equation (7.11c) for laminar flow with constant ($\rho$, $\beta$, $\mu$, $c_p$, $k$), using the same variables defined in Problem 7.8. Comment on parameters that arise.

7.10 Compute the Grashof number $Gr_L$ for buoyant meltwater flow up the vertical submerged side of an iceberg, if $L = 100$ m, $T_w = 0$°C freshwater, and $T_\infty = 8$°C seawater of 35 ppt salinity.

7.11 If transition to turbulence occurs at $Gr_x = 10^9$, and $T_w = 20$°C and $T_\infty = 0$°C, estimate the transition point $x$ for vertical free convection of (a) hydrogen, (b) liquid ammonia, (c) engine oil.

7.12 Find a cubic polynomial velocity profile approximation for vertical free convection by dropping the term $d\eta^4$ from Eq. (7.16) and dropping the requirement $\partial^2 u/\partial\eta^2 = 0$ at $\eta = 1$ from Eq. (7.17). This was the profile originally used by Eckert [8].

7.13 The following free-convection velocity profile shapes have been suggested to replace Eq. (7.18):

a) $u/u_0 = \eta - \eta^2$
b) $u/u_0 = \eta - \eta^3$
c) $u/u_0 = \eta e^{-6\eta}$
d) $u/u_0 = \sin(\pi\eta)$

With a sketch and a qualitative description, evaluate these shapes as to whether they are realistic approximations.

7.14 Use the following simplified velocity and temperature profiles for free convection on a vertical plate

$$u/u_0 \doteq \eta - 2\eta^2 + \eta^3,$$

$$\frac{T - T_\infty}{T_w - T_\infty} \doteq 1 - \eta$$

with the integral method of Section 7.2.1 to derive alternate formulas comparable to Eqs. (7.23)–(7.25).

7.15 Equally spaced vertical plates 8 in. high at 180°F are to be cooled in still air at 60°F. Estimate the minimum plate spacing to avoid interference between the boundary layers.

7.16 Figure P7.16 shows the velocity profiles on a vertical plate computed from an exact boundary layer theory by Ostrach [6]. Both coordinates are dimensionless. For the special case Pr = 0.01, compare Ostrach's results for (a) boundary layer thickness and (b) maximum velocity with the approximate integral results of Section 7.2.1. Also compare with Ostrach's computed heat transfer result $\text{Nu}_x = 0.057\text{Gr}_x^{1/4}$ for this case.

7.17 For the plates of Problem 7.15, assuming that their spacing is greater than the minimum, compute the total heat transfer rate from one plate if $b = 6$ in. Would it make any difference if the plates were rotated so that $b = 8$ in. and $L = 6$ in.? What would be the percent difference, and why?

7.18 A vertical plate is 30 cm wide with $T_w = 0$°C and is suspended in air at 40°C and 1 atm. What should be the vertical length $L$ of the plate if it is to dissipate 25 W of heat (a) from one side only or (b) from both sides together?

7.19 Using the concept of Grashof number being equivalent to Reynolds-number-squared, show that there is a direct analogy between the free-convection boundary layer thickness from Eq. (7.23) and the forced convection solution (6.13). Compare the two numerically if Pr = 1.

7.20 If, as an approximation, $\text{Nu}_x \propto \text{Gr}_x^n$, for what power $n$ does the heat transfer coefficient $h_x$ become constant? Is there any range of Grashof numbers for which this situation actually occurs in the Churchill-Chu general correlation, Eq. (7.35)?

7.21 Consider two identical plates of equal wall temperature placed such that one is in the free-convection wake of the other, as in Fig. P7.21. Will the upper plate have more or less heat transfer than the lower? Assuming laminar flow, make a simplified analysis to express the upper plate heat transfer as a fraction of the lower plate value.

7.22 A vertical plate 20 cm wide and 30 cm high is immersed in water at 20°C. What should be the plate wall temperature for the water to remove 300 W of heat (a) from one side or (b) from both sides together?

7.23 A copper plate at 70°C is 40 cm by 40 cm by 1 mm and hangs vertically

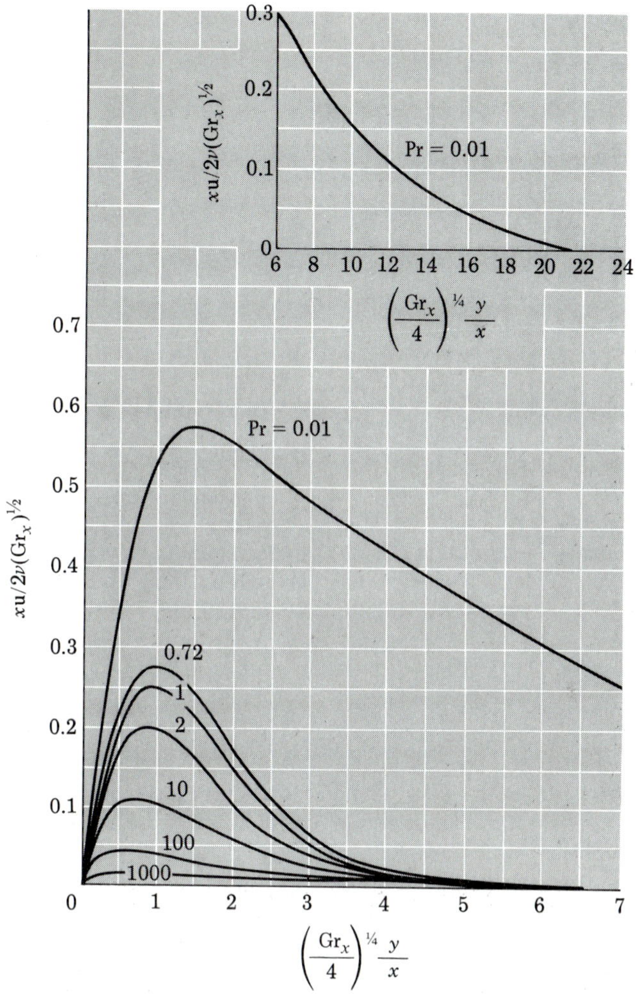


**Figure P7.16** After Ostrach [6].

in still air at 1 atm and 10°C. Estimate (a) the initial rate of heat loss and (b) the time for the plate to cool to 40°C. What complicates part (b)?

7.24 Repeat Problem 7.23 if the fluid is water.

7.25 A thin 5-cm-square plate hangs vertically in still air at 1 atm and 20°C. If electrically heated with 5 W of power approximating uniform-flux conditions, estimate the maximum plate wall temperature if heat is convected by (a) both sides and (b) only one side.

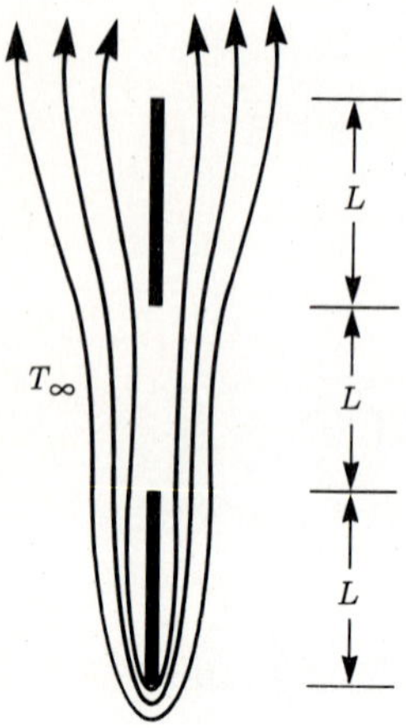


Figure P7.21

7.26 In Problems 7.23 and 7.24 the plate was lumpable, that is, uniform temperature $T_w(t)$, but the heat transfer coefficient was time-variable. In laminar free convection, formulas such as Eq. (7.28) suggest that $\bar{h} = C\ \Delta T^{1/4}$, where $C$ is a constant depending upon fluid properties and body size. Utilize this approximation to modify our lumped analysis, Eqs. (4.10) and (4.11), and show that the solution is of the form

$$\Delta T = T_w - T_\infty = \left(\frac{4}{Kt + 4/\Delta T_0^{1/4}}\right)^4,$$

where $K$ is another constant and $\Delta T_0$ is the initial temperature difference. Evaluate $K$ for the conditions of Problem 7.23 and repeat part (b) for that problem.

7.27 A 6-mm-thick yellow pine plywood wall 2 m wide and 1.5 m high separates two air enclosures at 40°C and 0°C, respectively. Combine free-convection and solid-conduction theory to estimate the heat loss through the wall in watts.

7.28 The free convection flow in Problem 7.27 is called a *conjugate boundary layer,* as illustrated in Fig. P7.28. The warm side of the plate is actually a cold wall, with downward convection, while the cold-air side represents hot-wall upward flow. The plate temperatures "float" between the two flows and vary with distance $x$ along the plate [60]. Make a simplified analysis based on constant plate surface temperatures and derive a formula for the total heat transfer between the hot and cold air.

7.29 A 50-cm-square plate hangs vertically in still fluid at 1 atm and 20°C. Estimate the required constant wall temperature if one side of the plate is to dissipate 140 W of heat when the fluid is (a) air and (b) water.

7.30 For the vertical iceberg surface of Problem 7.10, estimate the average

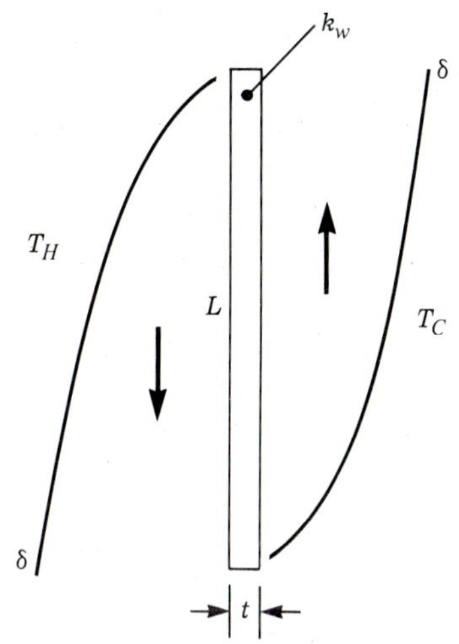


Figure P7.28

ice melting rate in m/day if the uniform composition formula (7.35) is valid and the latent heat of melting is 334 kJ/kg.

7.31 A fluid bath at 20°C is to be heated by immersing a vertical 20-cm-square thin electrically heated plate in the bath. What is the maximum electric power allowed if $T_w$ is not to exceed 120°C when the fluid is (a) engine oil and (b) mercury.

7.32 The answer to Problem 7.12 is $u/u_0 \doteq \eta - 2\eta^2 + \eta^3$. Consider this as a reward if you read ahead this far before attempting the problem. Now use this profile and $T(\eta)$ from Eq. (7.19) to repeat the analysis of Section 7.1.2 and show that the resulting Nusselt number is $\mathrm{Nu}_x \doteq \mathrm{Pr}^{1/2}\mathrm{Gr}_x^{1/4}/[15(\mathrm{Pr} + 20/21)]^{1/4}$. Compare with our version, Eq. (7.25). The above result is from Eckert [8].

7.33 A 22-cm-square circuit board, insulated on one side, contains 100 equally spaced computer chips, each of which dissipates 0.1 W of heat. When exposed to air at 20°C, as in Fig. P7.33, what will be the maximum chip surface temperature? What is the total board heat dissipation allowed if chip temperature is not to exceed 80°C?

7.34 For a vertical plate with uniform heat flux $q''_w$ an integral analysis is possible using the approximate profiles

$$u \doteq u_0\eta(1 - \eta)^2, \qquad \eta = y/\delta$$
$$T - T_\infty \doteq (T_w - T_\infty)(1 - \eta)^2,$$

as suggested by E. M. Sparrow (NACA TN 3508, July 1955). Show that substitution of these profiles into the integral momentum and energy

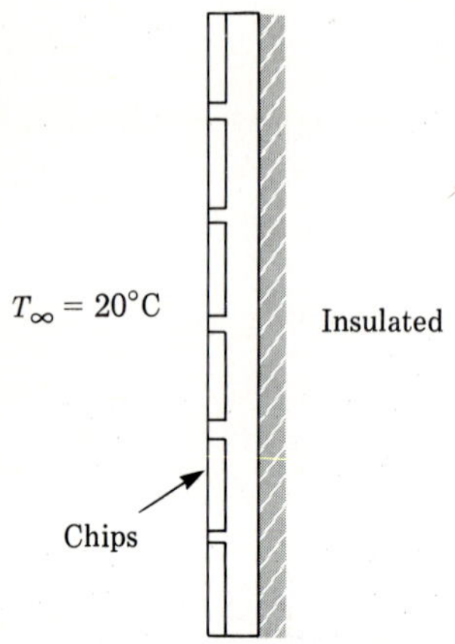


**Figure P7.33**

equations and assuming $u_0 = C_2x^n$ and $\delta = C_1x^m$ yield the result

$$\mathrm{Nu}_x \doteq \left(\frac{4}{45 + 36/\mathrm{Pr}}\right)^{1/5} \mathrm{Ra}_x^{1/5}.$$

Then use this result to derive a formula for the wall temperature variation $T_w(x)$ of the plate.

7.35 Repeat Example 7.6 if the cylinder is covered with a ¼-in. thickness of insulation with $k = 0.03$ Btu/hr · ft · °F.

7.36 In Problem 7.35 find the insulation thickness that will reduce the total heat loss to 5900 Btu/hr (10% of Example 7.6 loss). What nonlinear effect complicates an accurate analysis?

7.37 A horizontal heating element, suitable for an electric stove, is 1 cm in diameter and 50 cm long. Estimate its equilibrium surface temperature in still 20°C air when operating at its rated power of 100 W. Include radiation with $\varepsilon$ equal to (a) 0; (b) 0.8.

7.38 An insulated steam pipe is 12 cm in diameter and has a surface temperature of 70°C. It cuts through a cold room where the air temperature is 0°C, proceeding horizontally for 3 m and then vertically for 3 m. How much heat in watts is lost from the horizontal and vertical legs? If they are not comparable, explain why.

7.39 A horizontal 40 W fluorescent tube is 3.8 cm in diameter and 120 cm long and stands in still air at 1 atm and 20°C. If the equilibrium surface temperature is 40°C and radiation is neglected, what percent power is being dissipated by convection?

7.40 A square duct that is 15 cm on a side has a surface temperature of 50°C and is to be immersed in argon at 0°C. Three different orientations from the horizontal are shown in Fig. P7.40. Estimate the free convection heat

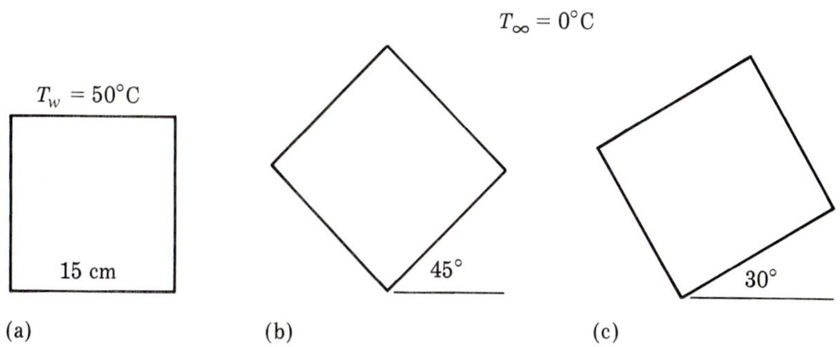


**Figure P7.40**

loss per meter of duct for these three cases and compare with an estimate based on Lienhard's "particle-travel" concept in Section 7.3.6.

7.41 Following up on Example 7.7, suppose the outer surface is at 10°C also and is undergoing free convection to a still atmosphere. Estimate the atmospheric temperature and comment on the result.

7.42 An immersion heater of the type used to heat water in a coffee cup is 1 cm in diameter, 10 cm long, and rated at 150 W. If the heater is horizontal and is immersed in a cup of water at 60°C, estimate the heater surface temperature.

7.43 The manufacturer of the immersion heater in Problem 7.42 says if you accidentally turn it on in still room air it will be Melt City. Estimate the heater temperature under these conditions.

7.44 A hockey rink is 200 ft by 85 ft and the ice surface is maintained at 20°F. If the arena air is still at 65°F and 1 atm, estimate the power required to keep the ice cool, in kW.

7.45 On a clear, still day, solar heating at a rate of 700 W/m$^2$ can make a dark absorbent roof hot enough to fry eggs, as they say. Consider the roof in Fig. P7.45, which is slanted at 45° and has a depth of 5 m into the paper. If convection losses are entirely by free convection, what is the maximum possible roof temperature if the ambient air is at 20°C?

7.46 A light bulb, approximated as a 7-cm-diameter sphere, has a surface temperature of 500°C when immersed in air at 20°C. If the surface emissivity is 0.6, estimate the total heat dissipated owing to both free convection and radiation.

7.47 Following up on Problem 6.100, suppose Newton runs outside with his hot sphere and there is no wind whatever. How long will it take the sphere to cool to 200°C under free-convection conditions?

7.48 Steam at 175°C flows through a thin, bare copper tube of diameter 2.54 cm and length 6 m. The heat transfer coefficient on the steam side is 95

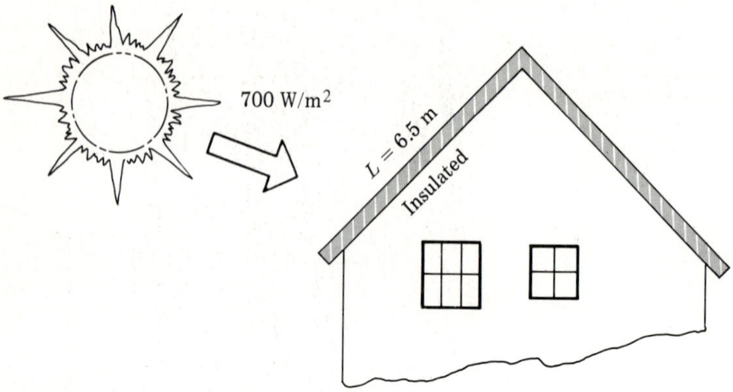


Figure P7.45

W/(m$^2 \cdot$ K). The outside is exposed to still air at 20°C. What will be the total heat loss of the steam and its temperature drop over the tube length if the flow rate is 25 kg/hr?

7.49 An electric hot plate is a 6-in. horizontal disk, immersed in still air at 1 atm and 60°F. Estimate the maximum allowable power in watts if the surface is not to exceed 450°F.

7.50 A sheet-metal horizontal heating duct is 32 cm wide and 20 cm high and passes through a cellar where air temperature is 10°C. If the duct surface is at 50°C and the duct length is 15 m, estimate the total heat loss.

7.51 For free convection on an array of vertical plates, the Bar-Cohen and Rohsenow optimization, Eqs. (7.43–7.44), was obtained from experimental data. Lacking such data, we might estimate the optimum condition to occur when the plate boundary layers exactly meet at the top, as in Fig. P7.51. Assuming laminar flow with negligible plate thickness, use Fig. P7.51 to compute the plate spacing, the Nusselt number $\overline{\mathrm{Nu}}_S$, and the total array heat transfer and compare with Section 7.3.7.

7.52 An electronic component dissipates heat through free convection on an array of equally spaced plates as in Fig. 7.8, with $L = 2$ cm, $b = 4$ cm, and a total width of 6 cm. If the plates are at 100°C and the air is at 20°C, estimate (a) optimum plate spacing, $S$; (b) total number of plates; and (c) total heat dissipated. Assume that $b/L$ is large so that Fig. 7.8 is valid. Neglect plate thickness.

7.53 A domestic baseboard heater consists of a small hot-water pipe passing through the center of equally spaced 6-in.-square vertical plate fins. If the plates are at 120°F and the air is at 60°F, compute (a) the optimum plate spacing; (b) the optimum number of plates per foot; and (c) the heat flux per foot of pipe. The plate thickness is 0.05 inch.

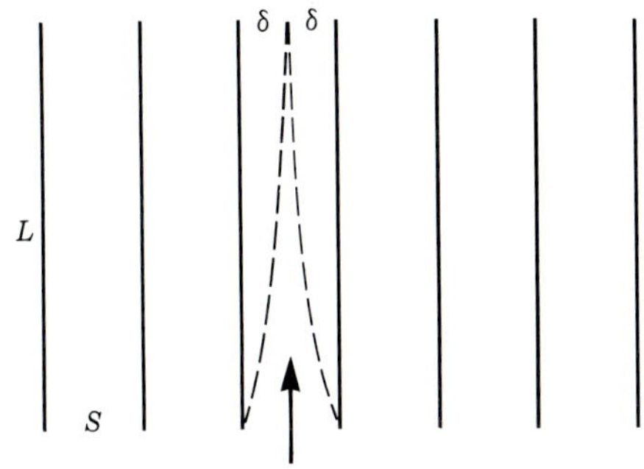


Figure P7.51

7.54 Repeat Example 7.7 if the enclosure spacing $L = 3$ cm. What is the percent increase in heat loss compared to $L = 8$ cm?

7.55 A horizontal enclosure is 1 m square, with a height or "gap" of 3 cm. The upper surface is at 20°C and the gap is filled with air. How hot must the lower surface be before convective motion ensues? What is the heat flux if $T_{lower} = 40°C$?

7.56 A storm window consists of two 80-cm-square vertical glass panes 5 mm thick separated by a 10-cm air gap. Inside room air is still at 65°C, and outside air is still at 0°C. Estimate the heat loss through the window. What percent resistance is the air gap compared to the total resistance of the system?

7.57 A vertical enclosure is formed by two plates 50 cm high and 40 cm wide, separated by a 5-cm gap. The plates are at 60°C and 20°C, respectively. If the gap is filled with air at 1 atm, estimate the heat flux between the plates.

7.58 The enclosure of Problem 7.57 will approach a pure-conduction condition if the air pressure is reduced. Estimate the maximum air pressure at which pure conduction ($\overline{\mathrm{Nu}}_L = 1$) occurs.

7.59 For opposing mixed convection past a vertical plate, using laminar boundary theory, find the ratio $\mathrm{Gr}_x/\mathrm{Re}_x^2$ for which the heat transfer becomes negligibly small, for Prandtl number equal to (a) 0.03; (b) 0.71; and (c) 1000. If we then reverse the flow to be *assisting,* what percent increase will we achieve over the nominal free-convection heat loss?

7.60 A thin vertical plate 30 cm high and 20 cm wide is immersed in a channel containing mercury at 40°C flowing gently parallel to the plate at $U =$ 2 cm/s. If the plate temperature is 10°C, estimate the heat transfer to the plate for (a) assisting flow and (b) opposing flow. For opposing flow, for what velocity $U$ will the heat transfer nearly vanish?

7.61 A hot-wire anemometer is a thin wire mounted horizontally between two needles as in Fig. P7.61. It measures forced-flow velocity $U$ by monitoring the wire heat loss and correlating with a forced-convection formula such

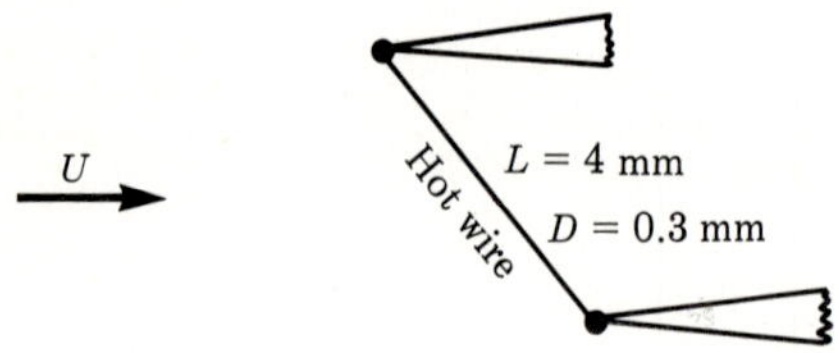


Figure P7.61

as Eq. (6.115) [7, p. 214]. Suppose that the fluid in Fig. P7.61 is air at 20°C and the wire is heated to $T_w$ = 235°C. For what flow velocity $U$ in cm/s will free- and forced-convection effects be comparable? What will be the wire heat loss at $U$ = 20 cm/s?

7.62 A 1-mm-diameter carbon steel sphere falls at terminal velocity through engine oil at 10°C. The sphere surface is at 30°C. Is this a mixed-convection problem? What is the heat loss from the sphere in milliwatts?

7.63 Water at 10°C flows upward at 35 cm/s past a thin vertical plate 50 cm wide and 30 cm high. The plate temperature is 90°C. Is this a mixed-convection flow? Estimate the total heat loss.

## Special Projects

7.64 Reynolds, Prandtl, Rayleigh, Nusselt, and Mach were all famous engineers or scientists who contributed greatly to heat transfer and fluid mechanics. They deserved to have dimensionless numbers named after them. But who was Grashof? Was he deserving? Report to the class on Grashof: The Man and the Myth. A good place to start is *Int. J. Heat and Mass Transfer,* vol. 15, 1972, pp. 562–563.

7.65 Square tiles, 15 by 15 by 1 cm, with physical properties approximating common brick, are to be cooled in air at 20°C until their center plane is at 150°C, as in Fig. P7.65. The tiles are initially at a uniform temperature of 400°C. It is desired to move these tiles along on a trolley through a distance no more than 30 m as part of a production line. What is the maximum trolley velocity $U$ allowable? Will the trolley motion cause a mixed-convection condition?

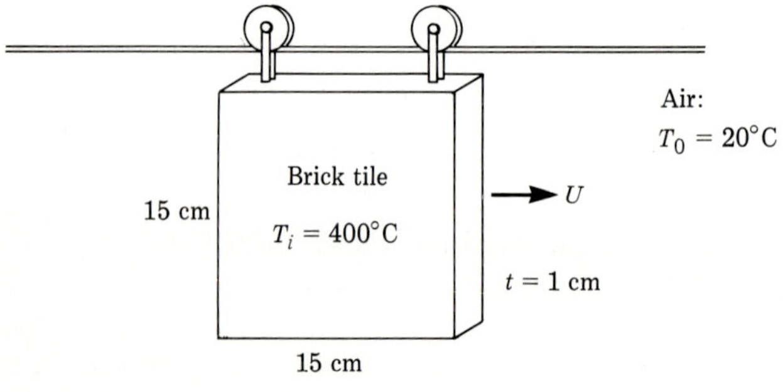


Figure P7.65

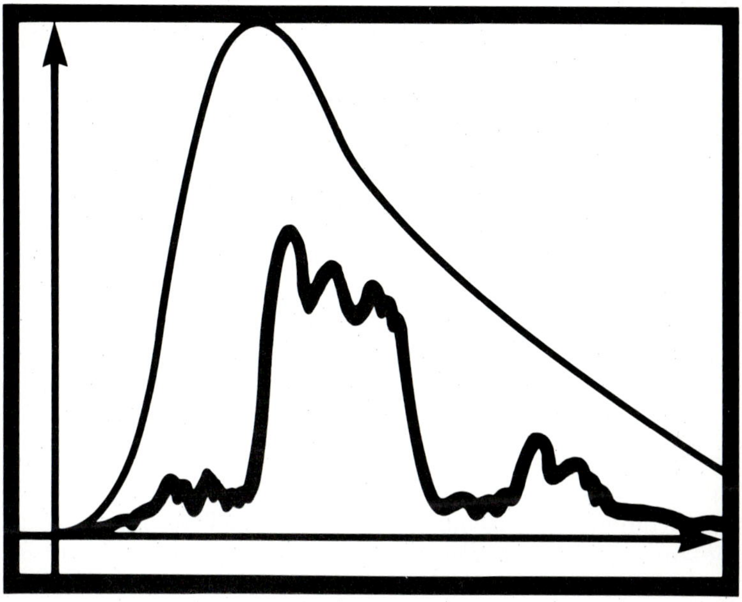

# Radiation

# 8

## 8.1 Introduction to Thermal Radiation Physics

In the preceding chapters we have been primarily concerned with the conduction/convection mode of heat transfer, in which the heat flux is caused by local temperature gradients in a solid or a fluid. Now we consider the details of the second mode, *thermal radiation,* which is not directly related to temperature gradients. Rather, thermal radiation is simply *emitted* by atomic excitation of any substance, and the radiant energy travels at the speed of light, even through a vacuum, until it strikes another substance where it may be absorbed, reflected, transmitted, or scattered.

Thus, unlike conduction and convection, radiant heat transfer does not require an intervening medium, and its properties can be described both by the classical electromagnetic wave theory and by the quantum theory of photons. Thermal radiation differs from other electromagnetic waves only in the mechanism of formation: Thermal waves are caused by temperature excitation, X-rays by electron bombardment of a metal, gamma rays by nuclear reactions, radio waves by excitation of crystal substances, and so on. All travel at the speed of light $c$ and all are characterized by a *spectrum* of wavelengths $\lambda$ and frequencies $\nu$ related by

$$c = \lambda \nu. \tag{8.1}$$

Light travels at its maximum speed in a vacuum, where $c = c_0 = 2.9979 \times 10^8$ m/s. When traveling through a transparent medium such as gases or water or glass or clear plastic, light travels at the speed

$$c = c_0/n, \tag{8.2}$$

where $n \geqslant 1$ is the *index of refraction* of the medium. Our primary concern here is with gases, where $n \doteq 1$, $c \doteq c_0$.

### 8.1.1 The Electromagnetic Wavelength Spectrum

Electromagnetic waves occur over an enormous range of wavelengths, from $10^{-16}$ m (cosmic rays) to $10^6$ m (electric power waves). A schematic of the complete electromagnetic wave *spectrum* is shown in Fig. 8.1. Thermal radiation is in the range $\lambda = 10^{-7}$ to $10^{-3}$ m, which is more conveniently described by the small length unit called the *micron:*

$$1 \text{ micron} = 1\ \mu\text{m} = 10^{-6}\ \text{m} = 10^4\ \text{Å}. \tag{8.3}$$

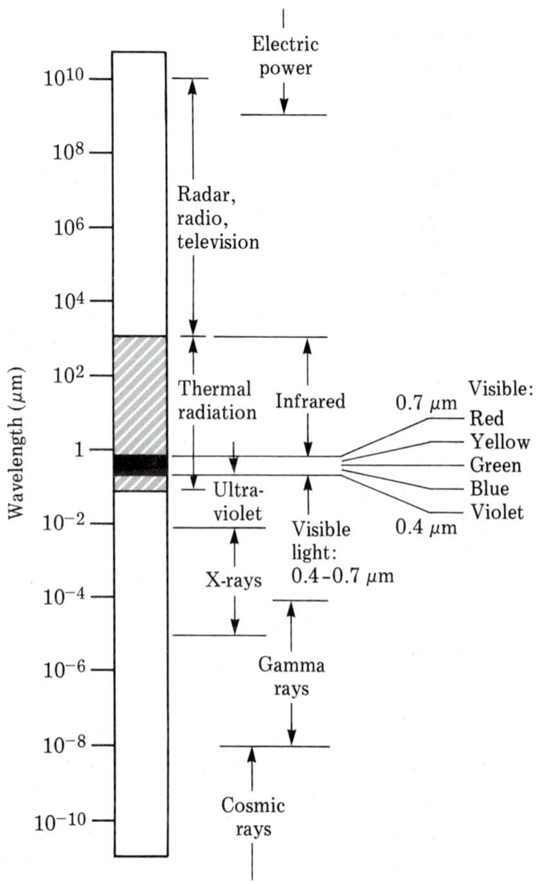


**Figure 8.1 The electromagnetic wave spectrum.**

In Fig. 8.1, then, the crosshatched thermal radiation region extends over wavelengths from 0.1 $\mu$m to 1000 $\mu$m. Outside of this region thermal radiation exists but usually has negligible energy.

Within the thermal radiation region in Fig. 8.1 is a thin black region, $0.4 \leq \lambda \leq 0.7$ $\mu$m, of *visible radiation,* where the human eye can detect radiation as colors of light. As we shall see, the eye can pick up radiation only from rather hot sources, $T > 800$K, such as the sun, lamp filaments, or molten metals. The "colors" we see on ordinary objects are actually reflected radiation from these hot sources. Colder bodies, $T < 800$K, emit radiation in the *infrared* range, $1 \leq \lambda \leq 1000$ $\mu$m, which can be detected only by special optical means.

The fact that the visible range is so narrow leads to some misleading beliefs about the effectiveness of a radiating surface, black surfaces being thought of as "good" and white as "poor." This is true only in the visible range: Both white and black surfaces are excellent radiation emitters and absorbers in the infrared range, where many practical engineering problems occur.

We now present some of the details of the physics and engineering analysis of thermal radiation. Radiation is a rich subject, and no single elementary chapter can do it full justice. There are at least six modern engineering textbooks devoted entirely to radiation heat transfer [1–6]. The reader is referred to these books for further details and more advanced discussions.

### 8.1.2 Planck's Blackbody Spectral Energy Distribution

The radiant energy emitted by any body at any temperature varies continuously over a range of wavelengths. We did not discuss this fact in Chapter 1, when we made preliminary statements about the radiation mode of heat transfer. Rather, we simply introduced an expression for integrated or "total" energy emitted by a radiating surface:

$$q''_{\text{rad}} = \varepsilon \sigma T^4, \tag{8.4a}$$

where $\sigma = 5.67 \times 10^{-8}\ \text{W}/(\text{m}^2 \cdot \text{K}^4)$ is the Stefan-Boltzmann constant and $\varepsilon$ is the dimensionless *emissivity,* which for a blackbody equals unity and for a real surface is a fraction that depends upon temperature and surface parameters: roughness, texture, color, material, degree of oxidation, and whether painted or otherwise coated. It is imperative that *absolute* temperature $T$ be used in Eq. (8.4a).

A *blackbody* is a perfect emitter and perfect absorber of radiant energy. It has unit emissivity, independent of wavelength or temperature or direction, and it emits the maximum possible energy at a given temperature. We denote total blackbody radiant energy by the symbol $E_b$:

$$E_b = q''_{\text{rad}}(\varepsilon = 1) = \sigma T^4. \tag{8.4}$$

A blackbody also absorbs all incoming thermal radiation, no matter what wavelength or temperature. It is thus a standard against which a real radiating surface can be compared, much like the Carnot cycle in thermodynamics.

Equation (8.4) was found experimentally by J. Stefan in 1879 and can be predicted by classical thermodynamics, as first shown by L.

Boltzmann in 1884. However, the spectral or wavelength distribution of blackbody radiant energy, $E_{b\lambda} = dE_b/d\lambda$, cannot be predicted by thermodynamics, yet it is extremely important for estimating real surface characteristics. Although it had been known since the end of the eighteenth century that radiant energy had a wavelength content and very accurate spectral data had been obtained by the end of the nineteenth century, it remained for Max Planck to complete the theory of blackbody radiation in a remarkable paper in 1901 [7]. Planck modeled a blackbody as a set of dipole oscillators whose energy, at a given frequency $\nu$, could only be a multiple of a discrete or "quantum" energy level $h\nu$, where $h$ is now called *Planck's constant.* After statistically averaging the oscillator energies, Planck arrived at the following famous expression for the wavelength energy spectrum or "emissive power" of a blackbody:

$$E_{b\lambda} = dE_b/d\lambda = \frac{C_1}{\lambda^5[\exp(C_2/\lambda T) - 1]}, \tag{8.5}$$

where $C_1$ and $C_2$ are constants that are related to the speed of light $c_0$, Planck's constant $h$, and Boltzmann's constant $\kappa$, as follows:

$$\begin{aligned} C_1 &= 2\pi h c_0^2 = 3.7415 \times 10^{-16}\ \mathrm{W \cdot m^2}, \\ C_2 &= h c_0/\kappa = 1.4388 \times 10^{-2}\ \mathrm{m \cdot K}. \end{aligned} \tag{8.6}$$

A complete list of these fundamental constants is given in Table 8.1. In deriving Eq. (8.5), Planck of course had thereby invented quantum

**Table 8.1** Fundamental radiation constants

| |
|---|
| 1. *Speed of light in a vacuum:* |
| $c_0 = 2.9979 \times 10^8$ m/s $= 9.8356 \times 10^8$ ft/s |
| 2. *Planck's constant:* |
| $h = 6.6257 \times 10^{-34}$ J · s $= 1.7444 \times 10^{-40}$ Btu · hr |
| 3. *Boltzmann's constant:* |
| $\kappa = 1.3805 \times 10^{-23}$ J/K $= 7.2694 \times 10^{-27}$ Btu/°R |
| 4. *Stefan-Boltzmann constant:* |
| $\sigma = 5.6696 \times 10^{-8}$ W/m² · K⁴ $= 1.7121 \times 10^{-9}$ Btu/hr · ft² · °R⁴ |
| 5. *Blackbody constants:* |
| $C_1 = 3.7415 \times 10^{-16}$ W · m² $= 1.3742 \times 10^{-14}$ Btu · ft²/hr |
| $C_2 = 1.4388 \times 10^{-2}$ m · K $= 8.4969 \times 10^{-2}$ ft · °R |
| $C_3 = 2897.8$ μm · K $= 5216.1$ μm · °R |

theory, which is the foundation of modern physics. A more complete description is in Planck's book [8].

The blackbody spectrum (8.5) is plotted versus λ for various body temperatures in Fig. 8.2. By inspection of the curves we immediately see the following characteristics:

1. The blackbody spectrum is continuous and skewed to the right, that is, a majority (about 75%) of the emitted energy lies to the right of the maximum. (Note the log scale on the abscissa.)

**Figure 8.2** Blackbody spectral emissive power for various surface temperatures, from Eq. (8.5).

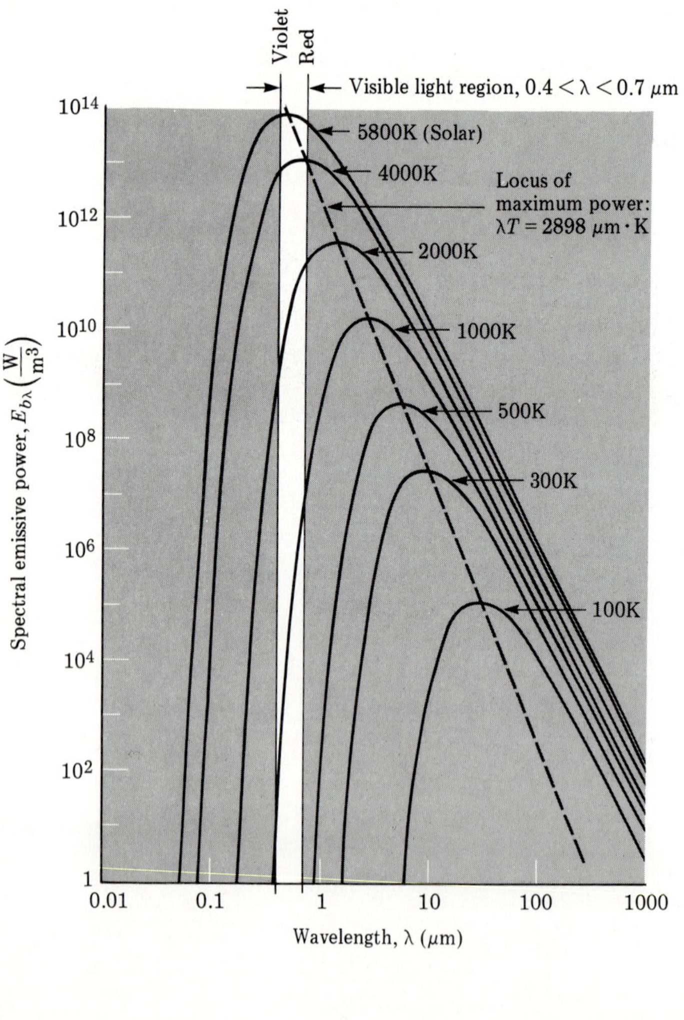

2. At all wavelengths the emissive power increases with increasing temperature.
3. As temperature increases, the emissive power becomes concentrated at shorter and shorter wavelengths.
4. Solar radiation ($T \doteq 5800$K) has its peak power in the middle of the visible light range. In contrast, for $T < 700$K, the emissive power is infrared and totally outside the visible region.
5. The maximum emissive power increases as $T^5$. However, the total emitted radiation (area under the curves) increases only as $T^4$. These two facts are probably not obvious from the curves.

### 8.1.3 Wien's Displacement Law

For any constant temperature, we may find the point of maximum emissive power by differentiating Eq. (8.5) with respect to $\lambda$ and setting the result equal to zero. We obtain

$$(\lambda T)_{\text{max power}} = C_3 = 2897.8\ \mu\text{m} \cdot \text{K}. \tag{8.7}$$

This locus of peaks, called *Wien's displacement law,* is plotted as a dashed line in Fig. 8.2. It was derived, using classical thermodynamics, by W. Wien in 1894, seven years before Planck gave the general spectrum analysis. For this we see that the solar radiation peak occurs at $\lambda = 2898/5800 = 0.5\ \mu$m, or near the middle of the visible range.

By substituting Eq. (8.7) into (8.5) we find that the maximum blackbody emissive power is given by

$$E_{b\lambda,\,\text{max}} = (1.2865 \times 10^{-5}\ \text{W/m}^3 \cdot \text{K}^5)T^5. \tag{8.8}$$

This verifies our contention in item 5 of Section 8.1.2.

The human eye begins to detect color from radiating bodies at an emissive power of about $10^6$ W/m$^3$. For red wavelengths, $\lambda \doteq 0.7\ \mu$m, this occurs when $T \doteq 950$K (1250°F), for which a blackbody would seem to glow dull red. For violet wavelengths, $\lambda \doteq 0.4\ \mu$m, $E_{b\lambda} = 10^6$ W/m$^3$ corresponds to $T \doteq 1500$K (2200°F), which is enough visible radiation to appear nearly "white" to the eye. Even though seemingly white hot, a 1500 K body has its emissive power peak at $\lambda \doteq 1.9\ \mu$m, far out in the infrared range.

### 8.1.4 The Stefan-Boltzmann Law

By definition in Eq. (8.5), the total radiant energy emitted by a blackbody is equal to the integral of $E_{b\lambda}$ with respect to $\lambda$ at constant temperature:

$$E_b = \int_0^\infty E_{b\lambda}(\lambda, T)d\lambda. \tag{8.9}$$

When $E_{b\lambda}$ is introduced from Eq. (8.5), the integral is exact and the result is

$$E_b = \sigma T^4, \qquad \sigma = \pi^4 C_1/15C_2^4 = 5.6696 \times 10^{-8}\ \mathrm{W/m^2 \cdot K^4}. \tag{8.10}$$

This is seen to be the classical Stefan-Boltzmann law (8.4) for blackbody total radiation. The Stefan-Boltzmann constant $\sigma$ is a combination of the spectrum constants $C_1$ and $C_2$. In this chapter we will use the notation $E_b$ rather than $q''_{\mathrm{rad}}$ to denote blackbody radiation.

The fact that $E_b$ increases as $T^4$ means of course that radiation becomes increasingly important at high temperatures and usually dominates over conduction and convection when the temperature is greater than about 1000K. For example, the equivalent radiation heat transfer coefficient, $h_r = E_b/\Delta T$, exceeds 300 W/m$^2 \cdot$ K at $T$ = 1100K.

### Example 8.1

At a wavelength of 0.7 $\mu$m, for what temperature is the blackbody emissive power equal to $10^8$ W/m$^3$? What is the total emitted energy at this temperature?

*Solution* To find the temperature, we set $E_{b\lambda} = 10^8$ W/m$^3$ and $\lambda = 0.7$ $\mu$m in the Planck spectrum relation (8.5):

$$10^8 = \frac{3.7415 \times 10^{-16}}{(0.7 \times 10^{-6})^5\{\exp[0.014388/(0.7 \times 10^{-6})T] - 1\}},$$

or

$$\exp(20{,}554/T) - 1 = 2.2262 \times 10^7.$$

Solve for

$$T = 1215\mathrm{K}. \quad [Ans.]$$

The blackbody total emitted energy at this temperature is

$$E_b = \sigma T^4 = (5.6696 \times 10^{-8})(1215)^4 = 124{,}000\ \mathrm{W/m^2}. \quad [Ans.]$$ ■

### 8.1.5 Emission in a Finite-Wavelength Band

Since the emission characteristics of real surfaces often vary with wavelength, it is useful to know the blackbody energy contained within a finite-wavelength band, say between $\lambda_1$ and $\lambda_2$. To do this, we first notice that Planck's spectrum (8.5) may be rearranged so that $E_{b\lambda}/T^5$ is a function of the single variable $\lambda T$:

$$E_{b\lambda}/T^5 = \frac{C_1}{\xi^5[\exp(C_2/\xi) - 1]}, \qquad \xi = \lambda T. \tag{8.11}$$

Note, however, that $E_{b\lambda}/T^5$ is not dimensionless.

Now the total emitted blackbody energy between 0 and $\lambda T$ is the integral of $E_{b\lambda}$ over that range:

$$\delta E_b(0 \to \lambda T) = \int_0^{\lambda T} E_{b\lambda}(\lambda, T)d\lambda,$$

and we have already seen that $\delta E_b(0 \to \infty) = \sigma T^4$ in Eq. (8.10). The ratio of these two is called the *fractional function,* that is, the fraction of total energy between zero and $(\lambda T)$:

$$f_e(\lambda, T) = \delta E_b(0 \to \lambda T)/(\sigma T^4). \tag{8.12}$$

By introducing $E_b$ from Eq. (8.11) and rewriting the integral, we find that $f_e$ also is a function only of $(\lambda T)$:

$$f_e(\lambda T) = \frac{15}{\pi^4}\int_\zeta^\infty \frac{\zeta^3 d\zeta}{\exp(\zeta) - 1}, \qquad \zeta = C_2/\lambda T.$$

Unfortunately, the integral is not known in closed form. But it may be evaluated as a rapidly converging series [9]:

$$f_e(\lambda T) = \frac{15}{\pi^4}\sum_{i=1}^{\infty} i^{-4} e^{-i\zeta}[(i\zeta)^3 + 3(i\zeta)^2 + 6(i\zeta) + 6], \tag{8.13}$$

where again $\zeta = C_2/\lambda T$. The series converges to four-decimal-place accuracy in six terms if $\zeta$ is unity or greater. The series is appropriate for entering in a programmable handheld calculator.

The emissive power from Eq. (8.11) and the fractional energy function from Eq. (8.13) are tabulated versus $\lambda T$ in Table 8.2 and plotted in Fig. 8.3. The linear scales in Fig. 8.3 give a clearer picture of the right-skewed power distribution than do the log-log scales in Fig. 8.2. Only 25% of the energy is contained in the left-side range from zero to the maximum-power wavelength. The 50% point is at $\lambda T = 0.004107$ m · K, far to the right of maximum power.

The energy emitted in any bandwidth from $\lambda_1$ to $\lambda_2$, at any temperature $T$, is computed by subtracting the fractional functions at

**Table 8.2** Blackbody radiation functions

| $\lambda T$ – m · K | Spectral Emissive Power: $E_{b\lambda}/T^5$ – W/m$^3$ · K$^5$ × $10^5$ | Fractional Energy Function $f_e$ |
|---|---|---|
| 0.0010 | 0.021107 | 0.000321 |
| 0.0015 | 0.336408 | 0.012849 |
| 0.0020 | 0.878836 | 0.066725 |
| 0.0025 | 1.216947 | 0.161347 |
| 0.002898** | 1.286495** | 0.250050 |
| 0.0030 | 1.282823 | 0.273218 |
| 0.0035 | 1.187304 | 0.382896 |
| 0.0040 | 1.029575 | 0.480852 |
| 0.004107*** | 0.993554 | 0.500000*** |
| 0.0045 | 0.864017 | 0.564292 |
| 0.0050 | 0.713877 | 0.633715 |
| 0.0055 | 0.586251 | 0.690873 |
| 0.0060 | 0.481105 | 0.737780 |
| 0.0065 | 0.395756 | 0.776312 |
| 0.0070 | 0.326887 | 0.808067 |
| 0.0075 | 0.271369 | 0.834360 |
| 0.0080 | 0.226525 | 0.856245 |
| 0.0085 | 0.190169 | 0.874563 |
| 0.0090 | 0.160557 | 0.889984 |
| 0.0095 | 0.136312 | 0.903039 |
| 0.0100 | 0.116354 | 0.914152 |
| 0.0110 | 0.086083 | 0.931841 |
| 0.0120 | 0.064901 | 0.945050 |
| 0.0130 | 0.049773 | 0.955090 |
| 0.0140 | 0.038763 | 0.962849 |
| 0.0150 | 0.030610 | 0.968932 |
| 0.0160 | 0.024477 | 0.973765 |
| 0.0170 | 0.019796 | 0.977651 |
| 0.0180 | 0.016176 | 0.980811 |
| 0.0190 | 0.013343 | 0.983404 |
| 0.0200 | 0.011102 | 0.985552 |
| 0.0250 | 0.004924 | 0.992164 |
| 0.0300 | 0.002502 | 0.995289 |
| 0.0350 | 0.001401 | 0.996952 |
| 0.0400 | 0.000844 | 0.997916 |
| 0.0450 | 0.000538 | 0.998512 |
| 0.0500 | 0.000359 | 0.998901 |

**Maximum emissive power

***50% fractional energy

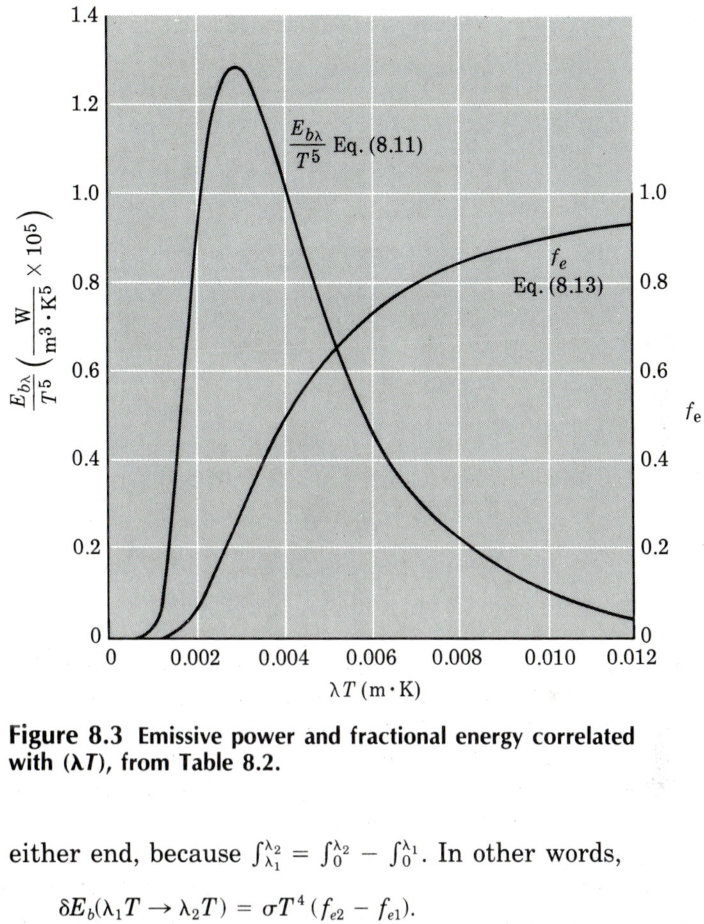


**Figure 8.3** **Emissive power and fractional energy correlated with ($\lambda T$), from Table 8.2.**

either end, because $\int_{\lambda_1}^{\lambda_2} = \int_0^{\lambda_2} - \int_0^{\lambda_1}$. In other words,

$$\delta E_b(\lambda_1 T \rightarrow \lambda_2 T) = \sigma T^4 (f_{e2} - f_{e1}). \qquad \textbf{(8.14)}$$

This artifice enables us to make many bandwidth calculations from a single table. More extensive tables are in [9–11].

## Example 8.2

The sun approximates a blackbody with a surface temperature of 5762K. Estimate the percentage of solar energy (a) in the visible range and at wavelengths (b) shorter than and (c) longer than the visible range.

*Solution* Assume the visible range is between $\lambda_1 = 0.4$ and $\lambda_2 = 0.7\ \mu$m. Then the appropriate values of ($\lambda T$) are

$$\lambda_1 T = (0.4 \times 10^{-6})(5762) = 0.00231 \text{ m} \cdot \text{K},$$

$$\lambda_2 T = (0.7 \times 10^{-6})(5762) = 0.00403 \text{ m} \cdot \text{K}.$$

By evaluating Eq. (8.13), or interpolating in Table 8.2, we find that

$$f_{e1} \doteq 0.123, \qquad f_{e2} \doteq 0.486.$$

Then the fraction of solar energy contained in the visible range is

$$f_{e2} - f_{e1} = 0.486 - 0.123 = 0.363 = 36.3\%. \quad [Ans. (a)]$$

The fraction of energy at wavelengths shorter than visible is

$$f_{e1} = 0.123 = 12.3\%. \quad [Ans. (b)]$$

The fraction of solar energy in the infrared range is

$$(1 - f_{e2}) = 0.514 = 51.4\%. \quad [Ans. (c)]$$

Approximately half of solar energy is infrared, about a third is visible, and a small amount is ultraviolet. ■

---

### 8.1.6 Brief Historical Outline

The development of the principles of radiation physics occurred primarily in the nineteenth century. Isaac Newton discovered the visible color spectrum in 1666 by passing solar light through a prism. But no one took serious notice of the thermal energy in the "invisible" or infrared spectrum until 1800, when Sir Frederick Herschel, passing a thermometer through the spectrum, found the bulk of the energy to fall in what he called the "invisible light" range. Herschel verified that such radiant energy, whether from the sun, a candle, a hot poker, or a fire, obeyed the same laws of reflection and refraction as visible light. In 1801, Johann Ritter found that there was also radiation in the ultraviolet region. Accuracy of radiant intensity data was greatly improved in 1829 by Leopoldi Nobili's invention of the thermopile. Nobili's friend Macedonio Melloni improved the device in 1831 by arranging the thermopile into a ring "multiplier" with great sensitivity. Between 1831 and 1854, Melloni reported radiation data for a great variety of materials.

In 1880 Samuel Langley developed a sensitive galvanometer that detected temperature differences as low as $10^{-5}$ °C. Using this device and a specially ruled refraction grating (142 lines/mm), Langley in 1883 reported the first accurate solar radiation spectrum, out to $\lambda = 2.8\ \mu$m. Today a unit of solar energy, cal/cm$^2$, is called the *langley* in his honor. The perfection of gratings by ruling machines now makes wavelength determination a routine experimental procedure.

By 1889, although many accurate radiation data were known, Langley remarked that "we know almost nothing about the relation between temperature and radiation." Classical theory had resulted in a relation between emissivity and absorptivity predicted by Gustav Kirchhoff in 1860 and the verification of Joseph Stefan's 1879 measurement of the fourth-power blackbody emission law in a theory by Ludwig Boltzmann in 1884. The wavelength spectrum failed to yield to classical theory, although Willy Wien in 1894 and 1896 and Lord Rayleigh in 1900 made important contributions. Finally, using very accurate blackbody spectrum data of Willy Wien and Ernst Pringsheim as a guide, Max Planck developed his quantum concept and in 1901 published the famous spectrum formula (8.5), which began modern theoretical physics [7].

Further details about the history of radiation research may be found in the interesting survey by E. S. Barr [12].

### 8.1.7 Directional Radiation Intensity

Blackbody radiation has no directional variation and is said to be *diffuse,* that is, the emitted radiation per unit area is the same in every direction. Real surfaces, however, do exhibit directional variations. Here we define some directional parameters as preparation for the real surface properties to be discussed in the next section.

Consider an element of emitting surface area $dA$ in the $xy$-plane, as shown in Fig. 8.4, surrounded by a hemisphere of some convenient radius $r$. Any given *direction* of radiation from $dA$ can be denoted by the two polar angles $\theta$ measured from the vertical ($z$-axis) and $\phi$ measured in the surface plane from some convenient datum line (the $x$-axis). As shown in Fig. 8.4, let $dA_n$ be the projection of $dA$ onto the hemisphere. Then we define the monochromatic *intensity* $I_\lambda$ of emitted radiation as the energy flux in the given direction ($\theta$, $\phi$) per unit normal area per unit solid angle about this direction:

$$I_\lambda = \frac{dq/dA_n}{d\omega} = I_\lambda(\theta, \phi, \lambda, T). \tag{8.15}$$

When defined in this manner, $I_\lambda$ is a finite quantity. Also, by definition,

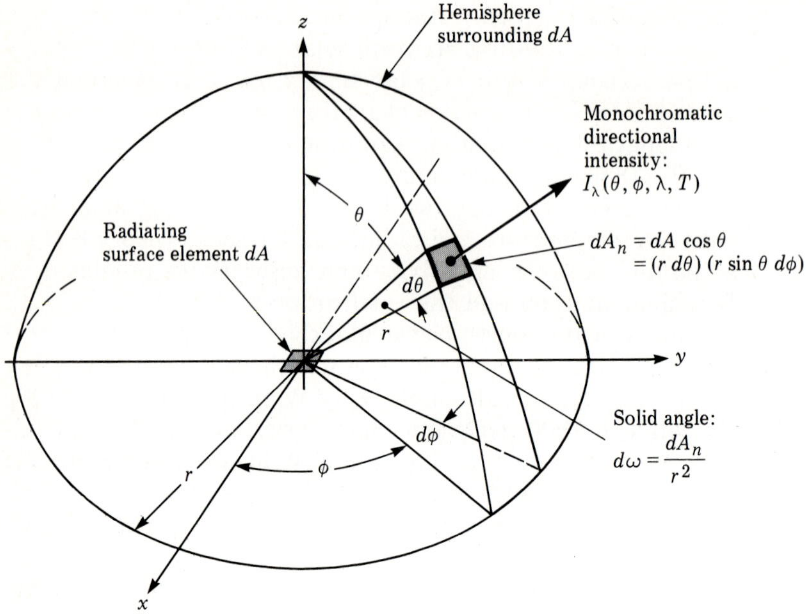


**Figure 8.4 Definition sketch showing directional angles for radiation emitted in a hemispherical region surrounding a surface element.**

the included solid angle $d\omega$ is given by

$$d\omega = dA_n/r^2. \tag{8.16}$$

In general, as denoted in Eq. (8.15), intensity $I_\lambda$ is a function of direction, wavelength, and temperature.

From the geometry of Fig. 8.4, we see that the normal area element $dA_n$ on the hemisphere is related to $dA$ by

$$dA_n = dA \cos\theta.$$

Replacing $dA_n$ in Eq. (8.15) and noting by definition that $dq/dA = dE_\lambda$, we obtain

$$E_\lambda = \int I_\lambda(\theta, \phi, \lambda, T)(\cos\theta)\, d\omega, \tag{8.17}$$

where the integration is over the entire hemisphere. For *black* radiation, $I_\lambda = I_{b\lambda}$ is independent of direction, so that

$$dE_{b\lambda}/d\omega = I_{b\lambda}(\lambda, T)\cos\theta. \tag{8.18}$$

This is called *Lambert's cosine law*. However, for real surfaces, we must retain $I_\lambda$ under the integral sign in Eq. (8.17).

Finally, note from Fig. 8.4 that the solid angle $d\omega$ is a function only of $\theta$ and $\phi$:

$$\begin{aligned} d\omega &= \frac{dA_n}{r^2} = \frac{(r d\theta) r (\sin\theta) d\phi}{r^2} \\ &= (\sin\theta)\, d\theta\, d\phi. \end{aligned} \tag{8.19}$$

Then the evaluation of Eq. (8.18) is a double integration over the two polar direction angles:

$$E_\lambda = \int_0^{2\pi} d\phi \left[ \int_0^{\pi/2} I_\lambda(\theta, \phi, \lambda, T)(\sin\theta)(\cos\theta) d\theta \right]. \tag{8.20}$$

This expression is quite general. If the monochromatic intensity $I_\lambda$ is known from experiment or theory, Eq. (8.20) may be used to evaluate the *monochromatic hemispherical emissive power, $E_\lambda$*, of any real surface. The modifier *monochromatic* means "at a given wavelength," and the term *hemispherical* means "integrated over the entire hemisphere above the surface."

A final integration over wavelength then yields the *total hemispherical emissive power, E*, of the real surface:

$$E(T) = \int_0^\infty E_\lambda(\lambda, T) d\lambda. \tag{8.21}$$

The term *total* means "integrated over all wavelengths."

Most natural surfaces are *isotropic,* that is, their intensity $I_\lambda$ does not depend on the circumferential angle $\phi$. (Artificial surfaces can be made *anisotropic* by, for example, aligning grooves or roughness elements along a particular surface axis.) For an isotropic surface, Eq. (8.20) becomes

$$E_\lambda = 2\pi \int_0^{\pi/2} I_\lambda(\theta, \lambda, T)(\sin\theta)(\cos\theta) d\theta. \tag{8.22}$$

Equation (8.21) applies again for the total power.

A *black* surface is not only isotropic but also diffuse, that is, its intensity $I_{b\lambda}$ varies neither with $\theta$ nor with $\phi$. Equation (8.22) becomes, for a black surface,

$$\begin{aligned} E_{b\lambda} &= 2\pi I_{b\lambda}(\lambda, T) \int_0^{\pi/2} (\sin\theta)(\cos\theta) d\theta \\ &= \pi I_{b\lambda}. \end{aligned} \tag{8.23}$$

Thus the blackbody intensity $I_{b\lambda}$ has exactly the same shape as the monochromatic emission spectrum $E_{b\lambda}(\lambda, T)$ from Planck's relation (8.5) but is smaller by a factor of $\pi$.

Similarly, after integration over wavelength, the total intensity is related to total emissive power by

$$E_b = \pi I_b = \sigma T^4. \tag{8.24}$$

These are ideal surface results. Of more practical interest is the application of relations such as Eq. (8.22) to the radiative properties of real surfaces, in Section 8.2.

## 8.2 Radiation Characteristics of Real Surfaces

In studying the properties of blackbody surfaces in Section 8.1, we constantly noted that real surfaces are not ideal and that they exhibit nonblack, nondiffuse, and anisotropic emission characteristics. Real surfaces also possess nonideal absorption, reflection, and transmission properties, all of which we define in this section. The net effect of this nonideal behavior is to make the *algebra* of practical radiation computation rather messy and laborious, tempting us strongly to make some rather gross simplifications.

First let us consider radiation emission, which is a function only of the surface characteristics and is not dependent upon external conditions.

### 8.2.1 The Hohlraum or Black Radiator

A *hohlraum* is a cavity containing a small hole, as in Fig. 8.5. It is a near-perfect simulation of a black surface. Incoming radiation entering the hole will incur multiple absorptions, and little or none will make its way back out the hole. The hole is "black," absorbing all incoming radiation. Conversely, if the cavity surface is isothermal and insulated, its emitted energy, after multiple reflections, has a diffuse character when it leaves the hole. The cavity thus simulates both a black emitter and a black absorber, regardless of its actual surface properties. Further details of hohlraum construction are given in [1, Section 2.5].

A hohlraum and a wavelength filter can be used to establish experimentally the directional and monochromatic properties of real surfaces. Let us discuss typical experimental results.

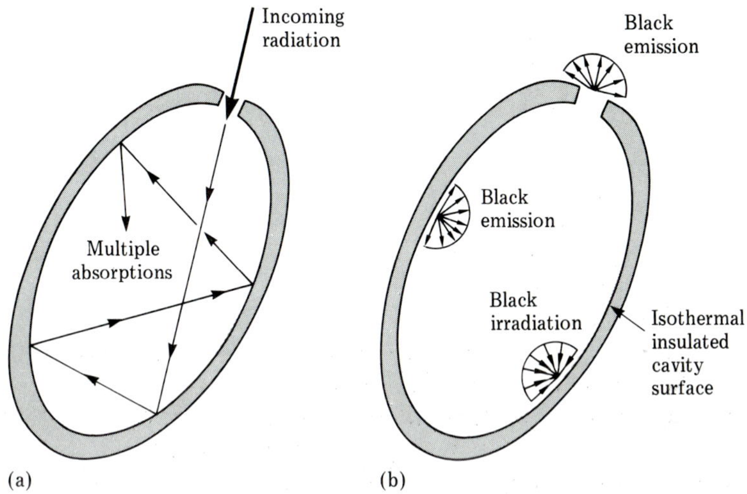


**Figure 8.5** A hohlraum simulates a blackbody: (a) a small hole absorbs all incoming radiation; (b) if the cavity surface is isothermal and insulated, all radiation through the hole and within the cavity is black.

### 8.2.2 Emissivity

No real surface is truly black. The (dimensionless) ratio of its actual (measured) emission to that of a blackbody is called the *emissivity* and is given the symbol $\varepsilon$, which we hope will not confuse the reader with roughness height (Chapter 6).

At its lowest or most specific level, the emissivity is a function of wavelength and both surface directions:

$$\varepsilon_{\lambda\theta\phi}(\lambda, \theta, \phi, T) = I_\lambda(\lambda, \theta, \phi, T)/I_{b\lambda}(\lambda, T), \tag{8.25}$$

where the subscripts on $\varepsilon$ indicate that the measurement is taken at a particular direction and wavelength.

Assume that the surface is isotropic (no $\phi$ variation). Then, by integrating Eq. (8.25) over all wavelengths, we obtain the *directional total emissivity:*

$$\varepsilon_\theta(\theta, T) = \frac{1}{I_{b\lambda}}\int_0^\infty \varepsilon_{\lambda\theta}(\lambda, \theta, T) I_{b\lambda}(\lambda, T)\,d\lambda, \tag{8.26}$$

where $I_{b\lambda} = E_{b\lambda}/\pi$ from Eq. (8.5). Similarly, by integrating Eq. (8.25) over all directions at a given wavelength, we obtain the *monochromatic hemispherical emissivity:*

$$\varepsilon_\lambda(\lambda, T) = 2\int_0^{\pi/2} \varepsilon_{\lambda\theta}(\lambda, \theta, T)\cos\theta(\sin\theta)d\theta. \quad \textbf{(8.27)}$$

Finally, by averaging over both wavelength and direction, we obtain the familiar overall or *total hemispherical emissivity:*

$$\varepsilon(T) = \frac{1}{E_b}\int_0^{\infty} \varepsilon_\lambda(\lambda, T)\, E_{b\lambda}(\lambda, T)d\lambda, \quad \textbf{(8.28)}$$

where $E_b = \sigma T^4$ from Eq. (8.10). In the next sections we discuss some experimental measurements of $\varepsilon_\theta$, $\varepsilon_\lambda$, and $\varepsilon$.

### 8.2.3 The Gray-Lambert Body Approximation

Most emissivity data reported in the literature are either averaged over direction ($\varepsilon_\lambda$) or over wavelength ($\varepsilon_\theta$). Relatively few direct measurements exist for total hemispherical emissivity, $\varepsilon$. Here, and in Appendix I, we illustrate only a few experimental results. For more comprehensive design data, see the text by McAdams [13], the monographs by Gubareff [14] and by Wood [15], and — most especially — the monumental compilation by Touloukian and co-workers [16], which contains over 4600 pages of thermal radiation data. Also see data summaries in the advanced texts on radiation [1–6].

Consider first the monochromatic hemispherical emissivity, $\varepsilon_\lambda$, sketched in Fig. 8.6(a). Whereas a blackbody has unit emissivity, $\varepsilon_\lambda = \varepsilon = 1.0$, real surface measurements show a pronounced wavelength variation, such as the wiggly line in Fig. 8.6a. This wiggly line will also vary with temperature, so that evaluation of total emissivity, $\varepsilon$, from Eq. (8.28) would require repeated and laborious numerical or graphical quadratures. A gross but useful simplification is to replace the wiggly line by an average emissivity, shown as a dashed line in Fig. 8.6(a). This is the *graybody approximation:*

$$\varepsilon_\lambda = \text{constant} = \varepsilon. \quad \textbf{(8.29)}$$

The effect of this simple assumption on emissive power is shown in Fig. 8.6(b). By neglecting the wiggles in $\varepsilon_\lambda$, we may miss some important peaks and valleys in $E_\lambda$. However, since the "black" curve shifts with temperature, there may be a temperature range where the wiggles are negligible and the graybody approximation is accurate. We illustrate in Example 8.3.

Next consider total directional emissivity, $\varepsilon_\theta$, sketched in Fig. 8.7. The blackbody has $\varepsilon_\theta = \varepsilon = 1.0$, independent of $\theta$ (diffuse surface).

Typical real surfaces shown in Fig. 8.7 are of two types. Nonmetallic surfaces (electrically nonconducting) have emissivity that is high at small angles and then drops off rather sharply to zero at $\theta = 90°$. Metals have emissivity that is small at low angles and then rises to a peak, followed by a sharp drop to zero at $\theta = 90°$. No real surface is diffuse (constant $\varepsilon_\theta$) at large angles from the normal. Nonmetals appear "brightest" viewed from the normal; metals are brightest from a side view. Consider a sphere heated into the red visible range. A blackbody sphere would appear uniformly red, a nonmetallic sphere

**Figure 8.6** Emission of ideal, gray, and real surfaces: (a) monochromatic hemispherical emissivity; (b) monochromatic emission powers corresponding to (a).

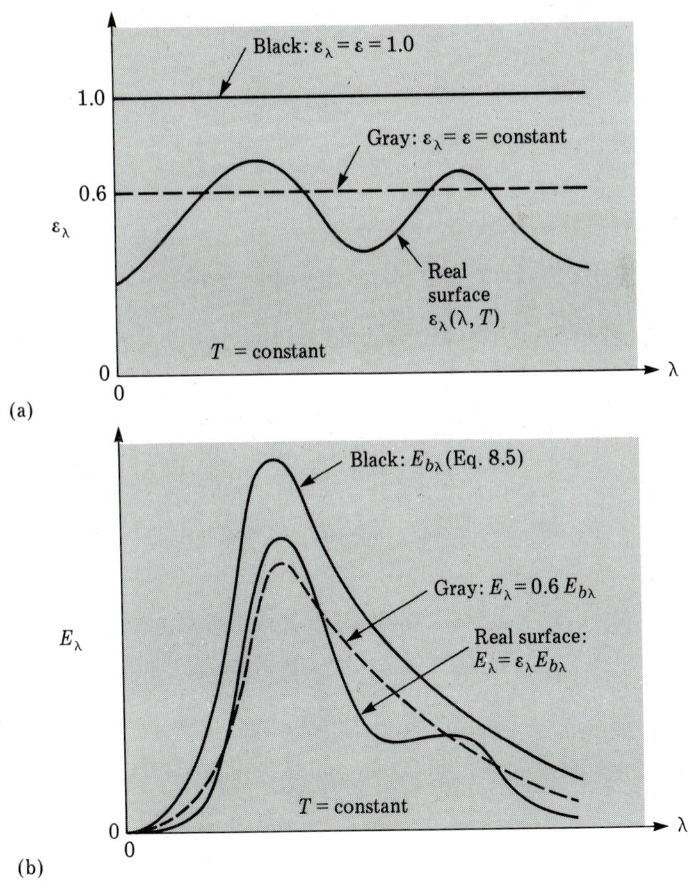

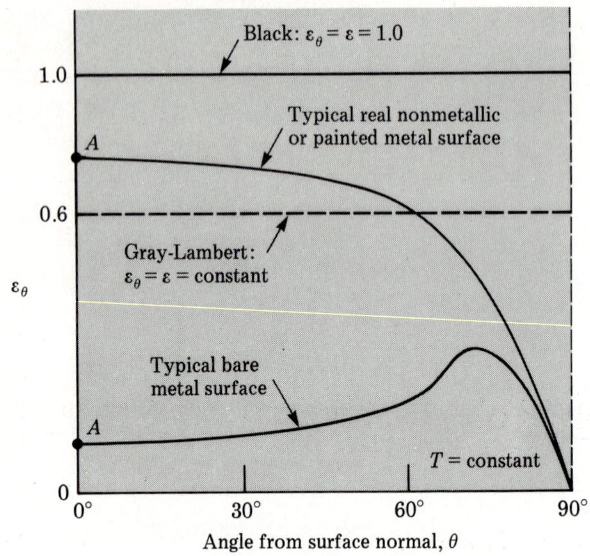


**Figure 8.7 Total directional emissivity of ideal, gray-lambert, and real surfaces. Most monochromatic data in the literature are "normal," measured at points *A* ($\theta$ = 0°).**

would have a red center, and a bare metal sphere would seem to have a red rim.

Experimenters often report only the *normal total emissivity,* $\varepsilon_\theta(0°)$, or points *A* in Fig. 8.7. This is only a rough estimate for the directional average, $\varepsilon$; it is too high for nonmetals and too low for metals. A semi-empirical correlation of the two values, taken from [17], is shown in Fig. 8.8.

For radiant interchange computations, it is common to make the *gray-lambert body* approximation, a dashed line in Fig. 8.7:

$$\varepsilon_\theta = \varepsilon = \text{constant}. \tag{8.30}$$

This is clearly a crude assumption, especially at large angles, but the advantage of being able to use the Lambert cosine law, Eq. (8.18), is significant.

### 8.2.4 Experimental Emissivity Data

Hundreds of emissivity data are reported in the literature [13–16]. Because of the strong effect of surface conditions (oxidation, roughness, and so on), all such data are subject to large uncertainty.

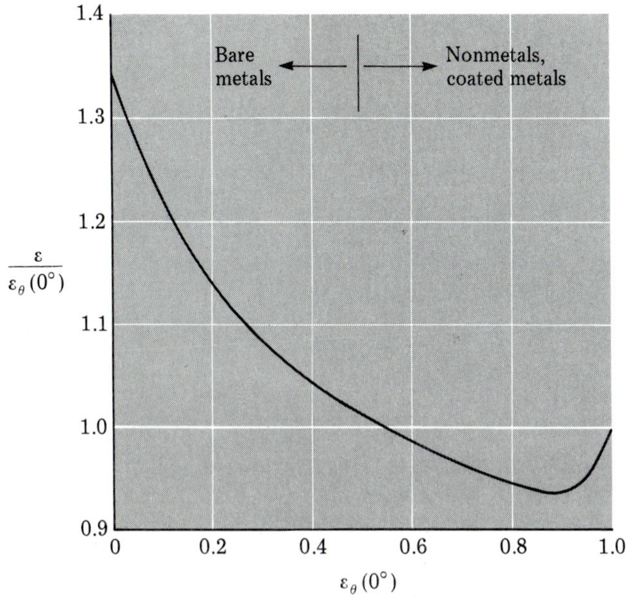


**Figure 8.8** Semi-empirical correlation between normal and hemispherical emissivity (From [17].)

The uncertainty is especially acute for metals, whose surfaces vary from micropolished to black-oxidized.

An example of metal data uncertainty is shown for copper in Fig. 8.9, taken from [16]. The raw data from 64 experiments are shown in Fig. 8.9(a). The same data are smoothed and interpreted in Fig. 8.9(b) by the authors of [16]. Oxidation and roughness are seen to be the main reasons for the scatter and uncertainty. Since surface conditions usually vary during the useful life of a given design, the heat transfer engineer must always keep in mind this large uncertainty in radiative property data.

The overall scatter in nonmetal data is usually less than that for metals, if only because there are no low values. A nonmetal emissivity might vary from 0.7 to 0.9, instead of the broad range 0.02 to 0.92 shown for copper in Fig. 8.9.

Monochromatic and directional emissivity data also show considerable uncertainty. Figure 8.10 gives data for normal monochromatic emissivity, $\varepsilon_{\lambda\theta}(0°)$ for various materials. We see two distinct types of wavelength behavior. All polished metals have emissivity that decreases with increasing wavelength. Thus, when integrated over wavelength, the total emissivity $\varepsilon$ of a metal increases with temperature, as in

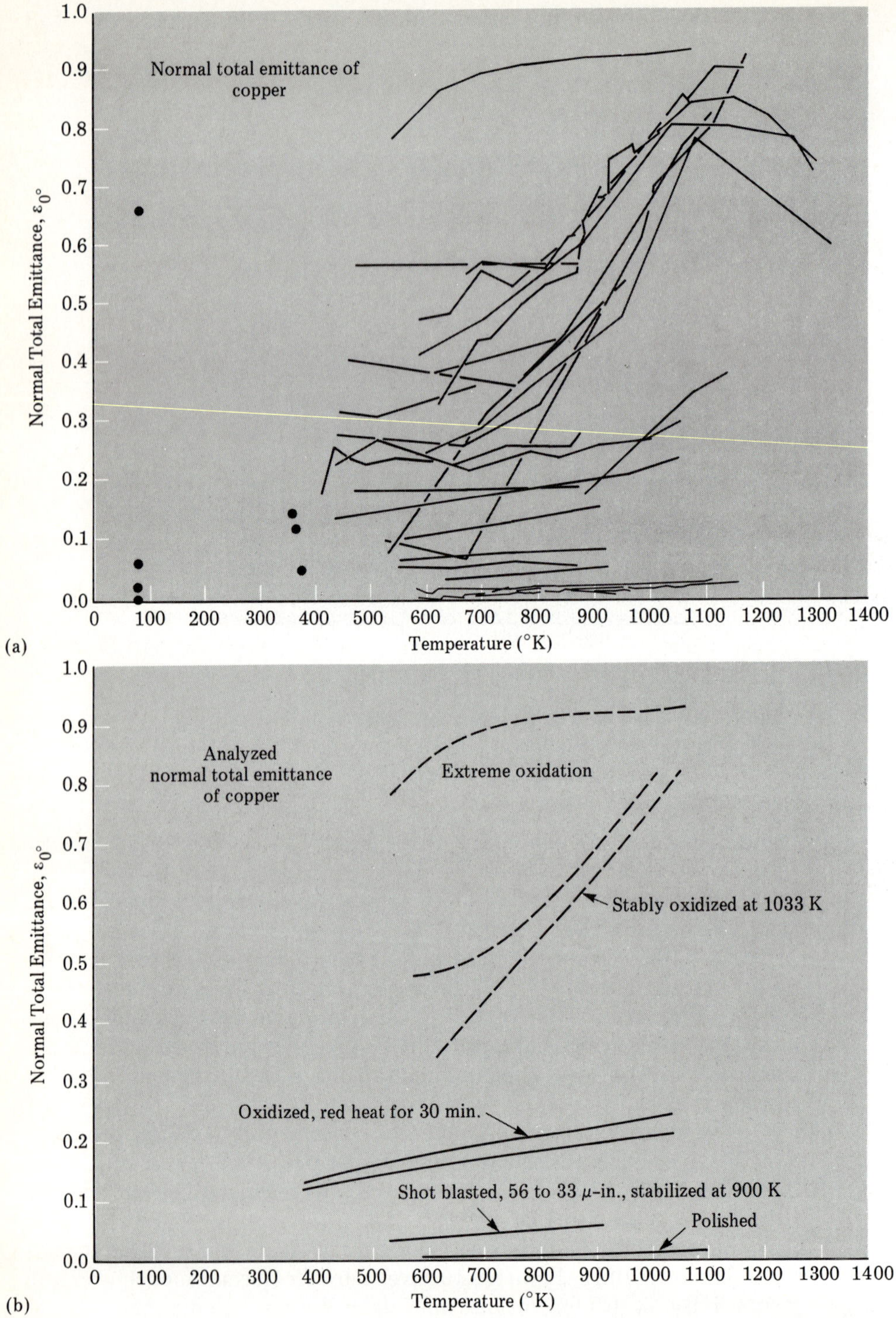


**Figure 8.9** **Illustration of uncertainty in emissivity data: (a) normal total emissivity of copper from 64 different experiments; (b) the same data as analyzed, interpreted, and smoothed by Touloukian and DeWitt [16].**

Fig. 8.9(b). This is because Wien's displacement law, Eq. (8.7), links low wavelength to high temperature.

Conversely, the emissivity of the nonmetals in Fig. 8.10 increases with wavelength; hence the integrated value, $\varepsilon$, will generally decrease with temperature.

These integrated effects for total hemispherical emissivity, $\varepsilon(T)$, are shown in Fig. 8.11 for various clean surfaces. The tendency for metal values to rise with temperature and for nonmetal values to drop is clear. These variations cause the total emitted radiation, $E = \varepsilon\sigma T^4$, to vary from the ideal fourth-power law. From Fig. 8.11, for example, in the temperature range 1000–1500K, $E \propto T^3$ for fire clay and $E \propto T^{4.3}$ for aluminum. Again we emphasize that all radiative property data are uncertain.

### 8.2.5 Approximate Integration with Fractional Functions

The determination of $\varepsilon(T)$ by integration of $\varepsilon_\lambda(\lambda, T)$ in Eq. (8.28) is quite laborious if we use the actual data in Fig. 8.10. An effective simplification is to approximate the curves for $\varepsilon_\lambda$ by a series of piecewise-

**Figure 8.10** Measured monochromatic normal emissivity for polished metals and clear nonmetal surfaces. (Data from many sources.)

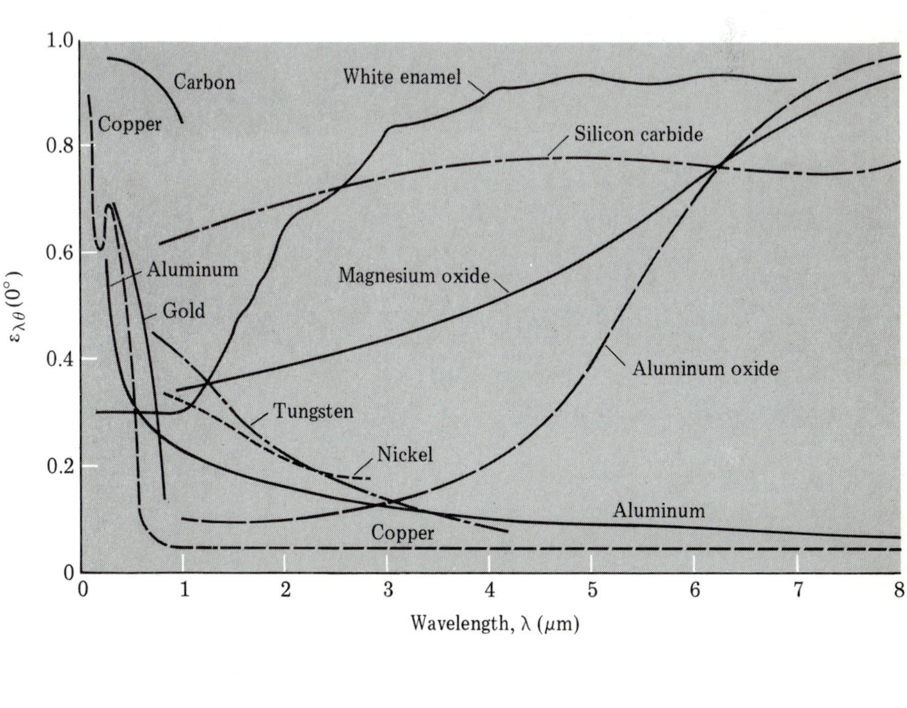

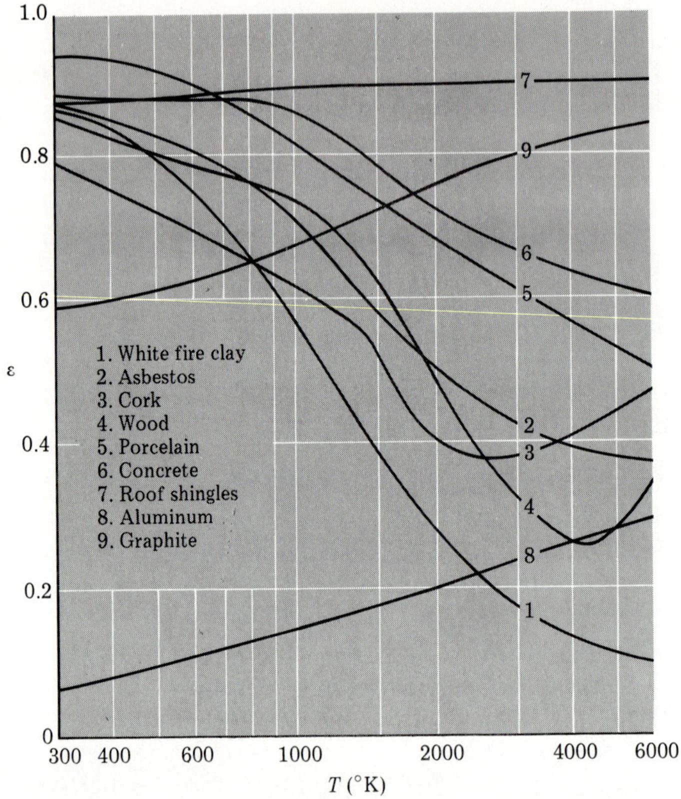


**Figure 8.11** **Measured total hemispherical emissivity of various clean surfaces, from [18]. Because of possible surface condition variations, these curves have a large uncertainty.**

constant values. This allows us to use the blackbody fractional functions $f_e$ from Eq. (8.12).

Figure 8.12 illustrates this approximation. We simulate the measured curve for $\varepsilon_\lambda$ by three piecewise constants: $\varepsilon_A$ from 0 to $\lambda_1$, $\varepsilon_B$ from $\lambda_1$ to $\lambda_2$, and $\varepsilon_C$ from $\lambda_2$ to infinity. The integral for total emissivity, Eq. (8.28), becomes

$$\begin{aligned}\varepsilon(T) &= \frac{1}{E_b}\int_0^\infty \varepsilon_\lambda(\lambda, T)E_{b\lambda}(\lambda, T)d\lambda \\ &\doteq \frac{1}{\sigma T^4}\left(\varepsilon_A\int_0^{\lambda_1} E_{b\lambda}d\lambda + \varepsilon_B\int_{\lambda_1}^{\lambda_2} E_{b\lambda}d\lambda + \varepsilon_C\int_{\lambda_2}^{\infty} E_{b\lambda}d\lambda\right) \qquad \textbf{(8.31)} \\ &= \varepsilon_A f_e(\lambda_1 T) + \varepsilon_B[f_e(\lambda_2 T) - f_e(\lambda_1 T)] + \varepsilon_C[1 - f_e(\lambda_2 T)].\end{aligned}$$

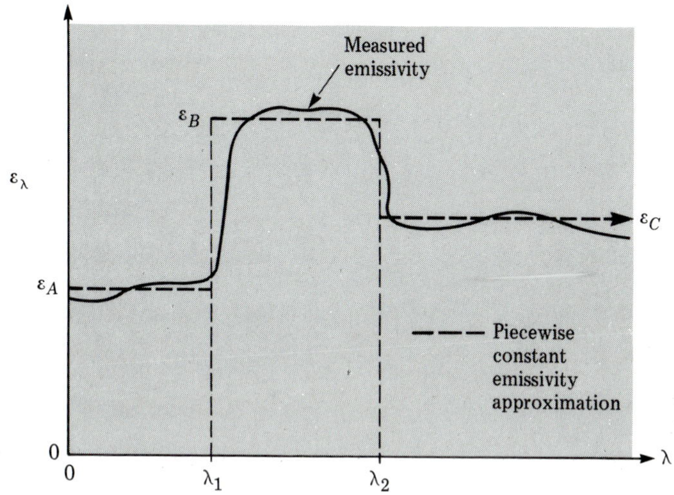


**Figure 8.12 Illustration of the approximate evaluation of total emissivity by piecewise use of blackbody fractional functions.**

It should be clear from this expression how to extend the method to more than three piecewise constants. The results are approximate but quite adequate for preliminary design of total emissive properties of a proposed surface.

## Example 8.3

The monochromatic emissivity of white fireclay may be approximated by $\varepsilon_\lambda \doteq 0.1$ for $0 \leq \lambda \leq 3$ $\mu$m and by $\varepsilon_\lambda \doteq 0.9$ for $\lambda > 3$ $\mu$m. Estimate the total emissivity of white fireclay at 1500K and compare with Fig. 8.11.

*Solution* This is a two-piece approximation, with $\varepsilon_A = 0.1$, $\lambda_1 = 3$ $\mu$m, $\varepsilon_B = 0.9$, and $\lambda_2 = \infty$. The argument of $f_e$ at the breakpoint is $(\lambda_1 T) = (3 \times 10^{-6}\text{ m})(1500\text{K}) = 0.0045$ m $\cdot$ K. From Table 8.2, $f_e(0.0045) = 0.5643$. Equation (8.31) applies with the third term dropped:

$$\varepsilon = \varepsilon_A f_e(0.0045) + \varepsilon_B[1 - f_e(0.0045)]$$

$$= 0.1(0.5643) + 0.9(1 - 0.5643) = 0.45. \quad [\textit{Ans.}]$$

From Fig. 8.11, the measured emissivity at 1500K is 0.39. By

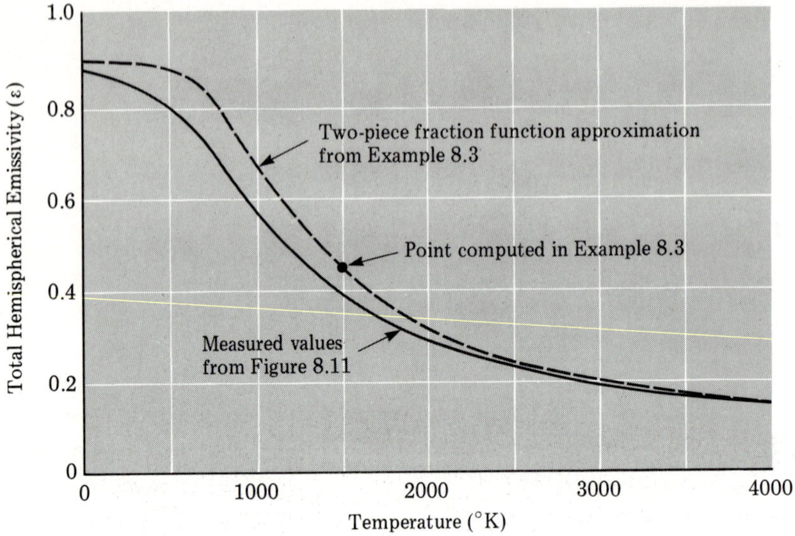


**Figure 8.13 Comparison of measured and computed emissivity of white fireclay.**

repeating the computation for various $T$ we may construct the entire $\varepsilon(T)$ curve as in Fig. 8.13. The agreement is reasonable. ■

## Example 8.4

An accurate curve-fit to data for the total directional emissivity of copper oxide reported by Schmidt and Eckert [17] is

$$\varepsilon_\theta \doteq \varepsilon_{0^\circ}\,[1 - (2\theta/\pi)^8], \tag{1}$$

where $\varepsilon_{0^\circ} \doteq 0.8$. Estimate the total hemispherical emissivity.

*Solution* One estimate is simply to enter Fig. 8.8 at $\varepsilon_{0^\circ} = 0.8$ and read $\varepsilon/\varepsilon_{0^\circ} \doteq 0.95$, whence $\varepsilon \doteq (0.8)(0.95) \doteq 0.76$. [*Ans.*]

A second method is to integrate the curve-fit (1) above using Eq. (8.26) for an isotropic surface:

$$\varepsilon = \frac{1}{\pi}\left(\int_0^{2\pi} d\phi\right)\left[\int_0^{\pi/2} \varepsilon_\theta \cos\theta(\sin\theta)d\theta\right]$$

$$\doteq 2\,\varepsilon_{0^\circ}\int_0^{\pi/2} [1 - (2\theta/\pi)^8]\cos\theta(\sin\theta)d\theta \tag{2}$$

$$= 2\,\varepsilon_{0^\circ}\,(0.500 - 0.025) = 0.95\,\varepsilon_{0^\circ}.$$

This gives exactly the same estimate:

$$\varepsilon \doteq (0.95)(0.8) = 0.76. \quad [Ans.]$$

It was from data such as these that Schmidt and Eckert constructed their semi-empirical correlation in Fig. 8.8. Equation (1) above is very similar to the nonmetal curve in Fig. 8.7. ■

---

### 8.2.6 Absorptivity, Reflectivity, and Transmissivity

The previous discussion has been devoted to the *emission* of thermal radiation from a surface. Similar concepts apply to radiation *received* by a surface from the surroundings.

Figure 8.14 shows the general case of a surface receiving radiation from the surroundings, usually called *irradiation*. We let the received radiant energy flux be denoted by $G$ if it is total and hemispherical (analogous to $E$), by $G_\lambda$ if it is at a specific wavelength, and by $G_{\lambda,\theta,\phi}$ if it is at a specific wavelength and direction. When the irradiation impinges on the surface, a portion of it is absorbed, a portion is reflected, and a portion is transmitted through. If the transmission is negligible, the surface is said to be *opaque* to irradiation. If a finite

**Figure 8.14** Schematic of a surface receiving irradiation and splitting it into absorbed, reflected, and transmitted portions.

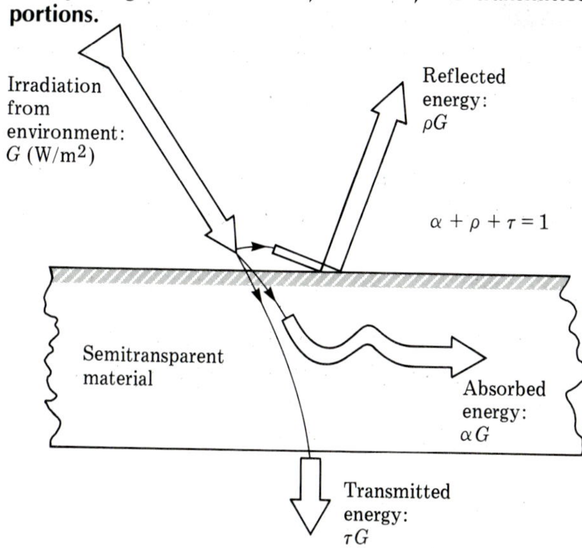

transmission occurs, the surface is termed *semitransparent*. Most surfaces are opaque; incident radiation is absorbed within a thin surface layer perhaps one or two microns thick. Examples of semitransparent materials are glass, clear plastic, rock salt, and water.

Figure 8.14 illustrates total hemispherical irradiation. We define the *absorptivity*, $\alpha$, as the ratio of absorbed to total irradiation. Similarly, *reflectivity*, $\rho$, is the ratio of reflected to total energy, and *transmissivity*, $\tau$, is the ratio of transmitted to total irradiation. Since there is no other path for the energy, it follows that the three energies must sum to total irradiation, or the three dimensionless coefficients or ratios must sum to unity at all levels:

$$\begin{aligned} \alpha + \rho + \tau &= 1, \\ \alpha_\lambda + \rho_\lambda + \tau_\lambda &= 1, \\ \alpha_{\lambda\theta\phi} + \rho_{\lambda\theta\phi} + \tau_{\lambda\theta\phi} &= 1. \end{aligned} \tag{8.32}$$

For an opaque material, $\tau \doteq 0$ at all levels. The Greek notation $(\alpha, \rho, \tau)$ is a strong custom in radiation theory. We hope the context will make it clear that we are not referring to thermal diffusivity, density, or shear stress.

Unlike $\varepsilon$, the quantities $(\alpha, \rho, \tau)$ are not strictly surface properties but depend also on the character of the irradiation. Generally, the irradiation $G$ is nonblack and nongray and depends upon wavelength, direction, and temperature $T_e$ of the irradiating body. Meanwhile, the receiving surface has its own temperature $T_w$ and its own nonblack, nongray spectral and directional characteristics. Thus, in the general case, $(\alpha, \rho, \tau)$ depend upon $T_e$, $T_w$, and a mixture of $(\lambda, \theta, \phi)$ effects from both the irradiating and the receiving surfaces. It is most tempting to assume that both bodies are black or gray.

By analogy with Eqs. (8.26)–(8.28), the integral forms for isotropic $(\alpha, \rho, \tau)$ are

$$\begin{aligned} \alpha_\lambda &= \frac{\int_0^{\pi/2} \alpha_{\lambda\theta}\, G_{\lambda,\,\theta} \cos\theta (\sin\theta)\, d\theta}{\int_0^{\pi/2} G_{\lambda,\,\theta} \cos\theta(\sin\theta)\, d\theta}, \\ \alpha &= \frac{\int_0^{\infty} \alpha_\lambda G_\lambda\, d\lambda}{\int_0^{\infty} G_\lambda\, d\lambda}, \end{aligned} \tag{8.33}$$

with exactly similar expressions for $(\rho_\lambda, \rho)$ and $(\tau_\lambda, \tau)$. The form of these equations makes it clear that the particular form of irradiation

$G$ strongly affects the hemispherical and total absorption, reflection, and transmission.

If the incoming irradiation is black or gray-lambert, Eqs. (8.33) reduce to

$$\alpha_\lambda = 2\int_0^{\pi/2} \alpha_{\lambda\theta} \cos\theta(\sin\theta)d\theta,$$
$$\alpha = \frac{1}{\sigma T_e^4}\int_0^{\infty} \alpha_\lambda E_{b\lambda}(\lambda, T_e)d\lambda. \tag{8.34}$$

Even with this simplification, the absorptivity is still a function of both $T_e$ and $T_w$. For example, the value of $\alpha$ for room-temperature paint subjected to black solar radiation is much lower than when the same paint receives black irradiation from a 500K source.

### 8.2.7 Specular and Diffuse Reflection

Reflected radiant energy may show strong directional effects, depending on both the incoming irradiation direction and the surface characteristics. There are two basic types of reflection, illustrated in Fig. 8.15. If the surface is rough compared to the incoming wavelength, $h > 2\lambda$, the incoming rays see a "mountain range" of projections and cavities and tend to reflect back in random directions. The result is *diffuse* or uniform hemispherical reflection, as shown in Fig. 8.15(a). Most commercial "as received" surfaces are nearly diffuse, having roughness heights of the order of 100 $\mu$m, compared to radiation wavelengths of 1–10 $\mu$m.

In the other extreme, Fig. 8.15(b), the surface is "smooth" compared to the incoming wavelength, $h < \lambda/10$, and the radiation reflects *specularly,* like a mirror or billiard ball. The outgoing ray leaves at the mirror image of the incoming angle. Since thermal radiation tends to have longer than visible wavelengths, we can state conclusively that thermal rays will reflect specularly off surfaces with a visible "mirror" finish.

A typical surface is likely to fall in the intermediate or "mixed-reflection" range, $0.5h < \lambda < 10h$, as in Fig. 8.15(c). Rays reflect in all directions, reaching a peak intensity near the mirror-image angle [19]. Sometimes there are also off-specular peaks at large wavelengths [20]. It is common in computing radiation interchange (Section 8.3) to assume that all surfaces are *diffuse*. The theory of mixed diffuse/specular interchange between surfaces is given, for example, in [2, Chapter 5].

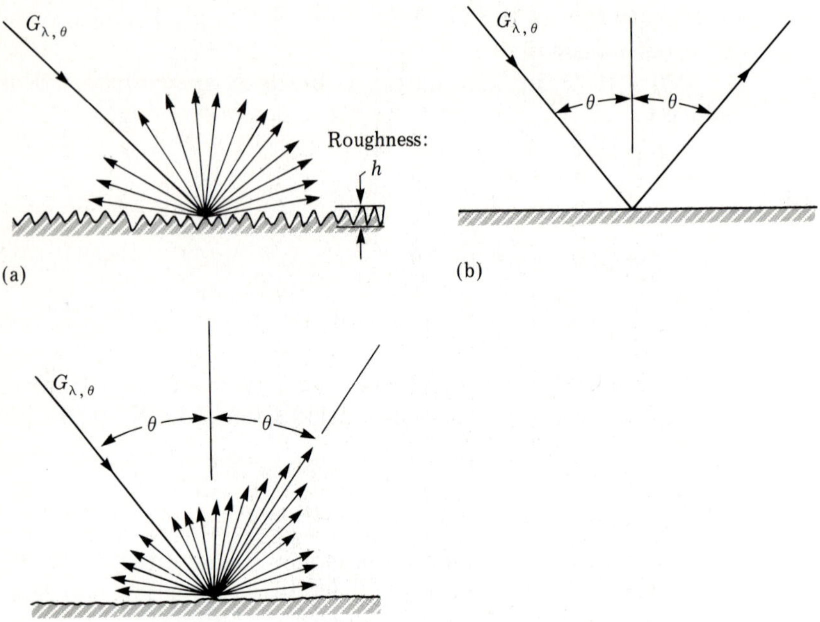


(a) (b) (c)

**Figure 8.15 Reflection characteristics of rough surfaces: (a) diffuse reflection, $\lambda < h/2$; (b) specular reflection, $\lambda > 10h$; (c) mixed reflection, $0.5h < \lambda < 10h$.**

### 8.2.8 Kirchhoff's Law

For a given real surface, we need data for four quantities: emissivity, absorptivity, reflectivity, and transmissivity. If two of the latter three are measured, the third may be computed from Eqs. (8.32). An additional relation may also be used under conditions to be outlined. This relation, now called *Kirchhoff's law,* was deduced with a thermodynamic argument by Gustav Kirchhoff in 1860. The derivation assumes that the surface is at temperature $T_s$ and is completely surrounded by black irradiation at the same temperature, $T_s$. Then, for equilibrium, the absorbed and emitted energies must be the same, or

$$\alpha_{\lambda\theta\phi}(T_s) = \varepsilon_{\lambda\theta\phi}(T_s). \tag{8.35}$$

This equality at the lowest level, that is, given wavelength, direction, and temperature, is quite general for any real surface. However, it fails if the emitted and received radiation are at different temperatures unless the surface is *gray-lambert* in character. The complete proof is given in [8].

If either the irradiation or the surface is *diffuse,* we may integrate Eq. (8.35) over direction to obtain a higher level of Kirchhoff's law:

$$\alpha_\lambda(T_s) = \varepsilon_\lambda(T_s). \tag{8.36}$$

Again, this fails if receiver and source are not at the same temperature.

Finally, if either the surface or the irradiation is *gray or black,* we may integrate again to the highest level of Kirchhoff's law:

$$\alpha(T_s) = \varepsilon(T_s). \tag{8.37}$$

Once again the relation fails if the temperatures are not the same. However, if the surface itself is gray-lambert, then $\alpha = \varepsilon$, independent of temperature.

There is of course a great temptation for engineers to assume the validity of Eq. (8.37) even for nongray surfaces receiving nongray irradiation. Then, for an opaque surface, only a single piece of the data is needed, say $\varepsilon(T)$, from which we compute $\alpha(T) \doteq \varepsilon(T)$ and $\rho(T) \doteq 1 - \varepsilon(T)$. However, the error in this approximation may be quite large, as the next example shows. See [1, pp. 59–65] for details.

### Example 8.5

A certain diffuse surface has $\varepsilon_\lambda = 0.8$ between 2 and 4 $\mu$m and zero absorptivity elsewhere. The surface is at 500K. Estimate its total hemispherical emissivity and absorptivity when receiving black radiation from a source at (a) 500K and (b) 2000K.

*Solution* The emissivity is dependent only upon *surface* temperature, $T_s = 500$K. Equation (8.31) applies, with $\varepsilon_A = \varepsilon_C = 0$, $\varepsilon_B = 0.8$, $\lambda_2 T = (4 \times 10^{-6}\ \text{m})(500\text{K}) = 0.002\ \text{m} \cdot \text{K}$, and $\lambda_1 T = (2 \times 10^{-6}\ \text{m})(500\text{K}) = 0.001\ \text{m} \cdot \text{K}$. Then, from Table 8.2,

$$\varepsilon(500\text{K}) = 0.8[f_e(0.002) - f_e(0.001)] = 0.8(0.06673 - 0.00032)$$

$$= 0.053. \quad [Ans.]$$

This result is independent of the temperature $T_e$ of the irradiation.

Since the surface is diffuse, Kirchhoff's law (8.36) applies, with $\alpha_\lambda(\lambda) = \varepsilon_\lambda(\lambda)$. Then, since the irradiation is black, the total absorptivity may be evaluated from Eq. (8.34):

$$\alpha(T_e) = \frac{1}{\sigma T_e^4}\int_{\lambda_1}^{\lambda_2} \alpha_\lambda E_{b\lambda}(\lambda, T_e)d\lambda$$
$$= 0.8[f_e(\lambda_2 T_e) - f_e(\lambda_1 T_e)]. \tag{1}$$

If $T_e = 500\text{K}$, the numerical values are the same as in the previous computation of emissivity $\varepsilon(500\text{K})$:

$$\alpha(500\text{K}) = 0.8(0.06673 - 0.00032) = 0.053. \quad [Ans.\ (a)]$$

We see that this verifies Kirchhoff's law in the total form (8.37) for black irradiation of a diffuse surface at equal temperatures.

If the source is at $T_e = 2000\text{K}$, however, all the numbers are quite different:

$$T_e = 2000\text{K}, \quad \lambda_2 T_e = 0.008 \text{ m} \cdot \text{K}, \quad \lambda_1 T_e = 0.004 \text{ m} \cdot \text{K},$$

$$\alpha(2000\text{K}) = 0.8[f_e(0.008) - f_e(0.004)] = 0.8(0.85625 - 0.48085)$$

$$= 0.300. \quad [Ans.\ (b)]$$

This is nearly 6 times greater than the emissivity at the given surface temperature of 500K. The surface is *nongray* and Eq. (8.37) cannot be stretched to apply accurately to unequal source and surface temperatures. ■

### Example 8.6

A white fireclay surface, assumed diffuse but not gray, is at $T_s = 1500\text{K}$ and completely surrounded by black walls at 800K. Estimate the net radiation heat flux from the clay surface.

*Solution* For black radiation toward a diffuse nongray surface, the net heat flux is emission minus absorption:

$$q''_w = \varepsilon(T_s)\sigma T_s^4 - \alpha(T_e)\sigma T_e^4. \quad \textbf{(1)}$$

From Fig. 8.11 for white fireclay at $T_s = 1500\text{K}$, read $\varepsilon(T_s) \doteq 0.39$. For these conditions, Kirchhoff's law holds at the highest level (8.37): $\alpha(T_e) = \varepsilon(T_e)$. Therefore use Fig. 8.11 again at $T_e = 800\text{K}$ to read $\alpha(T_e) \doteq 0.66$. Equation (1) then becomes

$$q''_{w,\,\text{net}} = (5.67 \times 10^{-8}\ \text{W/m}^2 \cdot \text{K}^4)[0.39(1500\text{K})^4 - 0.66(800\text{K})^4]$$

$$= 97{,}000\ \text{W/m}^2. \quad [Ans.]$$

Since fireclay is not strictly diffuse, for more accuracy one would have to measure its absorptivity at $T_e = 800\text{K}$. ■

### 8.2.9 Selective Surfaces

The designer often wishes to use a *selective* radiating surface to promote specific spectral or directional characteristics. If, for simplicity, we divide the spectrum into "low" versus "high" wavelengths, we obtain four different types of selective surfaces, as discussed in [21]. The basic types are shown in Fig. 8.16(a–d).

First is the *flat absorber* (Fig. 8.16a), which has high absorptivity over the entire spectrum. It is intended to capture all incoming irradiation. Good approximations to flat absorbers are black paints and enamels and black or dark oxides or black anodized coatings.

Second is the *flat reflector* (Fig. 8.16b), which has low absorptivity over the whole spectrum. Such a surface reflects most irradiation and is achieved by bare or polished metals and foils, electroplated or vacuum-deposited metal coatings, aluminized films, and metallic paints.

Third, and truly "selective," is the *solar absorber* (Fig. 8.16c), which absorbs solar (low-wavelength) irradiation but is highly reflective (with low emission) in the infrared range. Such surfaces are commonly

**Figure 8.16** Four basic types of selective radiating surfaces: (a) flat absorber; (b) flat reflector; (c) solar absorber; (d) solar reflector. (From [21].)

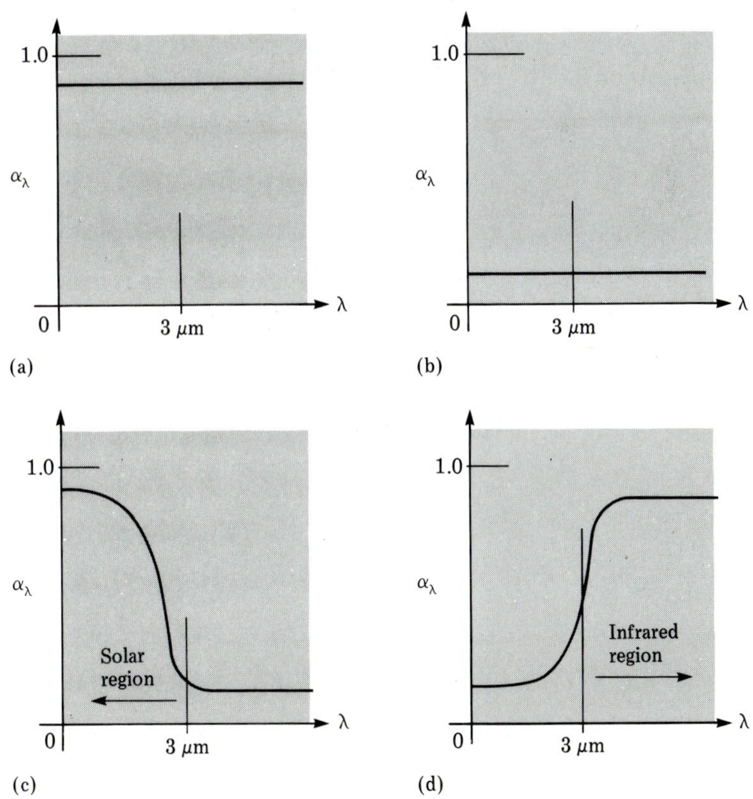

used in solar collector surfaces and for satellite temperature control. Typical designs use thin oxide coatings, sprayed or baked-on finishes, and vacuum-deposited films. Some examples are given in [21], [23], and [25].

Finally, the *solar reflector* (Fig. 8.16d) has high absorptivity and emission in the infrared range but reflects most of the low-wavelength (solar) irradiation. It is simulated by white or light-colored paints, enamels, or ceramic coatings or anodized aluminum. Such surfaces help cool exterior structures such as fuel tanks or cryogenic containers by reflecting solar irradiation and emitting low-temperature (infrared) heat from within.

The surfaces sketched in Fig. 8.16 are assumed to be opaque. Transparent or semitransparent materials can also be selective, with somewhat different design results. For example, thin window glass has characteristics similar to Fig. 8.16(d) (see [24]) but is a solar *transmitter,* that is, it has low absorptivity *and* reflectivity in the solar region. Thin glass thus *transmits* most solar radiation. But glass absorbs most infrared radiation and transmits little. This is the "greenhouse effect," wherein a glass enclosure captures most solar radiation and keeps the interior of the structure warm even in winter.

Further details about selective surfaces may be found in the texts by Edwards [23] and Siegel and Howell [1]. A very interesting "directionally selective" surface is given in [26].

## Example 8.7

Window glass less than 0.3 cm thick has a monochromatic transmissivity of 0.9 in the range 0.3 to 2.5 $\mu$m and nearly zero elsewhere. Estimate the total transmissivity of the window for (a) nearly black solar radiation at 5800K, and (b) black-room radiation at 300K.

*Solution* Given $\tau_\lambda = 0.9$ between $\lambda_1 = 0.3\ \mu$m and $\lambda_2 = 2.5\ \mu$m, and $\tau_\lambda = 0$ elsewhere. From Eq. (8.34), with $\tau$ replacing $\alpha$,

$$\tau = \frac{1}{\sigma T_e^4}\int_0^\infty \tau_\lambda E_{b\lambda}(\lambda, T_e)d\lambda = \frac{1}{\sigma T_e^4}\int_{\lambda_1}^{\lambda_2} (0.9)E_{b\lambda}(\lambda, T_e)d\lambda$$
$$= 0.9[f_e(\lambda_2 T_e) - f_e(\lambda_1 T_e)]. \quad (1)$$

a) For solar radiation, $T_e = 5800$K,

$\lambda_1 T_e = (0.3 \times 10^{-6}\ \text{m})(5800\text{K}) = 0.00174\ \text{m} \cdot \text{K}$,

$\lambda_2 T_e = (2.5 \times 10^{-6}\ \text{m})(5800\text{K}) = 0.0145\ \text{m} \cdot \text{K}$.

Then Eq. (1) becomes

$$\tau = 0.9[f_e(0.0145) - f_e(0.00174)] = 0.9(0.9661 - 0.0326)$$

$$= 0.840 \quad \text{(at 5800K)}. \quad [Ans.\ (a)]$$

Here we have evaluated the $f_e$ terms exactly from Eq. (8.13). Linear interpolation in Table 8.2 would give $\tau \doteq 0.834$, or less than 1% error.

b) At room temperature, $T_e = 300\text{K}$, $\lambda_1 T_e = 0.00009\ \text{m} \cdot \text{K}$ and $\lambda_2 T_e = 0.00075\ \text{m} \cdot \text{K}$. Then Eq. (1) becomes

$$\tau = 0.9[f_e(0.00075) - f_e(0.00009)] = 0.9(0.00001 - 0.00000)$$

$$\doteq 0 \quad \text{(at 300K)}. \quad [Ans.\ (b)]$$

Thus, for this slightly idealized window glass, 84% of the received solar irradiation is transmitted to the room, but essentially none of the room radiation is transmitted out. This is the greenhouse effect, common not only to greenhouses and glassed-in rooms but also quite evident if one leaves a parked car in the sun with the windows closed. ■

## 8.3 Radiation Shape Factors

A fundamental problem in radiation heat transfer is the computation of net radiant exchange between two surfaces. If suitable assumptions are made about the surface characteristics, the calculation reduces to the use of a purely geometric quantity called a *shape factor*. Once this factor is known between *two* surfaces, its use can be readily generalized to radiant exchange between multiple surfaces or enclosures.

From the extensive discussion of surface property effects in Section 8.2, the reader will realize that the assumption of a gray-diffuse surface is, at best, "reasonable" and more likely is only "fair" or even "poor." Yet this assumption is commonly made because of the tremendous simplification it allows in radiant-exchange algebra. To eschew the gray-diffuse assumption is to take on the burden of a complex calculation, often requiring a large computer program. The reader is warned, meanwhile, of the possible inaccuracy of the simple shape factor computations presented here.

### 8.3.1 The Concept of Radiosity

The computation of radiant exchange is concerned with the total energy streaming away from a given surface, as illustrated in Fig.

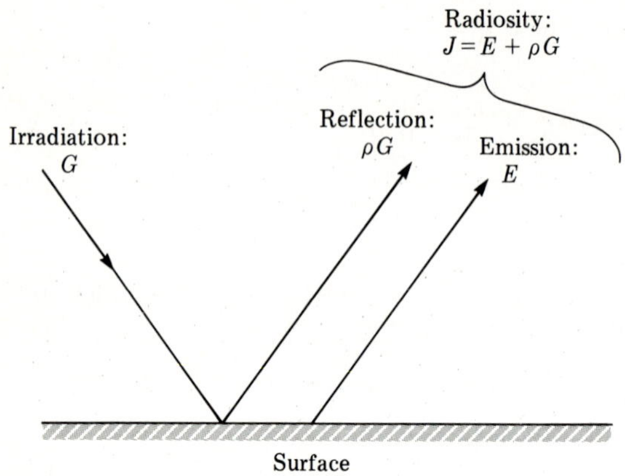


**Figure 8.17** **Sketch illustrating the radiosity concept.**

8.17. The surface is emitting its own energy at a rate $E$ and also reflecting any received irradiation $G$ at a rate $\rho G$, where $\rho$ is the reflectivity. The total energy leaving is called the *radiosity, J*:

$$J = E + \rho G = \varepsilon E_b + \rho G, \tag{8.38}$$

where $\varepsilon$ is the surface emissivity. Each surface $i$ has a radiosity $J_i$. If the radiation is nondiffuse, $J_i$ depends on the direction $(\theta, \phi)$ of radiation.

### 8.3.2 Definition of the Shape Factor

Now consider two infinitesimal surfaces, $dA_1$ and $dA_2$, exchanging radiation as in Fig. 8.18. The line connecting the two surfaces is of length $r$ and makes angles $\theta_1$ and $\theta_2$, respectively, with the surface normals $n_1$ and $n_2$.

Let $I_1$ be the net intensity of radiation leaving surface 1 in the direction $\theta_1$. Then the total energy intercepted by surface $dA_2$ is

$$dE_{1\to 2} = I_1\, dA_1 \cos\theta_1\, d\omega_1, \qquad d\omega_1 = dA_2 \cos\theta_2 / r^2.$$

Meanwhile, the total energy leaving surface 1 is

$$dE_1 = dA_1 \left[ 2\pi \int_0^{\pi/2} I_1(\theta) \sin\theta (\cos\theta) d\theta \right].$$

The ratio of these two is called the *shape factor* $dF_{1\to 2}$: the fraction of total radiation leaving surface 1 that is intercepted by surface 2:

$$dF_{1\to 2} = dE_{1\to 2} / dE_1. \tag{8.39}$$

Other commonly used names in the literature are *view factor, configuration factor,* and *angle factor.*

If $I_1$ is directionally dependent, the evaluation of $dF_{1\to 2}$ is laborious and its extension to finite areas even more burdensome. We simplify by assuming that both surfaces are *diffuse,* that is, that $I_1$ and $I_2$ are independent of $\theta_1$ and $\theta_2$. Then $I_1 = J_1/\pi$, $dE_1 = J_1\,dA_1$, and Eq. (8.39) becomes

$$dF_{dA_1\to dA_2} = \frac{\cos\theta_1\cos\theta_2}{\pi r^2}\,dA_2. \tag{8.40}$$

The intensity cancels out for diffuse surfaces, leaving a purely *geometric* factor that can be evaluated in advance before any system computations are begun.

If we repeat the analysis from surface 2 looking toward surface 1, we find the companion shape factor:

$$dF_{dA_2\to dA_1} = \frac{\cos\theta_2\cos\theta_1}{\pi r^2}\,dA_1. \tag{8.41}$$

This is exactly equivalent to reversing the subscripts 1 and 2, since both are diffuse and neither is preferred.

**Figure 8.18** Definition sketch for radiant interchange between two infinitesimal areas.

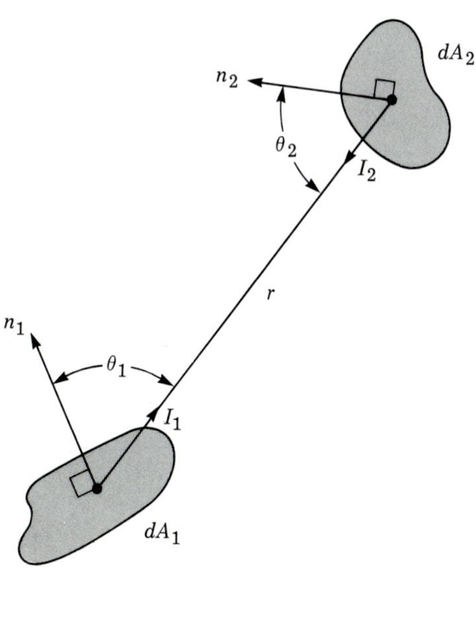

Comparing Eqs. (8.40) and (8.41), we deduce the *reciprocity rule* for infinitesimal diffuse areas:

$$dA_1 dF_{dA1 \to dA2} = dA_2\, dF_{dA2 \to dA1}. \tag{8.42}$$

This rule greatly simplifies the algebra of radiant interchange.

### 8.3.3 Finite-Area Shape Factors

The shape factor concept is readily extended to finite areas. Suppose that the radiosity $J_1$ is known all over surface $A_1$. Then we may integrate the energy radiated from $dA_1$ over $A_1$ and evaluate Eq. (8.39) with one area finite:

$$dF_{A1 \to dA2} = \frac{dA_2 \int J_1 \cos\theta_1(\cos\theta_2) dA_1/(\pi r^2)}{\int J_1\, dA_1}.$$

The integration is laborious if $J_1$ varies over the surface because of variations in surface temperature, emissivity, or reflectivity. We therefore assume uniform radiosity over both surfaces, for which Eq. (8.43) becomes

$$dF_{A1 \to dA2} = \frac{dA_2}{A_1} \int_{A1} \cos\theta_1(\cos\theta_2) dA_1/(\pi r^2), \tag{8.43}$$

which is purely geometric. Similarly, looking from $dA_2$ toward a finite $A_1$, we obtain

$$dF_{dA2 \to A1} = \int_{A1} \cos\theta_1(\cos\theta_2) dA_1/(\pi r^2). \tag{8.44}$$

By comparing Eqs. (8.43) and (8.44) we see that the reciprocity rule is also valid if one area is finite:

$$A_1\, dF_{A1 \to dA2} = dA_2\, dF_{dA2 \to A1}. \tag{8.45}$$

This requires that $J_1$ be uniform over the area $A_1$.

Finally, if both surfaces $A_1$ and $A_2$ are finite, the assumption of uniform radiosity yields the *finite* shape factor

$$A_1 F_{A1 \to A2} = \int_{A1} \int_{A2} \cos\theta_1(\cos\theta_2) dA_1\, dA_2/(\pi r^2). \tag{8.46}$$

The integral is unchanged if the subscripts are reversed. Thus there is reciprocity between finite areas:

$$A_1 F_{A1 \to A2} = A_2 F_{A2 \to A1}. \tag{8.47}$$

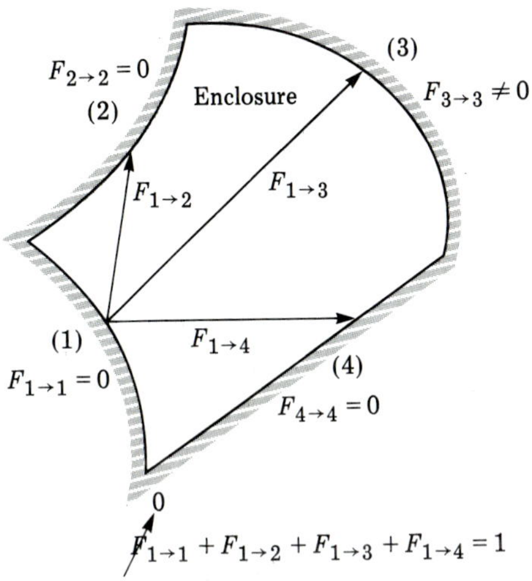


**Figure 8.19 Illustration of radiation within enclosures for diffuse surfaces with uniform radiosity. Surfaces 1, 2, and 4 cannot "see" themselves, unlike concave surface 3.**

This requires that both surfaces be diffuse and have uniform radiosity $J = \varepsilon E_b + \rho G$. If these conditions are met, the shape factors can be tabulated and the radiation streaming from one surface to another computed simply as

$$E_{1\to2} = A_1 F_{1\to2} J_1. \tag{8.48}$$

This enables us to set up multiple-interchange problems in a matrix or algebraic network form, to be discussed in Section 8.4.2.

### 8.3.4 The Summation Rule for Enclosures

In addition to the reciprocity rules (Eqs. 8.42, 8.45, and 8.48) for diffuse, uniform-radiosity surfaces, two additional rules are useful. The first is the concept of "seeing oneself." Figure 8.19 shows four different surfaces: Surfaces 1 and 2 are convex, surface 3 is concave, and surface 4 is planar. Only surface 3 can "see itself" and thus has a nonzero self-radiation shape factor. Thus a rule exists for self-radiation:

*Convex or plane surface:*

$$F_{i\to i} = 0,$$

*Concave surface:*

$$F_{i\to i} \neq 0.$$

The other rule concerns radiation in enclosures and states that the total radiation equals the sum of its parts. For any surface $i$ exchanging radiation with an enclosure consisting of $N$ surfaces, the *summation rule* states that

$$\sum_{j=1}^{N} F_{i\to j} = 1. \tag{8.49}$$

The enclosure in Fig. 8.19 illustrates this rule for the particular case of surface 1, which cannot "see itself." A second example would be for surface 3, which "sees itself":

$$F_{3\to 3} + F_{3\to 1} + F_{3\to 2} + F_{3\to 4} = 1. \tag{8.50}$$

This rule is needed to complete a typical set of algebraic relations for interchange in enclosures.

### 8.3.5 The Two-Surface Enclosure

A common configuration is illustrated in Fig. 8.20: a convex surface 1 completely surrounded by surface 2. There are four factors $F_{i\to j}$, $i, j = 1, 2$. They are related by two summation rules, one reciprocity

**Figure 8.20** Shape factors for a two-surface enclosure with diffuse, uniform-radiosity walls.

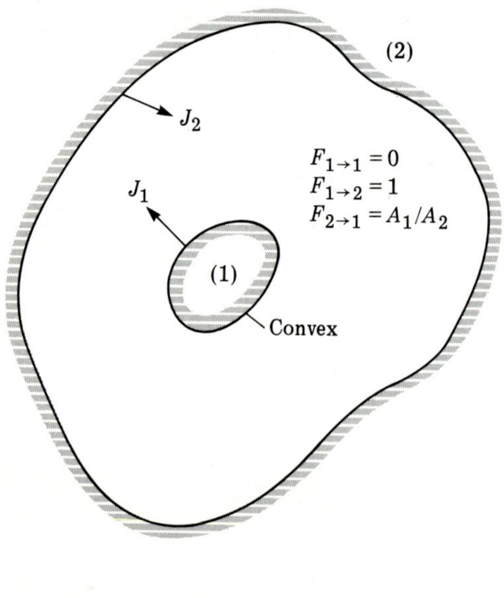

rule, and the fact that surface 1 cannot "see itself":

$$F_{1\to1} + F_{1\to2} = 1,$$
$$F_{2\to1} + F_{2\to2} = 1,$$
$$A_1F_{1\to2} = A_2F_{2\to1},$$
$$F_{1\to1} = 0.$$

By rearranging we find, almost by inspection, unique values for the three unknown shape factors:

$$F_{1\to2} = 1,$$
$$F_{2\to1} = A_1/A_2, \tag{8.51}$$
$$F_{2\to2} = 1 - (A_1/A_2).$$

If $A_1$ is much smaller than $A_2$, then $F_{2\to1} \doteq 0$ and $F_{2\to2} \doteq 1$, which are common approximations for a small body enclosed within a large room.

In general, for an enclosure with $N$ surfaces, there are $N^2$ shape factors, which may be displayed in a square array:

$$\begin{matrix} F_{1\to1} & F_{1\to2} & \cdots & F_{1\to N} \\ F_{2\to1} & F_{2\to2} & \cdots & F_{2\to N} \\ \vdots & \vdots & & \vdots \\ F_{N\to1} & F_{N\to2} & \cdots & F_{N\to N} \end{matrix}$$

These $N^2$ factors are not independent. Subtracting $N(N - 1)/2$ reciprocity relations and $N$ summation rules leaves only $N(N - 1)/2$ independent shape factors that need evaluation. If, further, all surfaces are convex or flat, the $N$ diagonal terms are all zero, leaving only $N(N - 3)/2$ independent factors. For example, if $N = 5$, or $5^2 = 25$ total factors, only $5(5 - 1)/2$, or 10, concave wall shape factors need direct computations, and only $5(5 - 3)/2 = 5$ independent flat wall factors are needed.

In all these computations of the shape factor, it is assumed that the medium between the two surfaces is either a vacuum or a "nonparticipating" gas, that is, a gas such as oxygen or nitrogen that causes negligible scatter and absorption of radiant energy. Further details on shape factor theory are given in a review article [28]. The problem of interchange between nondiffuse or specular surfaces is treated in [2, Chapter 5].

### Example 8.8

A 5-cm-diameter sphere is at 600°C and is near an infinite wall at 100°C. Both surfaces are black. Estimate the net radiant heat transfer between the two.

*Solution* First we need the shape factors, which we deduce by inspection (see the accompanying figure). If the sphere were totally enclosed by two walls, $F_{1\to2}$ would be unity, with 2 representing both walls. With only one wall, the sphere is half-enclosed and we see that $F_{1\to2} = 0.5$, $F_{2\to1} = 0$. The net heat radiated is

$$q_{1\to2} = E_{1\to2} - E_{2\to1} = A_1F_{1\to2}E_{b1} - A_2F_{2\to1}E_{b2}.$$

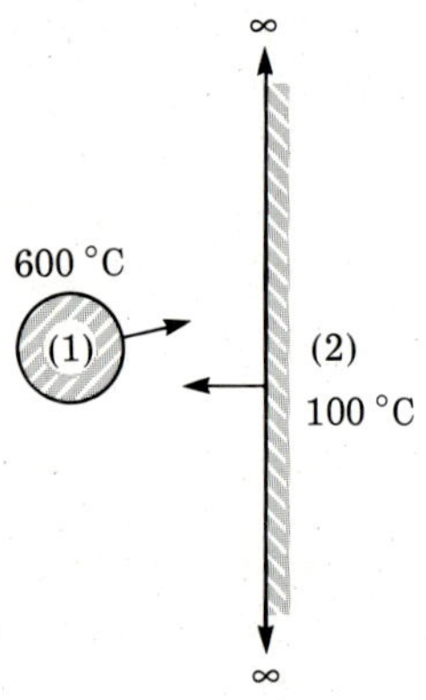


Even though $F_{2\to1} = 0$, the second term is not negligible, because $A_2 \to \infty$ and from Eq. (8.47) $A_1F_{1\to2} = A_2F_{2\to1}$. Thus

$$q_{1\to2} = A_1F_{1\to2}(E_{b1} - E_{b2}), \qquad E_b = \sigma T^4,$$

or

$$q_{1\to2} = \pi\,(0.05\text{ m})^2(0.5)(5.67 \times 10^{-8}\text{ W/m}^2\cdot\text{K}^4)[(873\text{K})^4 - (373\text{K})^4]$$

$$= 125\text{ W.} \quad [Ans.]$$

We must be careful always to use absolute temperature in radiation computations. ■

### 8.3.6 Analytical Evaluation of Shape Factors

The shape factors are defined for differential and finite surfaces by Eqs. (8.40), (8.43), (8.44), and (8.46), assuming diffuse surfaces with uniform radiosity. The integrals involved look deceptively simple but actually are generally quite difficult to evaluate. Fortunately, many solutions are reported in the literature [49]. Hamilton and Morgan [29] give extensive tables of shape factors, and the text by Siegel and Howell [1] lists 32 explicit formulas plus references to find 150 others. H. C. Hottel, in the text by McAdams [13], explains an intriguing "crossed-string analogy" method, which we use in Problem 8.51 in this text. The text by Jakob [30] discusses the experimental determination of shape factors. Any attempt to compute a shape factor by hand rapidly leads to a feeling of extreme gratitude to these authors for compiling accurate formulas and tables.

The following example illustrates a relatively simple shape factor result.

#### Example 8.9

Find a formula for the (diffuse) shape factor $dF_{1\to 2}$ for the configuration of Fig. 8.21. The differential area $dA_1$ is parallel to and on the centerline of the finite disk of radius $R_2$.

*Solution* The appropriate shape factor definition is Eq. (8.44) with subscripts reversed:

$$dF_{dA_1\to A_2} = \int_{A_2} \cos\theta_1(\cos\theta_2)dA_2/(\pi r^2). \tag{1}$$

A suitable differential area for surface 2 is the circular ring strip shown in Fig. 8.21, $dA_2 = 2\pi R dR$. All around this strip the angles $\theta_1$ and $\theta_2$ and the distance $r$ are constant:

$$\cos\theta_1 = \cos\theta_2 = L/(L^2 + R^2)^{1/2},$$

$$r^2 = L^2 + R^2.$$

Figure 8.21 should be inspected to verify these relations. Substituting into Eq. (1) above, we may integrate:

$$dF_{1\to 2} = \int_0^{R_2} \frac{L^2/(L^2 + R^2)}{\pi(L^2 + R^2)} 2\pi R dR$$

$$= R_2^2/(L^2 + R_2^2). \quad [Ans.]$$

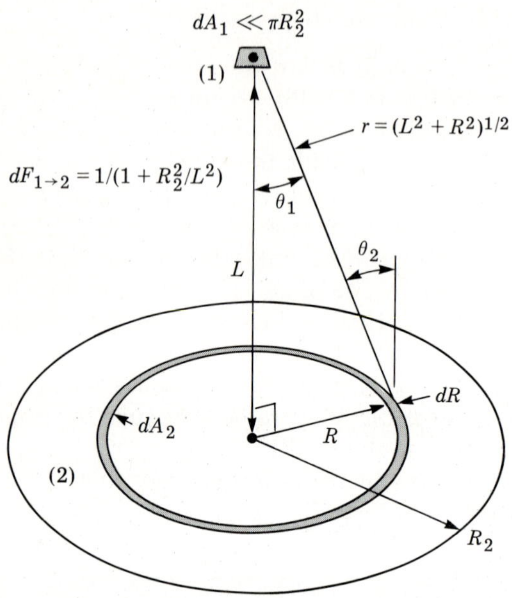


**Figure 8.21** **Shape factor evaluation for a small element $dA_1$ parallel to and on the centerline of a finite disk $R_2$. See Example 8.9 for the result.**

This result was unusually easy to obtain and therefore is an appropriate example for a writer of limited integration ability. To extend this result to compute the shape factor for two *finite* disks is quite difficult. (See Table 8.4.) ■

---

Some shape factors for two-dimensional geometries are listed in Table 8.3. The surfaces 1 and 2 are viewed on edge in the table and are assumed to have infinite length into the paper.

Table 8.4 lists some three-dimensional shape factors. The resulting formulas are considerably more complex than those in Table 8.3. To assist in routine computations, the first three formulas in Table 8.4 are plotted. Figure 8.22 shows the shape factor for parallel coaxial disks. Equal opposing parallel rectangles are plotted in Fig. 8.23, and Fig. 8.24 charts rectangles at 90° with a common edge. For other geometries, consult [49] and [29] especially.

**Table 8.3 Shape factors for two-dimensional geometries**

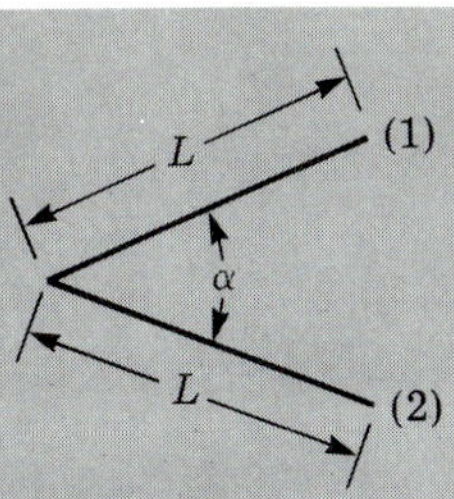


1. $F_{1\to2} = F_{2\to1} = 1 - \sin(\alpha/2)$

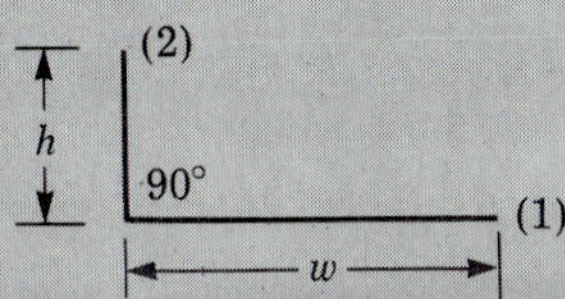


2. $F_{1\to2} = 0.5[1 + h/w - (1 + h^2/w^2)^{1/2}]$

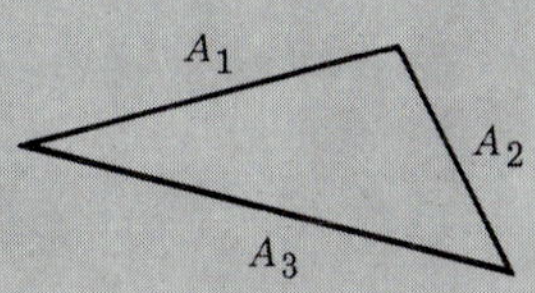


3. $F_{1\to2} = (A_1 + A_2 - A_3)/(2A_1)$

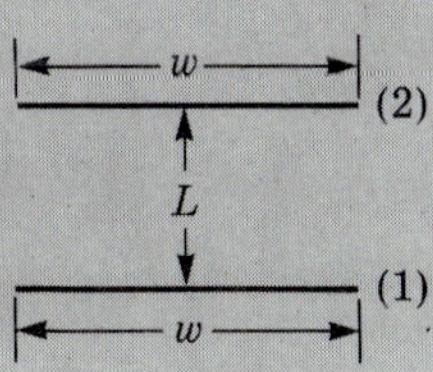


4. $F_{1\to2} = F_{2\to1} = (1 + L^2/w^2)^{1/2} - L/w$

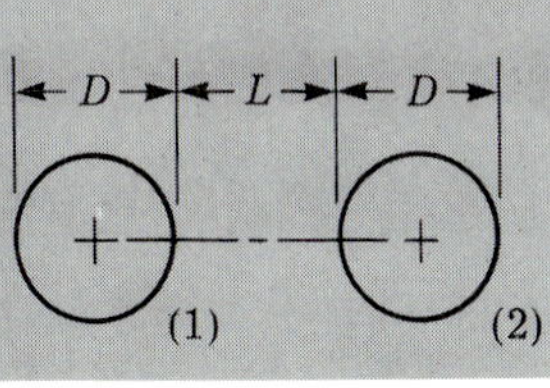


5. $F_{1\to2} = F_{2\to1}$

$$= \frac{1}{\pi}[(X^2 - 1)^{1/2} + \sin^{-1}(1/X) - X],$$

where $X = 1 + L/D$

### Example 8.10

The room in the accompanying figure is 20 ft by 20 ft wide and 9 ft high. The floor is at 100°F, the walls are at 60°F, and the ceiling is at 40°F. All surfaces are assumed black. Estimate the net radiation heat transfer (a) from floor to walls and (b) from floor to ceiling.

**Table 8.4** Shape factors for three-dimensional geometries

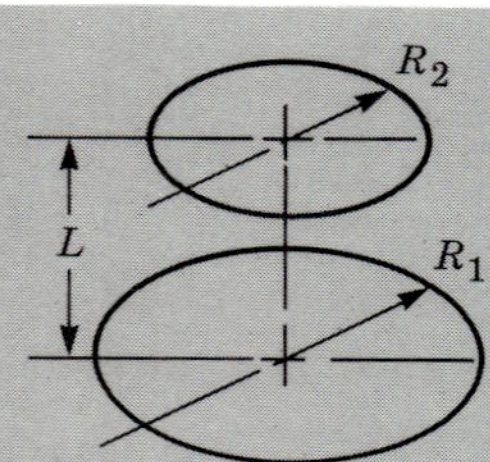


1. Let $Z_1 = R_1/L$, $Z_2 = R_2/L$, $X = 1 + (1 + Z_2^2)/Z_1^2$:

$$F_{1\to 2} = 0.5[X - (X^2 - 4Z_2^2/Z_1^2)^{1/2}]$$

(See Fig. 8.22.)

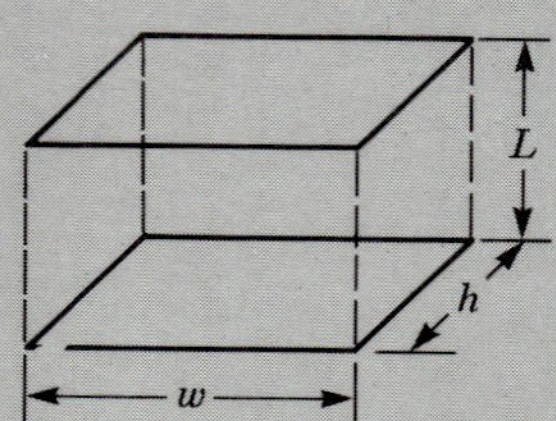


2. $$F_{1\to 2} = \frac{2}{\pi XY}\left\{\ln\left[\frac{(1 + X^2)(1 + Y^2)}{1 + X^2 + Y^2}\right]^{1/2} + Y\sqrt{1 + X^2}\tan^{-1}\frac{Y}{\sqrt{1 + X^2}} + X\sqrt{1 + Y^2}\tan^{-1}\frac{X}{\sqrt{1 + Y^2}} - X\tan^{-1}X - Y\tan^{-1}Y\right\}$$

where $X, = w/L$ and $Y = h/L$. (See Fig. 8.23.)

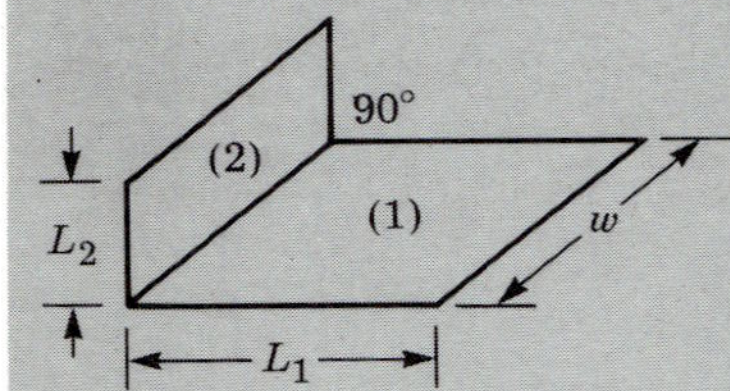


3. $$F_{1\to 2} = \frac{1}{\pi W}\left(W \tan^{-1}\frac{1}{W} + H \tan^{-1}\frac{1}{H} - \sqrt{H^2 + W^2}\tan^{-1}(H^2 + W^2)^{-1/2}\right.$$
$$\left. + \frac{1}{4}\ln\left\{\left[\frac{(1+W^2)(1+H^2)}{1+W^2+H^2}\right]\left[\frac{W^2(1+W^2+H^2)}{(1+W^2)(W^2+H^2)}\right]^{W^2}\left[\frac{H^2(1+H^2+W^2)}{(1+H^2)(H^2+W^2)}\right]^{H^2}\right\}\right),$$
where $H = L_2/w$ and $W = L_1/w$. (See Fig. 8.24.)

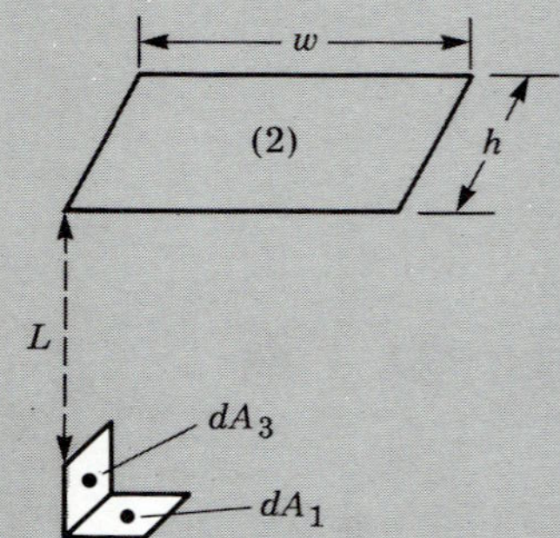


4. Corner differential element $dA_1$ parallel to rectangle:
$$F_{1\to 2} = \frac{1}{2\pi}\left(\frac{X}{\sqrt{1+X^2}}\tan^{-1}\frac{Y}{\sqrt{1+X^2}} + \frac{Y}{\sqrt{1+Y^2}}\tan^{-1}\frac{X}{\sqrt{1+Y^2}}\right),$$
where $X = w/L$ and $Y = h/L$.
Corner differential element $dA_3$ normal to rectangle:
$$F_{3\to 2} = \frac{1}{2\pi}\left(\tan^{-1}\frac{1}{Y} - \frac{Y}{\sqrt{X^2+Y^2}}\tan^{-1}\frac{1}{\sqrt{X^2+Y^2}}\right)$$
where $X = w/h$ and $Y = L/h$.

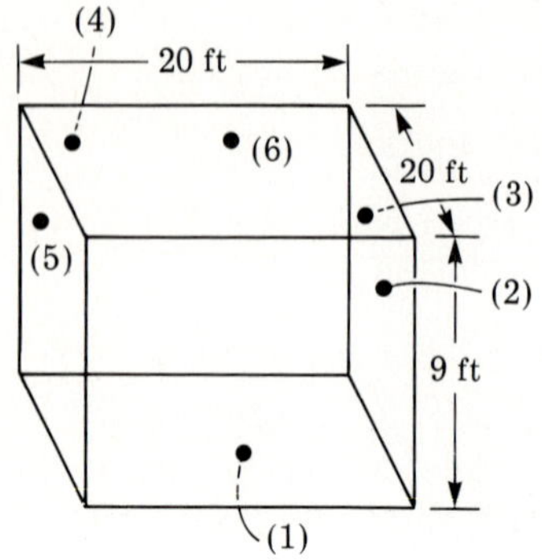


*Solution* Being square, the floor has an identical view of all four walls. Thus only two different shape factors are needed: floor to wall and floor to ceiling. For the view from floor to one wall, Fig. 8.24 applies, with $L_1/w = 1.0$ and $L_2/w = 9/20 = 0.45$. Then, from Fig. 8.24 or item 3 in Table 8.4:

$$F_{1\to2} = F_{1\to3} = F_{1\to4} = F_{1\to5} = 0.1374.$$

Actually, of course, one can read only two figures at best from Fig. 8.24. Since there is only one other surface — the ceiling — we could use the summation rule (8.49) to compute

$$F_{1\to6} = 1 - 4F_{1\to2} = 1 - 4(0.1374) = 0.4504.$$

Alternatively, the floor-ceiling configuration fits Fig. 8.23, with $h/L = w/L = 20/9$. From Fig. 8.23 or the formula in item 2 of Table 8.4, we compute

$$F_{1\to6} = 0.4504.$$

For black surfaces, the net energy radiated from floor to walls is

$$\begin{aligned} q_{1\to2,3,4,5} &= 4A_1F_{1\to2}(E_{b1} - E_{b2}) = 4A_1F_{1\to2}\sigma(T_1^4 - T_2^4) \\ &= 4(400\ \text{ft}^2)(0.1374)(1.7121 \times 10^{-9})[(100 + 460)^4 \\ &\quad - (60 + 460)^4] \\ &= 9500\ \text{Btu/hr} \quad (\text{floor to 4 walls}). \quad [\textit{Ans. (a)}] \end{aligned}$$

The net energy from floor to ceiling is

$$\begin{aligned} q_{1\to6} &= A_1F_{1\to6}\sigma(T_1^4 - T_6^4) \\ &= (400\ \text{ft}^2)(0.4504)(1.7121 \times 10^{-9})[(100 + 460)^4 - (40 + 460)^4] \\ &= 11{,}100\ \text{Btu/hr} \quad (\text{floor to ceiling}). \quad [\textit{Ans. (b)}] \end{aligned}$$

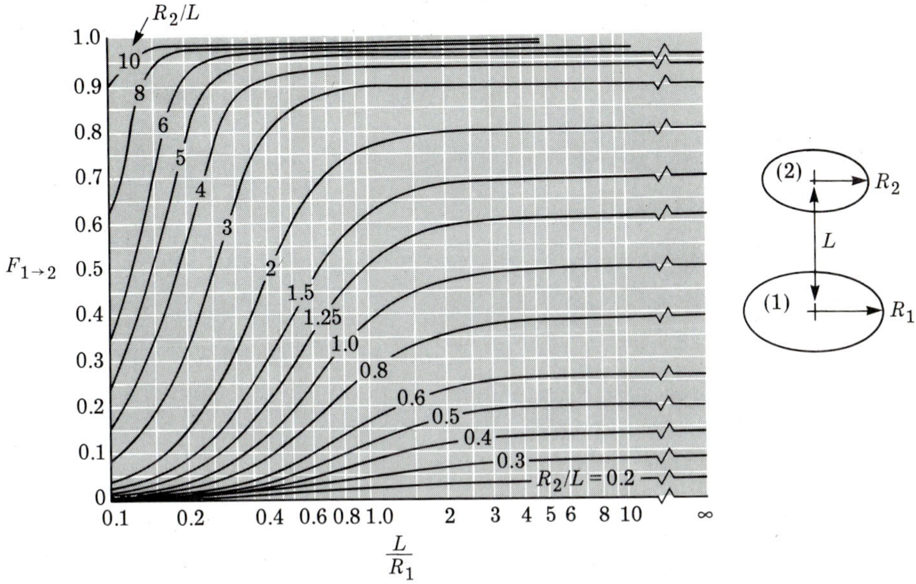


**Figure 8.22** Shape factor between two parallel coaxial disks. (From [29].)

**Figure 8.23** Shape factor for equal opposing parallel rectangles. (From [29].)

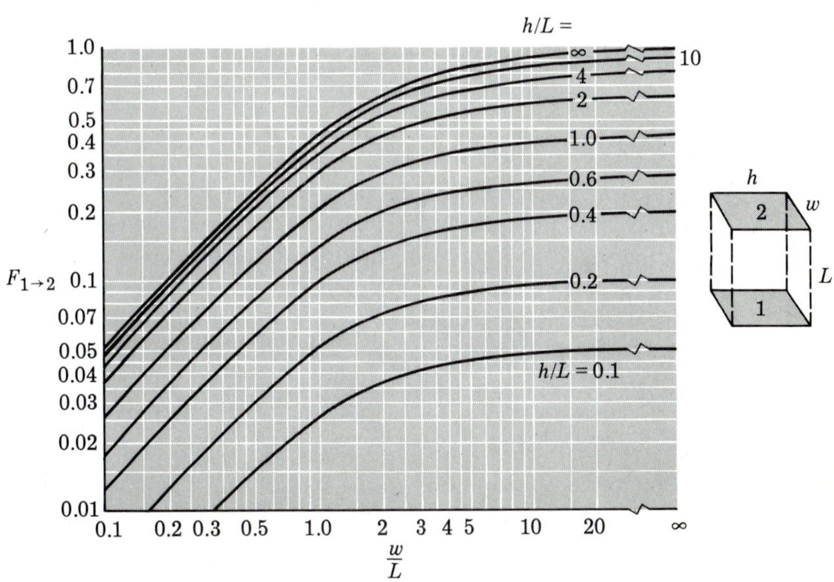

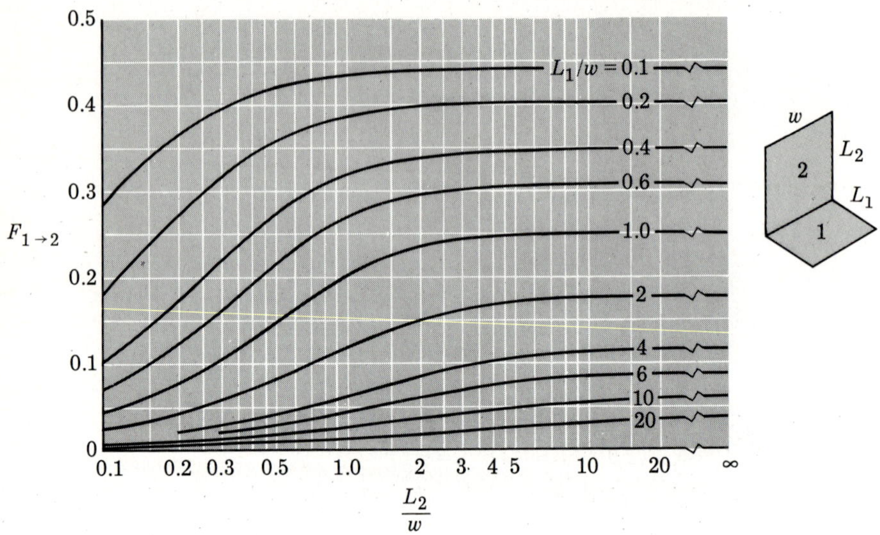


**Figure 8.24** Shape factor for rectangles at 90° with a common edge. (From [29].)

*Note:* The analysis is laborious without uniform radiosity and diffuse walls. ■

---

### 8.3.7 Shape Factor Algebra

Although the basic geometries listed in Table 8.4 are rather limited, they may be combined and manipulated into results for other geometries, using a technique known as *shape factor algebra*. The procedure involves use of the summation and reciprocity rules plus some shrewd maneuvers.

Suppose surface 1 looks toward several other surfaces (2, 3, 4, 5). Then its total radiant energy toward these surfaces must equal the sum of the various parts:

$$F_{1\to2,3,4,5} = F_{1\to2} + F_{1\to3} + F_{1\to4} + F_{1\to5}. \tag{8.52}$$

The opposite of this rule is *not* true because the surfaces (2, 3, 4, 5) do not have the same common area. That is,

$$F_{2,3,4,5\to1} \neq F_{2\to1} + F_{3\to1} + F_{4\to1} + F_{5\to1}. \tag{8.53}$$

Rather, use of the reciprocity rule $A_iF_{i\to j} = A_jF_{j\to i}$ would reveal the following rule for many surfaces looking toward a single one:

$$A_{2,3,4,5}F_{2,3,4,5\to 1} = A_2F_{2\to 1} + A_3F_{3\to 1} + A_4F_{4\to 1} + A_5F_{5\to 1}. \quad \textbf{(8.54)}$$

By use of Eqs. (8.52) and (8.54) we may generalize Table 8.4 into other geometries. For example, consider the basic rectangular corner geometry of Fig. 8.25(a), which has a known formula in item 3 of Table 8.4 and is plotted in Fig. 8.24.

Suppose we wish to compute the shape factor $F_{1\to 3}$ in Fig. 8.25(b), where the rectangles do *not* have a common edge. Then, from the summation rule (8.52), $F_{1\to 2,3} = F_{1\to 2} + F_{1\to 3}$. Rearranging, we have

$$F_{1\to 3} = F_{1\to 2,3} - F_{1\to 2}. \quad \textbf{(8.55)}$$

Since both terms on the right-hand side may be found from the basic geometry result of Fig. 8.24, the problem is solved. This result splits only one part of the basic geometry.

**Figure 8.25** **Development of complex shape factors from a single basic correlation: (a) basic geometry; (b) single-splitting process solved in Eq. (8.55); (c) double-splitting process solved in Eq. (8.58).**

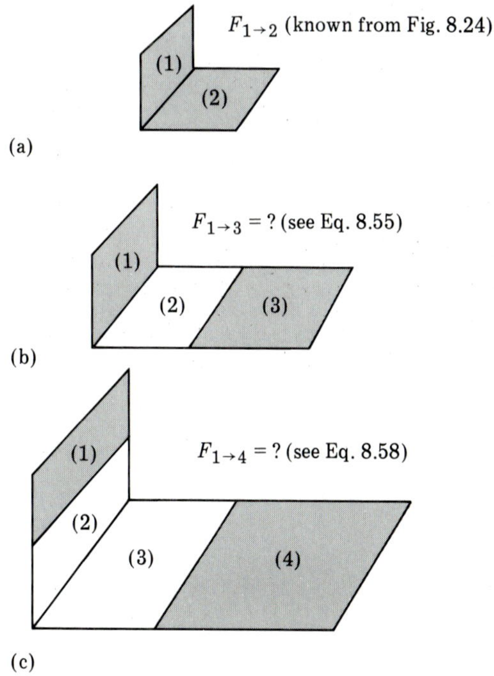

As a second example, Fig. 8.25(c) illustrates a double-splitting process. We wish to compute $F_{1\to 4}$, where the edges now are neither common nor in the corner. Yet the basic geometry, Fig. 8.25(a), still applies, as follows. From Eq. (8.54),

$$A_{1,2}F_{1,2\to 3,4} = A_1F_{1\to 3,4} + A_2F_{2\to 3,4} \qquad \textbf{(8.56)}$$
$$= A_1(F_{1\to 3} + F_{1\to 4}) + A_2F_{2\to 3,4}.$$

The factors $F_{1,2\to 3,4}$ and $F_{2\to 3,4}$ are in basic form but $F_{1\to 3}$ is not. Therefore, use the reciprocity rule and the single-splitting result from Eq. (8.55) to obtain

$$A_1F_{1\to 3} = A_3F_{3\to 1} = A_3(F_{3\to 2,1} - F_{3\to 2}). \qquad \textbf{(8.57)}$$

Using this to substitute for $F_{1\to 3}$ in Eq. (8.56) and rearranging, we obtain the desired result:

$$F_{1\to 4} = \frac{1}{A_1}(A_{1,2}F_{1,2\to 3,4} + A_3F_{3\to 2} - A_3F_{3\to 2,1} - A_2F_{2\to 3,4}). \qquad \textbf{(8.58)}$$

All shape factors on the right-hand side are in the standard form plotted in Fig. 8.24. Note the sharp increase in complexity of Eq. (8.58) compared to the single-splitting (8.55).

Hamilton and Morgan [29] show how this shape factor algebra process may be extended to compute the shape factor between rectangles in practically any orientation, using only Fig. 8.23 and 8.24 to develop the results. The algebra is quite complex, however. For difficult orientations of irregularly shaped surfaces, it is usually more efficient to compute the shape factor from a direct numerical simulation of the basic integral relation, Eq. (8.46).

### Example 8.11

For the internal surfaces of the right circular cylinder in the accompanying figure, compute $F_{1\to 3}$ and $F_{3\to 3}$.

*Solution* Neither result is given in our tables and figures. However, for the two parallel disks, Fig. 8.22 applies. For $L/R_1 = 10/4 = 2.5$ and $R_2/L = 4/10 = 0.4$, read from Fig. 8.22 or compute from item 1 of Table 8.4,

$$F_{1\to 2} = 0.1230$$

Then, from the summation rule (8.49), since $F_{1\to 1} = 0$,

$$F_{1\to 3} = 1 - F_{1\to 2} = 1 - 0.1230 = 0.8770. \quad [Ans.]$$

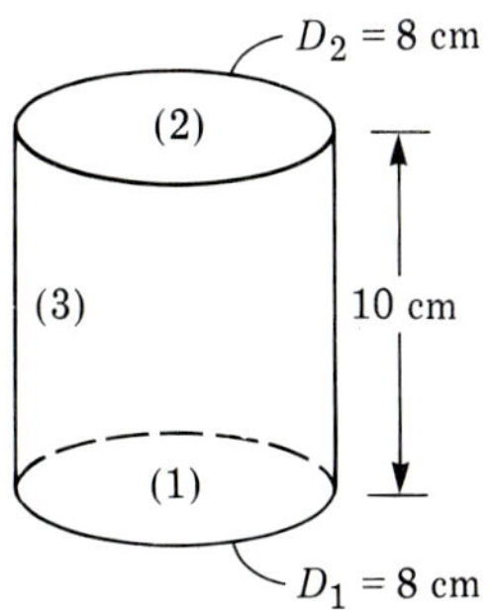


This result would be most difficult to obtain by integration.

For the cylindrical surface 3, the summation rule requires

$$F_{3\to3} + F_{3\to1} + F_{3\to2} = 1 = F_{3\to3} + 2F_{3\to1}, \qquad \textbf{(1)}$$

since the cylinder has identical views of the top and bottom disks. To obtain $F_{3\to1}$, use the reciprocity rule (8.47):

$$A_3F_{3\to1} = A_1F_{1\to3},$$

where $A_3 = 2\pi R_1 L = 80\pi\ \text{cm}^2$ and $A_1 = \pi R_1^2 = 16\pi\ \text{cm}^2$. Then

$$F_{3\to1} = (A_1/A_3)F_{1\to3} = \left(\frac{16\pi}{80\pi}\right)(0.8770) = 0.1754.$$

Finally, from Eq. (1) above,

$$F_{3\to3} = 1 - 2(0.1754) = 0.6492. \quad [Ans.]$$

The cylindrical surface is thus rather narcissistic. Here again a direct integration of $F_{3\to3}$ would be very laborious. ■

## Example 8.12

Find the shape factor between the small element $dA_1$ in the floor and the entire 6-m-by-6-m ceiling in the figure.

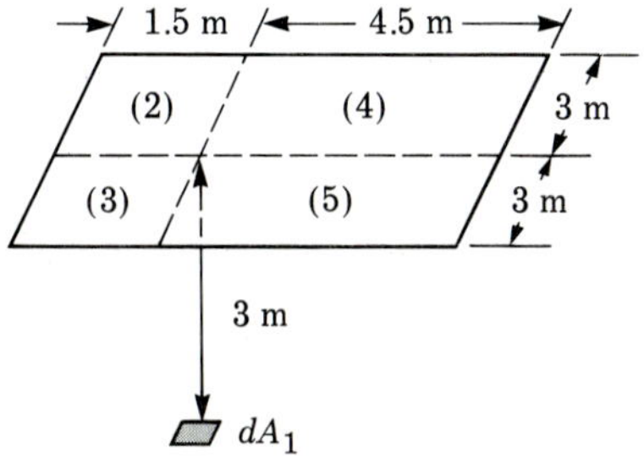

*Solution* We have no formula for the entire ceiling, but item 4 of Table 8.4 applies if we break the ceiling into four parts, as shown in the figure:

$$F_{1\to 2,3,4,5} = F_{1\to 2} + F_{1\to 3} + F_{1\to 4} + F_{1\to 5} \tag{1}$$

$$= 2F_{1\to 2} + 2F_{1\to 4} \quad \text{by symmetry.}$$

For subarea 2, $X = 3/3 = 1.0$ and $Y = 1.5/3 = 0.5$; hence from item 4 of Table 8.4 compute $F_{1\to 2} = 0.0902$. For subarea 4, $X = 4.5/3 = 1.5$ and $Y = 3/3 = 1.0$, from which we compute $F_{1\to 4} = 0.1588$. Then from Eq. (1) above, the desired total shape factor is

$$F_{1\to 2,3,4,5} = 2(0.0902) + 2(0.1588) = 0.498. \quad [Ans.]$$

Thus the ceiling captures approximately 50% of the radiation emitted by this particular floor element. ■

## 8.4 Radiation Between Multiple Gray Surfaces

Knowledge of the shape factor essentially completes the problem of blackbody interchange. Given $F_{1\to 2}$, the net radiation between any two blackbodies is given by

$$q(\text{net})_{1\to 2} = A_1 F_{1\to 2}\sigma(T_1^4 - T_2^4). \tag{8.59}$$

This result is independent of the presence of other surfaces (3, 4, 5, and so on), since a blackbody has zero reflection.

For nonblack surfaces, the net interchange must account for reflection of radiation from other surfaces. The algebraic relations are straightforward if we assume that all surfaces are gray, diffuse, and opaque:

$$\tau = 0, \quad \alpha = \varepsilon, \quad \rho = 1 - \varepsilon. \tag{8.60}$$

Knowledge of the emissivity $\varepsilon$ thus fully characterizes such an idealized surface. Interchange relations for real (nongray, nondiffuse) surfaces are far more complicated and are treated, for example, in [2].

Recalling the concept of radiosity $J$ and irradiation $G$ from Fig. 8.17, we can compute the net heat radiated from any surface $i$ as given by

$$q_i = A_i(J_i - G_i), \tag{8.61}$$

where

$$J_i = \varepsilon_i E_{bi} + \rho_i G_i.$$

We have assumed that $J_i$ and $G_i$ are uniform over the surface $A_i$. Eliminating $G_i$ in Eqs. (8.61) and replacing $\rho_i = 1 - \varepsilon_i$ from Eq. (8.60), we obtain a basic formulation for gray-body local net radiation:

$$q_i = \frac{\varepsilon_i A_i}{(1 - \varepsilon_i)}(E_{bi} - J_i). \tag{8.62}$$

We may interpret this either algebraically or electrically.

### 8.4.1 The Electrical Analogy

Gray-body relations such as Eq. (8.62) were given an electrical analogy in 1956 by Oppenheim [31], with $(E_{bi} - J_i)$ as a "driving potential" and $q_i$ as a "current":

$$q_i = \frac{E_{bi} - J_i}{R_i}, \qquad R_i = (1 - \varepsilon_i)/(\varepsilon_i A_i), \tag{8.63}$$

where $R_i$ represents the "radiation resistance" of the gray surface.

In a similar manner, the shape factor relation between surfaces $q_{\text{net}} = A_1 F_{1\to 2}(J_1 - J_2)$, can be interpreted as

$$q_{\text{net}} = \frac{J_1 - J_2}{R_{12}}, \qquad R_{12} = 1/(A_1 F_{1\to 2}), \tag{8.64}$$

where $R_{12}$ is a "shape" or "view" resistance between the two surfaces. These two concepts of surface and view resistances can be combined into an electrical "network" that models the radiation interchange between gray diffuse surfaces.

A simple example of the electric analogy is shown in Fig. 8.26 for two gray surfaces that "see" only each other. Surface 1 has "ideal" potential $E_{b1} = \sigma T_1^4$, but because of gray-body surface resistance it delivers only a lesser "surface" potential $J_1$. A similar gray-body surface loss occurs for surface 2. The heat flux "circuit" is completed by connecting the two surface potentials with a "shape" resistance $1/(A_1 F_{1\to 2})$. In terms of the overall ideal potential difference, the net radiation heat flux between the two is $\Delta E_b/\Sigma R$, or:

$$q_{\text{net},1\to 2} = \frac{E_{b1} - E_{b2}}{\dfrac{1 - \varepsilon_1}{\varepsilon_1 A_1} + \dfrac{1}{A_1 F_{1\to 2}} + \dfrac{1 - \varepsilon_2}{\varepsilon_2 A_2}}, \tag{8.65}$$

where $E_{b1} - E_{b2} = \sigma(T_1^4 - T_2^4)$. The electric analogy affords a rather simple formulation for the heat flux, relating it directly to the blackbody energies and minimizing the algebra.

Two examples of a three-surface enclosure problem are shown in

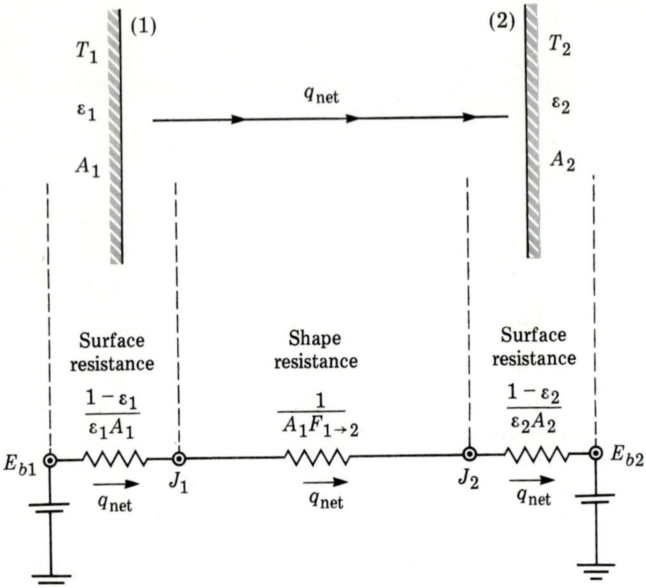


**Figure 8.26** Electric analogy for net heat flux between two gray surfaces that "see" only each other.

Fig. 8.27. It is assumed that none of the surfaces can "see themselves." In Fig. 8.27(a), all three surfaces have finite heat transfer. In Fig. 8.27(b), surface 3 is insulated and merely "reradiates" heat received from surfaces 1 and 2.

In Fig. 8.27(a), application of Kirchhoff's current laws to each node gives three simultaneous equations for the three unknown radiosities $J_1$, $J_2$, and $J_3$. Example 8.13 illustrates this procedure.

In Fig. 8.27(b), the insulated surface 3 has no ground potential $E_{b3}$ but simply "floats" at a potential $J_3$ and serves as a reradiating path for heat flux from 1 to 2. The shape resistances between $J_1$ and $J_2$ form a parallel system of net resistance

$$R_{\text{net}} = \frac{R_{12}(R_{13} + R_{23})}{R_{12} + R_{13} + R_{23}},$$

which when added to the surface resistances gives the net heat radiated from 1 to 2:

$$q_{\text{net},1\to 2} = \frac{E_{b1} - E_{b2}}{\dfrac{1 - \varepsilon_1}{\varepsilon_1 A_1} + R_{\text{net}} + \dfrac{1 - \varepsilon_2}{\varepsilon_2 A_2}}. \tag{8.66}$$

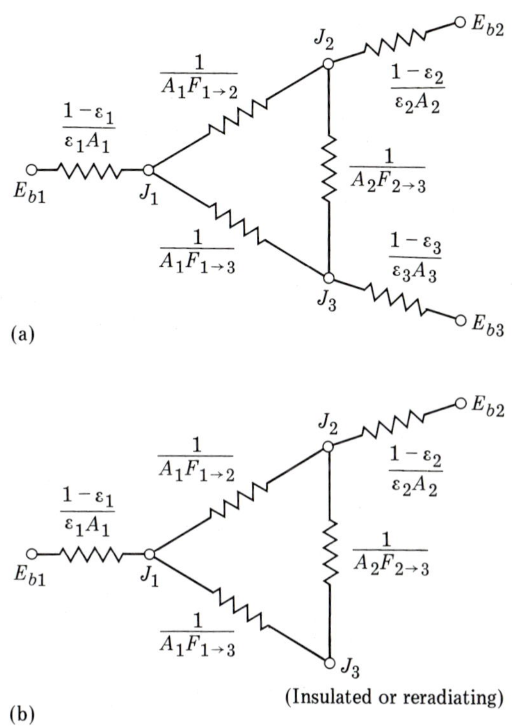


**Figure 8.27 Radiation network for a three-surface enclosure: (a) all three surfaces radiating net heat flux; (b) surface 3 insulated or reradiating. These networks assume that no surface can "see" itself.**

Again the straightforward simplicity of the electric analogy is demonstrated. It is assumed of course that the enclosure contains a vacuum or a nonparticipating gas.

In principle, the electric analogy may be extended to any number of surfaces forming an enclosure. Participating gases and nondiffuse or specular surfaces can also be accounted for. A discussion of these extensions to the analogy is given in the text by Holman [32].

## Example 8.13

The right triangular duct in the first figure is very long and has surface conditions $T_1 = 500\text{K}$, $\varepsilon_1 = 0.625$, $T_2 = 600\text{K}$, $\varepsilon_2 = 0.4$,

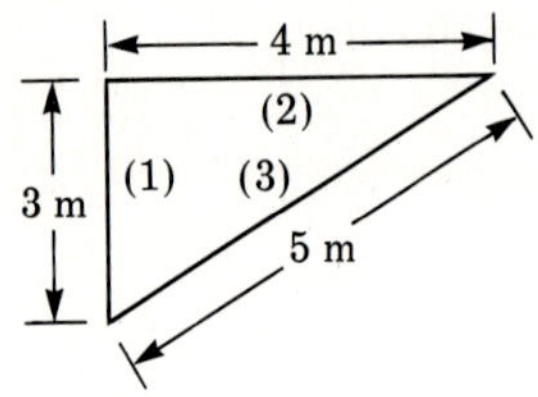


$T_3 = 800\text{K}$, and $\varepsilon_3 = 0.8$. Set up the electric network and compute the internal radiosities per unit width into the paper.

*Solution* This problem fits the network of Fig. 8.27(a). The necessary shape factors are obtained from item 3 of Table 8.3:

$$F_{1\to2} = (A_1 + A_2 - A_3)/2A_1 = (3 + 4 - 5)/2(3) = 0.333,$$

$$F_{1\to3} = (A_1 + A_3 - A_2)/2A_1 = (3 + 5 - 4)/2(3) = 0.667,$$

$$F_{2\to3} = (A_2 + A_3 - A_1)/2A_2 = (4 + 5 - 3)/2(4) = 0.75,$$

from which the shape resistances are

$$R_{12} = 1/A_1F_{1\to2} = 1/3(0.333) = 1.0\ \text{m}^{-2},$$

$$R_{13} = 1/A_1F_{1\to3} = 1/3(0.667) = 0.5\ \text{m}^{-2},$$

$$R_{23} = 1/A_2F_{2\to3} = 1/4(0.75) = 0.333\ \text{m}^{-2}.$$

Meanwhile, the three surface resistances are

$$R_1 = (1 - \varepsilon_1)/\varepsilon_1A_1 = (1 - 0.625)/0.625(3) = 0.2\ \text{m}^{-2},$$

$$R_2 = (1 - \varepsilon_2)/\varepsilon_2A_2 = (1 - 0.4)/0.4(4) = 0.375\ \text{m}^{-2},$$

$$R_3 = (1 - \varepsilon_3)/\varepsilon_3A_e = (1 - 0.8)/0.8(5) = 0.05\ \text{m}^{-2}.$$

The ideal wall potentials are

$$E_{b1} = \sigma T_1^4 = (5.67 \times 10^{-8})(500)^4$$

$$= 3544\ \text{W/m}^2,$$

$$E_{b2} = \sigma T_2^4 = 7348\ \text{W/m}^2,$$

$$E_{b3} = \sigma T_3^4 = 23{,}224\ \text{W/m}^2.$$

The complete numerical network is shown in the second figure. Setting the net "current" entering each junction $J_i$ equal to zero, we obtain the following algebraic relations:

$$(3544 - J_1)/0.2 + (J_2 - J_1)/1.0 + (J_3 - J_1)/0.5 = 0, \quad \textbf{(1)}$$

$$(7348 - J_2)/0.375 + (J_1 - J_2)/1.0 + (J_3 - J_2)/0.333 = 0, \quad \textbf{(2)}$$

$$(23{,}224 - J_3)/0.05 + (J_1 - J_3)/0.5 + (J_2 - J_3)/0.333 = 0. \quad \textbf{(3)}$$

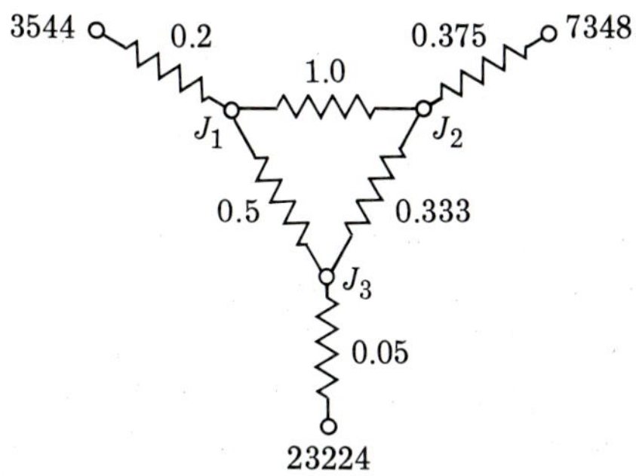


These may be rearranged into three simultaneous equations:

$$8J_1 - J_2 - 2J_3 = 17{,}720,$$

$$-J_1 + 6.667J_2 - 3J_3 = 19{,}595,$$

$$-2J_1 - 3J_2 + 25J_3 = 464{,}480.$$

These are well-conditioned relations. The solutions are

$J_1 = 9174\ \text{W/m}^2,$

$J_2 = 13{,}749\ \text{W/m}^2,$

$J_3 = 20{,}963\ \text{W/m}^2.$ [*Ans.*]

From Eqs. (1), (2), and (3), the net wall heat fluxes are $q_1 = (3544 - 9174)/0.2 = -28{,}150$ W, $q_2 = (7348 - 13{,}749)/0.375 = -17{,}070$ W, and $q_3 = (23{,}224 - 20{,}963)/0.05 = +45{,}220$ W. The sum of these is zero. ∎

### 8.4.2 Matrix Formulation

The electric analogy of the previous section is simple and gives good insight. However, it is laborious if the enclosure has many surfaces, say $N \geqslant 5$. For complex enclosures, it is more convenient to rewrite the "network" equations in matrix form for solution by computer.

For the three-surface enclosure of Fig. 8.27(a), for example, the "zero current" network relation for node 1 is

$$\frac{E_{b1} - J_1}{(1 - \varepsilon_1)/\varepsilon_1 A_1} + \frac{J_2 - J_1}{1/(A_1 F_{1\to 2})} + \frac{J_3 - J_1}{1/(A_1 F_{1\to 3})} = 0.$$

This may be rearranged into an algebraic equation for the unknown radiosities:

$$J_1 - (1 - \varepsilon_1)F_{1\to 2}\, J_2 - (1 - \varepsilon_1)F_{1\to 3}\, J_3 = \varepsilon_1 E_{b1}.$$

This illustrates the form of the basic relation: The "nodal" value $J_{i=1}$ has coefficient unity, while the other radiosities $J_{j\neq 1}$ have coefficients $-(1 - \varepsilon_i)F_{i\to j}$. Thus the general matrix formulation for each $i$th node is as follows:

$$J_i - (1 - \varepsilon_i) \sum_{j=1}^{N} F_{i\to j} J_j = \varepsilon_i E_{bi}. \tag{8.67}$$

Applying this relation for $i = 1, 2, 3, \ldots, N$ gives $N$ linear algebraic equations for the $N$ radiosities. Equations (8.67) are appropriate for surfaces with known temperatures, so that $E_{bi} = \sigma T_i^4$ is a known constant.

An alternative boundary condition is that of known surface heat flux, $q_i''$, instead of known temperature. In this case we can eliminate $E_{bi}$ in favor of $q_i$ from Eq. (8.63) and rearrange Eqs. (8.67) as follows:

$$J_i - \sum_{j=1}^{N} F_{i\to j} J_j = q_i''. \tag{8.68}$$

Again, applying this to each surface gives $N$ equations in the $N$ unknown radiosities. For mixed conditions, we use either Eqs. (8.67) or (8.68), depending on whether temperature or heat flux is known at the boundary.

If all surfaces have known temperature, we may rewrite the nodal relations directly in terms of the $N$ unknown heat fluxes, $q_i''$:

$$\frac{1}{\varepsilon_i} q_i'' - \sum_{j=1}^{N} \frac{1 - \varepsilon_j}{\varepsilon_j} F_{i\to j} q_j'' = E_{bi} - \sum_{j=1}^{N} F_{i\to j} E_{bj}, \tag{8.69}$$

for $i = 1, 2, 3, \ldots, N$. All three of these basic nodal relations (8.67)–(8.69) are of the matrix form

$$\mathbf{AX} = \mathbf{B},$$

where $\mathbf{X}$ is the column matrix of unknowns ($J_i$ or $q_i''$). Thus these linear algebraic equations can be solved by any convenient method for well-conditioned matrixes, such as matrix inversion, Gauss elimination, or Gauss-Seidel iteration as in Section 3.4.2. A FORTRAN program for matrix inversion is given in [33].

### 8.4.3 Radiation Shields

An important application of gray-body interchange theory is the *radiation shield,* which is generally intended to reduce the net radiation heat transfer between two surfaces. To make a shield, one inserts a thin, highly reflective (low-emissivity) sheet of material between the two surfaces. As a rule of thumb, such shields reduce $q_{\text{net}}$ by a factor of $1/(1 + n)$, where $n$ is the number of shields. A common application is the use of multilayer shields (50 per inch or more) in vacuum insulation. A shield is also effective in isolating a gas temperature measurement from the (unwanted) radiation from nearby hot or cold walls.

Figure 8.28 shows a schematic of a single radiation shield. We compare the heat transfer $q_{\text{net}}$ from surface 1 to 2 with and without the shield. The figure shows the electric analogy of the surfaces plus the shield. We see that the shield introduces two additional surface resistances plus a doubling of the shape resistance between the surfaces.

With the shield absent, the total resistance between surfaces 1 and 2 is

$$R_0 = \frac{1 - \varepsilon_1}{\varepsilon_1 A_1} + \frac{1}{A_1 F_{1\to 2}} + \frac{1 - \varepsilon_2}{\varepsilon_2 A_2}.$$

With the shield inserted, the total resistance is

$$R = R_0 + \frac{2(1 - \varepsilon_s)}{\varepsilon_s A_s} + \frac{1}{A_1 F_{1\to s}} + \frac{1}{A_s F_{s\to 2}} - \frac{1}{A_1 F_{1\to 2}}. \tag{8.70}$$

There is clearly a substantial increase in $R$, especially if $\varepsilon_s$ is small (highly reflective shield), and the net radiation heat flux drops as $q = q_0 R_0 / R$. The percentage drop depends on the relative values

**Figure 8.28 Radiation heat flux between two surfaces separated by a radiation shield.**

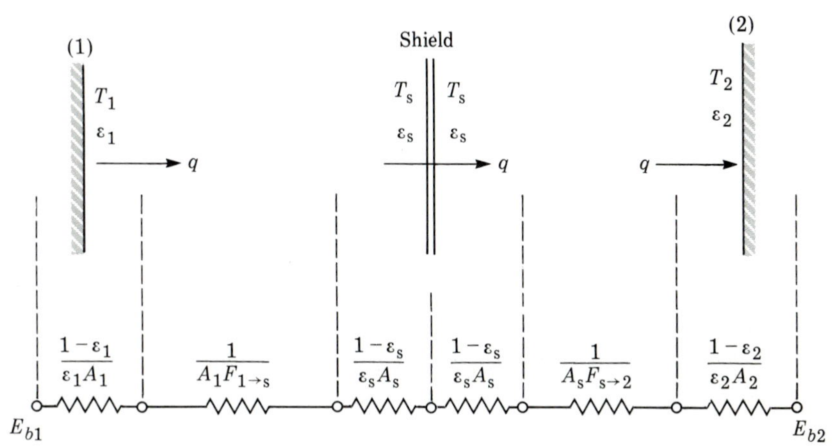

of ($\varepsilon_1$, $\varepsilon_2$, $\varepsilon_s$) and ($A_1$, $A_2$, $A_s$). To illustrate, suppose that $\varepsilon_1 = \varepsilon_2 = \varepsilon_s = \varepsilon$ (comparable shield and surface material) and $A_1 = A_2 = A_s$ (large, closely spaced rectangular surfaces). Further let $F_{1\to2} = F_{1\to s} = F_{s\to2} = F$. Then

$$R_0 = \frac{2(1-\varepsilon)}{\varepsilon A} + \frac{1}{AF},$$

$$R = R_0 + \frac{2(1-\varepsilon)}{\varepsilon A} + \frac{1}{AF} = 2R_0.$$

The single shield doubles the resistance, hence halves the net heat flux. If there are $n$ such shields,

$$R = R_0 + \frac{2n(1-\varepsilon)}{\varepsilon A} + \frac{n}{AF} = (1+n)R_0. \tag{8.71}$$

This proves our contention that the use of $n$ shields roughly reduces the heat flux by a factor of $1/(1+n)$.

## Example 8.14

As shown in the accompanying figure, a 2-mm-diameter thermocouple wire at 200°C is within a 10-cm-diameter duct whose walls are at 20°C. A thin shield, $r_s = 5$ mm, is proposed to reduce radiation heat loss from the wire. What shield emissivity $\varepsilon_s$ is required to reduce $q_1$ by 50%?

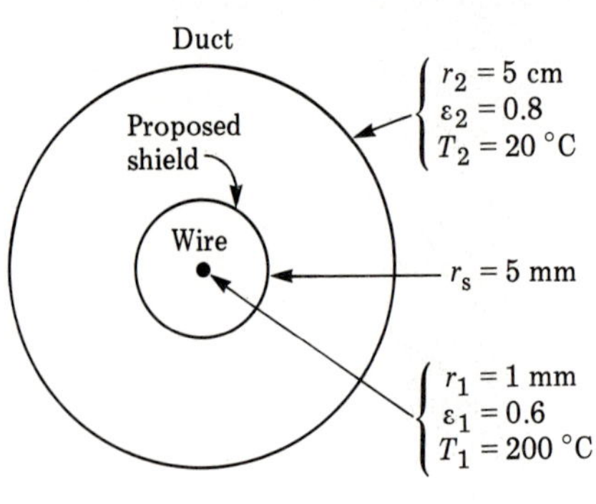


*Solution* This system fits the requirements of Fig. 8.28. Since the duct completely encloses the shield, which itself encloses the thermocouple, $F_{1\to s} = F_{s\to2} = F_{1\to2} = 1.0$. If the shield is absent, the total radiation resistance is

$$R_0A_1 = \frac{(1-\varepsilon_1)}{\varepsilon_1} + \frac{1}{F_{1\to 2}} + \frac{(1-\varepsilon_2)A_1}{\varepsilon_2 A_2}$$

$$= \frac{1-0.6}{0.6} + \frac{1}{1.0} + \frac{(1-0.8)(2\pi r_1 L)}{0.8(2\pi r_2 L)}$$

$$= 0.667 + 1.0 + 0.005 = 1.672,$$

where $L$ is the axial length of the wire. The ideal emissive powers are $E_{b1} = \sigma T_1^4 = (5.67 \times 10^{-8})(200 + 273)^4 = 2842\ \text{W/m}^2$ and $E_{b2} = \sigma T_2^4 = (5.67 \times 10^{-8})(20 + 273)^4 = 545\ \text{W/m}^2$. Then the radiation heat loss of the unshielded thermocouple is, from Eq. (8.65),

$$q_0/A_1 = \frac{(E_{b1} - E_{b2})}{(R_0A_1)} = \frac{(2842 - 545)}{1.672}$$

$$= 1374\ \text{W/m}^2.$$

We wish the shield to reduce this by 50%, to 687 W/m$^2$. With the shield inserted, the resistance from Eq. (8.70) is

$$RA_1 = R_0A_1 + [2(1-\varepsilon_s)/\varepsilon_s](A_1/A_s) + 1/F_{1\to s} + A_1/A_sF_{s\to 2} - 1/F_{1\to 2}$$

$$= 1.672 + [2(1-\varepsilon_s)/\varepsilon_s](1/5) + 1.0 + (1/5)/1.0 - 1.0$$

$$= 1.872 + 0.4(1-\varepsilon_s)/\varepsilon_s.$$

We wish the heat transfer with the shield to be

$$q/A_1 = (E_{b1} - E_{b2})/(RA_1) = 687$$

$$= (2842 - 545)/[1.872 + 0.4(1-\varepsilon_s)/\varepsilon_s].$$

Solving, we obtain $1.472 = 0.4(1-\varepsilon_s)/\varepsilon_s$, or

$\varepsilon_s = 0.4/(0.4 + 1.472) = 0.214.$ [*Ans.*] ■

## 8.5 Radiation Emission and Absorption by Gases

Throughout this chapter we have assumed, when computing radiation transfer between surfaces, that any intervening gas media are perfectly transparent or *nonparticipating,* that is, the gases neither emit nor absorb thermal radiation. This is an excellent approximation for gases with symmetric, diatomic molecules: $H_2$, $N_2$, and $O_2$, which emit negligible radiation until temperatures are high enough to ionize them. Similarly, the common monatomic gases, argon and helium, are nonparticipating.

Gases with nonsymmetric molecular structures, however, do emit significant radiation. For engineering calculations, the most important of these gases are $H_2O$, $CO_2$, $CO$, $SO_2$, and $NH_3$. Here we will concentrate only on the first two, water vapor and carbon dioxide. Data for other gases are given in [3] and [13], which have extensive treatments of gas radiation. The theoretical physics of gas radiation is explained in detail in [1]. There are also at least three excellent review articles about gas radiation [34–36].

Unlike solid surfaces, which emit radiation over a continuous spectrum approximating a blackbody or gray body (Fig. 8.6b), gases emit only in certain narrow-wavelength *bands* associated with their molecular energy levels. The most important emission bands for water vapor and carbon dioxide are as follows:

$H_2O$: 0.9–1.0, 1.1–1.2, 1.3–1.5, 1.8–2.0, 2.6–3.0, 5–8, and 16–22 $\mu$m

$CO_2$: 1.6–2.0, 2.5–3.0, 4–5, 8–10, and 15–20 $\mu$m.

Thus gas emission is not at all "gray," especially for thin layers, and thus the total emission or absorption of a gas volume depends on many variables: (a) gas temperature and partial pressure, (b) the shape of the gas volume, (c) whether other participating gases are also present, (d) the total pressure of the gas mixture, and (e) the temperature and properties of the enclosing surfaces. Edwards [36] especially warns against putting a single engineering number on gas emission, calling it the "gray-gas myth." For best engineering accuracy, Edwards recommends numerical integration of wavelength-band gas radiation data on, say, a programmable handheld calculator. Here we confine our treatment to a gas radiation correlation chart developed by H. C. Hottel and his co-workers about 40 years ago.

Consider monochromatic radiation passing through a thin, plane participating gas layer of thickness $L$. The radiation intensity $I_\lambda$ will decrease through any differential gas layer $dx$ by an amount $dI_\lambda = -\kappa_\lambda I_\lambda dx$, where $\kappa_\lambda$ is a gas absorption coefficient that depends on wavelength, temperature, and gas partial pressure. Integrating over finite thickness $L$ gives

$$I_\lambda(L) = I_{\lambda 0}\, e^{-\bar{\kappa}_\lambda L}, \tag{8.72}$$

where $\bar{\kappa}_\lambda$ is the value of $\kappa_\lambda$ averaged over the layer.

The fraction of total radiated energy absorbed by the layer can be defined as the monochromatic absorptivity of the layer:

$$\alpha_{\lambda,\text{gas}} = \frac{I_{\lambda 0} - I_\lambda(L)}{I_{\lambda 0}} = 1 - e^{-\bar{\kappa}_\lambda L}. \tag{8.73}$$

By Kirchhoff's law, assuming a diffuse gas, this also equals the gas layer emissivity at the same temperature, $\varepsilon_{\lambda,\text{gas}} = \alpha_{\lambda,\text{gas}}$.

From Eq. (8.28), then, the total emissivity of the gas layer is given by integration over the whole wavelength range:

$$\varepsilon_{\text{gas}} = \frac{1}{\sigma T_g^4} \int_0^\infty E_{b\lambda}(\lambda, T_g)\,(1 - e^{-\bar{\kappa}_\lambda L})\,d\lambda. \tag{8.74}$$

Since $\bar{\kappa}_\lambda$ is a function of many parameters, the integration can be quite laborious. Edwards [36] indicates a numerical procedure. Further, the results must depend on the *shape* of the radiating gas volume; to good approximation, this effect is accounted for by replacing the layer thickness $L$ in Eq. (8.74) by the *mean beam length*, $L_e$, which is a function only of gas volume shape.

Table 8.5 lists computed mean beam lengths for various gas volume

**Table 8.5** Mean beam length for various gas volume shapes (from [1])

| Gas Volume Geometry | Mean Beam Length, $L_e$ |
|---|---|
| Hemisphere radiating to the center of its base | $R$ |
| Sphere radiating to any point on its surface | $0.65D$ |
| Circular cylinder of semi-infinite height radiating to | |
| a) center of its base | a) $0.9D$ |
| b) its entire base | b) $0.65D$ |
| Circular cylinder of height equal to its base, radiating to | |
| a) center of its base | a) $0.71D$ |
| b) entire surface | b) $0.6D$ |
| Infinite circular cylinder radiating to its entire surface | $0.95D$ |
| Cube radiating to one face of side length $H$ | $0.6H$ |
| Infinite slab of thickness $H$ radiating to either face | $1.8H$ |
| An approximate formula for a gas volume of any shape | $\frac{3.6\,(\text{Volume})}{(\text{Surface area})}$ |

shapes, abstracted from a more extensive list in [1, pp. 570–571]. For shapes not tabulated, the geometric approximation $L_e \doteq 3.6$ (Gas volume)/(Gas surface area) gives results accurate to about $\pm 10\%$.

### 8.5.1 Gas Emissivity Measurements

During the 1930s, extensive measurements of the emissivity of various gases were performed by H. C. Hottel and co-workers. Their results for water vapor and carbon dioxide are summarized in a classic paper by Hottel and Egbert [37]. These workers found that, to good engineering approximation, the gas emissivity was a function of five parameters:

$$\varepsilon_{\text{gas}} \doteq \text{fcn}(P_{\text{gas}}L_e,\ T_{\text{gas}},\ P_{\text{gas}},\ P_T,\ PL \text{ of other emitting gases}). \tag{8.75}$$

The charts they developed to express this function for $H_2O$ and $CO_2$ are shown in Figs. 8.29–8.31.

As their "base" chart, Hottel and Egbert plotted $\varepsilon_{\text{gas}}$ versus $P_g L_e$ and $T_g$ for $P_g = 0$, $P_T = 1$ atm, with no other participating gases present. The base chart for water vapor is Fig. 8.29(a) and for carbon dioxide Fig. 8.30(a).

For values of $P_g \neq 0$ and $P_T \neq 1$ atm, a correction factor, $C_{\text{gas}}$, was developed to modify the "base" emissivity. The factor $C_{H_2O}$ is given by Fig. 8.29(b), and $C_{CO_2}$ is shown in Fig. 8.30(b).

Finally, if both $H_2O$ and $CO_2$ are contained in a single gas mixture, one sums the $H_2O$ contribution from Fig. 8.29 with the $CO_2$ result from Fig. 8.30 and then subtracts the "band overlap" correction $\Delta\varepsilon$ shown in Fig. 8.31. Thus the final estimate for emissivity of any mixture of $H_2O$, $CO_2$, and nonparticipating gases is given by

$$\varepsilon_{\text{gas}} = C_{H_2O}\varepsilon_{H_2O} + C_{CO_2}\varepsilon_{CO_2} - \Delta\varepsilon. \tag{8.76}$$

Example 8.15 illustrates the use of these correlation charts.

**Example 8.15**

A right circular cylinder 60 cm in diameter and 60 cm high contains a mixture of $N_2$, $H_2O$, and $CO_2$ such that $P_{N_2} = 1.0$ atm, $P_{H_2O} = 0.3$ atm, and $P_{CO_2} = 0.6$ atm. The mixture temperature is 1000°F. Estimate the gas emissivity.

*Solution* From Table 8.5, for a cylinder of equal diameter and height, $L_e = 0.6D = 0.6(60) = 36$ cm $= 1.18$ ft. The total pressure

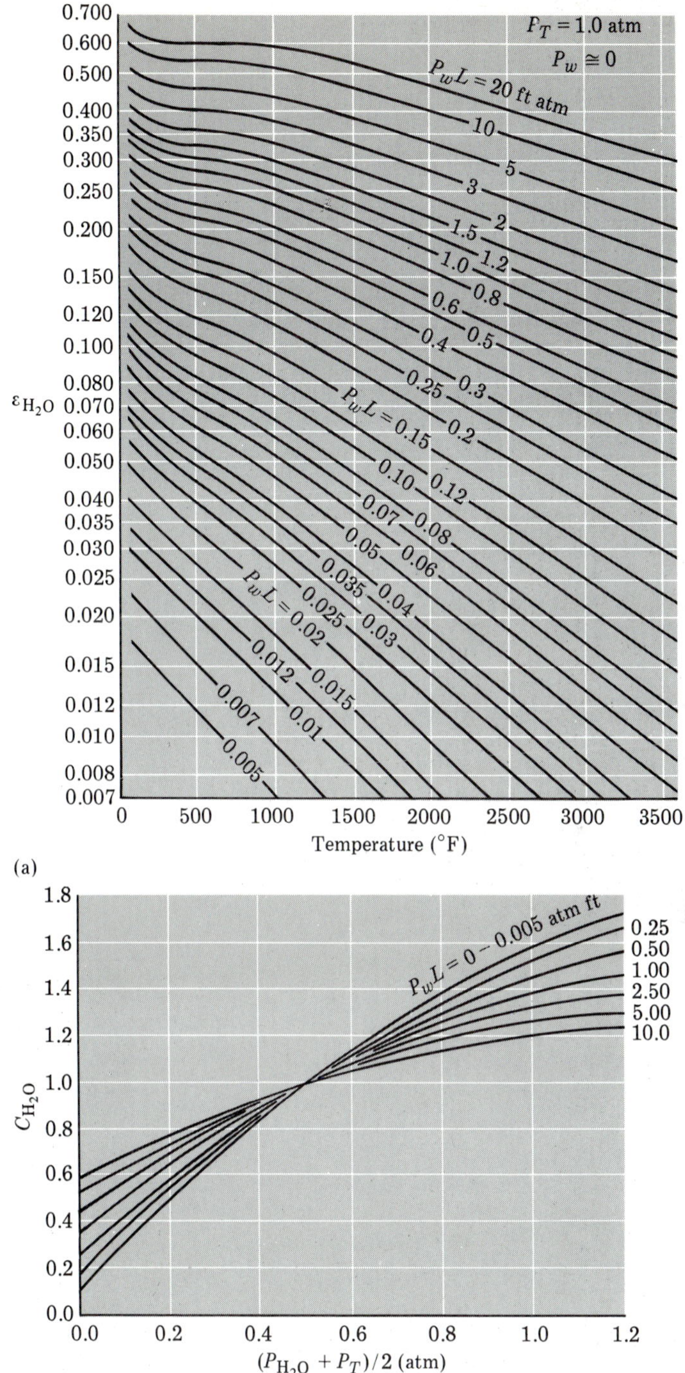


**Figure 8.29** Measured emissivity of water vapor by Hottel and Egbert [37]: (a) emissivity at near-zero water vapor pressure and 1 atm total pressure; (b) correction factor for different values of water vapor and total pressure.

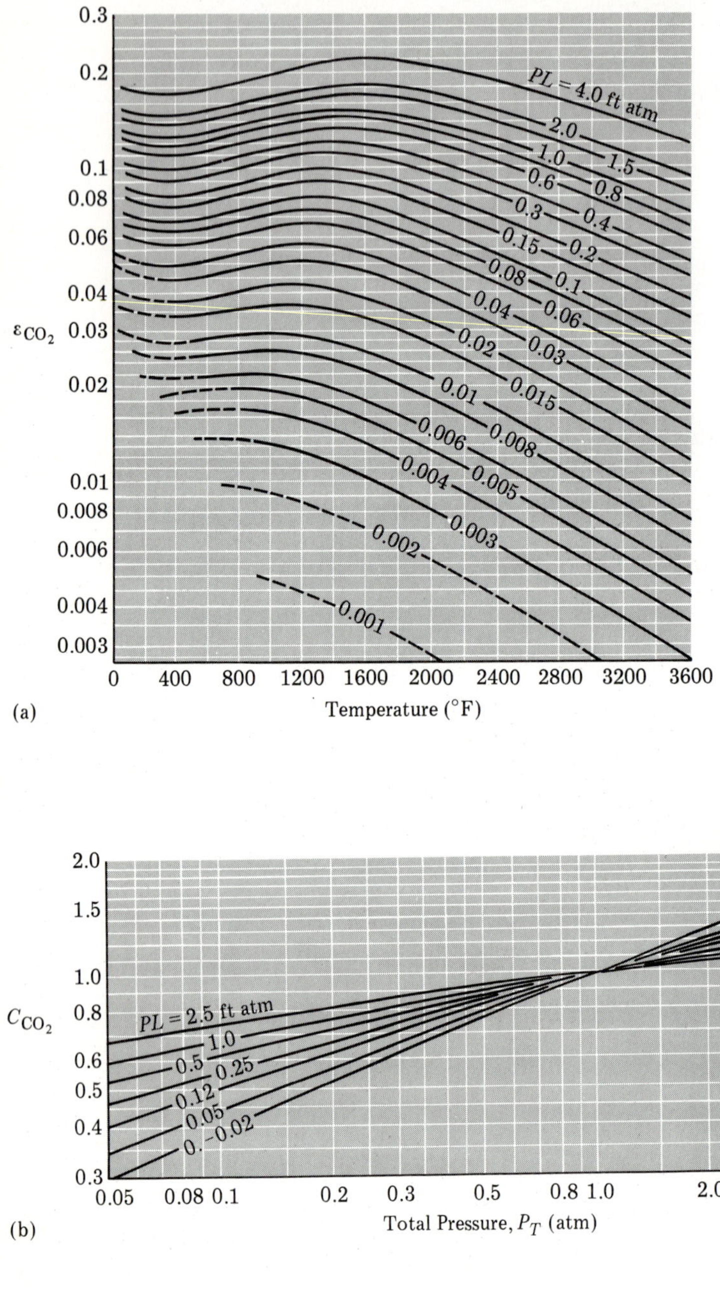


**Figure 8.30** Measured emissivity of carbon dioxide by Hottel and Egbert [37]: (a) emissivity at near-zero $CO_2$ vapor pressure and 1 atm total pressure; (b) correction factor for nonzero $CO_2$ pressure and various total pressures.

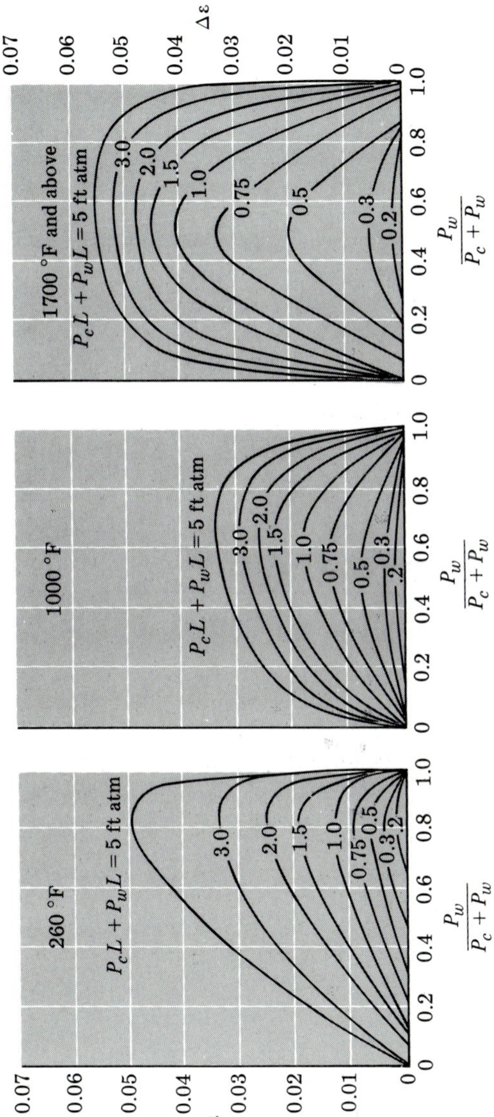


**Figure 8.31** Correction factor for gas emissivity, to be subtracted when both $H_2O$ and $CO_2$ vapor are present in a gas mixture, after Hottel and Egbert [37]. See Eq. (8.76) for its use.

is $P_T = 1.0 + 0.3 + 0.6 = 1.9$ atm. We will need all five charts from Figs. 8.29–8.31 to complete this problem.

For the "base" emissivity of $H_2O$, $P_{H_2O}L_e = (0.3)(1.18) = 0.35$ atm · ft, and $T_{gas} = 1000°F$. From Fig. 8.29(a) read $\varepsilon_{H_2O} \doteq 0.16$. Since $(P_{H_2O} + P_T)/2 = (0.3 + 1.9)/2 = 1.1$ atm, a correction is needed: From Fig 8.29(b) at $PL = 0.35$ atm · ft, read $C_{H_2O} \doteq 1.55$.

For the "base" emissivity of $CO_2$, compute $P_{CO_2}L_e = (0.6)(1.18) = 0.71$ atm · ft. From Fig. 8.30(a) at $T_{gas} = 1000°F$, read $\varepsilon_{CO_2} \doteq 0.13$. Again, since $P_T \neq 1$ atm, a pressure correction is needed. From Fig. 8.30(b) at $P_T = 1.9$ atm and $P_{CO_2}L_e = 0.71$ atm · ft, read $C_{CO_2} \doteq 1.1$.

Finally, since both $H_2O$ and $CO_2$ are present in the same mixture, a negative band-overlap correction is needed. Since $T_g = 1000°F$, the center chart in Fig. 8.31 applies. At $P_CL + P_WL = 0.71 + 0.35 = 1.06$ and $P_W/(P_C + P_W) = 0.3/(0.6 + 0.3) = 0.33$, read $\Delta\varepsilon \doteq 0.014$.

Our final emissivity estimate follows from Eq. (8.76):

$$\begin{aligned}\varepsilon_{gas} &= C_{H_2O}\,\varepsilon_{H_2O} + C_{CO_2}\,\varepsilon_{CO_2} - \Delta\varepsilon\\ &= (1.55)(0.16) + (1.1)(0.13) - 0.014\\ &\doteq 0.38. \quad [\textit{Ans.}]\end{aligned}$$

The uncertainty in this estimate is probably $\pm 10\%$. ■

---

### 8.5.2 Gas Absorption of Incoming Radiation

The charts in Figs. 8.29–8.31 allow evaluation of the effective emissivity of a gas volume at temperature $T_g$. The absorptivity of the same gas to incoming irradiation depends not only upon $T_g$ but also upon the source temperature, $T_s$. Hottel and co-workers, in [3] and [13], have developed a method of using the same charts to estimate absorptivity. Their suggested correlation formula is

$$\alpha_{gas} = C_{H_2O}\,\varepsilon'_{H_2O} + C_{CO_2}\,\varepsilon'_{CO_2} - \Delta\varepsilon', \tag{8.77}$$

where

$$\begin{aligned}\varepsilon'_{H_2O} &= (T_g/T_s)^{0.45}\varepsilon_{H_2O}(T_s, P_{H_2O}L_eT_s/T_g),\\ \varepsilon'_{CO_2} &= (T_g/T_s)^{0.65}\varepsilon_{CO_2}(T_s, P_{CO_2}L_eT_s/T_g),\\ \Delta\varepsilon' &= \Delta\varepsilon(T_s, P_{CO_2}L_e + P_{H_2O}L_e).\end{aligned}$$

The correction factors $C_{H_2O}$ and $C_{CO_2}$ are evaluated from $T_g$ and $PL_e$ without any modification from source temperature $T_s$.

If the gas is surrounded by nearly black enclosure walls at uniform temperature $T_s$, the net radiation from gas to wall is

$$q/A_w = \sigma[\varepsilon_{gas} T_g^4 - \alpha_{gas} T_s^4], \qquad (8.78)$$

where $\varepsilon_{gas}$ and $\alpha_{gas}$ are evaluated from Eqs. (8.76) and (8.77), respectively. Absolute temperatures are required for the power-law correlations in Eqs. (8.77).

### Example 8.16

For the hot gas of Example 8.15, suppose the wall temperature of the container is 260°F. Estimate the gas absorptivity and the net heat flux from gas to walls.

*Solution* Equation (8.77) applies, with $T_g$ = 1000°F = 811K and $T_s$ = 260°F = 400K. In Example 8.15 we already found $C_{H_2O}$ = 1.55 and $C_{CO_2}$ = 1.1, which are not affected by wall temperature. Evaluate $\varepsilon_{H_2O}$ from Fig. 8.29(a) at $T_s$ = 260°F and $P_{H_2O}L_eT_s/T_g$ = (0.35)(400)/(811) = 0.175. From the figure, read $\varepsilon_{H_2O} \doteq 0.14$. Then $\varepsilon'_{H_2O} = (811/400)^{0.45}(0.14) = 0.192$.

Similarly, at $T_s$ = 260°F and $P_{CO_2}L_eT_s/T_g$ = (0.71)(400/811) = 0.35, from Fig. 8.30(a) read $\varepsilon_{CO_2} \doteq 0.096$. Then the term $\varepsilon'_{CO_2} = (811/400)^{0.65}(0.096) = 0.152$.

The correction factor for both gases coexisting is read from Fig. 8.31 at $T_s$ = 260°F (the first chart) and at $P_CL_e + P_WL_e$ = 0.33, as in Example 8.15. Read $\Delta\varepsilon' \doteq 0.005$.

The gas absorptivity is now computed from Eq. (8.77):

$$\begin{aligned}\alpha_{gas} &= C_{H_2O}\,\varepsilon'_{H_2O} + C_{CO_2}\,\varepsilon'_{CO_2} - \Delta\varepsilon' \\ &= (1.55)(0.192) + (1.1)(0.152) - 0.005 = 0.46. \quad [Ans.]\end{aligned}$$

For a 60-cm-by-60-cm right circular cylinder, the wall area is $A_w = 2(\pi/4)(0.6\text{ m})^2 + \pi(0.6\text{ m})(0.6\text{ m}) = 1.70\text{ m}^2$. Finally, with $\varepsilon_{gas} \doteq 0.38$ from Example 8.15, the net radiation from gas to a nearly black wall is

$$\begin{aligned}q_{g\to w} &= \sigma A_w[\varepsilon_{gas}T_g^4 - \alpha_{gas}T_s^4] \\ &= (5.67 \times 10^{-8}\text{ W/m}^2\cdot\text{K}^4)(1.70\text{ m}^2)[0.38(811\text{K})^4 - 0.46(400\text{K})^4] \\ &= 14{,}700\text{ W}. \quad [Ans.]\end{aligned}$$

■

## 8.6 Solar Radiation

Although a small star by galactic standards, the sun produces an enormous amount of energy by earth standards. The sun approximates a blackbody at about 5762K and, being about $1.4 \times 10^9$ m in diameter, emits about $3.8 \times 10^{26}$ W of energy continuously. Of this energy, about half a billionth or $1.7 \times 10^{17}$ W impinges on the earth, warming the ground, oceans, and atmosphere and driving the plant photosynthesis process that maintains biological life. In contrast, the direct energy needs of humans are about $5 \times 10^{12}$ W, or 0.003% of the energy received from the sun. It seems a reasonable goal to try to extract our energy needs from solar irradiation.

Unfortunately, such large-scale utilization of solar energy is probably not feasible, for solar radiation is extremely diffuse, averaging about 1353 W/m$^2$ for a surface outside the atmosphere and normal to the sun's rays. Only about half of this is received at the earth's surface, the rest being lost by (a) absorption and scattering by air and dust molecules; (b) absorption by $H_2O$, CO, $CO_2$, and $O_3$ in the atmosphere; and (c) absorption and reflection by clouds. Thus, to meet civilization's total energy requirement of $5 \times 10^{12}$ W with ground-based solar equipment would require an area of 8000 mi$^2$ covered entirely with collection devices, assuming a collection efficiency of 60%. And recall that the sun rarely shines at night! Such massive projects are not planned.

The spectrum of solar radiation, as shown in Fig. 8.32, resembles that of a blackbody at about 5762K, but with much less intensity due to the great distance ($1.495 \times 10^{11}$ m) between the earth and the sun. The dashed curve in Fig. 8.32 approximates extraterrestrial radiation, with total energy of 1353 W/m$^2$. The solid curve represents typical measurements at the earth's surface, showing multiple absorption bands owing to atmospheric $O_3$, $H_2O$, and $CO_2$. In both cases the maximum spectral energy occurs at about 0.5 $\mu$m, which is in the visible range and agrees with Wien's law, Eq. (8.7), for $T = 5762$K.

The solar constant, or mean radiation outside the atmosphere, varies about $\pm 3.4\%$ throughout the year because of the earth's elliptical orbit. It is well fit by a cosine variation:

$$E_s'' \doteq 1353[1 + 0.034 \cos(2\pi n/365)] \text{ W/m}^2, \qquad \textbf{(8.79)}$$

where $n$ is the day of the year. This assumes that the receiving surface is normal to the sun's rays. If the surface has a fixed orientation — horizontal, say — its irradiation throughout the day will vary depending on the earth's rotation and the sun's declination. For details, see [38–42].

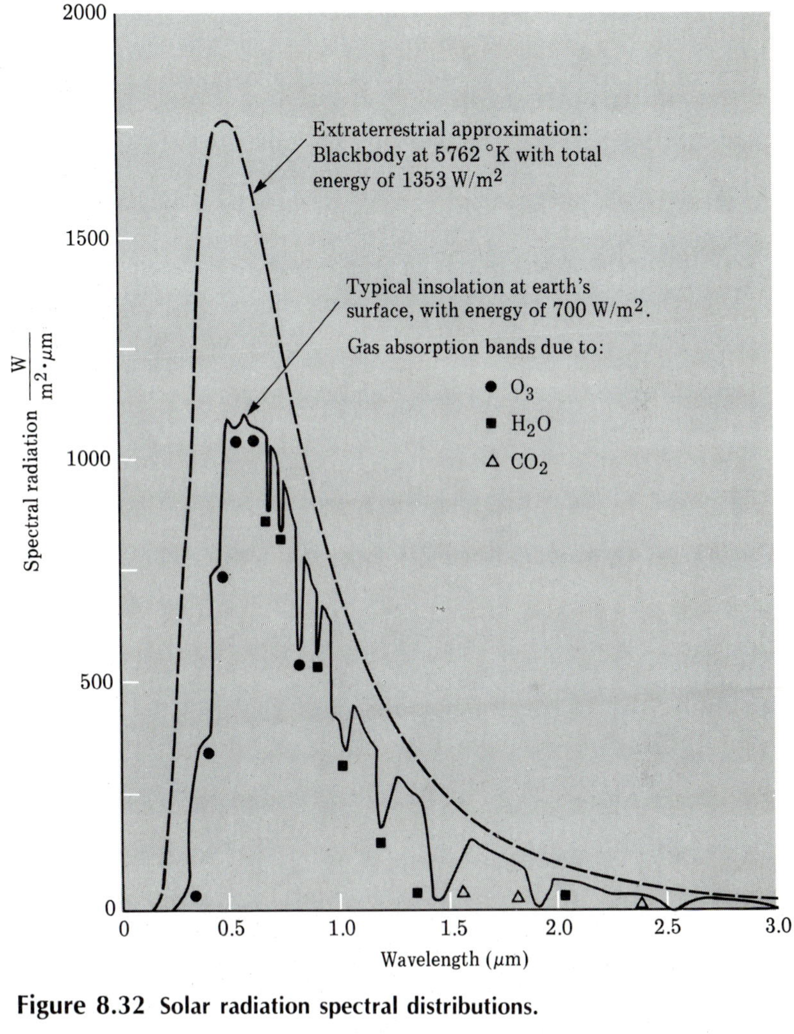


**Figure 8.32 Solar radiation spectral distributions.**

It is of interest to know the total energy per unit area received each day by a horizontal surface at the top of the earth's surface. Assuming no unusual solar activity, this daily energy varies only with latitude and day of the year; it is shown in Fig. 8.33. The curves are given in *langleys* per day, after Samuel Langley, who in 1883 made the first accurate measurements of the solar spectrum, which are reproduced in Fig. 8.32 (1 langley = 1 cal/cm$^2$ = 41,860 J/m$^2$ = 3.686 Btu/ft$^2$).

Examination of Fig. 8.33 reveals that substantial insolation is available to every region on earth — even the polar areas, except of course for their dark winters. But only a fraction of this extraterrestrial energy reaches the ground. This fraction can be as high as 85% on an extremely clear day with no haze or smog, or as low as 5% on a very cloudy day. Wetter regions, with more clouds, and industrialized areas, with more smog, receive less ground insolation than dry or desert areas at the same latitude.

Figure 8.34 shows the measured monthly average ground insolation for December and June in the United States, from [45]. December approximates the minimum condition and June the maximum. Mean annual solar energy may be estimated as the arithmetic mean of these two. We see that the wet and industrialized northeast and northwest states have lowest insolation, while the southwest states have the

**Figure 8.33** **Total daily solar energy received by a horizontal surface at the top of the earth's atmosphere, in langleys per day. (1 langley = 1 cal/cm²). (From [42].)**

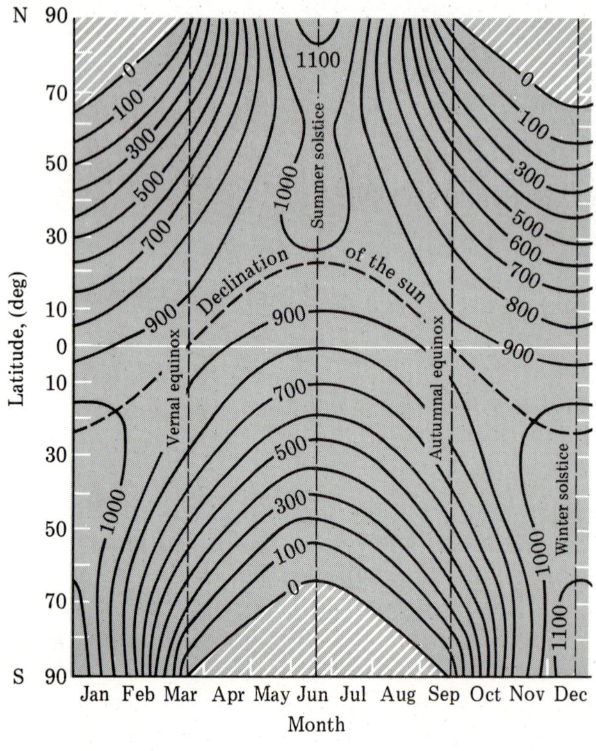

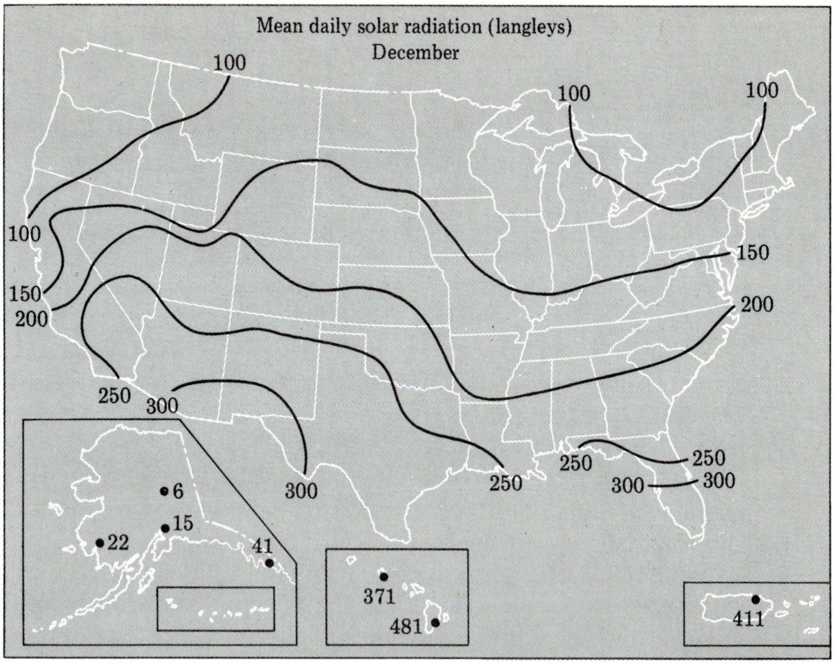


(a)

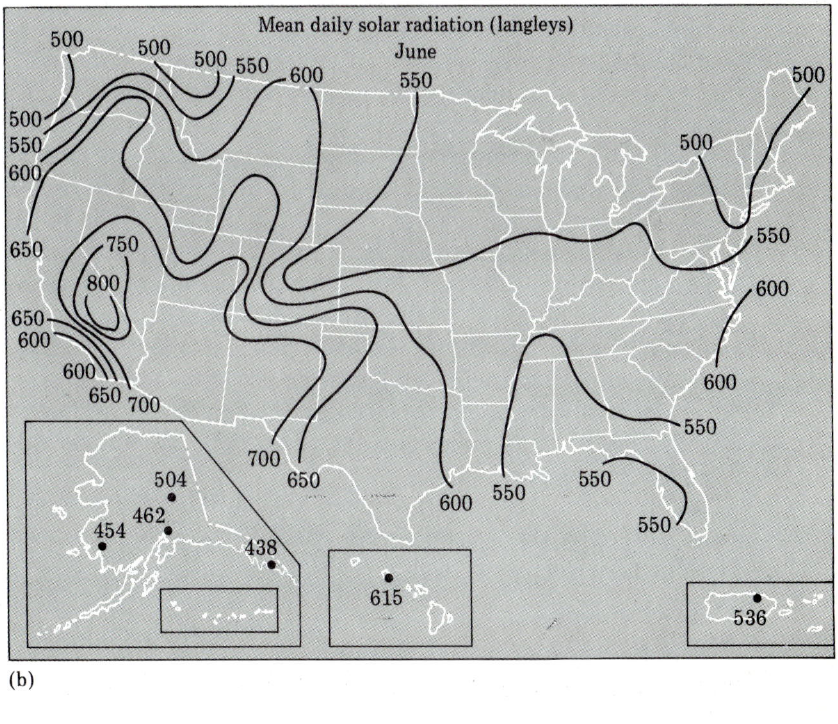


(b)

**Figure 8.34** Average measured daily insolation on a horizontal surface in the United States: (a) December; (b) June. (From [45].)

highest. Alaska is nearly dark in January but quite sunny in June. Hawaii, though near the equator, is relatively wet and cloudy. Reference [45] gives similar curves for all months of the year and is also an utter treasury of information about all United States climatic parameters: rainfall, cloudiness, snowfall, humidity, wind speeds, and so on. Reference [46] gives surface insolation curves for the entire globe.

Solar radiation received at the ground is partly *directed,* or arriving in a straight path from the sun, and partly *diffuse,* or scattered and reradiated by the atmosphere in such a way as to lose its directional characteristics. On a clear day, 80% of surface insolation is direct. On a cloudy day, nearly all solar radiation at the ground is diffuse and may amount to only 50–100 $W/m^2$. Solar energy may be "collected" for practical use either by *flat collectors,* which absorb the incident energy, or by *concentrating collectors* such as mirror arrays and parabolic or Fresnel lenses, which reflect the energy toward a small collector region. Concentrators allow much higher working fluid temperatures and thus higher thermodynamic efficiency, but they essentially collect only the directed solar energy. A recent discussion of advanced concepts in concentrator technology is given in [43].

The future of solar energy, as a quantitative contribution to world energy needs, is probably only moderate at best. One reason already given is its low intensity, so that very large collector areas are needed to yield significant energy. A second disadvantage is its intermittency: It is both periodic, because of the diurnal day/night cycle, and random, because of sporadic and often protracted cloudiness. Even the brightest areas of the American southwest "sunbelt" have 20% cloudiness, sometimes persisting for several days. Thus solar collector devices require either a costly fossil- or nuclear-fueled backup system or a costly and complex heat storage design. To date, neither solar-thermal nor solar-electric systems have achieved any wide use or clear economic viability. However, with continued research and development and government subsidy, solar systems may, by the year 2000, account for about 10% of our energy needs.

In addition to solar insolation, the sun is also responsible for several other types of geophysical energy that may be tapped by engineering devices:

1. wind energy,
2. hydroelectric energy from rainfall and river flow,
3. ocean thermal energy conversion (OTEC),
4. renewable combustible organic matter (wood),

5. ocean wave energy, and
6. microwave transmission from satellite solar collectors.

In all six cases, as with surface insolation collectors, very large areas are needed to generate significant energy. Wind energy designs are discussed in [44], and [48] treats the design of small and micro hydroelectric plants. Pilot plants for OTEC and ocean wave energy extraction are now being constructed. Wind and wood have long been sources of small-scale energy production.

There are many fine texts devoted to aspects of solar energy. The only book known to the writer that treats all phases of solar thermal and electric generation is [38]. The fundamentals of solar thermal processes are treated in [39] and [40], with practical design details given in [41], [42], and [47]. Being a source of what the media term "free" energy, the sun and its effects on the earth will continue to capture the imagination of both engineers and the public for many centuries to come.

### Example 8.17

The world's largest solar-electric pilot power plant was completed in 1982 in the Mojave Desert near Daggett, Calif. It covers 75 acres (1 acre = 4047 m$^2$) and uses hundreds of tracking mirrors to reflect direct solar insolation toward a tower-mounted steam generator. What is the overall efficiency of this plant, which delivers 10 MW of power?

*Solution* The total area available is 75 acres = 304,000 m$^2$. From Fig. 8.34, the Mojave Desert area yields about 800 langleys per day in June and about 300 in December. Assuming 14 hours of sunlight in June and 10 hours in December, the average intensity of insolation for this area is

*June:*

$$E_s'' = \frac{(800 \times 41{,}860\ \text{J/m}^2)}{(14 \times 3600\ \text{s})} = 664\ \text{W/m}^2;$$

*December:*

$$E_s'' = \frac{(300 \times 41{,}860\ \text{J/m}^2)}{(10 \times 3600\ \text{s})} = 349\ \text{W/m}^2.$$

The total average solar insolation available is thus

*June:*

$E_s = (664\ \text{W/m}^2)(304{,}000\ \text{m}^2) = 202\ \text{MW};$

*December:*

$E_s = (349\ \text{W/m}^2)(304{,}000\ \text{m}^2) = 106\ \text{MW}.$

The overall efficiency is the power delivered divided by the power available:

*June:*

$\eta = 10/202 = 5\%;$ [*Ans. (a)*]

*December:*

$\eta = 10/106 = 9\%.$ [*Ans. (b)*]

One reason for the low output is that the mirrors would interfere with one another if crowded together densely. As designed, the mirrors cover less than one-third of the 75 acres. Lacking significant storage, the plant is inoperable at night and on cloudy days. ■

---

## Summary

Being an electromagnetic wave phenomenon, thermal radiation is so completely different from conduction and convection that a whole new set of concepts must be defined. At high temperatures, radiation becomes the dominant mode of heat transfer.

The chapter begins with a brief description of the physics of electromagnetic radiation, the central result of which is Planck's blackbody spectral energy function. There follows a detailed discussion of the engineering properties of real (nonblack) surfaces: emissivity, absorptivity, reflectivity, transmissivity, and directional variations. Kirchhoff's law of reciprocity is defined and some special "selective" surfaces are described. The gray-body approximation is introduced.

The concept of surface radiosity is discussed, leading to the definition of the radiation shape factor between two surfaces. For diffuse surfaces at uniform temperatures, the resulting integrals are purely geometric and may be tabulated for a given geometry. Shape factor reciprocity and summation laws are defined, leading to an interesting type of algebra. Multiple-surface enclosures are then discussed, leading to the electric-analogy solution of simultaneous equations. A matrix algebra solution technique is also indicated for enclosures.

There is a brief discussion of the radiation and absorption characteristics of gases, which complicate the analysis of radiant interchange between solid surfaces. The chapter ends with a brief treatment of the physics of solar radiation and its use in heating, cooling, and the generation of electricity.

## References


1. R. Siegel and J. R. Howell, *Thermal Radiation Heat Transfer,* 2nd ed., McGraw-Hill, New York, 1980.
2. E. M. Sparrow and R. D. Cess, *Radiation Heat Transfer,* Brooks/Cole, Belmont, Calif., 1970.
3. H. C. Hottel and A. F. Sarofim, *Radiative Transfer,* McGraw-Hill, New York, 1967.
4. M. N. Ozisik, *Radiative Transfer and Interactions with Conduction and Convection,* Wiley, New York, 1973.
5. T. J. Love, *Radiative Heat Transfer,* Merrill, Columbus, Ohio, 1968.
6. J. A. Wiebelt, *Engineering Radiation Heat Transfer,* Holt, Rinehart and Winston, New York, 1966.
7. Max Planck, "Distribution of Energy in the Spectrum," *Ann. Physik,* vol. 4, no. 3, 1901, pp. 553–563.
8. Max Planck, *The Theory of Heat Radiation,* Dover, New York, 1959.
9. R. V. Dunkle, "Thermal Radiation Tables and Applications," *ASME Transactions,* vol. 76, 1954, pp. 549–552.
10. M. Czerny and A. Walther, *Tables of Fractional Functions for the Planck Distribution Law,* Springer Verlag, Berlin, 1961.
11. M. Pivovonsky and M. R. Nagel, *Tables of Blackbody Radiation Functions,* Macmillan, New York, 1961.
12. E. S. Barr, "Historical Survey of the Early Development of the Infrared Spectral Region," *Amer. J. Physics,* vol. 28, no. 1, 1960, pp. 42–54.
13. W. H. McAdams, *Heat Transmission,* McGraw-Hill, New York, 1954.
14. G. G. Gubareff, J. E. Janssen, and R. H. Torberg, *Thermal Radiation Properties Survey,* 2nd ed., Honeywell Research Center, Minneapolis, Minn., 1960.
15. W. D. Wood, H. W. Deem, and C. F. Lucks, *Thermal Radiative Properties,* Plenum Press, New York, 1964.
16. Y. S. Touloukian and D. P. DeWitt, *Thermal Radiation Properties,* vol. 7: Metallic Elements and Alloys, vol. 8: Nonmetallic Solids, and vol. 9: Coatings, from *Thermophysical Properties of Matter,* Y. S. Touloukian and C. Y. Ho (Eds.), IFI/Plenum, New York, 1970–1972.

17. E. Schmidt and E. Eckert, "Über die Richtungsverteilung der Wärmestrahlung," *Forsch. Gebiete Ingenieurw,* vol. 6, 1935, pp. 175–183.

18. W. Sieber, "Zusammensetzung der von Werk — und Baustoffen zurückgeworfenen Wärmestrahlung," *Z. Tech. Physik,* vol. 22, 1941, pp. 130–135.

19. R. C. Birkebak and E. R. G. Eckert, "Effects of Roughness of Metal Surfaces on Angular Distribution of Monochromatic Radiation," *J. Heat Transfer,* vol. 87, 1965, pp. 85–94.

20. K. E. Torrance and E. M. Sparrow, "Off-Specular Peaks in the Directional Distribution of Reflected Thermal Radiation," *J. Heat Transfer,* vol. 88, 1966, pp. 223–230.

21. W. G. Camack and D. K. Edwards, "Effect of Surface Thermal Radiation Characteristics on the Temperature Control Problem in Satellites," in *Surface Effects on Spacecraft Materials,* F. J. Clauss (Ed.), Wiley, New York, 1960, pp. 3–54.

22. R. B. Dunkle, "Thermal Radiation Characteristics of Surfaces," in *Theory and Fundamental Research in Heat Transfer,* J. A. Clark (Ed.), Pergamon, New York, 1963, pp. 1–31.

23. D. K. Edwards, *Radiation Heat Transfer Notes,* Hemisphere Publishing, New York, 1981.

24. R. Gardon, "The Emissivity of Transparent Materials," *J. Am. Ceramic Soc.,* vol. 39, no. 8, 1956, pp. 278–287.

25. R. L. Long, "A Review of Recent Air Force Research on Selective Solar Absorbers," *J. Engineering for Power,* vol. 87, July 1965, pp. 277–280.

26. M. Perlmutter and J. R. Howell, "A Strongly Directional Emitting and Absorbing Surface," *J. Heat Transfer,* vol. 85, no. 3, 1963, pp. 282–283.

27. C. H. Liebert, "Spectral Emittance of Aluminum Oxide and Zinc Oxide on Opaque Substrates," NASA TN D-3115, Nov. 1965.

28. E. M. Sparrow, "Radiation Heat Transfer Between Surfaces," *Advances in Heat Transfer,* vol. 2, 1965, pp. 400–452.

29. D. C. Hamilton and W. R. Morgan, "Radiant-Interchange Configuration Factors," NACA TN 2836, 1952.

30. M. Jakob, *Heat Transfer,* vol. 1, Wiley, New York, 1949.

31. A. K. Oppenheim, "Radiation Analysis by the Network Method," *ASME Transactions,* vol. 78, 1956, pp. 725–735.

32. J. P. Holman, *Heat Transfer,* 5th ed., McGraw-Hill, New York, 1981.

33. R. H. Pennington, *Introductory Computer Methods and Numerical Analysis,* 2nd ed., Macmillan, London, 1970.

34. C. L. Tien, "Thermal Radiation Properties of Gases," *Advances in Heat Transfer,* vol. 5, 1968, pp. 253–324.

35. R. D. Cess and S. N. Tiwari, "Infrared Radiative Energy Transfer in Gases," *Advances in Heat Transfer,* vol. 8, 1972, pp. 229–283.

36. D. K. Edwards, "Molecular Gas Band Radiation," *Advances in Heat Transfer,* vol. 12, 1976, pp. 115–193.

37. H. C. Hottel and R. B. Egbert, "Radiant Heat Transmission from Water Vapor," *AIChE Transactions,* vol. 38, 1942, pp. 531–568.

38. F. Kreith and J. F. Kreider, *Principles of Solar Engineering,* McGraw-Hill Hemisphere, New York, 1978.

39. E. E. Anderson, *Solar Energy Fundamentals for Designers and Engineers,* Addison-Wesley, Reading, Mass., 1982.

40. J. A. Duffie and W. A. Beckman, *Solar Engineering of Thermal Processes,* Wiley, New York, 1980.

41. M. J. Fisk and H. C. Anderson, *Introduction to Solar Technology,* Addison-Wesley, Reading, Mass., 1982.

42. J. F. Kreider and F. Kreith, *Solar Heating and Cooling,* Scripta Book Co./McGraw-Hill, New York, 1975.

43. L. Leibowitz and E. Hanseth, "Solar Thermal Technology — Outlook for the 80's," *Mechanical Engineering,* vol. 104, Jan. 1982, pp. 30–35.

44. J. Park, *The Wind Power Book,* Chesire Books, Palo Alto, Calif., 1981.

45. *Climatic Atlas of the United States,* U.S. Government Printing Office, Washington, D.C., 1968.

46. B. de Jong, *Net Radiation Received by a Horizontal Surface at the Earth,* Delft Univ. Press, Holland, 1973.

47. *Solar Heating and Cooling of Residential Buildings,* 2 vols., U.S. Dept. of Commerce, Washington, D.C., Oct. 1977.

48. R. Noyes (Ed.), *Small and Micro Hydroelectric Plants,* Noyes Data Corp., Park Ridge, N.J., 1980.

49. J. R. Howell, *Catalog of Radiation Configuration Factors,* McGraw-Hill, New York, 1982.

## Review Questions

1. What are the physical differences between the radiation and conduction heat transfer mechanisms?
2. How fast does radiation travel compared to conduction heat flux?
3. What is the wavelength range of visible light?
4. Which are longer, infrared or ultraviolet waves?
5. Why is absolute temperature necessary for radiation formulas?
6. What is the basic idea behind Planck's radiation theory?

7. Why do hot bodies glow and cold bodies do not?
8. What is Wien's displacement law?
9. What is the Stefan-Boltzmann law?
10. Define a blackbody; a gray body; a lambert body.
11. Who was Samuel Langley? Sir Frederick Herschel? Max Planck?
12. Define the monochromatic intensity of radiation.
13. What is Lambert's cosine law? Do real surfaces follow it?
14. Define diffuse and specular reflection.
15. What is hohlraum?
16. Define emissivity, absorptivity, transmissivity, and reflectivity. Simplify these for an opaque, gray-lambert body.
17. Distinguish between "normal" and "hemispherical" emissivity.
18. What is Kirchhoff's law? Why does it have three "levels"?
19. Define the radiation shape factor for diffuse surfaces.
20. What is radiosity? How does it relate to surface interchange?
21. What is the shape factor reciprocity law?
22. What is the electric analogy for radiation in enclosures?
23. What distinguishes gas radiation from blackbody radiation?
24. What is the solar constant?

## Problems

### Problem distribution by sections

**The Problem Assignments Are Organized as Follows:**

| Problems | Sections Covered | Topics Covered |
|---|---|---|
| 8.1 –8.18 | 8.1 | Thermal radiation physics |
| 8.19–8.31 | 8.2 | Real surface characteristics |
| 8.32–8.57 | 8.3 | Shape factors |
| 8.58–8.74 | 8.4 | Multiple surface enclosures |
| 8.75–8.80 | 8.5 | Gas absorption and emission |
| 8.81–8.88 | 8.6 | Solar radiation |
| 8.89–8.93 | All | Design Projects |

8.1 Calculate the maximum emissive power in $W/m^3$ and the wavelength at which it occurs in microns for a blackbody surface at (a) 300K and (b) 3000K.

8.2 The yellow lines in the visible spectrum have an approximate wavelength of 0.58 $\mu$m. Estimate (a) the frequency of yellow waves and (b) the temperature for which yellow waves would radiate at maximum blackbody emissive power.

8.3 If, as inferred in Fig. 8.2, the sun approximates a blackbody at 5800K, estimate its radiated energy in $W/m^2$. Compare this with the solar energy received by the earth's upper atmosphere, which is 1353 $W/m^2$. Can you account analytically for the vast difference in these two numbers?

8.4 From Planck's law show analytically that (a) $E_{b\lambda}$ increases with temperature for any constant $\lambda$ and (b) $E_{b\lambda}(\max)$ increases as $T^5$.

8.5 A small hole is cut in the side of a large spherical container whose inner surface is painted black and held at 1200°C. What will be the total radiated energy emitted from the hole? Find the wavelength $\lambda_1$ below which 20% of this power is emitted.

8.6 At $\lambda = 0.7$ $\mu$m, for what temperature will a blackbody have spectral emissive power of $10^6$ $W/m^3$?

8.7 For a wavelength of 0.4 $\mu$m, what blackbody temperature will cause a spectral emissive power of $10^9$ $W/m^3$?

8.8 What percentage of emissive power is in the wavelength range 0 to 1.2 $\mu$m for a blackbody at temperature (a) 500K and (b) 5000K?

8.9 Consider the four equally spaced infinite black parallel plates in Fig. P8.9. For the given temperatures $T_1$ and $T_4$, compute the total radiated energy $q''$ through the plates (a) for plates 1 and 4 standing alone; (b) for plates 1, 2, and 4 only; and (c) for all four plates standing together. Comment on the ratios of these three results.

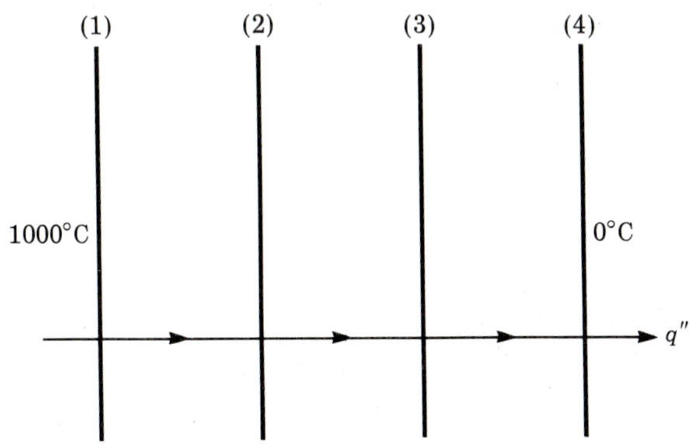


**Figure P8.9**

8.10 A light bulb filament, assumed black, is at 3200K. What percentage of its radiation is in the (a) visible and (b) infrared range?

8.11 A flame, assumed black, burns at 1500°F. What percentage of its emission is in the (a) visible and (b) infrared range?

8.12 In 1896, Willy Wien proposed an approximation to Planck's blackbody spectrum law, $E_{b\lambda} \doteq (C_1/\lambda^5)\exp(-C_2/\lambda T)$. For what wavelength range does this expression agree to within 10% of Planck's law? Sketch both Wien's and Planck's law on the same graph.

8.13 Rayleigh in 1900 and Jeans in 1905 independently proposed the blackbody approximation $E_{b\lambda} \doteq C_1T/(C_2/\lambda^4)$. For what wavelength range does this expression agree to within 10% of Planck's law? Sketch both the Rayleigh-Jeans law and Planck's law on the same graph.

8.14 A thin silica glass sheet transmits 92% of incident radiant energy between 0.3 and 2.8 $\mu$m and essentially zero otherwise. What percent of incident blackbody solar energy does it transmit?

8.15 Fused quartz transmits 90% of incident radiation in the range between 0.2 and 4.0 $\mu$m and is opaque otherwise. What percent of blackbody radiation at 300K does it transmit?

8.16 Show that the total radiation emitted by a black surface between the directions $\theta$ and 90° is given by

$$\delta E_b(\theta \rightarrow 90°) = E_b(1 - \sin^2\theta).$$

What fraction of total energy is emitted between 20° and 50°?

8.17 Consider two elements $dA_1$ and $dA_2$ near point 0 as shown in Fig. P8.17. Surface element $dA_0$ is black and held at 1800°C. For each element 1 and 2, compute (a) the solid angle subtended about point 0 and (b) the total radiated energy from point 0 striking the element.

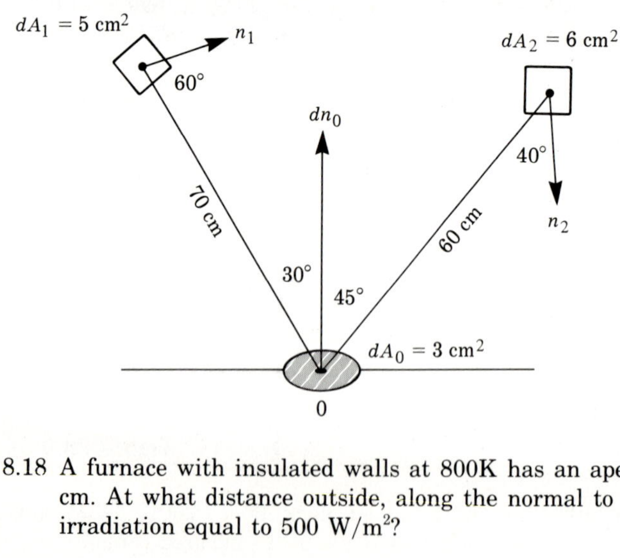


Figure P8.17

8.18 A furnace with insulated walls at 800K has an aperture of diameter 3 cm. At what distance outside, along the normal to the aperture, is the irradiation equal to 500 W/m²?

8.19 A diffuse surface has $\varepsilon_\lambda = 0.8$ for $0 \leqslant \lambda \leqslant 2\ \mu$m and $\varepsilon_\lambda = 0.2$ for $\lambda .>$ 2 m. Compute the total hemispherical emissivity of the surface at (a) 500K and (b) 5000K.

8.20 A directionally selective isotropic surface has an emissivity distribution shown in Fig. P8.20. These values are independent of $\lambda$. Compute the hemispherical emissivity and its ratio to normal emissivity and compare with Fig. 8.8.

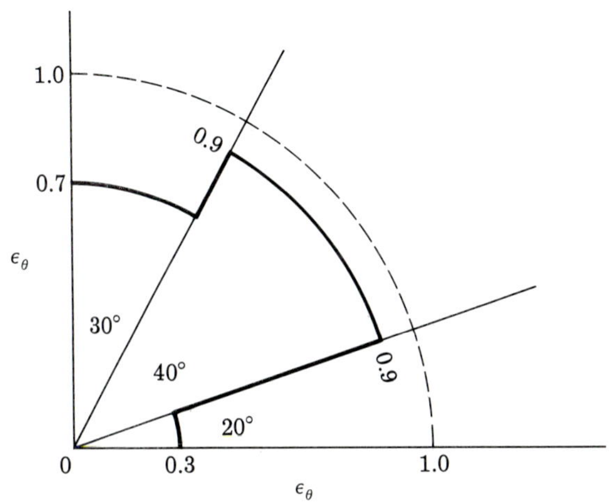


**Figure P8.20**

8.21 A painted isotropic surface has $\varepsilon_\theta = 0.9\ (\cos\ \theta)^{1/4}$, independent of $\lambda$. Compute its hemispherical emissivity and compare with the estimate given by Fig. 8.8.

8.22 The spectral emissivity of tungsten from Fig. 8.10 is approximated by $\varepsilon_\lambda = 0.4$ for $0 \leqslant \lambda \leqslant 1.5\ \mu$m and $\varepsilon_\lambda = 0.15$ for $\lambda > 1.5\ \mu$m. Estimate the total emissivity of tungsten at (a) 3000K and (b) 1500K.

8.23 Assuming that a light bulb filament is lumpable, has constant emissivity, and loses heat only by radiation, derive an expression for the time required to cool from $T_0$ to $T_1$ after the current is suddenly shut off. Apply your result to a tungsten filament (Appendix I) cooling from 3200K to 1000K if the filament diameter is 1 mm.

8.24 From Fig. 8.10, the spectral emissivity of magnesium oxide may be fit to a three-piece approximation: $\varepsilon_\lambda = 0.36$ for $\lambda < 2\ \mu$m, $\varepsilon_\lambda = 0.5$ for $2 \leqslant \lambda \leqslant 5\ \mu$m, and $\varepsilon_\lambda = 0.8$ for $\lambda > 5\ \mu$m. Estimate the total emissivity at 1250 K. Does the emissivity drop with temperature?

8.25 A panel mounted on a satellite approximates graphite from Fig. 8.11. It is insulated on the inside, and the outside is normal to solar radiation of 1353 W/m$^2$. Estimate the equilibrium temperature of the panel.

8.26 Repeat Problem 8.25 if the panel is made of aluminum as given in Fig. 8.11.

8.27 Given the spectral characteristics of surfaces $A$, $B$, and $C$ in Fig. P8.27, determine $\varepsilon$ for each at $T = 1000$K. Which surface is better for (a) a solar absorber and (b) a solar reflector?

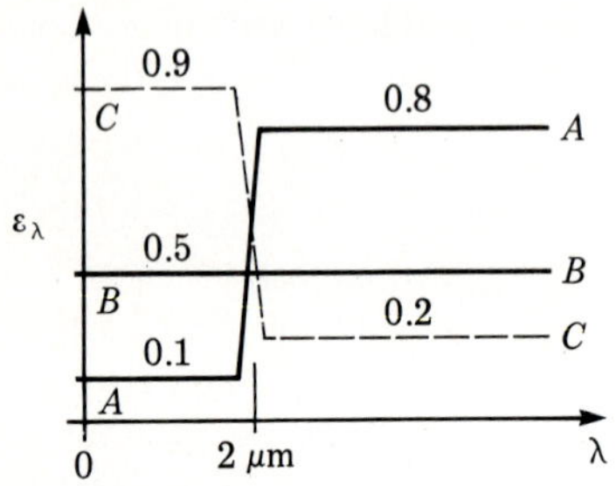


**Figure P8.27**

8.28 A shallow pan full of water, with bottom and sides insulated, is placed outside on a clear night, when the effective sky radiation temperature is 230K. The air temperature is 5°C and the convective heat transfer coefficient is 20 W/m$^2$ · K. The water surface emissivity is 0.8. Will the water freeze? What air temperature will ensure no freezing?

8.29 The hohlraum in Fig. 8.5 does simulate a blackbody, but the hole in the cavity must not be too large. If $A_w$ is the cavity surface area and $\varepsilon_w$ its emissivity, the effective emissivity of the hole is given by

$$\varepsilon_h \cong \frac{\varepsilon_w}{\varepsilon_w + (A_h/A_w)(1 - \varepsilon_w)}.$$

Suppose that the holhraum is a white fire clay cylinder as in Fig. P8.29. If the cavity properties are given by Fig. 8.11, what is the largest diameter hole that can be used and still ensure that $\varepsilon_h = 0.999$?

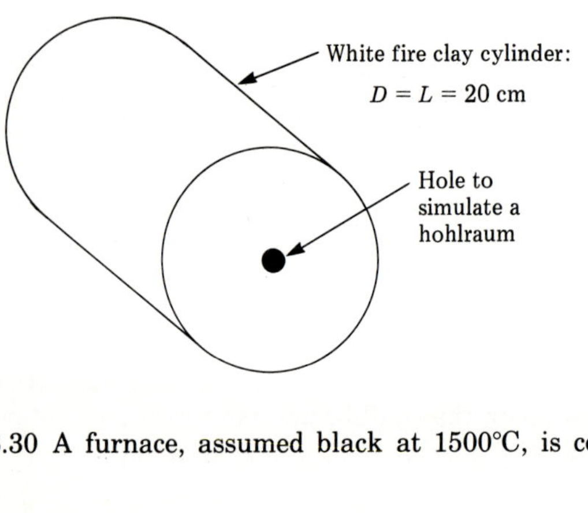


**Figure P8.29**

8.30 A furnace, assumed black at 1500°C, is connected by a window to an environment, assumed black, at 20° C. The window glass has the following spectral characteristics:

*For* $\lambda \leq 2\ \mu m$:

$\varepsilon_\lambda = 0.1,\ \tau_\lambda = 0.9,\ \rho_\lambda = 0.0;$

*For* $\lambda > 2\ \mu m$:

$\varepsilon_\lambda = 0.7,\ \tau_\lambda = 0.1,\ \rho_\lambda = 0.2.$

For steady-state conditions, neglecting conduction and convection, estimate (a) the window temperature and (b) the heat loss from the furnace through the window.

8.31 A 10-cm-diameter sphere has a diffuse surface with spectral characteristics $\varepsilon_\lambda = 0.3$ for $\lambda \leq 3\ \mu m$ and $\varepsilon_\lambda = 0.8$ for $\lambda > 3\ \mu m$. It is placed in a large furnace whose walls are at 500K. The furnace air is at 450K and the convective heat transfer coefficient is 18 $W/m^2 \cdot K$. At one instant while the sphere heats up, the net heat flux to the sphere is found to be 5000 $W/m^2$. Estimate the sphere surface temperature at this time.

***Note:* For Problems 8.32–8.57, assume that all surfaces are diffuse and have uniform radiosity.**

8.32 Determine the shape factors $F_{1\to2}$, $F_{2\to1}$, and $F_{2\to2}$ for the long, concentric cylinders in Fig. P8.32, where $R_2 = 3R_1$. Would the answers change if the inner cylinder were eccentric?

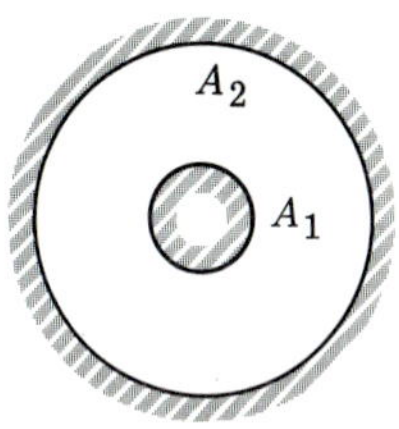


**Figure P8.32**

8.33 Repeat Problem 8.32 if the bodies in Fig. P8.32 are spheres.

8.34 In Fig. P8.34 a long cylinder $R_1$ is concentric to a long half-cylinder $R_2 = 4R_1$. Compute $F_{1\to2}$ and $F_{2\to1}$. Also compute $F_{2\to e}$, where $e$ denotes the environment.

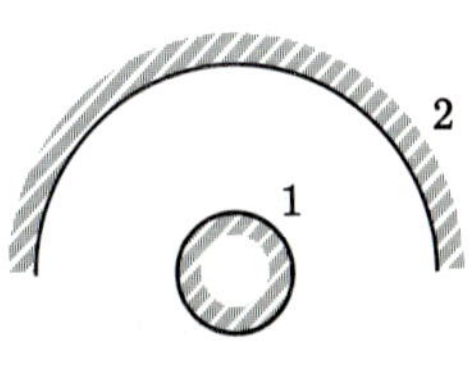


**Figure P8.34**

8.35 For the long semicircular duct of Fig. P8.35, compute $F_{1\to2}$, $F_{2\to1}$, and $F_{2\to2}$. Do not integrate.

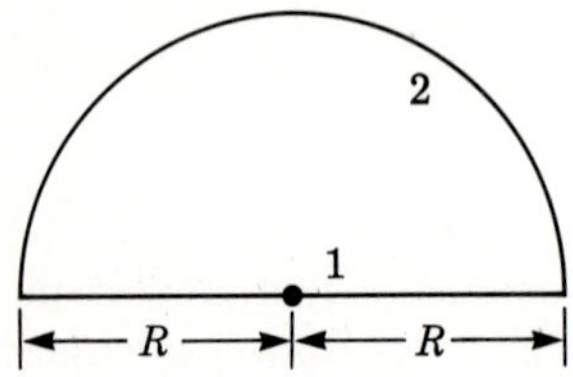


Figure P8.35

8.36 For the long triangular duct of Fig. P8.36, all surfaces are black, with $T_1 = 800°C$ and $T_2 = T_3 = 100°C$. What should be the angle $\theta$ so that only 40% of the radiant energy from surface 1 is captured by surface 3?

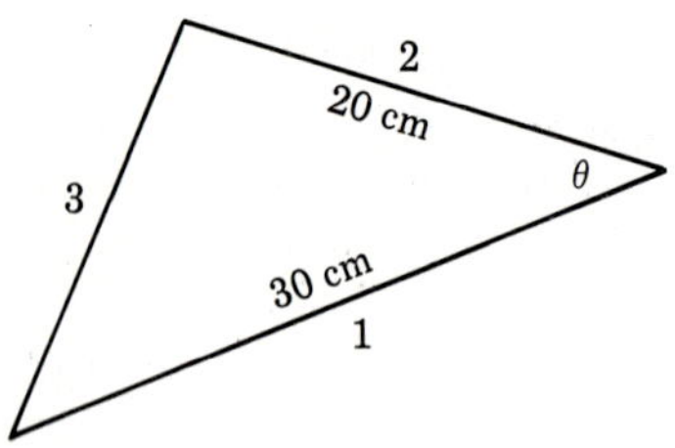


Figure P8.36

8.37 A long, black 16-cm-diameter steam pipe passes down the middle of a long, black 50-×-50-cm square duct. Pipe walls are at 200°C and duct walls are at 20°C. Determine (a) the radiation loss from pipe to duct, and (b) the radiation heat transfer coefficient, per meter of pipe length.

8.38 For the two-dimensional configuration of Fig. P8.38, determine $F_{1\to2}$, $F_{1\to3}$, $F_{1\to2,3}$, and $F_{1\to e}$, where $e$ is the environment.

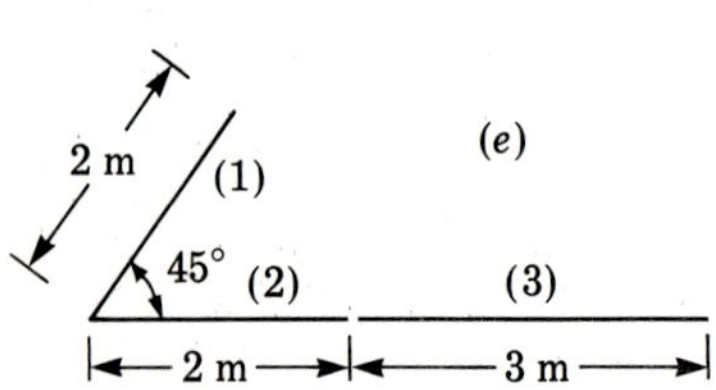


Figure P8.38

8.39 The conical shape in Fig. P8.39 has flat black surface 1 at 500°C and conical black surface 2 at 20°C. The bottom flat surface is open. What should be the height $H$ so that only 20% of the energy from surface 1 passes out the opening? What is the net energy received by surface 2 for this condition?

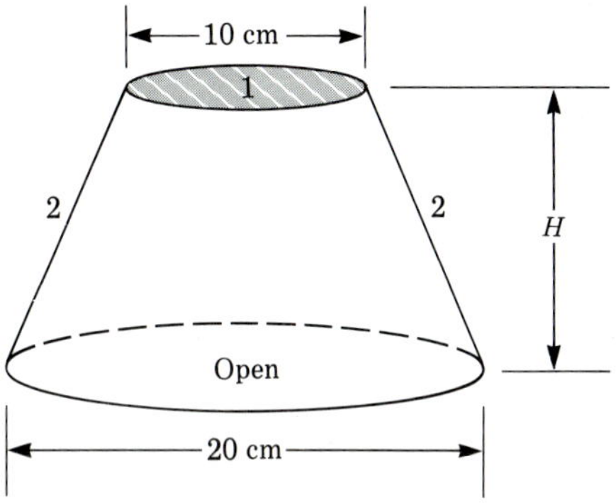


**Figure P8.39**

8.40 Compute the shape factors $F_{1\to 2}$ and $F_{2\to 1}$ if body 1 is a sphere of radius $R$ lying on body 2, which is a very large floor.

8.41 The long 60° vee-groove in Fig. P8.41 has black walls at 150°C. Compute the net heat loss per meter of groove length to a black environment at 20°C.

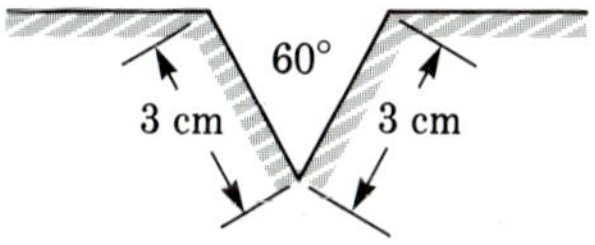


**Figure P8.41**

8.42 The long triangular duct in Fig. P8.42 has black sides with $T_1 = 300$°C and $T_2 = T_3 = 100$°C. Compute the net radiation between sides 1 and 2, per meter of duct length.

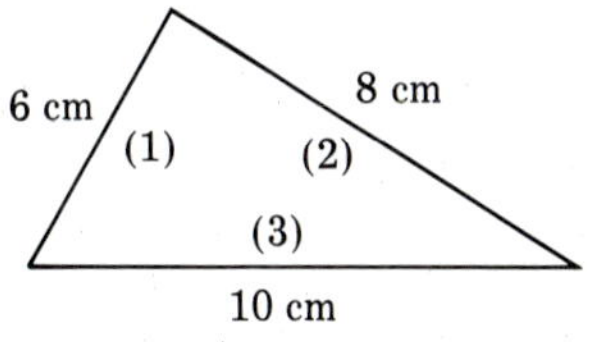


**Figure P8.42**

8.43 Two long, black 18-cm-diameter pipes are parallel and 8 cm apart at their closest point. Pipe 1 is at 300°C and pipe 2 at 20°C. Compute the net radiation between the pipes, per meter of pipe length.

8.44 A long, black rectangular duct has 34-cm top and bottom and 25-cm sides. The bottom is at 300°C and the other three surfaces are at 20°C. Compute the net radiation, per meter of duct length, from the bottom (a) to one side and (b) to the top.

8.45 For the two-dimensional right-angle geometry of Fig. P8.45, compute (a) $F_{2\to 3}$ and (b) $F_{1\to 4}$. Use algebra, not integration. Compare the result for $F_{1\to 4}$ with a computation using "point" areas $A_1$ and $A_4$ and a mean distance between the two.

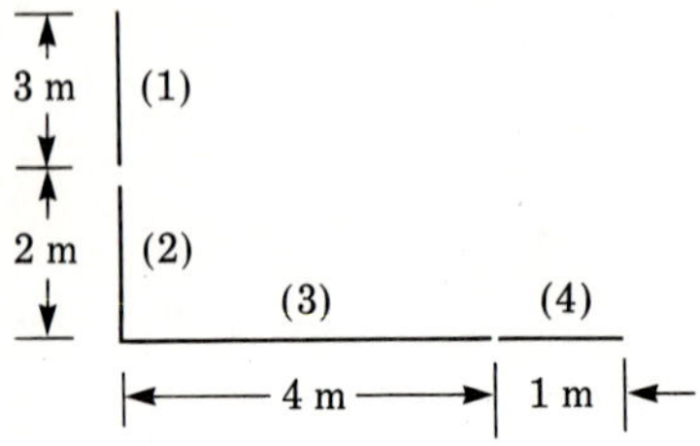


**Figure P8.45**

8.46 For the rectangular two-dimensional geometry of Fig. P8.46, determine the distance $H$ for which $F_{1\to 2} = F_{1\to 3}$.

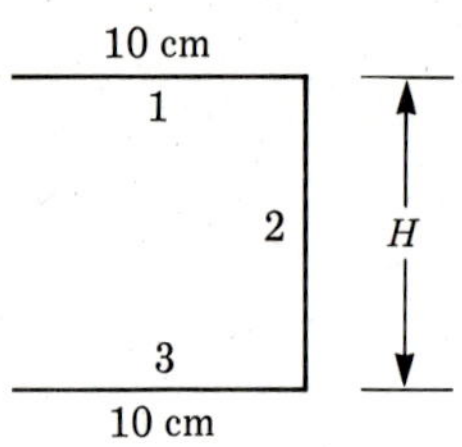


**Figure P8.46**

8.47 A black cubical room has sides of length 5 m, with the floor, walls, and ceiling all at 20°C except for a 20-by-20-cm radiant heater at 350°C in the corner of the ceiling. Estimate the net radiant heat transfer from this heater to (a) the floor and (b) the far wall.

8.48 A black, 50-cm-diameter disk at 100°C is concentric and parallel to an identical second disk at 20°C. They are brought together until the net radiation between the two is 74 W. How far apart will they be, in cm?

8.49 Repeat Problem 8.47 with the radiant heater in the center of the ceiling.

8.50 Two long black 10-cm-diameter cylinders are laid parallel to each other, with one surface at 400°C and the other at 20°C. What is the minimum distance between centerlines to ensure that the net radiant heat transfer between the two will be no more than 200 watts per meter of length?

8.51 Consider the two-dimensional plates $L_1$ and $L_2$ in Fig. P8.51. Pretend that "strings" are sketched across the ends, shown as dashed lines. Show that the shape factor is given by $F_{1\to 2} = (a + b - c - d)/2L_1$, where $a$ and $b$ are the "crossed-string" lengths and $c$ and $d$ are the "end-string" lengths. This is the *string method* of H. C. Hottel [13, p. 67] and can be extended to curved surfaces.

8.52 Repeat Problem 8.45 by the string method of Problem 8.51. Does the method work even if bodies 3 and 2 intersect?

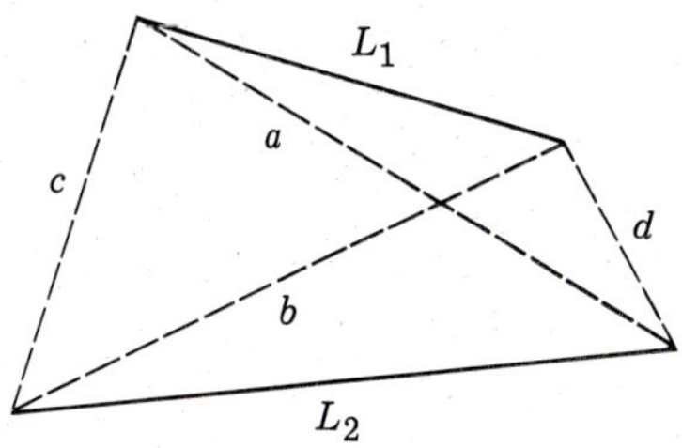


Figure P8.51

8.53 Compute $F_{1\rightarrow3}$ for the 90° geometry of Fig. P8.53 using shape factor algebra.

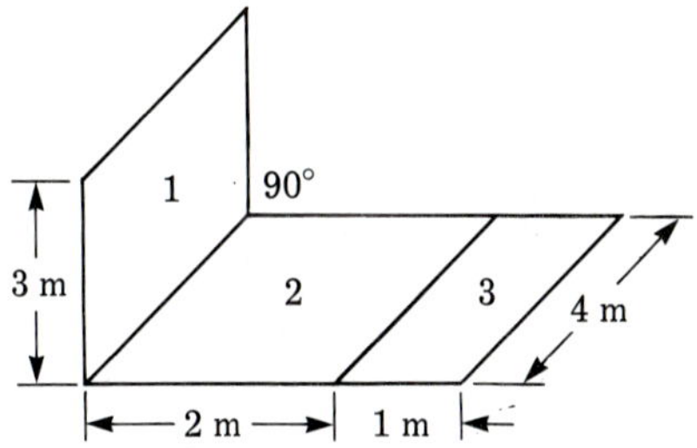


Figure P8.53

8.54 Using shape factor algebra, compute $F_{1\rightarrow4}$ for the geometry of Fig. P8.54.

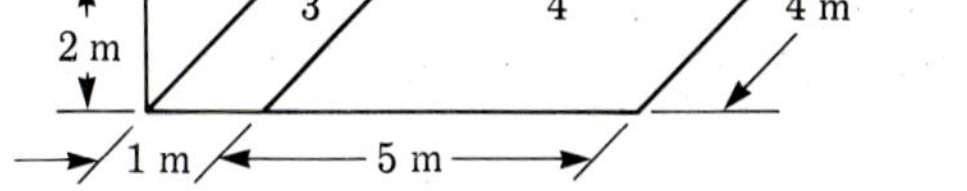


Figure P8.54

8.55 Using shape factor algebra, compute $F_{1\rightarrow4}$ for the geometry of Fig. P8.55.

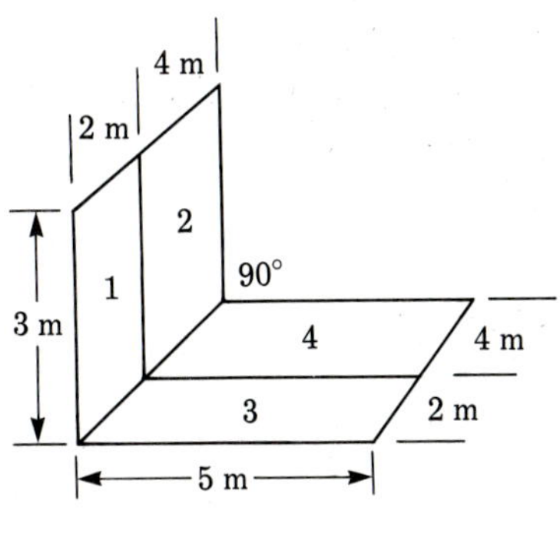


Figure P8.55

8.56 A right circular cylinder is 40 cm in diameter and 100 cm high. All inside surfaces are black, with the (flat) bottom at 100°C and all other surfaces at 20°C. Compute the net radiation heat transfer from the bottom to (a) the top and (b) the curved sides.

8.57 The open right circular cylinder in Fig. P8.57 is insulated and has its inside (black) surface heated electrically to 400°C. The environment (black) is at 20°C. Compute the electric power needed to maintain the surface temperature. Neglect convection.

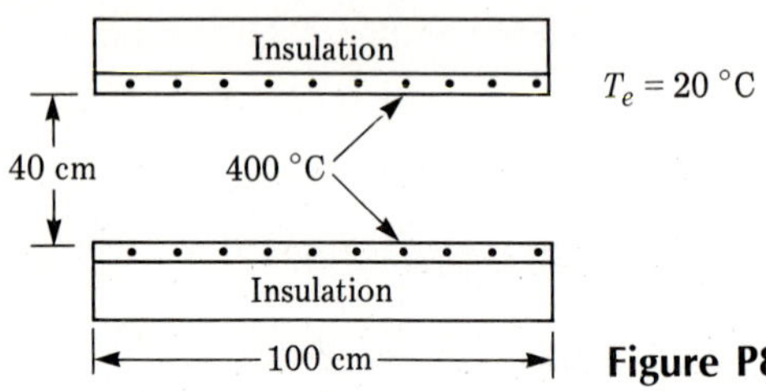


**Figure P8.57**

***Note:* In Problems 8.58–8.74, assume that all surfaces are gray and diffuse unless otherwise stated.**

8.58 Show for the geometry of Fig. P8.32 that, if the surfaces are gray with emissivities $\varepsilon_1$ and $\varepsilon_2$, respectively, the net radiation heat transfer is given by

$$q_{1\to 2} = \frac{A_1\sigma(T_1^4 - T_2^4)}{\dfrac{1}{\varepsilon_1} + \dfrac{A_1}{A_2}\left(\dfrac{1}{\varepsilon_2} - 1\right)}.$$

Evaluate this expression in the limit $R_2 \gg R_1$ and interpret.

8.59 For the semicircular duct of Fig. P8.35, assume that $R = 20$ cm, $T_1 = 200$°C, $\varepsilon_1 = 0.6$, $T_2 = 20$°C, and $\varepsilon_2 = 0.4$. Compute the net radiation from surface 1 to surface 2 per meter of duct.

8.60 A cryogenic container consists of two thin spherical shells of diameters 120 and 160 cm, with the clearance evacuated. The inner sphere is filled with liquid oxygen. The sphere surfaces have emissivity 0.15. Assume that the outer sphere is at ambient temperature of 20°C and the inner sphere is at the oxygen saturation temperature of 90K. If the latent heat of oxygen is 213,000 J/kg, estimate the rate of evaporation of the oxygen in kg/hr.

8.61 A thin cylinder 15 cm in diameter and 10 m long is filled with liquid methane at its saturation temperature of 111K. It is surrounded by a shielding cylinder 20 cm in diameter, and the space between is evacuated. The latent heat of vaporization of methane is 509 kJ/kg. Neglecting resistance of the pipe walls, estimate the rate of methane evaporation in kg/hr.

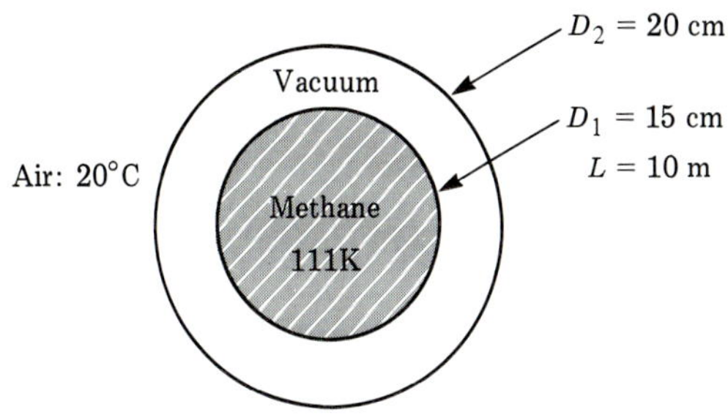


Figure P8.61

8.62 Two parallel concentric 40-cm-diameter disks are placed 15 cm apart. Their respective conditions are $T_1 = 250°C$, $\varepsilon_1 = 0.8$ and $T_2 = 50°C$, $\varepsilon_2 = 0.7$. Environmental radiation is negligible. Compute the net radiation between the disks.

8.63 Repeat Problem 8.62 if the disks are surrounded by a gray, adiabatic (reradiating) sphere of diameter 2 m and emissivity 0.6.

8.64 Repeat Problem 8.60 if an additional thin-foil spherical shield with $\varepsilon_3 = 0.15$ is placed midway in the clearance space.

8.65 The right circular conical container in Fig. P8.65 has its base (1) at 300°C, its curved surface (2) at 20°C, and its upper surface (3) at 100°C. All surfaces are oxidized steel ($\varepsilon = 0.6$). Compute the net radiant heat transfer from each surface. (*Note:* Since surface 2 sees itself, the electric analogy, (Fig. 8.27) fails, and we must use the matrix formulation, Eq. (8.67) or (8.69).)

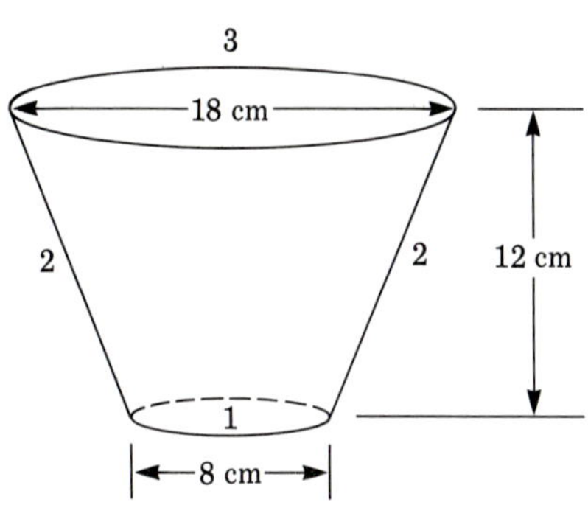


Figure P8.65

8.66 How close is a hole to a hohlraum? Consider a hole of diameter 1 cm and depth 5 cm drilled into a material whose emissivity is 0.6. Let the hole surface be heated to 400°C. Compute the radiation emitted from the hole and compare to the emission of a black disk of diameter 1 cm, at 400°C.

8.67 Generalize Problem 8.66 to a hole of diameter $D$, depth $L$, surface temperature $T_0$, and emissivity $\varepsilon_0$. Compute and plot, in the range $1 \leq L/D$

$\leqslant 10$, the ratio $q_{\text{hole}}/(A_{\text{hole}}\sigma T_0^4)$, which should approach unity for a true hohlraum.

8.68 The oven in Fig. P8.68 is 2 m deep into the paper and is used to bake a carbon-fiber cloth with an electric heater at the top. All vertical walls are reradiating. Take $\varepsilon_1 = 0.7$, $\varepsilon_2 = 0.9$, and $\varepsilon_3 = 0.8$. When 20 kW of power are supplied, the heater temperature is 650°C. Neglecting convection, what is the cloth temperature?

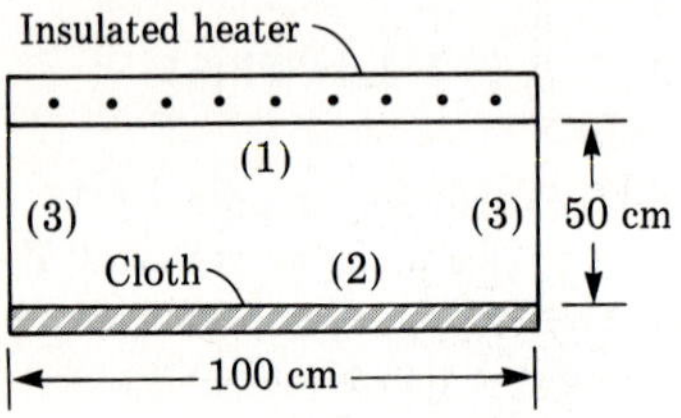


Figure P8.68

8.69 As shown in Fig. P8.69, a large disk $R_1$ faces a small disk, $R_2 = 24$ cm, within a large room at $T_3 = 15$°C. The backs of the disks are insulated, with $T_1 = 90$°C, $\varepsilon_1 = 0.8$, $T_2 = 25$°C, and $\varepsilon_2 = 0.6$. Neglecting convection, find the disk size $R_1$ required to maintain these steady-state conditions.

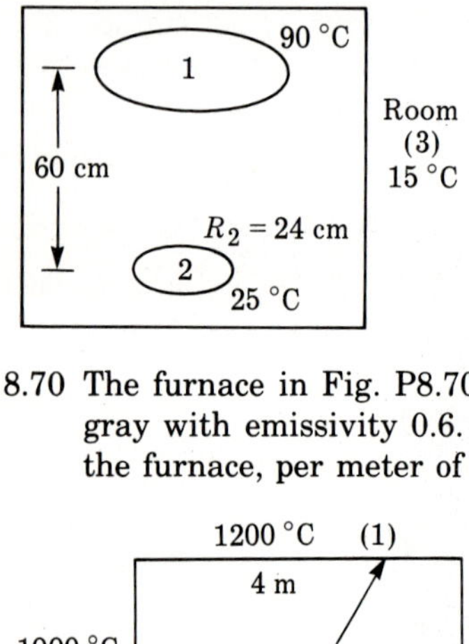


Figure P8.69

8.70 The furnace in Fig. P8.70 is very long into the paper. All surfaces are gray with emissivity 0.6. Find the radiant heat transfer to the floor of the furnace, per meter of length into the paper.

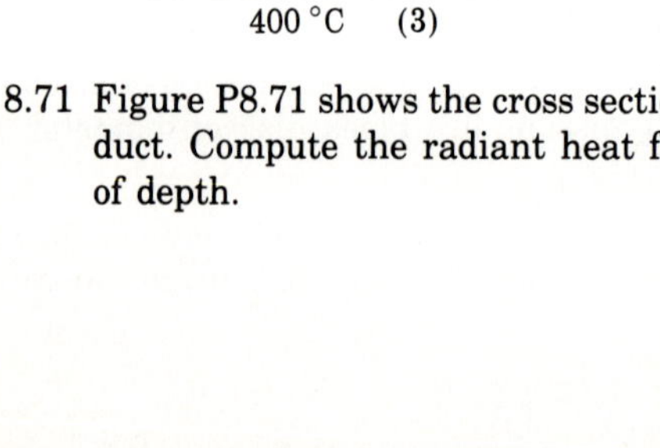


Figure P8.70

8.71 Figure P8.71 shows the cross section of a very long equilateral triangular duct. Compute the radiant heat flux to the floor of the duct, per meter of depth.

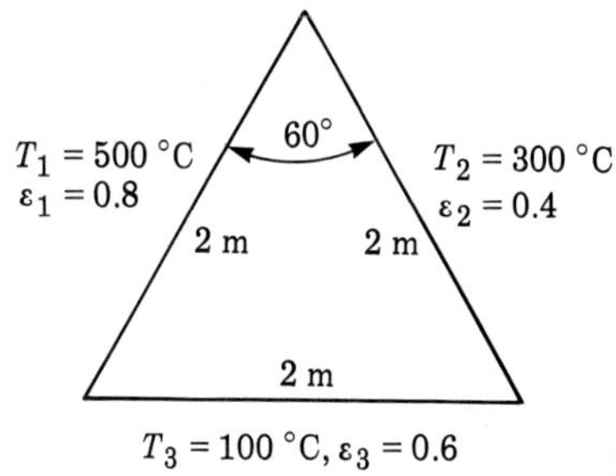


**Figure P8.71**

8.72 The two perpendicular planes in Fig. P8.72 are in a large room, $T_3$ = 300K. Surface 1 is a heater, insulated on the back side, with $T_1$ = 1200K and $\varepsilon_1$ = 0.8. Surface 2 is reradiating, with $\varepsilon_2$ = 0.6. Compute the power delivered by the heater and the temperature $T_2$.

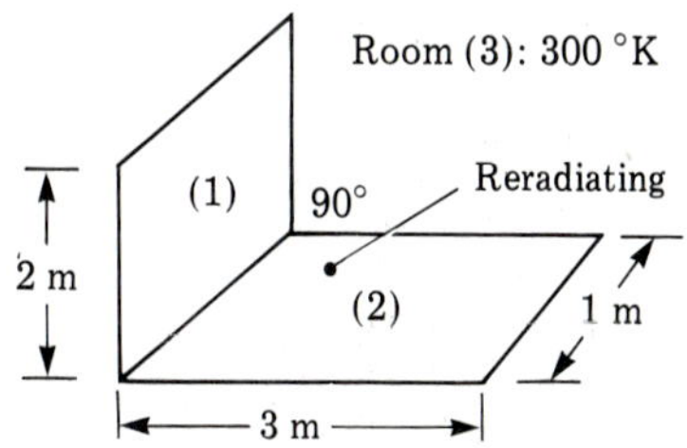


**Figure P8.72**

8.73 Two large, parallel, 5 × 5-m planes are separated by a 10-cm gap, with $T_1$ = 500°C, $\varepsilon_1$ = 0.8, $T_2$ = 100°C, and $\varepsilon_2$ = 0.7. Compute the radiant heat flux between the two (a) as described, and also when (b) one and (c) two shields with $\varepsilon$ = 0.2 are inserted in the gap.

8.74 For radiation between two nonblack bodies, some texts (such as [13]) define a "gray-body shape factor" Γ such that

$$q_{1\to 2}(\text{net}) = \Gamma_{1\to 2} A_1 (E_{b1} - E_{b2}),$$

where $\Gamma_{1\to 2}$ = fcn($\varepsilon_1$, $\varepsilon_2$, $A_1$, $A_2$, $F_{1\to 2}$). Use this concept to define $\Gamma_{1\to 2}$ for (a) the unshielded plate geometry of Problem 8.73, and (b) the concentric cylinders or spheres of Problem 8.32. Comment on the usefulness of the concept.

8.75 A 1-m-diameter sphere, with black walls at 500°F, contains 70% nitrogen and 30% water vapor at 1000°F. Compute the gas emissivity and absorptivity if the total pressure is 1 atm.

8.76 A right circular cylinder 50 cm high and 50 cm in diameter contains 50% $N_2$, 30% $CO_2$, and 20% $H_2O$ at a total pressure of 2 atm and temperature of 800°F. The container walls are black at 500°F. Estimate the gas emissivity and absorptivity.

8.77 A 1 × 1 × 1-m cubical enclosure contains 60% $N_2$, 25% $CO_2$, and 15% $H_2O$ at a total pressure of 50 kPa and a temperature of 700°F. The cube walls are black at 500°F. Estimate $\varepsilon_{gas}$ and $\alpha_{gas}$.

8.78 For the gas-filled sphere of Problem 8.75, estimate the radiant heat flux to the wall and the time required to cool the gas to 900°F.

8.79 For the gas-filled cube of Problem 8.77, estimate the radiant heat flux to the wall and the time required to cool the gas to 650°F.

8.80 A cylindrical furnace, with length and diameter 80 cm, contains 80% $N_2$ and 20% $CO_2$ at 1200°F. If the walls are black, how much cooling is required to maintain them at 600°F if $P$ = 1 atm?

8.81 Find a reference book and estimate the total number of daylight hours available annually at latitude (a) 0° (equator); (b) 23.5°N (Tropic of Cancer); (c) 41.5°N (Rhode Island); and 66°N (Arctic Circle).

8.82 Assume that $D_{sun} = 1.39 \times 10^9$ m, $D_{earth} = 1.27 \times 10^7$ m, and the mean distance from sun to earth is $1.49 \times 10^{11}$ m. If the sun is a blackbody at 5762K, verify that the solar constant is 1350 W/m$^2$.

8.83 Using the same concepts as in Problem 8.82, find the "solar constant" for (a) Mars, which is $2.28 \times 10^{11}$ m from the sun; (b) Mercury, which is $5.79 \times 10^{10}$ m away; and (c) Neptune, $4.5 \times 10^{12}$ m from the sun. Why do the results not depend upon the *sizes* of the planets?

8.84 On a clear day in Kingston, R.I., we may estimate that the atmosphere absorbs 90% of the solar radiation less than 0.4 $\mu$m, 30% between 0.4 and 2.0 $\mu$m, and 100% above 2 $\mu$m. What is the insolation received at the ground if the solar constant is 1350 W/m$^2$?

8.85 Using Fig. 8.34, estimate the total required area of flat plate solar collector at 50% efficiency, in both Yuma, Ariz., and Kingston, R.I., to heat 80 U.S. gallons of water each day from 50°F to 120°F in (a) December and (b) June.

8.86 A flat horizontal plate, insulated underneath, receives insolation of 700 W/m$^2$. If convection is negligible, estimate its equilibrium temperature if its surface is (a) white enamel and (b) silicon carbide. Use Fig. 8.10.

8.87 Repeat Problem 8.86 by accounting for a convection coefficient of 8 W/m$^2$ · K from surrounding air at 20°C.

8.88 Reference [43] describes a project being studied at the Jet Propulsion Laboratory, in which 60 12-m parabolic dish solar concentrators provide 1 MW electric power to a small community. Estimate the efficiency and comment on the potential of this concept.

### Special Projects

8.89 The approximate answers to Problem 8.73 are $q_{12}$ = 285 kW with no shields, 45 kW with one shield, and 24 kW with two shields. Consider this a reward if you are currently working that problem and have read this far ahead. It is proposed that we can reduce this heat flux still further by filling the spaces between walls with steam, which is an absorbing gas. Treat the walls as infinite parallel plates and the steam layers as reradiating bodies, with each layer at the mean temperature of its bounding walls. Using Fig. 8.29, make a parametric study of the effect of steam pressure on the heat flux with no shields. Plot the percent reduction versus steam pressure. What is the maximum reduction possible for the Hottel/Egbert data? Make a brief study of the effect of steam pressure with one shield and with two shields. Why is the heat flux reduction so much less? Explain how you might program this study on a personal computer.

8.90 For the 3 × 4 × 5-m rectangular room of Fig. P8.90, the ceiling (surface 1) is to be a radiant heater delivering 500 W/m$^2$ to the room. The floor (surface 2) and the walls (surfaces 3–6) are all at 20°C. It is desired to know what the effect of surface emissivity $\varepsilon$ (taken to be the same for all surfaces) will be on surface heat transfer and ceiling temperature. Since there are six sides, the electric analogy is cumbersome, and the matrix method of Eqs. (8.67) and (8.68) is recommended. Set up a computer program to solve for $q''_{2-6}$ and $T_1$ for values of $\varepsilon$ from 0.1 to 1.0 in increments of 0.1. Comment on the results.

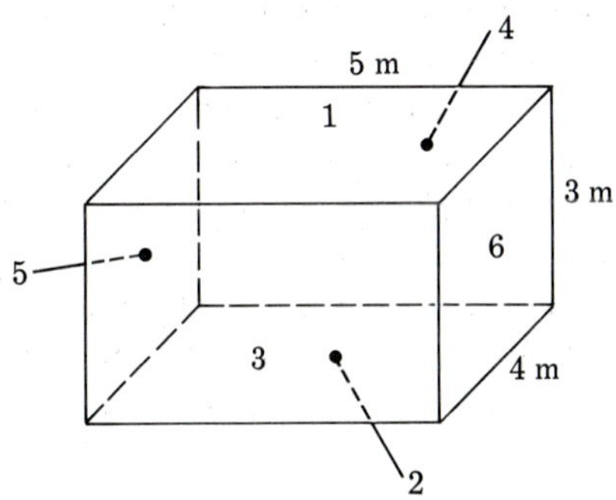


**Figure P8.90**

8.91 For the long quarter-circular duct of Fig. P8.91, assume the same emissivity $\varepsilon$ for all surfaces. Letting $\varepsilon$ vary from 0.1 to 1.0 in increments of 0.1, compute the radiant heat flux from each surface per meter of length. Plot the results and comment on the trends and the physics.

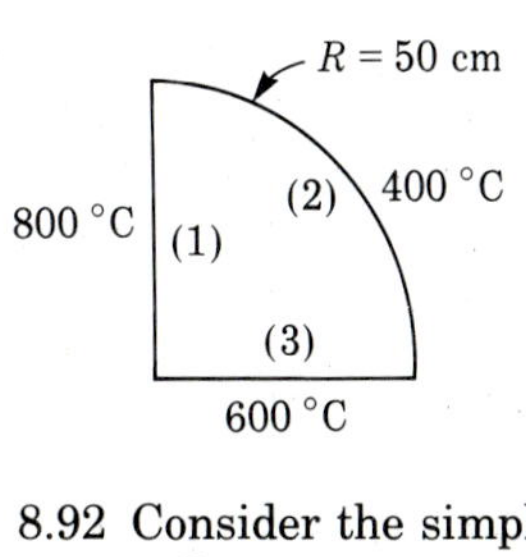


**Figure P8.91**

8.92 Consider the simplified flat plate solar collector in Fig. P8.92. The glass collector (2) is transparent to solar insolation (1) of 700 W/m$^2$ but opaque to reradiation from the cooler collector surface (3). Assume that the glass has the same emissivity $\varepsilon_2$ on both sides, uniform temperature $T_2$, and negligible conduction resistance. Assume constant environmental convection coefficient $h_e = 15\ \mathrm{W/(m^2 \cdot K)}$ and account for convection in the air gap by use of Eq. (7.47). Letting $\varepsilon_2$ vary from 0.5 to 1.0 in increments of 0.05, compute the heat transfer received by the collector and its temperature $T_3$. Is there an optimum condition for best collection? (Note that the glass has different radiosities $J_{2i}$ and $J_{2o}$ on its inside and outside, respectively.)

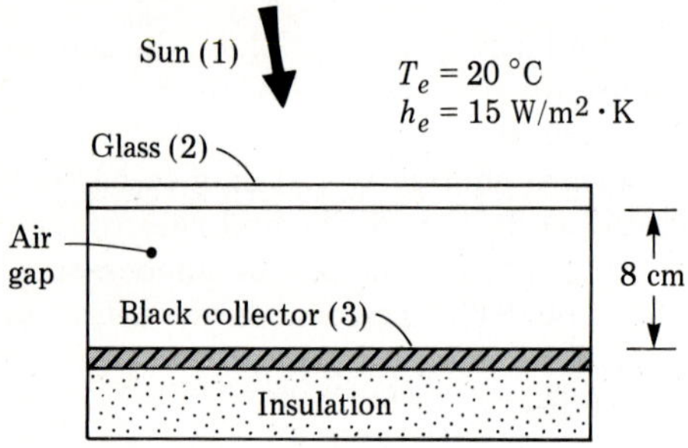


Figure P8.92

8.93 In Problem 8.66 we considered radiation from a hole 1 cm in diameter and 5 cm deep, with $\varepsilon = 0.6$. Taking the entire inner curved surface as a single surface 2, we computed that the hole would emit radiation as if $\varepsilon_{hole} = 0.953$. Now let us increase the accuracy of this estimate by letting the inner curved surface be divided into five parts, $J_2$ through $J_6$, as in Fig. P8.93. Using the matrix method, Eqs. (8.67), compute the radiosities and the total heat flux out of the hole and compare with an equivalent black hole. In computing the required shape factors it might be useful to consider the dashed lines (7–11) as "surfaces." If your computer program works especially well, try varying the inner emissivity to see the effect on hole heat flux. If you have trouble estimating the required shape factors, reduce the inner surface to two unknown radiosities $J_2$ and $J_3$ over 2.5-cm-long pieces. Also compare your result with Problem 8.29.

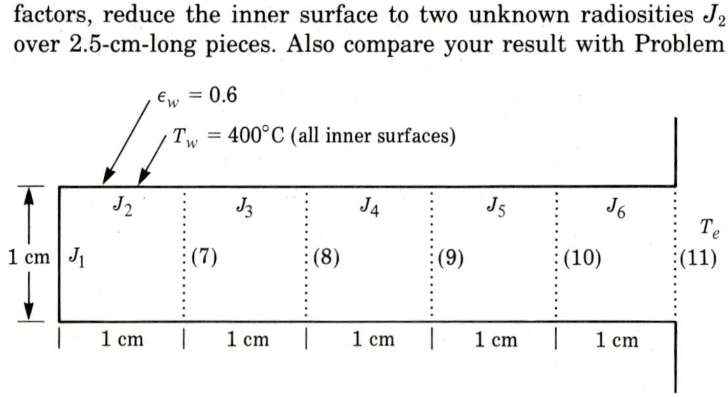


Figure P8.93

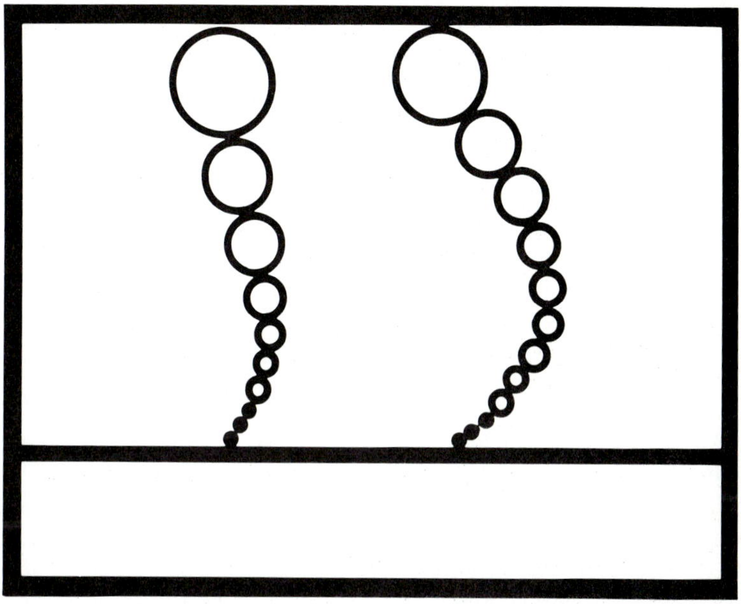

# Heat Transfer with Phase Changes

# 9

## 9.1 Introduction to Phase-Change Effects

In studying heat convection in Chapters 5, 6, and 7 we dealt only with heat transfer to single-phase fluids, that is, pure liquids or pure gases. Phase changes profoundly affect convection and generally require entirely new formulas to replace the forced- and free-convection analyses of Chapters 6 and 7. Three new parameters become dominant in phase-change problems:

1. the latent heat associated with the phase change;
2. the specific weight difference between vapor and liquid; and
3. the surface tension between the vapor/liquid interface.

We introduced the latent heat effect by assigning forced- and free-convection analyses of a melting iceberg in Problems 6.37 and 7.30, but the present chapter is our introduction to specific weight and surface tension effects.

Recall from thermodynamics that the pressure-temperature diagram of a pure substance looks like a Y, with three saturation lines fanning out from the triple point and separating the solid, liquid, and vapor regions. Thus there are six possible phase changes across these three lines:

1. vaporization line — liquid-to-vapor (boiling) and vapor-to-liquid (condensation);
2. fusion line — solid-to-liquid (melting) and liquid-to-solid (freezing);
3. sublimation line — solid-to-vapor and vapor-to-solid.

Of these six, condensation and boiling are the most important in heat transfer studies, since they cause some striking fluid mechanics effects and extremely large heat transfer coefficients. This chapter concentrates primarily on condensation and boiling. We will make brief mention of freezing and melting problems, which exhibit straightforward convection effects but some interesting moving-boundary conduction problems. We will not treat sublimation heat transfer, which can be important in ablation-type thermal protection designs.

Basically, there are two different types of condensation processes: (1) film condensation and (2) dropwise condensation. And there are two different types of boiling phenomena: (1) film boiling and (2) nucleate boiling. Film condensation and boiling are driven primarily by the density difference $(\rho_f - \rho_g)g$, where subscript $f$ denotes the liquid phase and subscript $g$ the gas phase. Dropwise condensation and nucleate boiling are governed by both density difference and surface tension. The chief energy exchange in all these processes is the latent heat of vaporization, $h_{fg}$, plus the vigorous mixing that occurs in processes such as nucleate boiling.

### 9.1.1 Dimensionless Parameters

Consider either a boiling or a condensation process. The heat transfer rate $q''$ will depend upon the temperature difference $\Delta T = |T_{\text{wall}} - T_{\text{sat}}|$, the density difference $g(\rho_f - \rho_g)$, the latent heat $h_{fg}$, surface tension $\Upsilon$, an appropriate length scale $L$, and liquid (or vapor) transport properties ($\rho$, $\mu$, $k$, $c_p$):

$$q'' = \text{fcn}(\Delta T, g\Delta\rho, h_{fg}, \Upsilon, L, \rho, \mu, k, c_p). \tag{9.1}$$

By straightforward dimensional analysis we find that the dimensionless heat transfer is a function of four dimensionless parameters:

$$\frac{q''L}{k\Delta T} = \text{fcn}\left(\frac{\rho g\Delta\rho L^3}{\mu^2}, \frac{c_p\Delta T}{h_{fg}}, \frac{\mu c_p}{k}, \frac{g\Delta\rho L^2}{\Upsilon}\right),$$

or

$$\text{Nu} = \text{fcn}(\text{Gr}, \text{Ja}, \text{Pr}, \text{Bo}). \tag{9.2}$$

All these parameters have names: the Nusselt number is a function of the Grashof number, the Jakob number, the Prandtl number, and the Bond number. We recognize Nu, Gr, and Pr of course from Chapters 5–7 on single-phase convection.

The Jakob number, $\text{Ja} = c_p\Delta T/h_{fg}$, is named for Max Jakob, who published many pioneering studies of both condensation and boiling between 1928 and 1942. The Bond number, $\text{Bo} = g\Delta\rho L^2/\Upsilon$, is the ratio of gravity force to surface tension force and arises in any problem where gravity and surface tension interact. These two parameters and the Prandtl number have only a slight effect on condensation and film boiling but are important in nucleate boiling. The primary parameter is the Grashof number, which will be rearranged in different forms to suit each phase-change process in the following sections. Note that $\text{Gr} = \rho g\Delta\rho L^3/\mu^2$ has the generic form of Eq. (5.31); the "thermal expansion" form $g\beta\Delta T L^3/\nu^2$ of Chapter 7 is irrelevant here.

We will begin our analyses by studying film condensation, which is very similar in form to the single-phase free-convection theories of Chapter 7, leading to the Nusselt number as a function of modified Grashof number. This chapter gives only a brief introduction to a subject that fills whole textbooks on boiling and condensation [1–4]. There are also two texts on flow problems in two-phase motion [5–6] and several fine review articles [7–15] on aspects of two-phase heat transfer.

## 9.2 Film Condensation

The theory of laminar film condensation is well developed and similar in many respects to the single-phase free-convection theories of Chapter 7. Except for cleaning up small details, the problem was solved in its entirety in a classic paper in 1916 by Nusselt [16]. We give the solution here for a vertical cold surface and indicate its extension to other geometries.

Figure 9.1 illustrates a cold vertical surface, $T_w < T_{sat}$, submerged in a saturated vapor. A liquid condensate film forms and grows downward with increasing thickness $\delta(x)$. The film velocity and temperature

**Figure 9.1** Schematic of the development of a condensed liquid film forming on a cold vertical surface.

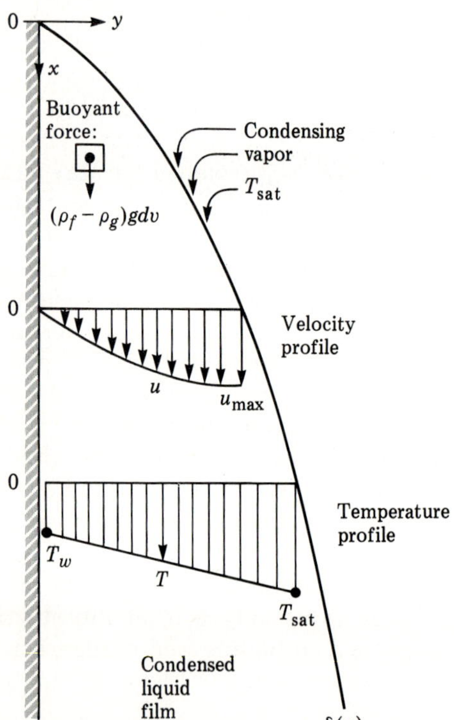

profiles would be approximately as shown. The film begins at $x = 0$, the top of the plate.

The film grows slowly, and the boundary layer approximations apply to the falling liquid. Any volume element $d\mathcal{V}$ in the film has a net weight $(\rho_f - \rho_g)g\,d\mathcal{V}$ relative to the surrounding vapor, as shown. The boundary layer continuity, $x$-momentum, and energy equations for laminar flow are, from Section 7.1,

$$\frac{\partial u}{\partial x} + \frac{\partial v}{\partial y} = 0, \tag{9.3a}$$

$$\rho_f\left(u\frac{\partial u}{\partial x} + v\frac{\partial u}{\partial y}\right) = (\rho_f - \rho_g)g + \mu_f\frac{\partial^2 u}{\partial y^2}, \tag{9.3b}$$

$$\rho_f c_{pf}\left(u\frac{\partial T}{\partial x} + v\frac{\partial T}{\partial y}\right) = k_f\frac{\partial^2 T}{\partial y^2} + \mu_f\left(\frac{\partial u}{\partial y}\right)^2. \tag{9.3c}$$

Because the film moves so slowly, inertia and dissipation are negligible. Nusselt neglected the convective acceleration term in Eq. (9.3b) plus both the convective temperature change and the viscous dissipation in Eq. (9.3c). This greatly simplifies the momentum and energy equations:

*Momentum:*

$$\frac{\partial^2 u}{\partial y^2} \doteq -(\rho_f - \rho_g)g/\mu_f; \tag{9.4a}$$

*Energy:*

$$\frac{\partial^2 T}{\partial y^2} \doteq 0. \tag{9.4b}$$

These can be solved directly for $u$ and $T$ without recourse to the continuity equation.

### 9.2.1 Nusselt's Solution for the Vertical Wall

Having derived Eqs. (9.4), Nusselt [16] adopted boundary conditions similar to the profile sketches in Fig. 9.1. He assumed that the velocity would be zero at the wall (no-slip) and that the shear stress would be nearly zero at the outer edge of the condensate. Thus,

$$u_{y=0} = 0; \qquad \left.\frac{\partial u}{\partial y}\right|_{y=\delta} = 0. \tag{9.5}$$

The condensate temperature would equal $T_w$ at the wall (no temperature jump) and be saturated at the outer edge:

$$T_{y=0} = T_w; \qquad T_{y=\delta} = T_{\text{sat}}. \tag{9.6}$$

The solution to Eq. (9.4a) subject to conditions (9.5) is, for constant properties ($\rho_f$, $\rho_g$, $\mu_f$),

$$u = u_{\max}[2y/\delta - y^2/\delta^2], \qquad u_{\max} = (\rho_f - \rho_g)\delta^2 g/2\mu_f, \tag{9.7}$$

where $\delta(x)$ is as yet undetermined. This parabolic shape is analogous to fully developed laminar flow between parallel plates or in a pipe, Eq. (6.58).

The solution to Eq. (9.4b) subject to conditions (9.6) is

$$T = T_w + (T_{\text{sat}} - T_w)y/\delta, \tag{9.8}$$

which is analogous to pure conduction in a slab of condensate.

We can use these solutions for $u$ and $T$ to estimate the validity of neglecting the convective terms in Eqs. (9.3). We find that the convective acceleration in (9.3b) is negligible if

$$\text{Ja} = c_p(T_{\text{sat}} - T_w)/h_{fg} \ll 1. \tag{9.9}$$

This is quite realistic, since the latent heat is usually much greater than the sensible heat of the condensate.

Evaluation of the convective energy change in Eq. (9.3c) shows that it is negligible if

$$\text{Re}_\delta \text{Ja}/\text{Pr}_f \ll 1, \qquad \text{Re}_\delta = \frac{2}{3}\rho_f u_{\max}\delta/\mu_f. \tag{9.10}$$

Assuming that Ja is small as required by Eq. (9.9), this condition requires small Reynolds numbers and Prandtl number of at least order unity. A more exact analysis by Sparrow and Gregg [18] corrects the Nusselt theory for large Jakob numbers (very cold walls) and small Prandtl numbers (liquid metals). Here we give only the basic Nusselt result.

To complete the analysis, we need an expression for $\delta(x)$, which is found by evaluating total mass flow and heat flux in the condensate layer. The mass flux per unit layer width is

$$\dot{m} = \int_0^\delta \rho_f u \, dy = \frac{2}{3}\rho_f u_{\max}\delta = \rho_f(\rho_f - \rho_g)g\delta^3/3\mu_f. \tag{9.11}$$

Meanwhile, the heat flux at the wall must equal the total change of enthalpy between vapor and condensate:

$$q_w'' = k_f \left.\frac{\partial T}{\partial y}\right|_{y=0} = \frac{d\dot{m}}{dx}\left[h_{fg} + \frac{1}{\dot{m}}\int_0^\delta \rho_f u c_{pf}(T_{\text{sat}} - T)\, dy\right],$$

or

$$q_w'' = k_f(T_{\text{sat}} - T_w)/\delta = \frac{d\dot{m}}{dx} h_{fg}\left(1 + \frac{3}{8}\text{Ja}\right), \tag{9.12}$$

where we introduced $u$ and $T$ from Eqs. (9.7) and (9.8). The rate of change of mass flux is found by differentiating Eq. (9.11):

$$\frac{d\dot{m}}{dx} = [\rho_f(\rho_f - \rho_g)g\delta^2/\mu_f]\frac{d\delta}{dx}. \tag{9.13}$$

If we combine Eqs. (9.12) and (9.13) and separate the variables, we obtain a first-order differential equation for layer thickness:

$$\delta^3 d\delta = \frac{k_f\mu_f(T_{\text{sat}} - T_w)}{\rho_f(\rho_f - \rho_g)gh_{fg}(1 + 0.375\text{Ja})}dx.$$

Integration yields the layer thickness varying as $x^{1/4}$:

$$\delta = \left(\frac{4k_f\mu_f(T_{\text{sat}} - T_w)x}{\rho_f(\rho_f - \rho_g)gh'_{fg}}\right)^{1/4}, \tag{9.14}$$

where $h'_{fg} = h_{fg}(1 + 0.68\text{Ja})$. The change from 0.375Ja to 0.68Ja was suggested by Rohsenow [19] as an improvement to the theory.

Finally, from Eq. (9.12), the local wall Nusselt number is

$$\text{Nu}_x = q''_w x/k_f(T_{\text{sat}} - T_w) = x/\delta = 0.707\left(\frac{\rho_f(\rho_f - \rho_g)gh'_{fg}x^3}{\mu_f k_f(T_{\text{sat}} - T_w)}\right)^{1/4}. \tag{9.15}$$

Since $h_x = q''_w/(T_{\text{sat}} - T_w) \propto x^{-1/4}$, it follows that the mean Nusselt number over a plate of length $L$ is one-third higher than the local trailing-edge Nusselt number:

$$\bar{h} = \frac{1}{L}\int_0^L h_x\,dx = \frac{4}{3}h_x(x = L),$$

or

$$\overline{\text{Nu}_L} = 0.943\left(\frac{\rho_f(\rho_f - \rho_g)gh'_{fg}L^3}{\mu_f k_f(T_{\text{sat}} - T_w)}\right)^{1/4}. \tag{9.16}$$

These are the basic results of Nusselt's theory of condensation on a vertical wall. The flow is assumed laminar.

Examining Eq. (9.15), we see that the wall heat transfer is proportional to the $\frac{3}{4}$ power of $(T_{\text{sat}} - T_w)$ and only the $\frac{1}{4}$ power of $h'_{fg}$. Values of $T_{\text{sat}}$, $h_{fg}$, and surface tension are given in Table 9.1 for various fluids at 1 atm pressure. We see that water has the highest surface tension and by far the highest latent heat.

These results are valid for a plate inclined at angle $\theta$ from the vertical, at least up to $\theta = 60°$, if $g$ is replaced by $g \cos\theta$. They are also valid for condensation on the outside of a vertical tube if the tube radius is large compared to the condensate thickness. For condensation *inside* a tube, however, large vapor velocities are induced

**Table 9.1** Properties of boiling and condensing liquids at 1 atm (from [17])

| Liquid | Saturation Temperature $T_{sat}$, K | Latent Heat $h_{fg}$, kJ/kg | Surface Tension $\gamma$, N/m at 25°C |
|---|---|---|---|
| Acetone | 329 | 519 | 0.0231 |
| Benzene | 353 | 391 | 0.0282 |
| Carbon tetra-chloride | 350 | 194 | 0.0263 |
| Ethanol | 351 | 847 | 0.0223 |
| Ethylene glycol | 470 | 800 | 0.0482 |
| Glycerine | 563 | 975 | 0.0630 |
| Mercury | 630 | 295 | 0.0484 |
| Methanol | 338 | 1103 | 0.0222 |
| Octane | 398 | 298 | 0.0211 |
| Propane | 231 | 428 | 0.0066 |
| Propylene glycol | 460 | 914 | 0.0363 |
| Refrigerant-11 | 297 | 180 | 0.0183 |
| Refrigerant-12 | 243 | 165 | 0.0089 |
| Refrigerant-21 | 232 | 233 | 0.0084 |
| Toluene | 384 | 363 | 0.0273 |
| Water | 373 | 2257 | 0.0720 |

along the tube axis and the heat flux is markedly changed by vapor shear on the condensate film. See [20] for further details.

### Example 9.1

Water at 1 atm condenses on one side of a vertical wall 20 cm high and 30 cm wide, held at 80°C. Estimate (a) $\delta(L)$; (b) $u_{\max}(L)$; (c) $\mathrm{Re}_\delta(L)$; and (d) $q_w$.

*Solution* At $T_{sat} = 100$°C, the required physical properties of water are $h_{fg} = 2257$ kJ/kg, $\rho_f = 958$ kg/m$^3$, $\rho_g = 0.6$ kg/m$^3$, $k_f = 0.68$ W/m · K, $\mu_f = 2.79 \times 10^{-4}$ kg/m · s, and $c_{pf} = 4.217$ kJ/kg · K. The Jakob number is

$$\mathrm{Ja} = \frac{c_{pf}(T_{sat} - T_w)}{h_{fg}} = \frac{(4.217)(100 - 80)}{2257} = 0.0374.$$

Then

$$h'_{fg} = h_{fg}(1 + 0.68\text{Ja}) = 2257[1 + 0.68(0.0374)] = 2314 \text{ kJ/kg}.$$

From Eq. (9.14), the film thickness at $x = L$ is

$$\delta(L) = \left[\frac{4(0.68 \text{ W/m}\cdot\text{K})(20\text{K})(2.79 \times 10^{-4} \text{ kg/m}\cdot\text{s})(0.2 \text{ m})}{(958 \text{ kg/m}^3)(958 - 0.6 \text{ kg/m}^3)(9.81 \text{ m/s}^2)(2{,}314{,}000 \text{ J/kg})}\right]^{1/4}$$

$$= 0.00011 \text{ m} = 0.11 \text{ mm}. \quad [\textit{Ans. (a)}]$$

This is extremely small compared to the plate length of 20 cm, and it justifies our use of boundary layer approximations. The maximum film velocity at the trailing edge, from Eq. (9.7), is

$$u_{\max}(L) = \frac{(\rho_f - \rho_g)g\delta^2}{2\mu_f} = \frac{(958 - 0.6)(9.81)(0.00011)^2}{2(2.79 \times 10^{-4})}$$

$$= 0.203 \text{ m/s}. \quad [\textit{Ans. (b)}]$$

This is a modest speed but removes a great deal of condensate. From Eq. (9.10), the film Reynolds number at $x = L$ is

$$\text{Re}_\delta = \frac{2\rho_f u_{\max}}{3\mu_f} = \frac{2(958)(0.203)(0.00011)}{3(2.79 \times 10^{-4})}$$

$$= 51. \quad [\textit{Ans. (c)}]$$

This is definitely laminar. Experiments show film transition to turbulence at about $\text{Re}_\delta = 450$.

From Eq. (9.16), the mean Nusselt number is

$$\overline{\text{Nu}}_L = 0.943\left[\frac{958(958 - 0.6)(9.81)(2{,}314{,}000)(0.2)^3}{(2.79 \times 10^{-4})(0.68)(20)}\right]^{1/4} = 2427,$$

from which

$$\bar{h}_L = \frac{\overline{\text{Nu}}_L k_f}{L} = \frac{(2427)(0.68)}{0.2} = 8250 \text{ W/m}^2\cdot\text{K}.$$

The total heat transfer is

$$q_w = \bar{h}_L A_w(T_{\text{sat}} - T_w) = 8250(0.2 \text{ m})(0.3 \text{ m})(20°)$$

$$= 9900 \text{ W}. \quad [\textit{Ans. (d)}]$$

This small plate, only 20 cm by 30 cm and only 20°C below $T_{\text{sat}}$, is absorbing 10 kW of heat and removing 15.4 kg/hr of condensate. ■

### 9.2.2 The Film Condensate Grashof Number

Although our dimensional analysis of the condensation process led to the Nusselt number being a function of three parameters (Gr, Ja, Pr) in Eq. (9.2) — assuming that Bo is not important — the actual analysis, Eq. (9.16), contains only a single parameter. This single quantity is a combination of Gr, Ja, and Pr and may be termed a "film condensate Grashof number":

$$\begin{aligned} \mathrm{Gr}_{\mathrm{film}} &= \frac{\rho\Delta\rho g h'_{fg} L^3}{\mu\, k\Delta T} = \frac{\rho\Delta\rho g L^3}{\mu^2}\frac{\mu c_p}{k}\frac{h'_{fg}}{c_p\Delta T} \\ &= \mathrm{Gr}_L \mathrm{Pr}\frac{1 + 0.68\mathrm{Ja}}{\mathrm{Ja}}, \end{aligned} \tag{9.17}$$

where ($\rho$, $\mu$, $k$, $c_p$) are evaluated for the liquid. Without Nusselt's analysis we would not have predicted that Gr, Ja, and Pr would combine in this manner. A very extensive set of experiments would have been required to investigate the separate parametric effects of Gr, Ja, and Pr, and in fact the results may not have revealed the simple form predicted by Eq. (9.16). This illustrates the importance of an inspired *analysis* — even if simplified — in predicting the form of a relationship and in guiding subsequent experiments.

Almost the identical form of modified Grashof number in Eq. (9.17) holds also for *film boiling,* except that ($\rho$, $\mu$, $k$, $c_p$) are evaluated for the vapor instead of the liquid. See Section 9.4.5 for further details.

### 9.2.3 Transition and Turbulence in a Vertical Film

The preceding solution by Nusselt is valid only for laminar flow and, like other viscous flows, breaks down into turbulence at a finite Reynolds number $\mathrm{Re}_\delta$. The definition of $\mathrm{Re}_\delta$ is given in Eq. (9.10) or may be rewritten in the form

$$\mathrm{Re}_\delta = \dot{m}/\mu_f, \qquad \dot{m} = \int_0^\delta \rho_f u\, dy. \tag{9.18}$$

If we assume that $\rho_g << \rho_f$, that is, far from the critical point, we can rewrite Eq. (9.16) in terms of the Reynolds number and a modified Nusselt number:

$$(\bar{h}_L/k_f)(\nu_f^2/g)^{1/3} = 0.924\,\mathrm{Re}_\delta^{-1/3}, \quad \nu_f = \mu_f/\rho_f. \tag{9.19}$$

Experiments [39] show that waves begin to form on a liquid film interface at about $\mathrm{Re}_\delta \doteq 6$. The flow is still laminar but the wall heat flux is increased over Nusselt's formula (9.19) and is approximated

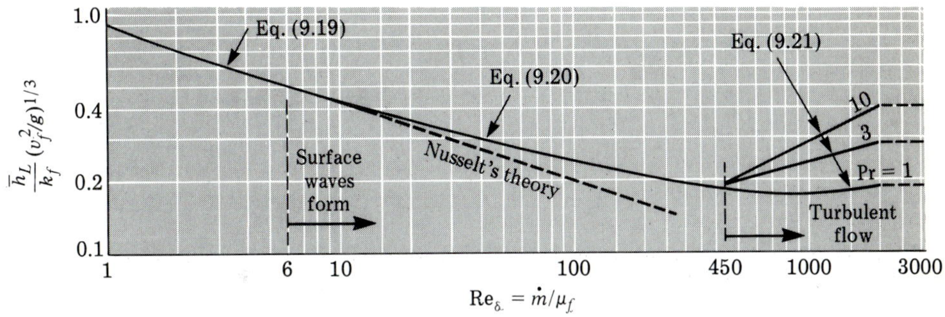


**Figure 9.2** **Theoretical and experimental correlations for mean heat transfer coefficient in film condensation on a vertical wall.**

by the correlation of Kutateladze [41]:

$$(\bar{h}_L/k_f)(\nu_f^2/g)^{1/3} \doteq (1.47\,\mathrm{Re}^{0.22} - 1.3/\mathrm{Re})^{-1}, \qquad 6 \leq \mathrm{Re} \leq 450. \tag{9.20}$$

Then, at Re $\doteq$ 450, transition to turbulence occurs in the film, and the heat flux begins to increase according to a correlation suggested by the experiments of Labuntsov [42]:

$$(\bar{h}_L/k_f)(\nu_f^2/g)^{1/3} \doteq \mathrm{Re}/[2188 + 41(\mathrm{Re}^{0.75} - 89.5)/\mathrm{Pr}^{1/2}], \qquad \mathrm{Re} > 450. \tag{9.21}$$

These relations are plotted in Fig. 9.2. Both theory and experiment [43] indicate that $\bar{h}_L$ is independent of Re if Re is greater than 2000.

## Example 9.2

Modify Example 9.1 to correct the total wall heat flux for surface wave instability in the film.

*Solution* The film kinematic viscosity is $\nu_f = \mu_f/\rho_f = 2.91 \times 10^{-7}\ \mathrm{m^2/s}$. The length scale $(\nu_f^2/g)^{1/3} = [(2.91 \times 10^{-7})/9.81]^{1/3} = 0.0205$ mm. In Example 9.1 we computed $\mathrm{Re}_\delta = 51$, which is greater than 6. Hence surface waves have formed on the laminar film, enhancing the heat flux. Equation (9.20) applies:

$$(\bar{h}_L/k_f)(\nu_f^2/g)^{1/3} = \frac{\bar{h}_L(0.0205 \times 10^{-3})}{0.68} = [1.47(51)^{0.22} - 1.3/51]^{-1} = 0.289$$

or

$$\bar{h}_L = 9570\ \mathrm{W/m^2 \cdot K}.$$

The total wall heat flux is

$$q_w = \bar{h}_L A_w(T_{sat} - T_w) = (9570)(0.2)(0.3)(20)$$
$$= 11{,}500 \text{ W.} \quad [Ans.]$$

This is about 16% more than the simple Nusselt theory predicts. ■

---

### 9.2.4 Laminar Condensation for Other Geometries

Nusselt's solution for a vertical wall can be extended to other shapes such as cylinders, spheres, and cones, if the local surface slope and possible axisymmetric effects are accounted for. Dhir and Lienhard [22] showed that the Nusselt formulas could be directly applied to such geometries if $g$ was replaced by a local effective acceleration of gravity,

$$g_{\text{eff}} = \frac{gx(R\cos\theta)^{4/3}}{\int_0^x (\cos\theta)^{1/3} R^{4/3}\, dx}, \tag{9.22}$$

where

$x$ = surface arc length in the condensate flow direction;

$\theta$ = angle between the surface flow and the vertical;

$R$ = local surface radius with respect to a vertical axis (used only for vertical axisymmetric bodies).

Dhir and Lienhard applied this formula to a variety of geometries. For an inclined plate, $\theta$ is constant and $R \to \infty$; hence $G_{\text{eff}} = g\cos\theta$, as mentioned earlier.

For a vertical tube, $\theta = 0°$ and $R$ is constant, so that $g_{\text{eff}} = g$ itself.

For a horizontal cylinder, $g_{\text{eff}}$ is variable, and an integration yields the following mean heat transfer formula:

$$\overline{\text{Nu}}_D = 0.729\left(\frac{\rho_f(\rho_f - \rho_g)gh'_{fg}D^3}{\mu_f k_f(T_{sat} - T_w)}\right)^{1/4}, \tag{9.23}$$

where $D$ is the cylinder diameter.

For a sphere, numerical integration is again needed, and the result reported by [22] is

$$\overline{\text{Nu}}_D = 0.815\left(\frac{\rho_f(\rho_f - \rho_g)gh'_{fg}D^3}{\mu_f k_f(T_{sat} - T_w)}\right)^{1/4}. \tag{9.24}$$

Both latter formulas are in good agreement with experiment. Other examples such as cones and disks are given in [22].

For a vertical bank of horizontal tubes, the heat flux to the lower tubes is reduced by the extra condensate falling on them. A conservative estimate of average heat transfer to a tube bank is given by replacing $D$ on both sides of Eq. (9.23) by $ND$, where $N$ is the number of horizontal tube rows.

### Example 9.3

A horizontal tube 2 m long is cooled by water flow inside so that its surface temperature is 60°C. It is desired to place the tube in saturated steam at 1 atm and condense 300 kg/hr. What is the proper diameter for the tube?

*Solution* Take the physical properties for saturated steam and water at 100°C from Example 9.1. The temperature difference is $(100 - 60) = 40$°C, so the Jakob number is

$$\text{Ja} = \frac{c_{pf}\Delta T}{h_{fg}} = \frac{(4.217)(40)}{2257} = 0.0747.$$

Then

$$h'_{fg} = h_{fg}(1 + 0.68\text{Ja}) = 2257[1 + 0.68(0.0747)]$$

$$= 2372 \text{ kJ/kg}.$$

The condensate flow rate times $h'_{fg}$ equals the wall heat flux:

$$q_w = \dot{m}h'_{fg} = (300/3600 \text{ kg/s})(2372 \text{ kJ/kg}) = 197.6 \text{ kW}.$$

This enables us to evaluate the average Nusselt number:

$$\overline{\text{Nu}}_D = \frac{(q_w/\pi DL\Delta T)D}{k_f} = \frac{197{,}600/\pi(2.0)(40)}{0.68} = 1156.$$

This Nusselt number must equal the value predicted by Eq. (9.23), for which everything is known except $D$:

$$\overline{\text{Nu}}_D = 1156 = 0.729\left[\frac{958(958 - 0.6)(9.81)(2{,}372{,}000)D^3}{(0.000279)(0.68)(40)}\right]^{1/4}.$$

Solve for $D = 0.132$ m. [*Ans.*]

The Reynolds number can be estimated from $\text{Re}_\delta = (\dot{m}/L\delta)\delta/\mu_f$

or

$$\text{Re}_\delta = \dot{m}/L\mu_f = \left[\frac{300/3600}{(2.0)(0.000279)}\right] = 149.$$

The film flow is laminar and probably has surface wave-enhanced heat flux. Our estimated size $D = 13.2$ cm is conservative. ■

## 9.3 Dropwise Condensation

Most commercial condenser designs result in film condensation. It has been known for fifty years, however, that if the cold surface is nonwetting to the condensate, *dropwise condensation* occurs and is much more efficient. A sketch of the process is shown in Fig. 9.3.

**Figure 9.3** Dropwise condensation on a nonwetting surface: (a) droplets grow and touch and coalesce; (b) when the merged droplet is large enough, it slides down, leaving a dry surface in its wake; (c) new droplets form at nucleation sites in the dry wake.

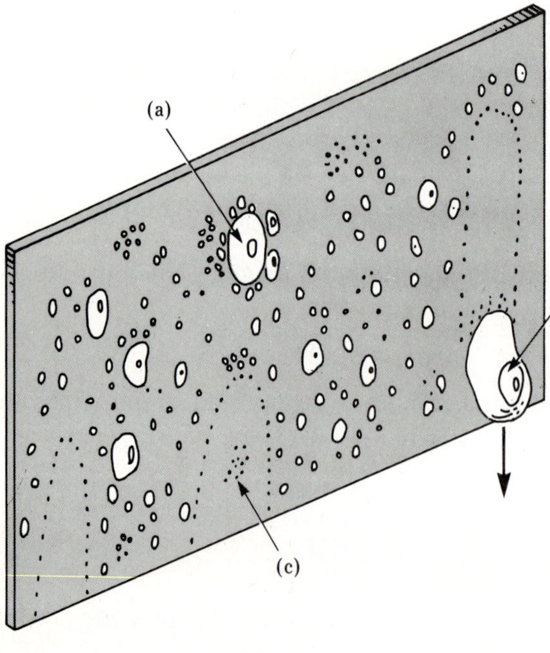

Small droplets form at nucleation sites on the surface. They grow, touch each other, and coalesce into large droplets. Eventually a merged droplet becomes so large that it falls, sliding down the surface and leaving a dry wake in its path. New small droplets form on the dry wake surface and the process repeats.

Dropwise condensation is one of the most intense heat transfer mechanisms known, and heat transfer rates as high as $10^6$ W/m$^2 \cdot$ K have been measured. The subject is reviewed in [14] and [23]. There are four methods of creating the nonwetting surface:

1. seeding the saturated vapor with a promoting chemical;
2. treating the surface with promoter chemicals such as butter or oil;
3. bonding a polymer such as Teflon to the surface; and
4. plating the surface with gold or other noble metals.

If droplet formation can be promoted, heat flux will be 5 to 10 times higher than for film condensation. The difficulty is that the surface ages in a few days, loses its nonwetting properties, and reverts to filmwise flows. This loss of efficiency seems especially frustrating with steam applications.

Some recent work with various types of gold-plating processes [24] seems to hold promise for future designs incorporating stable, nonwetting, dropwise condensers.

## 9.4 Pool Boiling

Boiling is at once the most common and the most perplexing of heat transfer phenomena. We have all seen fluids boil; yet we still cannot explain the process completely. This is not to say we haven't tried: [4] contains over 1000 pages of discussion of boiling, with many interesting insights. The published literature on boiling approaches 2000 papers. But we still cannot predict, in advance, the rate of vapor bubble formation or the total heat flux on an arbitrary hot surface. Let us discuss the regimes of boiling and some existing analyses.

We may divide boiling into two classes: (1) *pool boiling* from a hot surface into a stationary bulk of fluid and (2) *flow boiling* from a hot surface into a moving bulk fluid. This section treats pool boiling, which is akin to the process that occurs when we heat a pan of water on a stove. Recall your stove experiences. The pan of water is initially quite cold, far below the saturation temperature. It seems to take forever to heat up; you notice small bubbles forming on the bottom quite early, but these are merely air bubbles being driven out of solution. Then, when the water is hot but still below saturation, water vapor bubbles form in great numbers at the bottom, rise, and collapse in the cooler water above. This is *subcooled boiling* and can be quite vigorous. The rising bubbles quickly heat the remainder of the water in the pan to saturation, after which we see a "full rolling boil," as the cookbooks describe it. This is *saturated boiling* and is very energetic, yet the stove power is actually low enough, for safety reasons, that the pan remains far below the maximum heat flux attainable. A final fact, not evident by simply observing pan boiling, is that the bottom of the pan is hotter than the saturation value, $T_w > T_{sat}$: It takes an *excess temperature* to boil any liquid. Let us now put some numbers on saturated boiling.

### 9.4.1 Nukiyama's Famous Experiment

In 1934, S. Nukiyama [25] placed a horizontal Nichrome wire in saturated water at 1 atm and heated it electrically, using the wire resistance as a measure of its temperature. He noticed that boiling did not begin until the excess wire temperature, $\Delta T = T_w - T_{sat}$, was finite, about 5°C. He increased the power and found that $\Delta T$ rose very slowly while the power (heat flux of boiling) reached enormous levels, about $10^6$ W/m$^2$. Suddenly the wire temperature shot up and it melted (*burnout*). Nukiyama secured a platinum wire and tried again: The power level and wire temperature followed nearly the same curve, line *ab* in Fig. 9.4. When he tried to raise $q''_w$ above $10^6$ W/m$^2$, the platinum wire suddenly glowed white hot but did not melt. Cooling the wire caused it to move down an entirely different curve, *dc* in Fig. 9.4. When power was reduced below point *c*, the wire temperature shot *down* suddenly and then cooled along the original curve *ba* back to saturation.

Nukiyama had discovered the *pool boiling curve,* which has many different physical regimes. He realized that there was a piece of the curve, *bc*, missing, which could not be attained by power-controlled devices such as electric heating. Subsequent experiments with a temperature-controlled apparatus (superheated steam condensing on the inside of a tube) have established the complete shape of the curve for many liquids.

Figure 9.5 shows the approximate pool boiling curve for saturated water at 1 atm on a clean hot metal surface. The general shape of the curve is the same for all liquids, and there are four basic regimes:

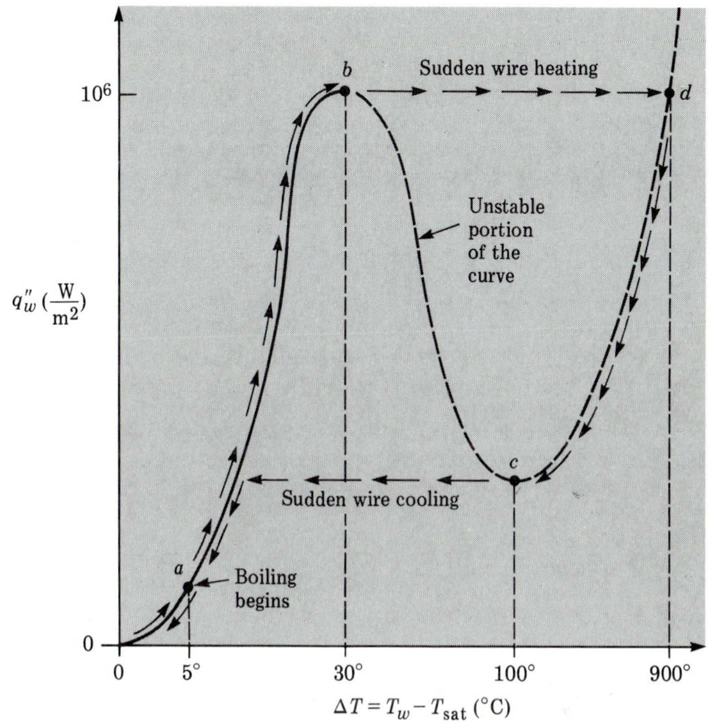


**Figure 9.4** Nukiyama's famous experiment in 1934 [25] establishing the pool boiling curve. The arrows follow his observations with the platinum wire in a pool of saturated water.

1. *Free convection* up to point $A$: $q''_w$ is predicted by the single-phase natural-convection theories of Chapter 7. There is no boiling.
2. *Nucleate boiling* regime between points $A$ and $C$: The first bubble forms at a nucleation site at $A$. At point $B$ many bubbles are rising at a great rate from many sites. Between $B$ and $C$ the bubbles rise in columns and jets of vapor, and $q''_w$ is approximately proportional to $\Delta T^3$. Maximum heat transfer coefficient occurs at point $H$ between $B$ and $C$, but $h$ is a rather irrelevant concept when heat flux is such a nonlinear function of temperature difference. The photograph in Fig. 9.6(a) illustrates intensive nucleate boiling at about point $H$ on the methanol curve.
   a) Point $C$ is the "critical" or maximum heat flux (sometimes called the "burnout" point). The bubble columns are so dense

and crowded that excess vapor begins to impede the heat transfer, which levels off to a peak.

3. The *transition boiling regime* is between points $C$ and $L$: Bubble columns become more and more crowded and periodically blanket the entire surface. The blanketing effect occurs more and more often until, at point $L$, it is continuous. Figure 9.6(b) shows transition boiling in methanol.
   a) Point $L$ is the Leidenfrost point, where vapor completely blankets the surface and heat flux is a minimum. It is named for J. G. Leidenfrost, who pointed out in 1756 that water globules on a very hot metal surface "dance about" and boil away very slowly, being insulated from the metal by a thin blanket of vapor.
4. The *film boiling regime* is above point $L$, where vapor film continuously blankets the surface and inhibits heat transfer. As wall

**Figure 9.5** The pool boiling curve for saturated water boiling at 1 atm pressure on a clean metal surface. Curves for other liquids are of similar shape but with different numerical values.

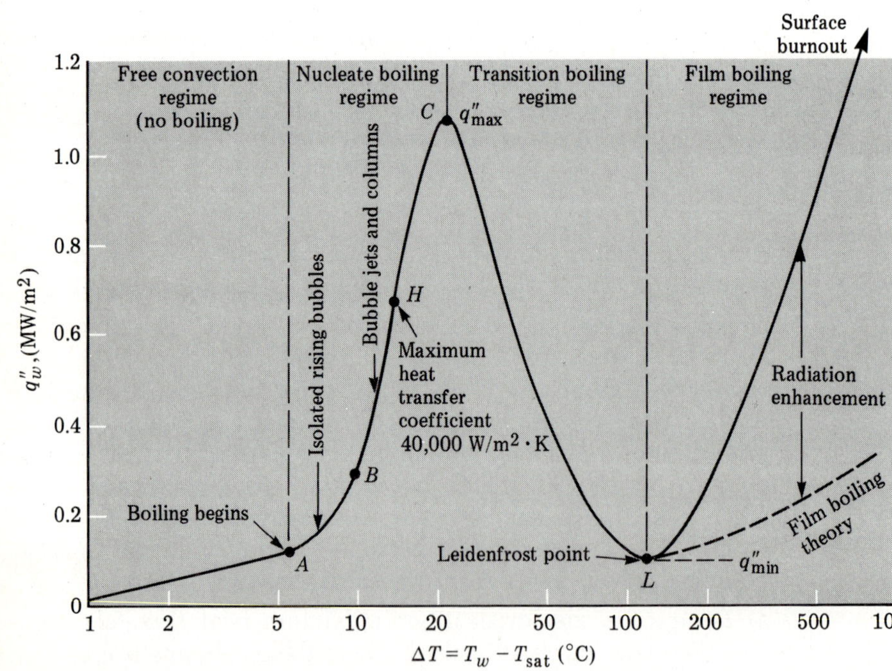

(a)

(b)

(c)

**Figure 9.6** Photographs of methanol pool boiling from [26]: (a) nucleate boiling, similar to point $H$ in Fig. 9.5, $\Delta T = 37°C$, $q''_w = 0.24\ MW/m^2$; (b) transition boiling about midway on the transition curve, $\Delta T = 62°C$, $q''_w = 0.22\ MW/m^2$ ($q''_{max} = 0.54\ MW/m^2$); (c) film boiling, $\Delta T = 82°C$, $q''_w = 0.04\ MW/m^2$ ($q''_{min} = 0.015\ MW/m^2$).

temperature increases, radiation across the vapor film to the liquid enhances the flux rate. The physics of film boiling is quite similar to film condensation (Section 9.2), and similar formulas apply. Figure 9.6(c) shows film boiling in a pool of saturated methanol.

We now briefly discuss heat transfer correlations for the various regimes and points. The state of the art is as follows:

1. Up to point $A$ in Fig. 9.5, there is excellent free-convection theory (Chapter 7) for pool boiling and also forced-convection theory (Chapter 6) for flow boiling.
2. The strong nucleate boiling range, between $B$ and $C$ in Fig. 9.5, is crucially dependent upon surface conditions — the availability and efficiency of vapor nucleation sites. The theory is weak and limited, and specific experiments are needed.
3. There is a reasonably accurate and general theory for predicting the critical heat flux at point $C$ in Fig. 9.5.
4. There is also a reasonable theory for predicting minimum heat flux at the Leidenfrost point $L$ in Fig. 9.5.
5. There is little theory in the transition region, and a practical design rarely operates in this range. If numbers are necessary, we can compute points $C$ and $L$ and draw a straight line.
6. Film boiling theory is quite adequate and may be supplemented by a semi-empirical radiation enhancement method.

Our treatment will be brief. Recall that there are whole textbooks on boiling [1–4]. Also, a student learning for the first time about boiling would greatly benefit from reading the masterful discussion of boiling in the undergraduate text by Lienhard [27], who is a specialist in the field.

### 9.4.2 Nucleate Boiling Correlations

The difficulty in analyzing nucleate boiling lies in predicting the distribution of surface nucleation sites and their rate of bubble production. The physics of nucleation has undergone intensive study [4, 11, 13], but its link to quantitative heat flux prediction is weak. The most widely accepted correlation dates back to 1952, when Rohsenow [28] postulated that excess wall temperature (or Jakob number, Ja $= c_p \Delta T / h_{fg}$) should vary with Prandtl number and a "bubble Reynolds number,"

$$\mathrm{Re}_b = G_b D_b / \mu_f, \qquad \textbf{(9.25)}$$

where $G_b = \dot{m}_b/A$ should be proportional to the amount of liquid boiled away: $G_b \propto q''_w/h_{fg}$. From static nucleation theory, the bubble diameter should be related to surface tension and density difference: $D_b \propto [\Upsilon/(\rho_f - \rho_g)g]^{1/2}$. Thus the bubble Reynolds number is really a dimensionless heat flux, $\text{Re}_b = (q''/\mu h_{fg})(\Upsilon/g\Delta\rho)^{1/2}$. Rohsenow correlated various nucleate boiling experiments and found that a power-law formula gave reasonable accuracy if an empirical, dimensionless constant $C_{sf}$ was included:

$$\frac{q''_w}{\mu_f h_{fg}}\left[\frac{\Upsilon}{g(\rho_f - \rho_g)}\right]^{1/2} \doteq (\text{Ja}/C_{sf})^3 \,\text{Pr}_f^{-n}, \qquad \text{Ja} = c_{pf}(T_w - T_{\text{sat}})/h_{fg}, \tag{9.26}$$

where fluid properties are evaluated at $T_{\text{sat}}$. The constant $C_{sf}$ and the exponent $n$ vary with the particular fluid and surface. Some values given by Rohsenow are listed in Table 9.2. The correlation is valid only for a "clean," relatively smooth surface, and there is considerable scatter. When $\Delta T$ is known, errors in $q''_w$ can be $\pm 100\%$; when $q''_w$ is known, the error in predicted $\Delta T$ can be $\pm 30\%$. In the absence of specific data, Eq. (9.26) is widely accepted as a design estimate.

Note that Eq. (9.26) predicts that $q''_w \propto \Delta T^3$, as expected, and also $q''_w \propto h_{fg}^{-2}$ for a given temperature difference. Since $h_{fg}$ drops off sharply with saturation pressure (or temperature), the heat flux in nucleate

**Table 9.2** Values of $C_{sf}$ and $n$ for use in Eq. (9.26)

| Surface and Fluid | $C_{sf}$ | $n$ |
|---|---|---|
| Water-copper | 0.013 | 3.0 |
| Water-brass | 0.006 | 3.0 |
| Water-nickel | 0.006 | 3.0 |
| Water-platinum | 0.013 | 3.0 |
| Benzene-chromium | 0.010 | 5.1 |
| $n$-pentane-chromium | 0.015 | 5.1 |
| Ethanol-chromium | 0.0027 | 5.1 |
| Isopropanol-copper | 0.0025 | 5.1 |
| $n$-butanol-copper | 0.0030 | 5.1 |
| Water-stainless steel: | | |
| mechanically polished | 0.013 | 5.1 |
| ground and polished | 0.0080 | 5.1 |

boiling rises significantly with saturation pressure of the liquid, as confirmed by experiment.

**Example 9.4**

Assume that Fig. 9.5 is for a copper or platinum surface. Use Eq. (9.26) to predict $q''_w$ at $\Delta T = 15°C$ and compare with Fig. 9.5. The liquid pressure is 1 atm.

*Solution* At 1 atm, $T_{sat} = 100°C$ and the properties from Example 9.1 apply again: $h_{fg} = 2257$ kJ/kg, $c_{pf} = 4.217$ kJ/kg · K, $\rho_f = 958$ kg/m$^3$, $\rho_g = 0.6$ kg/m$^3$, $\mu_f = 0.000279$ kg/m · s, and $\mathrm{Pr}_f = 1.73$. From Table 9.2 for water-platinum, $C_{sf} \doteq 0.013$ and $n \doteq 3.0$. The surface tension is $\Upsilon = 0.0589$ N/m. Equation (9.26) predicts

$$\frac{q''_w}{(0.000279)(2{,}257{,}000)}\left[\frac{0.0589}{(9.81)(958-0.6)}\right]^{1/2} \doteq \left[\frac{4.217(15°)}{2257(0.013)}\right]^3 (1.73)^{-3}.$$

Solve for

$$q''_w \doteq 490{,}000 \text{ W/m}^2 = 0.49 \text{ MW/m}^2. \quad [Ans.]$$

This compares well with Fig. 9.5 at $\Delta T = 15°C$, for which the heat flux is between 0.5 and 0.6 MW/m$^2$. Note that the formula requires no conversion factors if consistent SI units are used.

If the computation is repeated for a pressure of 26 atm, $T_{sat} = 500$K ($h_{fg} = 1825$, $c_{pf} = 4.59$, $\rho_f = 831$, $\rho_g = 13$, $\mu_f = 0.000113$, $\mathrm{Pr}_f = 1.28$, $\Upsilon = 0.0316$), the formula yields $q''_w \doteq 1.21$ MW/m$^2$, or about 2.5 times higher than at 1 atm. The experimental flux at 26 atm is about 1.7 MW/m$^2$ [28]. ■

### 9.4.3 Critical Heat Flux Prediction

The analysis of critical (maximum) heat flux, point *C* in Fig. 9.5, has been quite successful, and engineers can use the theory to avoid burnout in their boiling designs. The peak flux is nearly independent of surface conditions and only weakly dependent upon heater geometry (wires, plates, spheres, and so on). Working independently, S. S. Kutateladze in Russia in 1948 and N. Zuber in the United States in 1958 derived the same relation. Zuber's theory was quite elegant, postulating a Helmholtz instability along the vertical jet-column vapor/liquid interfaces. Kutateladze's attack was through dimensional analysis: He assumed that $q''_{max}$ varied only with latent heat, vapor density, mean bubble diameter, and surface tension:

$$q''_{max} = \text{fcn}(h_{fg}, \rho_g, D_b, \Upsilon), \quad D_b = [\Upsilon/g\Delta\rho]^{1/2}. \tag{9.27}$$

The reader should verify as an exercise that the only dimensionally consistent combination of these parameters is

$$q''_{max} = (\text{constant})h_{fg}\rho_g^{1/2} D_b^{-1/2} \Upsilon^{1/2}.$$

The constant is dimensionless and has an experimental value of about 0.15 (Zuber's theory predicted $C = \pi/24 = 0.131$). Thus we have the accepted correlation for peak boiling heat flux:

$$q''_{max} \doteq 0.15\, \rho_g^{1/2}\, h_{fg}[g(\rho_f - \rho_g)\Upsilon]^{1/4}. \tag{9.28}$$

Note that the formula is independent of fluid viscosity, conductivity, or specific heat.

We recommend Eq. (9.28) as a general preliminary design formula, but in fact it is strictly valid only for a heater surface size that is large compared to the mean bubble diameter, $D_b$. As shown by Lienhard [27, pp. 407–410], the formula needs a correction factor if the heater size is less than about $3D_b$. Examples of heater "size" are the body radius for spheres and horizontal cylinders, and the smallest width for a flat plate. We will not pursue small-heater corrections here.

The success of the Zuber/Kutateladze peak flux formula is illustrated in Fig. 9.7 for water. All common fluids show the same behavior as in Fig. 9.7(a): Their latent heat and surface tension drop monotonically with pressure and vanish at the thermodynamic critical point (where the liquid and vapor are indistinguishable). For water, as shown, $p_{crit} = 22.12$ MPa. When the data from Fig. 9.7(a) are entered into Eq. (9.28), the solid curve in Fig. 9.7(b) results and is in good agreement with a variety of peak flux water boiling data from [29]. As Cichelli and Bonilla [29] point out, peak flux rises with pressure up to about $\frac{1}{3}p_{crit}$, after which the decreases in $h_{fg}$ and $\Upsilon$ outweigh the increase in $\rho_g$, and $q''_{max}$ drops toward zero at $p_{crit}$. This is characteristic of all common liquids, so that a plot of $q''_{max}/q''_{max}(p_c/3)$ versus $p/p_c$ is a "universal" curve.

## Example 9.5

Continuing with our verification of the pool boiling curve in Fig. 9.5, predict the critical flux of water at 1 atm and compare with the figure.

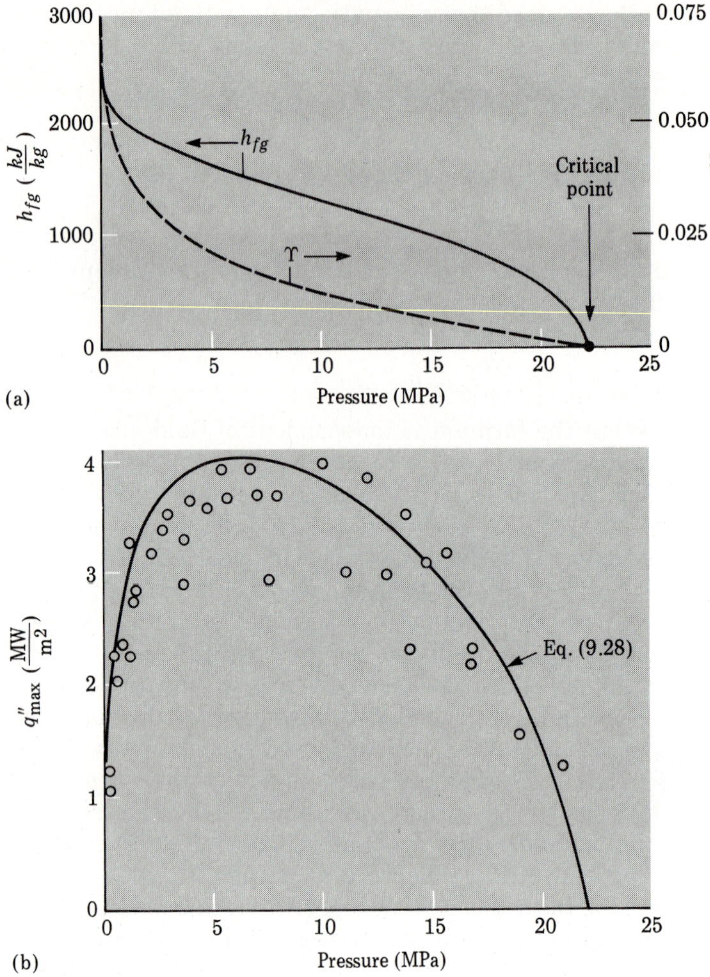


**Figure 9.7 Critical heat flux for water boiling at high pressure: (a) latent heat and surface tension of water; (b) comparison of Eq. (9.28) with data from [29].**

*Solution* At $T_{sat} = 100°C$ the relevant properties are $h_{fg} = 2257$ kJ/kg, $\rho_g = 0.596$ kg/m$^3$, $\rho_f = 958$ kg/m$^3$, $\Upsilon = 0.0589$ N/m. Then the Zuber/Kutateladze correlation (9.28) predicts

$$q''_{max} = 0.15(0.596)^{1/2}(2{,}257{,}000)[9.81(958 - 0.596)(0.0589)]^{1/4}$$

$$= 1{,}270{,}000 \text{ W/m}^2 = 1.27 \text{ MW/m}^2. \quad [Ans.]$$

In Fig. 9.5 the measured peak flux is about 1.1 MW/m$^2$, or about 13% less. This is the expected accuracy of the correlation. ■

### 9.4.4 Minimum Flux: The Leidenfrost Point

In addition to analyzing critical flux with his interfacial wave stability theory, Zuber [30] was also able to predict minimum flux at the Leidenfrost point $L$ in Fig. 9.5. His formula was modified by Berenson [31] to give better agreement with experiment. The recommended correlation is

$$q''_{\min} \doteq 0.09\, \rho_g\, h_{fg}[g(\rho_f - \rho_g)Y]^{1/4}(\rho_f + \rho_g)^{-1/2}. \tag{9.29}$$

At low pressures, the predicted ratio $q''_{\min}/q''_{\max} = 0.6(\rho_g/\rho_f)^{1/2}$, or from 0.02 to 0.05. Thus a high wall temperature gives a much less efficient boiling rate, and film boiling conditions are used in typical designs only where necessary, such as in chemical processing equipment.

When compared to a variety of fluids, Eq. (9.29) has a scatter of $\pm 50\%$ and is especially inaccurate at higher pressures, as shown in the data of [3].

**Example 9.6**

Predict the minimum (Leidenfrost) heat flux of water at 1 atm and compare with Fig. 9.5.

*Solution* Use the water properties from Example 9.5. Equation (9.29) yields

$$q''_{\min} = 0.09(0.596)(2{,}257{,}000)[9.81(958 - 0.6)0.0589]^{1/4}(958 + 0.6)^{-1/2}$$

$$= 19{,}000\ \text{W/m}^2. \quad [\textit{Ans.}]$$

Figure 9.5 is difficult to read, but point $L$ seems to be at approximately $q''_{\min} \doteq 30\ \text{MW/m}^2$, or 58% higher than our formula. ∎

### 9.4.5 Film Boiling

Film boiling is physically quite similar to film condensation, with the liquid and vapor roles reversed. In film boiling there is a vapor film or "reverse condensate" that flows upward into the saturated liquid. Nusselt's analysis from Section 9.2 should be approximately valid with $\rho_f$ and $\rho_g$ reversed. Two possible discrepancies in this analogy are (1) the boiling vapor layer is much thicker than a liquid film so that its temperature distribution is not linear and (2) the vapor/liquid interface is much more unstable than a liquid film. It is likely that the vapor layer flow is laminar, and it turns out that the role-reversed condensation formulas (9.23) and (9.24) are reasonably accurate for film boiling if the constants are reduced by 15%:

$$\overline{\mathrm{Nu}}_D \doteq C\left[\frac{\rho_g(\rho_f - \rho_g)gh'_{fg}D^3}{\mu_g k_g(T_w - T_{\mathrm{sat}})}\right]^{1/4} = \bar{h}D/k_g, \tag{9.30}$$

where $C \doteq 0.62$ for horizontal cylinders and 0.67 for spheres, and where $h'_{fg} \doteq h_{fg} + 0.4c_{pg}(T_w - T_{\mathrm{sat}})$.

If the wall temperature exceeds 300°C, radiation through the vapor layer can enhance film boiling, as sketched in Fig. 9.5. Bromley [32] suggests computing a radiation heat transfer coefficient,

$$h_{\mathrm{rad}} = \frac{\varepsilon_w \sigma(T_w^4 - T_{\mathrm{sat}}^4)}{(T_w - T_{\mathrm{sat}})} \tag{9.31}$$

and then combining this with the boiling coefficient $h_b$ from Eq. (9.30) to give an overall coefficient, $h$, as follows:

$$h \doteq h_b(h_b/h)^{1/3} + h_{\mathrm{rad}}. \tag{9.32}$$

Bromley points out that simple addition of $h_b$ and $h_{\mathrm{rad}}$ is not accurate because radiation increases the vapor layer thickness and reduces the overall heat transfer.

Film boiling is difficult to achieve in an electrically heated laboratory demonstration, because of the danger of burnout as in the "jump" *bd* of Fig. 9.4. But it can be done, as shown in Fig. 9.8, by using a fluid with low saturation temperature (R-11 in this instance) and providing sufficient thermal inertia and electric current control that the "jump" is partially aborted without surface burnout.

We should note that Nusselt's analysis, Eq. (9.16), is *not* accurate if role-reversed into vertical plate film boiling. The vertical vapor layer is too unstable for the analogy to hold. (See [27] for further details.)

### Example 9.7

Estimate the boiling heat flux in water at 1 atm when $\Delta T = 200$°C for a 5-cm-diameter horizontal cylinder.

*Solution* With $T_w = 573$K and $T_{\mathrm{sat}} = 373$K, the vapor layer spans a large temperature range and its properties should be evaluated at the mean value $(T_w + T_{\mathrm{sat}})/2 = 473$K: $\rho_g = 0.47$ kg/m$^3$, $c_{pg} = 1.97$ kJ/kg · K, $\mu_g = 1.61 \times 10^{-5}$ kg/m · s, and $k_g$

**Figure 9.8** Film boiling in a laboratory demonstration device. The fluid is Refrigerant-11 at 1 atm and $(T_w - T_{sat}) = 100°C$. The measured heat flux is 30 $kW/m^2$. (Courtesy of C.A.A. Scientific, Prime Mover Laboratory Systems.)

$= 0.032$ W/m · K. Equation (9.30) should be used with $C = 0.62$ for a cylinder. The effective latent heat is

$$h'_{fg} = h_{fg} + 0.4c_{pf}\Delta T = 2257 + 0.4(1.97)(200) = 2415 \text{ kJ/kg}.$$

Equation (9.30) then predicts

$$\overline{\text{Nu}}_D = 0.62\left[\frac{0.47(958 - 0.47)(9.81)(2{,}415{,}000)(0.05)^3}{(0.0000161)(0.032)(200)}\right]^{1/4} = 209,$$

or

$$\overline{h} = \frac{\overline{\text{Nu}}_D k_g}{D} = \frac{(209)(0.032)}{(0.05)} = 134 \text{ W/m}^2 \cdot \text{K}.$$

The wall is hot, so check for a radiation effect, with $\varepsilon_w = 1$:

$$h_{rad} = (1)(5.67 \times 10^{-8})(573^4 - 373^4)/200 = 25 \text{ W/m}^2 \cdot \text{K}.$$

Then, with $h_b = 134$ and $h_{\text{rad}} = 25$, Eq. (9.32) predicts, after iteration, that the overall $h = 153\ \text{W/m}^2 \cdot \text{K}$.

The final predicted heat flux is

$$q'' = h\Delta T = (153)(200) = 31{,}000\ \text{W/m}^2. \quad [Ans.]$$

Radiation enhancement is small at this temperature. Again, Fig. 9.5 is difficult to read but the prediction seems reasonable. ■

---

### 9.4.6 Effect of Gravity

Typical applications of pool boiling assume earth gravity, $g = 9.81\ \text{m/s}^2$. However, there are important cases of low gravity in aerospace applications and of high gravity in devices encountering centrifugal effects. Siegel [8] has reviewed low-gravity boiling effects, down to 0.01 of earth gravity. Our recommended formulas for critical flux (9.28), minimum flux (9.29), and film boiling (9.30) all indicate that $q''$ is proportional to $g^{1/4}$; Siegel concludes from available data that this is an accurate prediction.

On the other hand, the nucleate boiling correlation (9.26) predicts that $q'' \propto g^{1/2}$, whereas Siegel finds that the data show only an insignificant effect of $g$. This is another indication of the weak underpinnings of present theories of nucleate boiling.

### 9.4.7 Effect of Surface Roughness

Surface roughness has only a slight effect on critical flux, minimum flux, and film boiling. It has a striking effect on nucleate boiling, as shown from the experiments by Berenson [33] in Fig. 9.9. The figure is really rather shocking, exhibiting flux increases and excess temperature decreases of a factor of 10 as roughness increases. Clearly the roughness greatly enhances the number of nucleation sites available. And the power-law changes from $q'' \propto \Delta T^2$ for the smoothest surface up to $q'' \propto \Delta T^4$ for the roughest. Unfortunately, this strong enhancement degrades back to the smooth curve in a few days, as trapped vapor slowly leaks away from the roughness cavities. What is needed is a more stable vapor-trap design.

### 9.4.8 Nucleate Boiling Enhancement Surfaces

Although the performance of the rough surfaces in Fig. 9.9 decays with time, there are now commercially available special surfaces that

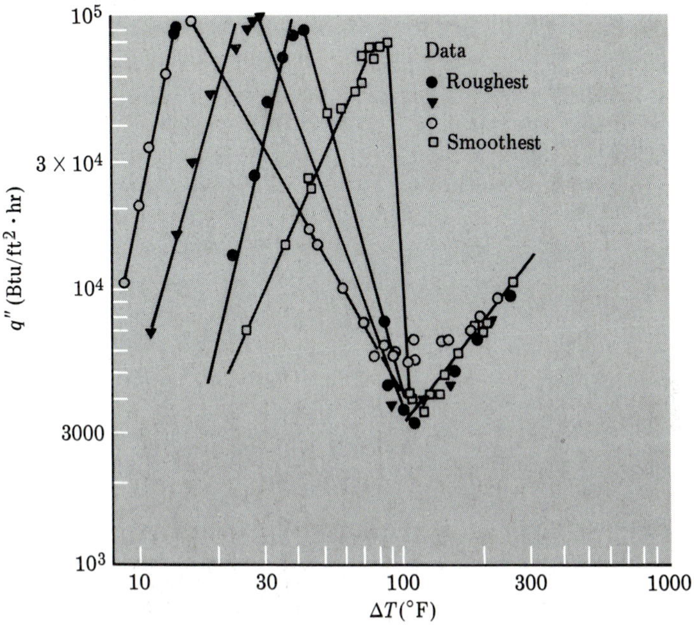


**Figure 9.9** Effect of surface roughness on pool boiling of pentane on a flat copper surface, after Berenson [33].

are stable and that continue to enhance nucleate boiling. Webb [34] reviewed the evolution of these new boiling surfaces and described 29 of them, some of which are proprietary and all of which are patented. We see in Fig. 9.10 that these special surfaces provide stable increases of up to a factor of 10 in the nucleate boiling and transition heat flux, compared to a plain metal surface. There is also a surprising factor-of-3 variation in critical flux, which is not predicted by the Zuber/Kutateladze formula (9.28).

These enhancement surfaces are of two types: very porous or mechanically formed to have deep cavities with a small opening. Surface tension at the narrow exit prevents gradual degassing of any vapor trapped in the cavity, an effect clearly seen in the "bubble nucleation" portion of the excellent educational fluid mechanics film on surface tension [35] familiar from undergraduate fluid mechanics courses. Webb [34] points out some start-up and tube-bundle difficulties with enhanced surfaces.

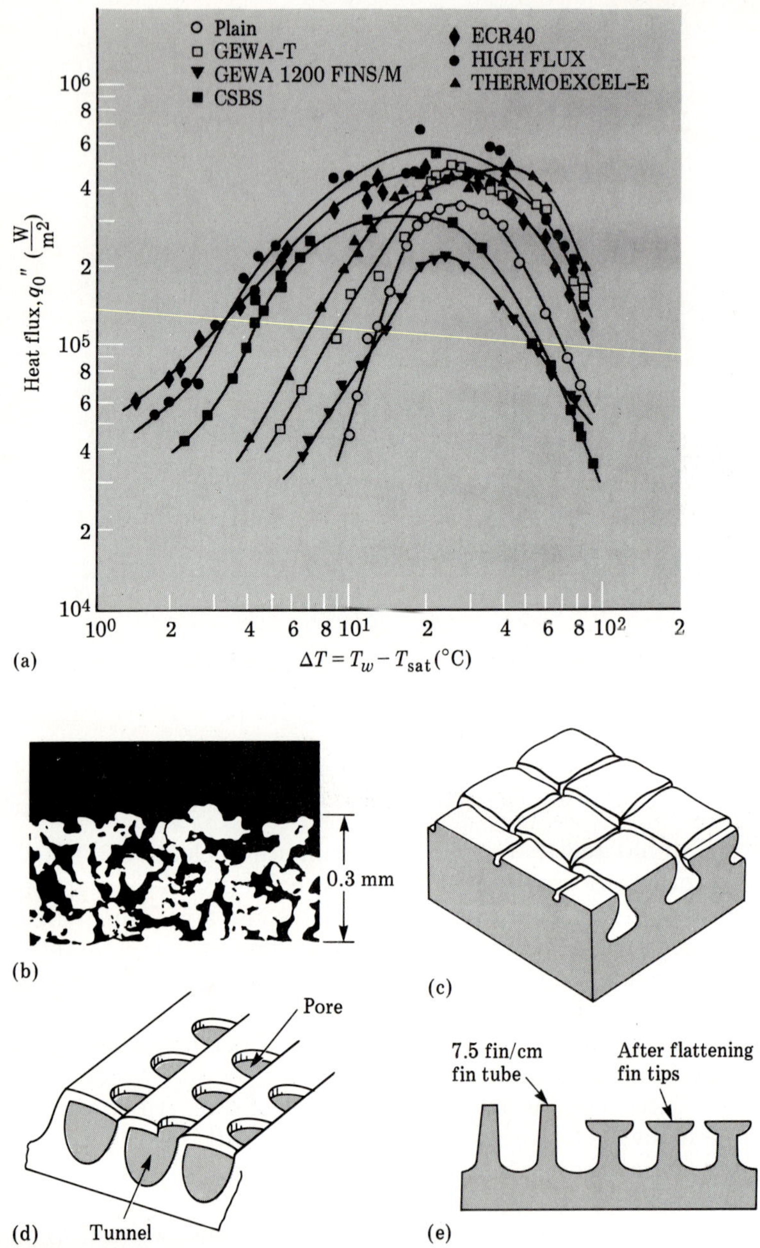


**Figure 9.10** Nucleate boiling enhancement surfaces, after Webb [34]; (a) comparison of single-tube pool boiling results for *p*-xylene at 1 atm; (b) High-Flux surface; (c) ECR-40; (d) Thermoexcel-E; (e) GEWA-T.

## 9.5 Flow Boiling

Section 9.4 gave some ideas for predicting the pool boiling curve of a saturated liquid. In that case, the only "flow" is the upward buoyant motion of bubbles and warm liquid near the hot surface. Now consider superimposing on that process a mean bulk velocity or *forced convection* past the hot surface. This would greatly complicate the overall motion, depending on how the vapor interacts with the forced flow. The geometry is also important: An *external* convection over a hot plate or tube bank is simpler to analyze than an *internal* flow in a hot duct, where the generated vapor must go faster and faster down the duct to satisfy mass conservation.

The general effect of forced convection on saturated boiling is shown in Fig. 9.11. The lowest line is the standard pool boiling curve (Fig. 9.5), which follows a natural convection correlation $q''_{\mathrm{NC}}$ until boiling begins at heat flux $q''_{\mathrm{i}}$.

If there is forced convection of bulk liquid past the hot surface, the heat flux follows a forced-convection correlation $q''_{\mathrm{FC}}$ until boiling begins at about the same point $\Delta T_{\mathrm{i}}$ as for pool boiling. The curves then turn upward to merge with the nucleate pool boiling correlation. The critical or peak flux increases, as shown, with forced-convection velocity.

To a zeroth approximation, one could simply add the forced convection and pool boiling predictions, $q''_{\mathrm{FC}} + q''_{\mathrm{PB}}$, to estimate overall heat flux in flow boiling. But Rohsenow and Bergles [15, pp. 13–46] suggest an interpolation:

$$q'' \doteq [\, q''^2_{\mathrm{FC}} + q''^2_{\mathrm{PB}} - q''^2_{\mathrm{i}}\,]^{1/2}, \tag{9.33}$$

for $\Delta T > \Delta T_{\mathrm{i}}$, $q'' > q''_{\mathrm{i}}$. This formula merges smoothly with $q''_{\mathrm{FC}}$ at $\Delta T_{\mathrm{i}}$ and also merges with $q''_{\mathrm{PB}}$ as we move higher up the nucleate boiling curve. For best results, $q''_{\mathrm{PB}}$ should be taken from nucleate boiling data for the given surface. If no data are available, we can use the Rohsenow correlation (9.26) to estimate $q''_{\mathrm{PB}}$.

Equation (9.33) gives best results for light to moderate vapor production, especially subcooled boiling. If the bulk liquid is subcooled, $T_b < T_{\mathrm{sat}}$, $q''_{\mathrm{FC}}$ should be based on $(T_w - T_b)$, while $q''_{\mathrm{PB}}$ remains based on $(T_w - T_{\mathrm{sat}})$. The interpolation (9.33) is less accurate for saturated-flow boiling but is used anyway as a design estimate.

Both forced convection and subcooling markedly increase the critical or peak heat flux in nucleate boiling. The writer is not aware of a general theory relating to subcooling, but Lienhard and Eichhorn [36] give an elegant analysis of the effect of saturated liquid velocity on peak flux in flow across horizontal cylinders.

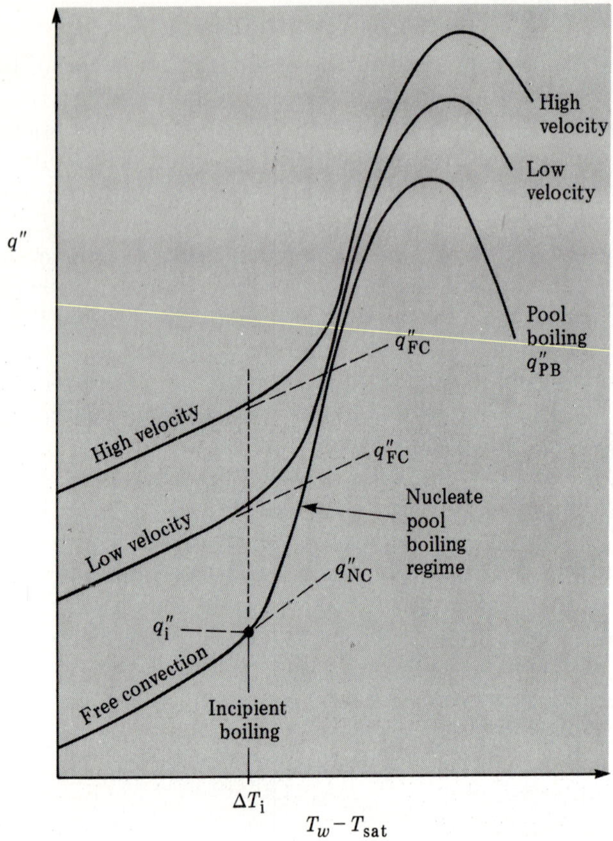


**Figure 9.11 Illustration of the effect of forced convection on external flow boiling.**

To illustrate these effects, we can fit peak flux data for flow boiling of saturated or subcooled water to the following dimensional formula:

$$q''_{max} \doteq 1.1(1 + 2.05V^{1/3}) + 0.077(T_{sat} - T_b)(1 + 1.4V^{2/3}), \tag{9.34}$$

with $q''_{max}$ in $MW/m^2$, $V$ in m/s, and $(T_{sat} - T_b)$ in °C. (We will assign as a problem the determination of the dimensionless parameters that may lurk in this formula.) Equation (9.34) is plotted in Fig. 9.12, showing a strong effect of velocity and an even more striking effect of subcooling. It seems that peak flux can be raised indefinitely in this manner, and experimental values as high as 35 $MW/m^2$ have been reported [4].

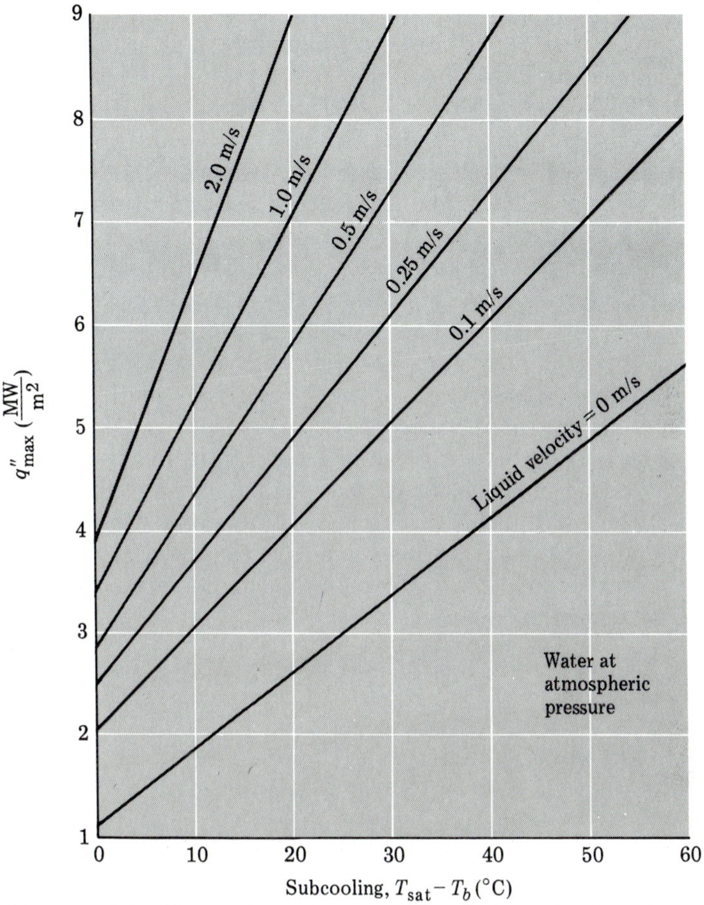


**Figure 9.12** Effect of forced-convection velocity and subcooling on measured peak flux of water, from Eq. (9.34). These data are strictly valid only for very small heaters such as fine wires.

The boiling process for internal tube flow is far more complex than for external flow and is beyond our scope. As a subcooled liquid enters a long, hot tube, it undergoes seven different types of two-phase flow pattern changes before emerging as a superheated vapor. For an excellent discussion of the internal flow boiling process, the reader is referred to the text by Collier [2].

## Summary

Phase-change problems are very common and quite different from our single-phase convection theories in Chapters 5–7. The convective motion is primarily induced by the density jump across the phase change, and the heat transfer is dominated by the latent heat. Interestingly, the latent heat does not even appear in the equations of motion. Rather, it arises in the boundary condition at the phase interface.

The theory of film condensation is well established, and solutions are possible for almost any geometry. Dropwise condensation is more efficient but difficult to maintain; hence most condensation applications involve film formation.

Boiling in a stationary "pool" of saturated liquid has been studied intensively for 50 years. We present theory and experiment to correlate the various regimes of pool boiling: preboiling convection, nucleate boiling, critical heat flux, the Leidenfrost point, and film boiling. Nucleate boiling is strongly affected by surface conditions, and several patented boiling enhancement surfaces are now available commercially.

Flow boiling occurs when the bulk fluid moves with a nonzero velocity past the hot surface. The nucleate boiling range is much the same in flow boiling, but the critical flux is increased markedly. Subcooling of the bulk fluid also causes a rise in critical flux. Internal flow boiling in tube flow is much more complicated hydrodynamically than an external flow process.

## References

1. L. S. Tong, *Boiling and Two Phase Flow,* Wiley, New York, 1965.
2. J. C. Collier, *Convective Boiling and Condensation,* McGraw-Hill, New York, 1972.
3. E. Hahne and U. Grigull, *Heat Transfer in Boiling,* Hemisphere/Academic Press, New York, 1977.
4. S. van Stralen and R. Cole, *Boiling Phenomena,* 2 vols., McGraw-Hill/Hemisphere, New York, 1979.
5. J. N. Ginoux, *Two Phase Flow and Heat Transfer,* McGraw-Hill/Hemisphere, New York, 1978.
6. S. I. Pai, *Two Phase Flows,* Vieweg and Sohn, Braunschweig, 1977.
7. G. Leppert and C. C. Pitts, "Boiling," *Advances in Heat Transfer,* vol. 1, 1964, pp. 185–266.
8. R. Siegel, "Effect of Reduced Gravity on Heat Transfer," *Advances in Heat Transfer,* vol. 4, 1967, pp. 143–228.
9. D. P. Jordan, "Film and Transition Boiling," *Advances in Heat Transfer,* vol. 5, 1968, pp. 55–128.

10. H. Merte, Jr., "Condensation Heat Transfer," *Advances in Heat Transfer,* vol. 9, 1973, pp. 181–272.

11. R. Cole, "Boiling Nucleation," *Advances in Heat Transfer,* vol. 10, 1974, pp. 85–166.

12. E. K. Kalinin, I. I. Berlin, and V. V. Kostyuk, "Film Boiling Heat Transfer," *Advances in Heat Transfer,* vol. 11, 1975, pp. 51–197.

13. G. S. Springer, "Homogeneous Nucleation," *Advances in Heat Transfer,* vol. 14, 1978, pp. 281–346.

14. W. M. Rohsenow and P. Griffith, "Condensation," in *Handbook of Heat Transfer,* McGraw-Hill, New York, 1973, chap. 12.

15. W. M. Rohsenow, "Boiling," in *Handbook of Heat Transfer,* McGraw-Hill, New York, 1973, chap. 13.

16. W. Nusselt, "Die Oberflachenkondensation des Wasserdampfes," *Z. Ver. Deut. Ing.,* vol. 60, 1916, pp. 541, 569.

17. R. E. Bolz and G. L. Tuve (Eds.), *CRC Handbook of Tables for Applied Engineering Science,* CRC Press, Boca Raton, Fla., 1979.

18. E. M. Sparrow and J. L. Gregg, "A Boundary Layer Treatment of Laminar Film Condensation," *J. Heat Transfer,* vol. 81, 1959, pp. 13–18.

19. W. M. Rohsenow, "Heat Transfer and Temperature Distribution in Laminar Film Condensation," *ASME Transactions,* vol. 78, 1956, pp. 1645–1648.

20. W. M. Rohsenow, J. M. Weber, and A. T. Ling, "Effect of Vapor Velocity on Laminar and Turbulent Film Condensation," *ASME Transactions,* vol. 78, 1956, pp. 1637–1644.

21. A. P. Colburn, "The Calculation of Condensation Where a Portion of the Condensate Layer Is in Turbulent Flow," *AIChE Transactions,* vol. 30, 1933, p. 187.

22. V. K. Dhir and J. H. Lienhard, "Laminar Film Condensation on Plane and Axisymmetric Bodies in Non-Uniform Gravity," *J. Heat Transfer,* vol. 93, no. 1, 1971, pp. 97–100.

23. I. T. Tanasawa, "Dropwise Condensation: The Way to Practical Applications," *Proc. Sixth Intl. Heat Transfer Conference,* Toronto, vol. 6, 1978, pp. 393–405.

24. D. W. Woodruff and J. W. Westwater, "Steam Condensation on Various Gold Surfaces," *J. Heat Transfer,* vol. 103, Nov. 1981, pp. 685–692.

25. S. Nukiyama, "The Maximum and Minimum Values of the Heat Transmitted from Metal to Boiling Water at Atmospheric Pressure," *J. Soc. Mech. Engrs. Japan,* vol. 37, 1934, pp. 367–374 (trans. in *Int. J. Heat Mass Transfer,* vol. 9, 1966, pp. 1419–1433).

26. J. W. Westwater and J. G. Santangelo, "Photographic Study of Boiling," *Industrial and Engineering Chemistry,* vol. 47, Aug. 1955, pp. 1605–1610.

27. J. H. Lienhard, *A Heat Transfer Textbook,* Prentice-Hall, Englewood Cliffs, N.J., 2nd ed., 1987.

28. W. M. Rohsenow, "A Method of Correlating Heat Transfer Data for Surface Boiling of Liquids," *ASME Transactions,* vol. 74, 1952, pp. 969–975.

29. M. T. Cichelli and C. F. Bonilla, "Heat Transfer to Liquids Boiling Under Pressure," *AIChE Transactions,* vol. 41, 1945, pp. 755–787.

30. N. Zuber, "On the Stability of Boiling Heat Transfer," *ASME Transactions,* vol. 80, 1958, pp. 711–720.

31. P. J. Berenson, "Film Boiling Heat Transfer for a Horizontal Surface," *J. Heat Transfer,* vol. 83, 1961, pp. 351–358.

32. A. L. Bromley, "Heat Transfer in Stable Film Boiling," *Chem. Engrg. Prog.,* vol. 46, 1950, pp. 221–227.

33. P. J. Berenson, "Experiments on Pool Boiling Heat Transfer," *Int. J. Heat Mass Transfer,* vol. 5, 1962, pp. 985–999.

34. R. L. Webb, "The Evolution of Enhanced Surface Geometries for Nucleate Boiling," *Heat Transfer Engrg.,* vol. 2, nos. 3–4, 1981, pp. 46–69.

35. "Surface Tension in Fluid Mechanics," National Committee for Fluid Mechanics, Film No. 21610, narrated by L. Trefethen.

36. J. H. Lienhard and R. Eichhorn, "Peak Boiling Heat Flux on Cylinders in a Crossflow," *Int. J. Heat Mass Transfer,* vol. 19, 1976, pp. 1135–1142.

37. J. W. Rose, "Dropwise Condensation Theory," *Int. J. Heat Mass Transfer,* vol. 24, 1981, pp. 191–194.

38. S. Levy, "Generalized Correlation of Boiling Heat Transfer," *J. Heat Transfer,* vol. 81, 1959, pp. 37–42.

39. S. J. Friedman and C. O. Miller, "Liquid Films in the Viscous Flow Region," *Industrial and Engineering Chemistry,* vol. 33, 1941, pp. 885–891.

40. V. K. Dhir, "Quasi-Steady Laminar Film Condensation of Steam on Copper Spheres," *J. Heat Transfer,* Aug. 1975, pp. 347–351.

41. S. S. Kutateladze, *Fundamentals of Heat Transfer,* Academic Press, New York, 1963.

42. D. A. Labuntsov, "Heat Transfer in Film Condensation of Pure Steam on Vertical Surfaces and Horizontal Tubes," *Teploenergetika,* Vol. 4, 1957, pp. 72–80.

43. S. J. Chen, "Turbulent Film Condensation on a Vertical Plate," *Heat Transfer 1986,* vol. 4, Hemisphere, New York, 1986, pp. 1601–1606.

44. G. A. Greene and T. F. Irvine, Jr., "Film Boiling of R-11 on Liquid Metal Surfaces," *Heat Transfer 1986,* vol. 4, Hemisphere, New York, 1986, pp. 2049–2054.

45. G. Meyer, E. S. Gaddis, and A. Vogelpohl, "Critical Heat Flux on a Cylinder of Large Diameter in a Crossflow," *Heat Transfer 1986,* vol. 5, Hemisphere, 1986, pp. 2125–2130.

## Review Questions

1. Name six different types of phase change.
2. What dimensionless parameters affect the phase-change Nusselt number? Which are the most important?
3. Define film condensation and discuss its characteristics.
4. In Nusselt's analysis of film condensation, how many parameters affect Nusselt number in the final formulation?
5. Can a film condensate achieve a turbulent-flow condition?
6. Does it make sense to talk about Reynolds number of a film?
7. Discuss the physics of dropwise condensation.
8. Sketch the pool boiling curve and discuss its regimes.
9. Discuss Nukiyama's experiment and the burnout phenomenon.
10. What is the Leidenfrost point?
11. Define a bubble Reynolds number for nucleate boiling.
12. How does the critical heat flux vary with fluid pressure?
13. Which regimes of boiling does surface tension affect?
14. How does radiation enhance film boiling?
15. How do boiling enhancement surfaces work?
16. Sketch a flow boiling curve for (a) a saturated liquid and (b) a subcooled liquid.

## Problems

### Problem distribution by sections

**The Problem Assignments are Organized as Follows:**

| Problems | Sections Covered | Topics Covered |
|---|---|---|
| 9.1–9.16 | 9.1, 9.2 | Film condensation |
| 9.17–9.18 | 9.3 | Dropwise condensation |
| 9.19–9.36 | 9.4 | Pool boiling |
| 9.37–9.42 | 9.5 | Flow boiling |
| 9.43–9.45 | All | Special Projects |

9.1 In film condensation, surface tension and specific heat have only a weak effect and $k$ occurs only in the conduction combination $k\Delta T$. Thus Eq. (9.1) reduces to

$$q'' = \text{fcn}(k\Delta T, g\Delta\rho, h_{fg}, L, \mu, \rho).$$

Rewrite this in nondimensional form. To what does your result reduce if $g\Delta\rho$ and $h_{fg}$ occur only in the combination $(g\Delta\rho h_{fg})$?

9.2 Using a length scale $L$, a temperature scale $(T_{sat} - T_w)$, and a velocity scale $(gL(\rho_f - \rho_g/\rho_f)^{1/2}$ as reference properties, nondimensionalize Eqs. (9.3) to find some dimensionless parameters that affect condensation. Why don't $q''_w$ and $h_{fg}$ appear directly? Do your dimensionless equations make it clear when convective terms can be neglected?

9.3 Equations (9.3) were solved exactly in [18] for film condensation. Write the exact boundary conditions for these equations and put them in nondimensional form.

9.4 One side of a 30-cm-high, 40-cm-wide vertical plate at 40°C is in contact with saturated steam at 1 atm. Compute the film thickness and maximum velocity at the bottom of the plate and the total heat flux to the plate.

9.5 A vertical plate 30 cm high and 30 cm wide is used to condense ethylene glycol vapor at 1 atm, as shown in Fig. P9.5. The other side of the plate is cooled by forced water flow such that the water heat transfer coefficient is 4000 W/(m$^2$ · K). Assume laminar film condensation with negligible plate conduction resistance. What water temperature $T_\infty$ and plate temperature $T_w$ are necessary to condense 50 kg/hr of the ethylene glycol? What is the film thickness as it leaves the plate?

9.6 A horizontal tube 3 cm in diameter and 60 cm long is placed in steam at 1 atm. How cold should the tube wall be if it is to condense 30 kg of water per hour?

9.7 A tube is 9 cm in diameter and 150 cm long and has its outside surface cooled to 50°C by water flow inside. If placed in steam at 1 atm, how

**Figure P9.5**

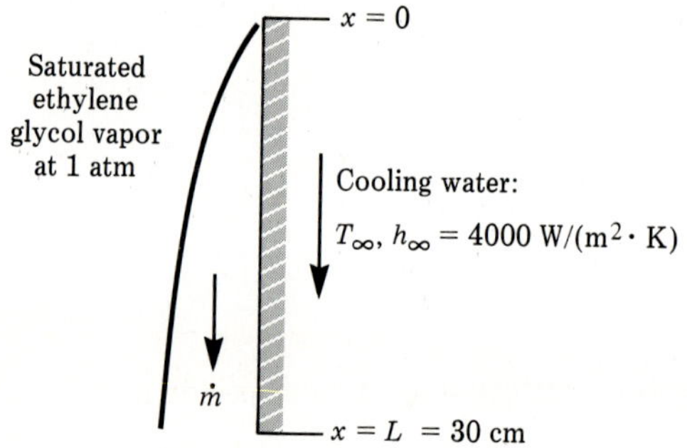

much water will it condense per hour if the tube is (a) vertical and (b) horizontal. What do you conclude?

9.8 It is desired to condense steam at 1 atm with copper tubes placed horizontally and cooled to 80°C. There is enough copper to make one 10-cm-diameter tube or 10 1-cm-diameter tubes. Which design would condense more water? Why? Does it matter if the ten-tube system is arrayed in a horizontal or vertical row?

9.9 Figure P9.9 shows the experimental data of V. K. Dhir [40] for the mean Nusselt number for steam at 1 atm condensing on a 1-inch diameter copper sphere. The abscissa is $(T_{sat} - T_w)$ as measured by a thermocouple. The Jakob number varied from 0.015 to 0.11 during the runs. Sketch in the theory for laminar sphere condensation on this chart and comment on the accuracy, for Ja = 0.015 and 0.11. Can you estimate from the data whether film turbulence might have been present?

9.10 It is desired to condense 2000 kg/hr of steam at 1 atm using a horizontal array of 3-cm-diameter, 70-cm-long tubes. If the tube wall temperature is 85°C, how many tubes are needed?

9.11 It is desired to condense ethylene glycol at 1 atm with a horizontal water-cooled tube 3 cm in diameter and 80 cm long. The water enters at 20°C at a rate of 1000 kg/hr. How much ethylene glycol is condensed?

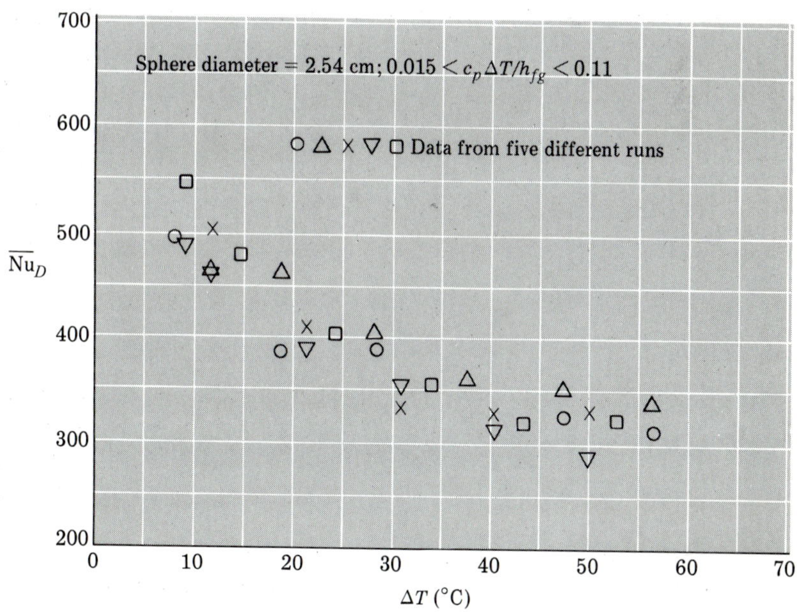


**Figure P9.9** (from [40])

9.12 A 50-cm-diameter sphere at a surface temperature of $-50$°C is placed in saturated R-12 vapor at 1 atm. How many kilograms of refrigerant will condense per hour?

9.13 A vertical plate 50 cm wide at 80°C is placed in steam with $T_{sat} = 100$°C. How high is the plate if transition to turbulence is occurring exactly at the trailing edge? How much steam is being condensed?

9.14 Condensation is slower when the liquid collects in a container and cannot run off, as in Fig. P9.14. Assume that the walls are insulated and the bottom has area $A$ and temperature $T_w$. Show, with a simplified heat balance across the condensate layer, that its growth rate is

$$\delta \doteq \left[\frac{2k_f t(T_{sat} - T_w)}{\rho_f h_{fg}}\right]^{1/2}.$$

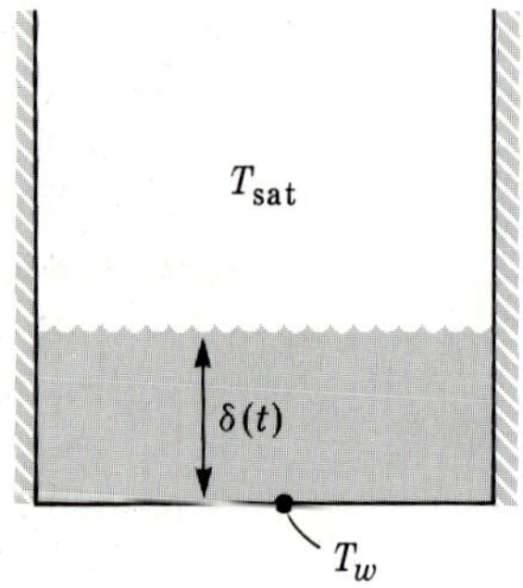


Figure P9.14

9.15 Make a Nusselt-type analysis of condensate forming on a vertical cone of half-angle $\theta$ as shown in Fig. P9.15. Show that the mean heat transfer coefficient is given by

$$\overline{h}_L \doteq \left[\frac{g \cos\theta\, \rho_f(\rho_f - \rho_g)k_f^3 h'_{fg}}{L\, \mu_f(T_{sat} - T_w)}\right]^{1/4}.$$

Compare this result with the prediction of the Dhir-Lienhard transformation, Eq. (9.22).

Figure P9.15

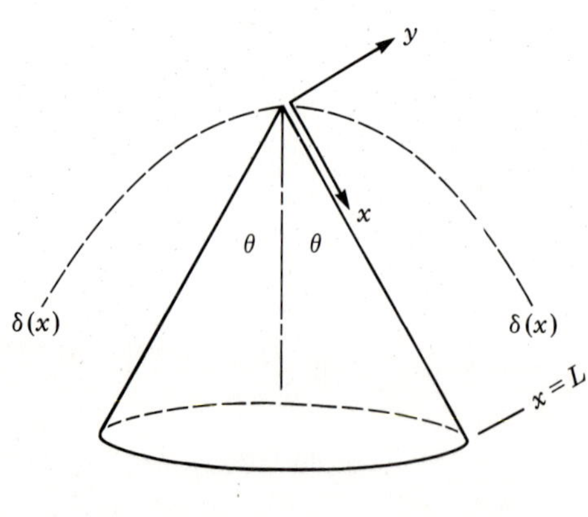

9.16 A power plant condenser consists of a bank of horizontal copper tubes 3 cm in diameter and 2 m long, cooled by water that enters each tube at 20°C and 2 m/s. It is desired to condense 10,000 kg/hr of steam at 100°C. How many tubes are needed?

9.17 In dropwise condensation, the surface is 90% covered with drops, and most of the heat transfer occurs for drop size less than 0.1 mm. Assuming that a drop is a hemisphere of liquid, estimate its heat transfer coefficient and compare with film condensation.

9.18 In an interesting theory of dropwise condensation, Rose [37] proposes a droplet size distribution,

$$A(r) \doteq \frac{1}{r}[0.871(r/r_0)^{1/2} - 1.39(r/r_0) + 1.296(r/r_0)^2],$$

where $A(r)dr$ is the fractional area covered by drops having base radius between $r$ and $r + dr$, and

$$r_0 = \frac{2\Upsilon T_{sat}}{\rho_f h_{fg}(T_{sat} - T_w)}$$

is the minimum drop size. The mean heat flux is then given by

$$q'' \doteq \int_{r_0}^{R} Q_{drop} A(r) dr,$$

where $Q_{drop} \doteq k_f(T_{sat} - T_w)/r$ is a gross simplification. The upper limit $R \doteq 1$ mm. For steam at 1 atm with $\Delta T = 5$°C, sketch the curve $A(r)$, evaluate $q''$, and compare to measured flux of 1.5 MW/m$^2$.

9.19 Show with a sketch how we might construct the complete pool boiling curve for any liquid at any pressure boiling on a clean metal surface. Label each part of the curve with the appropriate equation number.

9.20 Point $A$ in Fig. 9.5 is where boiling first begins. It may be found analytically as the intersection of free convection with the nucleate boiling correlation, Eq. (9.26), as in Fig. P9.20. Take the special case of a 2-mm-diameter platinum wire heating saturated water at 1 atm. Compute the heat flux and temperature difference at the intersection of these two curves and compare the results with point $A$ in Fig. 9.5.

9.21 Water at 1 atm boils in a copper pan whose bottom is at 110°C and has diameter 30 cm. Estimate the stove burner power required to maintain boiling and the evaporation rate in kg/hr. What bottom temperature causes the water to boil at maximum flux?

9.22 The following are nucleate-boiling data taken by J. N. Addoms (Sc.D. Thesis, M.I.T., 1948) for a 0.61-mm-diameter platinum wire immersed

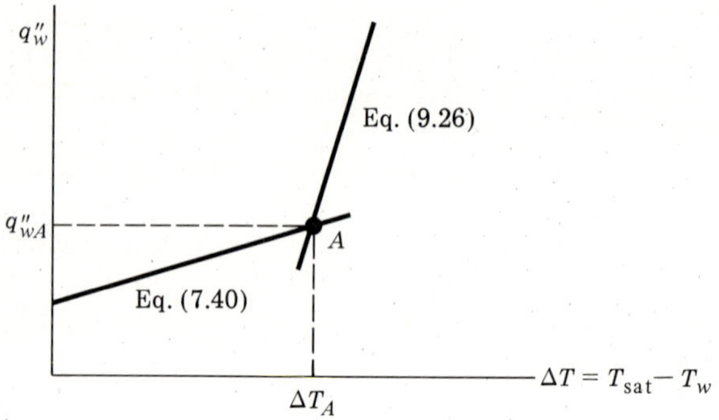


**Figure P9.20**

in saturated water at 5.31 MPa:

| $\Delta T$ (°C): | 3 | 6 | 10 | 16 |
|---|---|---|---|---|
| $q''_w$ (W/m$^2$): | $7.6 \times 10^4$ | $5.7 \times 10^5$ | $1.8 \times 10^6$ | $4.4 \times 10^6$ |

The highest heat flux corresponds to the critical (burnout) point. Compare the three lower flux values with the Rohsenow correlation, Eq. (9.26). What is the best-fit value of $C_{sf}$? Compare the burnout value with the Zuber/Kutateladze correlation, Eq. (9.28).

9.23 A chromium surface at 360K is placed in saturated ethanol at 1 atm. Estimate the boiling heat flux in W/m$^2$.

9.24 It is desired to boil saturated water at 1 atm using copper tubes 2 cm in diameter and 50 cm long. The tubes operate in the nucleate boiling range at 70% of maximum flux. How many tubes are required to boil 500 kg/hr? What total power is required? What will be the tube surface temperature?

9.25 In [26] the measured critical flux of methanol at 1 atm is 172,000 Btu/hr · ft$^2$. Compare this with the correlation (9.28).

9.26 In [4, p. 28], the maximum possible boiling heat flux in methanol is stated to be 1 MW/m$^2$ at a pressure of 2 MPa. Compare this with the Zuber/Kutateladze correlation (9.28). If you can't find $h_{fg}$ and Y of methanol at 2 MPa, scale what you have in the same manner as for water in Fig. 9.7(a).

9.27 In Fig. 9.5 the point of maximum heat transfer coefficient is shown to be somewhere along the nucleate-boiling curve. Is this fact predicted by the Rohsenow correlation, Eq. (9.26)? If not, what is the reason for the discrepancy?

9.28 A 7-cm-diameter copper sphere is immersed in saturated ethylene glycol at 1 atm. Estimate the total boiling heat transfer rate and the sphere surface temperature (a) at the burnout point and (b) at the Leidenfrost point.

9.29 As discussed in detail in Section 10.3 of the text by Lienhard [27], the Zuber/Kutateladze peak-flux correlation, Eq. (9.28), is accurate only for a heater surface of fairly large size. For small heaters of size $L$, such as spheres, wires, rods, and ribbons, Lienhard shows that peak flux depends upon a dimensionless parameter $\zeta$ combining $L$, $g$, the density difference $(\rho_f - \rho_g)$, and surface tension $\Upsilon$. What is the appropriate form of $\zeta$? Can you relate this parameter to the length of a *capillary wave* between fluids of different density [35]?

9.30 Levy [38] gives the following dimensional formula for boiling of water at high pressures:

$$q''(\mathrm{W/m^2}) \doteq 283.2[p\ (\mathrm{MPa})]^{4/3}[\Delta T\ (^\circ\mathrm{C})]^3,$$
$$0.7 < p < 14\ \mathrm{MPa}$$

Does this raw formula correspond to any of our dimensionless boiling correlations? If so, compare values at pressures of about 3 and 10 MPa in a suitable range of $\Delta T$.

9.31 A cylindrical copper pan 16 cm in diameter is filled with 6 cm of water at 20°C and placed on the stove with the burner set at 1500 W. Estimate the time required to boil away 1 cm of water.

9.32 The complete pool boiling curve may be reproduced in reverse by quenching a hot body in a saturated liquid, because, if the body is "lumpable" (Section 4.2), its cooling rate is proportional to heat flux. Using Fig. 9.5 as a guide, sketch the "quenching curve" $T_w(t)$ of a hot body placed in saturated water. Point out any important events, such as the Leidenfrost point.

9.33 Try out some numbers from Problem 9.32. A 3-cm copper sphere at 773K is suddenly thrust in saturated water at 1 atm. Estimate the time for the sphere to cool to (a) 473K; (b) 293K; and (c) 275K.

9.34 A copper bar 1 cm in diameter and 2 cm long is at 700°C when it is suddenly thrust into saturated water at 1 atm. Estimate the time for the bar temperature to decrease to 500°C.

9.35 Film boiling can occur on a flat horizontal surface, but Eq. (9.30) does not apply because a flat surface has no "diameter." Berenson [31] showed that Eq. (9.30) should be modified to replace $D$ by the capillary wavelength of the boiling fluid:

$$\lambda = \{\Upsilon/(g(\rho_f - \rho_g))\}^{1/2}.$$

Greene and Irvine [44] recently tested Berenson's idea by boiling a pool of R-11 on a very hot pool of molten lead, as in Fig. P9.35. They measured a heat flux of 80 kW/m$^2$ when the lead was 500°C hotter than the R-11 pool. Compare this result with Berenson's formula above. Use the R-11 properties from Table 9.1 and assume that other needed properties correspond to R-12 in Appendix F. The chemical formula for R-11 is $CCl_3F$.

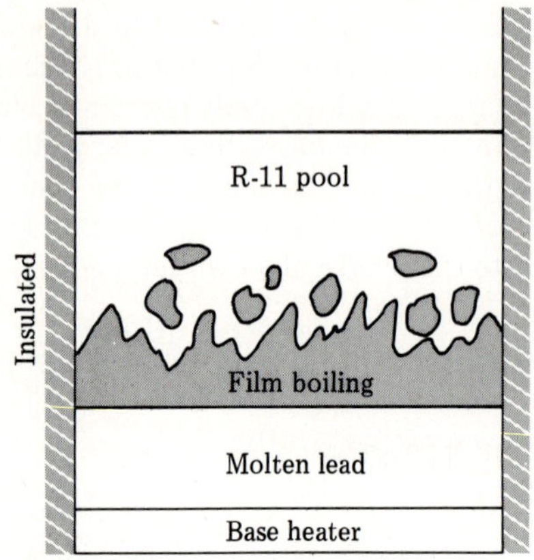


**Figure P9.35**

9.36 A 3-mm wire with emissivity 0.8 is immersed in water at 1 atm and 100°C and electrically heated to 600°C. Estimate the electric power required per meter of wire length.

9.37 Water at 80°C and 1 atm flows at 1 m/s past a 1-cm-diameter horizontal copper cylinder whose wall temperature is 112°C. Estimate the total heat flux, taking $\Delta T_i = 5°C$ and $q_i'' = 30\ kW/m^2$.

9.38 If all other parameters in Problem 9.37 are held constant, estimate the water temperature (subcooling) that would be needed for the total heat flux to be 600 $kW/m^2$.

9.39 For the conditions of Problem 9.37, estimate the maximum possible boiling heat flux and the wall temperature for which it occurs.

9.40 The data of Fig. 9.12 show the optimum effect of subcooling and forced convection on peak flux and are appropriate for a small heater size, such as a fine wire immersed in a stream. In tests in R-11 and R-113 with a larger heater (a 1-inch-diameter cylinder), Meyer et al. [45] show a less pronounced effect. In fact, their data show a *decrease* in peak flux at a low forced velocity of 0.25 m/s, and the measured local heat flux for this condition shows a region of *film boiling* on the lower front quadrant ($\theta$ = 270°–360° in Fig. 6.17). Can you explain physically and qualitatively why this is so?

9.41 Note in Eq. (9.34) that the leading coefficient, 1.1 $MW/m^2$, is the pool boiling peak flux for water from Fig. 9.5. With this in mind, can you suggest a way to nondimensionalize Eq. (9.34), using gas density, body length, and surface tension to nondimensionalize $V$? What quantities might best nondimensionalize $(T_{sat} - T_b)$?

9.42 According to data in [4, p. 8], the effect of subcooling peak flux at 1 atm is quite different for water versus carbon tetrachloride:

$H_2O$:

$q''_{max} \doteq (1.1\ MW/m^2)[1 + 0.089(T_{sat} - T_b)°C]$;

$CCl_4$:

$q''_{max} \doteq (0.3\ MW/m^2)[1 + 0.018(T_{sat} - T_b)°C]$.

Plot these in the range of subcooling up to 100°C. What accounts for the striking difference? Why is water so much better a coolant? Can the formulas be nondimensionalized?

## Special Projects

9.43 The complication to an analysis of *turbulent* film condensation on, say, a vertical plate is that the velocity and temperature profiles are assumed rather than found by direct integration as in Eqs. (9.7) and (9.8) for laminar condensation. Some typical profile shapes are shown in Fig. P9.43 for turbulent flow. You are asked to make a turbulent-flow analysis of this problem, resulting in a Nusselt number prediction, perhaps similar to Eq. (9.21). An acceptable velocity profile would be

$u/u_{max} \doteq (y/\delta)^{1/7}$,

and the wall shear could be related to $u_{max}$ by either the simple estimate

$\tau_w \doteq 0.002\rho_f u_{max}^2$

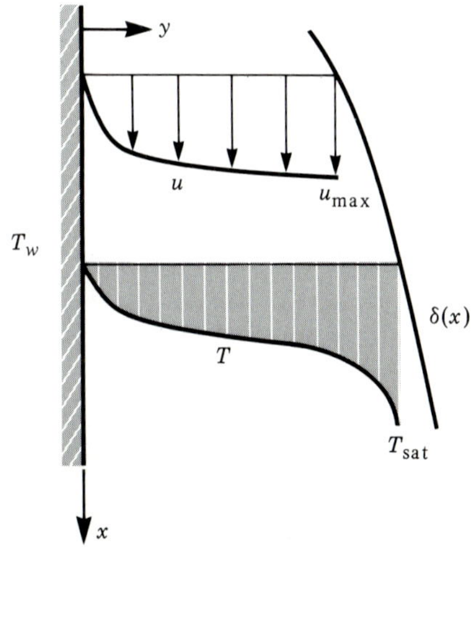


Figure P9.43

or by the more complicated formula in Eq. (7.33). Wall heat transfer could be estimated by the Reynolds analogy $\mathrm{St}_x \doteq c_f/(2\mathrm{Pr}^{2/3})$. Wall shear is related to the driving force $g(\rho_f - \rho_g)$ through Eq. (8.3b), replacing $\mu_f(\partial^2 u/\partial y^2)$ by a turbulent term $\partial\tau/\partial y$. Finally, the heat of vaporization is introduced by the enthalpy balance of Eq. (9.12), again converting from laminar to turbulent forms. Your results could also be compared with the more complex differential analysis of [43].

9.44 You are asked to make some preliminary calculations for a small fire-tube boiler. Combustion products, approximated as hot air at 800K, are pumped through an array of parallel horizontal 1-m-long copper tubes that pass through a pool of saturated water at 1 atm. Tube diameter may be from 1 to 10 cm, with smooth surfaces assumed. The hot air is at 1 atm and is provided by the furnace at a maximum rate of 3000 kg/hr. Maximum air pumping power available is 2000 W. Estimate the maximum rate of steam generation possible for such a system. Check to see that burnout does not occur in the tube entrances.

9.45 List some assumptions and methods for adapting our film condensation correlations to predicting the amount of water that will condense on a cold pipe in a warm, humid room. What difficulty most inhibits our direct use of the formulas of the present chapter? Why is the rate of condensation so much less than that predicted by Eq. (9.23)? Make sketches of the water vapor pressure and temperature distribution in and near the condensing water film. Cite at least two references for your conclusions.

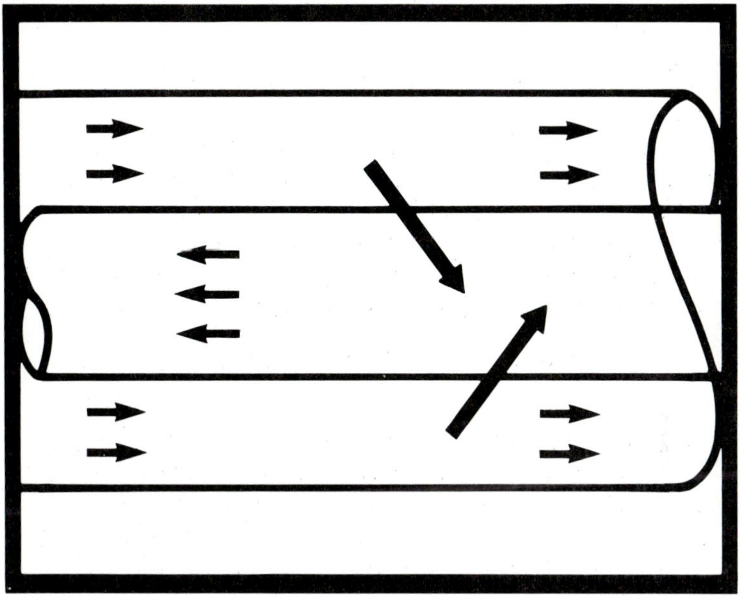

# Heat Exchangers

# 10

## 10.1 Introduction and Classification

In the previous chapters we have learned how to analyze conduction, convection, and radiation in various geometries and media. As engineers, we do not just sit around and enjoy this knowledge; we utilize it in design and development. The prototypical application of heat transfer analysis is to the design of a heat exchanger.

A *heat exchanger* is a device that efficiently transfers heat between a warmer substance and a colder substance. Although there are radiation exchangers in outer-space applications, normally the two exchanging substances are flowing fluids separated by a wall, as in Fig. 10.1. The heat transfer is a combination of convection on the "inner" and "outer" sides plus conduction through the wall. Both sides of the wall may be "fouled" by an unwanted deposit. The fluids may flow in the same direction (*parallel flow*) or in opposite directions (*counterflow*). This chapter reviews the characteristics of heat exchangers. One chapter is certainly not enough for one to become a specialist. As the reader may guess, there are many monographs devoted entirely to heat exchangers [1–7]. Most of the concepts used today were outlined in 1950 in a classic text by Kern [1], except the notion of heat exchanger "effectiveness," which was introduced by Kays and London in 1955 [3].

**Figure 10.1** **Sketch of a heat exchanger process: (a) definition of terms; (b) convection, fouling, and conduction resistances.**

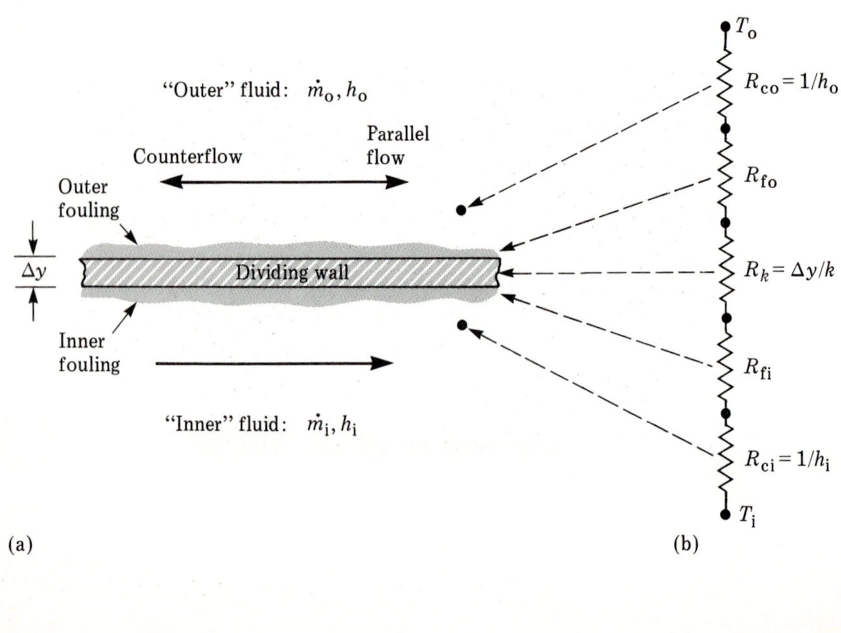

### 10.1.1 Classification Schemes

As pointed out in an interesting review by R. K. Shah [7, pp. 9–48], heat exchangers can be classified by (a) construction type, (b) flow arrangements, (c) heat transfer mechanisms, (d) surface density, and (e) whether or not the fluids are in contact. The first two are the most common classifications.

The flow arrangements may be either (a) single-pass or (b) multiple-pass. Figure 10.2 shows some examples. In single-pass designs, both fluids traverse the length of the system only once, as in Fig. 10.2(a). In multiple-pass exchangers, either or both fluids may zigzag back and forth across the system, like swimmers on a long race in a short pool. Figure 10.2(b) shows two tube passes and one outer or "shell" pass, while Fig. 10.2(c) has four tube passes and two shell passes. Note that while single-pass systems are either parallel flow or counterflow, multiple passes are a combination of both.

The fluids may also be arranged in crossflow, as in Figs. 10.2(d,e). Single passes are shown but multiple passes are possible. In Fig. 10.2(d) both fluids are "unmixed," that is, they flow through individual passages without intermingling. In Fig. 10.2(e) the left-to-right crossflow is "mixed" and the fluid can move about freely in an open shell area. If the shell area is large, crossflow will produce a much higher shell-side heat transfer coefficient than axial flow along the tube as in Fig. 10.2(a).

The second popular classification scheme is by the type of construction. There are four major types:

1. Tubular exchanger: (a) double-pipe, (b) shell-and-tube, (c) spiral tube.
2. Plate heat exchanger.
3. Extended surfaces: (a) plate-fin, (b) tube-fin.
4. Regenerators: (a) fixed-matrix, (b) rotary.

A simple double-pipe heat exchanger is sketched in Fig. 10.2(a). Both the inner and the outer pipe can be fluted or grooved to increase area, promote turbulence, and enhance heat transfer. The inner tube can be spiraled and densely packed within the shell.

The *shell-and-tube* geometry is very popular in industrial applications. A small example and a large example are shown in Fig. 10.3. The tube bundle can be densely packed and the shell flow baffled to crisscross over the tubes. Multiple passes may be utilized for both tube and shell flows.

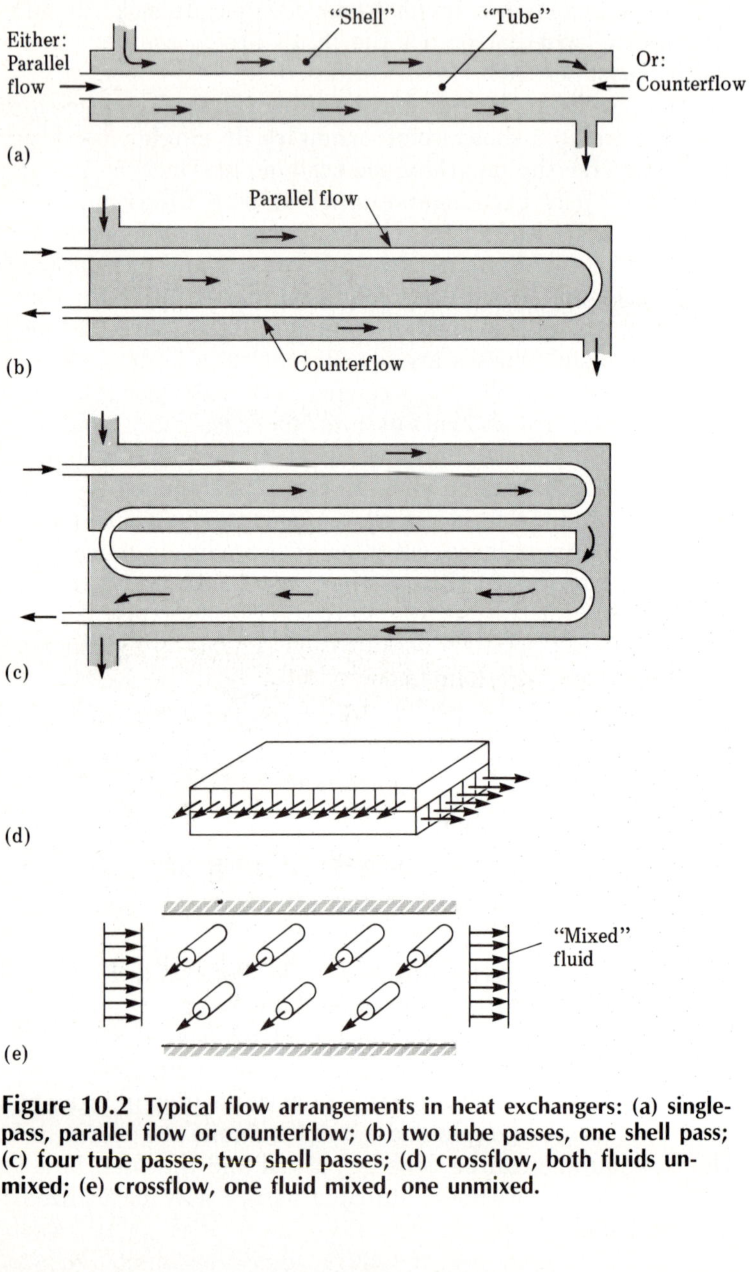


**Figure 10.2** Typical flow arrangements in heat exchangers: (a) single-pass, parallel flow or counterflow; (b) two tube passes, one shell pass; (c) four tube passes, two shell passes; (d) crossflow, both fluids unmixed; (e) crossflow, one fluid mixed, one unmixed.

(a)

(b)

**Figure 10.3** Illustration of shell and tube-bundle construction: (a) 6-in.-diameter shell, four tube passes, 76 tubes, segmental baffles (Courtesy of Young Radiator Co.); (b) 26-in.-diameter shell, two tube passes, 632 tubes, segmental baffles (Courtesy of Heat Transfer Research, Inc.).

Some popular plate baffle designs are shown in Fig. 10.4. Inducing crossflow over the bundle enhances the shell-side heat transfer coefficient. There are many shell-and-tube designs. In the late 1930s, the industry formed the Tubular Exchanger Manufacturers Association (TEMA), whose excellent and extensive handbook of standards is now in its sixth edition [8].

Figure 10.5 shows a commercial *plate heat exchanger.* The flows enter from the corners and pass up and down between the parallel plates, with the hot fluid occupying, say, the even-numbered passages and the cold fluid the odd passages. The plates are corrugated, knurled, or finned to promote heat transfer and inhibit fouling. As you can see, the plate exchanger is easy to disassemble and clean and is thus quite popular in the process industries.

Bare or "prime" surfaces such as tubes and plates have only a limited efficiency. To increase heat flux for a given exchanger size, *extended surfaces* such as fins, spines, or grooves are used. Figure 10.6 shows an external spine-fin surface made by wrapping a long piece of thin, multislitted sheet metal spirally around a tube. According to Abbott and co-workers [9, pp. 37–55] this design provides the least weight and size of any finned tubing presently used in air conditioner condensers and evaporators.

Figure 10.7 shows an array of internal fin-tube designs. Other examples of fin-tube geometries were shown earlier in Figs. 2.10 and 6.13. Such extended-surface products are relatively expensive but do increase the convection heat transfer coefficient.

All of the above designs — double-pipe, shell-and-tube, plate, and extended surface — are variations on a theme of hot and cold flowing fluids separated by a wall. A quite different geometry is the *regenerator,* in which the hot and cold fluids alternately occupy the same exchanger passages. The cellular exchanger or "matrix" serves as a heat storage element, being periodically primed by the hot fluid and then giving up its heat to the cold fluid. If the hot and cold fluids alternately enter a stationary exchanger, the design is called a *fixed matrix.* A contrasting design, called the *rotary regenerator,* has a circular matrix that rotates and exposes its surfaces alternately to the hot and then the cold fluid. Regenerators are popular in gas turbine and furnace preheater applications. The theory of periodic regeneration is discussed

**Figure 10.4 Shell-and-tube designs: (a) two tube passes, two shell passes; (b) single-segmental baffle; (c) double-segmental baffle; (d) disk and doughnut baffle; (e) orifice baffle.**

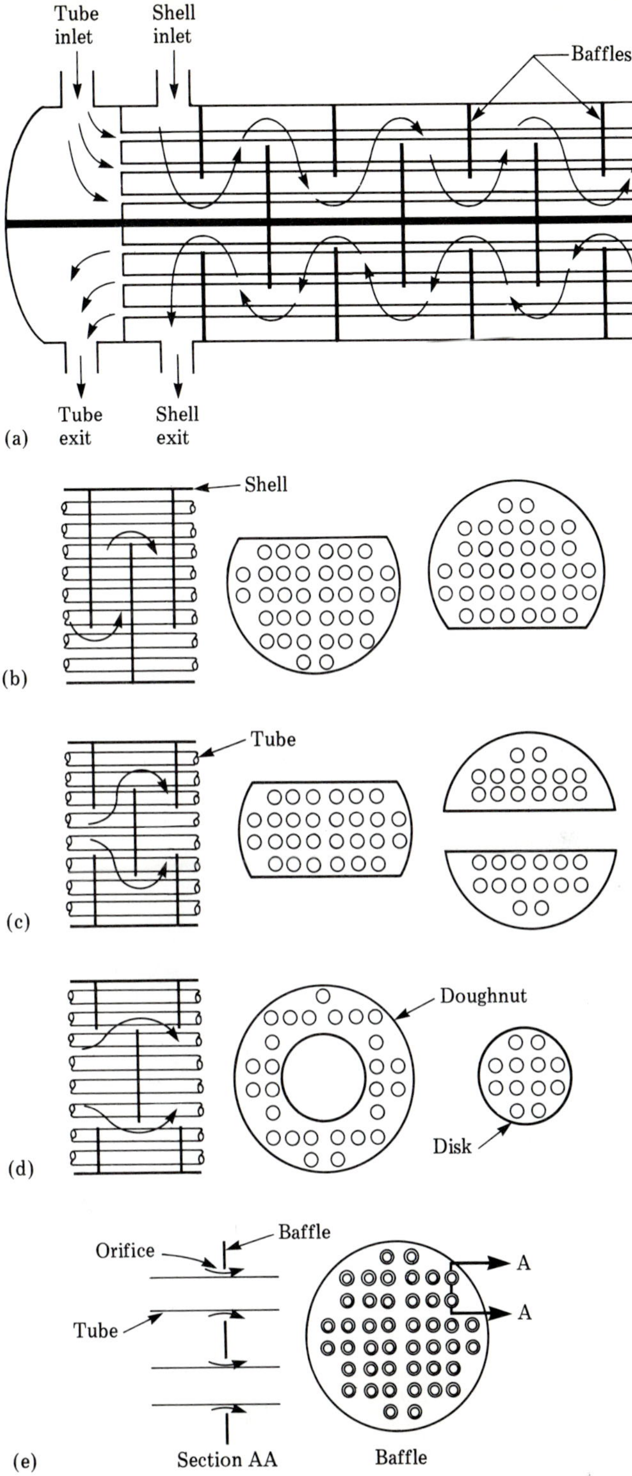
Tube inlet
Shell inlet
Baffles
Tube exit
Shell exit
(a)
Shell
(b)
Tube
(c)
Doughnut
Disk
(d)
Orifice
Baffle
Tube
A
A
Section AA
Baffle
(e)

by Shah [7, pp. 721–763]. Review articles on regenerators by H. Hausen and by P. Razelos can be found in [9].

A heat exchanger is said to be *compact* if it has a high "area density" or ratio of heat transfer surface to body volume. For example, a plain tube has an area density of $(\pi DL)/(\pi D^2L/4)$, or $(4/D)$. The text by Kays and London [3] is a classic compilation of data and design formulas for compact heat exchangers. In reviewing the history of compact designs, London [9, pp. 1–4] defines an exchanger to be compact if its area density is greater than 200 $ft^2/ft^3$ (656 $m^2/m^3$). This corresponds to a tube of diameter 0.24 in. (6.1 mm). As examples, London cites an 8 fin/in. automobile radiator (200 $ft^2/ft^3$), an aircraft plate-

**Figure 10.5** A plate heat exchanger. (Courtesy of A.P.V. Company, Inc.)

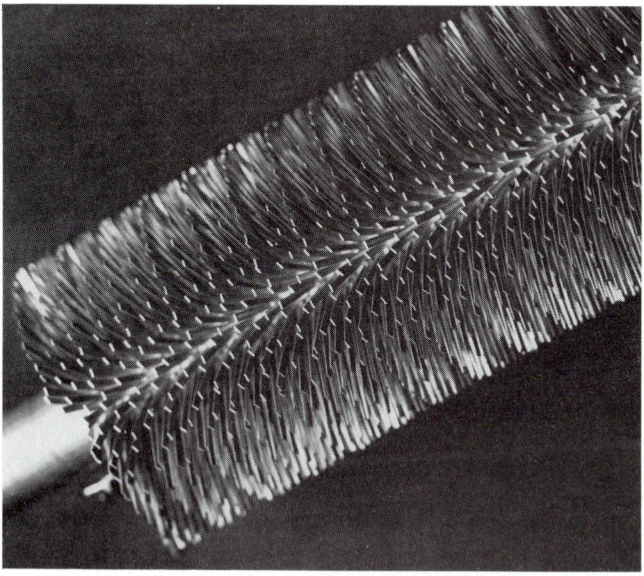

Figure 10.6 A tube covered with spine-fins for use in air-conditioner condensers. (Courtesy of Air Conditioning Business Division, General Electric Co.)

Figure 10.7 An array of internal fin-tube designs. (Courtesy of Forge-Fin Division, Noranda Metal Industries, Inc.)

fin exchanger (600 $ft^2/ft^3$), a glass-ceramic gas turbine exchanger (2000 $ft^2/ft^3$), and the human lung (6000 $ft^2/ft^3$). Reference [9] is devoted entirely to various aspects of compact heat exchanger design.

## 10.2 The Overall Heat Transfer Coefficient

A useful concept in exchanger analysis is the *overall heat transfer coefficient U*, which was discussed earlier in Section 2.4.6.

$$q = UA\,(T_h - T_c), \tag{10.1}$$

where $T_h$ and $T_c$ are the local hot and cold fluid temperatures, respectively, $A$ is the local surface area between fluids, and $q$ is the local heat transfer rate. Recall from Fig. 10.1 that $U$ will be a combination of convection, conduction, and fouling resistances.

In two-dimensional or plane heat flow, the area $A$ is the same for all sections, and we obtain from Fig. 10.1(b)

$$U = \frac{1}{\dfrac{1}{h_o} + R_{fo} + \dfrac{\Delta y}{k} + R_{fi} + \dfrac{1}{h_i}}, \tag{10.2}$$

where $R_{fo}$ and $R_{fi}$ are empirical "fouling coefficients," some of which are tabulated in Table 10.1.

In a tube or circular geometry, area varies with radius and we must be careful to cite whether $U$ refers to the inner or outer area. In terms of inner area, $A_i = 2\pi r_i L$,

$$U_i = \frac{1}{\dfrac{1}{h_i} + R_{fi} + \dfrac{r_i}{k}\ln\left(\dfrac{r_o}{r_i}\right) + \dfrac{r_i}{r_o}R_{fo} + \dfrac{r_i}{r_o h_o}}. \tag{10.3}$$

Alternatively, if we refer $U$ to the outer surface, $A_o = 2\pi r_o L$,

$$U_o = \frac{1}{\dfrac{r_o}{r_i h_i} + \dfrac{r_o}{r_i}R_{fi} + \dfrac{r_o}{k}\ln\left(\dfrac{r_o}{r_i}\right) + R_{fo} + \dfrac{1}{h_o}}. \tag{10.4}$$

The relation between these two is $U_i A_i = U_o A_o$.

The fouling factors suggested in Table 10.1 are only rough guidelines for preliminary design. In practice, of course, the fouling factor is zero for a new installation and grows with time as deposits form on the exchanger surface. The study of heat exchanger fouling is still one of the most frustrating and least understood problems in heat transfer

**Table 10.1** Approximate fouling factors, from [8]

| Fluid | $R_f$ – $m^2 \cdot$ °C/W | |
|---|---|---|
| Seawater, river water, boiler feedwater | 0.0001<br>0.0002 | (<50°C)<br>(>50°C) |
| Fuel oil | 0.0009 | |
| Quenching oil | 0.0007 | |
| Industrial air | 0.0004 | |
| Alcohol vapors | 0.0001 | |
| Steam (free of oil) | 0.0001 | |
| Refrigerant liquids | 0.0002 | |

[10]. See also [12] and the review by Collier and co-workers [7, pp. 999–1047].

Often the overall coefficient $U$ is dominated by the presence of a relatively low convection coefficient, usually on the gas side, as the following example illustrates.

## Example 10.1

A fire-tube boiler has hot air flowing with a heat transfer coefficient of 55 $W/m^2 \cdot K$ through 2-mm-thick steel tubes with an inside diameter of 3.8 cm. On the outside of the tubes is water boiling with a heat transfer coefficient of 7500 $W/m^2 \cdot K$. Estimate the coefficient $U_i$ and comment.

*Solution* The tube dimensions are $r_i = 0.019$ m and $r_o = 0.021$ m. For steel, take $k = 50$ W/m · K. From Table 10.1 for air take $R_{fi} = 0.0004$ and for water take $R_{fo} = 0.0002\ m^2 \cdot K/W$. Equation (10.3) is used for the inner version of $U$:

$$U_i = \left[\frac{1}{55} + 0.0004 + \frac{0.019}{50}\ln\left(\frac{0.021}{0.019}\right) + \frac{0.019}{0.021}(0.0002) + \frac{0.019}{0.021(7500)}\right]^{-1}$$

$$= [0.01818 + 0.0004 + 0.00004 + 0.00018 + 0.00012]^{-1}$$

$$= 53\ W/m^2 \cdot K. \quad [Ans.]$$

Alternatively, we could have computed

$$U_o = \frac{U_i r_i}{r_o} = \frac{(53)(0.019)}{0.021} = 47.8\ W/m^2 \cdot K.$$

Note that $U$ is dominated by and nearly equal to the air-side heat transfer coefficient $h_i = 55\ \text{W/m}^2 \cdot \text{K}$. ■

---

## 10.3 The Mean Temperature Difference

In a real heat exchanger, it is clear that the overall coefficient $U$ varies with position. In the shell-and-tube exchanger of Fig. 10.4(a), for example, $h_i$ is highest in the entrance region of the tubes (recall Figs. 6.5 and 6.14), while $h_o$ is highest between the baffles and lowest near the baffles. For best accuracy, then, one should compute the overall heat transfer between fluids by integrating Eq. (10.1) over the area:

$$q_{\text{total}} = \int U(T_h - T_c)\, dA. \tag{10.5}$$

For a complex design, numerical or graphical integration is required.

For preliminary design it is customary to take $U$ equal to a constant, average value, $\overline{U}$. This reduces Eq. (10.5) to the determination of an appropriate mean temperature difference between hot and cold fluids:

*If $U$ = constant:*

$$q_{\text{total}} = \overline{U} A_{\text{total}}\, \Delta T_{\text{mean}}. \tag{10.6}$$

We now show that, for a single-pass system, either parallel or counterflow, $\Delta T_{\text{mean}}$ is none other than our old friend the log-mean temperature difference from Eq. (6.69). For multipass systems, an additional correction factor is needed to relate $\Delta T_{\text{mean}}$ to the LMTD.

### 10.3.1 Single-Pass Exchangers: The LMTD

The derivation of the LMTD in Section 6.3.3 for tube flow assumed a constant wall temperature. But this is not necessary, as we now show. Schematic hot and cold temperature distributions are given in Fig. 10.8 for a single-pass parallel or counterflow heat exchanger. For any local part of the wall, $dA$, the local heat transfer can be related to either the exchanger wall or to the change in enthalpy of the flowing streams:

$$dq = U(T_h - T_c)dA = \dot{m}_c c_{pc}\, dT_c = -\dot{m}_h c_{ph}\, dT_h. \tag{10.7}$$

This is true for either parallel flow or counterflow. But note the striking difference in the temperature curves in Fig. 10.8. In parallel flow (Fig.

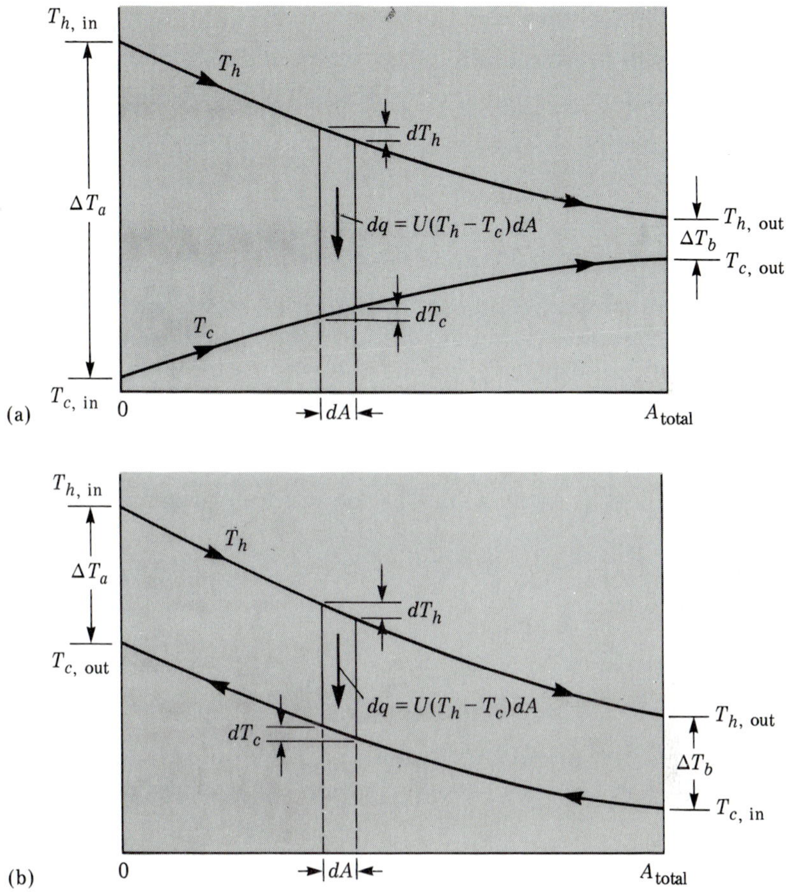


**Figure 10.8 Temperature distributions in a single-pass heat exchanger: (a) parallel flow; (b) counterflow.**

10.8a), the hot and cold fluid temperatures approach each other such that $T_{\text{c, out}}$ is always less than $T_{\text{h, out}}$. But in counterflow (Fig. 10.8b), the cold fluid can continue rising toward the hot fluid entrance such that, quite often, $T_{\text{c, out}}$ can be greater than $T_{\text{h, out}}$. Clearly the counterflow case is more effective for a given system size.

To continue analyzing Eq. (10.7) we make some assumptions:

1. $\dot{m}_{\text{c}}$ and $\dot{m}_{\text{h}}$ are constant (steady flow);
2. $c_{p\text{c}}$ and $c_{p\text{h}}$ are constant;

3. $U$ is constant;
4. there are no losses to the exchanger surroundings; and
5. axial conduction along the tubes is negligible.

The enthalpy changes in Eq. (10.7) may be integrated across the entire exchanger:

$$\begin{aligned} q &= C_c(T_{c,\,out} - T_{c,\,in}), \qquad C_c = \dot{m}_c c_{pc} \\ &= C_h(T_{h,\,in} - T_{h,\,out}), \qquad C_h = \dot{m}_h c_{ph}. \end{aligned} \tag{10.8}$$

The grouping $(\dot{m}c_p)$ occurs so often that we adopt the short notation $C$ for convenience to denote fluid *heat capacity rate*.

Now eliminate $dq$ among the various terms in Eq. (10.7):

$$dT_h - dT_c = d(\Delta T) = -dq\left(\frac{1}{C_h} + \frac{1}{C_c}\right) = -U\Delta T\, dA\left(\frac{1}{C_h} + \frac{1}{C_c}\right).$$

We may separate the variables and integrate over the whole exchanger:

$$\int \frac{d(\Delta T)}{\Delta T} = -U\left(\frac{1}{C_h} + \frac{1}{C_c}\right)\int dA,$$

or

$$\ln\left(\frac{\Delta T_b}{\Delta T_a}\right) = -U\left(\frac{1}{C_h} + \frac{1}{C_c}\right) A_{total}. \tag{10.9}$$

Finally, eliminate $C_h$ and $C_c$ in favor of $q$ from Eqs. (10.8). The result has the same form for parallel or counterflow:

$$q_{total} = UA_{total}\,\Delta T_{LM}, \tag{10.10}$$

where

$$\Delta T_{LM} = \frac{\Delta T_b - \Delta T_a}{\ln(\Delta T_b/\Delta T_a)} = \frac{\Delta T_a - \Delta T_b}{\ln(\Delta T_a/\Delta T_b)}.$$

This is the *log-mean temperature difference,* where $\Delta T_a$ and $\Delta T_b$ are the left and right fluid temperature differences in the exchanger, as labeled in Figs. 10.8(a,b). Note that, for parallel flow, $\Delta T_a = (T_{h,\,in} - T_{c,\,in})$ and $\Delta T_b = (T_{h,\,out} - T_{c,\,out})$. For counterflow, $\Delta T_a = (T_{h,\,in} - T_{c,\,out})$ and $\Delta T_b = (T_{h,\,out} - T_{c,\,in})$.

In one form of heat exchanger design problem, the fluid inlet and outlet temperatures are known. Then the required exchanger area is $A = q/(U\Delta T_{LM})$, and the configuration that produces a larger $\Delta T_{LM}$ (counterflow) will result in a smaller exchanger area. The following example illustrates.

**Example 10.2**

In a single-pass exchanger, the hot fluid enters at 300°F and leaves at 220°F, while the cold fluid enters at 75°F and exits at 200°F. Compute the LMTD for (a) parallel flow and (b) counterflow, and comment on the difference.

*Solution* For parallel flow, from Fig. 10.8(a),

$$\Delta T_a = T_{\text{h, in}} - T_{\text{c, in}} = 300 - 75 = 225°\text{F},$$

$$\Delta T_b = T_{\text{h, out}} - T_{\text{c, out}} = 220 - 200 = 20°\text{F}.$$

Then

$$\Delta T_{\text{LM}}(\text{parallel}) = \frac{225 - 20}{\ln(225/20)} = 84.7°\text{F}. \quad [Ans. (a)]$$

Meanwhile, for counterflow, from Fig. 10.8(b),

$$\Delta T_a = T_{\text{h, in}} - T_{\text{c, out}} = 300 - 200 = 100°\text{F},$$

$$\Delta T_b = T_{\text{h, out}} - T_{\text{c, in}} = 220 - 75 = 145°\text{F},$$

from which

$$\Delta T_{\text{LM}}(\text{counterflow}) = \frac{100 - 145}{\ln(100/145)} = 121.1°\text{F}. \quad [Ans. (b)]$$

This latter solution looks a little weird with the ratio of negative numbers, but your calculator will handle it perfectly.

For the same conditions in this case, the LMTD for counterflow is 43% higher, so a counterflow design will require only 1/1.43, or 70%, of the exchanger area needed for a parallel flow system. ■

### 10.3.2 Multiple Passes: The LMTD Correction Factor

The LMTD concept breaks down in multiple-pass shell-and-tube exchangers, such as Figs. 10.2(b,c), where exchange conditions are partly counterflow and partly parallel flow. The concept also needs revision for crossflow exchangers, such as Figs. 10.2(d,e), because the temperature distribution is two-dimensional and each passage sees different conditions. The analysis of these multipass and crossflow geometries was begun by Nusselt in 1911 and culminated in a classic summary paper by Bowman, Mueller, and Nagle in 1940 [11].

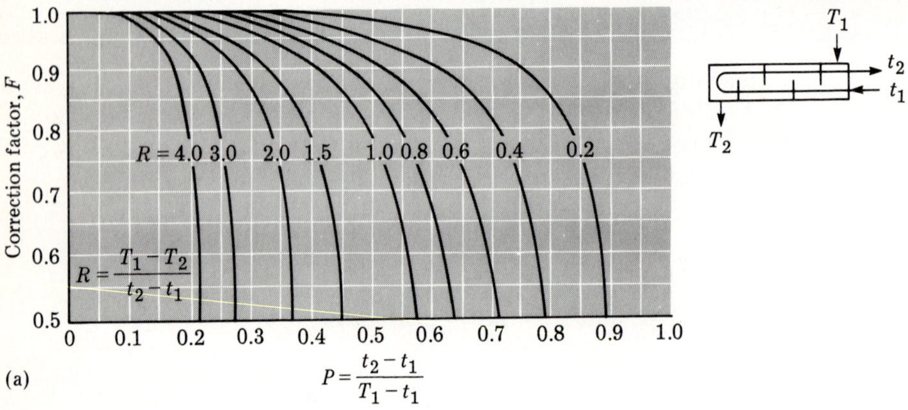
Correction factor, $F$
$R = 4.0$ 3.0 2.0 1.5 1.0 0.8 0.6 0.4 0.2
$R = \frac{T_1 - T_2}{t_2 - t_1}$
$P = \frac{t_2 - t_1}{T_1 - t_1}$
$T_1$
$t_2$
$t_1$
$T_2$

(a)

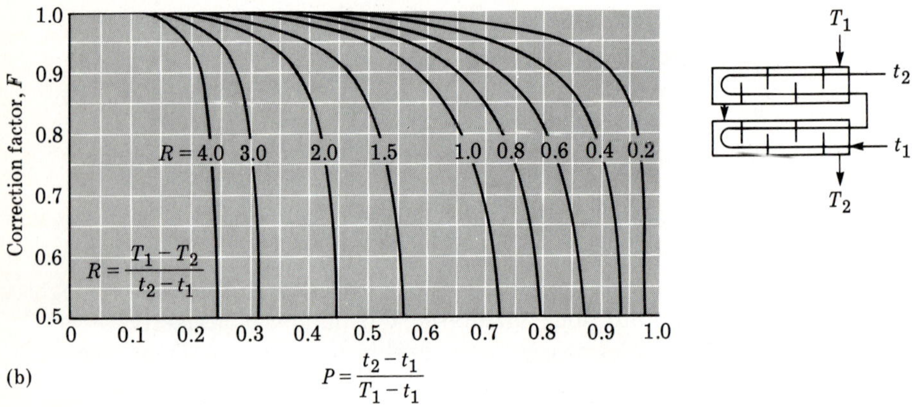
Correction factor, $F$
$R = 4.0$ 3.0 2.0 1.5 1.0 0.8 0.6 0.4 0.2
$R = \frac{T_1 - T_2}{t_2 - t_1}$
$P = \frac{t_2 - t_1}{T_1 - t_1}$
$T_1$
$t_2$
$t_1$
$T_2$

(b)

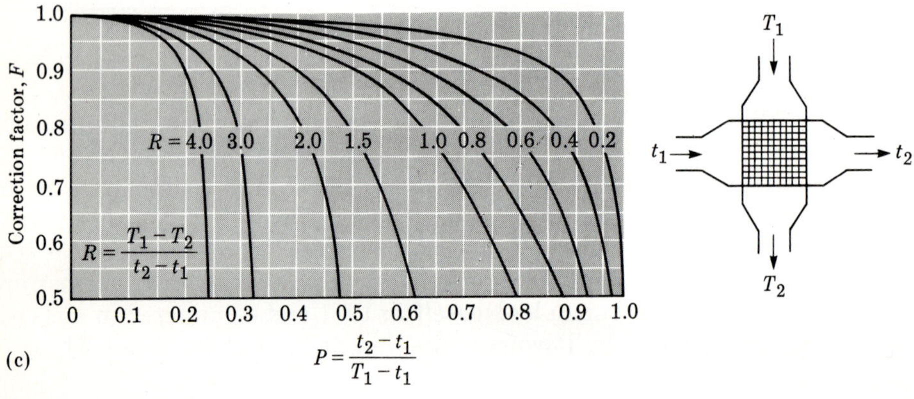
Correction factor, $F$
$R = 4.0$ 3.0 2.0 1.5 1.0 0.8 0.6 0.4 0.2
$R = \frac{T_1 - T_2}{t_2 - t_1}$
$P = \frac{t_2 - t_1}{T_1 - t_1}$
$T_1$
$t_1$
$t_2$
$T_2$

(c)

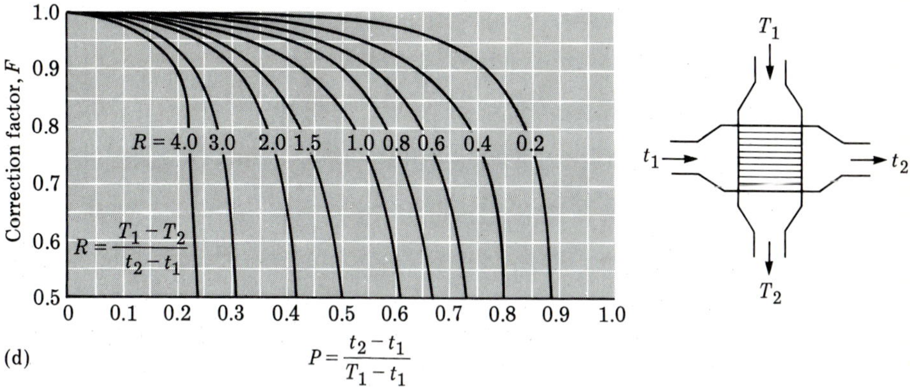


(d)

**Figure 10.9** **LMTD correction factors for heat exchangers: (a) one shell pass, any multiple of two tube passes; (b) two shell passes, any multiple of four tube passes; (c) crossflow, both fluids unmixed; (d) crossflow, one fluid mixed.**

Reference [11] shows that the correct mean temperature difference for a complex geometry is equal to the LMTD multiplied by a correction factor, $F$:

$$\Delta T_{\text{mean}} = F \Delta T_{\text{LM}}, \tag{10.11}$$

where $F \leqslant 1.0$. Two characteristic temperature ratios occur:

$$R = \frac{T_1 - T_2}{t_2 - t_1} = \frac{C(\text{tube})}{C(\text{shell})}, \quad \text{and} \quad P = \frac{t_2 - t_1}{T_1 - t_1}, \tag{10.12}$$

where $T$ and $t$ denote shell-side and tube-side temperatures, respectively, and 1 and 2 denote inlet and outlet, respectively. It does not matter which is the hot or cold fluid. In all cases the correction factor has the form

$$F = \text{fcn}(R, P, \text{geometry}).$$

For example, the analysis of a single shell pass and any even number (2, 4, 6, . . .) of tube passes (Fig. 10.9a) leads to the formula

$$F = \frac{B \ln[(1 - P)/(1 - RP)]}{(R - 1)\ln\left[\dfrac{2 - P(R + 1 - B)}{2 - P(R + 1 + B)}\right]}, \qquad B = (1 + R^2)^{1/2}. \tag{10.13}$$

This expression is plotted versus $P$ for various values of $R$ in Fig. 10.9(a). Three similar cases are also plotted in Fig. 10.9, as taken from [11]. Many additional cases (not shown here) are plotted in [8] and [11]. For all geometries, the correction factor $F$ begins to drop sharply — signaling a poor exchanger design — as the product $RP$ rises toward unity.

For all multipass designs, the LMTD is computed as if there were pure counterflow conditions:

$$\text{LMTD} = \frac{(T_1 - t_2) - (T_2 - t_1)}{\ln[(T_1 - t_2)/(T_2 - t_1)]}.$$

Note from Fig. 10.9 that, if either $R$ or $P$ approaches zero, $F$ approaches unity. This is the case where either the shell-side or tube-side fluid is at constant temperature, such as in a condenser or evaporator, as studied in Section 6.3.3. When this happens, $\Delta T_{\text{mean}}$ = LMTD regardless of exchanger geometry.

This type of approach, through the LMTD and the factor $F$, is a design problem of the "first kind": fluid temperatures are known and an estimate of $U$ is available. We evaluate LMTD and $F$ and thus obtain, for specified flow rates, the total heat transfer area, $A$, required for the design. Example 10.3 illustrates this procedure. The "second kind" of design problem is discussed in Section 10.4.

## Example 10.3

Water in two tube passes is used to cool oil in one shell pass. As shown in the figure, the water enters at 20°C and 3 kg/s and leaves at 70°C. The oil enters at 180°C and leaves at 80°C. Find (a) the oil flow rate, (b) the heat exchanged, and (c) the required exchanger area, if $U$ = 350 W/m$^2$ · K.

*Solution* For water at a mean of 45°C, take $c_{pc}$ = 4180 J/kg · °C. For (unused engine) oil at a mean of 130°C, take $C_{ph}$ = 2350

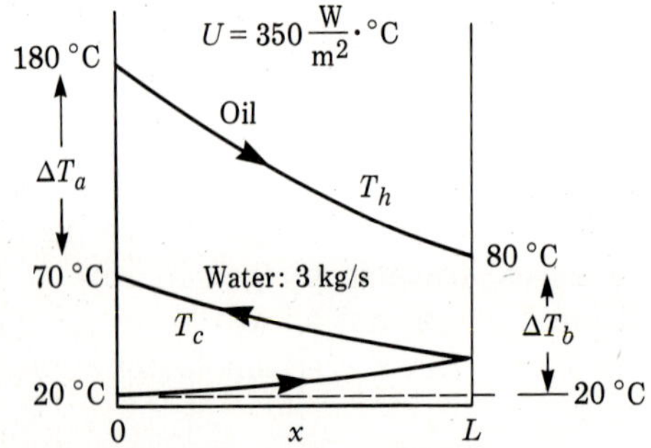

J/kg · °C. A heat balance as in Eq. (10.8) gives

$$q = \dot{m}_c c_{pc}(T_{c,\ out} - T_{c,\ in}) = \dot{m}_h c_{ph}(T_{h,\ in} - T_{h,\ out})$$

$$= (3\ \text{kg/s})(4180\ \text{J/kg}\cdot{}^\circ\text{C})(70 - 20^\circ\text{C})$$

$$= \dot{m}_h(2350\ \text{J/kg}\cdot{}^\circ\text{C})(180 - 80^\circ\text{C}) = 627{,}000\ \text{W}. \quad [Ans.\ (b)]$$

Solve for the oil flow rate, $\dot{m}_h = 2.67$ kg/s. [*Ans. (a)*]

With reference to the figure, $\Delta T_a = 180 - 70 = 110$°C and $\Delta T_b = 80 - 20 = 60$°C. Then, from Eq. (10.9),

$$\text{LMTD} = \Delta T_{\text{LM}} = \frac{110 - 60}{\ln(110/60)} = 82.5^\circ\text{C}.$$

The correction factor $F$ relates to Fig. 10.9(a). Oil is the shell side, so $T_1 = 180$°C, $T_2 = 80$°C, $t_1 = 20$°C, and $t_2 = 70$°C. Then the parameters $R$ and $P$ are

$$R = \frac{180 - 80}{70 - 20} = 2.0, \quad \text{and} \quad P = \frac{70 - 20}{180 - 20} = 0.3125.$$

From Fig. 10.9(a) or Eq. (10.13), $F = 0.86$. Then we may solve for the required heat exchanger area from Eq. (10.6):

$$q = 627{,}000\ \text{W} = UAF\Delta T_{\text{LM}} = (350\ \text{W/m}^2\cdot{}^\circ\text{C})A(0.86)(82.5^\circ\text{C}),$$

or

$$A = 25.3\ \text{m}^2. \quad [Ans.\ (c)]$$

We could apportion this area among the tubes if we knew their sizes. But if we don't know their sizes, how did we find out that $U = 350$ W/m$^2$ · K? This is a dilemma in practice: We cannot estimate exchanger area from known flow rates and temperatures unless we obtain an estimate of $U$ from an experienced engineer. ■

## Example 10.4

In Example 10.3 the area of 25.3 m$^2$ is equivalent to 100 tubes of 2 cm in diameter and 4 m long (50 tubes per pass). Neglecting tube thickness and fouling, estimate $U$ for this geometry if the tubes are bundled and baffled such that the oil approaches each tube at a crossflow velocity of 8 cm/s.

*Solution* The exact analysis of $h_i$ and $h_o$ in a complex heat exchanger geometry requires much experience and experimental data. We can make a preliminary estimate of $h_i$ using fully developed water flow in the tubes. For water at $t_{mean} = 45°C$, $Pr = 3.93$, $\rho = 990\ kg/m^3$, $\mu = 0.0006\ kg/m \cdot s$, and $k = 0.638\ W/m \cdot K$. The mass flow per tube is $\dot{m}_c/N = 3.0/50 = 0.06$ kg/s. Then the Reynolds number for water flow in a tube is

$$Re_{D,\,i} = \frac{4\dot{m}}{\pi D \mu} = \frac{4(0.06\ kg/s)}{\pi(0.02\ m)(0.0006\ kg/m \cdot s)}$$
$$= 6366.$$

For simplicity, use the Dittus-Boelter formula, Eq. (6.107), to estimate

$$Nu_{D,\,i} \doteq 0.027\ Re_{D,\,i}^{0.8}\ Pr_i^{1/3} = 0.027(6366)^{0.8}(3.93)^{1/3} = 47.1.$$

Then

$$h_i = \frac{Nu_{D,\,i} k_i}{D} = \frac{(47.1)(0.638)}{0.02} = 1500\ W/m^2 \cdot K.$$

On the shell side, for oil at $T_{mean} = 130°C$, $\rho = 825\ kg/m^3$, $Pr = 152$, $\mu = 0.0087\ kg/m \cdot s$, and $k = 0.134\ W/m \cdot K$. The outside Reynolds number is

$$Re_{D,\,o} = \frac{\rho_o V_o D}{\mu_o} = \frac{(825\ kg/m^3)(0.08\ m/s)(0.02\ m)}{0.0087\ kg/m \cdot s}$$
$$= 151.$$

The outside Nusselt number can be estimated from Eq. (6.115):

$$Nu_{D,\,o} \doteq 0.3 + \frac{0.62(151)^{1/2}(152)^{1/3}}{[1 + (0.4/152)^{2/3}]^{1/4}}\left[1 + \left(\frac{151}{282{,}000}\right)^{5/8}\right]^{4/5}$$
$$= 41.1.$$

Then

$$h_o = \frac{Nu_{D,\,o} k_o}{D} = \frac{(41.1)(0.134\ W/m \cdot K)}{0.02\ m} = 275\ W/m^2 \cdot K.$$

With tube conduction and fouling neglected, the overall coefficient of heat transfer is

$$U = [(1/h_i) + (1/h_o)]^{-1} = [(1/1500) + (1/275)]^{-1}$$
$$\doteq 232\ W/m^2 \cdot K. \quad [Ans.]$$

This is in the ballpark of our guess $U = 350\ W/m^2 \cdot K$ in Example 10.3 but is 34% less and implies that 50% more exchanger area is needed. The heat exchanger design process may converge slowly. ■

## 10.4 The Effectiveness/NTU Method

In Section 10.3 we considered the "first kind" of heat exchanger problem, where inlet and outlet temperatures are known, $U$ is known, and the exchanger area is desired. The solution was straightforward through the LMTD and $F$ factor.

Now consider the "second" kind of design problem. The inlet temperatures and mass flows are known, $U$ and $A$ are known, but the outlet temperatures are to be determined. By the LMTD method, one would have to guess the outlet temperatures, evaluate LMTD and $F$, and iterate until the heat exchange computations match.

A noniterative method to solve this second problem was devised by Kays and London in 1955 in the first edition of their book [3]. They defined a dimensionless parameter called the heat exchanger *effectiveness*, $\varepsilon$:

$$\varepsilon = \frac{\text{Actual heat transfer}}{\text{Maximum possible heat transfer}}. \tag{10.14}$$

Examining Fig. 10.8, we deduce that the maximum possible heat transfer would occur if one fluid — the hot fluid, say — had a massive flow rate and remained at its inlet temperature $T_{\text{h, in}}$ throughout. If the exchanger area were large, the cold fluid temperature would rise until, far downstream, it would approach $T_{\text{h, in}}$.

Alternatively, suppose the cold fluid were flowing at a colossal rate and maintained its temperature at $T_{\text{c, in}}$. Then the hot fluid, flowing at a lower rate $C_{\min} = (\dot{m}c_p)_{\min}$, would cool down in a large exchanger until it nearly reached $T_{\text{c, in}}$.

In either case, the maximum heat transfer possible is

$$q_{\max} = C_{\min}(T_{\text{h, in}} - T_{\text{c, in}}), \tag{10.15}$$

where $C_{\min}$ is the smaller of the two capacity rates $C_\text{c}$ and $C_\text{h}$. The effectiveness may be defined for either the hot or cold fluid:

$$\varepsilon = \frac{C_\text{h}(T_{\text{h,in}} - T_{\text{h,out}})}{C_{\min}(T_{\text{h,in}} - T_{\text{c,in}})} = \frac{C_\text{c}(T_{\text{c,out}} - T_{\text{c,in}})}{C_{\min}(T_{\text{h,in}} - T_{\text{c,in}})}. \tag{10.16}$$

We choose whichever form suits our needs. In either case, there is only one unknown: $T_{\text{h,out}}$ or $T_{\text{c,out}}$.

Kays and London [3] showed that, for any heat exchanger geometry, $\varepsilon$ is a function solely of the heat capacities of the fluids and $UA$.

Let us illustrate with a parallel flow exchanger (Fig. 10.8a) for which $C_{\min} = C_\text{h}$ and $C_{\max} = C_\text{c}$. Then, from the first form of Eq. (10.16),

$$\varepsilon = \frac{T_{\text{h, in}} - T_{\text{h, out}}}{T_{\text{h, in}} - T_{\text{c, in}}}. \tag{10.17}$$

Further, from a heat balance on hot and cold fluid,

$$\frac{T_{\text{c, out}} - T_{\text{c, in}}}{T_{\text{h, in}} - T_{\text{h, out}}} = \frac{C_{\min}}{C_{\max}}. \tag{10.18}$$

Our temperature difference solution, Eq. (10.9), may be rewritten in the form

$$\ln\left(\frac{T_{\text{h, out}} - T_{\text{c, out}}}{T_{\text{h, in}} - T_{\text{c, in}}}\right) = -\frac{UA}{C_{\min}}\left(1 + \frac{C_{\min}}{C_{\max}}\right). \tag{10.19}$$

Now, using Eqs. (10.17) and (10.18), manipulate the argument of the logarithm into an effectiveness form:

$$\begin{aligned}\frac{T_{\text{h, out}} - T_{\text{c, out}}}{T_{\text{h, in}} - T_{\text{c, in}}} &= \frac{(T_{\text{h, out}} - T_{\text{h, in}}) + (T_{\text{h, in}} - T_{\text{c, in}}) - \left(\dfrac{C_{\min}}{C_{\max}}\right)(T_{\text{h, in}} - T_{\text{h, out}})}{(T_{\text{h, in}} - T_{\text{c, in}})} \\ &= -\varepsilon + 1 - (C_{\min}/C_{\max})\varepsilon.\end{aligned} \tag{10.20}$$

Substitute this back into the logarithm in Eq. (10.19) and solve for the effectiveness for a parallel flow exchanger:

$$\varepsilon = \frac{1 - \exp[-\text{NTU}(1 + C_{\min}/C_{\max})]}{1 + C_{\min}/C_{\max}}, \tag{10.21}$$

where $\text{NTU} = UA/C_{\min}$ is a dimensionless parameter called the *number of transfer units.*

Repeating this analysis with the cold fluid as the "minimum" fluid gives exactly the same relation, Eq. (10.21). Kays and London [3] show that, for any heat exchanger,

$$\varepsilon = \text{fcn}\left(\text{NTU}, \frac{C_{\min}}{C_{\max}}, \text{geometry}\right).$$

Some of their analytical results are summarized in Table 10.2. These formulas are reasonably amenable to evaluation on a handheld calculator, although many extensive plots are given in [3] and [8].

Typical effectiveness versus NTU variations are sketched in Fig. 10.10 for several heat exchanger designs. Maximum effectiveness occurs at $C_{\min}/C_{\max} = 0$ and is the same for all exchangers of any design:

$$C_{\min}/C_{\max} = 0:$$
$$\varepsilon = \varepsilon_{\max} = 1 - e^{-\text{NTU}}. \tag{10.22}$$

Minimum effectiveness occurs when the capacity rates are equal, $C_{\min}/C_{\max} = 1.0$, and the numerical values in Fig. 10.10 are different for each design. Most effective of all, for a given NTU, is the counterflow exchanger (curve a), followed closely by the crossflow exchanger with both fluids unmixed (curve b). The other designs shown in Fig. 10.10 are less effective at large $C_{min}/C_{max}$. Poorest of all, as might be expected from the discussion of Fig. 10.8(a), is the parallel flow design.

For values of $C_{min}/C_{max}$ between zero and unity, the upper and any specific lower curve in Fig. 10.10 can be interpolated linearly to about $\pm 5\%$ error. For increased accuracy, one should use the exact formulas from Table 10.2.

Figure 10.10 (and Table 10.2) is the solution to the "second kind" of design problem. Given inlet temperatures, heat capacities, and $UA$, one can enter the chart, find $\varepsilon$, and compute outlet temperatures from Eqs. (10.16). Note that the solid curve ($C_{min}/C_{max} = 0$) is appropriate for condensers, evaporators, and cases where the capacity rates are greatly different.

**Table 10.2** Effectiveness relations for various exchangers, from [3]

Let $N$ = NTU = $UA/C_{min}$, $r = C_{min}/C_{max}$

| | |
|---|---|
| 1. Single-pass parallel flow | $\varepsilon = \dfrac{1 - \exp[-N(1 + r)]}{1 + r}$ |
| 2. Single-pass counterflow | $\varepsilon = \dfrac{1 - \exp[-N(1 - r)]}{1 - r\exp[-N(1 - r)]}$ |
| *Shell-and-tube:* | |
| 3. One shell pass, (2, 4, 6, . . .) tube passes | $\varepsilon = 2\left[1 + r + B\dfrac{1 + \exp(-NB)}{1 - \exp(-NB)}\right]^{-1}$, where $B = (1 + r^2)^{1/2}$ |
| 4. Two shell passes, (4, 8, 12, . . .) tube passes | $\varepsilon = \left[\left(\dfrac{1 - \varepsilon^* r}{1 - \varepsilon^*}\right)^2 - 1\right]\left[\left(\dfrac{1 - \varepsilon^* r}{1 - \varepsilon^*}\right)^2 - r\right]^{-1}$, where $\varepsilon^*$ is computed from Eq. (3) above, evaluated at $r$ and $N/2$. |
| *Crossflow:* | |
| 5. Both fluids mixed | $\varepsilon = N\left[\dfrac{N}{1 - \exp(-N)} + \dfrac{rN}{1 - \exp(-rN)} - 1\right]^{-1}$ |
| 6. Both fluids unmixed | $\varepsilon \doteq 1 - \exp\{[\exp(-rN^{0.78}) - 1]N^{0.22}/r\}$ |
| 7. $C_{max}$ mixed, $C_{min}$ unmixed | $\varepsilon = \dfrac{1}{r}\{1 - \exp[-r(1 - e^{-N})]\}$ |
| 8. $C_{max}$ unmixed, $C_{min}$ mixed | $\varepsilon = 1 - \exp\left[-\dfrac{1}{r}(1 - e^{-Nr})\right]$ |

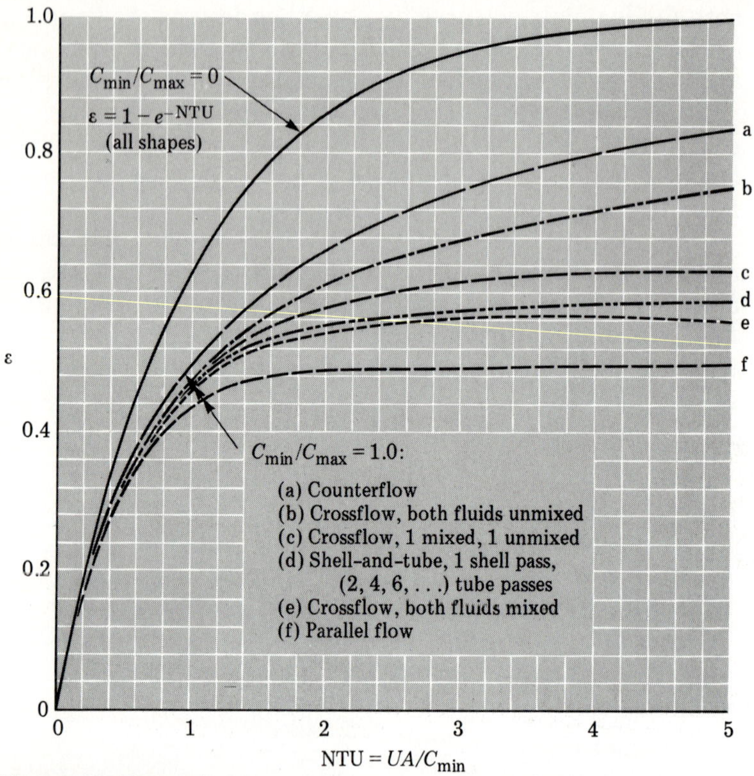


**Figure 10.10** Maximum ($C_{min}/C_{max} = 0$) and minimum ($C_{min}/C_{max} = 1$) effectiveness of various heat exchangers. For intermediate $C_{min}/C_{max}$, interpolate linearly or use Table 10.2 for greater accuracy.

### Example 10.5

To illustrate the NTU method, suppose we work Example 10.3 backward by assuming known area but unknown outlet temperatures. Then we are given oil in one shell pass and water in two tube passes, with $U = 350\ \text{W/m}^2 \cdot {}^\circ\text{C}$ and $A = 25.3\ \text{m}^2$. The oil enters at 180°C and the water enters at 20°C. Find the outlet temperatures if $\dot{m}_{\text{oil}} = 2.67$ kg/s and $\dot{m}_{\text{water}} = 3$ kg/s.

*Solution* Recall the specific heats: $c_{ph} = 2350\ \text{J/kg} \cdot {}^\circ\text{C}$ for the oil and $c_{pc} = 4180\ \text{J/kg} \cdot {}^\circ\text{C}$ for the water. Then

$$C_c = (3\ \text{kg/s})(4180\ \text{J/kg} \cdot {}^\circ\text{C}) = 12{,}540\ \text{W/}^\circ\text{C} = C_{\max},$$

$$C_h = (2.67\ \text{kg/s})(2350\ \text{J/kg} \cdot {}^\circ\text{C}) = 6270\ \text{W/}^\circ\text{C} = C_{\min},$$

and

$$C_{min}/C_{max} = 6270/12{,}540 = 0.5.$$

Meanwhile,

$$\text{NTU} = \frac{UA}{C_{min}} = \frac{(350\ \text{W/m}^2\cdot{}^\circ\text{C})(25.3\ \text{m}^2)}{6270\ \text{W/}^\circ\text{C}}$$

$$= 1.41.$$

Enter Fig. 10.10 at NTU = 1.41 for the shell-and-tube case: $\varepsilon_{max} \doteq 0.75$ and $\varepsilon_{min} = 0.53$ (curve d). Interpolate linearly to $C_{min}/C_{max} = 0.5$:

$$\varepsilon \doteq (0.75 + 0.53)/2 \doteq 0.64.$$

Or use the exact formula, Eq. (3) of Table 10.2:

$$B = [1 + (0.5)^2]^{1/2} = 1.118,$$

$$\varepsilon = 2\left\{1 + 0.5 + (1.118)\frac{1 + \exp[-1.41(1.118)]}{1 - \exp[-1.41(1.118)]}\right\}^{-1} = 0.625.$$

The difference in these two estimates is only 2.4%. From the definition of effectiveness, Eqs. (10.16),

$$\varepsilon = 0.625 = \frac{(6270)(180^\circ\text{C} - T_{\text{h, out}})}{(6270)(180^\circ\text{C} - 20^\circ\text{C})} = \frac{(12{,}540)(T_{\text{c, out}} - 20^\circ\text{C})}{(6270)(180^\circ\text{C} - 20^\circ\text{C})}.$$

Solve for

$T_{\text{h, out}} = 80.0^\circ\text{C}$, [*Ans.*]

$T_{\text{c, out}} = 70.0^\circ\text{C}$. [*Ans.*]

These exactly match our given exit conditions in Example 10.3. Stop and imagine how relatively laborious it would have been to *guess* the outlet temperatures, evaluate the LMTD, read Fig. 10.9(a) for $F$, compare $q = UAF(\text{LMTD})$ to the fluid enthalpy change, and then repeat with a new guess. The NTU method of direct solution is a considerable advantage. ■

## Example 10.6

Air at 15°C enters a crossflow heat exchanger mixed and is heated by water entering an unmixed tube bank at 95°C. Their capacity rates are equal, $C_{air} = C_{water} = 2000$ W/°C, and $U = 70$ W/m$^2\cdot$°C. Using the NTU method, determine the required exchanger area in m$^2$ to heat the air to 60°C.

*Solution* Since $C_{min}/C_{max} = 1$, we are at minimum-effectiveness conditions — curve c in Fig. 10.10. The effectiveness is known:

$$\varepsilon = \frac{T_{c,\,out} - T_{c,\,in}}{T_{h,\,in} - T_{c,\,in}} = \frac{60 - 15}{95 - 15} = 0.5625.$$

From curve c of Fig. 10.10, read NTU $\doteq$ 1.8. For more accuracy, use Eq. (7) or (8) of Table 10.2 to compute NTU = 1.7526. Then, by definition,

$$\text{NTU} = 1.7526 = UA/C_{min} = (70\ \text{W/m}^2 \cdot {}^\circ\text{C})A/(2000\ \text{W}/{}^\circ\text{C}).$$

Solve for

$$A = 50\ \text{m}^2. \quad [\mathit{Ans.}]$$

This problem could also be solved by the LMTD method. First we would have to make a heat balance to compute the water exit temperature of 50°C. Then LMTD = 35°C, $R$ = 1.0, and $P$ = 0.56. From Fig. 10.9(d) read $F \doteq 0.73$. The area then follows from $q = UAF(\text{LMTD})$, giving $A \doteq 50\ \text{m}^2$. ■

## 10.5 Heat Exchanger Design

The LMTD and NTU methods of exchanger analysis are very useful concepts, but they do not normally result in complete "designs" of a system. In particular, they are based on three rather stringent assumptions:

1. that $U$ is constant through the exchanger;
2. that $U$ may be predicted from forced-convection correlations; and
3. that we are pleased with what we are designing.

In practice, $U$ is rarely constant along the exchanger surface. The closest to a constant-$U$ case is a double-pipe parallel or counterflow system (Fig. 10.2a) or a compact exchanger with straight, uniform flow passages (Fig. 10.2d).

Prediction of $U$ from theory and dimensional analysis can result in large uncertainty, except perhaps for circular double-pipe designs. Fortunately, there is a great deal of published data from compact exchanger designs, especially in [3]. Accurate correlations are rather lacking for large, complex, baffled-shell designs (Fig. 10.3). Manufacturers commonly piece together extensive test data into a large computer program for predicting the overall heat transfer coefficient of complicated geometries. If the spatial variation of $U$ can be predicted, it is clear from Eq. (10.9) that the proper "constant" $U$ to be used in the LMTD and NTU methods is the area-average:

$$\overline{U} = \frac{1}{A} \int U dA. \tag{10.23}$$

Even with this refinement, the final estimate can be subject to considerable uncertainty.

And we may not be pleased with our design. We may not even like the geometry we chose. While the average engineer may only be called upon to design a small, moderately efficient compact exchanger, specialized manufacturers require an extremely detailed design process. One need only glance over the 58 case studies in Kern's classic text [1] or the multiple configurations in the TEMA standards [8] to realize that effective designs do not spring forward overnight. There is constant redesign and reconfiguring to optimize the system among heat transfer, pressure drop, total size and volume, cost, reliability, and maintenance.

J. Taborek [13] gives an eloquent discussion of the shell-and-tube exchanger design process. Among the points made by Taborek are the following:

1. Cost questions relate not only to the fixed and operating costs of the device but also to the cost of the engineering design effort and the cost of a poor design, that is, the penalty that results if the expected performance is not achieved.
2. A given design can be either heat-transfer-limited or pressure-drop-limited. In turbulent flow, heat transfer rises approximately as $\dot{m}^{0.8}$, while pressure drop rises as $\dot{m}^{1.8}$ and pumping power as $\dot{m}^{2.8}$. It is thus easy to evolve a design that bumps up against a maximum power limit; such a design is usually a poor one and should be reconfigured. Further, if a design is severely heat-transfer-limited, it is usually well worth the effort to redesign for heat transfer improvements.
3. A crucial decision is the allocation of one fluid to the shell side and one fluid to the tube side. Generally the tube side should receive the fluid that is at higher pressure, is more corrosive, or has a high fouling tendency. The shell side should receive the more viscous fluid or the fluid that presents more pressure-drop difficulties. Finned tubes should be viewed with caution if heavy fouling is anticipated.
4. Even the best available heat transfer correlations result in an overall uncertainty of $\pm 30\%$ in $U$. The designer must be ready to compensate for such uncertainty.

5. If a design results in a correction factor $F$ that falls below the "elbows" of the curves in Fig. 10.9, it is almost always a poor design and can be improved.
6. Make sketches of temperature and flow distributions as you progress with the design. It is amazing how much you can learn from a simple graph. Figure 1 of Taborek's paper is a classic example that will strike any reader with its power to illuminate pressure-drop limits.
7. Commercial heat exchanger designs are making increasing use of massive computer programs. Taborek predicts that future designs may be performed entirely by computers, including plotter-generated shop drawings. This is an ominous prospect, since no human being can completely check such results. It will also lead to a decrease in the number of experienced human designers.

Perhaps a good compromise in future heat exchanger work will be a computer-aided design (CAD) backed up by experienced engineers who can choose proper parameters and options. The reader is encouraged to study the use of computers to organize and optimize design procedures. An excellent reference for the heat exchanger case is the text by Stoecker [14], which treats computer modeling, simulation, curve-fitting, optimization, search procedures, and dynamic/geometric/linear programming techniques. All of Stoecker's examples relate to heat transfer problems.

## Summary

Being an almost universal task for heat transfer engineers, the study of heat exchangers is an ideal application of the theory we learned in earlier chapters. The purpose of the present chapter has been to acquaint us with some concepts used in heat exchanger analysis.

The first thing we learned is that there are a great many different heat exchanger geometries. The design details can be obtained only from specialized monographs [1–8] and from the industrial literature. Next we saw that the determination of overall heat transfer coefficient for a complex exchanger is a formidable problem: The theory is limited, and experimental correlations are not widely published.

If, however, an estimate of average $U$ *can* be obtained, two useful concepts enable us to complete the analysis. The log-mean temperature difference (LMTD) and its geometrical correction factor, $F$, enable us to estimate exchanger area from known inlet and outlet conditions. Alternatively, the concept of effectiveness, $\varepsilon$, and number of transfer units, NTU, allow computation of outlet temperatures when the exchanger area and inlet temperatures are known.

Here we only introduced concepts, not actual designs. As Stoecker remarks [14], "Engineering students become proficient in solving homework problems which require 45 min., whereas most professional engineering problems require weeks or months to complete." Heat exchanger design is a professional problem.

## References


1. D. Q. Kern, *Process Heat Transfer,* McGraw-Hill, New York, 1950.
2. W. Hryniszak, *Heat Exchangers,* Butterworths, London, 1958.
3. W. Kays and A. L. London, *Compact Heat Exchangers,* 3rd ed., McGraw-Hill, New York, 1984.
4. A. P. Fraas and M. N. Özisik, *Heat Exchanger Design*, Wiley, New York, 1965.
5. N. Afgan and E. U. Schlünder, *Heat Exchangers: Design and Theory Sourcebook*, McGraw-Hill/Scripta, Washington, D.C., 1974.
6. D. Chisholm (Ed.), *Developments in Heat Exchanger Technology*, Applied Science Publishers, London, 1980.
7. S. Kakac, A. E. Bergles, and F. Mayinger (Eds.), *Heat Exchangers*, McGraw-Hill/Hemisphere, New York, 1981.
8. *Standards of the Tubular Exchanger Manufacturers Association*, 6th ed., TEMA, New York, 1978.
9. R. K. Shah, C. F. McDonald, and C. P. Howard (Eds.), *Compact Heat Exchangers,* ASME Symposium Volume HTD-10, 1980.
10. J. Taborek, T. Aoki, R. Ritter, J. Pallen, and J. Knudsen, "Fouling: The Major Unresolved Problem in Heat Transfer," *Chem. Eng. Prog.*, vol. 68, no. 2, 1972, p. 59; vol. 68, no. 7, 1972, p. 69.
11. R. A. Bowman, A. C. Mueller, and W. M. Nagle, "Mean Temperature Difference in Design," *ASME Transactions*, vol. 62, 1940, pp. 283–294.
12. J. M. Chenoweth and M. Impagliazzo (Eds.), *Fouling in Heat Exchange Equipment*, ASME Symposium Volume HTD-17, 1981.
13. J. Taborek, "Evolution of Heat Exchanger Design Techniques," *Heat Transfer Engineering*, vol. 1, no. 1, 1979, pp. 15–29.
14. W. F. Stoecker, *Design of Thermal Systems,* McGraw-Hill, New York, 1980.
15. E. U. Schlünder (Ed.), *Heat Exchanger Design Handbook*, 5 vols., Hemisphere Publishing, New York, 1982.

16. S. Kakac, R. K. Shah, and A. E. Bergles, *Low Reynolds Number Flow Heat Exchangers,* Hemisphere, New York, 1983.
17. J. Taborek, G. F. Hewitt, and N. Afgan, *Heat Exchangers: Theory and Practice*, Hemisphere, New York, 1983.
18. J. P. Gupta, *Fundamentals of Heat Exchanger and Pressure Vessel Technology*, Hemisphere, New York, 1986.

## Review Questions

1. Name five different types of heat exchangers.
2. Define the concept of a "compact" exchanger.
3. Define the overall heat transfer coefficient, $U$.
4. How does "fouling" enter into a heat exchanger analysis?
5. What is "mixed" versus "unmixed" fluid?
6. What is a regenerator? Name two applications.
7. What is the TEMA?
8. Name three different baffle designs for an exchanger.
9. What is the log-mean temperature difference?
10. Why do multipass heat exchangers need a correction factor for the LMTD?
11. Describe the "first" and "second" types of heat exchanger design problem.
12. For a given geometry, how does one estimate $U$?
13. Define effectiveness and number of transfer units.
14. Why is the parallel-flow case less effective by its very nature than a counterflow design?
15. Name some basic weaknesses of the LMTD and NTU approaches to heat exchanger design.
16. Name some parameters that one tries to optimize or otherwise match in a real heat exchanger design.
17. Does the CAD concept have a future in heat exchanger design?

## Problems

### Problem distribution by sections

| The Problem Assignments are Organized as Follows | | |
|---|---|---|
| Problems | Sections Covered | Topics Covered |
| 10.1–10.5 | 10.1 | Classification |
| 10.6–10.9 | 10.2 | Overall coefficient |
| 10.10–10.20 | 10.3 | Mean temperature difference |
| 10.21–10.34 | 10.4 | Effectiveness and NTU |
| 10.35–10.40 | All | Special Projects |

10.1 What type of heat exchanger is the truck radiator in Fig. 1.1?

10.2 Sketch the tube-pass patterns in the heat exchanger of Fig. 10.3(a).

10.3 Figure P10.3 shows a heat exchanger constructed of corrugated plate fins yielding triangular passages. If the triangles are equilateral, what is the maximum plate spacing $H$ for which the exchanger can be called "compact"?

10.4 What plate spacing in Fig. 10.5 would result in a smooth-plate area density of 200 $ft^2/ft^3$?

10.5 In Fig. 10.6 the tube diameter is 0.375 in., the fin length is 0.438 in., the fin width is 0.030 in., the fin thickness is 0.0065 in., and the fins are wrapped spirally with a pitch of 0.050 in. What is the area density? Is it a compact surface?

10.6 A heat exchanger 2 m long consists of a 4-cm-diameter pipe inside a 6-cm-diameter pipe. Water at an average of 40°C flows through the smaller pipe at 0.016 $m^3/s$. Unused engine oil at an average of 150°C flows through the annular region at 0.010 $m^3/s$. Neglect tube conduction re-

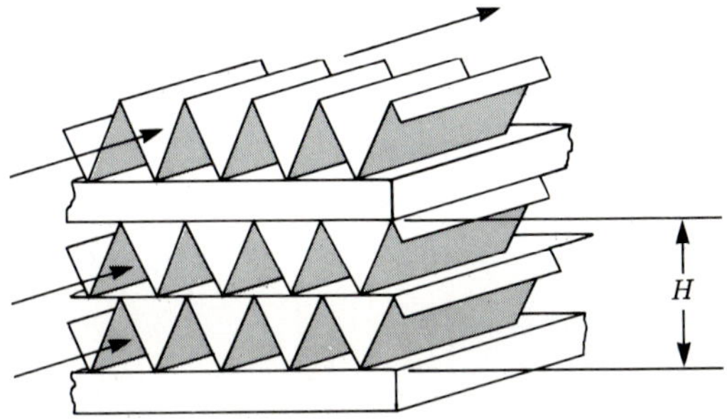


Figure P10.3

sistance and estimate the coefficient $U$ if the surface is (a) clean and (b) fouled.

10.7 Data from Kays and London [3] for the equilateral triangle passages of Fig. P10.3 result in the following approximate formulas for laminar and turbulent friction and heat transfer:

a) Laminar, Re < 1000:

$$c_f \doteq 13.5/\text{Re}$$

$$\text{St Pr}^{2/3} \doteq 2.7/\text{Re}$$

b) Turbulent, Re > 2000:

$$c_f \doteq 0.0654\ \text{Re}^{-0.24}$$

$$\text{St Pr}^{2/3} \doteq 0.00523\ \text{Re}^{-0.063}$$

where St, $c_f$, and Re are based on average velocity and hydraulic diameter of the passages. Suppose that the plate spacing is $H = 12$ mm and the passage length is $L = 200$ mm. Let one layer of 100 passages contain carbon dioxide at 1 atm and 40°C at a total flow rate of 160 kg/hr. The other layer of 100 passages contains helium at 1 atm and 0°C at a flow rate of 45 kg/hr. Neglecting plate conduction, estimate the overall heat transfer coefficient $U$ for clean surfaces. Also estimate the total pressure drop for each gas.

10.8 Air at an average temperature of 30°C cools ethylene glycol at an average temperature of 120°C in the parallel plate geometry of Fig. P10.8. The plates are 1 cm apart, 1 m wide, and 1 m long. The air flows at 0.2 m$^3$/s and the liquid at 0.05 m$^3$/s. Estimate $U$ if the plate surfaces are (a) clean and (b) fouled.

10.9 For the crossflow geometry of Fig. 10.2(d), assume that the passages are 1 cm by 1 cm by 1 m long. Air at an average velocity of 22 m/s and 300°C heats water at an average velocity of 1 m/s and 30°C. Estimate $U$ for clean surfaces.

10.10 In a double-pipe heat exchanger, water enters at 10°C and leaves at 60°C, cooling unused engine oil, which enters at 120°C and leaves at 65°C. Compute the LMTD for (a) parallel flow and (b) counterflow, and comment. Suppose that the oil flow rate is 0.5 kg/s, and $U = 225$ W/m$^3$ · °C. Compute (a) the mass flow of water, and the exchanger area required for (b) parallel flow and (c) counterflow.

10.11 Water at 90°C enters a thin 2.5-cm-diameter copper tube at a rate of 300

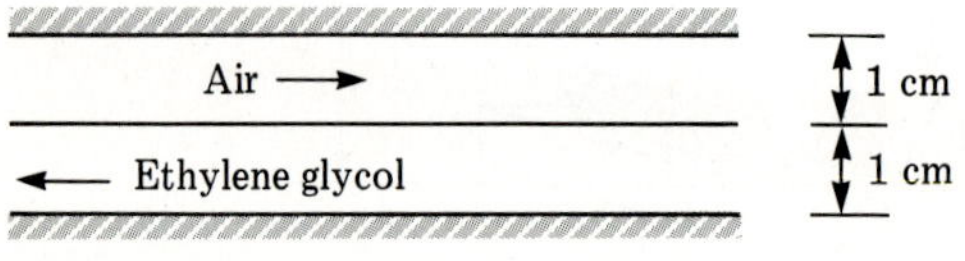


**Figure P10.8**

kg/hr. It is coiled by air at 10°C flowing across the 1-m-long tube, as in Fig. P10.11. Using convection theory, estimate (a) the overall heat transfer coefficient $U$ and (b) the exit temperature $t_2$ of the water.

10.12 A double-pipe counterflow heat exchanger has inner and outer diameters of 3 and 6 cm, respectively. Ethylene glycol, flowing in the outer annular region at 1.1 kg/s, is to be cooled from 100°C to 80°C by water, which enters the inner pipe at 5°C and leaves at 55°C. Estimating $U$ from convection theory, compute (a) the water mass flow rate and (b) the required pipe length.

10.13 A double-pipe heat exchanger is to cool argon from 200°C to 120°C, flowing at 100 kg/hr, using air, which enters at 0°C and flows at 45 kg/hr. If $U = 55\ \mathrm{W/(m^2 \cdot {}^\circ C)}$, compute the required surface area between the two fluids.

10.14 Oil ($c_p = 2150\ \mathrm{J/kg \cdot {}^\circ C}$) flows at 2.5 kg/s and is cooled in two tube passes from 100°C to 80°C by air that enters one shell pass at 15°C and 2.0 kg/s. If $U = 140\ \mathrm{W/m^2 \cdot {}^\circ C}$, use the LMTD/$F$ method to estimate the required exchanger area.

10.15 Air enters the shell of a double-pipe heat exchanger at 0.15 kg/s and 300°C and leaves at 200°C. Cooling water enters the 2-cm-diameter inner tube at 20°C and 0.09 kg/s. If $U = 275\ \mathrm{W/m^2 \cdot {}^\circ C}$, estimate the required pipe length for (a) parallel flow and (b) counterflow.

10.16 Saturated steam at 1 atm is condensed in a heat exchanger with one shell pass and two tube passes. Cooling water enters the tubes at 10°C and 1.2 kg/s and leaves at 40°C. If $U = 1200\ \mathrm{W/m^2 \cdot {}^\circ C}$, compute (a) the required heat exchanger area and (b) the amount of steam condensed per hour.

10.17 Water, flowing at 0.6 kg/s on the shell side of a two-shell-pass, four-tube-pass heat exchanger, is to be cooled from 20°C to 8°C by a brine solution which enters the tubes at 0°C and leaves at 6°C. If $U = 450\ \mathrm{W/(m^2 \cdot {}^\circ C)}$, estimate the required heat exchanger surface area.

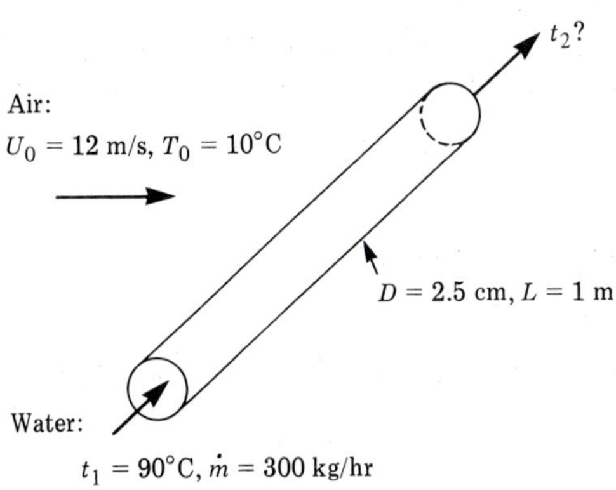


**Figure P10.11**

10.18 A crossflow exchanger uses mixed air at 4 kg/s that enters at 15°C and leaves at 45°C. The air cools unmixed water, which enters at 95°C and 0.7 kg/s. If $U = 150\ \text{W/m}^2 \cdot {}^\circ\text{C}$, what exchanger area is required?

10.19 Cold water at 4 kg/s is to be heated from 20°C to 45°C on the tube side of an exchanger. There are 50 tubes per pass of diameter 2 cm. On the shell side is hot water that enters at 2.5 kg/s and 90°C and makes one pass. If $U = 1100\ \text{W/m}^2 \cdot {}^\circ\text{C}$ and the maximum allowable tube length is 2 m, compute (a) the proper number of tube passes and (b) the proper tube length.

10.20 A single-pass crossflow heat exchanger similar to Fig. 10.2(d) uses air entering at 0.16 kg/s and 350°C and leaving at 230°C to heat water entering the other level at 0.25 kg/s and 20°C. The top and bottom of the exchanger are insulated. Each small passage in the exchanger is 1 cm by 1 cm and of length $L$. Using smooth-duct forced convection correlations to estimate $U$, with negligible wall conduction resistance, find the proper length $L$ for the exchanger. [*Hint:* $L$ is of the order of 1 meter.]

10.21 Derive the effectiveness of a counterflow exchanger, Eq. (2) of Table 10.2, by assuming $C_{min}$ is the hot fluid.

10.22 The effectiveness and LMTD/$F$ methods are really two sides of the same coin, namely, heat balances combined with integration of a variable temperature difference between two fluids. Prove this by showing that Eq. (10.13) for one shell pass and two tube passes is exactly equivalent to its effectiveness twin, item 3 of Table 10.2. Take the special case $R = 0.4$, $T_1 = 100°\text{C}$, and $t_1 = 0°\text{C}$, varying $t_2$ from, say, 10°C to 70°C with the requirement that $T_2 = 100 - 0.4t_2$ to keep $R$ constant. Compute $F$ from Eq. (10.13) and use this to compute NTU and hence compare directly with the effectiveness computed from Table 10.2.

10.23 Figure 10.10 contains a curve e labeled "crossflow, both fluids mixed." Does this make sense? What might such a system look like? Why does it have such a low effectiveness relative to the crossflow with both fluids unmixed? For a numerical example, consider Problem 10.8, for which the answer to part (a) was $U \doteq 87\ \text{W/(m}^2 \cdot {}^\circ\text{C)}$. Suppose that the air and glycol are in *crossflow* rather than the counterflow conditions shown in Fig. P10.8. If the air enters at 20°C and the ethylene glycol enters at 150°C, use the effectiveness method to compute (a) the air outlet temperature; (b) the glycol outlet temperature; and (c) the total heat transferred.

10.24 Show that the maximum effectiveness relation, Eq. (10.22), and the exponential temperature distribution of Eq. (6.54) are essentially equivalent formulas.

10.25 As was demonstrated in Problem 10.22, the LMTD/$F$ and effectiveness methods are equivalent concepts and either can be used to solve any heat exchanger problem. In Problem 10.14 we used the LMTD/$F$ method to compute $A \doteq 18\ \text{m}^2$. Now solve the same problem by the effectiveness method, for which the unknown will be the NTU of the exchanger.

10.26 A one-shell-pass, two-tube-pass exchanger uses water entering the shell at 1.2 kg/s and 30°C. Oil ($c_p = 2100$ J/kg · °C) enters the tubes at 2.0 kg/s and 120°C. If $U = 275\ \text{W/m}^2 \cdot °\text{C}$ and $A = 20\ \text{m}^2$, estimate the outlet water and oil temperatures.

10.27 In Problem 10.26, if the exchanger area can be expanded without limit, what is the maximum possible outlet water temperature and minimum oil outlet temperature? What is the value of $F$ for this case?

10.28 A crossflow exchanger with both fluids unmixed uses air entering at 300°C and 0.6 kg/s to heat water entering at 20°C. If $U = 180\ \text{W/m}^2 \cdot °\text{C}$ and $A = 7\ \text{m}^2$, what is the proper mimimum water flow rate to keep the exit water temperature below 90°C?

10.29 A crossflow exchanger preheats air for a boiler, using exhaust gases (air). The exchanger area is 12 m$^2$, with cold air entering at 25°C and 0.6 kg/s, and exhaust air entering at 400°C and 0.8 kg/s. If $U = 60\ \text{W/m}^2 \cdot °\text{C}$, compute the outlet air temperatures assuming that both fluids are unmixed.

10.30 Unused engine oil enters the double-pipe heat exchanger of Fig. P10.30 at 180°C and 0.005 kg/s. It is cooled by water that enters the inner pipe at 20°C and 0.001 kg/s. The pipes are 8 m long. Using convection correlations to estimate $U$, compute the outlet oil and water temperatures and the total heat transferred, assuming smooth walls with no fouling.

10.31 A counterflow exchanger uses hot air that enters at 0.8 kg/s and 250°C and leaves at 100°C. The air heats engine oil that enters at 35°C and leaves at 110°C. If $U = 85\ \text{W/m}^2 \cdot °\text{C}$, estimate the required exchanger area by the NTU method. What will be the exit oil temperature if the area is increased to 25 m$^2$?

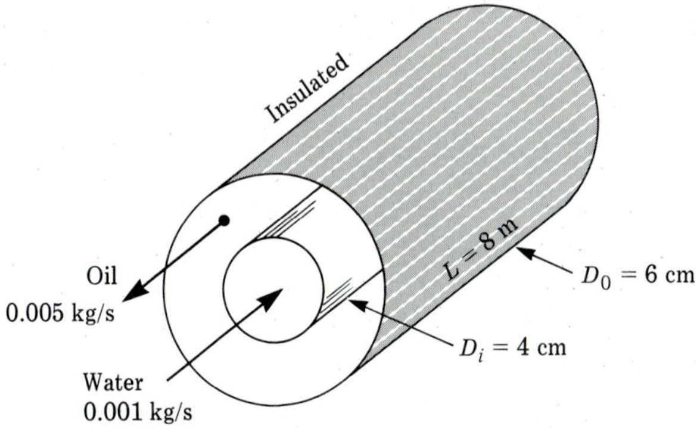


**Figure P10.30**

10.32 Oil is used to heat water in two shell passes and four tube passes. The oil ($c_p$ = 2100 J/kg · °C) enters at 2 kg/s and 150°C, while the water enters at 1.5 kg/s and 20°C. If $U$ = 700 W/m$^2$ · °C for an exchanger area of 15 m$^2$, estimate the outlet temperature of the water.

10.33 A single-pass shell-and-tube exchanger is designed to condense 18 kg/hr of saturated steam at 1 atm. Cooling water enters the tubes at 25°C and leaves at 60°C. If $U$ = 1800 W/m$^2$ · °C, what is the proper exchanger area?

10.34 It is desired to condense 1000 kg/hr of saturated steam at 100°C in the shell of a single-pass exchanger. The tubes are 2 cm in diameter and 3 m long and have a total fouling resistance of 0.0003 m$^2$ · °C/W. A total of 20 kg/s of water at 20°C is available to feed the tubes. If tube conduction is negligible and $U$ is estimated from condensation and forced-convection theory, how many tubes are needed?

### Special Projects

10.35 Figure P10.35 is a schematic of a heat exchanger with three shell passes and six tube passes, which is equivalent to a series of three exchangers with one shell pass and two tube passes. Its performance is therefore computed by applying the basic 1:2 exchanger formulas three times. For the LMTD method, with a given value of $R$ and $F$, Bowman [11] recommends that the value of $P$ for the 3:6 exchanger is given by

$$P_{36} = \frac{X^3 - 1}{X^3 - R}, \qquad X = (1 - P_{12}R)/(1 - P_{12}), \tag{1}$$

where $P_{12}$ is the value of $P$ that matches $R$ and $F$ in the basic 1:2

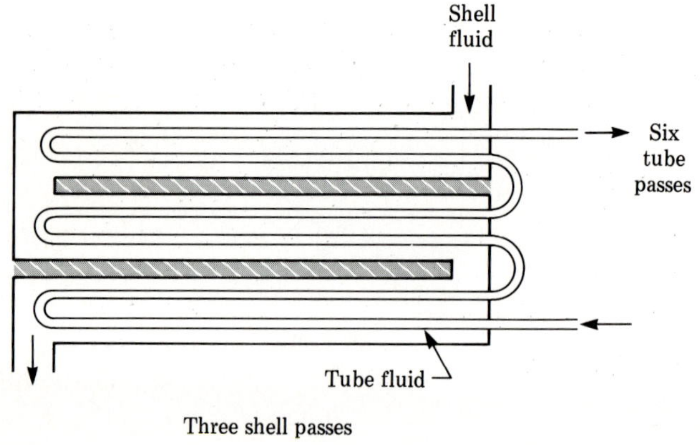


Figure P10.35

configuration, Eq. (10.13). For the effectiveness method, Kays and London [3] recommend the formula

$$\varepsilon_{36} = \frac{Y^3 - 1}{Y^3 - r}, \qquad Y = (1 - \varepsilon_{12} r)/(1 - \varepsilon_{12}), \tag{2}$$

where $r = C_{\min}/C_{\max}$ and $\varepsilon_{12}$ is the effectiveness of a 1:2 exchanger evaluated from item 3 of Table 10.2 at $r$ and $N/3$. First, show how Eqs. (1) and (2) above are derived. Then write a computer program to evaluate them and plot an LMTD chart similar to Fig. 10.9(b) and an effectiveness plot similar to Fig. 10.10 for this 3:6 exchanger. Discuss whether the 3:6 performance is better in any sense than the 1:2 configuration.

10.36 The LMTD and effectiveness formulas used in this chapter are based on simplifying assumptions: (a) constant $U$ independent of position and (b) constant fluid properties ($\rho$, $\mu$, $k$, $c_p$). Both assumptions may be inaccurate in practice. Figure P10.36 shows an unmixed crossflow exchanger similar to Fig. 10.2(d). Each layer has $N$ passages of square size, $h$ by $h$, and length $L = Nh$. Break the exchanger up into $N^2$ elements of size $h$ by $h$, with height equal to $2h$, the exchanger depth. Assume thin walls of negligible thermal resistance, with the top and bottom and closed sides insulated so that heat flows only between walls containing fluid on each side. Derive a heat balance for each arbitrary element $(I, J)$ in Fig. P10.36, with local heat transfer coefficients based on convection theory and local properties based on local mean fluid temperature. Combine these elements into a computer program that solves for the total heat transfer and outlet temperatures of the two fluids. For a numerical example, take $h = 1$ cm and $N = 50$, with fluid inlet data given by Problem 10.20: $T_{hi} = 350$°C and $\dot{m}_h = 0.16$ kg/s (air), $T_{ci} = 20$°C and $\dot{m}_c = 0.25$ kg/s (water). Use your computer program to calculate the outlet air and water temperatures and the overall heat transfer and compare with a simpler estimate based on the effectiveness formula, item 6 of Table 10.2.

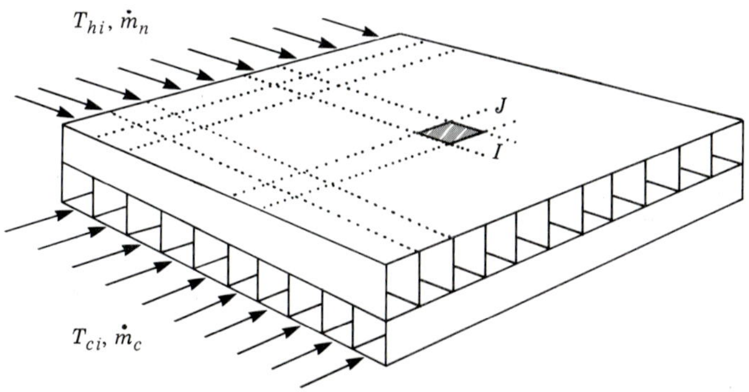


**Figure P10.36**

10.37 Figure P10.37 shows a compact heat exchanger made by stacking together a number of the two-layer crossflow exchangers of Fig. P10.36. Can this system be analyzed by the effectiveness method? Let the exchanger be a cube of side length 60 cm, with 1 cm by 1 cm passages. It uses exhaust gases at 330°C and 2.5 kg/s to preheat a fuel-air mixture entering at 10°C. For simplicity, model both gases as pure air. Compute (perhaps from a PC program) and plot the gas outlet temperatures as the fuel-air mass flow varies from 0.5 to 10 kg/s. Estimate $U$ from noncircular duct convection theory.

10.38 Attack the compact crossflow exchanger of Fig. P10.37 using the same computer-element technique proposed in Prob. 10.36. Now the element ($I$, $J$) will be $2Mh$ high, where $M$ is the number of two-layer systems stacked together. Is it permissible to assume a vertical periodicity, where, for example, each hot-fluid part of the element sees the same cold-fluid temperature above and below? Make a heat balance for this element and develop a computer program that computes fluid outlet temperatures and total heat transfer. Apply your program to the air-air data of Problem 10.37. If time is short, use only the single fuel-air mass flow rate of 5.0 kg/s, but plotting the results versus mass flow rate is very instructive.

10.39 Consider the crossflow heat exchanger shown in Fig. P10.39. Cooling water at 10°C is fed through an array of in-line 1-cm-diameter thin copper tubes to cool air that enters the open section at 230°C, 1.1 atm, and 2 kg/s. It is desired that the exit air temperature be as low as possible subject to the following constraints:

a) The size of the exchanger is fixed.
b) The maximum water flow rate is 70 kg/s.
c) The maximum water pumping power is 3 kW.
d) The maximum allowable air pressure drop is 20 kPa.

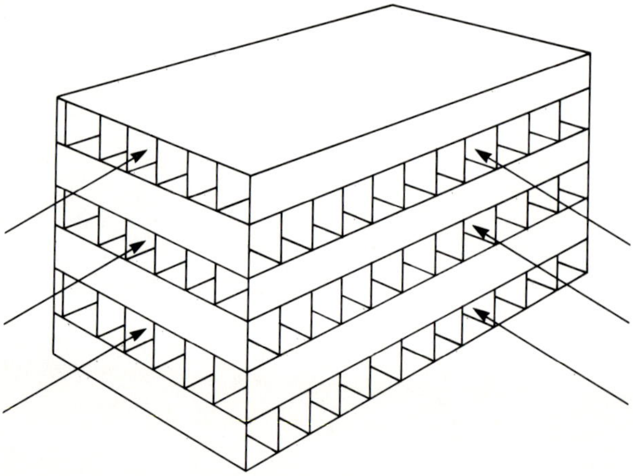

**Figure P10.37**

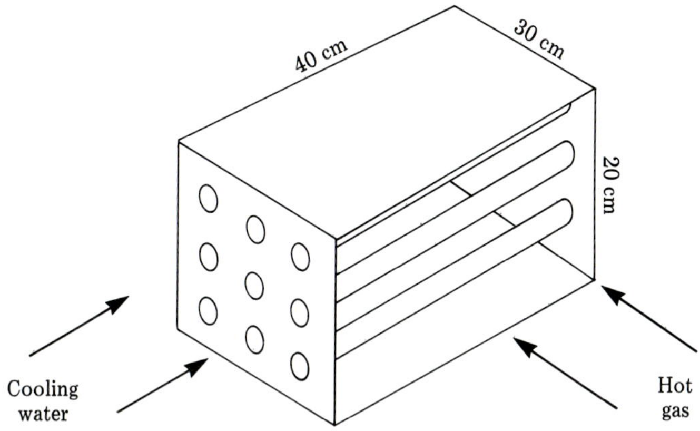


**Figure P10.39**

Any number of in-line tubes may be used, with the caution that large arrays will block the air flow and increase the pressure drop. Neglect fouling and tube conduction resistance. Use in-line array correlations from Section 6.7.1.

10.40 For a different kind of design problem, modify Problem 10.39 so that the hot gas is saturated steam at 1 atm. It is desired to maximize the amount of steam condensed on the cold-water tubes subject to the constraints of Problem 10.39. The steam flow rate is unknown and will balance the amount of steam condensed. Use condensation correlations to estimate the steam-side heat transfer.

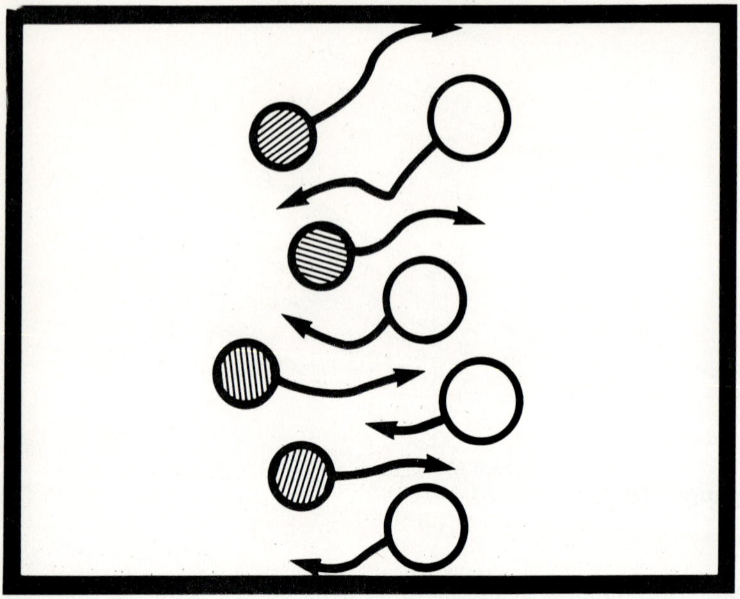

# Mass Transfer

11

## 11.1 Introduction

The phenomenon of mass transfer is familiar in daily life. A drop of dye in water spreads, dilutes, and soon colors the water in the entire container. An odor in one corner of a room is soon noticeable throughout the room. A wet surface soon dries out, even if there is no ambient air motion. Helium in a glass container seems to leak because it is migrating by mass transfer through the solid walls. All of these are examples of the *diffusion* of mass due to differences in concentration of substances in a mixture. Mass transfer tends toward making a mixture uniform, just as heat transfer tries to make the temperature uniform.

The term *mass transfer* is well accepted in the literature, although *mass diffusion* is also used. We may define the process as follows:

> *Mass transfer is a movement of particles of a given species through a mixture as a result of a gradient in the concentration of that species.*

We should not confuse this molecular process with the bulk *mass flow* such as $\dot{m} = \rho A V$ through a duct. In this chapter we must distinguish between the bulk or average flow of a mixture and the individual flows of its various species.

Many industrial operations involve mass transfer. Examples in the aerospace industry involve ablative, film, and transpiration cooling of vehicles and rocket engines. Air and water pollution processes are diffusion controlled. Bioengineering designs such as blood oxygenators, respirators, and artificial kidneys involve mass transfer. The chemical industries have perhaps the strongest mass transfer emphasis: separation, extraction, gas absorption, distillation, adsorption, crystallization, ore and isotope production, drying, humidification, chromatography, and sublimation all depend strongly upon mass transfer processes.

The mass transfer literature is enormous, including several advanced textbooks [1–3]. There are texts that emphasize chemical engineering [4–6] or mechanical engineering [7–8]. Specialized texts treat chemical reactions [9–10] and mathematical techniques [11]. Then, since there is a strong analogy among mass transfer, heat transfer, and fluid mechanics, a number of texts are devoted to treating all three topics together [12–16]. This chapter is clearly only a brief introduction to a broad subject.

### 11.1.1 Molecular Consideration

The analysis of mass transfer is firmly rooted in the kinetic theory of gases and liquids [17]. Consider a gas consisting of two types of molecules, A and B. Molecules have a random spectrum of velocities and directions and tend to collide with each other in random fashion. If an imaginary plane separates a region with high A concentration from one with fewer A molecules, more molecules from the A-rich area will cross that plane into the A-poor region because of the greater numbers involved in the random motion. This is mass transfer — migration of a given species from high to low concentration, in close analogy to the flow of heat from regions of high to low temperature.

A typical binary diffusion process is schematically illustrated in Fig. 11.1. The eye notes that the left side of the container is rich in the dark A molecules, while the right side contains mostly light B molecules. Mass transfer results even if the container mixture has no ordered motion, with A molecules moving to the right and B to the left at respective rates $j_{Ax}$ and $j_{Bx}$ (kg/s·m$^2$). If the process continues indefinitely, the final result will be a uniform concentration of both A and B, in this case about 50% each.

The kinetic theory correctly predicts that the mass transfer rate $j_{Ax}$ is proportional to the *gradient* of the concentration of A. This is called *Fick's law* and will be discussed in detail in Section 11.3.

Diffusion occurs in all phases of a substance, mass transfer rates

**Figure 11.1** **Illustration of mass transfer due to unequal concentrations of two gases in a container. At large times, in the absence of gas sources, $\omega_A$ and $\omega_B$ should approach uniformity.**

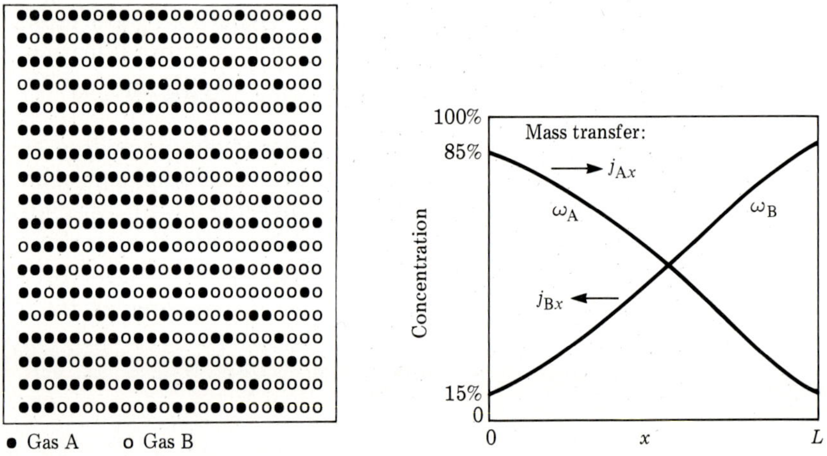

in gases being somewhat higher than in liquids. Solids may also diffuse into each other [18, 23] but usually at much smaller rates. There are also cases of gases or liquids diffusing through solids, such as helium migrating through the walls of a container.

As in heat transfer, mass transfer may be greatly enhanced by convection, especially if the bulk flow is turbulent. We then speak of concepts such as *eddy diffusion,* representing the mean mixing effect of turbulent eddies. Convective mass transfer is strongly analogous to convective heat transfer, and many of the same formulas are used.

Although many important applications involve multicomponent mixtures [4], this chapter will primarily emphasize *binary diffusion* in a mixture of two fluids, A and B.

## 11.2 Mass Transfer Definitions

Mass transfer concerns mixtures of two or more species. We therefore need rules for defining and summing species properties and relating them to average or bulk parameters. Unfortunately, this has led to an excessive number of different approaches to the subject. Here we define *mass* and *molar* averages and emphasize the former in most cases. There are still other averaging techniques such as the number of particles or number-density concepts [7]. The reader is advised to study the following definitions carefully to avoid confusion with the many different forms of mass transfer formulas in this chapter and in the literature.

### 11.2.1 Mass Averages

Consider a mixture of $n$ fluids, with individual total masses $m_1$, $m_2$, $m_3$, . . . , $m_n$, contained in a volume $v$. The bulk density of this mixture would be defined by

$$\rho = m/v, \qquad \text{where } m = \sum_{i=1}^{n} m_i. \tag{11.1}$$

In similar manner the density of any particular species is

$$\rho_i = m_i/v, \tag{11.2}$$

so the bulk density may also be defined as a summation:

$$\rho = \sum_{i=1}^{n} \rho_i. \tag{11.3}$$

Finally, the *mass fraction* of any species $i$ is convenient:

$$\omega_i = m_i/m = \rho_i/\rho. \tag{11.4}$$

It follows that the sum of the mass fractions is unity:

$$\sum_{i=1}^{n} \omega_i = 1.$$

Flow parameters are also related to mass averages. In particular, each species in a flowing mixture moves with a different velocity $\mathbf{V}_i$ and different mass flow rate $\rho_i \mathbf{V}_i$. We define the bulk mass flow per unit area as

$$\mathbf{G} = \rho \mathbf{V} = \sum_{i=1}^{n} \rho_i \mathbf{V}_i. \tag{11.5}$$

The notation $G = \dot{m}''$ was introduced for convection in Eq. (6.46). The bulk velocity $\mathbf{V}$ is thus defined as

$$\mathbf{V} = \frac{1}{\rho} \sum_{i=1}^{n} \rho_i \mathbf{V}_i = \sum_{i=1}^{n} \omega_i \mathbf{V}_i. \tag{11.6}$$

Since $\mathbf{V}$ is weighted by the constituent mass fractions, it is called a *mass-averaged* velocity. This concept is illustrated in Fig. 11.2 for the $x$ direction. It is the *difference* $(u_i - u)$, which is called the diffusion

**Figure 11.2 Illustration of individual, bulk, and diffusion velocities of a binary mixture of fluids A and B.**

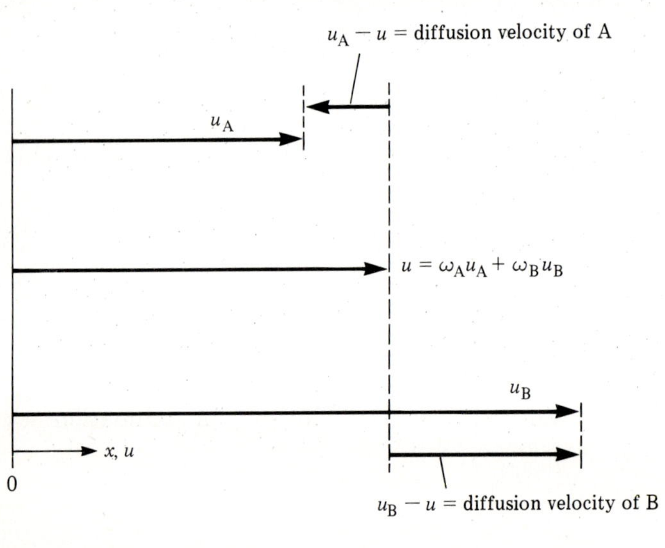

velocity of component $i$. Of more interest is the *diffusion mass flux* of species $i$:

$$j_{ix} = \rho_i (u_i - u)$$

or (11.7)

$$\mathbf{j}_i = \rho_i (\mathbf{V}_i - \mathbf{V}).$$

This concept is also shown in Fig. 11.2. It is $j_{ix}$, which is proportional to the $x$-gradient of the concentration of $i$. It follows from Eq. (11.5) that the net diffusion mass flux is zero:

$$\sum_{i=1}^{n} \mathbf{j}_i = 0.$$

However, because of differing species densities, the net diffusion velocity will not generally be zero and is not of particular interest anyway.

Note that bulk velocity as defined by Eq. (11.6) is consistent with single-component and uniform-mixture flows, as used in Chapters 1–10. When we referred there to the "velocity" of air, which is a complex but uniform mixture, we were using the mass-average definition from Eq. (11.6). Finally, note that $\mathbf{V}$ and $\mathbf{V}_i$ are velocities relative to a fixed coordinate system.

### 11.2.2 Molar Averages

A second approach, popular in the chemical engineering literature, is to average over the number of mols of a given species of molecular weight $M_i$:

$$N_i = m_i / M_i. \tag{11.8}$$

Then the *molar density* of species $i$ is defined as

$$C_i = N_i / v = \rho_i / M_i \tag{11.9}$$

with units of (kmols/m$^3$). Thus the bulk molar density of the mixture is

$$C = \sum_{i=1}^{n} C_i = \frac{N}{v}, \qquad N = \sum_{i=1}^{n} N_i.$$

For averaging purposes it is convenient to define a *mole fraction* $x_i$ of a species:

$$x_i = C_i / C, \qquad \text{where} \sum_{i=1}^{n} x_i = 1. \tag{11.10}$$

Then the mean molecular weight of a mixture is given by

$$M = \sum_{i=1}^{n} x_i M_i. \tag{11.11}$$

Similarly, the *molar average velocity* is

$$\mathbf{V}^* = \sum_{i=1}^{n} x_i \mathbf{V}_i, \tag{11.12}$$

and the *molar diffusion flux* is given by

$$\mathbf{J}_i^* = C_i(\mathbf{V}_i - V^*) \tag{11.13}$$

with units of (kmols/s·m$^2$). As with mass diffusion, $\Sigma \mathbf{J}_i^* = 0$.

Note that $V^*$ does not equal the mass-average $V$ in general and would be awkward in a mass-balance equation. However, $V^*$ and $V$ would be nearly equal in very dilute concentrations or in mixtures of nearly equal molecular weight.

We may convert from mass-fraction to mole-fraction as follows:

$$\omega_i = \frac{x_i M_i}{\sum_j x_j M_j}; \qquad x_i = \frac{\omega_i / M_i}{\sum_j \omega_j / M_j}. \tag{11.14}$$

These formulas are useful when handling data averaged in different ways.

### 11.2.3 Ideal Gas Simplifications

In a mixture of ideal gases, each species $i$ has a *partial pressure* $p_i$ that satisfies the ideal state law:

$$p_i = m_i R_i T = N_i \overline{R} T \tag{11.15}$$

where $\overline{R} = 8310\ \mathrm{m^2/(s^2 \cdot K)}$ is the universal gas constant and $R_i = \overline{R}/M_i$. Dalton's law states that the mixture pressure is the sum of the partial pressures and the mixture satisfies an overall ideal gas law:

$$p = \sum_i p_i = \sum_i m_i R_i \frac{T}{v} = \frac{m}{v} RT. \tag{11.16}$$

Comparing (11.15) and (11.16), we see that

$$p_i / p = x_i.$$

The partial pressure ratio equals the mole fraction. Also,

$$R = \sum_i \omega_i R_i. \tag{11.17}$$

We will use these relations to describe the local concentration of a mixture of gases undergoing convection and diffusion.

### Example 11.1

Air can be approximated as a mixture of 22% oxygen, 77% nitrogen, and 1% argon by weight. What are its mole fractions? What are the mixture gas constant and molecular weight?

*Solution* The phrase "by weight" means that we are given the mass fractions of the following mixture:

$$\omega_{O_2} = 0.22, \qquad \omega_{N_2} = 0.77, \qquad \omega_{Ar} = 0.01.$$

The molecular weights of these gases are in Appendix K:

$$M_{O_2} = 32.00, \qquad M_{N_2} = 28.02, \qquad M_{Ar} = 39.94.$$

The species gas constants follow from $R_i = \overline{R}/M_i$:

$$R_{O_2} = 259.7, \qquad R_{N_2} = 296.2, \qquad R_{Ar} = 208.1\,\frac{\text{m}^2}{\text{s}^2\cdot\text{K}}.$$

The mole fractions are computed from Eq. (11.14). First,

$$\sum \omega_j/M_j = 0.22/32.00 + 0.77/28.02 + 0.01/39.94 = 0.03461.$$

Then

$$\left.\begin{aligned} x_{O_2} &= \frac{0.22/32.00}{0.03461} = 0.1987 \\ x_{N_2} &= \frac{0.77/28.02}{0.03461} = 0.7941 \\ x_{Ar} &= \frac{0.01/39.94}{0.03461} = \underline{0.0072} \end{aligned}\right\} \quad [Ans.]$$

$$\sum: 1.0000 \quad \text{(check)}$$

The gas constant is computed from Eq. (11.17):

$$R = (0.22)(259.7) + (0.77)(296.2) + (0.01)(208.1) = 287.5\,\frac{\text{m}^2}{\text{s}^2\cdot\text{K}} \qquad [Ans.]$$

and, from Eq. (11.11),

$$M = (0.1987)(32.00) + (0.7941)(28.02) + (0.0072)(39.94) = 28.90 \qquad [Ans.]$$

As a check, compute $RM = (287.5)(28.90) = 8310 = \overline{R}$ (check). ∎

## 11.3 Fick's Law of Diffusion

As predicted by molecular theory, the diffusion mass flow is proportional to the gradient of the concentration of the species. This concept was first proposed by Adolph Fick in 1855 [19]. Fick's law may be written in vector form, using either mass or molar averages:

$$\begin{aligned} &\text{a) mass:} \quad \mathbf{j}_A = -\rho D_{AB}\nabla\omega_A = \rho_A(\mathbf{V}_A - \mathbf{V}), \\ &\text{b) molar:} \quad \mathbf{J}_A^* = -CD_{AB}\nabla x_A = C_A(\mathbf{V}_A^* - \mathbf{V}^*), \end{aligned} \tag{11.18}$$

for species A diffusing through medium B. The quantity $D_{AB}$ has units ($m^2/s$) and is called the *molecular diffusion coefficient* (or, alternatively, the *mass diffusivity*). It is a property of temperature and pressure and the mixture components A and B. The last terms on the right of (11.18) are to remind us that diffusion is proportional to the difference between species velocity and mixture velocity.

There is a strong analogy between Fick's law and Fourier's law for heat conduction:

$$\mathbf{q}'' = -k\nabla T, \tag{11.19}$$

although the units of $D_{AB}$ and $k$ are not the same. There is a weak analogy with the newtonian shear law and simple shear flows such as boundary layers:

$$\tau_{xy} = \mu\frac{\partial u}{\partial y}.$$

The shear stress $\tau$ is a tensor, not a vector, and the gradient $\partial u/\partial y$ is part of a more complex formulation that is not related to the vector operator $\nabla$. We cautioned against a shear/conduction analogy in Section 1.3.6, and we caution in this chapter against a shear/diffusion analogy.

By summing over the species in Eqs. (11.18) and recalling that $\Sigma\mathbf{j}_i = \Sigma\mathbf{J}_i^* = 0$ and $\Sigma\omega_i = \Sigma x_i = 1$, we find that

$$D_{AB} = D_{BA}. \tag{11.20}$$

There is reciprocity: Only a single binary diffusion coefficient needs to be measured for two given species.

This text considers only concentration-gradient diffusion. Gradients in temperature and pressure may also cause mass transfer, and concentration gradients may cause heat transfer (the Dufour effect). These effects are usually small and will be neglected here.

### 11.3.1 Dimensionless Diffusivity Ratios

The molecular diffusion coefficient $D_{AB}$ has the same units as kinematic viscosity, $\nu = \mu/\rho$, and thermal diffusivity, $\alpha = k/\rho c_p$.

Their ratios have been defined and named and are very useful dimensionless fluid properties:

$$\begin{aligned} &\text{a) Prandtl number:} && \mathrm{Pr} = \nu/\alpha = \mu c_p/k, \\ &\text{b) Schmidt number:} && \mathrm{Sc} = \nu/D_{\mathrm{AB}}, \\ &\text{c) Lewis number:} && \mathrm{Le} = \alpha/D_{\mathrm{AB}} = \mathrm{Sc}/\mathrm{Pr}. \end{aligned} \tag{11.21}$$

We saw in Chapter 6 that the Prandtl number correlates convective heat transfer data. In an exactly analogous manner the Schmidt number correlates convective mass transfer data. Meanwhile, the Lewis number is useful in problems involving both convective heat and mass transfer.

Table 11.1 lists some values of Schmidt and Lewis numbers for gases in air and liquids in water for dilute mixtures at standard conditions. We see that gases have Schmidt and Lewis numbers near unity, indicating that momentum, heat, and mass transfer effects are of comparable magnitude. Liquids, however, have large Sc and Le, indicating small diffusion rates compared to momentum and heat transfer.

Solids may have extremely small diffusion rates. For example, for aluminum diffusing into copper, [18] lists a diffusion coefficient $D_{\mathrm{AB}} \doteq 1.3 \times 10^{-34}\ \mathrm{m^2/s}$.

### 11.3.2 Diffusivity Variations in a Binary Mixture

Experimentally, the diffusion coefficient is a function of temperature, pressure, and the relative concentrations of the components. Primary data sources are [20] and [21].

For binary gas mixtures, $D_{\mathrm{AB}}$ is nearly independent of concentration but increases sharply with temperature and decreases nearly inversely with the pressure. If the diffusivity is known at pressure $p_0$ and temperature $T_0$, a suggested curve-fit expression is

$$D_{\mathrm{AB}}(p, T) \doteq D_{\mathrm{AB}}(p_0, T_0)(T/T_0)^m (p/p_0)^{-n}, \tag{11.22}$$

where $m \doteq 1.8$ and $n \doteq 1.0$. Appendix L lists values of $m$ for various gas mixtures. The value $n = 1.0$ is appropriate for monatomic and diatomic gases, but more complex gases such as methane may show differences. For methanol vapor diffusing in nitrogen, for example, $n \doteq 0.7$ [20].

Figure 11.3 shows data for mass diffusivity of three dilute gases in air. The agreement of the power law (11.22) with the measurements is excellent. Note that Eq. (11.22) requires absolute temperature for evaluation.

The diffusivity of liquid mixtures is a strong function of concen-

**Table 11.1** Mass diffusivities of dilute mixtures at 20°C and 1 atm

| I. Dilute Gases in Air | | | |
|---|---|---|---|
| **Gas** | $\mathcal{D}_{AB}$(m²/s) | **Sc** | **Le** |
| $H_2$ | $7.0 \times 10^{-5}$ | 0.21 | 0.30 |
| $O_2$ | $2.0 \times 10^{-5}$ | 0.74 | 1.0 |
| $N_2$ | $2.0 \times 10^{-5}$ | 0.74 | 1.0 |
| He | $7.1 \times 10^{-5}$ | 0.21 | 0.30 |
| Ar | $2.0 \times 10^{-5}$ | 0.74 | 1.0 |
| $CO_2$ | $1.6 \times 10^{-5}$ | 0.94 | 1.3 |
| CO | $2.0 \times 10^{-5}$ | 0.72 | 1.0 |
| $H_2O$ | $2.5 \times 10^{-5}$ | 0.60 | 0.84 |
| $SO_2$ | $1.2 \times 10^{-5}$ | 1.25 | 1.75 |
| NO | $1.7 \times 10^{-5}$ | 0.88 | 1.2 |
| Acetone | $1.2 \times 10^{-5}$ | 1.2 | 1.7 |
| Ammonia | $2.4 \times 10^{-5}$ | 0.63 | 0.88 |
| Benzene | $0.9 \times 10^{-5}$ | 1.7 | 2.3 |
| Ethanol | $1.2 \times 10^{-5}$ | 1.2 | 1.7 |
| Methanol | $1.5 \times 10^{-5}$ | 0.98 | 1.4 |
| Naphthalene | $0.6 \times 10^{-5}$ | 2.5 | 3.5 |

| II. Dilute Liquid Solutions in Water | | | |
|---|---|---|---|
| **Liquid** | $\mathcal{D}_{AB}$(m²/s) | **Sc** | **Le** |
| $H_2$ | $5.4 \times 10^{-9}$ | 190 | 27 |
| $O_2$ | $2.1 \times 10^{-9}$ | 490 | 70 |
| $N_2$ | $1.8 \times 10^{-9}$ | 570 | 80 |
| $CO_2$ | $1.9 \times 10^{-9}$ | 540 | 75 |
| $Cl_2$ | $1.2 \times 10^{-9}$ | 850 | 120 |
| NaCl | $1.4 \times 10^{-9}$ | 730 | 100 |
| Ammonia | $1.8 \times 10^{-9}$ | 570 | 80 |
| Glycerol | $0.9 \times 10^{-9}$ | 1150 | 160 |
| Ethanol | $1.1 \times 10^{-9}$ | 930 | 130 |
| Methanol | $1.3 \times 10^{-9}$ | 790 | 110 |

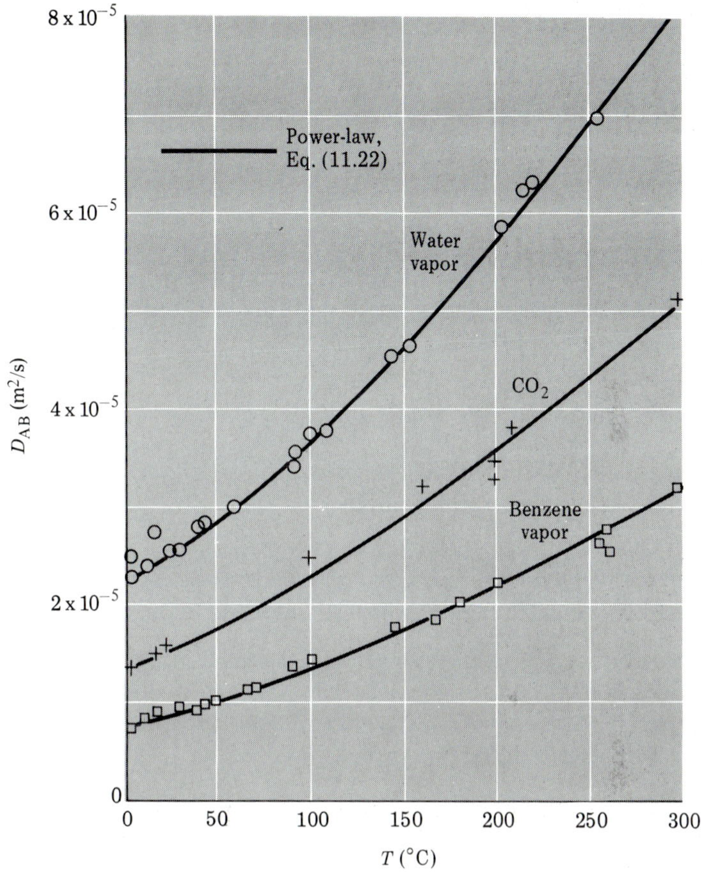


**Figure 11.3** Mass diffusivities of three dilute gases in air at 1 atmosphere.

tration and usually increases with temperature with little pressure dependence. There is an interesting review of diffusivity variations in Chapter 16 of [12].

### 11.3.3 Flux Relations for Binary Mixtures

The examples in this chapter will concern a binary mixture of two fluids, A and B. We list here some useful relations for binary mass transfer.

First, the total diffusion flux vanishes:

$$\mathbf{j}_A + \mathbf{j}_B = 0, \qquad \mathbf{J}_A^* + \mathbf{J}_B^* = 0. \tag{11.23}$$

Second, the average velocities are given by

$$\mathbf{V} = \omega_A \mathbf{V}_A + \omega_B \mathbf{V}_B, \qquad \mathbf{V}^* = x_A \mathbf{V}_A^* + x_B \mathbf{V}_B^*. \tag{11.24}$$

Finally, the absolute mass flow or molar flow of a given species is the sum of the diffusion flux plus the convective contribution:

$$\mathbf{G}_A = \rho_A \mathbf{V}_A = \mathbf{j}_A + \rho_A \mathbf{V} \qquad (\text{kg/s·m}^2)$$

$$\mathbf{G}_A^* = C_A \mathbf{V}_A^* = \mathbf{J}_A^* + C_A \mathbf{V}^* \qquad (\text{kmols/s·m}^2)$$

These may be converted to an interesting form involving mass and molar fractions by substituting for diffusion flux from Eq. (11.18) and for bulk velocity from Eq. (11.24):

$$\begin{aligned} \mathbf{G}_A &= -\rho D_{AB} \nabla \omega_A + \omega_A(\mathbf{G}_A + \mathbf{G}_B), \\ \mathbf{G}_A^* &= -C D_{AB} \nabla x_A + x_A(\mathbf{G}_A^* + \mathbf{G}_B^*). \end{aligned} \tag{11.25}$$

A special case will be diffusion through a *stagnant medium,* for which average velocities vanish and $\mathbf{G}_A = -\mathbf{G}_B$.

### Example 11.2

Helium is diffusing through the 4-mm-thick glass walls of a container. The helium concentration on the inside surface is $C_{A1} = 0.2$ kmols/m$^3$, and the outside surface has $C_{A2} = 0$. Estimate the flow of helium.

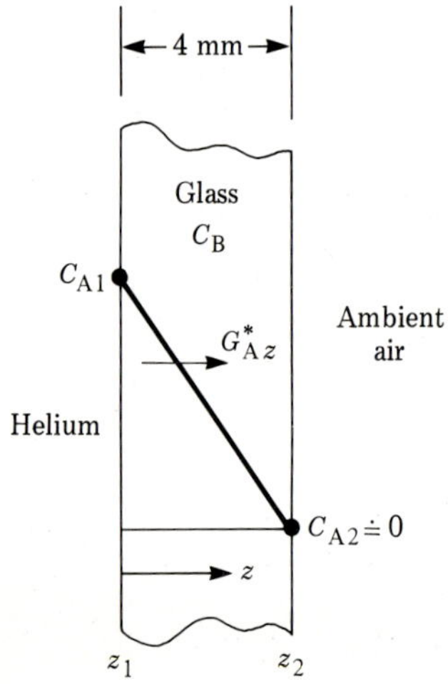


Figure E11.2

*Solution* For helium, $M_A = 4.003$; for glass, $M_B = 60$ and $\rho_B = 2600\ \text{kg/m}^3$. Thus $C_B = \rho_B/M_B = 2600/60 \doteq 43\ \text{kmols/m}^3$. Since $x_A = C_A/C_B = 0.2/43 = 0.0046$, the second or "convection" term in Eq. (11.25) will be neglected. From Appendix M the diffusivity for helium in glass is approximately $2 \times 10^{-12}\ \text{m}^2/\text{s}$. From Eq. (11.25) the diffusion rate on a molar basis is

$$G^*_{Az} \doteq -CD_{AB}\frac{dx_A}{dz} \doteq -D_{AB}\frac{dC_A}{dz} \doteq -D_{AB}\frac{C_{A2} - C_{A1}}{z_2 - z_1}.$$

For the given numerical values,

$$G^*_{Az} = -(2 \times 10^{-12})\left(\frac{0 - 0.2}{0.004 - 0}\right) = 1 \times 10^{-10}\ \text{kmols/s} \cdot \text{m}^2.$$

The mass diffusion rate is

$$G_{Az} = M_A G^*_{Az} = (4.003)(1 \times 10^{-10}) = 4 \times 10^{-10}\ \text{kg/s} \cdot \text{m}^2. \quad (\textit{Ans.})$$

If the container area is 1 $\text{m}^2$, it would take 29 days for one gram of helium to diffuse out of the container. ■

## 11.4 The Differential Equation of Species Conservation

In a flowing and diffusing mixture the number of molecules of species A is conserved except where mass of A is created by a chemical reaction occurring in the flow. We may therefore develop a species-conservation equation for each component.

Figure 11.4 shows flow of species A into and out of a two-dimensional control volume of unit depth in the $z$ direction. The net mass flow in must equal the rate of storage of A plus its generation within the volume:

$$\rho_A u_A\, dy + \rho_A v_A\, dx - \left[\rho_A u_A + \frac{\partial}{\partial x}(\rho_A u_A)\, dx\right] dy - \left[\rho_A v_A + \frac{\partial}{\partial y}(\rho_A v_A)\, dy\right] dx = \frac{\partial \rho_A}{\partial t}\, dx\, dy + \dot{m}'''_A\, dx\, dy.$$

After cancellation and rearrangement we obtain one form of the species conservation relation for two-dimensional flow:

$$\frac{\partial \rho_A}{\partial t} + \frac{\partial}{\partial x}(\rho_A u_A) + \frac{\partial}{\partial y}(\rho_A v_A) = \dot{m}'''_A. \tag{11.26}$$

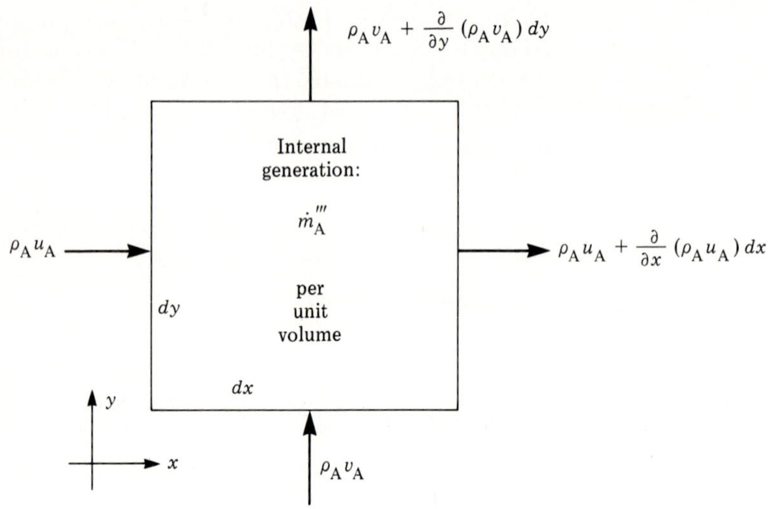


**Figure 11.4 Differential control volume for derivation of the equation of conservation of species A.**

A similar differential equation holds for species B, C, D, etc. If we sum Eqs. (11.26) over all species in the mixture and note that $\Sigma \dot{m}_i''' = 0$ because reactions merely transfer mass among species, the result is

$$\frac{\partial \rho}{\partial t} + \frac{\partial}{\partial x}(\rho u) + \frac{\partial}{\partial y}(\rho v) = 0. \tag{11.27}$$

This is the two-dimensional *continuity equation* for the entire mixture, using mass-averaged velocities. Its incompressible form was discussed in Eq. (5.6) of Chapter 5. Thus species conservation in mass transfer is an individual component of the overall mixture mass balance.

Equation (11.26) is inconvenient because it uses the species velocities $(u_A, v_A)$. We convert the relation to bulk velocities $(u, v)$ by using Eq. (11.18):

$$\frac{\partial \rho_A}{\partial t} + \frac{\partial}{\partial x}\left(\rho_A u - \rho D_{AB}\frac{\partial \rho_A}{\partial x}\right) + \frac{\partial}{\partial y}\left(\rho_A v - \rho D_{AB}\frac{\partial \rho_A}{\partial y}\right) = \dot{m}_A'''. \tag{11.28}$$

This is the general species conservation or *mass diffusion equation* for any species A in a flowing mixture. It is a second-order nonlinear partial differential equation, to be solved in conjunction with the continuity, momentum, and energy equations for the mixture.

The equivalent molar form of the diffusion equation is

$$\frac{\partial C_A}{\partial t} + \frac{\partial}{\partial x}\left(C_A u^* - CD_{AB}\frac{\partial x_A}{\partial x}\right) + \frac{\partial}{\partial y}\left(C_A v^* - CD_{AB}\frac{\partial C_A}{\partial y}\right) = \dot{n}_A''',$$

where $\dot{n}_A'''$ is rate of generation of species A by reactions (kmols/s·m$^3$). We see that this relation is usually awkward because it uses the molar-averaged velocities, which are not applicable to the momentum equation.

### 11.4.1 Incompressible Bulk Flow

Equation (11.28) is complex and nonlinear, and solutions can usually be obtained only by numerical modeling on a digital computer [22]. We simplify the diffusion equation by assuming that the bulk density is constant, even though $\rho_A$, $\rho_B$, etc. may vary. The result is

$$\frac{\partial \rho_A}{\partial t} + u\frac{\partial \rho_A}{\partial x} + v\frac{\partial \rho_A}{\partial y} = \frac{\partial}{\partial x}\left(D_{AB}\frac{\partial \rho_A}{\partial x}\right) + \frac{\partial}{\partial y}\left(D_{AB}\frac{\partial \rho_A}{\partial y}\right) + \dot{m}_A'''.$$

A further simplification is to take the diffusion coefficient $D_{AB}$ as constant:

$$\frac{\partial \rho_A}{\partial t} + u\frac{\partial \rho_A}{\partial x} + v\frac{\partial \rho_A}{\partial y} = D_{AB}\left(\frac{\partial^2 \rho_A}{\partial x^2} + \frac{\partial^2 \rho_A}{\partial y^2}\right) + \dot{m}_A'''. \tag{11.29}$$

This is the *incompressible diffusion equation* for any species A. It has exactly the same form as the mixture energy equation (5.23) if we neglect viscous dissipation:

$$\frac{\partial T}{\partial t} + u\frac{\partial T}{\partial x} + v\frac{\partial T}{\partial y} = \alpha\left(\frac{\partial^2 T}{\partial x^2} + \frac{\partial^2 T}{\partial y^2}\right) + \frac{\dot{q}}{\rho c_p}, \tag{11.30}$$

where $\dot{q}$ is the rate of energy generation per unit volume. Thus if the boundary conditions are the same, solutions for mass transfer are found by analogy to energy solutions.

### 11.4.2 Boundary Conditions

The mass diffusion equation (11.29) is a second-order "boundary-value" problem and thus requires information about the species density $\rho_A$ on every boundary of the flow. The three most common types of boundary conditions are

1. Solid wall or stationary interface: known concentration or concentration-gradient.

2. Interface between two diffusing immiscible fluids: equal concentrations on each side, equal diffusion fluxes on each side.
3. An inlet or outlet of the flow: known distribution of concentration.

Also, in the solution to Eq. (11.29) the velocities $u$ and $v$ would be known functions in forced-motion problems, as computed from the continuity and momentum equations. The reaction mass addition $\dot{m}'''_A$ is also assumed to be known or related functionally to the concentrations. We will not treat problems dealing with chemical reactions in this chapter (see, for example, Refs. [9] and [10]).

For transient mass transfer problems we must also specify a complete *initial condition* of the values of $\rho_A\ (x, y, z, t = 0)$ at every point within the boundaries. The same requirement was necessary for the transient heat transfer problems of Chapter 4.

### 11.4.3 The Heat and Mass Transfer Analogy

Since the diffusion equation (11.29) and the energy equation (11.30) are mathematically identical and have similar boundary conditions, a great many heat transfer solutions from Chapters 1–7 may be rewritten and applied immediately to mass transfer problems. This is done by replacing energy variables by their exactly analogous mass transfer variables. Table 11.2 illustrates the substitution.

It should be stressed that the analogy fails if the boundary and initial conditions for the mass transfer problem are not exactly the same as those for the associated heat transfer problem. A number of examples will be shown in the next sections.

**Table 11.2** Heat transfer and analogous mass transfer variables

| Heat Transfer Variable | Analogous Mass Transfer Variable |
|---|---|
| Temperature, $T$ | Species density, $\rho_A$ |
| Thermal diffusivity, $\alpha = k/\rho c_p$ | Mass diffusivity, $D_{AB}$ |
| Bulk velocities, $u$, $v$ | Bulk velocities, $u$, $v$ |
| Heat generation, $\dfrac{\dot{q}}{\rho c_p}$ | Reaction rate, $\dot{m}_A$ |
| Prandtl number, Pr | Schmidt number, Sc |

## 11.5 Mass Transfer without Convection or Reactions

For mass transfer in a stationary medium, with no bulk flow or chemical reactions, the mass diffusion equation (11.29) reduces to

$$\frac{\partial \rho_A}{\partial t} = D_{AB} \nabla^2 \rho_A \tag{11.31a}$$

or, in molar form,

$$\frac{\partial C_A}{\partial t} = D_{AB} \nabla^2 C_A. \tag{11.31b}$$

These are mathematically identical to the unsteady heat conduction equation from Chapter 4:

$$\frac{\partial T}{\partial t} = \alpha \nabla^2 T. \tag{4.1}$$

Thus a whole array of "conduction" solutions from Chapters 2–4 are available for use in mass transfer problems, especially if the boundary conditions specify known concentrations.

Table 11.3 shows four typical "conduction" solutions for mass transfer in a stationary medium. The first three are analogous to the one-dimensional conduction problems in Table 2.1, and the fourth is taken from the transient semi-infinite slab solution from Section 4.3.1. The table gives concentration distributions $C(x)$ and the total mass transfer rate $n_x$ (kmols/s). If the concentrations are replaced by species density $\rho_{A1}$, $\rho_{A2}$, etc., then $n_x$ represents the diffusion rate in kg/s.

In a similar manner, all of the conduction shape factor solutions from Table 3.1 are valid for mass transfer if the temperatures $T_1$ and $T_2$ are replaced by surface concentrations $C_1$ and $C_2$. Equation (3.46) would be replaced by

$$n_x = D_{AB} S (C_1 - C_2) \qquad \text{(kmols/s)}, \tag{11.32}$$

where the shape factor $S$ is purely geometric and given by Table 3.1.

Transient conduction problems involving surface *convection,* such as the Heisler-chart problems of Section 4.4 and Appendix J, are also analogous to mass transfer problems. For mass transfer the Biot number $hL/k$ must be replaced by the parameter $h_m L/D_{AB}$, where $h_m$ is the mass transfer coefficient to be defined in Section 11.6.1.

### Example 11.3

A thick mild steel slab has a uniform carbon concentration of 0.5% by weight and is suddenly subjected to a carbon environment

**Table 11.3 Solutions for mass transfer in a stationary medium**

| | Concentration Distribution | Mass Transfer Rate |
|---|---|---|
| 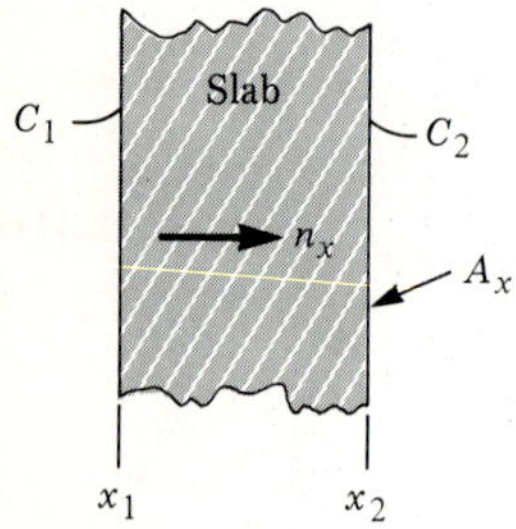  | $\dfrac{C - C_1}{C_2 - C_1} = \dfrac{x}{L}$ | $n_x = D_{AB} A_x \left( \dfrac{C_1 - C_2}{x_2 - x_1} \right)$ |
| 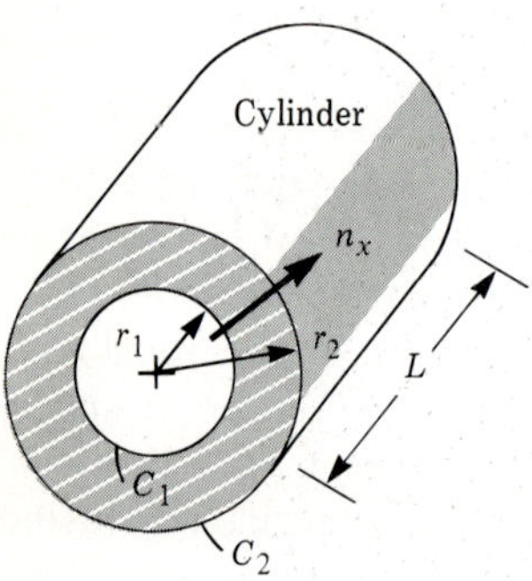  | $\dfrac{C - C_1}{C_2 - C_1} = \dfrac{\ln(r/r_1)}{\ln(r_2/r_1)}$ | $n_x = \dfrac{2\pi L D_{AB}}{\ln(r_2/r_1)} (C_1 - C_2)$ |
| 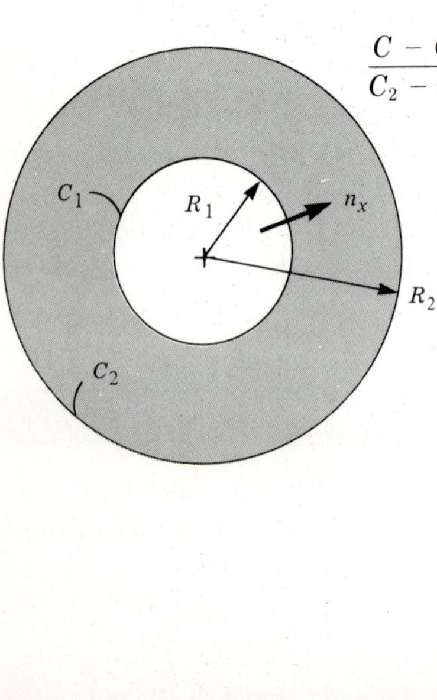  | $\dfrac{C - C_1}{C_2 - C_1} = \dfrac{R - R_1}{R_2 - R_1} \left( \dfrac{R_2}{R} \right)$ | $n_x = \dfrac{4\pi R_1 R_2 D_{AB}}{R_2 - R_1} (C_1 - C_2)$ |

**Table 11.3 (continued)**

| | Concentration Distribution | Mass Transfer Rate |
|---|---|---|
| 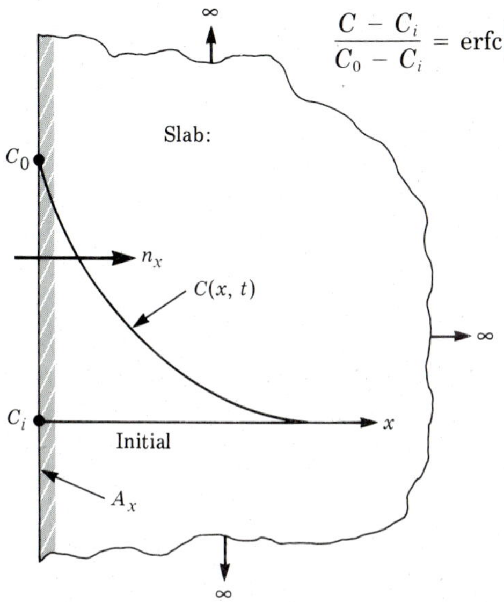 | $\dfrac{C - C_i}{C_0 - C_i} = \operatorname{erfc}\left(\dfrac{x}{2(D_{AB}\,t)^{1/2}}\right)$ | $n_x\Big\vert_{x=0} = A_x\left(\dfrac{D_{AB}}{\pi t}\right)^{1/2}(C_0 - C_i)$ |



such that the surface concentration jumps to 1.3%. How long will it take the carbon concentration at a point 1 mm inside the surface to rise to 0.7% if the diffusion coefficient is $6 \times 10^{-10}$ m$^2$/s?

*Solution* This problem fits the geometry of the fourth case in Table 11.2. We are given the mass fractions:

$$\omega_0 = 0.013, \qquad \omega_i = 0.005, \qquad \omega(x = 1\text{ mm}) = 0.007.$$

The concentration formula applies with $C$ converted to mass fractions:

$$\frac{\omega - \omega_i}{\omega_0 - \omega_i} = \frac{0.007 - 0.005}{0.013 - 0.005} = 0.25 = \operatorname{erfc}(\eta).$$

From Appendix A, erfc $\doteq$ 0.25 at $\eta \doteq 0.813$. Thus

$$\frac{x/2}{(D_{AB}t)^{1/2}} = 0.813 = \frac{0.001/2}{(6 \times 10^{-10}\,t)^{1/2}}.$$

Solve for $t$ = 630 s = 10.5 min. [*Ans.*] ■

### 11.5.1 Equimolar Counterdiffusion

An interesting special case of steady diffusion is the concentration distribution which occurs between two large reservoirs, one primarily containing gas A and the other gas B, as shown in Figure 11.5. If the reservoirs are at constant temperature $T$ and pressure $p$, the concentrations in each reservoir will remain constant. Gas A will diffuse toward reservoir B and vice versa.

In steady diffusion between two uniform concentrations there can be no net molar flux in the connecting duct between reservoirs:

$$G^*_{Az} + G^*_{Bz} = 0$$

or

$$-CD_{AB}\frac{dx_A}{dz} = CD_{AB}\frac{dx_B}{dz}. \tag{11.33}$$

Since we assume constant mixture concentration and $D_{AB}$, it follows that

$$\frac{dx_A}{dz} = -\frac{dx_B}{dz}$$

or

$$x_{A1} - x_{A2} = x_{B2} - x_{B1}. \tag{11.34}$$

**Figure 11.5** Illustration of steady equimolar counterdiffusion between two reservoirs.

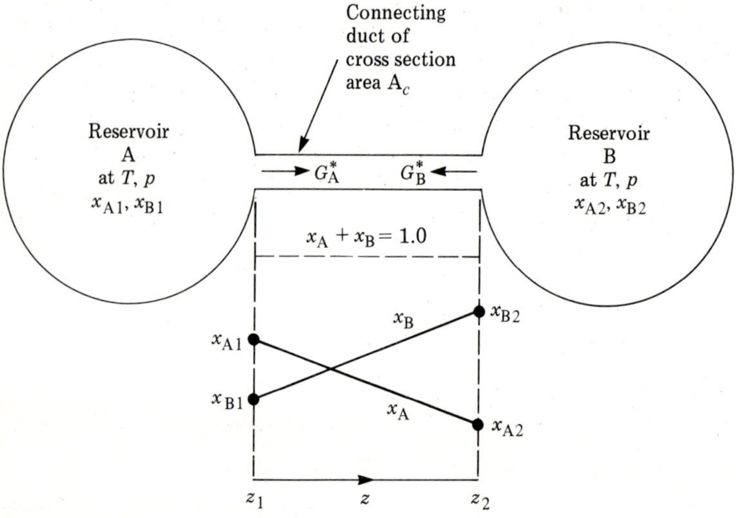

For ideal gases, $x_A = p_A/p$ and $x_B = p_B/p$, and since the total pressure $p = p_A + p_B$ is constant, Eq. (11.34) is equivalent to

$$\frac{dp_A}{dz} = -\frac{dp_B}{dz}$$

or

$$p_{A1} - p_{A2} = p_{B2} - p_{B1}. \tag{11.35}$$

This situation is called *equimolar counterdiffusion;* the mole fractions and partial pressures vary linearly through the connecting duct, and the molar fluxes are equal and opposite. The mass flows through the duct, however, are not equal:

$$\begin{aligned} G_{Az} &= -D_{AB}\frac{d\rho_A}{dz} = \frac{D_{AB}}{R_A T}\frac{p_{A1} - p_{A2}}{z_2 - z_1}, \\ G_{Bz} &= -D_{AB}\frac{d\rho_B}{dz} = \frac{D_{AB}}{R_B T}\frac{p_{B1} - p_{B2}}{z_2 - z_1}. \end{aligned} \tag{11.36}$$

Thus although the molar-average velocity $V_z^*$ is zero for this type of diffusion, the mass-average $V_z$ is not, and there is a slow but finite average motion of the mixture toward the gas with a lower molecular weight.

**Example 11.4**

A large tank of ethanol at 1 atm and 20°C is vented to atmospheric air through a tube 1 cm in diameter and 40 cm long. The concentration of air in the methanol tank is negligible. At what rate is the ethanol leaking out? At what rate is the air intruding into the tank?

*Solution* Following Fig. 11.5, let A be the ethanol and B be air. From Appendix K for ideal gases we find

$$M_A = 46.07, \qquad R_A = 180 \text{ J/kg·K};$$

$$M_B = 28.96, \qquad R_B = 287 \text{ J/kg·K}.$$

From Table 11.1 for ethanol/air, read $D_{AB} = 1.2 \times 10^{-5}$ m²/s. We assume that $p_{B1} = p_{A2} \doteq 0$; hence $p_{A1} = p_{B2} \doteq 1$ atm. The molar flux of ethanol is given by

$$\begin{aligned} G^*_{Az} &= \frac{D_{AB}}{\overline{R}T}\frac{p_{A1} - p_{A2}}{z_2 - z_1} = \frac{1.2 \times 10^{-5}}{(8310)(293)}\left(\frac{101{,}350 - 0}{0.4 - 0}\right) \\ &= 1.25 \times 10^{-6} \text{ kmols/m}^2\text{·s} = -G^*_{Bz}. \end{aligned}$$

Then the mass flow of ethanol into the atmosphere is

$$\dot{m}_{Az} = A_c M_A G^*_{Az} = \frac{\pi}{4}(0.01)^2 (46.07)(1.25 \times 10^{-6})$$

$$= 4.52 \times 10^{-9}\ \text{kg/s} = 0.0163\ \text{g/hr.} \quad [Ans.]$$

Similarly, the mass flow of air into the tank is

$$\dot{m}_{Bz} = A_c M_B G^*_{Bz} = -2.84 \times 10^{-9}\ \text{kg/s} = 0.0102\ \text{g/hr.} \quad [Ans.]$$

The air intrusion will gradually contaminate the tank gas. For example, if a 1 m$^3$ tank initially contains 100% ethanol or 1.92 kg, it will contain 1% air after about 78 days. ■

### 11.5.2 Diffusion from an Open Container

Consider the steady state evaporation and diffusion of liquid A through a stagnant container of gas B, as shown in Fig. 11.6(a). Vapor A rises through the container at molar rate $G^*_{Az}$ and is carried away by a gentle breeze of gas B that does not cause any turbulence in the

**Figure 11.6** **Evaporation and diffusion of species A through a stagnant gas B: (a) geometry of the container; (b) concentration distributions for various $x_{A0}$.**

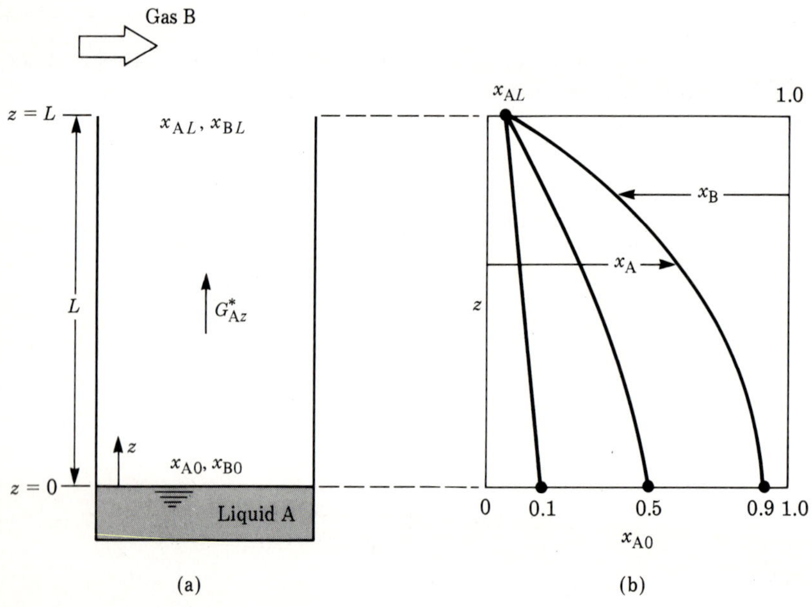

container. Let the liquid surface be ($z = 0$) and the container top be ($z = L$).

The mixture of gases A and B in the container is assumed to be at constant temperature $T$ and total pressure $p = p_A + p_B$. Then the boundary conditions on vapor A are as follows:

a) at $z = 0$: $p_{A0}$ = vapor pressure of A at $(T, p)$,

b) at $z = L$: $p_{AL}$ = partial pressure of A in the outer ambient gas.

We are to solve for the concentration $x_A(z)$ or its equivalent, the partial pressure $p_A(z)$ for steady diffusion. Three examples of the concentration distribution are shown in Fig. 11.6(b). If $x_{A0}$ is very small, the distribution $x_A(z)$ is linear. As $x_{A0}$ increases, $x_A(z)$ becomes more nonlinear.

An interesting facet to this problem is that liquid A acts as a *barrier* to the downward flux of gas B, whereas vapor A can both diffuse and flow upward. Thus if one neglects the very small rate at which gas B might dissolve into liquid A, the downward molar flux $G^*_{Bz}$ must be zero, and the flux of A upward from Eq. (11.25) reduces to

$$G^*_{Az} = -CD_{AB}\frac{dx_A}{dz} + x_A\,(G^*_{Az} + 0)$$

or

$$G^*_{Az} = \frac{-CD_{AB}}{1 - x_A}\frac{dx_A}{dz} = \text{const.} \tag{11.37}$$

This upward flux is constant because of the assumed steady flow conditions.

In terms of partial pressure $p_A/p = x_A$, Eq. (11.37) may be rewritten as

$$G^*_{Az} = \frac{-CD_{AB}}{p - p_A}\frac{dp_A}{dz} = \text{const.}, \tag{11.38}$$

since we have assumed constant $C$ and $D_{AB}$. Equation (11.38) may be integrated once to obtain

$$\ln(p - p_A) = C_1 z + C_2,$$

where $C_1$ and $C_2$ may be evaluated from the boundary conditions $p_A(x = 0) = p_{A0}$ and $p_A(x = L) = p_{AL}$. The final result is

$$\frac{p - p_A}{p - p_{A0}} = \left(\frac{p - p_{AL}}{p - p_{A0}}\right)^{z/L}$$

or $\qquad$ (11.39)

$$\frac{1 - x_A}{1 - x_{A0}} = \left(\frac{1 - x_{AL}}{1 - x_{A0}}\right)^{z/L}.$$

Examples of this distribution are shown in Fig. 11.6(b) for three different surface mole fractions $x_{A0}$. For very small $x_{A0} << 1$ the denominator $(p - p_a) \doteq p$ in Eq. (11.38), resulting in a linear pressure distribution $p_A = C_1 z + C_2$.

By substitution of $p_A(z)$ from (11.39) into (11.38) we find that the total molar and mass fluxes of vapor A upward are

$$G^*_{Az} = \frac{CD_{AB}}{L} \ln\left(\frac{p - p_{AL}}{p - p_{A0}}\right)$$

and **(11.40)**

$$\dot{m}_{Az} \doteq \frac{pA_c D_{AB}}{R_A TL} \ln\left(\frac{p - p_{AL}}{p - p_{A0}}\right).$$

where $A_c$ is the cross section area of the container.

## Example 11.5

Estimate the mass loss of benzene at 20°C from a container 30 cm high and 20 cm in diameter, diffusing into dry air at 20°C as shown. What is the benzene partial pressure at the midpoint of the container?

*Solution* From page D-206 of [24] the vapor pressure of benzene ($C_6H_6$) at 20°C is approximately 75 mm Hg. The ambient air is assumed to be free of benzene. Thus our initial data are

$$p = 1 \text{ atm} = 101325 \text{ Pa},$$

$$p_{A0} = 75 \text{ mm Hg} = 9986 \text{ Pa},$$

$$p_{AL} = 0 \text{ Pa},$$

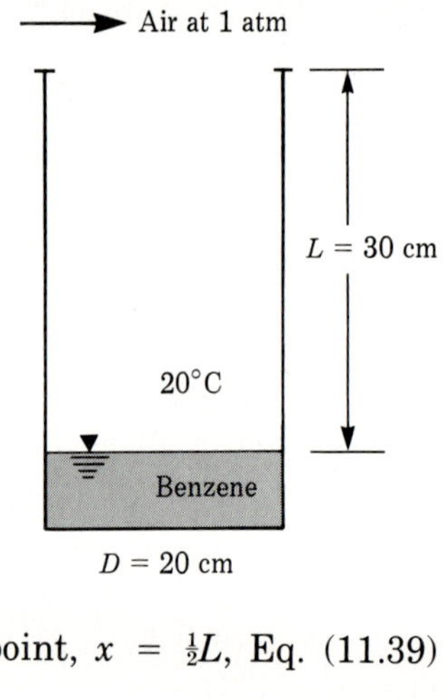


where gas A is benzene. At the midpoint, $x = \frac{1}{2}L$, Eq. (11.39) predicts that

$$\frac{p - p_A}{p - p_{A0}} = \frac{101325 - p_A}{101325 - 9986} = \left(\frac{101325 - 0}{101325 - 9986}\right)^{1/2} = 1.0533,$$

from which we solve for $p_A(\frac{1}{2}L) = 5122$ Pa. [*Ans.*]

From Table 11.1 for benzene in air, $D_{AB} = 0.9 \times 10^{-5}$ m²/s. The molecular weight of benzene is 78.11; hence its gas constant is $R_A = 8310/78.11 = 106.4$ m²/s²·K. Equation (11.40) then predicts a benzene mass flux of

$$\dot{m}_{Az} = \frac{(101325)\,\pi\,(0.1)^2\,(0.9 \times 10^{-5})}{(106.4)\,(293)\,(0.30)} \ln\left(\frac{101325 - 0}{101325 - 9986}\right)$$
$$= 3.18 \times 10^{-7}\ \text{kg/s} = 1.14\ \text{g/hr.} \quad [\textit{Ans.}]$$

This is a substantial loss rate that would be much larger if the air were allowed to *convect* past the liquid (Section 11.6).

The distribution $p_A(z)$ is nearly linear, since $x_{A0}$ is small (0.0986). A linear estimate of $p_A(\frac{1}{2}L)$ would be $9986/2 = 4993$ Pa, only 2.5% less than our actual estimate. ■

---

## 11.6 Mass Transfer with Forced Convection

Mass transfer in forced convection is similar and often exactly analogous to convection heat transfer as studied in Chapter 6. The recommended correlation formulas are very similar.

Consider steady boundary layer flow past a flat plate with non-uniform temperature and concentration profiles, as in Fig. 11.7. For incompressible flow with constant diffusivities the energy and species conservation equations are mathematically identical:

a) energy: $$u\frac{\partial T}{\partial x} + v\frac{\partial T}{\partial y} = \alpha\frac{\partial^2 T}{\partial y^2},$$

b) species: $$u\frac{\partial \rho_A}{\partial x} + v\frac{\partial \rho_A}{\partial y} = D_{AB}\frac{\partial^2 \rho_A}{\partial y^2},$$

where we have neglected heat and mass sources in the fluid. The stream has known values of velocity $U_\infty$, temperature $T_\infty$, and concentration or species density $\rho_{A\infty}$. The wall has known velocity $u = 0$ and known temperature $T_w$ and species density $\rho_{Aw}$. The temperature difference $(T_w - T_\infty)$ causes heat transfer $q''_w$ at the wall. Similarly, the density difference $(\rho_{Aw} - \rho_{A\infty})$ causes wall mass transfer $G_{Aw}$.

The temperature profiles $T(x, y)$ will be similar in shape to the concentration profiles $C(x, y)$ or $\rho_A(x, y)$. The boundary layer thicknesses

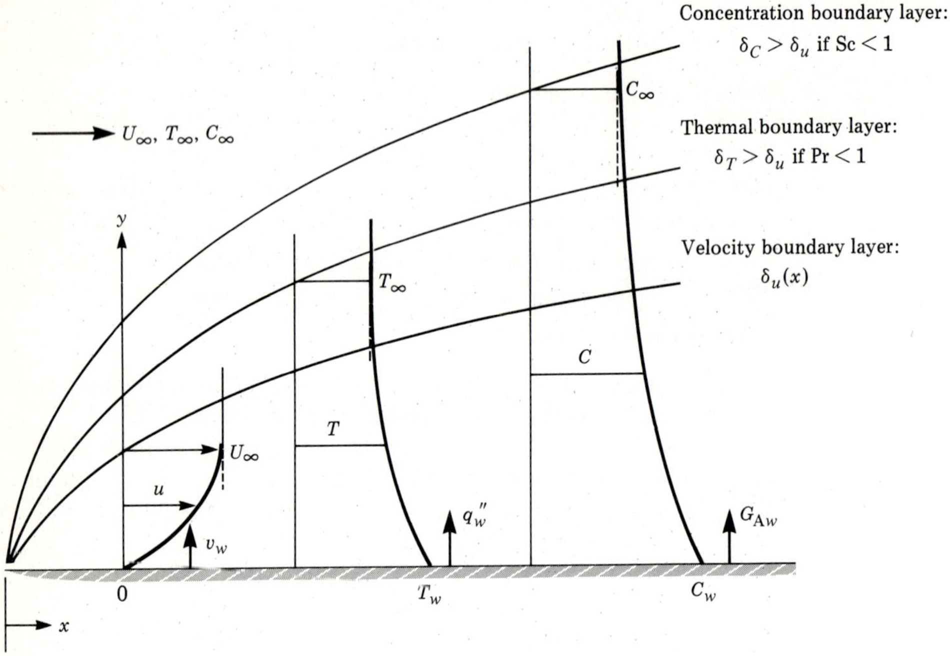


**Figure 11.7 Forced convection heat and mass transfer in flow past a flat plate. The boundary layer thicknesses are exaggerated. Compare this figure with Fig. 6.1.**

$\delta_T(x)$ and $\delta_C(x)$ will grow in a similar manner and will be thicker than $\delta_u(x)$ if the Prandtl and Schmidt numbers, respectively, are less than unity.

### 11.6.1 The Mass Transfer Coefficient

Recall from Eq. (5.1b) that the heat transfer coefficient $h$ in external (boundary layer) flow is defined by

$$q_w'' = h(T_w - T_\infty)$$

and that for a given convection problem the dimensionless heat transfer behavior is given by

$$hL/k = \text{fcn}(UL/\nu, \nu/\alpha, \text{geometry})$$

or

$$\text{Nu}_L = \text{fcn}(\text{Re}_L, \text{Pr}, \text{geometry}).$$

Many applications were given in Chapter 6. In an exactly similar manner the *mass transfer coefficient* $h_m$ is defined in either molar or mass diffusion form:

$$G_{Aw} = h_m(\rho_{Aw} - \rho_{A\infty})$$

or (11.41)

$$G^*_{Aw} = h_m(C_w - C_\infty).$$

We see that $h_m$ has dimensions of velocity, or (m/s). By analogy with heat transfer it will correlate with flow parameters in the following dimensionless form:

$$h_m L/D_{AB} = \text{fcn}(UL/\nu, \nu/D_{AB}, \text{geometry})$$

or (11.42)

$$\text{Sh}_L = \text{fcn}(\text{Re}_L, \text{Sc}, \text{geometry}),$$

where $h_m L/D_{AB}$ is called the *Sherwood number* and is exactly analogous to the Nusselt number.

### 11.6.2 The Heat and Mass Transfer Analogy

Let us restate the analogy that we discussed in Section 11.4.3 in the form of a rule of thumb:

> *If a convection mass transfer problem has the same geometry, flow pattern, and boundary conditions as a known heat transfer problem, then the heat transfer correlations may be converted to mass transfer by replacing h by $h_m$, Pr by Sc, and Nu by Sh for either laminar or turbulent flow.*

This analogy is widely used in engineering convection mass transfer computations. However, there is one note of caution: Mass transfer often causes a slight "blowing" effect at the wall, $v_w \neq 0$ (see Fig. 11.7). This blowing, though not large, may alter the mass transfer correlation and may not be negligible. We will discuss wall blowing in Section 11.6.4.

### 11.6.3 Flat Plate Mass Transfer

Consider flow of a uniform stream parallel to a flat plate, as in Fig. 11.7. If we neglect the wall blowing effect, the heat transfer formulas developed in Section 6.2 are valid for mass transfer by replacing Pr by Sc and Nu by Sh. For laminar flow, $\text{Re}_x < 500{,}000$, we obtain

a) local value: $\text{Sh}_x = h_m x/D_{AB} = 0.332\text{Re}_x^{1/2}\text{Sc}^{1/3}$, (11.43)

b) average value: $\overline{\text{Sh}_L} = \overline{h_m} L/D_{AB} = 0.664\text{Re}_L^{1/2}\text{Sc}^{1/3}$.

For turbulent flow with Schmidt numbers not too far from unity we can use the power law formulas from Eq. (6.91):

$$\begin{aligned}
&\text{a) local value:} && \mathrm{Sh}_x = h_m x/D_{\mathrm{AB}} = 0.0296\mathrm{Re}_x^{4/5}\mathrm{Sc}^{1/3},\\
&\text{b) average value:} && \overline{\mathrm{Sh}_L} = \overline{h}_m L/D_{\mathrm{AB}} = 0.037\mathrm{Re}_L^{4/5}\mathrm{Sc}^{1/3}.
\end{aligned} \tag{11.44}$$

For liquids, which usually have high Schmidt numbers (Table 11.1), it is more accurate to modify Eq. (6.92):

$$\begin{aligned}
&\text{a) local value:} && \frac{\mathrm{Sh}_x}{\mathrm{Re}_x\mathrm{Sc}} = \frac{h_m}{U_\infty} = \frac{c_f/2}{1 + 12.7(\mathrm{Sc}^{2/3} - 1)(c_f/2)^{1/2}},\\
&\text{b) average value:} && \overline{\mathrm{Sh}_L} = 1.15\mathrm{Sh}_x(x = L),
\end{aligned} \tag{11.45}$$

where the turbulent skin friction coefficient $c_f$ is computed from Eq. (6.89).

## Example 11.6

Dry air at 30°C flows at 4.5 m/s past a 1 m by 1 m wet plate, also at 30°C. Neglecting wall blowing, estimate the water loss rate from the plate in grams per hour.

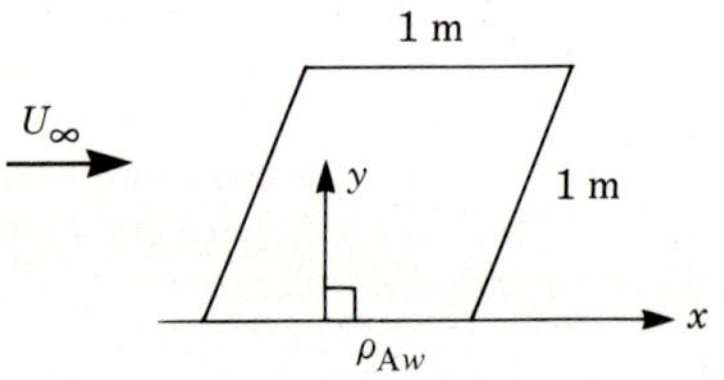


*Solution* For water vapor diffusing into air we can find most of the required physical properties in the Appendices:

$$\nu = 1.60 \times 10^{-5}\ \mathrm{m^2/s}, \qquad D_{\mathrm{AB}} = 2.66 \times 10^{-5}\ \mathrm{m^2/s}, \qquad \mathrm{Sc} = 0.603.$$

The molecular weight of water is 18.02; hence $R_{\mathrm{H20}} = 461\ \mathrm{m^2/(s^2 \cdot K)}$. From the Steam Tables [25] the vapor pressure of water at 30°C is 4242 Pa. Let water vapor be species A. Then the water density at the surface is

$$\rho_{\mathrm{A}w} = p_{\mathrm{H20}}/(R_{\mathrm{H20}}T) = \frac{4242}{(461)(303)} = 0.0304\ \mathrm{kg/m^3}.$$

Finally, since the air stream is "dry," we assume that $\rho_{\mathrm{A}\infty} = 0$. The Reynolds number of the flow is

$$\mathrm{Re}_L = U_\infty L/\nu = (4.5)(1.0)/1.6 \times 10^{-5} = 281{,}000.$$

Since this is less than 500,000, we assume that the flow is laminar, and Eq. (11.43b) applies:

$$\overline{\mathrm{Sh}}_L = 0.664(281{,}000)^{1/2}(0.603)^{1/3} = 297 = \overline{h}_m L/D_{\mathrm{AB}}$$

or

$$\overline{h}_m = 297(2.66 \times 10^{-5})/1.0 = 0.00791 \text{ m/s}.$$

Then the mass loss per unit area from the plate follows from the definition of $h_m$ from Eq. (11.41):

$$\begin{aligned}\overline{G_{\mathrm{A}w}} &= \overline{h}_m(\rho_{\mathrm{A}w} - \rho_{\mathrm{A}\infty}) = 0.00791(0.0304 - 0)\\ &= 2.4 \times 10^{-4} \text{ kg/(s·m}^2).\end{aligned}$$

Since the plate area is 1.0 m², we may convert this readily to total mass loss:

$$\dot{m}_{\mathrm{A}w} = 860 \text{ g/hr.} \quad [Ans.]$$

We will estimate the wall blowing effect, if any, for this problem in Example 11.7. ■

### 11.6.4 The Effect of Wall Blowing

In illustrating flat plate convective mass transfer, Fig. 11.7 showed, without much useful discussion, a wall blowing effect, that is, a nonzero wall normal velocity $v_w$ caused by the diffusion. We encountered this blowing effect in evaporation from a container (Section 11.5.2).

The basic reason for the blowing is that the wall acts as a barrier to the stream fluid B, so the mass flux of species B there is essentially zero: $G_{\mathrm{B}w} = 0$. Then from Eq. (11.25) the flux of wall fluid A is

$$G_{\mathrm{A}w} = -D_{\mathrm{AB}}\left(\frac{\partial \rho_{\mathrm{A}}}{\partial y}\right)_{y=0} + \omega_{\mathrm{A}}(G_{\mathrm{A}w} + 0) = \rho_{\mathrm{A}w} v_w,$$

where we have set the total flux equal to wall density times a wall velocity. Solving for $v_w$, we obtain

$$v_w = \frac{-D_{\mathrm{AB}}}{\rho - \rho_{\mathrm{A}w}}\left(\frac{\partial \rho_{\mathrm{A}}}{\partial y}\right)_{y=0} = \frac{-D_{\mathrm{AB}}}{1 - \omega_{\mathrm{A}w}}\left(\frac{\partial \omega_{\mathrm{A}}}{\partial y}\right)_{y=0}. \tag{11.46}$$

The wall velocity is small but finite and is proportional to the local concentration gradient of wall species A.

Generally speaking, the effect of wall blowing is to reduce skin friction and wall heat transfer and therefore, by analogy, to reduce the wall mass transfer. This is true for both laminar and turbulent flow. Extensive computations of laminar flat plate flow with blowing have been reported by Hartnett and Eckert [26] and others. The reduction in wall mass transfer (or heat transfer) is shown in Fig. 11.8 for gases. There is a weak effect of Schmidt number and a strong effect of the *blowing parameter:*

$$\zeta = (v_w/U_\infty)\,\mathrm{Re}_x^{1/2}. \tag{11.47}$$

The curves drop off nearly linearly with $\zeta$, and a reasonable curve-fit expression is

$$\mathrm{Sh}_x/\mathrm{Re}_x^{1/2} \doteq 0.332\mathrm{Sc}^{1/3} - 0.5\zeta. \tag{11.48}$$

At $\zeta \doteq 0.62$, both $\mathrm{Nu}_x$ and $\mathrm{Sh}_x$ have dropped to zero, and the boundary layer is said to have been entirely "blown off." However, in most

**Figure 11.8** Effect of wall blowing on convective heat and mass transfer in laminar flat plate flow (from data in [26]).

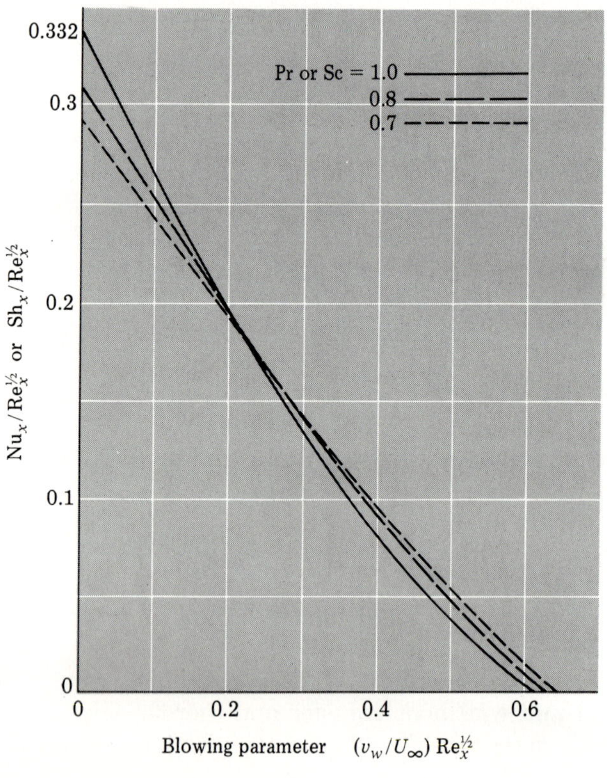

convective mass transfer problems we are operating in the range of $\zeta = 0$–$0.2$.

The blowing effect can be estimated iteratively by first computing $G_{Aw}$ assuming that $\zeta = 0$ and using this value to estimate $v_w$ from Eq. (11.46) and $\mathrm{Sh}_x$ from Eq. (11.48). We illustrate with the following example.

**Example 11.7**

Repeat Example 11.6 to estimate the effect of wall blowing.

*Solution* Assume that the mixture density equals the density

$$\rho = p_\infty / R_B T_\infty = \frac{101{,}325}{(287)(303)} = 1.165 \text{ kg/m}^3.$$

Our "no blowing" computation of wall mass transfer in Example 11.6 gave the result

$$G_{Aw}(\zeta = 0) = 2.4 \times 10^{-4} \text{ kg/(s·m}^2).$$

Then our first estimate of wall velocity is, from Eq. (11.46),

$$v_w \doteq \frac{G_{Aw}(\zeta = 0)}{\rho - \rho_{Aw}} = \frac{2.4 \times 10^{-4}}{1.165 - 0.0304} = 2.12 \times 10^{-4} \text{ m/s},$$

from which

$$\zeta = (v_w/U_\infty)\mathrm{Re}_L^{1/2} = \frac{2.12 \times 10^{-4}}{4.5}(281{,}000)^{1/2} = 0.025.$$

Figure 11.8 indicates that this blowing effect is slight. From Eq. (11.48),

$$\mathrm{Sh}_L/\mathrm{Re}_L^{1/2} \doteq 0.332(0.603)^{1/3} - 0.5(0.025) = 0.268$$

or

$$\overline{\mathrm{Sh}_L} = 2\mathrm{Sh}_L \mathrm{Re}_L^{1/2} = 2(0.268)(281{,}000)^{1/2} = 284.$$

This is 4% less than our previous estimate of $\overline{\mathrm{Sh}_L} = 297$. As in Example 11.6, we then compute $\overline{h_m} = 0.00755$ m/s and $\overline{G_{Aw}} = 2.30 \times 10^{-4}$ kg/(s·m²). The corrected mass transfer rate is thus

$$\dot{m}_{Aw} = 825 \text{ g/hr.} \quad [Ans.]$$

Since the blowing reduction is only 4%, there is no need to iterate this estimate further. ■

Wall blowing also has a significant effect upon turbulent flat plate mass transfer. There is a very good discussion in the text by Kays and Crawford [8], who recommend that the turbulent Sherwood number be modified as follows:

$$\mathrm{Sh}_x = 0.0296\mathrm{Re}_x^{4/5}\mathrm{Sc}^{1/3}\left|\frac{\ln(1+\beta)}{\beta}\right|, \tag{11.49}$$

where

$$\beta = \frac{\rho_{\mathrm{A}w} - \rho_{\mathrm{A}\infty}}{\rho - \rho_{\mathrm{A}w}} = \frac{\omega_{\mathrm{A}w} - \omega_{\mathrm{A}\infty}}{1 - \omega_{\mathrm{A}w}}.$$

The dimensionless ratio $\beta$ is a blowing parameter with much the same meaning as the laminar parameter $\zeta$ in Eq. (11.47). The same correction factor may be applied with reasonable accuracy to the average value:

$$\overline{\mathrm{Sh}_L} = 0.037\mathrm{Re}_L^{4/5}\mathrm{Sc}^{1/3}\left|\frac{\ln(1+\beta)}{\beta}\right|.$$

This blowing effect is not large unless the concentration of vapor A at the wall is large, for example, $\omega_{\mathrm{A}w} \geq 0.2$.

## Example 11.8

Repeat Example 11.7 for $U = 45$ m/s and $T_w = 70$°C.

*Solution* If the stream velocity is ten times higher, $\mathrm{Re}_L = 2.81 \times 10^6$, or turbulent flow. Then if we neglect blowing, Eq. (11.44) applies:

$$\overline{\mathrm{Sh}_L} = 0.037(2.81 \times 10^6)^{4/5}(0.603)^{1/3} = 4510$$

or

$$\overline{h_m} = \overline{\mathrm{Sh}_L} D_{\mathrm{AB}}/L = (4510)(2.66 \times 10^{-5})/1.0 = 0.120 \text{ m/s}.$$

At the new wall temperature of 70°C the vapor pressure of water is 31.16 kPa, or

$$\rho_{\mathrm{A}w} = p_{\mathrm{A}w}/R_{\mathrm{A}}T = \frac{31{,}160}{(461)(343)} = 0.197 \text{ kg/m}^3.$$

Then the mass loss of water, assuming zero blowing, is

$$\overline{G_{\mathrm{A}w}} = \overline{h_m}(\rho_{\mathrm{A}w} - \rho_{\mathrm{A}\infty}) = 0.120(0.197 - 0) = 0.0236 \text{ kg/(s·m}^2)$$

or

$$\dot{m}_{\mathrm{A}w} = (0.0236)(1 \text{ m}^2)(3600 \text{ s/hr})(1000 \text{ g/kg}) = 85{,}100 \text{ g/hr}.$$

To correct for wall blowing, we need the mass fractions of water and air. We are given the molar fractions:

$$x_{\mathrm{Aw}} = p_{\mathrm{A}w}/p = \frac{31{,}160}{101{,}350} = 0.307, \qquad x_{\mathrm{B}w} = 1 - x_{\mathrm{A}w} = 0.693.$$

Then from Eq. (11.14) we compute, for $M_{\mathrm{A}} = 18.02$ and $M_{\mathrm{B}} = 28.96$,

$$\omega_{\mathrm{A}w} = \frac{0.307(18.02)}{(0.307)(18.02) + (0.693)(28.96)} = 0.216.$$

For dry air, assume that $\omega_{\mathrm{A}\infty} = 0$. Then the blowing parameter from Eq. (11.49) is

$$\beta = \frac{0.216 - 0.0}{1.0 - 0.216} = 0.276.$$

The blowing correction is

$$\frac{\ln(1 + \beta)}{\beta} = \frac{\ln(1.276)}{0.276} = 0.883.$$

Then an accurate estimate of the water loss rate is

$$\dot{m}_{\mathrm{A}w} = (85{,}100)(0.883) = 75{,}000 \text{ g/hr}.$$

This is 90 times larger than the laminar flow example (Eq. 11.7). The increase is partly due to turbulence and partly due to the higher wall concentration of water vapor. Note that the blowing correction is 12%. ■

---

### 11.6.5 Other Geometries

In convective mass transfer near a solid surface we may neglect wall blowing if the wall mass fraction of the diffusing substance is small in comparison to unity. Then by the analogy of Section 11.6.2 we merely modify the heat transfer correlation for the same geometry by substituting Sc for Pr and Sh for Nu. We do not change the Reynolds number.

For example, in crossflow past a cylinder the mass transfer is given by modifying Eq. (6.115):

$$\overline{\mathrm{Sh}_D} \doteq 0.3 + \frac{0.62\mathrm{Re}_D^{1/2}\,\mathrm{Sc}^{1/3}}{\{1 + (0.4/\mathrm{Sc})^{2/3}\}^{1/4}}\left[1 + \left(\frac{\mathrm{Re}_D}{282{,}000}\right)^{5/8}\right]^{4/5} \tag{11.50}$$

valid for $\mathrm{Re}_D\mathrm{Sc} > 0.2$. Figure 6.18 may be readily interpreted as a plot of Sherwood versus Reynolds number for various Schmidt numbers.

Similarly, convective mass transfer of a sphere in a uniform stream may be modified from Eq. (6.119):

$$\overline{\mathrm{Sh}_D} \doteq 2 + (0.4\mathrm{Re}_D^{1/2} + 0.06\mathrm{Re}_D^{2/3})\mathrm{Sc}^{0.4} \qquad \textbf{(11.51)}$$

valid for $3.5 \leq \mathrm{Re}_D \leq 8 \times 10^4$ and $0.6 \leq \mathrm{Sc} \leq 380$.

Internal flows may also be modified with the convection analogy. For laminar flow ($\mathrm{Re}_D \leq 2000$) in the entrance of a circular tube at high Schmidt numbers we may modify Hausen's correlation, Eq. (6.75):

$$\mathrm{Sc} > 10: \quad \overline{\mathrm{Sh}_D} \doteq 3.66 + \frac{0.0668/\Lambda}{1 + 0.04/\Lambda^{2/3}}, \qquad \textbf{(11.52)}$$

where $\Lambda = (L/D)/(\mathrm{Re}_D\mathrm{Sc})$. For lower Schmidt numbers (gases) we may read Figure 6.5, interpreting Nu as Sh and Pr as Sc.

For fully developed turbulent mass transfer in pipe flow we may modify the Dittus-Boelter formula (6.107):

$$\overline{\mathrm{Sh}_D} \doteq 0.027\mathrm{Re}_D^{4/5}\mathrm{Sc}^{1/3}, \qquad \textbf{(11.53)}$$

valid for Sc near unity (gases). For higher Schmidt numbers we may modify the Petukhov-Kirillov formula (6.106):

$$\frac{\overline{\mathrm{Sh}_D}}{\mathrm{Re}_D\mathrm{Sc}} = \frac{\overline{h_m}}{V} = \frac{c_f/2}{1 + 12.7(\mathrm{Sc}^{2/3} - 1)(c_f/2)^{1/2}}, \qquad \textbf{(11.54)}$$

valid for $0.6 \leq \mathrm{Sc} \leq 1000$. The Reynolds number in Eqs. (11.53) and (11.54) should be turbulent (>4000).

It should be pointed out that the turbulent-flow analogy between heat and mass transfer is *not* exact, but only a first approximation. There is a good discussion of the possible inaccuracy of this analogy in the text by Spalding [27, Section 2.4]. Spalding recommends that better accuracy is always obtained in turbulent mass transfer if the blowing correction factor $\ln(1 + \beta)/\beta$ from Eq. (11.49) is included in the computation.

**Example 11.9**

A sphere of solid naphthalene has a diameter of 3 cm and is immersed in a stream of dry air at 20°C and $U = 6$ m/s. The naphthalene sublimes directly to gas, which has a vapor pressure of 7 Pa at 20°C. Estimate the mass loss rate in g/hr.

*Solution* Let species A be the naphthalene gas. Its molecular weight is 128.2 and, from Table 11.1, $D_{AB} = 0.6 \times 10^{-5}$ m²/s and Sc = 2.5. For air at 20°C, $\nu_B = 1.5 \times 10^{-5}$ m$^2$/s. The wall density of the naphthalene gas is approximately

$$\rho_{Aw} = \frac{p_{Aw}M_A}{\overline{R}T} = \frac{(7.0)(128.2)}{(8310)(293)} = 0.000369 \text{ kg/m}^3.$$

The Reynolds number of the sphere is

$$\text{Re}_D = UD/\nu = \frac{(6.0)(0.03)}{1.5 \times 10^{-5}} = 12{,}000.$$

Then from Eq. (11.51) the mean Sherwood number is

$$\overline{\text{Sh}_D} = 2 + (0.4(12{,}000)^{1/2} + 0.06(12{,}000)^{2/3})(2.5)^{0.4} = 111$$

or

$$\overline{h_m} = \overline{\text{Sh}_D}D_{AB}/D = 111(0.6 \times 10^{-5})/(0.03) = 0.022 \text{ m/s}.$$

Then the total mass loss of the sphere is

$$\begin{aligned}\dot{m}_{Aw} &= A_w\overline{h_m}(\rho_{Aw} - \rho_{A\infty}) = \pi(0.03)^2(0.0222)(0.000369 - 0)\\ &= 2.3 \times 10^{-8} \text{ kg/s} = 0.083 \text{ g/hr.} \quad [\textit{Ans.}]\end{aligned}$$

This is such a small mass loss that the blowing correction $\ln(1 + \beta)/\beta$ is negligible. The sphere would shrink to $D$ = 2.99 cm in about two hours and would disappear entirely in about one week. (The density of naphthalene is 1078 kg/m$^3$.) ■

---

### 11.6.6 The Naphthalene Sublimation Technique

In Example 11.9 we used the heat transfer analogy to estimate the diffusion mass loss of a naphthalene sphere in an air stream. The reverse of this procedure is now a useful heat transfer experimental technique. The experimental body shape is coated with solid naphthalene, which sublimes into the air stream. Using the heat/mass analogy, we then convert the mass loss data to heat transfer coefficients and Nusselt numbers for that body.

Some early users of the technique were Winding and Chaney [29], who measured the mass loss of naphthalene cylinders exposed to an air crossflow. When converted to Nusselt numbers, their data are comparable to the direct heat transfer measurements shown in Fig. 6.17. The density of naphthalene is 1078 kg/m$^3$, and its Schmidt number is 2.5 over a wide temperature range. The partial pressure of naphthalene gas at the surface is a function only of the surface temperature:

$$p_{Aw}\,(\text{pascals}) \doteq \exp(31.49 - 8673/T(K)). \tag{11.55}$$

If the air is isothermal, the naphthalene simulates "constant wall temperature", and the mass loss can be measured to high accuracy with modern balances that are accurate to $10^{-5}$ g. There is a good discussion of the technique on pp. 179–184 of [27]. Naphthalene sublimation is now so widely accepted that journal articles report "heat transfer experiments" in their title when in fact the data were obtained from naphthalene [30].

## Example 11.10

A long tube has a solid naphthalene inner lining of diameter 5 cm subjected to an air flow of 80 $m^3/h$ at 60°C. After one hour the inner diameter is found to be 5.15 cm. Estimate the Sherwood number and the Nusselt number of the equivalent air heat transfer problem and compare with Eq. (6.107).

*Solution* The average air velocity is

$$V = Q/A = \frac{80/3600}{\frac{\pi}{4}(0.05)^2} = 11.32\ \text{m/s}.$$

At 60°C, $\nu_{air} = 1.88 \times 10^{-5}\ m^2/s$, and since Sc = 2.5 for naphthalene, $D_{AB} = \nu/\text{Sc} = 0.75 \times 10^{-5}\ m^2/s$. The Reynolds number is

$$\text{Re}_D = VD/\nu = (11.32)(0.05)/1.88 \times 10^{-5} = 30{,}100.$$

For a unit length of tube, set the mass loss of naphthalene equal to the diffusion loss defined by Eq. (11.41):

$$m_{Aw} = \rho_{solid}\, L\, \Delta A_{solid} = A_w h_m(\rho_{Aw} - \rho_{A\infty}). \tag{1}$$

At 60°C = 333 K the naphthalene partial pressure is, from Eq. (11.55),

$$p_{Aw} = \exp(31.49 - 8673/333) = 235\ \text{Pa}.$$

Assuming an ideal gas, the gas density is therefore

$$\rho_{Aw} = \frac{(235)(128.2)}{(8310)(333)} = 0.0109\ \text{kg/m}^3.$$

Since the sublimation rate is so small, we assume that $\rho_{A\infty} = 0$. The density of the solid is 1078 kg/m$^3$, and we can assume an average diameter of 5.075 cm when computing the wall area. Equation (1) becomes

$$1078(1.0)\left(\frac{\pi}{4}\right)\{(0.0515)^2 - (0.05)^2\} = \pi(0.05075)(1.0)(0.0109 - 0)h_m,$$

which we solve for $h_m = 74.2$ m/hr $= 0.0206$ m/s. Then the experimental Sherwood number is

$$\text{Sh}_D = h_m D/D_{AB} = \frac{0.0206(0.05075)}{0.75 \times 10^{-5}} = 139. \quad [Ans.]$$

But this is *not* equal to the air Nusselt number because the Prandtl number of air is less than 2.5. It is necessary to convert, assuming that the one-third power of Sc expresses the Schmidt number effect accurately. Thus

$$\text{Nu}_D(\text{air}) \doteq 139(0.7/2.5)^{1/3} = 91. \quad [Ans.]$$

Spalding [27, p. 183] criticizes this conversion as inaccurate and a weakness of the naphthalene sublimation technique.

At $\text{Re}_D = 30{,}100$ and $\text{Pr} = 0.7$ the Dittus-Boelter formula (6.107) predicts that

$$\text{Nu}_D(\text{air}) = 0.027(30{,}100)^{0.8}(0.7)^{1/3} = 92.$$

The good agreement with the experimental value is somewhat misleading, since both rely upon a correlation $\text{Pr}^{1/3}$ or $\text{Sc}^{1/3}$. ■

## Summary

This chapter introduces the subject of molecular diffusion or mass transfer, with applications primarily to a mixture of two fluids, without chemical reactions. The concept is related to molecular collisions between particles of different species. The macroscopic result is Fick's law of diffusion, which is mathematically identical to Fourier's law of conduction.

The properties of mixtures of fluids are defined and lead to the concepts of mass and molar averages, along with ideal-gas simplifications. The diffusion coefficient is defined and tabulated for various binary mixtures. A new dimensionless parameter called the Schmidt number is defined; it is similar in character to the Prandtl number for heat transfer.

The basic differential equation of species conservation is derived, including convection effects, and found to be mathematically similar to the convection energy equation. Thus many known solutions to heat transfer problems can be converted to mass transfer by substituting concentration for temperature and diffusion coefficient for thermal diffusivity. For convective diffusion a mass transfer coefficient is defined with its associated Sherwood number. They are found to be mathematically similar to the heat transfer coefficient with its associated Nusselt number. One new complication is that mass transfer causes "wall blowing," which may have a significant effect on the transfer rates. The chapter ends with a number of practical applications to various geometries.

## References

1. E. L. Cussler, *Diffusion: Mass Transfer in Fluid Systems,* Cambridge University Press, New York, 1984.
2. R. E. Cunningham and R. J. Williams, *Diffusion in Gases and Porous Media,* Plenum Press, New York, 1980.
3. H. Tyrrell and K. Harris, *Diffusion in Liquids: A Theoretical and Experimental Study,* Butterworth, London, 1983.
4. R. E. Treybal, *Mass-Transfer Operations,* 3rd ed., McGraw-Hill, New York, 1980.
5. A. H. P. Skelland, *Diffusional Mass Transfer,* Wiley, New York, 1974.
6. T. K. Sherwood, R. L. Pigford, and C. R. Wilke, *Mass Transfer,* McGraw-Hill, New York, 1975.
7. E. R. G. Eckert and R. M. Drake, *Analysis of Heat and Mass Transfer,* McGraw-Hill, New York, 1972.
8. W. M. Kays and M. E. Crawford, *Convective Heat and Mass Transfer,* 2nd ed., McGraw-Hill, New York, 1980.
9. G. Astarita, *Mass Transfer with Chemical Reaction,* Elsevier, Amsterdam, 1967.
10. P. V. Danckwerts, *Gas-Liquid Reactions,* McGraw-Hill, New York, 1970.
11. J. Crank, *The Mathematics of Diffusion,* 2nd ed., Clarendon Press, Oxford, 1975.
12. R. B. Bird, W. E. Stewart, and E. N. Lightfoot, *Transport Phenomena,* Wiley, New York, 1960.
13. J. C. Slattery, *Momentum, Energy, and Mass Transfer in Continua,* McGraw-Hill, New York, 1972.

14. D. K. Edwards, V. E. Denny, and A. F. Mills, *Transfer Processes,* McGraw-Hill, New York, 1976.

15. W. M. Rohsenow and H. Y. Choi, *Heat, Mass, and Momentum Transfer,* Prentice-Hall, Englewood Cliffs, N.J., 1961.

16. C. O. Bennett and J. E. Myers, *Momentum, Heat, and Mass Transfer,* 2nd ed., McGraw-Hill, New York, 1974.

17. J. O. Hirschfelder, C. F. Curtiss, and R. B. Bird, *Molecular Theory of Gases and Liquids,* Wiley, New York, 1954.

18. R. M. Barrer, *Diffusion in and through Solids,* Macmillan, New York, 1941.

19. A. Fick, "On Liquid Diffusion," *Philosophical Magazine,* vol 4, no. 10, 1855, pp. 30–39.

20. N. B. Vargaftik, *Handbook of Physical Properties of Liquids and Gases,* 2nd ed., Hemisphere, New York, 1975.

21. R. C. Reid, J. M. Prausnitz, and T. K. Sherwood, *The Properties of Gases and Liquids,* 3rd ed., McGraw-Hill, New York, 1977.

22. S. V. Patankar, *Numerical Heat Transfer and Fluid Flow,* McGraw-Hill, New York, 1980.

23. W. Jost, *Diffusion in Solids, Liquids, Gases,* Revised Edition, Academic Press, New York, 1960.

24. R. C. Weast (Ed.), *CRC Handbook of Chemistry and Physics,* 66th edition, CRC Press, Boca Raton, Fla., 1985.

25. J. H. Keenan et al., *Steam Tables,* Wiley, New York, 1978.

26. J. P. Hartnett and E. R. G. Eckert, "Mass Transfer Cooling in a Laminar Boundary Layer with Constant Fluid Properties," *ASME Transactions,* vol. 79, 1957, pp. 247–254.

27. D. B. Spalding, *Convective Mass Transfer,* Edward Arnold Ltd., London, 1963.

28. L. Haar, J. S. Gallagher, and G. S. Kell, *NBS/NRC Steam Tables,* Hemisphere, New York, 1984.

29. C. C. Winding and A. J. Cheney, "Mass and Heat Transfer in Tube Banks," *Ind. Eng. Chem.,* vol. 40, 1948, pp. 1087–1093.

30. E. M. Sparrow and G. M. Chrysler, "Turbulent Flow and Heat Transfer in Bends of Circular Cross Section: I, Heat Transfer Experiments," *J. Heat Transfer,* vol. 108, Feb. 1986, pp. 40–47.

31. E. N. Fuller, P. D. Schettler, and J. C. Giddings, "A New Method for Prediction of Binary Gas-Phase Diffusion Coefficients," *Ind. Eng. Chem.,* vol. 58, May 1966, pp. 19–27.

32. Robert H. Perry (Ed.), *Chemical Engineer's Handbook,* 6th ed., McGraw-Hill, New York, 1984.

## Review Questions

1. Define mass transfer. How does it relate to molecular theory?
2. What is the "diffusion velocity" of a particular species?
3. What are mass averages versus molar averages?
4. How does the partial pressure of a gas relate to the mass fraction or molar fraction of that species?
5. Define Fick's law. How does it relate to Fourier's law?
6. Define the Schmidt number and Lewis number. Which one has an effect that is very similar to that of the Prandtl number?
7. What is the species conservation equation? How does it relate to the equation of conservation of energy?
8. What are typical boundary conditions in diffusion problems?
9. What is the heat/mass transfer analogy?
10. Can any of the conduction analyses of Chapters 2–4 be applied to mass transfer problems? Why?
11. Does stream flow increase the evaporation rate from a liquid surface? What is the mechanism?
12. Define and sketch a concentration boundary layer.
13. Define the mass transfer coefficient and the Sherwood number. Is there a mass transfer parameter equivalent to the Stanton number? To the Nusselt number?
14. Why does wall blowing occur in mass transfer? How does it affect the mass transfer rate?
15. What is the naphthalene sublimation technique?

## Problems

**Problem distribution by sections**

**The Problem Assignments Are Organized as Follows**

| Problems | Sections Covered | Topics Covered |
|---|---|---|
| 11.1–11.2 | 11.1 | Introduction |
| 11.3–11.7 | 11.2 | Properties of mixtures |
| 11.8–11.10 | 11.3 | Fick's law, diffusivities |
| 11.11–11.16 | 11.4 | The diffusion equation |
| 11.17–11.24 | 11.5 | Diffusion without convection |
| 11.25–11.26 | 11.5 | Equimolar counterdiffusion |

| | | |
|---|---|---|
| 11.27–11.34 | 11.5 | Diffusion from a container |
| 11.35–11.38 | 11.6 | Flat plate mass transfer |
| 11.39–11.41 | 11.6 | Wall blowing effects |
| 11.42–11.54 | 11.6 | Other geometries |
| 11.55–11.60 | 11.6 | Naphthalene sublimation |

11.1 Given the fact that diffusion in gases is a result of random molecular collisions, use the sketch in Fig. 11.1 to explain why the rate of diffusion should be proportional to the gradient of the gas concentration.

11.2 The simplest molecular theory of diffusion [12, p. 510] leads to the following expression for diffusivity:

$$D_{AB} = \frac{1}{3}u_m\lambda,$$

where $\lambda$ is the mean free path and $u_m$ is the mean molecular velocity. Is this formula dimensionally correct? What does it imply about the variation of $D_{AB}$ with temperature and pressure?

11.3 Consider a mixture of 40% carbon dioxide and 60% nitrogen gas by weight at 30°C with a density of 1.5 kg/m$^3$. Compute (a) the mole fractions; (b) the molar densities; (c) the mixture molecular weight; and (d) the pressure.

11.4 A gas mixture at 50°C consists of oxygen, nitrogen, and helium at equal partial pressures of 40 kPa. What are (a) the mass fractions; (b) the mixture molecular weight; and (c) the density?

11.5 Carbon dioxide and chlorine at partial pressures of 0.4 atm each are combined with 0.2 atm of an unknown gas. If the mixture temperature is 80°C and the density is 1.787 kg/m$^3$, what are the molecular weight and mass fraction of the unknown gas?

11.6 Derive Eqs. (11.14).

11.7 A gas mixture at 20°C and 1 atm consists of 50% helium, 30% oxygen, and 20% nitrogen by weight. Their respective velocities in the $x$ direction are $u_1 = -2$ m/s, $u_2 = 8$ m/s, and $u_3 = -7$ m/s. What are the mixture density and mass-averaged $x$ velocity? What are the respective mass transfer rates $j_{ix}$?

11.8 In the absence of experimental data the diffusion coefficient for a binary gas mixture may be estimated from the correlation formula of Fuller et al. [31]:

$$D_{AB}\,(\mathrm{m^2/s}) \doteq \frac{0.01T^{7/4}\,(M_A^{-1} + M_B^{-1})^{1/2}}{p(V_A^{1/3} + V_B^{1/3})^2} \tag{1}$$

with $T$ in °K and $p$ in pascals. The quantities $V_A$ and $V_B$ are the *molecular diffusion volumes* of the gases and are tabulated in Appendix K. Use this formula to predict the diffusivity of argon in hydrogen at 90°C and 100 atm and compare with the measured value of $1.0 \times 10^{-6}$ m$^2$/s [20].

11.9 Use Fuller's formula (1) in Problem 11.8 to predict the diffusion coefficients of (a) benzene vapor and (b) carbon dioxide in air at 300°C and 1 atm. Compare the results with Fig. 11.3.

11.10 Nitrogen at 20°C and 15 atm is confined in a rubber sphere of diameter 40 cm and thickness 3 mm. The nitrogen concentration at the inner sphere surface is 0.03 kmols/m$^3$, and the diffusion coefficient for $N_2$ through rubber is $1.5 \times 10^{-8}$ m$^2$/s [18]. Estimate the pressure in the sphere after four hours.

11.11 Using a differential control volume similar to Fig. 11.4, derive the molar form of the two-dimensional unsteady conservation of species relation (Eq. 11.28).

11.12 A hollow cylinder of inner radius $r_1$ and outer radius $r_2$ contains a gaseous species of uniform concentration $C_i$, as shown in Fig. P11.12. At time $t = 0$ the inner surface concentration is suddenly raised to $C_0$. There are no chemical reactions. Set up the differential equation and boundary conditions for the resulting distribution $C(r, t)$ in the cylinder. Solve for the distribution $C(r)$ after a long period of time.

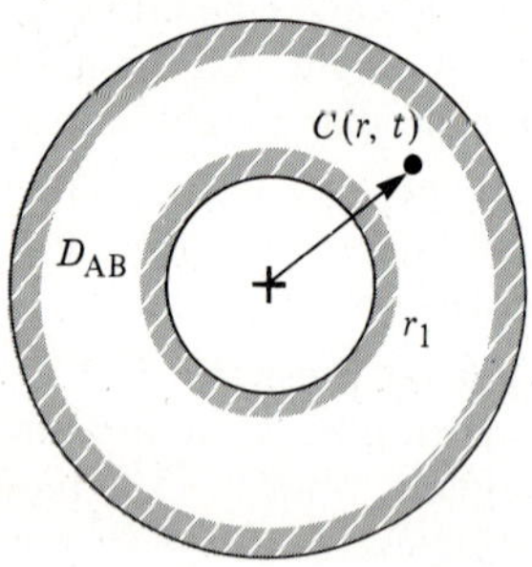


Figure P11.12

11.13 Repeat Problem 11.12 if the geometry is a hollow sphere instead of a cylinder.

11.14 Reconsider either Problem 11.12 or Problem 11.13 and assume that there are also uniform chemical reactions $\dot{m}_A$ throughout the solid. Find the equation for $C(r, t)$ and the final distribution $C(r, t \to \infty)$. You may choose whether to take the geometry as a cylinder or a sphere.

11.15 A Pyrex tube 1 m long, with inner and outer diameters of 6 cm and 8 cm, respectively, is filled with helium so that the inner surface concentration is 0.6 kmols/m$^3$. The outer concentration is negligible. If the diffusivity of helium in Pyrex is $2 \times 10^{-12}$ m$^2$/s, how many grams of helium will diffuse through the glass in one day?

11.16 Write out the three-dimensional incompressible form of the species conservation equation with constant $D_{AB}$ and compare with the three-dimensional form of the energy equation (5.23) for laminar flow. Under what conditions is the heat/mass transfer analogy valid?

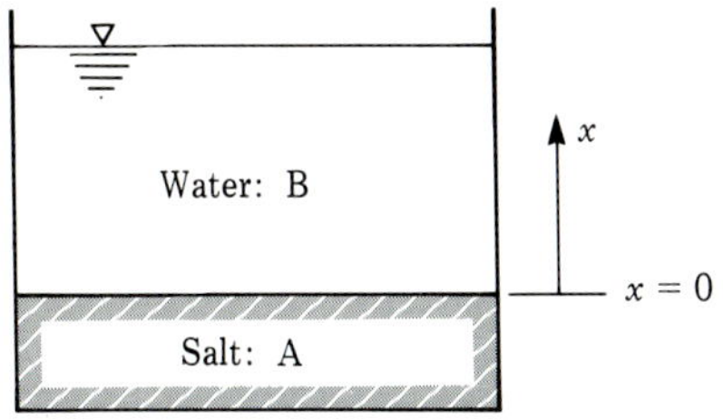


Figure P11.17

11.17 Fresh water is suddenly overlaid on a block of salt as in Fig. P11.17. The solubility of salt in water is 39.12% [24, p. B-125], so the salt density at the water interface ($x = 0$) immediately takes on the value $\rho_A(0) = 0.3912(998) = 390$ kg/m$^3$. Take the salt diffusivity from Table 11.1 and assume that the water is stagnant. How long will it take the salt concentration 1 cm above the interface to reach 100 kg/m$^3$? If the interface area is 1 m$^2$, how many kilograms of salt will have dissolved away in one day?

11.18 A 1-cm-diameter sphere of solid naphthalene is suspended in still air at 20°C and 1 atm. The sphere density is 1078 kg/m$^3$, and its partial pressure at the surface is 7 Pa (see Eq. 11.55). Take the diffusivity from Table 11.1. What is the initial rate of mass loss of naphthalene vapor? Derive a formula for the total time $t_0$ required to diffuse away the entire sphere.

11.19 A 1-mm droplet of water is suspended on a wire in still air at 20°C and 1 atm. Take the diffusivity of water vapor from Table 11.1. How long will it take the entire droplet to evaporate?

11.20 Oxygen is diffusing through the conical rubber stopper in Fig. P11.20. Its concentration $C_2 = 0$ and $C_1 = 0.04$ kmols/m$^3$, and $D_{AB} = 2.1 \times 10^{-8}$ m$^2$/s. Derive a formula for the diffusion mass flow $\dot{m}_{O_2,x}$ through the stopper and find its numerical value for this case. Does the mass flow change if we reverse $C_1$ and $C_2$?

Figure P11.20

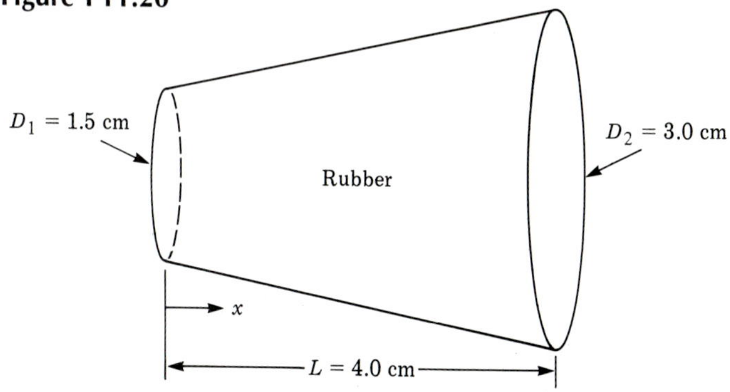

11.21 Modify Example 11.3 as follows: Let the steel slab have a *finite* thickness of 2 mm, with initial carbon concentration of 0.5%. If both sides of the slab are suddenly subjected to concentrations of 1.3%, how long will it take the center of the slab to rise to 0.7% carbon? Should your result be less or greater than the 630 s computed in Example 11.3? [*Hint:* This case corresponds to a modified Biot number of infinity in Table 4.1.]

11.22 Consider the interface between gas A and a liquid B in which it is weakly soluble, as in Fig. P11.22. The concentration of dissolved gas at the interface may be estimated from *Henry's law* [4, p. 279]:

$$x_A(\text{interface}) = p_A(\text{gas})/H,$$

where $p_A$ is the gas partial pressure at the interface and $H$ is Henry's constant, for which experimental values are tabulated in [32]. For oxygen dissolving in water at 20°C, $H$ = 40,100 atm per mole fraction. If the oxygen pressure is 21 kPa and the water is initially oxygen-free, how long will it take the water 1 cm below the interface to develop an oxygen mole fraction of $2 \times 10^{-6}$?

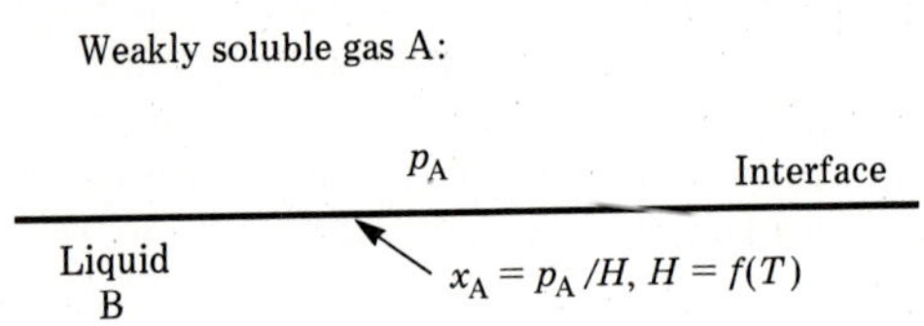


Figure P11.22

11.23 A 1-mm-diameter oxygen-free water droplet is suspended in pure oxygen at 1 atm and 20°C. If the surface concentration is estimated from Henry's law in Problem 11.22, how long will it take the oxygen mole fraction at the droplet center to reach $2 \times 10^{-5}$?

11.24 A solid naphthalene cylinder 1 cm in diameter and 40 cm long is suspended 5 cm below the ceiling in air at 20°C and 1 atm, as in Fig. P11.24. Take the naphthalene properties from Problem 11.18. Estimate (a) the initial naphthalene mass loss rate and (b) the time for the cylinder to sublime away completely.

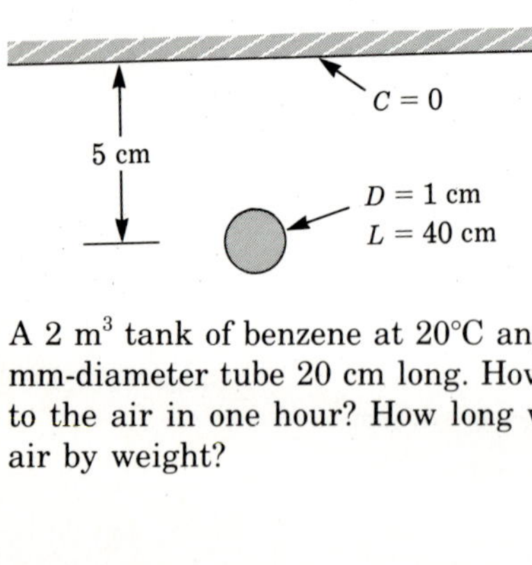


Figure P11.24

11.25 A 2 $m^3$ tank of benzene at 20°C and 1 atm is vented to air through a 4-mm-diameter tube 20 cm long. How many grams of benzene will be lost to the air in one hour? How long will it take the tank to contain 0.1% air by weight?

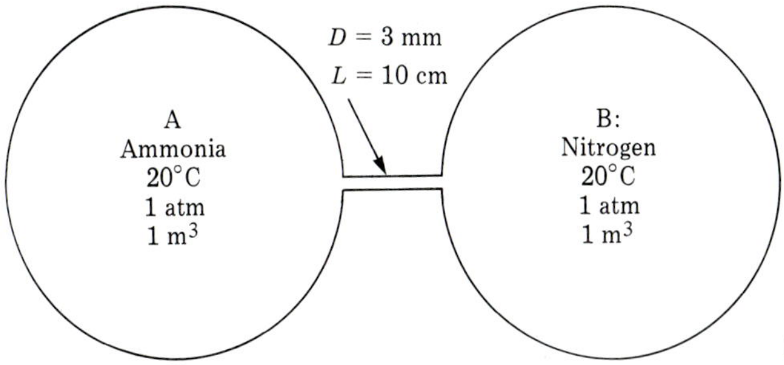


**Figure P11.26**

11.26 A tank of ammonia is connected to a tank of nitrogen by a short tube as in Fig. P11.26. Estimate the diffusivity from Fuller's formula, Problem 11.8. If each tank is initially pure, how many days will it take for the nitrogen tank to contain 1% ammonia by weight?

11.27 A test tube is 2 cm in diameter and 10 cm high and contains 1 mm of water at 30°C. The ambient air is dry at 1 atm and 30°C. How long will it take the water to evaporate completely from the tube?

11.28 Repeat Problem 11.27 if the ambient air has a relative humidity of 50%.

11.29 A shallow pan 2 cm high and 30 cm in diameter is placed in dry air at 20°C and 1 atm. The pan contains 1 mm of benzene at 20°C, for which $p_v$ = 75 mm Hg and $\rho$ = 874 kg/m$^3$. How long will it take the benzene to evaporate completely?

11.30 As a dramatic illustration of the "barrier" effect of diffusion on the evaporation process, Treybal [4, p. 21] states that (a) water at 25°C will evaporate at 3.3 kg/(s·m$^2$) into a complete vacuum and (b) if overlaid with a 1-mm-thick layer of air at 1 atm, the water will evaporate 600 times slower. How is estimate (a) obtained? Can you verify part (b)?

11.31 Consider air (gas B) in a closed container and initially dry. At time $t$ = 0, liquid A in the bottom of the container begins to diffuse into the stagnant air. Is the theory of Section 11.5.2 valid as a quasi-static ap-

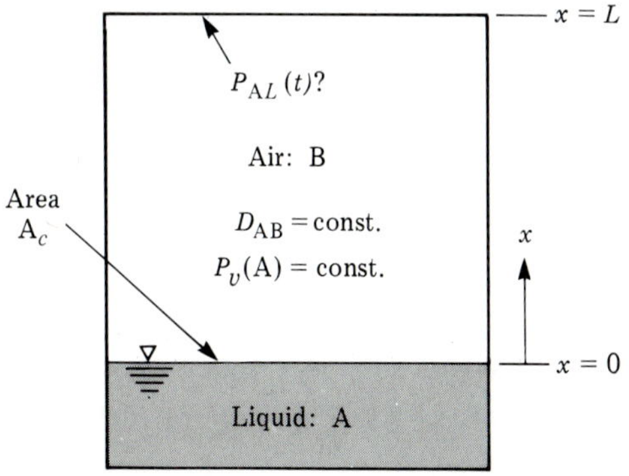


**Figure P11.31**

proximation? Assuming that the vapor pressure $p_v(A)$ is much smaller than the total pressure, develop an approximate theory for predicting the partial pressure $p_{AL}(t)$ at the top of the container. See Fig. P11.31 for the geometry.

11.32 The effect of convection on evaporation is large. When atmospheric air at 100°F and 30% relative humidity blows over a pan of water at 10 mi/hr, the estimated evaporation rate is 31 mm/day. What stagnant-air thickness $L$ in Fig. 11.6 would give the same evaporation rate?

11.33 Toluene ($C_7H_8$) liquid has a density of 862 kg/m$^3$ and a vapor pressure of 40 mm Hg at 32°C. When placed in a 1.5-cm-diameter test tube 7 cm high in air at 32°C, 0.093 cm$^3$ are observed to evaporate in five hours. Estimate the diffusivity of toluene vapor in air. What are some uncertainties in this measurement scheme?

11.34 A recent TV commercial derides alcohol for use in cosmetics because "it evaporates fast." Yet from Table 11.1 the diffusivities of methanol and ethanol vapor are modest, smaller than water vapor. Is alcohol's evaporation prowess a myth? If not, what accounts for the high rates?

11.35 Assume that the water pan in Problem 11.32 has a diameter of 70 cm and estimate the evaporation rate from flat plate convection mass transfer theory. If your result does not equal 31 mm/day, account for the discrepancy.

11.36 Worden's Pond in South Kingstown, Rhode Island, has a diameter of two statute miles. If dry air at 20°C and 1 atm blows over the pond at 8 mi/hr, estimate the evaporation rate in m$^3$/day. Neglect wall blowing.

11.37 Modify Example 11.5 so that the benzene liquid is in a shallow pan of diameter 20 cm and the air flows over the pan at 5 m/s. Estimate the mass loss in g/hr.

11.38 Methanol liquid at 20°C has a vapor pressure of 93 mm Hg and is placed in a shallow pan 1 m wide and 1 m long. It is desired to evaporate 3000 g/hr of the methanol by blowing over the pan with air at 20°C and 1 atm. What air velocity is required? Neglect blowing.

11.39 Modify Problem 11.36 to account for wall blowing. What water/air temperature is required for the blowing effect to be 10%?

11.40 Repeat Problem 11.38 by accounting for wall blowing.

11.41 Assuming a transition Reynolds number of $5 \times 10^5$, develop a formula for mean Sherwood number in flat plate flow at moderate turbulent Reynolds numbers where a significant portion of the flow is laminar.

11.42 Modify Problem 11.18 by letting the air flow past the sphere at a velocity of 2.5 m/s. Compute (a) the initial sphere loss and (b) the time for the sphere to disappear.

11.43 Modify Problem 11.24 by removing the ceiling and letting the air move in crossflow past the naphthalene cylinder at 4.5 m/s. Estimate the time for the cylinder to sublime away completely.

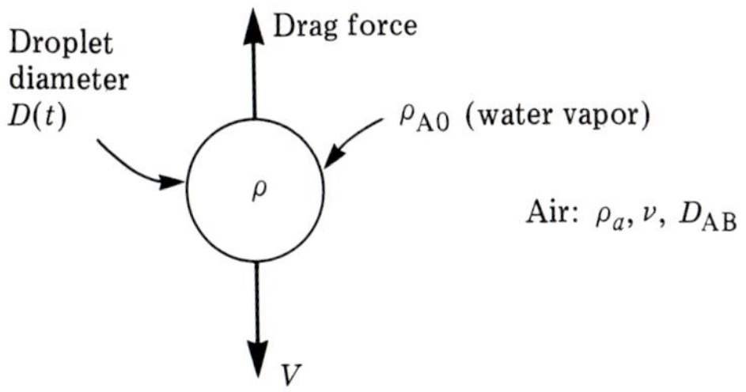


Figure P11.44

11.44 Can rain evaporate before it hits the ground? Consider a water droplet falling in terminal velocity in air of 50% relative humidity, as in Fig. P11.44. If the Reynolds number $VD/\nu_{\text{air}}$ is unity or less, the drag force is given by Stokes' law, $F = 3\pi\mu_{\text{air}}VD$. Assume that the water vapor pressure at the droplet surface, $p_{A0}$, is constant and that $\text{Re}_D \leq 1.0$. Derive expressions for (a) the fall velocity and (b) the distance $L$ a droplet of initial diameter $D_0$ will fall before it evaporates completely. Neglect the variation of air properties with altitude.

11.45 Apply the theory of Problem 11.44 to a water droplet of 0.5 mm diameter falling in dry air at 20°C and 1 atm. How far will the droplet fall before evaporating?

11.46 Warm, dry air supplied to heat a room is to be humidified by passing it at 8 m/s over a saturated wet drum 30 cm in diameter and 1 m long. The system is at 35°C. Estimate the amount of water added to the air in kg/hr. This rate greatly increases with temperature. What is the principal mechanism?

11.47 Approximate the human arm as a cylinder 9 cm in diameter and 50 cm long, with a surface temperature of 35°C. If exposed to a wind of 8 m/s in air at 20°C and 1 atm, estimate the heat loss of the arm in watts if it is (a) dry; (b) wet.

11.48 Air humidity can be measured by a *wet-bulb thermometer*, shown in Fig. P11.48. Liquid evaporates from the wet surface, and the enthalpy of

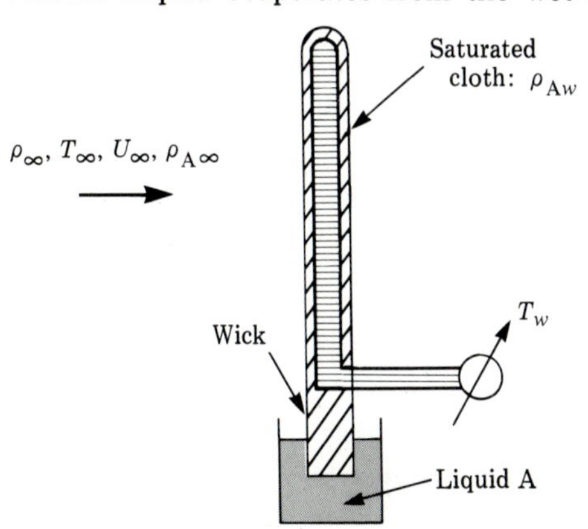


Figure P11.48

vaporization is balanced by heat transfer from the (warmer) gas stream. If radiation is negligible and the heat and mass transfer are proportional to $\mathrm{Pr}^{1/3}$ and $\mathrm{Sc}^{1/3}$, respectively, show that the wet-bulb temperature is approximately

$$T_w = T_\infty - \frac{(\rho_{Aw} - \rho_{A\infty})\, h_{fgA}}{\rho_\infty c_{p\infty}} \mathrm{Le}^{-2/3},$$

where $h_{fgA}$ is the heat of vaporization of liquid A, with values tabulated in Table 9.1. If, for example, $T_w$ is measured, we may compute the stream species density $\rho_{A\infty}$.

11.49 Suppose that a wet-bulb thermometer registers 15°C when the temperature of the air stream is 20°C at 1 atm. Using the theory of Problem 11.48, determine the relative humidity of the air stream.

11.50 Suppose the wet-bulb thermometer of Fig. P11.48 uses methanol as the liquid and registers a temperature $T_w$ = 20°C. If the vapor pressure of methanol is 93 mm Hg and the air stream is dry, estimate the temperature of the air stream. Neglect radiation.

11.51 The air temperature in Problem 11.50 is relatively high. Modify this problem by accounting for radiation exchange, assumed to be black, between the air stream and the wet bulb.

11.52 As shown in Fig. P11.52, dry air at 20°C and 1 atm is used to evaporate carbon tetrachloride that wets the inside of a tube 1 m long and 3 cm in diameter. The air flow rate is 2 kg/hr. Estimate (a) the exit $CCl_4$ density $\rho_{B2}$ and (b) the evaporation rate in kg/hr. The vapor pressure of $CCl_4$ at 20°C is 86 mm Hg.

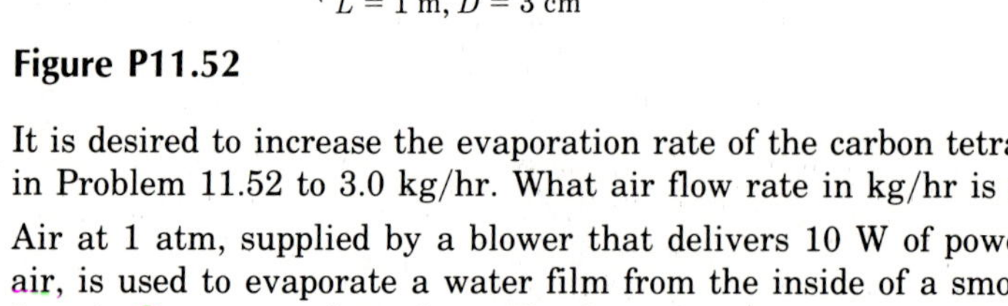


**Figure P11.52**

11.53 It is desired to increase the evaporation rate of the carbon tetrachloride in Problem 11.52 to 3.0 kg/hr. What air flow rate in kg/hr is needed?

11.54 Air at 1 atm, supplied by a blower that delivers 10 W of power to the air, is used to evaporate a water film from the inside of a smooth tube 5 cm in diameter and 4 m long. The desired evaporation rate is 10 kg/hr. If the system is isothermal, what is the appropriate temperature in °C?

11.55 Dry air at 20°C and 1 atm flows at 2.6 m/s past a solid naphthalene sphere of diameter 1 cm. After two hours the sphere diameter has decreased by 60 $\mu$m. Estimate (a) the mean Sherwood number and (b) the equivalent air-convection Nusselt number of this flow.

11.56 In the experiment of Problem 11.55 the naphthalene sphere is found to disappear entirely after 117 hours. Use this information to estimate the mean Sherwood number during the experiment. How does the result compare with Eq. (11.51)?

11.57 Air flows at 4.5 m/s across a solid naphthalene cylinder 1 cm in diameter and 40 cm long. The system is at 40°C and 1 atm. The measured cylinder mass loss is 1.4 g in 30 minutes. What is the mean Sherwood number? What is the equivalent mean Nusselt number for air convection heat transfer?

11.58 Air at 20°C and 1 atm flows at 10 m/s past a thin flat plate coated with solid naphthalene. The measured naphthalene mass loss is 2.5 g in 20 minutes. Estimate the mean Sherwood number and explain why the greatest mass loss is (a) at the leading edge and (b) in the center of the plate, which is 60 cm long and 50 cm wide.

11.59 Suppose that the air speed in Problem 11.58 is raised to 350 m/s and the plate is insulated. What new parameters complicate the mass transfer problem? Estimate how long it takes under these new conditions for the plate to lose 2.5 g of naphthalene.

11.60 A naphthalene sublimation experiment, using air at 20°C and 1 atm, is used to measure heat transfer in tube flow at a Reynolds number of 15,000. The tube is 3 cm in diameter and 90 cm long. An accurate weight cannot be determined until the mass loss is 2.5 grams. How long should the experimenter wait?

# Appendixes

# Appendix A
# Mathematical Tables

## Complementary error function, from Eq. (4.26)

| $\beta$ | erfc ($\beta$) | $\beta$ | erfc ($\beta$) |
|---|---|---|---|
| 0 | 1.0 | 1.1 | 0.11980 |
| 0.05 | 0.94363 | 1.2 | 0.08969 |
| 0.1 | 0.88754 | 1.3 | 0.06599 |
| 0.15 | 0.83200 | 1.4 | 0.04772 |
| 0.2 | 0.77730 | 1.5 | 0.03390 |
| 0.25 | 0.72367 | 1.6 | 0.02365 |
| 0.3 | 0.67137 | 1.7 | 0.01621 |
| 0.35 | 0.62062 | 1.8 | 0.01091 |
| 0.4 | 0.57161 | 1.9 | 0.00721 |
| 0.5 | 0.47950 | 2.0 | 0.00468 |
| 0.6 | 0.39615 | 2.5 | 0.000407 |
| 0.7 | 0.32220 | 3.0 | 0.0000221 |
| 0.8 | 0.25790 | 3.5 | 0.00000074 |
| 0.9 | 0.20309 | 4.0 | 0.00000001 |
| 1.0 | 0.15730 | $\infty$ | 0.0 |

## The first four roots of Eq. (4.38)

| $\beta_i \tan(\beta_i) = h_o L/k$ | | | | |
|---|---|---|---|---|
| $h_o L/k$ | $\beta_1$ | $\beta_2$ | $\beta_3$ | $\beta_4$ |
| 0 | 0 | 3.1416 | 6.2832 | 9.4248 |
| 0.001 | 0.0316 | 3.1419 | 6.2833 | 9.4249 |
| 0.002 | 0.0447 | 3.1422 | 6.2835 | 9.4250 |
| 0.004 | 0.0632 | 3.1429 | 6.2838 | 9.4252 |
| 0.006 | 0.0774 | 3.1435 | 6.2841 | 9.4254 |
| 0.008 | 0.0893 | 3.1441 | 6.2845 | 9.4256 |
| 0.01 | 0.0998 | 3.1448 | 6.2848 | 9.4258 |

The first four roots of Eq. (4.38) *continued*

$\beta_i \tan(\beta_i) = h_o L/k$

| $h_o L/k$ | $\beta_1$ | $\beta_2$ | $\beta_3$ | $\beta_4$ |
|---|---|---|---|---|
| 0.02 | 0.1410 | 3.1479 | 6.2864 | 9.4269 |
| 0.04 | 0.1987 | 3.1543 | 6.2895 | 9.4290 |
| 0.06 | 0.2425 | 3.1606 | 6.2927 | 9.4311 |
| 0.08 | 0.2791 | 3.1668 | 6.2959 | 9.4333 |
| 0.1 | 0.3111 | 3.1731 | 6.2991 | 9.4354 |
| 0.2 | 0.4328 | 3.2039 | 6.3148 | 9.4459 |
| 0.3 | 0.5218 | 3.2341 | 6.3305 | 9.4565 |
| 0.4 | 0.5932 | 3.2636 | 6.3461 | 9.4670 |
| 0.5 | 0.6533 | 3.2923 | 6.3616 | 9.4775 |
| 0.6 | 0.7051 | 3.3204 | 6.3770 | 9.4879 |
| 0.7 | 0.7506 | 3.3477 | 6.3923 | 9.4983 |
| 0.8 | 0.7910 | 3.3744 | 6.4074 | 9.5087 |
| 0.9 | 0.8274 | 3.4003 | 6.4224 | 9.5190 |
| 1.0 | 0.8603 | 3.4256 | 6.4373 | 9.5293 |
| 1.5 | 0.9882 | 3.5422 | 6.5097 | 9.5801 |
| 2.0 | 1.0769 | 3.6436 | 6.5783 | 9.6296 |
| 3.0 | 1.1925 | 3.8088 | 6.7040 | 9.7240 |
| 4.0 | 1.2646 | 3.9352 | 6.8140 | 9.8119 |
| 5.0 | 1.3138 | 4.0336 | 6.9096 | 9.8928 |
| 6.0 | 1.3496 | 4.1116 | 6.9924 | 9.9667 |
| 7.0 | 1.3766 | 4.1746 | 7.0640 | 10.0339 |
| 8.0 | 1.3978 | 4.2264 | 7.1263 | 10.0949 |
| 9.0 | 1.4149 | 4.2694 | 7.1806 | 10.1502 |
| 10.0 | 1.4289 | 4.3058 | 7.2281 | 10.2003 |
| 15.0 | 1.4729 | 4.4255 | 7.3959 | 10.3898 |
| 20.0 | 1.4961 | 4.4915 | 7.4954 | 10.5117 |
| 30.0 | 1.5202 | 4.5615 | 7.6057 | 10.6543 |
| 40.0 | 1.5325 | 4.5979 | 7.6647 | 10.7334 |
| 50.0 | 1.5400 | 4.6202 | 7.7012 | 10.7832 |
| 60.0 | 1.5451 | 4.6353 | 7.7259 | 10.8172 |
| 80.0 | 1.5514 | 4.6543 | 7.7573 | 10.8606 |
| 100.0 | 1.5552 | 4.6658 | 7.7764 | 10.8871 |
| $\infty$ | 1.5708 | 4.7124 | 7.8540 | 10.9956 |

Bessel functions of the first kind

| $x$ | $J_0(x)$ | $J_1(x)$ | $x$ | $J_0(x)$ | $J_1(x)$ |
|---|---|---|---|---|---|
| 0.0 | 1.0 | 0.0 | 6.0 | 0.15065 | −0.27668 |
| 0.2 | 0.99003 | 0.09950 | 6.2 | 0.20175 | −0.23292 |
| 0.4 | 0.96040 | 0.19603 | 6.4 | 0.24331 | −0.18164 |
| 0.6 | 0.91200 | 0.28670 | 6.6 | 0.27404 | −0.12498 |
| 0.8 | 0.84629 | 0.36884 | 6.8 | 0.29310 | −0.06522 |
| 1.0 | 0.76520 | 0.44005 | 7.0 | 0.30008 | −0.00468 |
| 1.2 | 0.67113 | 0.49829 | 7.2 | 0.29507 | 0.05432 |
| 1.4 | 0.56686 | 0.54195 | 7.4 | 0.27860 | 0.10963 |
| 1.6 | 0.45540 | 0.56990 | 7.6 | 0.25160 | 0.15921 |
| 1.8 | 0.33999 | 0.58152 | 7.8 | 0.21541 | 0.20136 |
| 2.0 | 0.22389 | 0.57672 | 8.0 | 0.17165 | 0.23464 |
| 2.2 | 0.11036 | 0.55596 | 8.2 | 0.12222 | 0.25800 |
| 2.4 | 0.00251 | 0.52019 | 8.4 | 0.06916 | 0.27079 |
| 2.6 | −0.09681 | 0.47082 | 8.6 | 0.01462 | 0.27275 |
| 2.8 | −0.18504 | 0.40970 | 8.8 | −0.03923 | 0.26407 |
| 3.0 | −0.26005 | 0.33906 | 9.0 | −0.09033 | 0.24531 |
| 3.2 | −0.32019 | 0.26134 | 9.2 | −0.13675 | 0.21741 |
| 3.4 | −0.36430 | 0.17923 | 9.4 | −0.17677 | 0.18163 |
| 3.6 | −0.39177 | 0.09547 | 9.6 | −0.20898 | 0.13952 |
| 3.8 | −0.40256 | 0.01282 | 9.8 | −0.23228 | 0.09284 |
| 4.0 | −0.39715 | −0.06604 | 10.0 | −0.24594 | 0.04347 |
| 4.2 | −0.37656 | −0.13865 | 10.2 | −0.24962 | −0.00662 |
| 4.4 | −0.34226 | −0.20278 | 10.4 | −0.24337 | −0.05547 |
| 4.6 | −0.29614 | −0.25655 | 10.6 | −0.22764 | −0.10123 |
| 4.8 | −0.24043 | −0.29850 | 10.8 | −0.20320 | −0.14217 |
| 5.0 | −0.17760 | −0.32760 | 11.0 | −0.17119 | −0.17679 |
| 5.2 | −0.11029 | −0.34322 | 11.2 | −0.13299 | −0.20385 |
| 5.4 | −0.04121 | −0.34534 | 11.4 | −0.09021 | −0.22245 |
| 5.6 | 0.02697 | −0.33433 | 11.6 | −0.04462 | −0.23200 |
| 5.8 | 0.09170 | −0.31103 | 11.8 | 0.00197 | −0.23229 |
|  |  |  | 12.0 | 0.04769 | −0.22345 |

# Appendix B
# Conversion Factors

| | |
|---|---|
| Acceleration | 1 $ft/s^2$ = 0.3048 $m/s^2$ |
| Area | 1 $ft^2$ = 0.092903 $m^2$ |
| Density | 1 $lb_m/ft^3$ = 16.0185 $kg/m^3$<br>1 $slug/ft^3$ = 515.38 $kg/m^3$ |
| Energy | 1 Btu = 1055.056 J<br>1 cal = 4.1868 J<br>1 ft $lb_f$ = 1.35582 J |
| Force | 1 $lb_f$ = 4.448222 N |
| Heat transfer rate | 1 Btu/hr = 0.29307 W |
| Heat flux | 1 Btu/hr · $ft^2$ = 3.15459 $W/m^2$ |
| Heat generation rate | 1 Btu/hr · $ft^3$ = 10.3497 $W/m^3$ |
| Heat transfer coefficient | 1 Btu/hr · $ft^2$ · °F = 5.67826 $W/m^2$ · K |
| Kinematic viscosity | 1 $ft^2/s$ = 0.092903 $m^2/s$ |
| Length | 1 ft = 0.3048 m<br>1 in. = 2.54 cm<br>1 mi = 1609.344 m |
| Latent heat | 1 $Btu/lb_m$ = 2326.0 J/kg |
| Mass | 1 $lb_m$ = 0.4535924 kg<br>1 slug = 14.5939 kg |
| Mass flow rate | 1 $lb_m/s$ = 0.4535924 kg/s |
| Power | 1 hp = 745.7 W<br>1 ft · $lb_f/s$ = 1.35582 W<br>1 Btu/hr = 0.29307 W |
| Pressure | 1 $lb_f/in.^2$ = 6894.76 $N/m^2$<br>1 $lb_f/ft^2$ = 47.880 $N/m^2$<br>1 bar = $10^5$ $N/m^2$<br>1 atm = 101,325 $N/m^2$ |
| Specific heat | 1 $Btu/lb_m$ · °F = 4186.8 J/kg · K |
| Temperature | $T$(°R) = 1.8 $T$(K)<br>$T$(°F) = 1.8 $T$(°C) + 32<br>$T$(°R) = $T$(°F) + 459.67<br>$T$(K) = $T$(°C) + 273.15 |
| Thermal conductivity | 1 Btu/hr · ft · °F = 1.7307 W/m · K |
| Thermal diffusivity | 1 $ft^2/s$ = 0.092903 $m^2/s$ |
| Thermal resistance | 1 hr · °F/Btu = 1.8956 K/W |
| Velocity | 1 ft/s = 0.3048 m/s |
| Viscosity | 1 $lb_m/ft$ · s = 1.4882 kg/m · s<br>1 slug/ft · s = 47.88 kg/m · s |
| Volume | 1 $ft^3$ = 0.028317 $m^3$<br>1 gal (U.S.) = 0.0037854 $m^3$ |
| Volume flow rate | 1 $ft^3/s$ = 0.028317 $m^3/s$<br>1 gal/min = 6.309 × $10^{-5}$ $m^3/s$ |

# Appendix C
# Properties of Metallic Solids

| Material | Melting Point (K) | Properties at 20°C | | | Thermal Conductivity, $k$ (W/m · K) | | | | | |
|---|---|---|---|---|---|---|---|---|---|---|
| | | $\rho$ (kg/m$^3$) | $c_p$ (J/kg · K) | $\alpha$ (m$^2$/s) | 0°C | 200°C | 400°C | 600°C | 800°C | 1000°C |
| Aluminum | | | | | | | | | | |
| Pure | 933 | 2702 | 902 | $9.7 \times 10^{-5}$ | 236 | 238 | 228 | 215 | | |
| 2024-T6 | 775 | 2780 | 880 | $6.9 \times 10^{-5}$ | 171 | 186 | | | | |
| Beryllium | 1550 | 1850 | 1809 | $6.3 \times 10^{-5}$ | 218 | 144 | 118 | 100 | 83 | 69 |
| Boron | 2573 | 2500 | 1090 | $1.1 \times 10^{-5}$ | 31.7 | 18.3 | 10.0 | 7.4 | 5.9 | 4.8 |
| Cadmium | 594 | 8650 | 231 | $5.2 \times 10^{-5}$ | 104 | 99 | | | | |
| Chromium | 2118 | 7160 | 451 | $2.9 \times 10^{-5}$ | 95 | 86 | 77 | 69 | 64 | 62 |
| Cobalt | 1765 | 8862 | 419 | $2.7 \times 10^{-5}$ | 104 | 77 | | | | |
| Copper | | | | | | | | | | |
| Pure | 1356 | 8933 | 385 | $11.6 \times 10^{-5}$ | 401 | 389 | 378 | 366 | 352 | 336 |
| Bronze | 1293 | 8800 | 420 | $1.4 \times 10^{-5}$ | 49 | 54 | 61 | | | |
| Brass | 1188 | 8526 | 382 | $3.4 \times 10^{-5}$ | 110 | 142 | 151 | | | |
| Constantan | 1493 | 8921 | 410 | $6.3 \times 10^{-6}$ | 22 | 26 | | | | |
| Germanium | 1211 | 5360 | 318 | $3.6 \times 10^{-5}$ | 67 | 40 | 19 | 17 | | |
| Gold | 1336 | 19300 | 129 | $12.7 \times 10^{-5}$ | 318 | 309 | 299 | 286 | 273 | 254 |
| Iron | | | | | | | | | | |
| Pure | 1810 | 7870 | 440 | $2.3 \times 10^{-5}$ | 83 | 66 | 53 | 41 | 32 | 30 |
| Wrought | | 7850 | 460 | $1.6 \times 10^{-5}$ | 59 | 52 | 45 | 33 | | |
| Armco | | 7870 | 447 | $2.1 \times 10^{-5}$ | 75 | 62 | 49 | 38 | 29 | 29 |
| Cast (4% C) | | 7272 | 420 | $1.7 \times 10^{-5}$ | 52 | 40 | 34 | 24 | 21 | |
| Steels | | | | | | | | | | |
| 1% carbon | | 7801 | 473 | $1.2 \times 10^{-5}$ | 43 | 42 | 36 | 29 | 28 | |
| 1% chrome | | 7913 | 448 | $1.7 \times 10^{-5}$ | 62 | 52 | 42 | 36 | 33 | |
| 304 stainless | | 7900 | 477 | $4.0 \times 10^{-6}$ | 14 | 18 | 21 | 24 | 26 | |
| 347 stainless | | 7978 | 480 | $3.9 \times 10^{-6}$ | 14 | 17 | 19 | 23 | 25 | 26 |

| | | | | | | | | | | |
|---|---|---|---|---|---|---|---|---|---|---|
| Lead | 601 | 11340 | 129 | $2.4 \times 10^{-5}$ | 36 | 33 | | | | |
| Lithium | 454 | 534 | 3560 | $4.0 \times 10^{-5}$ | 79 | | | | | |
| Magnesium | 923 | 1740 | 1017 | $8.8 \times 10^{-5}$ | 157 | 151 | 147 | 145 | | |
| Manganese | 1517 | 7430 | 477 | $2.2 \times 10^{-5}$ | 7.7 | | | | | |
| Molybdenum | 2883 | 10240 | 255 | $5.3 \times 10^{-5}$ | 139 | 131 | 123 | 116 | 109 | 103 |
| Nickel | | | | | | | | | | |
| Pure | 1726 | 8900 | 442 | $2.3 \times 10^{-5}$ | 94 | 74 | 65 | 69 | 73 | 78 |
| Nichrome (80% Ni, 20% Cr) | 1672 | 8360 | 430 | $3.5 \times 10^{-6}$ | 12 | 15 | 18 | 23 | | |
| Inconel-X | 1665 | 8510 | 439 | $3.1 \times 10^{-6}$ | 11 | 14 | 18 | 22 | 25 | 29 |
| Platinum | | | | | | | | | | |
| Pure | 2042 | 21450 | 130 | $2.6 \times 10^{-5}$ | 72 | 72 | 74 | 76 | 80 | 84 |
| 60% Pt, 40% Rh | 1800 | 16600 | 162 | $1.7 \times 10^{-5}$ | 46 | 54 | 62 | 66 | 70 | 74 |
| Rhenium | 3453 | 21100 | 138 | $1.7 \times 10^{-5}$ | 49 | 45 | 44 | 44 | 45 | 46 |
| Rhodium | 2233 | 12450 | 247 | $4.9 \times 10^{-5}$ | 151 | 141 | 132 | 125 | 119 | 113 |
| Silicon | 1685 | 2330 | 691 | $9.5 \times 10^{-5}$ | 168 | 93 | 59 | 41 | 30 | 25 |
| Silver | 1234 | 10500 | 235 | $17.3 \times 10^{-5}$ | 428 | 415 | 399 | 383 | 368 | 355 |
| Sodium | 371 | 971 | 1220 | $11.2 \times 10^{-5}$ | 135 | | | | | |
| Tin | 505 | 5750 | 227 | $5.1 \times 10^{-5}$ | 68 | 60 | | | | |
| Titanium | 1953 | 4500 | 510 | $9.6 \times 10^{-6}$ | 22 | 20 | 19 | 20 | 21 | 23 |
| Tungsten | 3653 | 19300 | 133 | $6.9 \times 10^{-5}$ | 182 | 152 | 134 | 125 | 118 | 114 |
| Uranium | 1406 | 19070 | 116 | $1.2 \times 10^{-5}$ | 27 | 30 | 35 | 39 | 45 | |
| Vanadium | 2192 | 6100 | 486 | $1.1 \times 10^{-5}$ | 31 | 33 | 35 | 37 | 40 | 44 |
| Zinc | 693 | 7140 | 388 | $4.4 \times 10^{-5}$ | 122 | 112 | 102 | | | |
| Zirconium | 2125 | 6570 | 280 | $1.2 \times 10^{-5}$ | 23 | 21 | 21 | 22 | 24 | 26 |

# Appendix D
# Properties of Nonmetallic Solids†

| Material | Melting Point (K) | Properties at 20°C | | | Thermal Conductivity, $k$ (W/m · K) | | | | | |
|---|---|---|---|---|---|---|---|---|---|---|
| | | $\rho$ (kg/m$^3$) | $c_p$ (J/kg · K) | $\alpha$ (m$^2$/s) | 0°C | 200°C | 400°C | 600°C | 800°C | 1000°C |
| Aluminum oxide | 2323 | 3970 | 765 | $1.3 \times 10^{-5}$ | 40. | 22 | 13 | 9.3 | 7.3 | 6.2 |
| Asbestos | | 383 | 816 | $3.6 \times 10^{-6}$ | 0.11 | | | | | |
| Beryllium oxide | 2725 | 3000 | 1030 | $9.2 \times 10^{-5}$ | 302. | 159 | 93 | 60 | 41 | 30 |
| Bricks | | | | | | | | | | |
| Common | | 1600 | 840 | $5.2 \times 10^{-7}$ | 0.7 | | | | | |
| Chrome | | 3000 | 840 | $8.7 \times 10^{-7}$ | 2.2 | 2.3 | 2.4 | 2.4 | 2.1 | |
| Fireclay | | 2000 | 960 | $5.2 \times 10^{-7}$ | 1.0 | 1.0 | 1.0 | 1.1 | 1.1 | 1.1 |
| Magnesite | | | 1130 | | 4.0 | 3.6 | 2.8 | 2.4 | 2.1 | 1.8 |
| Masonry | | 1700 | 837 | $4.6 \times 10^{-7}$ | 0.66 | | | | | |
| Silica | | 1900 | | | 1.1 | | | | | |
| Carbon | 1500 | 1950 | | | 1.6 | 1.9 | 2.2 | 2.4 | 2.6 | 2.8 |
| Cement mortar | | 1860 | 780 | $6.2 \times 10^{-7}$ | 0.9 | | | | | |
| Concrete | | 2300 | 880 | $4.9 \times 10^{-7}$ | 1.0 | | | | | |
| Coal | | 1370 | 1260 | $1.4 \times 10^{-7}$ | 0.24 | | | | | |
| Diamonds | | | | | | | | | | |
| Type I | | | | | 1000. | 300. | | | | |
| Type IIa | | 3500 | 510 | $1.4 \times 10^{-3}$ | 2650. | 1300. | | | | |
| Type IIb | | | | | 1510. | 780. | | | | |
| Earths | | | | | | | | | | |
| Clay | | 1500 | 880 | $1.1 \times 10^{-6}$ | 1.4 | | | | | |
| Diatomaceous | | | | | 1.3 | | | | | |
| Sand | | 1500 | 800 | $2.5 \times 10^{-7}$ | 0.3 | | | | | |

| | | | | | | | | | |
|---|---|---|---|---|---|---|---|---|---|
| Glasses | | | | | | | | | |
| Corning 7740 | 2700 | | | 1.1 | 1.3 | 1.5 | 1.9 | | |
| Pyroceram | 2600 | 810 | $1.9 \times 10^{-6}$ | 4.1 | 3.6 | 3.2 | 3.0 | 2.9 | 2.8 |
| Window | 2700 | 800 | $3.9 \times 10^{-7}$ | 0.84 | | | | | |
| Ice | 920 | 2000 | $1.2 \times 10^{-6}$ | 2.2 | | | | | |
| Insulations | | | | | | | | | |
| Cork, granular | 45–120 | 1900 | $2\text{–}5 \times 10^{-7}$ | 0.045 | | | | | |
| Corkboard | 160 | 1900 | $1.4 \times 10^{-7}$ | 0.043 | | | | | |
| Cellulose, loose | 45 | | | 0.038 | | | | | |
| Feltboard | 50–125 | | | 0.035 | | | | | |
| Glass fiber | 220 | | | 0.035 | | | | | |
| Glass wool | 40 | 700 | $1.4 \times 10^{-6}$ | 0.038 | | | | | |
| Kapok | | | | 0.035 | | | | | |
| Magnesia, 85% | 270 | | | 0.065 | 0.085 | | | | |
| Polystyrene | 50 | | | 0.025 | | | | | |
| Rubber, foam | 70 | | | 0.030 | | | | | |
| Rock wool | 160 | | | 0.040 | | | | | |
| Sawdust | | | | 0.059 | | | | | |
| Vermiculite, loose | 80 | | | 0.058 | | | | | |
| Magnesium oxide | | | | 53. | 29. | 18. | 12. | 8.8 | 7.3 |
| Plaster, gypsum | 1600 | 1000 | $3.8 \times 10^{-7}$ | 0.5 | | | | | |
| Quartz, fused | | | | 1.3 | 1.6 | 1.9 | 2.3 | 3.2 | 3.8 |
| Rocks | | | | | | | | | |
| Granite | 2640 | 800 | $1.4 \times 10^{-6}$ | 3.0 | | | | | |
| Limestone | 2400 | 860 | $1.0 \times 10^{-6}$ | 2.0 | | | | | |
| Marble | 2650 | 1000 | $1.0 \times 10^{-6}$ | 2.7 | | | | | |
| Sandstone | 2200 | 740 | $1.7 \times 10^{-6}$ | 2.8 | | | | | |
| Rubber, hard | 1170 | 2000 | $6.8 \times 10^{-8}$ | 0.16 | | | | | |
| Skin, human | | | | 0.37 | | | | | |
| Snow | | | | | | | | | |
| Loose | 110 | | | 0.05 | | | | | |
| Packed | 500 | | | 0.19 | | | | | |
| Teflon | 2200 | | | 0.35 | | | | | |

| Material | Melting Point (K) | Properties at 20°C $\rho$ (kg/m$^3$) | $c_p$ (J/kg · K) | $\alpha$ (m$^2$/s) | Thermal Conductivity, $k$ (W/m · K) 0°C | 200°C | 400°C | 600°C | 800°C | 1000°C |
|---|---|---|---|---|---|---|---|---|---|---|
| Woods | | | | | | | | | | |
| Balsa | | 140 | | | 0.055 | | | | | |
| Cypress | | 460 | | | 0.097 | | | | | |
| Fir | | 420 | 2700 | $9.7 \times 10^{-8}$ | 0.11 | | | | | |
| Maple or oak | | 600 | 2400 | $1.2 \times 10^{-7}$ | 0.17 | | | | | |
| Pine, yellow | | 640 | 2800 | $8.4 \times 10^{-8}$ | 0.15 | | | | | |
| Pine, white | | 440 | | | 0.11 | | | | | |
| Plywood | | 550 | 1200 | $1.8 \times 10^{-7}$ | 0.12 | | | | | |
| Wool | | 200 | | | 0.038 | | | | | |

†The properties of commercial nonmetallic materials are subject to large uncertainty (up to 50%) because of variations in composition and construction.

# Appendix E
# Properties of Liquid Metals

| $T$ (K) | $\rho$ (kg/m$^3$) | $c_p$ (J/kg · K) | $k$ (W/m · K) | $\alpha$ (m$^2$/s) | $\mu$ (kg/m · s) | $\nu$ (m$^2$/s) | Pr | $g\beta/\nu^2$ (m$^{-3}$ · K$^{-1}$) |
|---|---|---|---|---|---|---|---|---|
| *Bismuth* | | | | | | | | |
| 545† | 10069 | 143 | 16.8 | $1.17 \times 10^{-5}$ | $1.75 \times 10^{-3}$ | $1.74 \times 10^{-7}$ | 0.0148 | $43 \times 10^{8}$ |
| 600 | 9997 | 145 | 16.4 | $1.13 \times 10^{-5}$ | $1.61 \times 10^{-3}$ | $1.61 \times 10^{-7}$ | 0.0142 | $49 \times 10^{8}$ |
| 700 | 9867 | 150 | 15.6 | $1.06 \times 10^{-5}$ | $1.34 \times 10^{-3}$ | $1.36 \times 10^{-7}$ | 0.0128 | $68 \times 10^{8}$ |
| 800 | 9752 | 154 | 15.6 | $1.04 \times 10^{-5}$ | $1.12 \times 10^{-3}$ | $1.15 \times 10^{-7}$ | 0.0111 | $88 \times 10^{8}$ |
| 900 | 9636 | 159 | 15.6 | $1.02 \times 10^{-5}$ | $0.96 \times 10^{-3}$ | $0.99 \times 10^{-7}$ | 0.0098 | $126 \times 10^{8}$ |
| 1000 | 9510 | 163 | 15.6 | $1.01 \times 10^{-5}$ | $0.83 \times 10^{-3}$ | $0.87 \times 10^{-7}$ | 0.0087 | $177 \times 10^{8}$ |
| *Lead* | | | | | | | | |
| 601† | 10588 | 161 | 15.5 | $0.91 \times 10^{-5}$ | $2.62 \times 10^{-3}$ | $2.47 \times 10^{-7}$ | 0.0272 | $145 \times 10^{8}$ |
| 700 | 10476 | 157 | 17.4 | $1.06 \times 10^{-5}$ | $2.15 \times 10^{-3}$ | $2.05 \times 10^{-7}$ | 0.0194 | $257 \times 10^{8}$ |
| 800 | 10359 | 153 | 19.0 | $1.20 \times 10^{-5}$ | $2.05 \times 10^{-3}$ | $1.98 \times 10^{-7}$ | 0.0165 | $289 \times 10^{8}$ |
| 900 | 10237 | 149 | 20.3 | $1.33 \times 10^{-5}$ | $1.54 \times 10^{-3}$ | $1.50 \times 10^{-7}$ | 0.0113 | $528 \times 10^{8}$ |
| 1000 | 10111 | 145 | 21.5 | $1.47 \times 10^{-5}$ | $1.32 \times 10^{-3}$ | $1.30 \times 10^{-7}$ | 0.0089 | $736 \times 10^{8}$ |
| *Mercury* | | | | | | | | |
| 234† | 13723 | 142 | 7.3 | $3.8 \times 10^{-6}$ | $2.00 \times 10^{-3}$ | $1.46 \times 10^{-7}$ | 0.0389 | $82 \times 10^{9}$ |
| 273 | 13628 | 140 | 8.2 | $4.3 \times 10^{-6}$ | $1.69 \times 10^{-3}$ | $1.24 \times 10^{-7}$ | 0.0289 | $115 \times 10^{9}$ |
| 300 | 13562 | 139 | 8.9 | $4.7 \times 10^{-6}$ | $1.51 \times 10^{-3}$ | $1.11 \times 10^{-7}$ | 0.0237 | $143 \times 10^{9}$ |
| 350 | 13441 | 138 | 10.0 | $5.4 \times 10^{-6}$ | $1.31 \times 10^{-3}$ | $0.98 \times 10^{-7}$ | 0.0181 | $185 \times 10^{9}$ |
| 400 | 13320 | 137 | 11.0 | $6.1 \times 10^{-6}$ | $1.18 \times 10^{-3}$ | $0.89 \times 10^{-7}$ | 0.0147 | $227 \times 10^{9}$ |
| 500 | 13081 | 136 | 12.7 | $7.1 \times 10^{-6}$ | $1.02 \times 10^{-3}$ | $0.78 \times 10^{-7}$ | 0.0109 | $292 \times 10^{9}$ |
| 600 | 12816 | 134 | 14.2 | $8.3 \times 10^{-6}$ | $0.84 \times 10^{-3}$ | $0.66 \times 10^{-7}$ | 0.0080 | $480 \times 10^{9}$ |

| $T$ (K) | $\rho$ (kg/m³) | $c_p$ (J/kg · K) | $k$ (W/m · K) | $\alpha$ (m²/s) | $\mu$ (kg/m · s) | $\nu$ (m²/s) | Pr | $g\beta/\nu^2$ ($m^{-3} \cdot K^{-1}$) |
|---|---|---|---|---|---|---|---|---|
| *Lithium* | | | | | | | | |
| 454† | 512 | 4190 | 43 | $2.0 \times 10^{-5}$ | $6.1 \times 10^{-4}$ | $1.18 \times 10^{-6}$ | 0.059 | $134 \times 10^{7}$ |
| 500 | 508 | 4190 | 44 | $2.1 \times 10^{-5}$ | $5.9 \times 10^{-4}$ | $1.17 \times 10^{-6}$ | 0.056 | $136 \times 10^{7}$ |
| 600 | 498 | 4190 | 48 | $2.3 \times 10^{-5}$ | $5.7 \times 10^{-4}$ | $1.14 \times 10^{-6}$ | 0.050 | $143 \times 10^{7}$ |
| 700 | 489 | 4190 | 51 | $2.4 \times 10^{-5}$ | $5.4 \times 10^{-4}$ | $1.11 \times 10^{-6}$ | 0.045 | $151 \times 10^{7}$ |
| 800 | 480 | 4190 | 54 | $2.7 \times 10^{-5}$ | $5.2 \times 10^{-4}$ | $1.08 \times 10^{-6}$ | 0.040 | $160 \times 10^{7}$ |
| 900 | 471 | 4190 | 57 | $2.9 \times 10^{-5}$ | $4.9 \times 10^{-4}$ | $1.05 \times 10^{-6}$ | 0.036 | $169 \times 10^{7}$ |
| 1000 | 462 | 4190 | 60 | $3.1 \times 10^{-5}$ | $4.7 \times 10^{-4}$ | $1.02 \times 10^{-6}$ | 0.033 | $179 \times 10^{7}$ |
| *Potassium* | | | | | | | | |
| 337† | 827 | 802 | 55 | $8.3 \times 10^{-5}$ | $4.7 \times 10^{-4}$ | $5.6 \times 10^{-7}$ | 0.0068 | $86 \times 10^{8}$ |
| 400 | 812 | 798 | 52 | $8.0 \times 10^{-5}$ | $3.9 \times 10^{-4}$ | $4.9 \times 10^{-7}$ | 0.0061 | $119 \times 10^{8}$ |
| 500 | 789 | 790 | 48 | $7.7 \times 10^{-5}$ | $3.0 \times 10^{-4}$ | $3.8 \times 10^{-7}$ | 0.0050 | $199 \times 10^{8}$ |
| 600 | 766 | 783 | 44 | $7.3 \times 10^{-5}$ | $2.3 \times 10^{-4}$ | $3.0 \times 10^{-7}$ | 0.0041 | $331 \times 10^{8}$ |
| 700 | 742 | 775 | 40 | $7.0 \times 10^{-5}$ | $1.8 \times 10^{-4}$ | $2.4 \times 10^{-7}$ | 0.0034 | $550 \times 10^{8}$ |
| 800 | 718 | 767 | 37 | $6.7 \times 10^{-5}$ | $1.6 \times 10^{-4}$ | $2.2 \times 10^{-7}$ | 0.0033 | $683 \times 10^{8}$ |
| 900 | 693 | 760 | 34 | $6.5 \times 10^{-5}$ | $1.4 \times 10^{-4}$ | $2.0 \times 10^{-7}$ | 0.0032 | $840 \times 10^{8}$ |
| 1000 | 669 | 752 | 31 | $6.2 \times 10^{-5}$ | $1.3 \times 10^{-4}$ | $1.9 \times 10^{-7}$ | 0.0030 | $1040 \times 10^{8}$ |
| *Sodium* | | | | | | | | |
| 371† | 929 | 1382 | 88 | $6.9 \times 10^{-5}$ | $7.0 \times 10^{-4}$ | $7.5 \times 10^{-7}$ | 0.0110 | $51 \times 10^{8}$ |
| 400 | 922 | 1371 | 87 | $6.9 \times 10^{-5}$ | $6.1 \times 10^{-4}$ | $6.7 \times 10^{-7}$ | 0.0097 | $66 \times 10^{8}$ |
| 500 | 896 | 1334 | 82 | $6.8 \times 10^{-5}$ | $4.1 \times 10^{-4}$ | $4.6 \times 10^{-7}$ | 0.0067 | $145 \times 10^{8}$ |
| 600 | 871 | 1309 | 76 | $6.7 \times 10^{-5}$ | $3.2 \times 10^{-4}$ | $3.6 \times 10^{-7}$ | 0.0054 | $238 \times 10^{8}$ |
| 700 | 846 | 1284 | 72 | $6.6 \times 10^{-5}$ | $2.6 \times 10^{-4}$ | $3.0 \times 10^{-7}$ | 0.0046 | $356 \times 10^{8}$ |
| 800 | 822 | 1259 | 67 | $6.5 \times 10^{-5}$ | $2.1 \times 10^{-4}$ | $2.6 \times 10^{-7}$ | 0.0040 | $507 \times 10^{8}$ |
| 900 | 797 | 1256 | 63 | $6.2 \times 10^{-5}$ | $1.9 \times 10^{-4}$ | $2.4 \times 10^{-7}$ | 0.0039 | $604 \times 10^{8}$ |
| 1000 | 773 | 1256 | 58 | $6.0 \times 10^{-5}$ | $1.8 \times 10^{-4}$ | $2.3 \times 10^{-7}$ | 0.0038 | $708 \times 10^{8}$ |

†Melting point

# Appendix F
# Properties of Saturated Liquids

| $T$ (°C) | $\rho$ (kg/m$^3$) | $c_p$ (J/kg · K) | $k$ (W/m · K) | $\alpha$ (m$^2$/s) | $\mu$ (kg/m · s) | $\nu$ (m$^2$/s) | Pr | $g\beta/\nu^2$ (m$^{-3}$K$^{-1}$) |
|---|---|---|---|---|---|---|---|---|
| *Ammonia* | | | | | | | | |
| −40 | 692 | 4467 | 0.546 | $1.78 \times 10^{-7}$ | $2.81 \times 10^{-4}$ | $4.06 \times 10^{-7}$ | 2.28 | $1.05 \times 10^{11}$ |
| −20 | 667 | 4509 | 0.546 | $1.82 \times 10^{-7}$ | $2.54 \times 10^{-4}$ | $3.81 \times 10^{-7}$ | 2.09 | $1.31 \times 10^{11}$ |
| 0 | 640 | 4635 | 0.540 | $1.82 \times 10^{-7}$ | $2.39 \times 10^{-4}$ | $3.73 \times 10^{-7}$ | 2.05 | $1.51 \times 10^{11}$ |
| 20 | 612 | 4798 | 0.521 | $1.78 \times 10^{-7}$ | $2.20 \times 10^{-4}$ | $3.59 \times 10^{-7}$ | 2.02 | $1.81 \times 10^{11}$ |
| 40 | 581 | 4999 | 0.493 | $1.70 \times 10^{-7}$ | $1.98 \times 10^{-4}$ | $3.40 \times 10^{-7}$ | 2.00 | $2.34 \times 10^{11}$ |
| *Ethyl Alcohol* ($C_2H_6O$) | | | | | | | | |
| −40 | 823 | 2037 | 0.186 | $1.11 \times 10^{-7}$ | $4.81 \times 10^{-3}$ | $5.84 \times 10^{-6}$ | 52.7 | $0.29 \times 10^{9}$ |
| −20 | 815 | 2124 | 0.179 | $1.03 \times 10^{-7}$ | $2.83 \times 10^{-3}$ | $3.47 \times 10^{-6}$ | 33.6 | $0.84 \times 10^{9}$ |
| 0 | 806 | 2249 | 0.174 | $0.960 \times 10^{-7}$ | $1.77 \times 10^{-3}$ | $2.20 \times 10^{-6}$ | 22.9 | $2.12 \times 10^{9}$ |
| 20 | 789 | 2395 | 0.168 | $0.889 \times 10^{-7}$ | $1.20 \times 10^{-3}$ | $1.52 \times 10^{-6}$ | 17.0 | $4.54 \times 10^{9}$ |
| 40 | 772 | 2572 | 0.162 | $0.816 \times 10^{-7}$ | $0.834 \times 10^{-3}$ | $1.08 \times 10^{-6}$ | 13.2 | $9.31 \times 10^{9}$ |
| 60 | 755 | 2781 | 0.156 | $0.743 \times 10^{-7}$ | $0.592 \times 10^{-3}$ | $0.784 \times 10^{-6}$ | 10.6 | $18.1 \times 10^{9}$ |
| 80 | 738 | 3026 | 0.150 | $0.672 \times 10^{-7}$ | $0.430 \times 10^{-3}$ | $0.583 \times 10^{-6}$ | 8.7 | $33.5 \times 10^{9}$ |
| *Ethylene Glycol* ($C_2H_6O_2$) | | | | | | | | |
| 0 | 1131 | 2295 | 0.254 | $9.79 \times 10^{-8}$ | $65.1 \times 10^{-3}$ | $57.5 \times 10^{-6}$ | 588 | $0.0192 \times 10^{8}$ |
| 20 | 1117 | 2386 | 0.257 | $9.64 \times 10^{-8}$ | $21.4 \times 10^{-3}$ | $19.2 \times 10^{-6}$ | 199 | $0.173 \times 10^{8}$ |
| 40 | 1101 | 2476 | 0.259 | $9.50 \times 10^{-8}$ | $9.57 \times 10^{-3}$ | $8.69 \times 10^{-6}$ | 91 | $0.844 \times 10^{8}$ |
| 60 | 1088 | 2565 | 0.262 | $9.39 \times 10^{-8}$ | $5.17 \times 10^{-3}$ | $4.75 \times 10^{-6}$ | 51 | $2.82 \times 10^{8}$ |
| 80 | 1078 | 2656 | 0.265 | $9.26 \times 10^{-8}$ | $3.21 \times 10^{-3}$ | $2.98 \times 10^{-6}$ | 32 | $7.18 \times 10^{8}$ |
| 100 | 1059 | 2750 | 0.267 | $9.17 \times 10^{-8}$ | $2.15 \times 10^{-3}$ | $2.03 \times 10^{-6}$ | 22 | $15.5 \times 10^{8}$ |

| $T$ (°C) | $\rho$ (kg/m³) | $c_p$ (J/kg · K) | $k$ (W/m · K) | $\alpha$ (m²/s) | $\mu$ (kg/m · s) | $\nu$ (m²/s) | Pr | $g\beta/\nu^2$ (m⁻³K⁻¹) |
|---|---|---|---|---|---|---|---|---|
| *Freon-12 Refrigerant* ($CCl_2F_2$) | | | | | | | | |
| −40 | 1515 | 885 | 0.069 | $5.14 \times 10^{-8}$ | $4.24 \times 10^{-4}$ | $2.80 \times 10^{-7}$ | 5.4 | $2.52 \times 10^{11}$ |
| −20 | 1457 | 907 | 0.071 | $5.38 \times 10^{-8}$ | $3.43 \times 10^{-4}$ | $2.35 \times 10^{-7}$ | 4.4 | $3.73 \times 10^{11}$ |
| 0 | 1393 | 935 | 0.073 | $5.59 \times 10^{-8}$ | $2.98 \times 10^{-4}$ | $2.14 \times 10^{-7}$ | 3.8 | $5.04 \times 10^{11}$ |
| 20 | 1327 | 966 | 0.073 | $5.66 \times 10^{-8}$ | $2.62 \times 10^{-4}$ | $1.97 \times 10^{-7}$ | 3.5 | $6.54 \times 10^{11}$ |
| 40 | 1254 | 1002 | 0.069 | $5.46 \times 10^{-8}$ | $2.40 \times 10^{-4}$ | $1.91 \times 10^{-7}$ | 3.5 | $8.64 \times 10^{11}$ |
| *Glycerin* | | | | | | | | |
| −20 | 1288 | 2143 | 0.282 | $1.02 \times 10^{-7}$ | 134. | $104 \times 10^{-3}$ | $1020 \times 10^{3}$ | 0.42 |
| 0 | 1276 | 2261 | 0.284 | $0.98 \times 10^{-7}$ | 12.1 | $9.5 \times 10^{-3}$ | $96 \times 10^{3}$ | 50 |
| 20 | 1264 | 2386 | 0.287 | $0.95 \times 10^{-7}$ | 1.49 | $1.2 \times 10^{-3}$ | $12.4 \times 10^{3}$ | 3200 |
| 40 | 1252 | 2513 | 0.290 | $0.92 \times 10^{-7}$ | 0.27 | $0.2 \times 10^{-3}$ | $2.3 \times 10^{3}$ | 101000 |
| *Unused Engine Oil* | | | | | | | | |
| 0 | 899 | 1796 | 0.147 | $9.11 \times 10^{-8}$ | $3850 \times 10^{-3}$ | $4280 \times 10^{-6}$ | 47100 | 350 |
| 20 | 888 | 1880 | 0.145 | $8.72 \times 10^{-8}$ | $800 \times 10^{-3}$ | $901 \times 10^{-6}$ | 10400 | 7900 |
| 40 | 876 | 1964 | 0.144 | $8.34 \times 10^{-8}$ | $212 \times 10^{-3}$ | $242 \times 10^{-6}$ | 2870 | 111000 |
| 60 | 864 | 2047 | 0.140 | $8.00 \times 10^{-8}$ | $72.5 \times 10^{-3}$ | $83.9 \times 10^{-6}$ | 1050 | 939000 |
| 80 | 852 | 2131 | 0.138 | $7.69 \times 10^{-8}$ | $32.0 \times 10^{-3}$ | $37.5 \times 10^{-6}$ | 490 | $4.77 \times 10^{6}$ |
| 100 | 840 | 2219 | 0.137 | $7.38 \times 10^{-8}$ | $17.1 \times 10^{-3}$ | $20.3 \times 10^{-6}$ | 276 | $16.5 \times 10^{6}$ |
| 120 | 829 | 2307 | 0.135 | $7.10 \times 10^{-8}$ | $10.2 \times 10^{-3}$ | $12.4 \times 10^{-6}$ | 175 | $44.8 \times 10^{6}$ |
| 140 | 817 | 2395 | 0.133 | $6.86 \times 10^{-8}$ | $6.53 \times 10^{-3}$ | $8.0 \times 10^{-6}$ | 116 | $109. \times 10^{6}$ |
| 160 | 806 | 2483 | 0.132 | $6.63 \times 10^{-8}$ | $4.49 \times 10^{-3}$ | $5.6 \times 10^{-6}$ | 84 | $226. \times 10^{6}$ |

| *Water* ($T$ in °K) | | | | | | | | |
|---|---|---|---|---|---|---|---|---|
| 273.2 | 1000 | 4205 | 0.564 | $1.34 \times 10^{-7}$ | $1.79 \times 10^{-3}$ | $1.79 \times 10^{-6}$ | 13.4 | $-21 \times 10^{7}$ |
| 280 | 1000 | 4197 | 0.582 | $1.39 \times 10^{-7}$ | $1.44 \times 10^{-3}$ | $1.44 \times 10^{-6}$ | 10.4 | $+22 \times 10^{7}$ |
| 300 | 997 | 4177 | 0.608 | $1.46 \times 10^{-7}$ | $0.857 \times 10^{-3}$ | $0.86 \times 10^{-6}$ | 5.88 | $366 \times 10^{7}$ |
| 320 | 989 | 4176 | 0.637 | $1.54 \times 10^{-7}$ | $0.579 \times 10^{-3}$ | $0.59 \times 10^{-6}$ | 3.79 | $1250 \times 10^{7}$ |
| 340 | 980 | 4187 | 0.659 | $1.61 \times 10^{-7}$ | $0.423 \times 10^{-3}$ | $0.43 \times 10^{-6}$ | 2.69 | $2980 \times 10^{7}$ |
| 360 | 967 | 4204 | 0.674 | $1.66 \times 10^{-7}$ | $0.320 \times 10^{-3}$ | $0.33 \times 10^{-6}$ | 2.00 | $6250 \times 10^{7}$ |
| 373.2 | 958 | 4220 | 0.681 | $1.68 \times 10^{-7}$ | $0.282 \times 10^{-3}$ | $0.29 \times 10^{-6}$ | 1.75 | $8500 \times 10^{7}$ |
| 400 | 937 | 4241 | 0.686 | $1.73 \times 10^{-7}$ | $0.219 \times 10^{-3}$ | $0.23 \times 10^{-6}$ | 1.35 | $16100 \times 10^{7}$ |
| 450 | 890 | 4419 | 0.673 | $1.71 \times 10^{-7}$ | $0.153 \times 10^{-3}$ | $0.17 \times 10^{-6}$ | 1.01 | $40200 \times 10^{7}$ |
| 500 | 832 | 4647 | 0.635 | $1.64 \times 10^{-7}$ | $0.118 \times 10^{-3}$ | $0.14 \times 10^{-6}$ | 0.86 | $77100 \times 10^{7}$ |
| 550 | 756 | 5272 | 0.571 | $1.43 \times 10^{-7}$ | $0.095 \times 10^{-3}$ | $0.13 \times 10^{-6}$ | 0.88 | $144000 \times 10^{7}$ |
| 600 | 650 | 6691 | 0.481 | $1.11 \times 10^{-7}$ | $0.076 \times 10^{-3}$ | $0.12 \times 10^{-6}$ | 1.05 | $295000 \times 10^{7}$ |
| 647.3† | 315 | — | — | — | — | — | — | — |

†Critical point

*Source:* J. T. R. Watson, R. S. Basu, and J. V. Sengers, "An Improved Representative Equation for the Dynamic Viscosity of Water Substance," *J. Phys. & Chem. Ref. Prop.*, vol. 9, no. 4, 1980, pp. 1255–1290.

# Appendix G
# Properties of Gases at 1 atm

| $T$ (K) | $\rho$ (kg/m$^3$) | $c_p$ (J/kg · K) | $k$ (W/m · K) | $\alpha$ (m$^2$/s) | $\mu$ (kg/m · s) | $\nu$ (m$^2$/s) | Pr | $g\beta/\nu^2$ (m$^{-3}$K$^{-1}$) |
|---|---|---|---|---|---|---|---|---|
| *Air* | | | | | | | | |
| 200 | 1.766 | 1003 | 0.0181 | $1.02 \times 10^{-5}$ | $1.34 \times 10^{-5}$ | $0.76 \times 10^{-5}$ | 0.740 | $85700 \times 10^4$ |
| 250 | 1.413 | 1003 | 0.0223 | $1.57 \times 10^{-5}$ | $1.61 \times 10^{-5}$ | $1.14 \times 10^{-5}$ | 0.724 | $30200 \times 10^4$ |
| 300 | 1.177 | 1005 | 0.0261 | $2.21 \times 10^{-5}$ | $1.85 \times 10^{-5}$ | $1.57 \times 10^{-5}$ | 0.712 | $13300 \times 10^4$ |
| 350 | 1.009 | 1008 | 0.0297 | $2.92 \times 10^{-5}$ | $2.08 \times 10^{-5}$ | $2.06 \times 10^{-5}$ | 0.706 | $6600 \times 10^4$ |
| 400 | 0.883 | 1013 | 0.0331 | $3.70 \times 10^{-5}$ | $2.29 \times 10^{-5}$ | $2.60 \times 10^{-5}$ | 0.703 | $3630 \times 10^4$ |
| 450 | 0.785 | 1020 | 0.0363 | $4.54 \times 10^{-5}$ | $2.49 \times 10^{-5}$ | $3.18 \times 10^{-5}$ | 0.700 | $2160 \times 10^4$ |
| 500 | 0.706 | 1029 | 0.0395 | $5.44 \times 10^{-5}$ | $2.68 \times 10^{-5}$ | $3.80 \times 10^{-5}$ | 0.699 | $1360 \times 10^4$ |
| 550 | 0.642 | 1039 | 0.0426 | $6.39 \times 10^{-5}$ | $2.86 \times 10^{-5}$ | $4.45 \times 10^{-5}$ | 0.698 | $900 \times 10^4$ |
| 600 | 0.589 | 1051 | 0.0456 | $7.37 \times 10^{-5}$ | $3.03 \times 10^{-5}$ | $5.15 \times 10^{-5}$ | 0.698 | $616 \times 10^4$ |
| 700 | 0.504 | 1075 | 0.0513 | $9.46 \times 10^{-5}$ | $3.35 \times 10^{-5}$ | $6.64 \times 10^{-5}$ | 0.702 | $318 \times 10^4$ |
| 800 | 0.441 | 1099 | 0.0569 | $11.7 \times 10^{-5}$ | $3.64 \times 10^{-5}$ | $8.25 \times 10^{-5}$ | 0.704 | $180 \times 10^4$ |
| 900 | 0.392 | 1120 | 0.0625 | $14.2 \times 10^{-5}$ | $3.92 \times 10^{-5}$ | $9.99 \times 10^{-5}$ | 0.705 | $109 \times 10^4$ |
| 1000 | 0.353 | 1141 | 0.0672 | $16.7 \times 10^{-5}$ | $4.18 \times 10^{-5}$ | $11.8 \times 10^{-5}$ | 0.709 | $70 \times 10^4$ |
| 1200 | 0.294 | 1175 | 0.0759 | $22.2 \times 10^{-5}$ | $4.65 \times 10^{-5}$ | $15.8 \times 10^{-5}$ | 0.720 | $33 \times 10^4$ |
| 1400 | 0.252 | 1201 | 0.0835 | $27.6 \times 10^{-5}$ | $5.09 \times 10^{-5}$ | $20.2 \times 10^{-5}$ | 0.732 | $17.2 \times 10^4$ |
| 1600 | 0.221 | 1240 | 0.0904 | $33.0 \times 10^{-5}$ | $5.49 \times 10^{-5}$ | $24.9 \times 10^{-5}$ | 0.753 | $9.9 \times 10^4$ |
| 1800 | 0.196 | 1276 | 0.0970 | $38.8 \times 10^{-5}$ | $5.87 \times 10^{-5}$ | $29.9 \times 10^{-5}$ | 0.772 | $6.1 \times 10^4$ |
| 2000 | 0.177 | 1327 | 0.1032 | $44.1 \times 10^{-5}$ | $6.23 \times 10^{-5}$ | $35.3 \times 10^{-5}$ | 0.801 | $3.9 \times 10^4$ |
| *Ammonia* ($NH_3$) | | | | | | | | |
| 200 | 1.038 | 2199 | 0.0153 | $0.67 \times 10^{-5}$ | $6.89 \times 10^{-6}$ | $0.66 \times 10^{-5}$ | 0.990 | $113000 \times 10^4$ |
| 250 | 0.831 | 2248 | 0.0197 | $1.05 \times 10^{-5}$ | $8.53 \times 10^{-6}$ | $1.03 \times 10^{-5}$ | 0.973 | $37000 \times 10^4$ |
| 300 | 0.692 | 2298 | 0.0246 | $1.55 \times 10^{-5}$ | $10.27 \times 10^{-6}$ | $1.48 \times 10^{-5}$ | 0.959 | $14900 \times 10^4$ |

| | | | | | | | | |
|---|---|---|---|---|---|---|---|---|
| 350 | 0.593 | 2349 | 0.0302 | $2.17 \times 10^{-5}$ | $12.06 \times 10^{-6}$ | $2.03 \times 10^{-5}$ | 0.938 | $6800 \times 10^{4}$ |
| 400 | 0.519 | 2402 | 0.0364 | $2.92 \times 10^{-5}$ | $13.90 \times 10^{-6}$ | $2.68 \times 10^{-5}$ | 0.917 | $3400 \times 10^{4}$ |
| 450 | 0.461 | 2455 | 0.0433 | $3.82 \times 10^{-5}$ | $15.76 \times 10^{-6}$ | $3.42 \times 10^{-5}$ | 0.894 | $1860 \times 10^{4}$ |
| 500 | 0.415 | 2507 | 0.0506 | $4.86 \times 10^{-5}$ | $17.63 \times 10^{-6}$ | $4.25 \times 10^{-5}$ | 0.873 | $1090 \times 10^{4}$ |
| 550 | 0.378 | 2559 | 0.0580 | $6.00 \times 10^{-5}$ | $19.5 \times 10^{-6}$ | $5.16 \times 10^{-5}$ | 0.860 | $670 \times 10^{4}$ |
| 600 | 0.346 | 2611 | 0.0656 | $7.26 \times 10^{-5}$ | $21.4 \times 10^{-6}$ | $6.18 \times 10^{-5}$ | 0.852 | $430 \times 10^{4}$ |
| 700 | 0.297 | 2710 | 0.0811 | $10.1 \times 10^{-5}$ | $25.1 \times 10^{-6}$ | $8.45 \times 10^{-5}$ | 0.839 | $196 \times 10^{4}$ |
| 800 | 0.260 | 2810 | 0.0977 | $13.4 \times 10^{-5}$ | $28.8 \times 10^{-6}$ | $11.1 \times 10^{-5}$ | 0.828 | $100 \times 10^{4}$ |
| 900 | 0.231 | 2907 | 0.1146 | $17.1 \times 10^{-5}$ | $32.4 \times 10^{-6}$ | $14.0 \times 10^{-5}$ | 0.822 | $56 \times 10^{4}$ |
| 1000 | 0.208 | 3001 | 0.1317 | $21.1 \times 10^{-5}$ | $35.9 \times 10^{-6}$ | $17.3 \times 10^{-5}$ | 0.818 | $33 \times 10^{4}$ |
| *Argon* | | | | | | | | |
| 200 | 2.435 | 523.6 | 0.0124 | $0.98 \times 10^{-5}$ | $1.60 \times 10^{-5}$ | $0.66 \times 10^{-5}$ | 0.674 | $113000 \times 10^{4}$ |
| 250 | 1.948 | 522.2 | 0.0152 | $1.49 \times 10^{-5}$ | $1.95 \times 10^{-5}$ | $1.00 \times 10^{-5}$ | 0.672 | $39200 \times 10^{4}$ |
| 300 | 1.623 | 521.6 | 0.0177 | $2.09 \times 10^{-5}$ | $2.27 \times 10^{-5}$ | $1.40 \times 10^{-5}$ | 0.669 | $16700 \times 10^{4}$ |
| 350 | 1.392 | 521.2 | 0.0201 | $2.78 \times 10^{-5}$ | $2.57 \times 10^{-5}$ | $1.85 \times 10^{-5}$ | 0.666 | $8200 \times 10^{4}$ |
| 400 | 1.218 | 521.0 | 0.0223 | $3.52 \times 10^{-5}$ | $2.85 \times 10^{-5}$ | $2.34 \times 10^{-5}$ | 0.665 | $4480 \times 10^{4}$ |
| 450 | 1.082 | 520.9 | 0.0244 | $4.33 \times 10^{-5}$ | $3.12 \times 10^{-5}$ | $2.88 \times 10^{-5}$ | 0.665 | $2630 \times 10^{4}$ |
| 500 | 0.974 | 520.8 | 0.0264 | $5.20 \times 10^{-5}$ | $3.37 \times 10^{-5}$ | $3.45 \times 10^{-5}$ | 0.664 | $1640 \times 10^{4}$ |
| 550 | 0.886 | 520.7 | 0.0283 | $6.14 \times 10^{-5}$ | $3.60 \times 10^{-5}$ | $4.07 \times 10^{-5}$ | 0.662 | $1080 \times 10^{4}$ |
| 600 | 0.812 | 520.6 | 0.0301 | $7.12 \times 10^{-5}$ | $3.83 \times 10^{-5}$ | $4.72 \times 10^{-5}$ | 0.662 | $730 \times 10^{4}$ |
| 700 | 0.696 | 520.6 | 0.0336 | $9.28 \times 10^{-5}$ | $4.25 \times 10^{-5}$ | $6.11 \times 10^{-5}$ | 0.658 | $375 \times 10^{4}$ |
| 800 | 0.609 | 520.5 | 0.0369 | $11.6 \times 10^{-5}$ | $4.64 \times 10^{-5}$ | $7.62 \times 10^{-5}$ | 0.655 | $211 \times 10^{4}$ |
| 900 | 0.541 | 520.5 | 0.0398 | $14.1 \times 10^{-5}$ | $5.01 \times 10^{-5}$ | $9.26 \times 10^{-5}$ | 0.654 | $127 \times 10^{4}$ |
| 1000 | 0.487 | 520.5 | 0.0427 | $16.8 \times 10^{-5}$ | $5.35 \times 10^{-5}$ | $11.0 \times 10^{-5}$ | 0.652 | $81 \times 10^{4}$ |
| 1200 | 0.406 | 520.5 | 0.0481 | $22.8 \times 10^{-5}$ | $5.99 \times 10^{-5}$ | $14.8 \times 10^{-5}$ | 0.648 | $38 \times 10^{4}$ |
| 1400 | 0.348 | 520.4 | 0.0535 | $29.6 \times 10^{-5}$ | $6.56 \times 10^{-5}$ | $18.9 \times 10^{-5}$ | 0.638 | $19.7 \times 10^{4}$ |
| 1600 | 0.304 | 520.4 | 0.0588 | $37.1 \times 10^{-5}$ | $7.10 \times 10^{-5}$ | $23.2 \times 10^{-5}$ | 0.628 | $11.3 \times 10^{4}$ |
| 1800 | 0.271 | 520.4 | 0.0641 | $45.5 \times 10^{-5}$ | $7.60 \times 10^{-5}$ | $28.1 \times 10^{-5}$ | 0.617 | $6.9 \times 10^{4}$ |
| 2000 | 0.244 | 520.4 | 0.0692 | $54.6 \times 10^{-5}$ | $8.07 \times 10^{-5}$ | $33.1 \times 10^{-5}$ | 0.607 | $4.5 \times 10^{4}$ |

| $T$ (K) | $\rho$ (kg/m$^3$) | $c_p$ (J/kg · K) | $k$ (W/m · K) | $\alpha$ (m$^2$/s) | $\mu$ (kg/m · s) | $\nu$ (m$^2$/s) | Pr | $g\beta/\nu^2$ (m$^{-3}$K$^{-1}$) |
|---|---|---|---|---|---|---|---|---|
| *Carbon Dioxide* ($CO_2$) | | | | | | | | |
| 200 | 2.683 | 759 | 0.0095 | $0.47 \times 10^{-5}$ | $1.02 \times 10^{-5}$ | $0.38 \times 10^{-5}$ | 0.814 | $338000 \times 10^4$ |
| 250 | 2.146 | 806 | 0.0129 | $0.75 \times 10^{-5}$ | $1.26 \times 10^{-5}$ | $0.59 \times 10^{-5}$ | 0.790 | $113000 \times 10^4$ |
| 300 | 1.789 | 852 | 0.0166 | $1.09 \times 10^{-5}$ | $1.50 \times 10^{-5}$ | $0.84 \times 10^{-5}$ | 0.768 | $46500 \times 10^4$ |
| 350 | 1.533 | 897 | 0.0205 | $1.49 \times 10^{-5}$ | $1.73 \times 10^{-5}$ | $1.13 \times 10^{-5}$ | 0.755 | $22100 \times 10^4$ |
| 400 | 1.341 | 939 | 0.0244 | $1.94 \times 10^{-5}$ | $1.94 \times 10^{-5}$ | $1.45 \times 10^{-5}$ | 0.747 | $16900 \times 10^4$ |
| 450 | 1.192 | 979 | 0.0283 | $2.43 \times 10^{-5}$ | $2.15 \times 10^{-5}$ | $1.80 \times 10^{-5}$ | 0.743 | $6700 \times 10^4$ |
| 500 | 1.073 | 1017 | 0.0323 | $2.96 \times 10^{-5}$ | $2.35 \times 10^{-5}$ | $2.19 \times 10^{-5}$ | 0.740 | $4100 \times 10^4$ |
| 550 | 0.976 | 1049 | 0.0363 | $3.55 \times 10^{-5}$ | $2.54 \times 10^{-5}$ | $2.60 \times 10^{-5}$ | 0.734 | $2630 \times 10^4$ |
| 600 | 0.894 | 1077 | 0.0403 | $4.18 \times 10^{-5}$ | $2.72 \times 10^{-5}$ | $3.04 \times 10^{-5}$ | 0.727 | $1770 \times 10^4$ |
| 700 | 0.767 | 1126 | 0.0487 | $5.64 \times 10^{-5}$ | $3.06 \times 10^{-5}$ | $3.99 \times 10^{-5}$ | 0.708 | $880 \times 10^4$ |
| 800 | 0.671 | 1169 | 0.0560 | $7.14 \times 10^{-5}$ | $3.39 \times 10^{-5}$ | $5.05 \times 10^{-5}$ | 0.708 | $480 \times 10^4$ |
| 900 | 0.596 | 1205 | 0.0621 | $8.65 \times 10^{-5}$ | $3.69 \times 10^{-5}$ | $6.19 \times 10^{-5}$ | 0.716 | $284 \times 10^4$ |
| 1000 | 0.537 | 1235 | 0.0680 | $10.25 \times 10^{-5}$ | $3.97 \times 10^{-5}$ | $7.40 \times 10^{-5}$ | 0.721 | $179 \times 10^4$ |
| 1200 | 0.447 | 1283 | 0.0780 | $13.6 \times 10^{-5}$ | $4.49 \times 10^{-5}$ | $10.04 \times 10^{-5}$ | 0.739 | $81 \times 10^4$ |
| 1400 | 0.383 | 1315 | 0.0867 | $17.2 \times 10^{-5}$ | $4.97 \times 10^{-5}$ | $13.0 \times 10^{-5}$ | 0.754 | $42 \times 10^4$ |
| *Carbon Monoxide* (CO) | | | | | | | | |
| 200 | 1.708 | 1045 | 0.0175 | $0.98 \times 10^{-5}$ | $1.27 \times 10^{-5}$ | $0.75 \times 10^{-5}$ | 0.763 | $88100 \times 10^4$ |
| 250 | 1.366 | 1048 | 0.0214 | $1.50 \times 10^{-5}$ | $1.54 \times 10^{-5}$ | $1.13 \times 10^{-5}$ | 0.753 | $30900 \times 10^4$ |
| 300 | 1.138 | 1051 | 0.0252 | $2.11 \times 10^{-5}$ | $1.78 \times 10^{-5}$ | $1.56 \times 10^{-5}$ | 0.743 | $13400 \times 10^4$ |
| 350 | 0.976 | 1056 | 0.0288 | $2.80 \times 10^{-5}$ | $2.01 \times 10^{-5}$ | $2.05 \times 10^{-5}$ | 0.735 | $6640 \times 10^4$ |
| 400 | 0.854 | 1060 | 0.0323 | $3.57 \times 10^{-5}$ | $2.21 \times 10^{-5}$ | $2.59 \times 10^{-5}$ | 0.727 | $3650 \times 10^4$ |
| 450 | 0.759 | 1065 | 0.0355 | $4.39 \times 10^{-5}$ | $2.41 \times 10^{-5}$ | $3.18 \times 10^{-5}$ | 0.723 | $2160 \times 10^4$ |

| | | | | | | | | |
|---|---|---|---|---|---|---|---|---|
| 500 | 0.683 | 1071 | 0.0386 | $5.28 \times 10^{-5}$ | $2.60 \times 10^{-5}$ | $3.80 \times 10^{-5}$ | 0.720 | $1360 \times 10^{4}$ |
| 550 | 0.621 | 1077 | 0.0416 | $6.22 \times 10^{-5}$ | $2.77 \times 10^{-5}$ | $4.46 \times 10^{-5}$ | 0.717 | $896 \times 10^{4}$ |
| 600 | 0.569 | 1084 | 0.0444 | $7.20 \times 10^{-5}$ | $2.94 \times 10^{-5}$ | $5.17 \times 10^{-5}$ | 0.718 | $613 \times 10^{4}$ |
| 700 | 0.488 | 1099 | 0.0497 | $9.27 \times 10^{-5}$ | $3.25 \times 10^{-5}$ | $6.66 \times 10^{-5}$ | 0.718 | $316 \times 10^{4}$ |
| 800 | 0.427 | 1114 | 0.0549 | $11.5 \times 10^{-5}$ | $3.54 \times 10^{-5}$ | $8.29 \times 10^{-5}$ | 0.718 | $178 \times 10^{4}$ |
| 900 | 0.379 | 1128 | 0.0596 | $13.9 \times 10^{-5}$ | $3.81 \times 10^{-5}$ | $10.04 \times 10^{-5}$ | 0.721 | $108 \times 10^{4}$ |
| 1000 | 0.342 | 1142 | 0.0644 | $16.5 \times 10^{-5}$ | $4.06 \times 10^{-5}$ | $11.9 \times 10^{-5}$ | 0.720 | $69 \times 10^{4}$ |
| 1100 | 0.310 | 1155 | 0.0692 | $19.3 \times 10^{-5}$ | $4.30 \times 10^{-5}$ | $13.9 \times 10^{-5}$ | 0.718 | $47 \times 10^{4}$ |
| 1200 | 0.285 | 1168 | 0.0738 | $22.2 \times 10^{-5}$ | $4.53 \times 10^{-5}$ | $15.9 \times 10^{-5}$ | 0.717 | $32 \times 10^{4}$ |
| *Helium* | | | | | | | | |
| 200 | 0.2440 | 5197 | 0.115 | $0.91 \times 10^{-4}$ | $1.50 \times 10^{-5}$ | $0.61 \times 10^{-4}$ | 0.676 | 13200000 |
| 250 | 0.1952 | 5197 | 0.134 | $1.54 \times 10^{-4}$ | $1.75 \times 10^{-5}$ | $0.90 \times 10^{-4}$ | 0.680 | 4880000 |
| 300 | 0.1627 | 5197 | 0.150 | $1.77 \times 10^{-4}$ | $1.99 \times 10^{-5}$ | $1.22 \times 10^{-4}$ | 0.690 | 2190000 |
| 350 | 0.1394 | 5197 | 0.165 | $2.28 \times 10^{-4}$ | $2.21 \times 10^{-5}$ | $1.59 \times 10^{-4}$ | 0.698 | 1110000 |
| 400 | 0.1220 | 5197 | 0.180 | $2.83 \times 10^{-4}$ | $2.43 \times 10^{-5}$ | $1.99 \times 10^{-4}$ | 0.703 | 619000 |
| 450 | 0.1085 | 5197 | 0.195 | $3.45 \times 10^{-4}$ | $2.63 \times 10^{-5}$ | $2.43 \times 10^{-4}$ | 0.702 | 370000 |
| 500 | 0.0976 | 5197 | 0.211 | $4.17 \times 10^{-4}$ | $2.83 \times 10^{-5}$ | $2.90 \times 10^{-4}$ | 0.695 | 234000 |
| 550 | 0.0887 | 5197 | 0.229 | $4.97 \times 10^{-4}$ | $3.02 \times 10^{-5}$ | $3.40 \times 10^{-4}$ | 0.684 | 154000 |
| 600 | 0.0813 | 5197 | 0.247 | $5.84 \times 10^{-4}$ | $3.20 \times 10^{-5}$ | $3.93 \times 10^{-4}$ | 0.673 | 106000 |
| 700 | 0.0697 | 5197 | 0.278 | $7.67 \times 10^{-4}$ | $3.55 \times 10^{-5}$ | $5.09 \times 10^{-4}$ | 0.663 | 54100 |
| 800 | 0.0610 | 5197 | 0.307 | $9.58 \times 10^{-4}$ | $3.88 \times 10^{-5}$ | $6.37 \times 10^{-4}$ | 0.657 | 30200 |
| 900 | 0.0542 | 5197 | 0.335 | $11.9 \times 10^{-4}$ | $4.20 \times 10^{-5}$ | $7.75 \times 10^{-4}$ | 0.652 | 18200 |
| 1000 | 0.0488 | 5197 | 0.363 | $14.3 \times 10^{-4}$ | $4.50 \times 10^{-5}$ | $9.23 \times 10^{-4}$ | 0.645 | 11500 |
| 1200 | 0.0407 | 5197 | 0.416 | $19.7 \times 10^{-4}$ | $5.08 \times 10^{-5}$ | $12.5 \times 10^{-4}$ | 0.635 | 5240 |
| 1400 | 0.0349 | 5197 | 0.469 | $25.9 \times 10^{-4}$ | $5.61 \times 10^{-5}$ | $16.1 \times 10^{-4}$ | 0.622 | 2700 |
| 1600 | 0.0305 | 5197 | 0.521 | $32.9 \times 10^{-4}$ | $6.10 \times 10^{-5}$ | $20.0 \times 10^{-4}$ | 0.608 | 1530 |
| 1800 | 0.0271 | 5197 | 0.570 | $40.4 \times 10^{-4}$ | $6.57 \times 10^{-5}$ | $24.2 \times 10^{-4}$ | 0.599 | 930 |
| 2000 | 0.0244 | 5197 | 0.620 | $48.9 \times 10^{-4}$ | $7.00 \times 10^{-5}$ | $28.7 \times 10^{-4}$ | 0.587 | 595 |

| $T$ (K) | $\rho$ (kg/m$^3$) | $c_p$ (J/kg · K) | $k$ (W/m · K) | $\alpha$ (m$^2$/s) | $\mu$ (kg/m · s) | $\nu$ (m$^2$/s) | Pr | $g\beta/\nu^2$ (m$^{-3}$K$^{-1}$) |
|---|---|---|---|---|---|---|---|---|
| *Hydrogen* | | | | | | | | |
| 200 | 0.1229 | 13540 | 0.128 | $0.77 \times 10^{-4}$ | $0.68 \times 10^{-5}$ | $0.55 \times 10^{-4}$ | 0.717 | 16100000 |
| 250 | 0.0983 | 14070 | 0.156 | $1.13 \times 10^{-4}$ | $0.79 \times 10^{-5}$ | $0.80 \times 10^{-4}$ | 0.713 | 6070000 |
| 300 | 0.0819 | 14320 | 0.182 | $1.55 \times 10^{-4}$ | $0.89 \times 10^{-5}$ | $1.09 \times 10^{-4}$ | 0.705 | 2750000 |
| 350 | 0.0702 | 14420 | 0.203 | $2.01 \times 10^{-4}$ | $0.99 \times 10^{-5}$ | $1.42 \times 10^{-4}$ | 0.705 | 1400000 |
| 400 | 0.0614 | 14480 | 0.221 | $2.49 \times 10^{-4}$ | $1.09 \times 10^{-5}$ | $1.78 \times 10^{-4}$ | 0.714 | 778000 |
| 450 | 0.0546 | 14500 | 0.239 | $3.02 \times 10^{-4}$ | $1.18 \times 10^{-5}$ | $2.17 \times 10^{-4}$ | 0.719 | 464000 |
| 500 | 0.0492 | 14510 | 0.256 | $3.59 \times 10^{-4}$ | $1.27 \times 10^{-5}$ | $2.59 \times 10^{-4}$ | 0.721 | 292000 |
| 550 | 0.0447 | 14520 | 0.274 | $4.22 \times 10^{-4}$ | $1.36 \times 10^{-5}$ | $3.04 \times 10^{-4}$ | 0.722 | 193000 |
| 600 | 0.0410 | 14540 | 0.291 | $4.89 \times 10^{-4}$ | $1.45 \times 10^{-5}$ | $3.54 \times 10^{-4}$ | 0.724 | 130000 |
| 700 | 0.0351 | 14610 | 0.325 | $6.34 \times 10^{-4}$ | $1.61 \times 10^{-5}$ | $4.59 \times 10^{-4}$ | 0.724 | 66600 |
| 800 | 0.0307 | 14710 | 0.360 | $7.97 \times 10^{-4}$ | $1.77 \times 10^{-5}$ | $5.76 \times 10^{-4}$ | 0.723 | 36900 |
| 900 | 0.0273 | 14840 | 0.394 | $10.8 \times 10^{-4}$ | $1.92 \times 10^{-5}$ | $7.03 \times 10^{-4}$ | 0.723 | 22000 |
| 1000 | 0.0246 | 14990 | 0.428 | $11.6 \times 10^{-4}$ | $2.07 \times 10^{-5}$ | $8.42 \times 10^{-4}$ | 0.724 | 13800 |
| 1200 | 0.0205 | 15370 | 0.495 | $15.7 \times 10^{-4}$ | $2.36 \times 10^{-5}$ | $11.5 \times 10^{-4}$ | 0.733 | 6150 |
| *Nitrogen* | | | | | | | | |
| 200 | 1.708 | 1043 | 0.0183 | $1.02 \times 10^{-5}$ | $1.29 \times 10^{-5}$ | $0.75 \times 10^{-5}$ | 0.734 | $86500 \times 10^4$ |
| 250 | 1.367 | 1042 | 0.0222 | $1.56 \times 10^{-5}$ | $1.55 \times 10^{-5}$ | $1.13 \times 10^{-5}$ | 0.725 | $30700 \times 10^4$ |
| 300 | 1.139 | 1040 | 0.0260 | $2.19 \times 10^{-5}$ | $1.79 \times 10^{-5}$ | $1.57 \times 10^{-5}$ | 0.715 | $13300 \times 10^4$ |
| 350 | 0.967 | 1041 | 0.0294 | $2.92 \times 10^{-5}$ | $2.01 \times 10^{-5}$ | $2.08 \times 10^{-5}$ | 0.711 | $6500 \times 10^4$ |
| 400 | 0.854 | 1045 | 0.0325 | $3.64 \times 10^{-5}$ | $2.21 \times 10^{-5}$ | $2.59 \times 10^{-5}$ | 0.710 | $3650 \times 10^4$ |
| 450 | 0.759 | 1050 | 0.0356 | $4.47 \times 10^{-5}$ | $2.41 \times 10^{-5}$ | $3.17 \times 10^{-5}$ | 0.709 | $2170 \times 10^4$ |
| 500 | 0.683 | 1057 | 0.0387 | $5.36 \times 10^{-5}$ | $2.59 \times 10^{-5}$ | $3.79 \times 10^{-5}$ | 0.708 | $1370 \times 10^4$ |

| | | | | | | | | |
|---|---|---|---|---|---|---|---|---|
| 550 | 0.621 | 1065 | 0.0414 | $6.26 \times 10^{-5}$ | $2.76 \times 10^{-5}$ | $4.45 \times 10^{-5}$ | 0.711 | $900 \times 10^{4}$ |
| 600 | 0.569 | 1075 | 0.0441 | $7.20 \times 10^{-5}$ | $2.93 \times 10^{-5}$ | $5.14 \times 10^{-5}$ | 0.713 | $620 \times 10^{4}$ |
| 700 | 0.488 | 1098 | 0.0493 | $9.20 \times 10^{-5}$ | $3.24 \times 10^{-5}$ | $6.63 \times 10^{-5}$ | 0.720 | $319 \times 10^{4}$ |
| 800 | 0.427 | 1122 | 0.0541 | $11.3 \times 10^{-5}$ | $3.52 \times 10^{-5}$ | $8.24 \times 10^{-5}$ | 0.730 | $181 \times 10^{4}$ |
| 900 | 0.380 | 1146 | 0.0587 | $13.5 \times 10^{-5}$ | $3.79 \times 10^{-5}$ | $9.97 \times 10^{-5}$ | 0.739 | $110 \times 10^{4}$ |
| 1000 | 0.342 | 1168 | 0.0631 | $15.8 \times 10^{-5}$ | $4.04 \times 10^{-5}$ | $11.8 \times 10^{-5}$ | 0.747 | $70 \times 10^{4}$ |
| 1200 | 0.285 | 1205 | 0.0713 | $20.8 \times 10^{-5}$ | $4.50 \times 10^{-5}$ | $15.8 \times 10^{-5}$ | 0.761 | $33 \times 10^{4}$ |
| 1400 | 0.244 | 1233 | 0.0797 | $26.5 \times 10^{-5}$ | $4.92 \times 10^{-5}$ | $20.2 \times 10^{-5}$ | 0.761 | $17 \times 10^{4}$ |
| *Oxygen* | | | | | | | | |
| 200 | 1.951 | 906 | 0.0182 | $1.03 \times 10^{-5}$ | $1.47 \times 10^{-5}$ | $0.75 \times 10^{-5}$ | 0.728 | $87000 \times 10^{4}$ |
| 250 | 1.561 | 914 | 0.0225 | $1.58 \times 10^{-5}$ | $1.78 \times 10^{-5}$ | $1.14 \times 10^{-5}$ | 0.721 | $30300 \times 10^{4}$ |
| 300 | 1.301 | 920 | 0.0267 | $2.23 \times 10^{-5}$ | $2.07 \times 10^{-5}$ | $1.59 \times 10^{-5}$ | 0.711 | $12900 \times 10^{4}$ |
| 350 | 1.115 | 929 | 0.0306 | $2.95 \times 10^{-5}$ | $2.34 \times 10^{-5}$ | $2.10 \times 10^{-5}$ | 0.710 | $6380 \times 10^{4}$ |
| 400 | 0.976 | 942 | 0.0342 | $3.72 \times 10^{-5}$ | $2.59 \times 10^{-5}$ | $2.65 \times 10^{-5}$ | 0.713 | $3480 \times 10^{4}$ |
| 450 | 0.867 | 956 | 0.0377 | $4.55 \times 10^{-5}$ | $2.83 \times 10^{-5}$ | $3.26 \times 10^{-5}$ | 0.717 | $2050 \times 10^{4}$ |
| 500 | 0.780 | 971 | 0.0412 | $5.44 \times 10^{-5}$ | $3.05 \times 10^{-5}$ | $3.91 \times 10^{-5}$ | 0.720 | $1280 \times 10^{4}$ |
| 550 | 0.709 | 987 | 0.0447 | $6.38 \times 10^{-5}$ | $3.27 \times 10^{-5}$ | $4.61 \times 10^{-5}$ | 0.722 | $840 \times 10^{4}$ |
| 600 | 0.650 | 1003 | 0.0480 | $7.36 \times 10^{-5}$ | $3.47 \times 10^{-5}$ | $5.34 \times 10^{-5}$ | 0.725 | $574 \times 10^{4}$ |
| 700 | 0.557 | 1032 | 0.0544 | $9.46 \times 10^{-5}$ | $3.85 \times 10^{-5}$ | $6.91 \times 10^{-5}$ | 0.730 | $294 \times 10^{4}$ |
| 800 | 0.488 | 1054 | 0.0603 | $11.7 \times 10^{-5}$ | $4.21 \times 10^{-5}$ | $8.63 \times 10^{-5}$ | 0.736 | $165 \times 10^{4}$ |
| 900 | 0.434 | 1074 | 0.0661 | $14.2 \times 10^{-5}$ | $4.54 \times 10^{-5}$ | $10.5 \times 10^{-5}$ | 0.738 | $99 \times 10^{4}$ |
| 1000 | 0.390 | 1091 | 0.0717 | $16.8 \times 10^{-5}$ | $4.85 \times 10^{-5}$ | $12.4 \times 10^{-5}$ | 0.738 | $63 \times 10^{4}$ |
| 1200 | 0.325 | 1116 | 0.0821 | $22.6 \times 10^{-5}$ | $5.42 \times 10^{-5}$ | $16.7 \times 10^{-5}$ | 0.737 | $29 \times 10^{4}$ |
| 1400 | 0.278 | 1136 | 0.0921 | $29.1 \times 10^{-5}$ | $5.95 \times 10^{-5}$ | $21.3 \times 10^{-5}$ | 0.734 | $15 \times 10^{4}$ |

| $T$ (K) | $\rho$ ($kg/m^3$) | $c_p$ ($J/kg \cdot K$) | $k$ ($W/m \cdot K$) | $\alpha$ ($m^2/s$) | $\mu$ ($kg/m \cdot s$) | $\nu$ ($m^2/s$) | Pr | $g\beta/\nu^2$ ($m^{-3}K^{-1}$) |
|---|---|---|---|---|---|---|---|---|
| *Water Vapor (Steam)* | | | | | | | | |
| 300 | †0.0253 | 2041 | 0.0181 | †$35.1 \times 10^{-5}$ | $0.91 \times 10^{-5}$ | †$36.1 \times 10^{-5}$ | 1.03 | †$25 \times 10^4$ |
| 350 | †0.258 | 2037 | 0.0222 | †$4.22 \times 10^{-5}$ | $1.12 \times 10^{-5}$ | †$4.33 \times 10^{-5}$ | 1.02 | †$1490 \times 10^4$ |
| 400 | 0.555 | 2000 | 0.0264 | $2.38 \times 10^{-5}$ | $1.32 \times 10^{-5}$ | $2.38 \times 10^{-5}$ | 1.00 | $4330 \times 10^4$ |
| 450 | 0.491 | 1968 | 0.0307 | $3.17 \times 10^{-5}$ | $1.52 \times 10^{-5}$ | $3.10 \times 10^{-5}$ | 0.98 | $2270 \times 10^4$ |
| 500 | 0.441 | 1977 | 0.0357 | $4.09 \times 10^{-5}$ | $1.73 \times 10^{-5}$ | $3.92 \times 10^{-5}$ | 0.96 | $1280 \times 10^4$ |
| 550 | 0.401 | 1994 | 0.0411 | $5.15 \times 10^{-5}$ | $1.93 \times 10^{-5}$ | $4.82 \times 10^{-5}$ | 0.94 | $770 \times 10^4$ |
| 600 | 0.367 | 2022 | 0.0464 | $6.25 \times 10^{-5}$ | $2.13 \times 10^{-5}$ | $5.82 \times 10^{-5}$ | 0.93 | $480 \times 10^4$ |
| 700 | 0.314 | 2083 | 0.0572 | $8.74 \times 10^{-5}$ | $2.54 \times 10^{-5}$ | $8.09 \times 10^{-5}$ | 0.93 | $214 \times 10^4$ |
| 800 | 0.275 | 2148 | 0.0686 | $11.6 \times 10^{-5}$ | $2.95 \times 10^{-5}$ | $10.7 \times 10^{-5}$ | 0.92 | $106 \times 10^4$ |
| 900 | 0.244 | 2217 | 0.078 | $14.4 \times 10^{-5}$ | $3.36 \times 10^{-5}$ | $13.7 \times 10^{-5}$ | 0.95 | $58 \times 10^4$ |
| 1000 | 0.220 | 2288 | 0.087 | $17.3 \times 10^{-5}$ | $3.76 \times 10^{-5}$ | $17.1 \times 10^{-5}$ | 0.99 | $33 \times 10^4$ |

†At saturation pressure (less than 1 atm)

# Appendix H
# Surface Tension of Selected Liquids

**Surface tension of water in contact with air or water vapor**

| $T$ (°C) | $\Upsilon$ (N/m) | $T$ (°C) | $\Upsilon$ (N/m) |
|---|---|---|---|
| 0 | 0.0757 | 200 | 0.0377 |
| 10 | 0.0742 | 210 | 0.0354 |
| 20 | 0.0727 | 220 | 0.0331 |
| 30 | 0.0712 | 230 | 0.0308 |
| 40 | 0.0696 | 240 | 0.0284 |
| 50 | 0.0680 | 250 | 0.0261 |
| 60 | 0.0662 | 260 | 0.0237 |
| 70 | 0.0645 | 270 | 0.0214 |
| 80 | 0.0627 | 280 | 0.0190 |
| 90 | 0.0608 | 290 | 0.0167 |
| 100 | 0.0589 | 300 | 0.0144 |
| 110 | 0.0570 | 310 | 0.0121 |
| 120 | 0.0550 | 320 | 0.0099 |
| 130 | 0.0529 | 330 | 0.0077 |
| 140 | 0.0509 | 340 | 0.0056 |
| 150 | 0.0488 | 350 | 0.0037 |
| 160 | 0.0466 | 360 | 0.0019 |
| 170 | 0.0444 | 370 | 0.0004 |
| 180 | 0.0422 | 374† | 0.0 |
| 190 | 0.0400 | | |

†Critical point

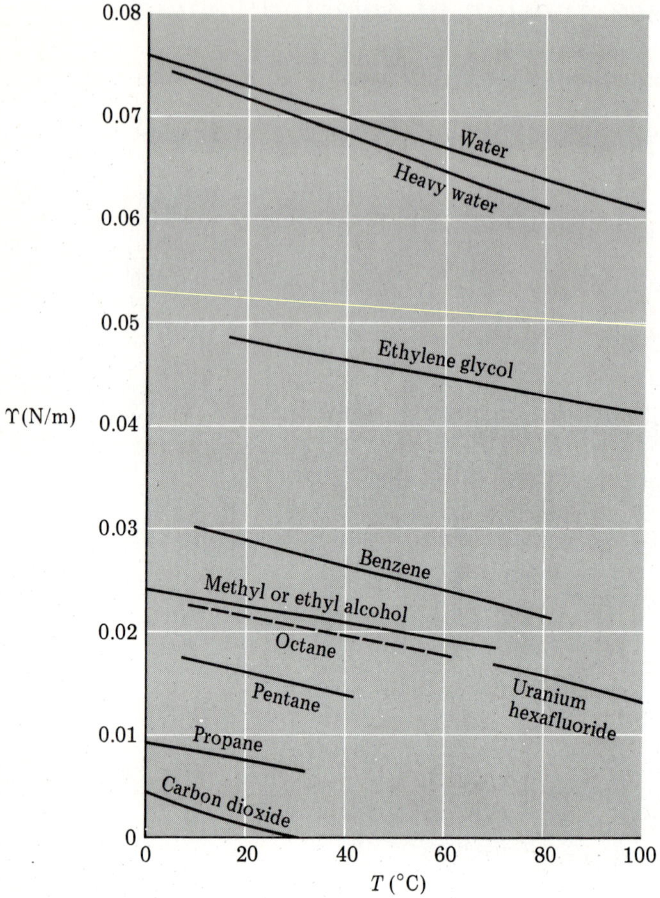


**Figure H.1** Surface tension of selected liquids in contact with their own vapor.

**Linear curve-fit near room temperatures**

$\Upsilon(\text{mN/m}) \doteq a - bT(°\text{C})$

| Liquid | $a$ (mN/m) | $b$ (mN/m · °C) |
|---|---|---|
| Acetone | 26.26 | 0.1120 |
| Alcohol | | |
| Butyl | 27.18 | 0.0898 |
| Ethyl | 24.05 | 0.0832 |
| Isopropyl | 22.90 | 0.0789 |
| Methyl | 24.00 | 0.0773 |
| Propyl | 25.26 | 0.0777 |
| Ammonia | 23.41 | 0.2993 |
| Benzene | 31.49 | 0.1303 |
| Butane | 14.87 | 0.1206 |
| Carbon dioxide | 4.34 | 0.16 |
| Carbon tetrachloride | 29.49 | 0.1224 |
| Ethylene glycol | 50.21 | 0.0890 |
| Mercury | 490.6 | 0.2049 |
| Nitrous oxide | 5.09 | 0.2032 |
| Octane | 23.52 | 0.0951 |
| *p*-xylene | 30.69 | 0.1074 |
| Pentane | 18.25 | 0.1102 |
| Propane | 9.22 | 0.0874 |
| Sulfur dioxide | 26.58 | 0.1948 |
| Toluene | 30.90 | 0.1189 |
| Uranium hexafluoride | 25.5 | 0.124 |
| Water | 75.83 | 0.1477 |
| Heavy water ($D_2O$) | 74.64 | 0.1616 |

*Source:* J. J. Jasper, "The Surface Tension of Pure Liquid Compounds," *J. Physical and Chemical Reference Data*, vol. 1, no. 4, 1972, pp. 841–1009.

# Appendix I
# Average Emissivity of Selected Surfaces

| | Temperature (K)† | Emissivity ($\varepsilon$) |
|---|---|---|
| *Metals* | | |
| Aluminum | | |
| Polished | 300–900 | 0.04–0.06 |
| Commercial sheet | 400 | 0.09 |
| Heavily oxidized | 400–800 | 0.20–0.33 |
| Anodized | 300 | 0.8 |
| Bismuth, bright | 350 | 0.34 |
| Brass | | |
| Highly polished | 500–650 | 0.03–0.04 |
| Polished | 350 | 0.09 |
| Dull plate | 300–600 | 0.22 |
| Oxidized | 450–800 | 0.6 |
| Chromium, polished | 300–1400 | 0.08–0.40 |
| Copper | | |
| Highly polished | 300 | 0.02 |
| Polished | 300–500 | 0.04–0.05 |
| Commercial sheet | 300 | 0.15 |
| Oxidized | 600–1000 | 0.5–0.8 |
| Black oxidized | 300 | 0.78 |
| Gold | | |
| Highly polished | 300–1000 | 0.03–0.06 |
| Bright foil | 300 | 0.07 |
| Iron | | |
| Highly polished | 300–500 | 0.05–0.07 |
| Cast iron | 300 | 0.44 |
| Wrought iron | 300–500 | 0.28 |
| Rusted | 300 | 0.61 |
| Oxidized | 500–900 | 0.64–0.78 |
| Lead | | |
| Polished | 300–500 | 0.06–0.08 |
| Unoxidized, rough | 300 | 0.43 |
| Oxidized | 300 | 0.63 |
| Magnesium, polished | 300–500 | 0.07–0.13 |
| Mercury | 300–400 | 0.09–0.12 |
| Molybdenum | | |
| Polished | 300–2000 | 0.05–0.21 |
| Oxidized | 600–800 | 0.80–0.82 |

†Linear interpolation is recommended over the given temperature range.

| | Temperature (K)† | Emissivity ($\varepsilon$) |
|---|---|---|
| *Metals* | | |
| Nickel | | |
| Polished | 500–1200 | 0.07–0.17 |
| Oxidized | 450–1000 | 0.37–0.57 |
| Platinum, polished | 500–1500 | 0.06–0.18 |
| Silver, polished | 300–1000 | 0.02–0.07 |
| Stainless steel | | |
| Polished | 300–1000 | 0.17–0.30 |
| Lightly oxidized | 600–1000 | 0.30–0.40 |
| Highly oxidized | 600–1000 | 0.70–0.80 |
| Steel | | |
| Polished sheet | 300–500 | 0.08–0.14 |
| Commercial sheet | 500–1200 | 0.20–0.32 |
| Heavily oxidized | 300 | 0.81 |
| Tin, polished | 300 | 0.05 |
| Tungsten | | |
| Polished | 300–2500 | 0.03–0.29 |
| Filament | 3500 | 0.39 |
| Zinc | | |
| Polished | 300–800 | 0.02–0.05 |
| Oxidized | 300 | 0.25 |
| *Nonmetals* | | |
| Alumina | 800–1400 | 0.65–0.45 |
| Aluminum oxide | 600–1500 | 0.69–0.41 |
| Asbestos | 300 | 0.96 |
| Asphalt pavement | 300 | 0.85–0.93 |
| Brick | | |
| Common | 300 | 0.93–0.96 |
| Fireclay | 1200 | 0.75 |
| Carbon filament | 2000 | 0.53 |
| Cloth | 300 | 0.75–0.90 |
| Concrete | 300 | 0.88–0.94 |
| Glass | | |
| Window | 300 | 0.90–0.95 |
| Pyrex | 300–1200 | 0.82–0.62 |
| Pyroceram | 300–1500 | 0.85–0.57 |
| Ice | 273 | 0.95–0.99 |
| Magnesium oxide | 400–800 | 0.69–0.55 |

| | Temperature (K)† | Emissivity ($\varepsilon$) |
|---|---|---|
| *Nonmetals* | | |
| Masonry | 300 | 0.80 |
| Paints | | |
| Aluminum | 300 | 0.40–0.50 |
| Black, lacquer, shiny | 300 | 0.88 |
| Oils, all colors | 300 | 0.92–0.96 |
| White acrylic | 300 | 0.90 |
| White enamel | 300 | 0.90 |
| Red primer | 300 | 0.93 |
| Paper, white | 300 | 0.90 |
| Plaster, white | 300 | 0.93 |
| Porcelain, glazed | 300 | 0.92 |
| Quartz, rough, fused | 300 | 0.93 |
| Rubber | | |
| Soft | 300 | 0.86 |
| Hard | 300 | 0.93 |
| Sand | 300 | 0.90 |
| Silicon carbide | 600–1500 | 0.87–0.85 |
| Skin, human | 300 | 0.95 |
| Snow | 273 | 0.80–0.90 |
| Soil, earth | 300 | 0.93–0.96 |
| Soot | 300–500 | 0.95 |
| Teflon | 300–500 | 0.85–0.92 |
| Water, deep | 273–373 | 0.95–0.96 |
| Wood | | |
| Beech | 300 | 0.94 |
| Oak | 300 | 0.90 |

# Appendix J
# The Heisler/Gröber Charts

Traditionally the Heisler/Gröber charts have been used to solve the problem of sudden immersion of a plate, cylinder, or sphere into a fluid. The geometries are given by Fig. 4.6. The fluid is at conditions ($h_o$, $T_o$) and the body is initially at uniform temperature $T_i$, with constants $k$ and $\alpha$.

Figures J.1(a), J.2(a), and J.3(a) show the decay of center point temperature $T_c$ as computed from Eq. (4.49) for the plate, cylinder, and sphere, respectively. Figures J.1(b), J.2(b), and J.3(b) show the temperature at other points in the three bodies as computed from Eqs. (4.50). These six charts were given by Heisler in a now famous paper: M. P. Heisler, "Temperature Charts for Induction and Constant Temperature Heating," *ASME Transactions*, vol. 69, 1947, pp. 227–236.

Figures J.1(c), J.2(c), and J.3(c) show the total heat transfer of the three bodies versus time, as computed from Eqs. (4.52). They are taken from the text by H. Gröber, S. Erk, and U. Grigull, *Fundamentals of Heat Transfer*, McGraw-Hill, New York, 1961.

You can see that it is difficult to read these charts to better than 10% accuracy. We recommend instead in Chapter 4 that you use the original equations (4.49), (4.50), and (4.52), since they are easily evaluated on a handheld calculator.

(a)

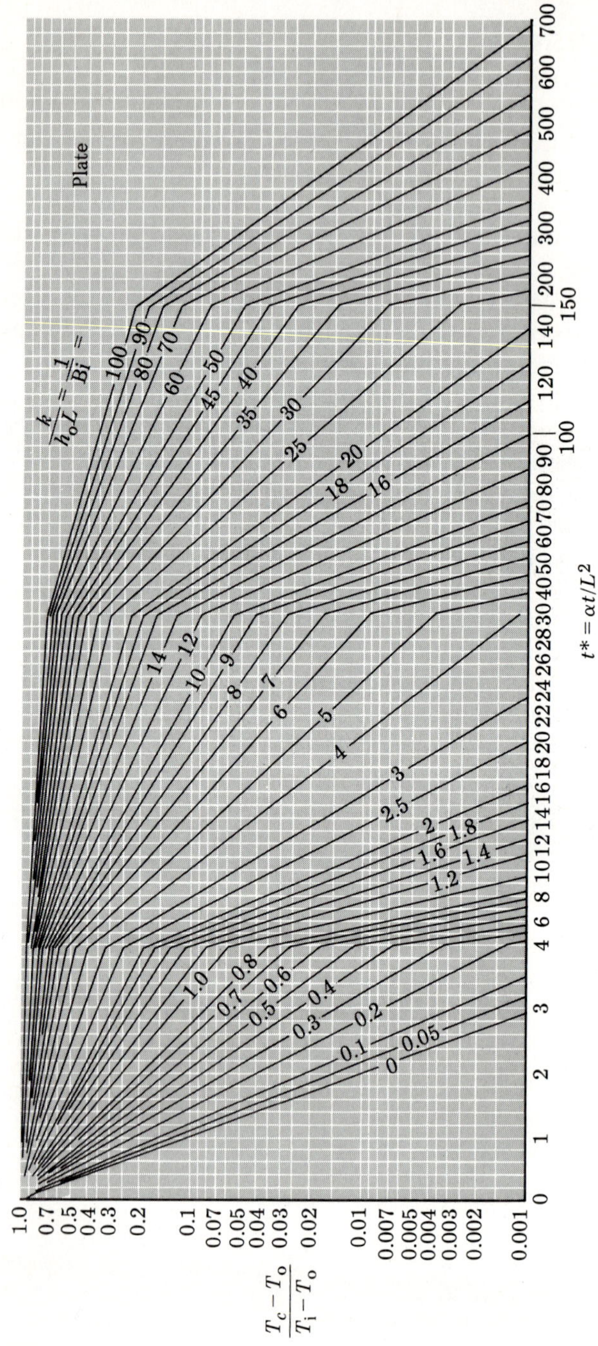
Plate
$\frac{k}{h_o L} = \frac{1}{Bi} =$
$\frac{T_c - T_o}{T_i - T_o}$
$t^* = \alpha t/L^2$
100
90
80
70
60
50
45
40
35
30
25
20
18
16
14
12
10
9
8
7
6
5
4
3
2.5
2
1.8
1.6
1.4
1.2
1.0
0.8
0.7
0.6
0.5
0.4
0.3
0.2
0.1
0.05
0

(b)

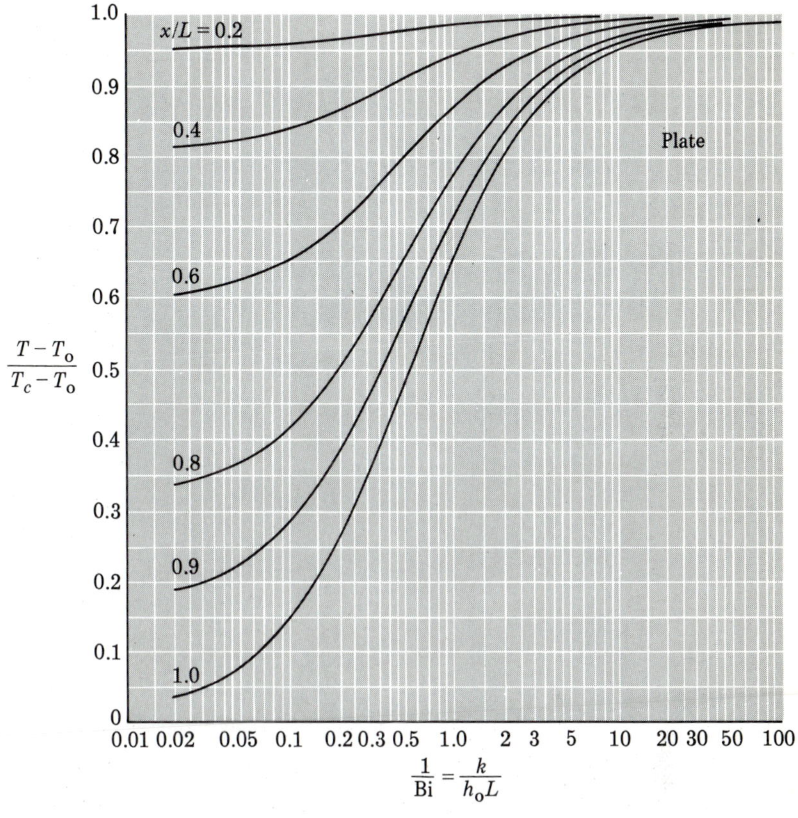


(c)

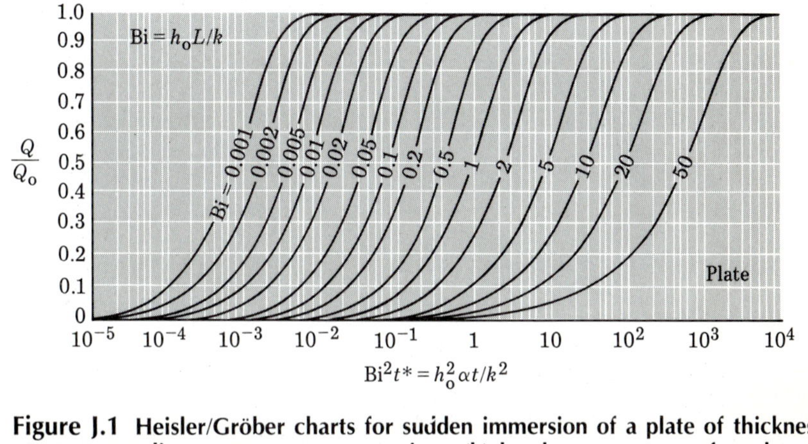


**Figure J.1** Heisler/Gröber charts for sudden immersion of a plate of thickness $2L$: (a) centerline temperature versus time; (b) local temperature referred to the centerline; (c) total heat transfer.

(a)

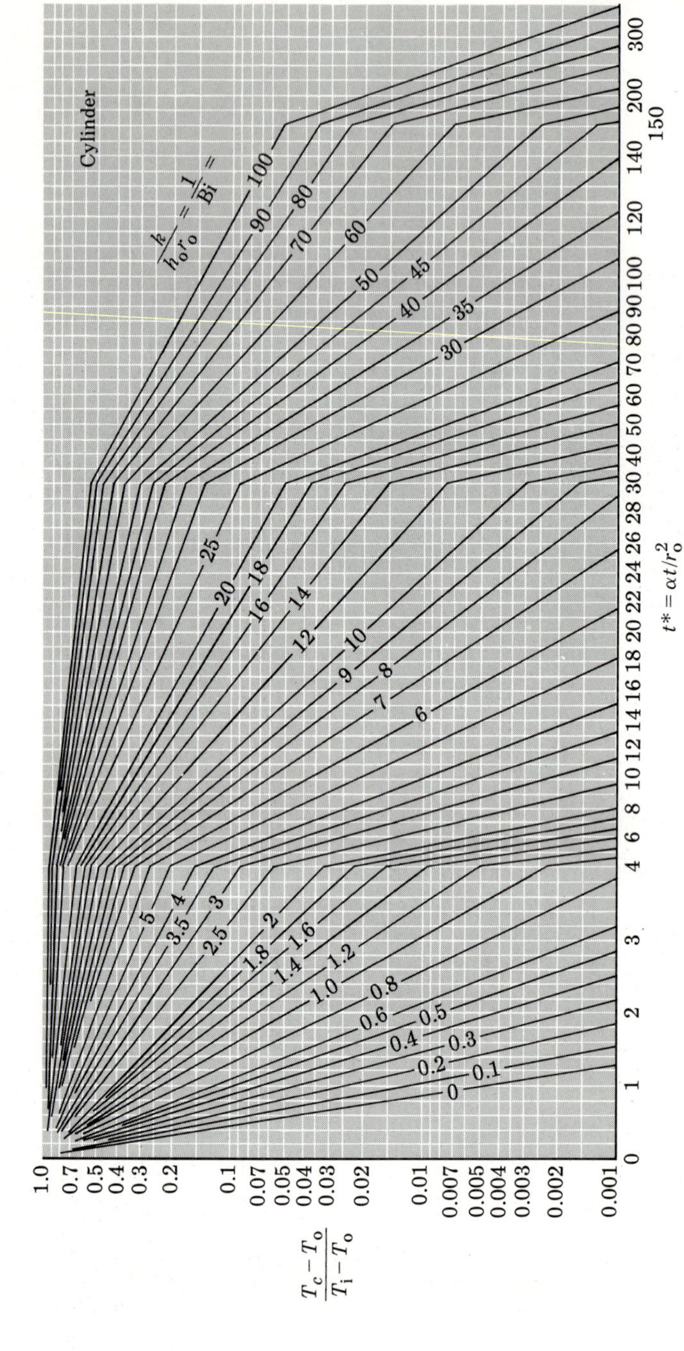

(b)

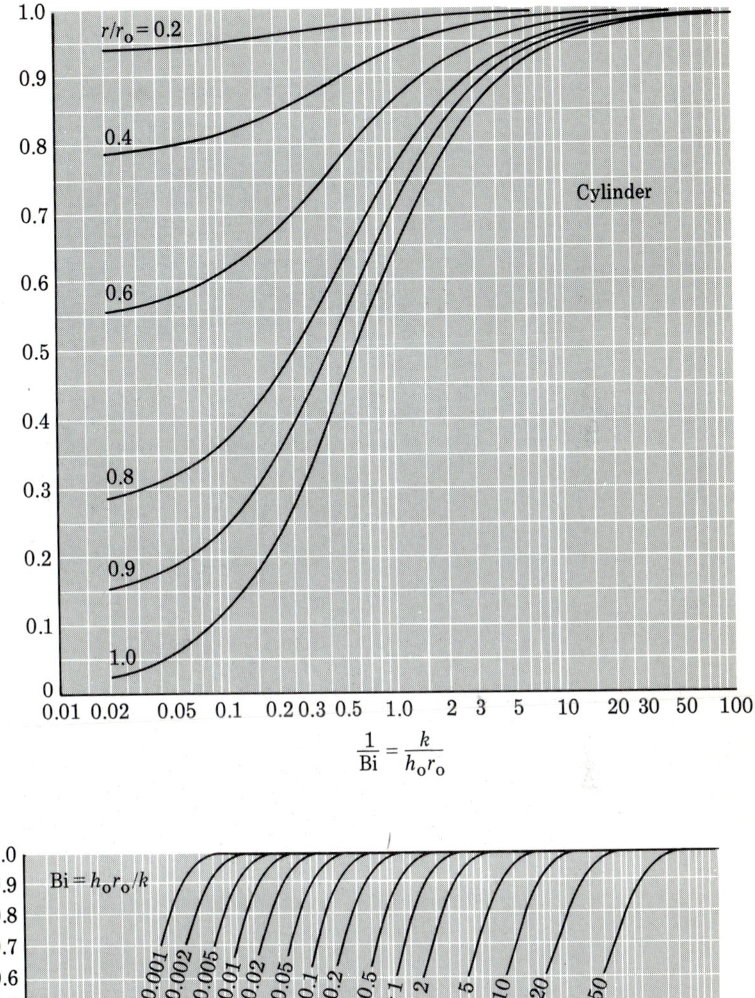


(c)

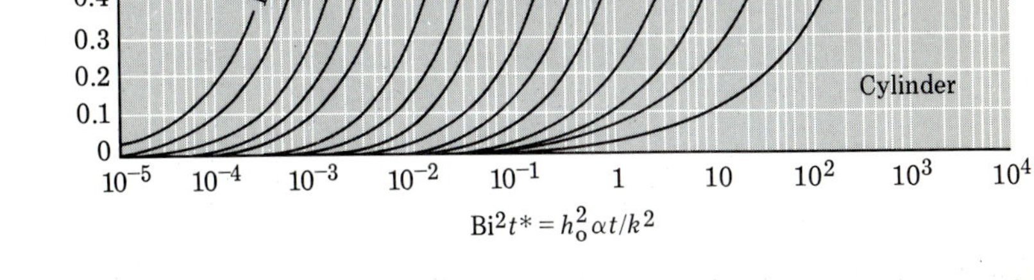


**Figure J.2** Heisler/Gröber charts for sudden immersion of a cylinder of radius $r_o$: (a) centerline temperature versus time; (b) local temperature referred to the centerline; (c) total heat transfer.

(a)

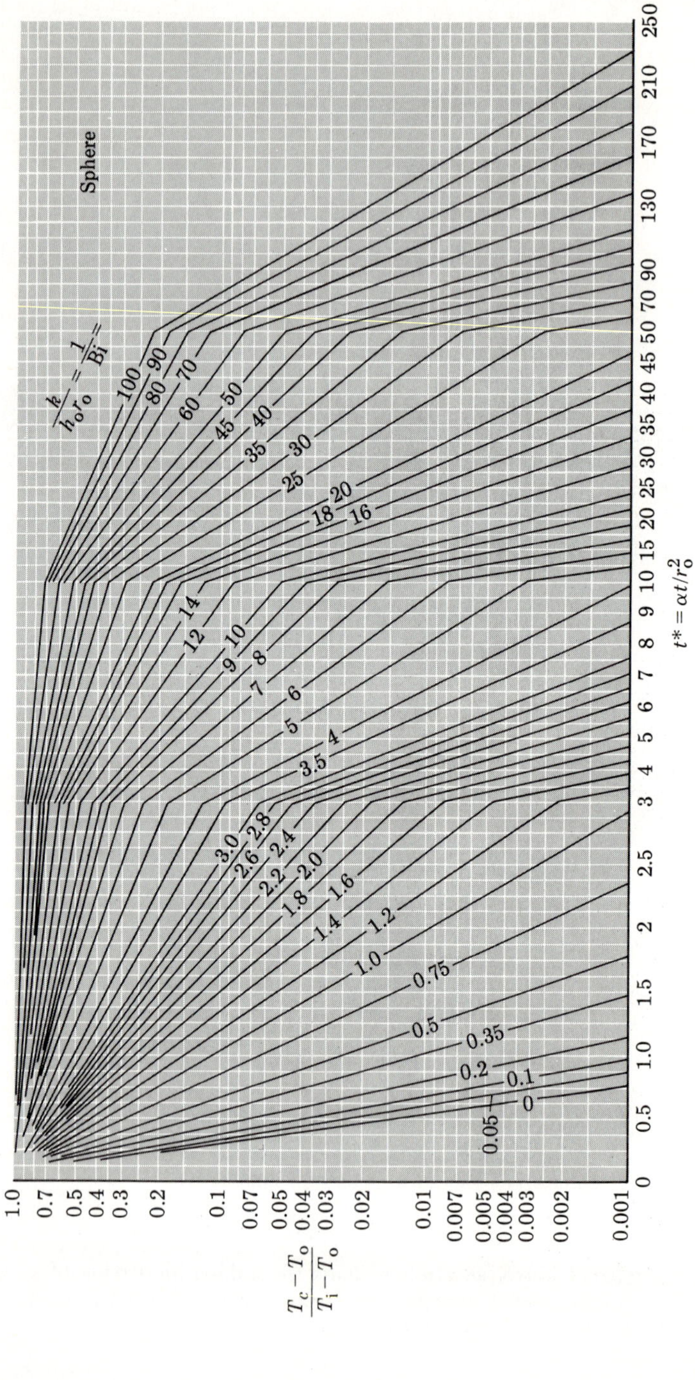

(b)

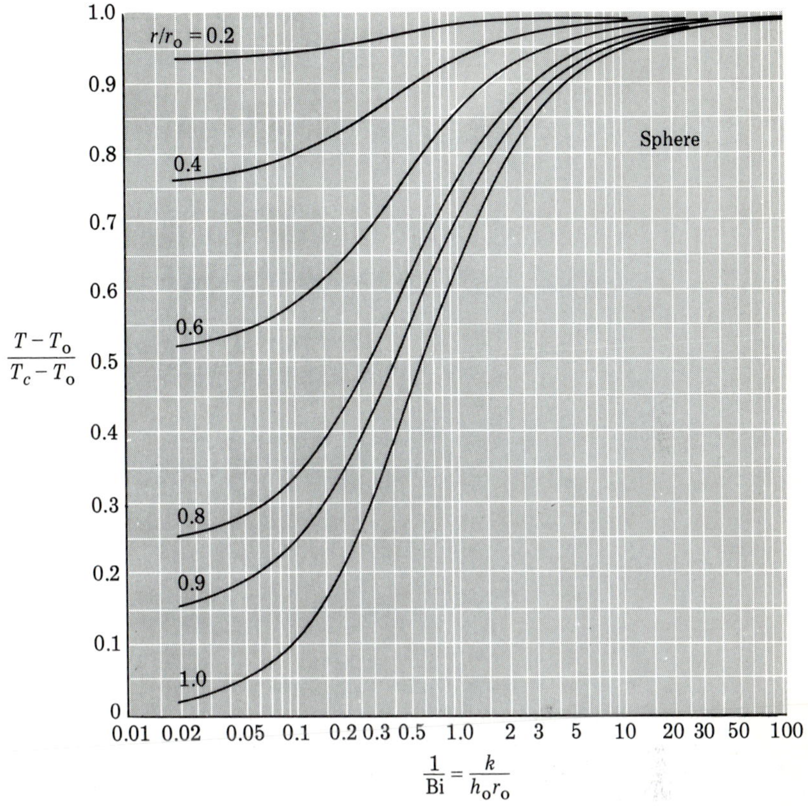


(c)

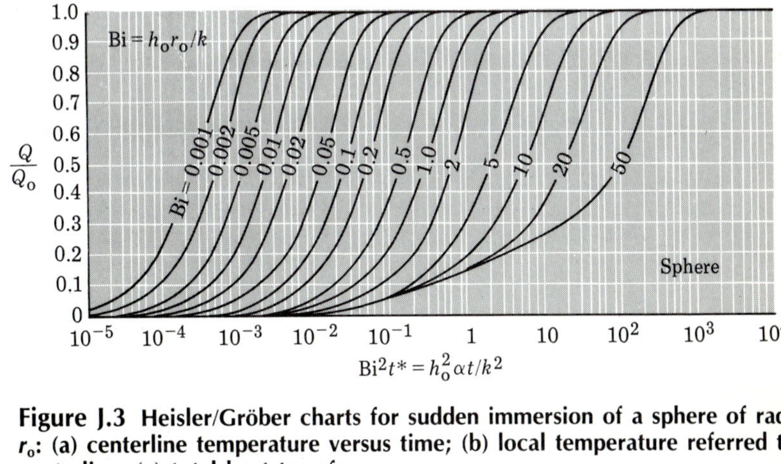


**Figure J.3** Heisler/Gröber charts for sudden immersion of a sphere of radius $r_o$: (a) centerline temperature versus time; (b) local temperature referred to the centerline; (c) total heat transfer.

# Appendix K
# Properties of Gases at 20°C and 1 atm

| Gas | Molecular Weight, *M* | Molecular† Volume, *V* | Viscosity $\mu$, kg/(m · s) |
|---|---|---|---|
| Air | 39.94 | 16.1 | 2.24 E-5 |
| Air | 28.96 | 20.1 | 1.80 E-5 |
| He | 4.003 | 2.9 | 1.97 E-5 |
| $H_2$ | 2.016 | 7.1 | 9.05 E-6 |
| $N_2$ | 28.02 | 17.9 | 1.76 E-5 |
| $O_2$ | 32.00 | 16.6 | 2.02 E-5 |
| $Cl_2$ | 70.91 | 38. | 1.33 E-5 |
| Ne | 20.18 | 5.6 | 3.11 E-5 |
| $O_2$ | 4.032 | 6.7 | — |
| $H_2O$ | 18.02 | 12.7 | 1.02 E-5 |
| $CO_2$ | 44.01 | 26.9 | 1.48 E-5 |
| CO | 28.01 | 18.9 | 1.82 E-5 |
| $N_2O$ | 44.02 | 36. | 1.45 E-5 |
| NO | 30.01 | 18. | 1.90 E-5 |
| $SO_2$ | 64.06 | 41.1 | 1.25 E-5 |
| $CH_4$ | 16.04 | 23. | 1.34 E-5 |
| $NH_3$ | 17.03 | 14.9 | 1.00 E-5 |
| $CCl_4$ | 153.83 | 92. | 1.25 E-5 |
| $C_6H_6$ | 78.11 | 83. | 7.53 E-6 |
| $C_2H_6O$ | 46.07 | 50. | 8.4 E-6 |
| $CH_4O$ | 32.03 | 32. | 9.5 E-6 |
| $C_3H_6O$ | 58.05 | 40. | 7.2 E-6 |
| $C_2H_4$ | 28.05 | 30. | 1.01 E-5 |
| $C_{10}H_8$ | 128.16 | 180. | — |

†For use in estimating mass diffusivity from Fuller's empirical correlation, Problem 11.8.

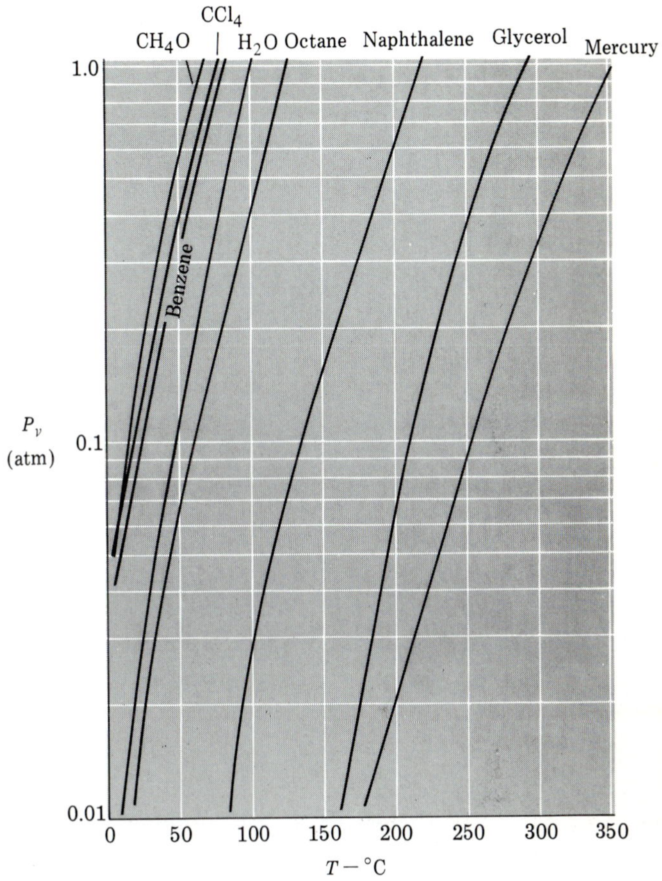
CCl4
CH4O
H2O
Octane
Naphthalene
Glycerol
Mercury
Benzene
1.0
0.1
0.01
Pv
(atm)
0
50
100
150
200
250
300
350
T − °C

# Appendix L
# Power-law Temperature Variation of Gas Diffusivity

| Gas Mixture | $D_{AB}$(273K, 1 atm), cm$^2$/s | Exponent $m$, Eq. (11.22) |
|---|---|---|
| Air-$H_2$ | 0.61 | 1.75 |
| Air-$O_2$ | 0.178 | 1.75 |
| Air-$N_2$ | 0.175 | 1.90 |
| Air-He | 0.61 | 1.69 |
| Air-Ar | 0.177 | 1.75 |
| Air-$Cl_2$ | 0.103 | 1.75 |
| Air-$H_2O$ | 0.23 | 1.74 |
| Air-CO | 0.19 | 1.75 |
| Air-$CO_2$ | 0.142 | 1.70 |
| Air-$SO_2$ | 0.108 | 1.75 |
| Air-NO | 0.170 | 1.75 |
| Air-acetone | 0.109 | 1.75 |
| Air-ammonia | 0.201 | 1.75 |
| Air-benzene | 0.0783 | 1.89 |
| Air-ethanol | 0.105 | 1.77 |
| Air-ethylene | 0.144 | 1.75 |
| Air-$CCl_4$ | 0.07 | 1.75 |
| Air-gasoline | 0.079 | 1.94 |
| Air-methane | 0.190 | 1.75 |
| Air-methanol | 0.134 | 1.77 |
| Air-napthalene | 0.054 | 1.75 |
| Air-octane | 0.0544 | 1.60 |
| Air-toluene | 0.075 | 1.76 |
| $H_2O$-$CO_2$ | 0.146 | 1.84 |
| $H_2O$-$H_2$ | 0.734 | 1.82 |
| $H_2O$-$O_2$ | 0.240 | 1.73 |
| Ar-$H_2$ | 0.715 | 1.89 |
| Ar-He | 0.638 | 1.75 |
| Ar-Ne | 0.28 | 1.73 |
| $CO_2$-CO | 0.137 | 1.86 |
| $CO_2$-He | 0.494 | 1.80 |
| $CO_2$-$O_2$ | 0.138 | 1.80 |

*Source:* N. B. Vargaftik, *Handbook of Physical Properties of Liquids and Gases*, Hemisphere, New York, 1975.

# Appendix M
# Molecular Diffusivity of Gases, Liquids, and Solids at 20°C and 1 atm

**M-1 Gases[1, 2]**

| Mixture | $D_{AB}$ ($m^2/s$) | Mixture | $D_{AB}$ ($m^2/s$) |
|---|---|---|---|
| $Ar\text{-}N_2$ | 1.9 E-5 | $H_2\text{-}CO_2$ | 6.2 E-5 |
| $H_2\text{-}N_2$ | 7.7 E-5 | $H_2O\text{-}CO_2$ | 1.7 E-5 |
| $CO_2\text{-}N_2$ | 1.6 E-5 | $He\text{-}CO_2$ | 5.8 E-5 |
| $O_2\text{-}N_2$ | 2.1 E-5 | $CO\text{-}CO_2$ | 1.5 E-5 |
| $H_2O\text{-}N_2$ | 2.3 E-5 | $NO\text{-}CO_2$ | 1.6 E-5 |
| $He\text{-}N_2$ | 7.0 E-5 | | |
| $CO\text{-}N_2$ | 2.1 E-5 | $O_2\text{-}H_2$ | 8.1 E-5 |
| | | $Ar\text{-}H_2$ | 7.7 E-5 |
| $H_2\text{-}O_2$ | 7.9 E-5 | $H_2O\text{-}H_2$ | 8.5 E-5 |
| $CO_2\text{-}O_2$ | 1.6 E-5 | $He\text{-}H_2$ | 1.4 E-4 |
| Ammonia-$O_2$ | 2.5 E-5 | | |
| Benzene-$O_2$ | 0.8 E-5 | He-Ar | 7.1 E-5 |
| $H_2O\text{-}O_2$ | 2.7 E-5 | Ne-Ar | 3.1 E-5 |

*Note:* For dilute gases in air and dilute liquids in water, see Table 11.1.

**M-2 Liquids[2]**

| Solution | $D_{AB}$ ($m^2/s$) |
|---|---|
| $CO_2$ in ethanol | 3.4 E-9 |
| Chloroform in ethanol | 1.2 E-9 |
| Phenol in ethanol | 0.8 E-9 |
| Acetic acid in benzene | 1.9 E-9 |
| Bromine in benzene | 2.0 E-9 |
| $CCl_4$ in benzene | 2.0 E-9 |
| Chloroform in benzene | 2.1 E-9 |
| Iodine in benzene | 1.9 E-9 |
| Phenol in benzene | 1.5 E-9 |
| Aniline in methanol | 1.5 E-9 |
| $CCl_4$ in methanol | 1.7 E-9 |
| Chloroform in methanol | 2.1 E-9 |

| M-3 Solids[3,4] | |
|---|---|
| **Substances** | $D_{AB}$ ($m^2/s$) |
| Aluminum in copper | 1.3 E-34 |
| Cadmium in copper | 2.7 E-19 |
| Lead in gold | 3.5 E-4 |
| Mercury in lead | 2.5 E-19 |
| Hydrogen in iron | 2.6 E-13 |
| Helium in glass ($SiO_2$) | 2.0 E-12 |
| Helium in Pyrex | 4.5 E-15 |
| Oxygen in rubber | 2.1 E-10 |
| Nitrogen in rubber | 1.5 E-10 |
| Carbon dioxide in rubber | 1.1 E-10 |
| Hydrogen in rubber | 8.5 E-10 |

*Sources:*

1. N. B. Vargaftik, *Handbook of Physical Properties of Liquids and Gases*, 2nd ed., Hemisphere, New York, 1975.

2. R. C. Reid, J. M. Prausnitz, and T. K. Sherwood, *The Properties of Liquids and Gases*, 3rd ed., McGraw-Hill, New York, 1980.

3. R. M. Barrer, *Diffusion in and through Solids*, Macmillan, New York, 1941.

4. W. Jost, *Diffusion in Solids, Liquids, Gases,* revised edition, Academic Press, New York, 1960.

# Appendix N
# Answers to Selected Even-Numbered Problems

## Chapter 1

1.8 $k = 0.1$ W/(m · K)
1.10 180 L/day
1.14 0.033 W/(m · K)
1.16 160°F
1.20 $T^{1/2}$ is too small for all gases
1.22 $k_B = 43$ W/(m · K)
1.24 604°F
1.28 $h = 13$ W/(m$^2$ · K)
1.30 (a) 7.0 m/s; (b) 1.7 m/s
1.32 (a) 85%; (b) 15%
1.36 39°C
1.38 192°C
1.40 $q'' = 8.3$ W/m$^2$; $T_i = 18.6$°C

## Chapter 2

2.4 $-k_b \left.\frac{\partial T}{\partial n}\right|_b = \varepsilon_b \sigma(T_b^4 - T_\infty^4)$
2.10 $q''_{net} = 26400$ W/m$^2$; $\partial T/\partial t = 0.115$ K/s at center
2.14 (a) 18.1 W/m · K; (b) 23.0; (c) 15.8
2.16 $T_1 = 683$ K, $T_4 = 250$K
2.18 *k* increases approximately exponentially with *T*
2.20 (a) 710 W/m$^2$; (b) 1900 W/m$^2$; wind chill factor = −110°C
2.22 (a) 276 W; (b) $20/month
2.24 76°C
2.28 q(one-dimensional) = 15600 W/m
2.30 8.3 W/m$^2$
2.32 (a) 24kW; (b) 398°C
2.34 $q_{bare} = 389000$ Btu/hr; (a) 6250 Btu/hr; (b) 11500 Btu/hr
2.38 0.0256 Btu/(hr · ft · °F)
2.40 $q = 9.3 \times 10^7$ J/day, Cost = $2.58 per day
2.44 300 ft$^2$ glass add loss of 85000 Btu/hr or 19 gal/day
2.48 522 kW/m$^3$
2.50 $q_{bare} = 89$ Btu/hr, $q_{insulated} = 112$ Btu/hr
2.52 $k_{min} = 0.012$ W/m · K; if $k = 0.035$, $q_{max} = 13.8$ W/m
2.54 (a) 0.060 in.; (b) 0.050 in.
2.58 $q = 25.2$ W, $T_c = 911$°C
2.66 (a) 224°C; (b) 95%; (c) 1.6 W
2.70 $k_X = 72$ W/m · K
2.72 (a) 296°F; (b) 90°F
2.80 $k_{min} = 1.27$ W/m · K
2.82 $L = 225$ mm; $\eta = 53\%$; $q = 12$ W

## Chapter 3

3.12 $T_c = 70$°C
3.16 $T = T_o \frac{\sinh(\pi x/H)}{\sinh(\pi L/H)} \sin(\pi y/H)$
3.18 $T = T_o \frac{\sinh(2\pi y/L)}{\sinh(2\pi H/L)} \sin(2\pi x/L)$
3.20 $\frac{T - T_o}{T_1 - T_o} = \frac{\ln(r/r_o)}{\ln(r_1/r_o)}$
3.22 10.66°C
3.24 $T(L/2, L/2) = 0.16234\, T_o$
3.26 $q = 4170$ W/m
3.28 $q = 11900$ Btu/hr
3.30 $q = 203000$ Btu/hr
3.34 $H = 1.15$ m
3.36 $q = 32000$ Btu/hr
3.40 $q = 21000$ W
3.46 $T_o = \frac{1}{3 + Bi}\left[T_2 + T_3 + \frac{1}{2}(T_1 + T_4) + BiT_\infty + \frac{3\dot{q}(\Delta x)^2}{8k}\right]$

3.48 $(2 + 2\beta)T_{m,n} \doteq T_{m+1,n} + \beta(1 + \eta)T_{m,n+1} + \beta(1 - \eta)T_{m,n-1}$ where $\beta = (r_n\Delta 0/\Delta r)^2$ and $\eta = \Delta r/r_n$

3.50 $T_1 = T_3 \doteq 24.0°$, $T_2 = T_4 \doteq 7.5°$

3.56 $T_1 = 68.75°C$, $T_2 = 118.75°C$, $T_3 = 56.25°C$ $T_4 = 106.25°C$

3.62 Three-point formula: $\frac{dT}{dx}$ at $(x = 1.0) \doteq -844.1°C/m$, error of 20°C/m if $T_1$ is off by 1°C.

## Chapter 4

4.2 $\frac{\partial\theta}{\partial n^*} = B(\theta^4 - 1)$ where $B = (\sigma F_{12}T_\infty^3 L/k)$ is equivalent to a Biot number

4.4 t = 760 s

4.6 Steady state T = 210°F everywhere.

4.10 t = 250 s

4.14 $D_{max} = 1.64$ mm

4.16 t = 8 hr

4.20 Thermocouple oscillates ±2.7°C with a phase lag of 74 degrees (1.3 radians).

4.24 V = 1.35 cm/s, $T_{final} = 88°C$

4.26 Setting back at night, option (a), saves 28%.

4.30 T(hot slab) = 188°F, T(cold slab) = 12°F

4.32 T(hot slab) = 66°C, T(cold slab) = 34°C

4.34 t = 121 s (no radiation)

4.36 t = 2.8 s

4.38 $q_o'' = 120$ kW/m$^2$

4.40 (a) 37 s; (b) 21 s

4.44 $h_o = 185$ Btu/(hr · ft$^2$ · °F)

4.46 V = 1.06 cm/s

4.50 t = 220 s

4.52 $T_c = 210°F$

4.54 t = 1.94 hr

4.56 t = 1.90 hr

4.58 $T_{edge} = 170°C$

4.60 t = 2.2 hr

4.64 $T_m^{j+1} = (1 - 2\sigma)T_m^j + \sigma(1 + \zeta)T_{m+1}^j + \sigma(1 - \zeta)T_{m-1}^j$ where $\sigma = \alpha\Delta t/\Delta r^2$ and $\zeta = \Delta r/2r_m$

4.66 $T_{m,n}^{j+1} = (1 - 4\sigma)T_{m,n}^j + (1 + \zeta)T_{m,n+1}^j + \sigma(1 - \zeta)T_{m,n-1}^j + \sigma(T_{m+1,n}^j + T_{m-1,n}^j)$, $\sigma = \alpha\Delta t/\Delta r^2$ and $\zeta = \Delta r/2r_{m,n}$

## Chapter 5

5.4 Velocity no more than (a) 9.3 ft/s; (b) 144 ft/s; (c) 1090 ft/s.

5.12 Equation (5.8) is unchanged by variable density.

5.14 (a) 68800 W/m$^2$ · K; (b) 66 kW

5.18 Pr = 1170

5.20 (a) $1.9 \times 10^9$; (b) $2.8 \times 10^{11}$; (c) $6.7 \times 10^6$

5.24 $h = 250$ W/m$^2$ · K, $q_w'' = 2500$ W/m$^2$

5.26 $\overline{h} = 102$ W/m$^2$ · K

5.30 U = 28 m/s

## Chapter 6

6.6 $u = Uy/H$

$$T = T_1 + (T_2 - T_1)y/H + \frac{\mu U^2}{2k}(y/H - y^2/H^2)$$

6.12 $U_o = 21$ m/s

6.14 $U \doteq 0.0007$ m/s (very low Reynolds number)

6.16 $T_o = 27°C$

6.22 $\theta/x = c_f \doteq (2 - \pi/2)^{1/2}/Re_x^{1/2}$
6.28 (a) 40°C; (b) 43.3°C
6.30 $x = 930$ m, LMTD $= 94.9$°C
6.34 $\Delta p = 25$ Pa, $T_o = 6.6$°C
6.42 $L \doteq 3200$ ft
6.44 $\dot{m} = 1.57$ kg/hr
6.46 $T_{mo} \doteq 20$°C
6.48 at $y^+ = 3.04$ (slightly thinner than the sublayer)
6.54 $q_w = 203$ kW
6.56 30 Btu/hr · ft$^2$ · °F
6.60 $t \doteq 20$ hr
6.64 $U \doteq 290$ ft/s
6.68 $T_w(L) = 37$°C
6.70 $T_w(x = 6 \text{ in.}) = 109$°F
6.74 $T_w = 78$°C, $\Delta p = 890$ Pa
6.76 $\dot{m} = 2.92$ lbm/s, $T_{mo} = 124$°F
6.82 $L \doteq 57$ cm
6.84 $\dot{m}_{max} = 0.4$ kg/s with $\Delta p =$ 660 Pa
6.88 $D = 10.2$ cm
6.90 (a) 15,700 Btu/hr; (b) 6.2 lbf
6.92 $U = 24$ m/s
6.94 $q_w = 2.0$ W
6.96 $q_w \doteq 90$ kW, $F \doteq 2$ kN ($Re_D$ too large)
6.100 $U = 5.5$ m/s
6.104 (a) 10900 W/m$^2$ · K; (b) 515 MW/m
6.106 $T_{aw} = 730$°R, $q''_w(x = 2 \text{ ft}) =$ 8600 Btu/hr · ft$^2$

## Chapter 7

7.4 (a) $1.6 \times 10^{11}$; (b) $2.0 \times 10^{12}$; (c) $1.5 \times 10^{14}$
7.8 Only $Gr_L$ appears.
71.0 $Gr_L \doteq 1.1 \times 10^{17}$
7.12 $u/u_o \doteq \eta - 2\eta^2 + \eta^3$
7.18 (a) 50 cm; (b) 20 cm
7.20 Above $Gr_x = 10^{12}$, $h_x$ is approximately constant.
7.22 According to Eq. (7.35), (a) 30°C; (b) 26°C
7.24 (a) 20,000 W; (b) 1.5 s
7.30 2.4 m/day
7.36 6.8 in.
7.38 500 W (horizontal), 460 W (vertical)
7.42 $T_w = 92$°C
7.44 25 kW (83,000 Btu/hr)
7.46 250 W
7.48 $q \doteq 400$ W, $\Delta T \doteq 28$°C
7.50 2800 W
7.54 $q = 330$ W
7.56 $q = 33$ W, $R_{gap} = 40\%$ of total resistance
7.58 at approximately 8000 Pa
7.60 4900 W (aiding), 4500 W (opposing)
7.62 $q = 33$ mW

## Chapter 8

8.6 955K
8.8 (a) negligible; (b) 74%
8.10 (a) 10.5%; (b) 89%
8.14 87%
8.18 10.2 cm
8.20 $\varepsilon = 0.78$
8.22 (a) 0.29; (b) 0.18
8.24 $\varepsilon = 0.55$ at 1250K
8.26 $T \doteq 670$K
8.28 the water will freeze unless $T_{air} = 6.3$°C
8.30 (a) 1300K; (b) 230 kW/m$^2$
8.32 $F_{1-2} = 1$, $F_{2-1} = 1/3$, $F_{2-2} =$ 2/3
8.34 $F_{1-2} = 1/2$, $F_{2-1} = 1/4$, $F_{2-e} =$ 3/4

8.38 $F_{1-2} = 0.617$, $F_{1-3} = 0.169$, $F_{1-2,3} = 0.786$, $F_{1-e} = 0.214$
8.40 $F_{1-2} = 0.5$, $F_{2-1} \doteq 0$
8.42 200 W/m
8.44 (a) 480 W/m; (b) 980 W/m
8.48 15 cm
8.50 28 cm
8.52 $F_{2-3} = 0.382$, $F_{1-4} = 0.041$
8.54 $F_{1-4} = 0.104$
8.56 (a) 3.2 W; (b) 82.4W
8.60 3.4 kg/hr
8.62 $q_{1-2} = 165$ W
8.64 1.7 kg/hr
8.66 0.872 W (95% of black emission)
8.68 $T_2 = 521°C$
8.70 $q_3 = 426$ kW/m
8.72 $q_1 = 185$ W, $T_2 = 384°C$
8.78 $q_{gas\text{-}wall} = 25$ kW
8.80 13 kW
8.84 904 $W/m^2$
8.86 (a) 325 K; (b) 357 W

## Chapter 9

9.4 $\delta = 0.16$ mm, $u_{max} = 0.42$ m/s, $q = 41$ kW
9.6 $T_w = 60°C$
9.8 one tube: 72 kg/hr; ten tubes: 128 kg/hr
9.10 118 tubes
9.12 200 kg/hr
9.16 113 tubes
9.22 $C_{sf} \doteq 0.012$
9.24 13 tubes at $T_w = 118°C$
9.30 Formula agrees with Eq. (9.26).
9.36 1400 W
9.38 approximate subcooling of 60°C

## Chapter 10

10.4 1.5 mm
10.6 (a) 2100 $W/m^2 \cdot K$; (b) 800 $W/m^2 \cdot K$
10.8 (a) 87 $W/m^2 \cdot K$; (b) 82 $W/m^2 \cdot K$
10.10 $\dot{m}_{water} = 0.288$ kg/s; parallel flow: LMTD = 36°C, A = 7.4 $m^2$; counterflow: LMTD = 57.5°C, A = 4.65 $m^2$.
10.12 (a) 0.282 kg/s; (b) 21 m
10.14 18 $m^2$
10.16 (a) 1.7 $m^2$; (b) 240 kg/hr
10.18 20 $m^2$
10.26 water: 71°C; oil: 72°C
10.28 0.46 kg/s
10.32 86°C
10.34 29 tubes

## Chapter 11

11.4 (a) $\omega_{O2} = 0.4998$, $\omega_{N2} = 0.4377$, $\omega_{He} = 0.0625$; (b) $M = 21.34$; (c) 0.9536 $kg/m^3$
11.8 $D_{AB} \doteq 1.09 \times 10^{-6}$ $m^2/s$
11.10 14.2 atm
11.12 As $t \to \infty$, $\dfrac{C(r) - C_o}{C_i - C_o} = \dfrac{\ln(r/r_1)}{\ln(r_2/r_1)}$
11.18 0.0005 kg/hr, $t_o = \rho_{Napth} r_o^2 / 2\rho_{Aw} D_{AB} \doteq 70$ days
11.20 $\dot{m} = 1.32$ g/hr
11.22 8.7 hr
11.24 (a) 0.0067 g/hr; (b) 280 days
11.28 67 hr
11.32 2.16 mm
11.36 0.005 m/day
11.38 5.5 m/s
11.42 (a) 0.0095 g/hr; (b) 5 days
11.50 167°C
11.58 $\overline{Sh} = 2000$
11.60 1.5 hr

# Index